OXYGEN TRANSPORT TO TISSUE XVI

ADVANCES IN EXPERIMENTAL MEDICINE AND BIOLOGY

Recent Volumes in this Series

OXYGEN TRANSPORT TO TISSUE XVI

Edited by

Michael C. Hogan
Odile Mathieu-Costello
David C. Poole
and Peter D. Wagner
University of California, San Diego
La Jolla, California

SPRINGER SCIENCE+BUSINESS MEDIA, LLC

Library of Congress Cataloging-in-Publication Data

Oxygen transport to tissue XVI / edited by Michael C. Hogan ... [et
al.].
      p.   cm. -- (Advances in experimental medicine and biology ; v.
361)
    "Proceedings of the 21st annual meeting of the International
Society on Oxygen Transport to Tissue, held August 14-18, 1993, in
San Diego, Calif."--T.p. verso.
    Includes bibliographical references and index.
    ISBN 978-1-4613-5763-6   ISBN 978-1-4615-1875-4 (eBook)
    DOI 10.1007/978-1-4615-1875-4
    1. Tissue respiration--Congresses.   I. Hogan, Michael C., 1954-
. II. International Society on Oxygen Transport to Tissue. Meeting
(21st : 1993 : San Diego) III. Title: Oxygen transport to tissue
16. IV. Series.
QP121.A10992 1994
599'.012--dc20   94-44823
                                                        CIP

Proceedings of the 21st annual meeting of the International Society on Oxygen Transport to Tissue, held August 14–18, 1993, in San Diego, California

ISBN 978-1-4613-5763-6

© 1994 Springer Science+Business Media New York
Originally published by Plenum Press in 1994
Softcover reprint of the hardcover 1st edition 1994

INTERNATIONAL SOCIETY ON OXYGEN TRANSPORT TO TISSUE
1992-1993

Officers

President:	P.D. Wagner, U.S.A.
Presidents-Elect:	C. Ince, Netherlands and
	K. Akpir, Turkey
Past President:	P. Vaupel, Germany
Secretary:	A. Hudetz, Hungary
Treasurer:	S.N. Cain, U.S.A.

Executive Committee

D.F. Bruley, U.S.A.	D.T. Delpy, United Kingdom
A. Eke, Hungary	T. Goldstick, U.S.A.
K. Groebe, Germany	D. Maguire, Australia
H. Metzger, Germany	K. Rakusan, Canada
M. Tamura, Japan	Z. Turek, The Netherlands
D.F. Wilson, U.S.A.	R. Zander, Germany

SAN DIEGO MEETING, AUGUST 14-18, 1993

Organizing Committee

D. Bebout	T. Davisson
M. Hogan	P. Keipert
J. Lessem	O. Mathieu-Costello
D. Poole	H. Wagner
P.D. Wagner	

SPONSORS

We are most grateful for the financial support for the 1993 ISOTT Meeting received from the following:

Alliance Pharmaceutical Corp.
Boehringer Ingelheim Gmbh
Eppendorf-Netheler-Hinz Gmbh
Hospex Fiberoptics
Medical Systems Corp.
Puritan-Bennett Corp.
Triton Technology, Inc.
MedGraphics, Corp.

PREFACE

Since its inception in 1973, The International Society on Oxygen Transport to Tissue
(ISOTT) has provided a unique forum to facilitate and encourage scientific interaction and
debate. Welcoming scientists and clinicians from a broad spectrum of disciplines, each with
their own particular skills and expertise, ISOTT unites them under the common theme of
oxygen transport. The successful blend of scientific presentations and informal discussion
which characterizes ISOTT is epitomized best by the many fundamental discoveries and
technical advancements which it has spawned. The breadth and strengths of The Society's
scientific base promotes the rapid progression of ideas from theoretical concepts to rigorous
scientific testing and often, ultimately to the clinical arena. Each publication of the ISOTT
proceedings has been recognized by Science Citation Index listing and the papers frequently
establish scientific precedents and become considered as standard works in their respective
fields.

The 21st ISOTT Meeting was held in San Diego from August 14th through August 18th,
1993. The San Diego Meeting attracted about 150 registrants and 40 accompanying persons.
Ten state-of-the-art lectures were presented by international experts in O_2 transport and there
were in addition two symposia - one dealing with assessment of tissue hypoxia and the other
with functional heterogeneity in different organ systems. There were 100 free
communications, consisting of posters accompanied by an abbreviated oral summary.

All manuscripts were reviewed by the Editors for form and content, but as is customary for
the ISOTT proceedings, rigorous scientific peer review was not undertaken.

The editors congratulate Dr. D.A. Benaron, from Stanford, for the honor of being selected
as the 1993 Melvin Knisely Award Winner for his outstanding contributions in the field of
research on optical imaging in the brain. We also wish to thank Peter Keipert, Eddie
Bebout, Harrieth Wagner, Tania Davisson and Julie Lessem for their tireless help in
mounting the meeting that gave rise to this book.

We look forward to the continued expansion of ISOTT and to the next meeting to be held in
summer 1994 in Istanbul, Turkey.

Michael C. Hogan

Odile Mathieu-Costello

David C. Poole

Peter D. Wagner

March 1994

CONTENTS

SYSTEMIC OXYGEN TRANSPORT

<u>Abstracts:</u>

HEART

LUNG

Abstracts:

KIDNEY & GUT

TUMOR

INDICES

LOCAL PLASMA CONVECTION CAN BE IMPORTANT FOR OXYGEN RELEASE IN TISSUE CAPILLARIES

C. Bos, L. Hoofd, and Z. Turek

Department of Physiology
University of Nijmegen
P.O. 9101
6500HB Nijmegen
The Netherlands

INTRODUCTION

Most of the research on oxygen transport from the red blood cells (RBC) in the capillaries to the tissue has been focused on diffusion. Convection is considered important with respect to the movement of blood. However, convection can also play a role in small-scale transport of oxygen. Mixing of the plasma between RBCs can possibly influence the local oxygen flux from the RBCs into the plasma and from the plasma into the tissue. To investigate this, a model of fluid flow is needed which has to be coupled to the mass transport. One was developed by Aroesty and Gross (1970). Their model of fluid dynamics was similar to that of others (for instance Bugliarello and Hsiao, 1970), but they were the first to couple it to oxygen transport. They concluded that the influence of local plasma mixing was negligible.

Their conclusions were based on calculations assuming constant oxygen concentrations along the boundaries of the gaps between RBCs. On the other hand, it has been shown that steep oxygen gradients along the boundaries are predicted by diffusional models (Groebe and Tews, 1989; Groebe, 1990; Hoofd, 1992). Therefore, in this presentation the model of Aroesty and Gross was extended to investigate the effect of plasma mixing on oxygen transport using more realistic boundary conditions.

Oxygen Transport to Tissue XVI
Edited by M.C. Hogan *et al.*, Plenum Press, New York, 1994

MODEL FORMULATION

The model is developed for a cylindrical coordinate system $\vec{r} = (r, \phi, z)$. Because of symmetry, the equations are independent of the angle ϕ. The RBCs are taken to be cylinders and exactly fit in the capillary. The system is defined in accordance with Aroesty and Gross and is shown in figure 1. The radius of the capillary is R, the gap length is 2 L, and the velocity of the RBCs is v_{RBC}. The equations and boundary conditions are set up for stationary RBCs and a wall moving in the opposite direction to the RBCs, but with the same velocity. In this way the fluid motion is determined relative to the RBCs.

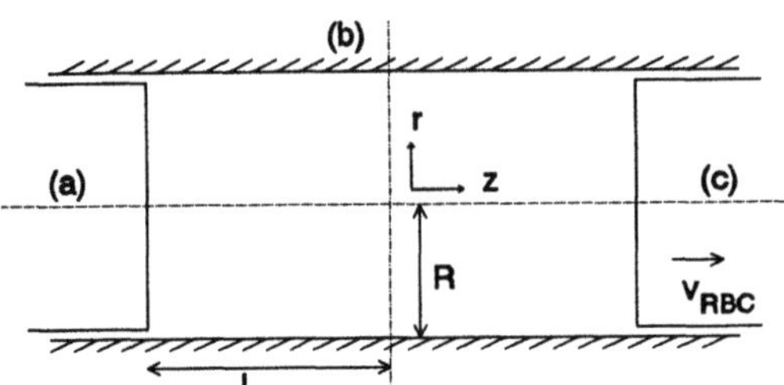

Figure 1. Model system for the calculation of plasma dynamics and concentration gradients. The upstream side of the gap is (a), (b) is the capillary side, and (c) is the downstream side.

The equations describing the fluid mechanics are the equation of continuity and the equation of motion (Bird et al., 1960). The former is derived from the mass balance and the latter from the momentum balance. The plasma fluid is taken to be incompressible, and as is valid for Newtonian fluids, of a constant dynamic viscosity μ. The Reynolds number is low, which leads to the assumption of Stokes flow. The equations are

$$\begin{cases} \vec{\nabla}p = \mu\nabla^2\vec{v} \\ (\vec{\nabla}\cdot\vec{v}) = 0 \end{cases}$$

where the second is the equation of continuity, p is the pressure, $\vec{v}$ is the velocity, $\vec{\nabla}$ is the gradient operator, ∇^2 is the Laplace operator and the dot denotes the inner product of two vectors. These equations are not easily solved, therefore a new variable is introduced: the stream function ψ. The velocity components are expressed as derivatives of ψ in such a way that the equation of continuity is satisfied. The solution lines with ψ a constant are the streamlines.

When the movement of the plasma is incorporated into the transport of oxygen, the equation of oxygen transport under the condition of steady state becomes

$$(\vec{v} \cdot \vec{\nabla} c) = D\nabla^2 c$$

where c is the concentration of oxygen and D the corresponding diffusion coefficient. In fact the calculations can be done for dimensionless concentrations $c^* = (c-c_0)/c_1$, since c_0 and c_1 are irrelevant for this investigation. The dimensionless constant governing the mixing is the Peclet number defined here as $Pe = (v_{RBC} \cdot L)/D$, which is zero when there is no mixing at all and is an increasing function of mixing. The equations are solved by finite difference approximation with an 11×11 grid for the convection in one quadrant of the gap, and therefore a 20×20 grid for the concentration in the whole gap.

To investigate the influence of mixing on the oxygen transport, concentration profiles have to be calculated. Aroesty and Gross used fixed values of 0 for the dimensionless concentration along the RBCs and of 1 along the capillary border. This gives exactly the same results for oxygen release from the RBCs at boundary concentrations of 0 at the capillary border and 1 at the RBC borders.

The boundary concentrations above were meant to get a first impression of the effect of mixing on oxygen transport. It can easily be seen that these conditions are unrealistic, as there is a singularity at the corners of the boundaries. More realistic boundary conditions can be found from studies where oxygen profiles have been calculated by means of diffusional models. The profiles used here are parabolic second order equations resulting in profiles comparable to those in literature. The equations are

$$\begin{cases} c^* = 1-a_1\left(\dfrac{r}{R}\right)^2 & \text{upstream RBC, } z=-L \\[2mm] c^* = 1-a_2-a_1\left(\dfrac{r}{R}\right)^2 & \text{downstream RBC, } z=L \\[2mm] c^* = \dfrac{a_3+a_4}{4}\left(\dfrac{z}{L}\right)^2-\dfrac{a_2}{2}\left(\dfrac{z}{L}\right)+\dfrac{a_3-a_4}{4} & \text{capillary, } r=R,-R \\[2mm] a_3 = 2-2a_1-a_2 \quad ; \quad a_4 = 2\sqrt{(a_1-1)(a_1+a_2-1)} \end{cases} \tag{1}$$

where a_1 is similar to the concentration drop from the centre of the RBC to the border, and a_2 is the concentration drop between two successive RBCs. The highest concentration is always set to 1, and the lowest is always 0. These equations are coupled such that a continuous concentration profile is generated without singularities, which is probably more realistic.

RESULTS AND DISCUSSION

Since the amount of mixing depends on the Pe, an estimation of the range for Pe is needed to be able to interpret the calculations. Aroesty and Gross showed some calculations for $R=L$, so with an RBC volume of 61 $(\mu m)^3$ (Altman et al., 1958) and a hematocrit of 40% this results in a radius $R = 2.44$ μm. For this radius and $(\alpha F/\wp)$ as reported by Hoofd et al. (1990), $v_{RBC}/D = (\alpha F/\wp)/(\pi R^2) = 0.565\ (\mu m)^{-1}$ and $Pe = L \cdot v_{RBC}/D = 1.4$. For lower hematocrits down to 20% the Peclet number increases up to 3.7. For heavily working rat heart muscle a v_{RBC}/D of approximately 5 times the value for resting heart can be reached, leading to $Pe = 18$ for a hematocrit of 20%. To cover a reasonable range the Peclet numbers used here are 1, 2 and 10.

Aroesty and Gross (1970) also used Peclet numbers of up to 10. They introduced a parameter to investigate the effect of mixing on the local mass transfer at the capillary wall by means of the ratio $\dot{m} = (\partial c/\partial r)_{Pe}/(\partial c/\partial r)_{Pe=0}$ or in words

$$\dot{m} = \frac{\text{Local mass transfer rate (convection + diffusion)}}{\text{Local mass transfer rate (diffusion only)}}$$

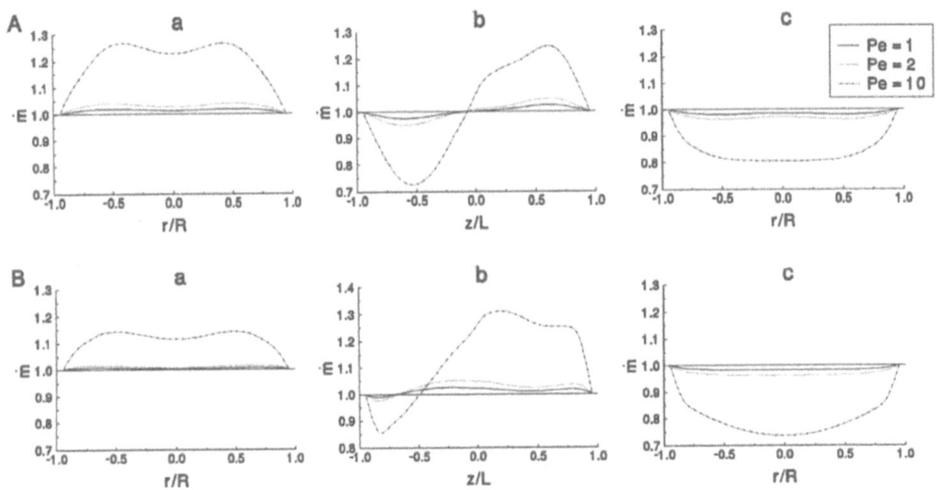

Figure 2. Local mass transfer ratios for different Peclet numbers and boundary concentrations with R/L = 1.0. The boundary conditions as chosen by Aroesty and Gross result in the A figures, and the B figures are for boundary concentrations derived for a 40% hematocrit in a model based on diffusion without mixing. The (a) graphs are for the upstream sides of the gap, the (b) graphs the capillary wall side, and the downstream side of the gap is shown in (c). The ratio $\dot{m}$ is shown for three Peclet numbers.

Graphs of $\dot{m}$ can be seen in figure 2, and the results of Aroesty and Gross are shown in picture 2Ab. They found that convection increased the mass transfer at the downstream side ($z > 0$) of the capillary wall, but that the mass transfer at the upstream side ($z < 0$) was decreased. Since, even at $Pe = 10$, the net effect on $\dot{m}$ is almost zero, they concluded that the effect of mixing was insignificant. Here we add pictures of $\dot{m}$ at the RBC walls (pictures 2Aa and 2Ac). It is clear from these pictures, that mixing increases the contribution of the upstream RBC and decreases that of the downstream erythrocyte. This already implies that non-symmetric boundary concentrations may cause the effect of mixing to become apparent.

4

The gradient ratio ṁ was also calculated for boundary concentrations as estimated from a diffusional model with point-like sources at a hematocrit of 40% (fig. 2B) (Hoofd et al., 1992). Under these conditions ṁ was markedly changed and the average effect on ṁ was far from zero. This might indicate an increased mass transfer. Therefore the change in flux at the capillary border was calculated (table 1). This can be done by calculation of

$$\varphi(Pe) = \frac{\int_{-L}^{L} J\,dz\big|_{Pe}}{\int_{-L}^{L} J\,dz\big|_{Pe=0}} = \frac{\int_{-L}^{L}\left(\frac{\partial c}{\partial r}\right)dz\big|_{Pe}}{\int_{-L}^{L}\left(\frac{\partial c}{\partial r}\right)dz\big|_{Pe=0}} \tag{2}$$

where J is the oxygen flux and φ is the ratio of flux with and without mixing. This gives an idea of the overall flux through the capillary wall, whereas ṁ gives merely an idea of change in the local flux at the capillary wall.

Table 1. Effect of mixing on the flux at the capillary wall

	a_1	a_2	R/L	Pe (rest)	Pe (work)	$\varphi(1)$	$\varphi(2)$	$\varphi(10)$
Aroesty and Gross	0.00	0.00	1.0	1.4	6.8	1.00	1.00	1.00
40% Hematocrit	0.34	0.18	1.0	1.4	6.8	1.01	1.03	1.14
20% Hematocrit	0.23	0.12	0.4	3.7	18	1.01	1.02	1.17

a_1 and a_2 from equation (1), and $\varphi(\ )$ from equation (2).

In table 1 it can be seen that, indeed, φ at the capillary wall is 1.00 for the boundary concentrations as chosen by Aroesty and Gross. This indicates that the flux is not altered by mixing. The change is even less than could be expected from the integrated ṁ. This is because the gradients are relatively small except for the outer sides near the RBCs, and particularly there the effects cancel each other. For the boundary conditions as estimated from a 40% hematocrit, the increase of flux by mixing is significant. Although here the increase in flux at low Peclet numbers for a hematocrit of 20% is less than for 40%, actually the increase might be more significant since the Peclet number is higher for the 20% case.

Although the oxygen flux definitely increases in these calculations, extrapolation to physiological situations is not straightforward. The use of a fixed concentration profile at the gap borders disregards any process in the RBCs and the tissue. However, since the mixing does change the local mass transfer, it will also change the profile at the borders. Therefore, to assess the importance of the flux and flux changes a more sophisticated model is needed that does take into account the RBCs and the tissue in realistic situations.

SUMMARY

Aroesty and Gross investigated the effect of mixing in gaps between two successive RBCs on the local mass transfer at the capillary wall. They found that the effect was insignificant. In this presentation it is shown that their conclusion results from their choice of boundary conditions. When boundary concentrations are chosen which are more similar to those in diffusional models the flux at the capillary wall increases significantly when mixing increases. The calculations presented indicate that mixing of plasma may enhance oxygen transport, although it is impossible to assess the physiological importance. To be able to investigate that, a model has to be developed that includes oxygen transport in both the RBCs and tissue, the oxygen release in the RBCs, and the oxygen consumption in the tissue.

REFERENCES

Altman, P.L., Gibson, J.F., and Wang, C.C., 1958, "Handbook of Respiration", Eds. Dittmar, D.S. and Grebe, R.M., Saunders company, Philadelphia and London

Aroesty, J., and Gross, J.F., 1970, Convection and diffusion in the microcirculation, *Microvasc. Res.* 2:247-267

Bird, R.B., Stewart, W.E., and Lightfoot, E.N., 1960, "Transport Phenomena", Wiley & Sons, New-York.

Bugliarello, G., and Hsiao, G.C., 1970, A mathematical model of the flow in axial plasmatic gaps of the smaller vessels, *Biorheology* 7:5-36

Groebe, K., and Thews, G., 1989, Effects of red cell spacing and red cell movement upon oxygen release under conditions of maximally working skeletal muscle, in: "Oxygen transport to tissue XI", *Adv. Exp. Med. Biol.*, 248:175-185

Groebe, K., 1990, A versatile model of steady state O_2 supply to tissue. Application to skeletal muscle. *Biophys. J.* 57:485-498

Hoofd, L., Olders, J., and Turek, Z., 1990, Oxygen pressures calculated in a tissue volume with parallel capillaries, in: "Oxygen transport to tissue XII", *Adv. Exp. Med. Biol.*, 277:21-29

Hoofd, L., 1992, Updating the Krogh model - assumptions and extensions. In: "Oxygen transport in biological systems", Soc. Exper. Biol. Seminaries Series 51, S. Eggington and H.F. Ross, eds., Cambridge University Press, Cambridge, pp.197-229

Hoofd, L., Bos, C., Turek, Z., 1992, Modelling erythrocytes as point-like O_2 sources in a Kroghian cylinder model, in: "Oxygen transport to tissue XV", Plenum Press, New York and London

FORMULATION AND REALIZATION OF A MULTICOMPARTMENTAL MODEL FOR O$_2$-CO$_2$ COUPLED TRANSPORT IN THE MICROCIRCULATION

Guo-Fan Ye

Biomedical Engineering and Science Institute
Drexel University
Philadelphia, PA 19104, USA

INTRODUCTION

A multicompartmental model for gas transport in the microcirculation (M.C.M) is a lumped form of the corresponding distributed models. This idea can be understood by the sketch shown in Fig.1. The microcirculation consists of a series of arteriole beds, capillary beds, venule beds and the surrounding tissue, with an oxygen partial pressure distribution and a carbon dioxide partial pressure distribution in both the radial and longitudinal directions. Distributed models, based on either the modified Krogh cylinder or a non-Kroghian geometry, usually simulate the radial and axial distributions in one vessel segment or within a local region of the microcirculation. If every vessel group is lumped into a compartment with a unique oxygen partial pressure and a unique carbon dioxide partial pressure, and the tissue is also lumped into one compartment (or more compartments, cf. Ye and Silverton, 1994), then a multicompartmental model is constructed. The model's block diagram in Fig.1 shows that axial convection of both blood flow and gas occurs along the sequential vessel compartments, and that simultaneous gas diffusions take place between the tissue compartment and every vessel compartment. Thus the global relationship of gas transport and blood supply is quite clear for a M.C.M. It is also easier for a M.C.M to be linked to a global multielement-lumped model for the whole cardiovascular system. Moreover, a compartmental model is described by algebraic equations (for steady-state) or by ordinary differential equations (for time-varying state), whereas a distributed model is described by ordinary differential equations (for one-dimension distributed, steady state) or by partial differential equations (otherwise). Thus, it is much simpler to mathematically solve a compartmental model than the corresponding distributed model. Therefore, multicompartmental modeling is a powerful tool for depicting the global properties of gas transport in the microcirculation-tissue system.

Oxygen Transport to Tissue XVI
Edited by M.C. Hogan *et al.*, Plenum Press, New York, 1994

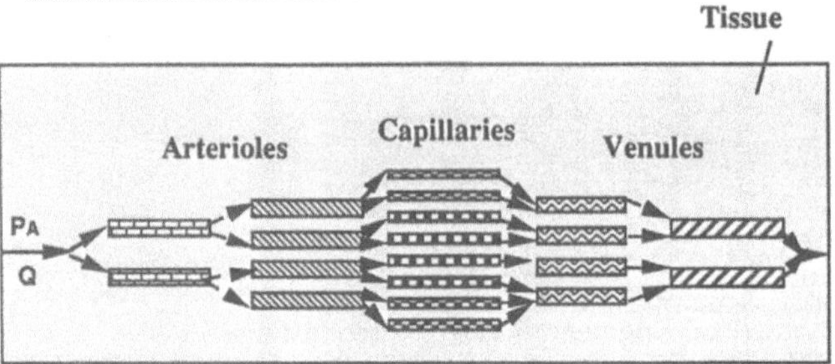

MULTICOMPARTMENTAL MODEL

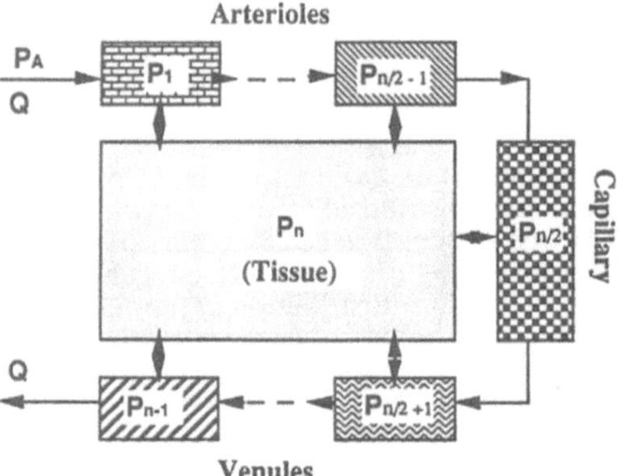

Figure 1. A sketch of the microcirculation-tissue system and a block diagram of the multicompartmental model for O2-CO2 coupled transport in the microcirculation. P represents the compartmental partial pressure of O2 or CO2, and Q represents the blood flow rate. The number of total compartments in this model is n, and n is even.

A few M.C.Ms for O2 transport or for O2-CO2 coupled transport have appeared in the literature. Roth and Wade (1986) presented a steady state model for O2-CO2 coupled transport in the skeletal muscle microcirculation. Sharan et al. (1989) proposed a steady state model for O2-only transport in the sheep brain. Since a M.C.M is the lumped form of the corresponding distributed models, the compartmental governing equations and the relevant lumped parameters of a M.C.M should be derived by space-averaging the governing equations and related parameters in the corresponding distributed models and then summing them up for each vessel group or tissue group. However, none of the existing M.C.Ms has been rigorously carried out in this way. Consequently, the predictions of these earlier models contained some apparent discrepancies from the experimental findings (cf. Table 2). Therefore, there exists a lack of a systematic formulation for the M.C.Ms.

In addition, the influence of the interaction between O2 and CO2 dissociations, i.e. Bohr effect and Haldane effect, on the model predictions needs further investigation.

Moreover, no time-varying M.C.M for the microcirculation is available in the literature.

This paper describes a work directed toward solving these problems (Ye, 1992; Ye et al., 1992a, 1992b, 1993). The work contains a systematic formulation for the M.C.M, and the computer realization for three M.C.Ms of various types including those for both the O2-only transport and the O2-CO2 coupled transport, and those for both the steady state and time-varying behavior. Results of comparisons are briefly reported in this paper.

METHODS

Derivation of Compartmental Formulation

The distributed governing equation, in a modified Krogh model, for the blood in a vessel segment is (cf. Reneau et al, 1969)

$$\alpha_{b,g}\frac{\partial p_{b,g}}{\partial \tau} = \alpha_{b,g}\frac{D_{b,g}}{r}\frac{\partial}{\partial r}(r\frac{\partial p_{b,g}}{\partial r}) - v\frac{\partial c_{b,g}}{\partial z} \qquad g=0,1 \qquad (1)$$

where $p_b=p_b(r,z,\tau)$ is the equivalent gas partial pressure in the blood, r, z, and τ are the radial, axial, and time coordinates, respectively, v is axial blood velocity at the given point, the subscript, g, indicates the gas species: g=0 refers to oxygen and g=1 refers to carbon

dioxide,. D_b is the gas diffusion coefficient and α_b is the gas solubility in blood, and c_b is the total concentration of gas in blood, which contains the part dissolved in plasma and the part bound with hemoglobin.

To facilitate the space averaging of Eq.1, two main assumptions were made on basis of experimental findings and predictions of distributed models:

a) the radial p_{O2} profile is near-parabolic for the homogeneous form of the blood (Reneau et al., 1967, 1969; Lagerland & Low, 1991; Schacterle et al., 1991);

b) the axial p_{O2} profile is linear for arterioles and the capillaries where $p_{O2}>20$ mmHg (Reneau et al., 1967, 1969; Gleighman et al., 1962; Popel & Gross, 1979; Schacterle et al., 1991), and is nonlinear for the capillaries where $p_{O2}<20$ mmHg (Reneau, et al., 1967, 1969). A logarithmic function was generated for the latter through data-fitting.

A cross-sectional average was conducted first. That is to apply the following operation to every term, X, on both sides of Eq.1:

$$\frac{2}{r_0^2}\int_0^{r_0} r X \, dr$$

where r_0 is inner radius of the vessel segment. Then the axial averaging was performed:

$$\frac{1}{L} \int_0^L X\, dz$$

where L is length of the vessel segment.

The modified Adair formula (Sharan et al., 1989) or Easton formula (Buerk & Bridges, 1986) was used for expressing the Bohr effect, and the Kelman (1966, 1967) equations were adopted for the Haldane effect.

Two specific factors were incorporated in determining the vessel-tissue diffusion conductance E: the radial distribution of gas partial pressure in blood, and the difference between the metabolism of vessel wall and that of tissue.

After the radial and axial averaging, governing equations for all vessels of a group were summed, which resulted in the final compartmental equation for this vessel group. The compartmental governing equations for all vessel groups were obtained in the same way. In these compartmental equations, all lumped terms, such as the lumped diffusion, lumped convection of partial pressure, lumped convection of dissociation and lumped countercurrent exchange, have been completely determined.

Evaluation of Lumped Terms

The following differences were estimated for the new model with respect to the earlier models in the literature: the difference in values of lumped convection, the difference in values of lumped dissociation, and the difference in values of vessel-tissue diffusion conductance. For the last item, the influences both of radial distribution of partial pressure and of vessel wall metabolism were evaluated separately.

Computer Realization of Three Simulations in Different Types

    a) O2-only transport, steady state: for the sheep brain microcirculation;
    b) O2-CO2 coupled transport, steady state: for the cat brain microcirculation;
    c) O2-CO2 coupled transport, time-varying: for the human brain microcirculation.

Evaluation of Model's Predictions

Two types of comparison in the behavior of the resulting compartmental gas distribution were performed for the new model: a comparison with respect to available experimental findings, and a comparison with respect to previous models in the literature. The latter included a comparison and contrast of the simulation for the sheep brain versus the Sharan et al (1989) model and a comparison and contrast of the simulation for the cat brain versus the Roth and Wade (1986) model.

In addition, the effect of O2-CO2 interaction and the effect of incorporation of both the radial distribution of gas partial pressure and the vessel wall metabolism upon the resulting model's predictions were assessed, respectively.

The model's predictions for various types of hypoxia and for the transient process of hypoxia were also analyzed. A particular attention was paid to the interesting phenomena observed in those predictions.

RESULTS

Table 1 lists the results of the evaluation of differences in values of the lumped terms between the new model and earlier models in the literature. As we can see, differences for all lumped terms are significant except for the cross-sectionally averaged dissociation terms.

Table 1. Evaluation of lumped terms in the compartmental governing equations

Lumped Term	New Model (N)	Earlier Model (E)	% Difference (E-N)/N
Convective Concentration	$c_{i\text{-}1}^{<r>}(z{=}L) - c_i^{<r>}(z{=}L)$	$C_{i\text{-}1} - C_i$	up to - 50%
Cross-Section Averaged	$S^{<r>}(p_{O_2}(r))$ for O_2	$S(p_{O2}^{<r>})$ for O_2	small
Dissociation Terms	$(Sp_{CO_2})^{<r>}$ for CO_2	$S^{<r>}p_{CO2}^{<r>}$ for CO_2	even smaller
Vessel-Tissue Diffusion Conductance	considers radial distribution	not considered	22% to 79% for capillary
	Mwall = 0.1 Mtissue	Mwall = Mtissue	-33%, thick arteriole
	together	together	-19% to +40%, arteriole

Note: c: gas concentration; C: compartmental concentration; S: oxygen saturation; p: partial pressure; M: metabolic rate; r: radial coordinate; z=L: at the end of a segment; superscript <r>: cross sectional average; subscript i: compartment number

Table 2. A comparison between experimental findings and model predictions

Experiment Finding about Distribution Behavior of O2	Experiment from which the Finding Came	Prediction of Earlier Models	Prediction of New Model
Pc > Pv in normal cases	Ivanov et al. 1982, rabbit brain Gleichmann et al. 1962, dog brain	always Pc ≈ Pv	Pc > Pv
Existence of Pt > Pv	Davies & Bronk 1957, cat cortex Gleichmann et al. 1962, dog brain Buerk et al. 1991, cat retina	always Pt ≤ Pv	can predict Pt > Pv
Existence of Pv-large >Pv-small	Pittman & Duling 1977, hamster cheek pouch Swain & Pittman 1984, hamster retractor muscle	always Pv-large ≤ Pv-small	can predict Pv-large > Pv-small
Pt = 38.7 mmHg	Nair et al. 1975, cat cortex		Pt =35.1 mmHg for the cat brain

Note: P: compartmental partial pressure; subscript: c: capillary; v: venule; t: tissue; v-large: larger venule; v-small: smaller venule

Table 2 summarizes three apparent differences in compartmental Po2 distribution behavior observed from the contrast between predictions of the sheep model and those of the Sharan et al (1989) model (cf. Ye et al. 1993 for details). The available experimental findings are also listed in Table 2 for a further comparison. The new model's predictions are supported by the experimental findings. Similar observations and the same conclusion can be yielded from a contrast between the cat model and the Roth and Wade (1986) model.

Fig.2 compares the compartmental Po2 and O2-Hb saturation distributions of the O2-

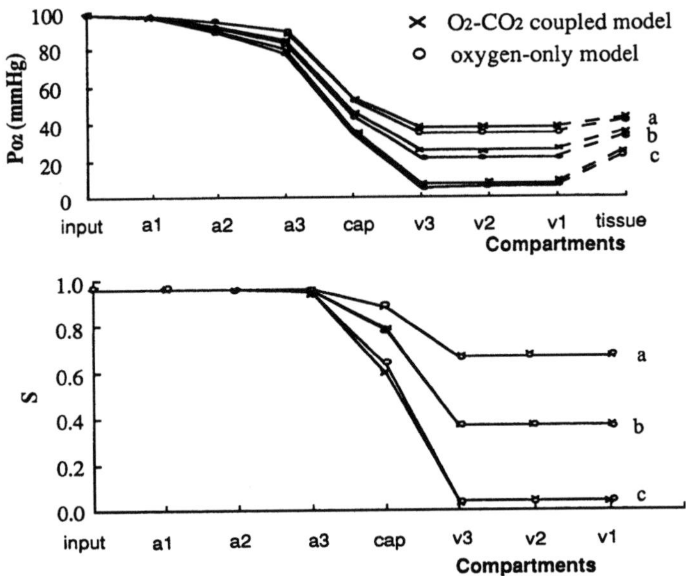

Fig.2. Compartmental Po2 and O2-Hb saturation (S) distributions predicted by the O2-CO2 coupled model and by the oxygen-only model for three different blood flow rate Q values (2.0, 1.0 and 0.65 times of the normal Q value for case a, b and c, respectively). a1, a2 and a3 represent arteriole compartments, cap represents the capillary compartment, v3, v2 and v1 represent venule compartments, tissue represents the tissue compartment, and input represents the input artery.

CO2 coupled model with those of the O2-only model under three different values of blood flow rate. While very little difference appears in the compartmental O2-Hb saturation distribution curves of the two models, a 20% difference can be observed in the Po2 values between the two models under a lower blood flow value. This comparison was performed in the cat brain model. Another comparison in the cat model demonstrated that the influence of incorporating the radial distribution of gas partial pressure and the wall metabolism upon compartmental Po2, Pco2 and pH distributions is significant.

The time-varying human brain model predicted transient changes of compartmental Po2, Pco2, S and pH distributions during the ischemic hypoxia process and hypoxic hypoxia process. Fig.3 depicts the transient response in Po2 and O2-Hb saturation distributions among each of the model compartments after a sudden reduction of cerebral blood flow rate Q (from Qnormal down to 0.7 Qnormal) at the moment of $\tau=0$. Another simulation of the model for an acute hypoxic hypoxia process after a sudden reduction of the input arterial PAO2 suggested an interesting damped oscillation phenomenon in Po2, Pco2, S and pH curves of all compartments before the distribution reached its final equilibrium.

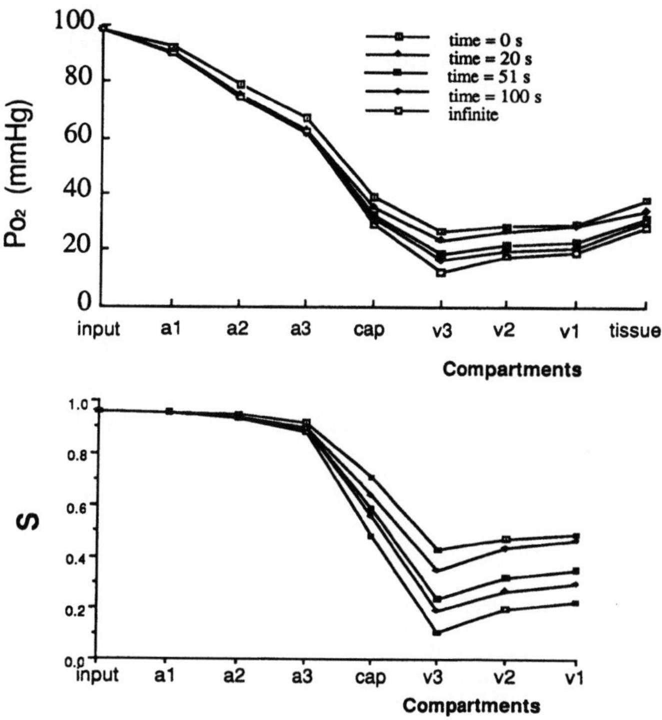

Fig.3. The transient response of compartmental Po2 and O2-Hb saturation (S) distributions after a sudden reduction of the   cerebral flow rate Q at the moment of t=0 (from Qnormal down to 0.7 Qnormal). In the input artery, PAO2=98.5 mmHg, PACO2=40 mmHg, and Qnormal=12.5 ml/sec.

A simulation to the rebreathing experiment by Hampson et al (1990) under normocapnic hypoxia and hypocapnic hypoxia was done in order to validate the reasonableness of the time-varying human brain model. The resulting behavior of the tissue Po_2 predicted by the model showed a similarity in tendency with the cerebral cytochrome a,a3 oxidation data provided by the experiment.

CONCLUSIONS

This work derived a systematic formulation for the multicompartmental model for O_2-CO_2 coupled transport in the microcirculation-tissue system through space-averaging and summing up governing equations of the corresponding distributed models. The model based on this formulation is able to predict diverse compartmental distributions of Po_2, Pco_2, oxygen-hemoglobin saturation and pH values for both the normal and various hypoxic conditions, and for both the steady state and time-varying behavior. Significant differences in the behavior of the resulting predictions were found between this new model and the earlier models in the literature. The predictions of the new model are supported by available experimental findings, and are more reasonable from a point of view of physiology. Results also indicated that coupling O_2-CO_2 interaction into an oxygen transport model is necessary. A time-varying simulation of the new multicompartmental model was performed. No similar work has been found in the literature. Predictions of the time-varying simulation for the hypoxic hypoxia suggested the probable oscillation phenomenon in the human body under some specific circumstance. The applications of this new M.C.M was extended to the study of additional microcirculation structures (Ye and Silverton, 1994).

ACKNOWLEDGMENTS

The author is grateful to Dr. D. Jaron, Dr. T. W. Moore and Dr. S. E. Dubin of Biomedical Engineering and Science Institute at Drexel University for their encouragement and advices. The author also appreciate Dr. D. G. Buerk of Ophthalmology Department at University of Pennsylvania for his comments and suggestions. This research was supported in part by the National Science Foundation under grant No. BCS 9022060 and in part by the Naval Air Development Center under contract No. N 62269-90-C-0242.

REFERENCES

Buerk, D.G., and Bridges, E.W., 1986, A simplified algorithm for computing the variation in oxyhemoglobin saturation with PH, PCO2, T and DPG, *Chem. Eng. Comm.* 47:113-124.

Buerk, D. G., Shonat, R.D., and Riva, C.E., 1991, Quantifying oxygen losses from retinal arterioles, in "Proceedings of 5th World Congress for Microcirculation", p.11.

Davies, P.W. and Bronk, D.W., 1957, Oxygen tension in mammalian brain, *Federation Proceedings* 16: 689-692.

Gleighmann, U., Ingvar, D. H., Liibbers, D. W., Siesjo, B. K., and Thews, G., 1962, Tissue PO2 and PCO2 of the cerebral cortex, related to blood gas tensions, *Acta Physiol. Scand.* 55: 127-138.

Hampson, N.B., Camporesi, E.M., Stolp, B.W., Moon, R.E., Shook, J.E., Griebel, J.A., and Piantadosi, C.A., 1990, Cerebral oxygen availability by NIR spectroscopy during transient hypoxia in humans, *J. Appl. Physiol.* 69:907-913.

Ivanov, K. P., Derry, A. N., Vovenko, E. P., Samilov, M. O., and Semionov, D. G., 1982, Direct measurements of oxygen tension at the surface of arterioles, capillaries and venules of the cerebral cortex, *Pflugers Archv.* 393: 118-120.

Lagerlund, T. D., and Low, P. A., 1991, Axial diffusion and Michaelis-Menten Kinetics in oxygen delivery in rat peripheral nerve, *Am. J. Physiol.* 260: R430-R440.

Nair, P., Whalen, W.J., and Buerk, D., 1975, PO2 of cat cerebral cortex: response to breathing N_2 and 100% O_2, *Microvas. Res.* 9: 158-165.

Pittman, R. N., and Duling, B. R., 1977, The determination of oxygen availability in the microcirculation, in "Oxygen and Physiological Function," F.F.Jobsis, ed., pp. 133-147, Professional Information Library, Dallas.

Popel, A. S., and Gross, J. F., 1979, Analysis of oxygen diffusion from arteriolar network, *Am. J. Physiol.* 237: H681-H689.

Reneau, D. D., Jr., Duane, D. F., and Knisely, M. H., 1967, A mathematical simulation of oxygen release, diffusion, and consumption in the capillaries and tissue of the human brain, in "Chemical Engineering in Medicine and Biology," D. Hershey, ed., pp. 135-241, Plenum Press, New York.

Reneau, D. D., Jr., Bruley, D. F., and Knisely, M. H., 1969, A digital simulation of transient oxygen transport in capillary-tissue systems, cerebral grey matter, *AIChE Journal* 15: 916-925.

Roth, A. C., and Wade, K., 1986, The effects of transport in the microcirculation: a two gas species model, *Microvasc. Res.* 32: 64-83.

Schacterle, R. S., Adams, J. M., and Ribando, R. J., 1991, A theoretical model of gas transport between arterioles and tissue, *Microvasc. Res.* 41: 210-228.

Sharan, M., Jones, M. D., Jr, Koehler, R. C., Traystman, R.J., and Popel, A. S., 1989, A Compartmental model for oxygen transport in brain microcirculation, *Ann. Biomed. Eng.* 17: 13-38.

Swain, D.P., and Pittman, R.N., 1984, Oxygen exchange in the microcirculation of hamster retractor muscle, Abstract, *Microvas. Res.*, 27: 266.

Ye, G-F., 1992, "Theoretical Basis and Computer Realization of a Compartmental Model for O2-CO2 Coupled Transport in the Microcirculation", Ph.D. dissertation, Drexel University.

Ye, G-F., Moore, T. W., and Jaron, D., 1992a, A compartmental model of oxygen transport derived from a distributed model: treatment of convective and oxygen dissociation properties, in "Proceedings of IEEE 18th Annual Northeast Conference on Bioengineering," W. J. Ohley, ed., pp. 83-84, Rhode Island.

Ye, G-F., Moore, T. W., Buerk, D.G., and Jaron, D., 1992b, Coupling of oxygen and carbon dioxide transport in a compartmental model of cat brain, in "Program and Abstracts of the Biomedical Engineering Society Third Annual Fall Meeting", H2.5. Salt Lake City, Utah. [Abstract]

Ye, G-F, Moore T.W., and Jaron, D., 1993, Contributions of oxygen dissociation and convection to the behavior of a compartmental oxygen transport model, *Microvasc. Res.* 46: 1-18.

Ye, G-F., and Silverton, S. F., 1994, Computer-modeling of oxygen supply to cartilage: addition of a compartmental model, in "Oxygen Transport to Tissue XVI", M. C. Hogan, O. Mathieu-Costello, D. C. Poole, and P. D. Wagner, eds, Plenum Publishing Co., New York.

ANALYSIS OF TISSUE DIFFUSIVITY USING MATHEMATICAL MODELS

A. Dutta and A. S. Popel

Department of Biomedical Engineering
School of Medicine
Johns Hopkins University
Baltimore, MD 21205

INTRODUCTION

It has been suggested that there is considerable heterogeneity of diffusion within cells and that oxygen diffuses unevenly across tissue, with more rapid diffusion in intracellular channels of high oxygen solubility (Longmuir, 1980). These oxygen channels are thought to be membranous structures within the cell with high lipid content, such as mitochondria and the sarcoplasmic reticulum in striated muscle. There is evidence from comparative physiology that higher maximal oxygen uptake capacity in aerobic animals is achieved by greater mitochondrial content (Weibel et al., 1992). This, combined with the fact that endurance training preferentially increases the volume of subsarcolemmal mitochondria and lipid droplets (Hoppeler and Billeter, 1991), suggests that higher lipid content and nonuniform mitochondrial distribution play an important role in cellular oxygen transport.

Most previous measurements of oxygen diffusion coefficient (DO_2) in muscle have been done at room temperature and extrapolated to 37°C using a temperature coefficient of 2.5%/°C (e.g., Ellsworth and Pittman, 1984). A temperature coefficient is determined by fitting an exponential equation to the data using, $DO_2 = C \exp(bT)$, where C is a constant, b is the temperature coefficient for diffusion having units of %/°C, and T is temperature (°C). Recently, Bentley et al. (1993) measured the diffusion coefficient of oxygen (DO_2) in the hamster retractor muscle directly at 37°C using a metabolic depressant to inhibit muscle oxygen consumption. These measurements have since been repeated in a hyperbaric chamber and extended up to 41°C (Bentley and Pittman, 1993). They found a temperature coefficient of 4.6%/°C between 11°C and 41°C and a DO_2 of 2.41×10^{-5} cm^2/s measured directly at 37°C. This value of DO_2 is much higher than the previously reported value based on measurements at 23°C and extrapolated to 37°C using a temperature coefficient of 2.5%/°C (Ellsworth and Pittman, 1984). Bentley et al. (1993) also found that DO_2 exhibited a marked increase between 23°C and 37°C and hypothesized that this could result from an alteration in the diffusion pathway above 23°C, possibly the result of a phase transition in the lipid membranes. Thus the conjecture that lipid

Oxygen Transport to Tissue XVI
Edited by M.C. Hogan *et al.*, Plenum Press, New York, 1994

membranes offer a low resistance pathway for oxygen diffusion from plasma membrane to the respiring mitochondria needs to be investigated thoroughly in light of the recent experimental data.

In this work, we will address the issue of intracellular heterogeneity by analyzing the experimental data of oxygen diffusion in hamster muscles as a function of temperature using available theoretical models of diffusion in two-phase media (Tai and Chang, 1974; Stroeve, 1977). These models will be utilized to predict the oxygen permeability, or Krogh coefficient, of cytosol and lipid as a function of temperature. Morphological data for hamster skeletal muscles (Sullivan and Pittman, 1987) will be used to evaluate the lipid contents of muscles. Furthermore, intracellular permeabilities will be calculated taking extracellular volume into account and the predictions will be compared to the available experimental data. It will be shown that the available models fail to predict the experimental values of tissue diffusivity using *in vitro* lipid permeability and a reasonable value for cytosol permeability. The results indicate that *in vivo* lipid permeability could be significantly higher than *in vitro* values and that both lipid and cytosol permeability appear to increase with increasing lipid content of muscles. In addition, we will show that experimental values of intracellular diffusivity are inconsistent with experimental tissue diffusivity values.

METHODS

Living tissues are heterogeneous; they consist of cells and extracellular spaces. Further, there are intracellular heterogeneities, for example, those caused by discrete oxygen consumption by mitochondria. These heterogeneities may affect the distribution of oxygen in the tissue. Several mathematical models are available for permeability coefficients (product of diffusivity and solubility) in heterogeneous tissue.

Tai and Chang (1974) proposed a model of heterogeneous tissue comprised of plane layers of cellular and extracellular material with different diffusion characteristics. In their model, it is assumed that cellular and extracellular media can be characterized by permeabilities P_d and P_c, respectively, and that the cells consume oxygen at a constant rate m (zero-order kinetics). If ϕ is the volume fraction of the cellular material, then volume-averaged tissue oxygen consumption is $M=\phi m$. The effective permeability of a layer of heterogeneous medium is defined as the permeability of a homogeneous medium with a uniform consumption rate M such that, for a given PO_2 difference at the boundaries, the oxygen fluxes at the corresponding boundaries are equal. For parallel and series arrangements of cellular and extracellular layers in the model, the effective permeabilities are, respectively,

$$P_e = \phi\, P_d + (1-\phi)\, P_c \tag{1}$$

$$P_e = \frac{P_d\, P_c}{(\phi\, P_c + (1-\phi)\, P_d)} \tag{2}$$

Note that P_e is independent of thickness of the tissue slice and consumption rate and depends only on permeabilities and on the volume fraction of the cellular space. The series arrangement tends to have more obstructions to the transport of oxygen from one surface to the other than real tissue whereas the parallel arrangement tends to have less. Although this model was originally developed to look at the effect of extracellular space in tissue on its diffusion properties, we will use this model to examine intracellular heterogeneities by considering a cell to be composed of aqueous cytosol and mitochondria and other lipid structures.

Stroeve (1977) developed a model of oxygen diffusion in heterogeneous tissue considering the tissue as consisting of spherical cells randomly dispersed in the continuous extracellular phase. Oxygen is consumed in the cells but not in the continuous phase. Following Maxwell's classical approach to deriving transport properties of a heterogeneous material by solving the diffusion equation analytically in and around spherical cells that follow zero-order kinetics and then volume averaging the results, he obtained an expression of the form

$$\frac{P_e}{P_c} = \frac{P_d + 2P_c - 2\phi\,(P_c - P_d)}{P_d + 2P_c + \phi\,(P_c - P_d)} \tag{3}$$

where P_d and P_c are the permeabilities of the cellular and extracellular phases respectively, and ϕ is the volume fraction of the cellular phase. It can be shown that the effective permeability predicted by this model lies between those predicted by the series and parallel models of Tai and Chang (1974).

Equations (1) to (3) can be used, as a first approximation, for estimating the effective permeability of a muscle cell if the permeabilities of the cytosol and lipids are known, as well as the volume density of mitochondria and other lipid structures. Since the volume fraction of extracellular material confined between muscle cells in skeletal muscle is very small (Hoppeler and Billeter, 1991), the tissue can be considered to be composed entirely of cytosol and mitochondria and other lipid structures. Then the effective permeabilities of a muscle cell and the tissue are equivalent. We will utilize these models to compute effective permeability of a muscle cell by considering the cell to be comprised of mitochondria and other lipids being suspended in an aqueous phase of cytosol. These models will also be used to calculate lipid and cytosol permeabilities when one of them is known and the other is computed by matching model predictions to experimental data on tissue permeability. Later, we will examine the effect of extracellular space in muscle on tissue and intracellular permeabilities.

As a first approximation, all the membranous structures in a · cell, e.g., mitochondria, sarcoplasmic reticulum, Golgi apparatus, plasma membrane, etc., will be assumed to be composed of lipids only. Thus, the total lipid content of a tissue can be computed from available morphological data for skeletal muscles (Hoppeler et al., 1987; Sullivan and Pittman, 1987). We will discuss later the implications of a smaller lipid content because of proteins present in the membranes. For the initial calculations, the oxygen permeability in cytosol will be computed as a fraction of the permeability in water utilizing a model developed by Stroeve (1975) and using a realistic value for protein concentration in the cytosol. The temperature dependence of oxygen permeability in water is available in the literature (Altman and Dittmer, 1971). Although different membranes present in the tissue could have different lipid compositions and hence very different physical characteristics, for the purpose of this study all membranous structures will be assigned a single value of oxygen permeability for dimyristoylphosphatidylcholine (DMPC), which is the dominant phospholipid present in various membranes (Fiehn et al., 1971). Literature values of oxygen permeability (Windrem and Plachy, 1980), diffusivity (Fischkoff and Vanderkooi, 1975; Subczynski and Hyde, 1981) and solubility (Smotkin et al., 1991) of the phospholipid at different temperatures will be used to calculate the effective permeability of tissue as a function of temperature.

However, since *in vivo* values of oxygen permeability in cytosol and in membrane lipids are unknown and could deviate significantly from their *in vitro* values, it might be too restrictive to use *in vitro* values of lipid permeability and a constant fraction of oxygen permeability in water for cytosol permeability. Therefore the calculation strategy we will employ is as follows. First, we will use *in vitro* values of lipid permeability and a reasonable value of cytosol permeability to calculate effective permeability of tissue at different temperatures. Second, we will keep lipid permeability at its *in vitro* value and extract values of cytosol permeability at different temperatures by matching model predictions to the experimental data for hamster retractor muscle (Bentley et al., 1993; Bentley and Pittman, 1993). Alternatively, we will use a realistic value for cytosol permeability and will calculate lipid permeability values predicted by different models by again matching model predictions to experimental tissue permeability values. We will also do these calculations for hamster sartorius and soleus muscles using experimental data of Ellsworth and Pittman (1984) and Sullivan and Pittman (1987). Sensitivity analysis will be done by perturbing the *in vitro* lipid permeability values and by varying the lipid fraction of a muscle. Finally, intracellular permeabilities in hamster retractor muscle as a function of temperature will be computed using various models and by employing realistic values of extracellular volume fraction and extracellular permeability. Similar calculations will be performed for hamster sartorius and soleus muscles and again a sensitivity study will be undertaken to investigate the effect of extracellular volume fraction and permeability on the calculated values for intracellular permeabilities.

RESULTS

Calculation of Tissue Permeability

For the initial calculations, oxygen permeability in cytosol was computed using a physiological value for protein volume fraction and bound water of 30% (Darnell et al., 1986) and according to a model of protein solution permeability to oxygen (Stroeve, 1975). This gave an upper bound for cytosol permeability which was 60% of the water permeability value. The lipid permeability was taken from an *in vitro* study of Windrem and Plachy (1980). Sullivan and Pittman (1987) reported the volume density of mitochondria and lipid droplets in three hamster muscles and these values were used as lipid volume fractions. The predicted oxygen permeability of tissue from the models were compared to the experimental values for hamster retractor muscle (Bentley et al., 1993; Bentley and Pittman, 1993). As shown in Figure 1(a), at low temperatures the effective permeability obtained from experiments falls between the values calculated from series and parallel models. However, at higher temperatures (including normal body temperatures), the experimental values of permeability fall outside these bounds. Since the sarcoplasmic reticulum and other membrane structures can contribute a significant amount of lipid (about 15%) to the total intracellular lipid content (Hoppeler et al., 1987), calculations were performed for the case of 30% lipid content. The higher lipid content results in higher permeability according to model predictions (Figure 1(b)), but the experimental data cannot be reconciled based on any single model. One intriguing observation of the model predictions is that at lower temperatures, the experimental values are closer to those predicted from a series model whereas at higher temperatures, the experimental values tend to be closer to those predicted from a parallel model which offers less resistance to oxygen transport. This trend appears to support the hypothesis that at higher temperatures membrane lipids offer a low resistance pathway for oxygen diffusion to tissue.

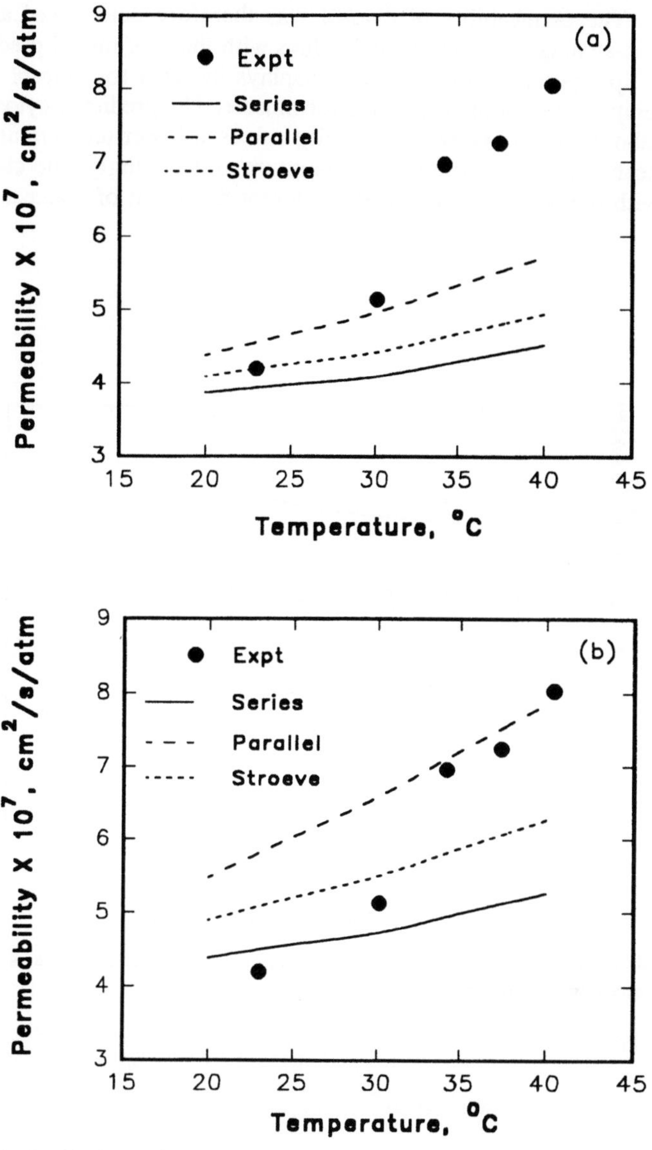

Figure 1. Calculated values of tissue permeability in hamster retractor muscle as a function of temperature using three models. (a) Lipid fraction = 13%. (b) Lipid fraction = 30%.

Calculation of Cytosol Permeability

The parameters we used to predict tissue permeability as a function of temperature are the cytosol permeability, lipid permeability, and the lipid content of tissue. Among these, the cytosol permeability has not been measured experimentally to our knowledge and the use of a fixed fraction of water permeability values at all temperatures may not be satisfactory. The permeability of cytosol was therefore determined as a function of temperature by matching the experimental values with those of model predictions using *in vitro* values of lipid permeability. Figure 2 displays the data for cytosol permeability as a function of temperature for hamster retractor muscle. The predicted cytosol permeability changes from about 50% of water permeability values at lower temperature to about 90% of water permeability values at higher temperature. Interestingly, the change in cytosol permeability with temperature is distinctly different from that of water and is suggestive

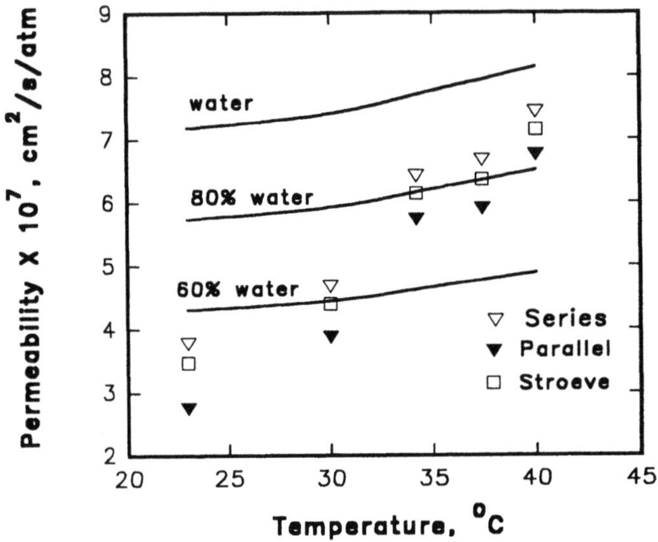

Figure 2. Calculated values of cytosol permeability in hamster retractor muscle as a function of temperature according to three models.

of a phase transition. It is likely, however, that this apparent phase transition comes from a phase transition in the lipid phase and model assumptions. Nevertheless, it is concluded from this analysis that the temperature coefficient for cytosol is different from that of water and that a fixed fraction of water permeability values can not be used for cytosol at all temperatures. When the lipid permeabilities were altered to within ±50% of their *in vitro* values, it was found not to have a major effect on calculated cytosol permeability values.

Ellsworth and Pittman (1984) have measured the diffusion coefficient of oxygen in hamster soleus, retractor, and sartorius muscles at 23°C. We used these values along with solubility data for frog sartorius muscle from Mahler et al. (1985) to arrive at permeability coefficients for these muscles. When the cytosol permeability for retractor muscle was used along with *in vitro* values for lipid permeability, the permeability data for sartorius and soleus muscles at 23°C could not be explained and values of 39% and 84% of water permeability were necessary to reconcile the experimental values of muscle permeability (Table 1). This clearly indicates that the cytosol permeability is not invariant across muscles and may indeed span a wide range of values from glycolytic muscle (sartorius) to oxidative muscle (soleus). Cytosol permeability appears to correlate with lipid content of a muscle and there may be physicochemical differences in cytosol composition and properties among muscles of different fiber type.

Table 1. Calculated values of cytosol permeability as a fraction of water permeability at 23°C for three hamster muscles

Muscle	Lipid fraction	Series	Stroeve	Parallel
Sartorius	9%	0.39	0.36	0.27
Retractor	13%	0.53	0.48	0.39
Soleus	19%	0.84	0.78	0.71

Calculation of Lipid Permeability

The *in vitro* value of lipid permeability we used in the previous calculations may not be valid *in vivo*. There may be differences in lipid permeability across muscles of different fiber types. Also, cytosol permeability values approaching those of water at high temperatures may not be realistic. The molecular weight of protein is several orders of magnitude larger than that of the oxygen molecule and as a consequence the protein molecules in cytosol present a steric hindrance to the diffusion of oxygen, since the mobility of the protein molecules is much smaller than that of the oxygen molecules. Using a physiological value for protein volume fraction and bound water (Darnell et al., 1986) and utilizing the model of protein solution permeability to oxygen (Stroeve, 1975), we arrived at an upper bound for cytosol permeability which was 60% of the water permeability values. Lipid permeability values were obtained using this value of cytosol permeability by matching the model predictions with those of experimental tissue permeability values (Bentley et al., 1993; Bentley and Pittman, 1993). Figure 3 displays the lipid permeability of hamster retractor muscle as a function of temperature as predicted by three different models. The solid line in the figure represents the *in vitro* values of lipid permeability from Windrem and Plachy (1980). At higher temperatures the series model and Stroeve model failed to predict the experimental permeability values even with very high lipid permeability values. Therefore, we conclude that models representative of a series arrangement of lipid and cytosol or lipid dispersed randomly in cytosol cannot explain the experimental values of permeability at higher temperatures. When a model using a parallel arrangement of lipid and cytosol is used, this model predicts a value of lipid permeability which is 71% higher than *in vitro* values. The fact that the series model and Stroeve model could not predict the experimental permeability values whereas the parallel model could, suggests that random, discrete arrangements of mitochondria cannot account for enhanced oxygen permeability observed in tissue at higher temperatures. Since the parallel arrangement of lipid and cytosol can be construed as representative of mitochondrial connectivity, it is possible that mitochondria *in vivo* may be connected to some extent. The same value of lipid permeability cannot explain the experimental values for sartorius and soleus muscles (Table 2) at 23°C. For sartorius muscle, the lipid permeability is found to be lower than the *in vitro* value, whereas for soleus muscle it can be as high as four times that of the *in vitro* value. Lipid permeability appears to increase with lipid content of muscle and as we found for cytosol permeability, different muscles exhibit a range of values for lipid permeability depending on the type of muscle fiber.

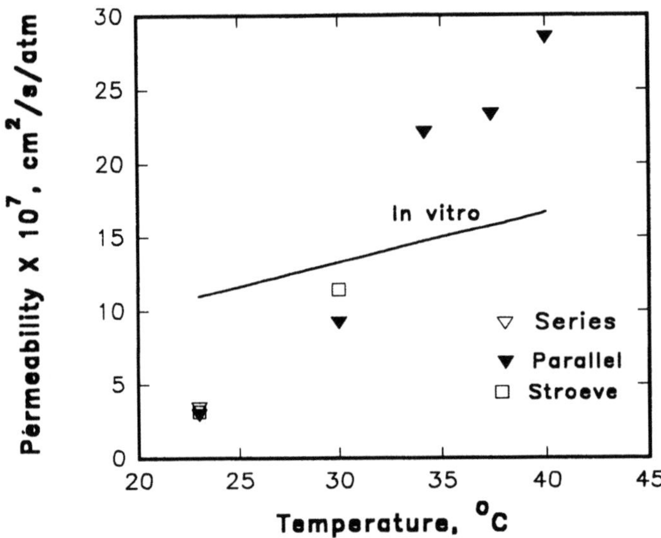

Figure 3. Predicted lipid permeability in hamster retractor muscle as a function of temperature for three models. *In vitro* lipid permeability values are from Windrem and Plachy (1980).

Table 2. Calculated values of lipid permeability as a fraction of *in vitro* lipid permeability at 23°C for three hamster muscles

Muscle	Lipid fraction	Series	Stroeve	Parallel
Sartorius	9%	0.05	-	-
Retractor	13%	0.24	0.23	0.22
Soleus	19%	-	4.48	1.23

Calculation of Intracellular Permeability

The extracellular volume fraction in skeletal muscle is small and it has been customary to neglect the interstitial space and assume that the entire muscle consists of muscle fiber (Hoppeler et al., 1987). Morphometric data indicate that this may lead to an error of 10% or less (Hoppeler et al., 1987). Tai and Chang (1974) and Stroeve (1977) used an extracellular volume fraction of 20% to 30% to calculate the ratio of extracellular to cellular permeability.

The intracellular diffusion coefficient of oxygen has been estimated to be 1.76×10^{-6} cm²/s at 30°C (Jones and Kennedy, 1986) and 3×10^{-6} cm²/s at room

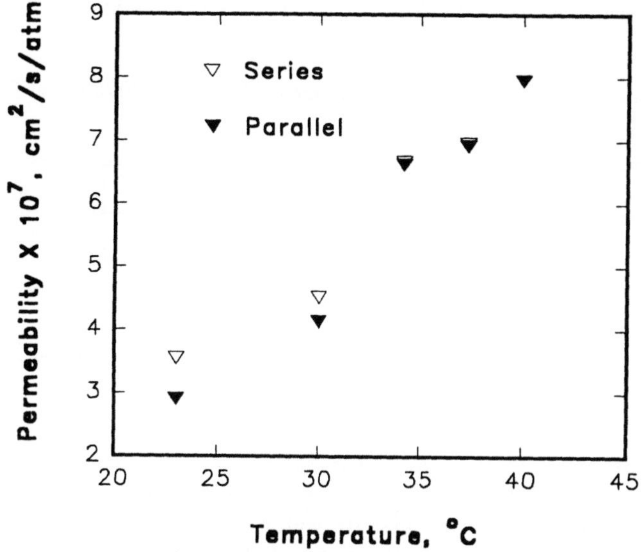

Figure 4. Predicted values of intracellular permeability in hamster retractor muscle as a function of temperature assuming extracellular volume fraction of 30% and extracellular permeability equal to that of water.

temperature (Rumsey et al., 1990) for rat cardiac myocytes. These values are an order of magnitude lower than experimentally measured tissue diffusivity values (Bentley et al., 1993; Mahler et al., 1985; Kawashiro et al., 1975). Since Rumsey et al. (1990) attributed this to high extracellular diffusivity, it is interesting to question whether the extracellular volume of tissue can account for the high tissue diffusivity values. We decided to calculate a lower bound for intracellular permeability by assuming a high extracellular volume fraction (30%) and extracellular permeability equal to that of water. In reality, the extracellular volume is lower, around 10% (Hoppeler et al., 1987), and its permeability is likely to be similar to those of physiological salt solutions, which is lower than that of water. The intracellular permeability was calculated by matching the experimental permeability values for retractor muscle with predictions based on the series and parallel models of Tai and Chang (1974). Stroeve's (1977) model should not be applicable to this situation of high cellular volume fraction since his model was derived for low values of cellular volume fraction.

Figure 4 depicts the calculated intracellular permeability for hamster retractor muscle as a function of temperature. The calculated values of intracellular permeabilities give rise to a diffusion coefficient which is an order of magnitude higher than experimentally reported values. For example, at 30°C the calculated intracellular permeability is about 4×10^{-7} cm^2/s/atm, which, using a water solubility value, gives an intracellular diffusion coefficient of 1.5×10^{-5} cm^2/s, compared to a value of 1.76×10^{-6} cm^2/s reported by Jones and Kennedy (1986) for cardiac myocytes. When intracellular permeabilities are calculated for hamster sartorius and soleus muscles from experimental data of Ellsworth and Pittman (1984) (Table 3), it is seen that intracellular permeability tends to increase with lipid content of muscle. Using solubility values of water at

Table 3. Calculated values of intracellular permeability (cm^2/s/atm) at 23°C for three muscles

Muscle	Lipid fraction	Series	Parallel
Sartorius	9%	2.39×10^{-7}	1.19×10^{-7}
Retractor	13%	3.57×10^{-7}	2.92×10^{-7}
Soleus	19%	6.55×10^{-7}	6.53×10^{-7}

23°C, the calculated intracellular diffusion coefficient can vary from 8.5×10^{-6} cm^2/s for sartorius muscle to 2.2×10^{-5} cm^2/s for soleus muscle. The calculated values for intracellular permeabilities are lower bounds since an extracellular volume fraction of 30% and extracellular permeability of that of water are probably higher than their actual physiological values. Therefore, it is concluded that our calculated intracellular permeability values are inconsistent with experimentally available values and that these experimental values of intracellular permeability are too low to explain the experimental values of tissue diffusivity.

Calculation of Muscle Solubility

The experimental values that have been used in this work were for the oxygen diffusion coefficient measurements in hamster skeletal muscles (Ellsworth and Pittman, 1984; Bentley et al., 1993; Bentley and Pittman, 1993). Solubility data for frog sartorius muscle measured by Mahler et al. (1985) were used to calculate the tissue permeability at different temperatures. No experimentally measured values of solubility are available for hamster retractor, sartorius, and soleus muscles at different temperatures. Since these muscles differ in their lipid content, it is possible that they might exhibit different solubility values. In our previous calculations, it was found that cytosol permeability and lipid permeability values were not invariant across these muscles and indeed they span a range of values. If one considers the muscle to be a composite of cytosol and lipid, then muscle solubility can be computed by volume fraction weighting of cytosol and lipid solubilities. Also, since muscle solubility is known (Mahler et al., 1985) and lipid solubility is available in the literature (Smotkin et al., 1991), cytosol solubility can be computed as a function of temperature. We attempted to calculate cytosol and lipid solubilities in different muscles as a function of temperature using this approach.

Solubility data for frog sartorius muscle (Mahler et al., 1985) and a lipid volume fraction of 10% were used along with lipid solubility data (Smotkin et al., 1991) to calculate cytosol solubility as a function of temperature. The calculated values for cytosol solubilities were unrealistically high, ranging from 124% of water solubility values at lower temperature to 88% of water solubility values at higher temperature. Since cytosol contains large molecular weight protein molecules, its solubility is expected to be much lower than that of water. The cytosol solubilities calculated for different muscles at 23°C showed the trend of higher solubility values with higher lipid content of muscle. Following our previous strategy, when the cytosol solubility was limited to 60% of water solubility values and the lipid solubility was calculated as a function of temperature by matching model predictions to the data of Mahler et al. (1985), we found several disturbing trends. First, the calculated lipid solubility values were two to five times higher than the experimentally measured values of Smotkin et al. (1991). Second, the lipid solubility values went down with temperature which is against the widely accepted notion of lipid phase transition resulting in higher oxygen solubility. Thus, we could not successfully compute oxygen solubilities in different muscles with different lipid content and had to use the data of Mahler et al. (1985).

DISCUSSION

In the present study, we have addressed the issue of intracellular heterogeneity of oxygen diffusion through the analysis of experimental data in hamster skeletal muscles. We utilized available theoretical models of diffusion in two-phase media to predict oxygen permeability of tissue, cytosol, and lipid as a function of temperature. It is shown that the available models cannot predict the experimental values of tissue diffusivity using *in vitro* lipid permeability values and a realistic value for cytosol permeability. When cytosol permeability was computed as a function of temperature, it exhibited temperature coefficients that were distinctly different from those of water. A single value of cytosol permeability failed to explain the permeability data for different muscles. It is also shown that intracellular lipid permeabilities may be two to four times their *in vitro* values and that lipid permeability appears to correlate with lipid content of muscle.

The analysis of intracellular permeability indicates that the oxygen permeability of a cell has to be an order of magnitude higher than the *in vitro* value (Jones and Kennedy, 1986; Rumsey et al., 1991) for the predicted muscle permeability to be consistent with the experimental data of Bentley et al. (1993). This conclusion was reached after variations of other parameters within physiological ranges of values failed to account for the differences between the predicted and observed oxygen permeability of tissue.

It is concluded that both lipid and cytosol permeabilities and hence, tissue permeabilities may be different among different muscles and one should exercise caution when data from one muscle are used to calculate or extrapolate values in other muscles. Oxidative muscles, by virtue of their higher mitochondrial content are likely to have higher oxygen permeabilities. It is conceivable that muscles with very high mitochondrial content (up to 40%), such as diaphragm and cardiac muscles, may exhibit an oxygen diffusion coefficient which is significantly higher than commonly accepted values. These results warrant measurements of tissue oxygen diffusivity directly at 37°C for various muscles, especially muscles with high lipid content.

In reality, the lipid contents of mitochondria, sarcoplasmic reticulum, plasma membrane, Golgi apparatus, etc., could be very different from each other and their physical properties (e.g., diffusivity, solubility, and phase transition temperature) would be dependent on their respective lipid and protein contents. The simple models used in this work did not account for these variations. Also, there is experimental evidence of mitochondrial clustering around capillaries and sarcolemma in response to endurance training, especially in highly oxidative muscle fibers (Weibel et al., 1992). In some muscles, e.g., diaphragm, mitochondria could form a reticulum (Kayar et al., 1988; Skulachev, 1990). The highly interconnected transverse portions of this network could, by virtue of the high solubility of oxygen in their lipid-rich membranes, facilitate oxygen diffusion within muscle fibers. Therefore, in the future it will be necessary to develop a model that takes into account realistic mitochondrial distribution and physical properties of lipid structures and cytosol to fully understand the pathway of oxygen transport through intracellular channels.

ACKNOWLEDGMENTS

The authors thank Drs. R. N. Pittman, T. B. Bentley, and S. R. Kayar for helpful comments and suggestions. This study was supported by NIH grant HL-18292 and a Research Fellowship award from American Heart Association, Maryland Affiliate, Inc.

REFERENCES

Altman, P. L. and Dittmer, D. S., Eds., 1971, *Respiration and Circulation,* Federation of American Societies for Experimental Biology, Bethesda, MD.

Bentley, T. B., Meng, H., and Pittman, R. N., 1993, Temperature dependence of oxygen diffusion in hamster retractor muscle. *Am. J. Physiol.,* 264:H1825-H1830.

Bentley, T. B. and Pittman, R. N., 1993, Influence of temperature on oxygen diffusion in hamster retractor muscle. Submitted to *Am. J. Physiol.*

Darnell, J. E., Lodish, H. F., and Baltimore, D., 1986, *Molecular Cell Biology,* W. H. Freeman, New York.

Ellsworth, M. L. and Pittman, R. N., 1984, Heterogeneity of oxygen diffusion through hamster striated muscle. *Am. J. Physiol.,* 246:H161-H167.

Fiehn, W., Peter, J. B., Mead, J. F., and Gan-Elepano, M., 1971, Lipids and fatty acids of sarcolemma, sarcoplasmic reticulum, and mitochondria from rat skeletal muscle. *J. Biol. Chem.,* 246, 18:5617-5620.

Fischkoff, S. and Vanderkooi, J. M., 1975, Oxygen diffusion in biological and artificial membranes determined by the fluorochrome pyrene. *J. Gen. Physiol.,* 65:663-676.

Hoppeler, H. and Billeter, R., 1991, Conditions for oxygen and substrate transport in muscles in exercising mammals. *J. Exp. Biol.,* 160:263-283.

Hoppeler, H., Kayar, S. R., Claassen, H., Uhlmann, E., and Karas, R. H., 1987, Adaptive variation in the mammalian respiratory system in relation to energetic demand: III. Skeletal muscles: setting the demand for oxygen. *Resp. Physiol.,* 69:27-46.

Jones, D. P. and Kennedy, F. G., 1986, Analysis of intracellular oxygenation of isolated adult cardiac myocytes. *Am. J. Physiol.,* 250:C384-C390.

Kawashiro, T., Nusse, W., and Scheid, P., 1975, Determination of diffusivity of oxygen and carbon dioxide in respiring tissue: results in rat skeletal muscle. *Pflugers Arch.,* 359:231-251.

Kayar, S. R., Hoppeler, H., Mermod, L., and Weibel, E. R., 1988, Mitochondrial size and shape in equine skeletal muscle: A three-dimensional reconstruction study. *The Anatomical Record,* 222:333-339.

Longmuir, I. S., 1980, Channels of oxygen transport from blood to mitochondria. *Adv. Physiol. Sci.,* 25:19-22.

Mahler, M., Louy, C., Homsher, E., and Peskoff, A., 1985, Reappraisal of diffusion, solubility, and consumption of oxygen in frog skeletal muscle, with applications to muscle energy balance. *J. Gen. Physiol.,* 86:105-134.

Rumsey, W. L., Schlosser, C., Nuutinen, E. M., Robiolio, M. and Wilson, D. F., 1990, Cellular energetics and the oxygen dependence of respiration in cardiac myocytes isolated from adult rat. *J. Biol. Chem.,* 265:15392-15399.

Skulachev, V. P., 1990, Power transmission along biological membranes. *J. Membrane Biol.*, 114:97-112.

Smotkin, E. S., Moy, F. T., and Plachy, W. Z., 1991, Dioxygen solubility in aqueous phosphatidylcholine dispersions. *Biochim. et Biophys. Acta.*, 1061:33-38.

Stroeve, P., 1975, On the diffusion of gases in protein solutions. *Ind. Eng. Chem. Fundam.*, 14, 2:140-141.

Stroeve, P., 1977, Diffusion with irreversible chemical reaction in heterogeneous media:application to oxygen transport in respiring tissue. *J. Theor. Biol.*, 64:237-251.

Subczynski, W. K. and Hyde, J. S., 1981, The diffusion-concentration product of oxygen in lipid bilayers using the spin-label T_1 method. *Biochim. et Biophys. Acta.*, 643:283-291.

Sullivan, S. M. and Pittman, R. N., 1987, Relationship between mitochondrial volume density and capillarity in hamster muscles. *Am. J. Physiol.*, 252:H149-H155.

Tai, R. C. and Chang, H. K., 1974, Oxygen transport in heterogeneous tissue. *J. Theor. Biol.*, 43:265-276.

Weibel, E. R., Taylor, C. R., and Hoppeler, H., 1992, Variations in function and design: Testing symmorphosis in the respiratory system. *Resp. Physiol.*, 87:325-348.

Windrem, D. A. and Plachy, W. Z., 1980, The diffusion-solubility of oxygen in lipid bilayers. *Biochim. et Biophys. Acta.*, 600:655-665.

COMPUTER - MODELING OF OXYGEN SUPPLY TO CARTILAGE: ADDITION OF A COMPARTMENTAL MODEL

Guo-Fan Ye[1] and Susan F. Silverton[2]

Biomedical Engineering & Science Institute, Drexel University[1]
Department of Medicine, University of Pennsylvania[2]
Philadelphia, PA 19104-6002 USA

INTRODUCTION

Our previous studies have focussed on the architecture of the avian growth plate and the oxygen consumption of growth plate chondrocytes in order to develop an appropriate computer model for estimating chondrocyte anoxia (Haselgrove et al., 1993). Initially, we used two models: the Krogh cylinder (Silverton et al., 1989), and a second model with similar geometry utilizing a complex oxygen consumption as a function of oxygen concentration (Silverton et al., 1990). For this purpose, we divided the growth plate into two anatomical regions; the region of resting-proliferating chondrocytes and the region of hypertrophic chondrocytes. We modeled the two growth plate regions separately and ignored the transition zone. We also used a two dimensional analysis assuming that the major flow of oxygen was radial rather than axial. To extend our model, we have now used

Oxygen Transport to Tissue XVI
Edited by M.C. Hogan *et al.*, Plenum Press, New York, 1994

a compartmental model originally developed for modeling the oxygen and carbon dioxide distribution in the microvasculature of the brain (Ye et al., 1993). With this model we have been able to evaluate the contribution of the microvacular structure to oxygen supply of the resting and hypertrophic regions of the growth cartilage and to estimate oxygen and carbon dioxide partial pressure variations in the growth plate.

The preceding models made two assumptions which simplified our initial investigations: e.g. that there was no drop in oxygen tension axially in the supplying vessel and, secondly, that carbon dioxide tensions in the tissue did not alter the oxygen available.

The compartmental model, however, requires more detail about the arterioles and venules supplying the growth plate cartilage. Thus, the definition of the architectural elements of the growth plate in resting and hypertrophic regions of the growth plate have required extensive amplification to provide the parameters for the compartmental model. The compartments comprise arterioles and venules as well as sinusoidal spaces and tissue spaces representing hypertrophic and resting cartilage regions. We have analyzed the relative frequency and comparative cross-sectional areas of arterioles, venules and sinusoidal structures in the resting and hypertrophic areas of the avian growth cartilage.

Values for carbon dioxide partial pressure in the growth plate were also estimated by this model and then calculated for comparison from literature information on the pH of the resting and hypertrophic regions (Howell et al., 1969). The determination of the bone nutrient artery oxygen concentration and blood flow rates in the bone marrow and growth plate were available from the literature (Davis et al., 1990) and from previous microsphere determinations of blood flow in the pig (Silverton et al., 1989).

The compartmental model showed that oxygen tension in the cartilage tissue compartments was more responsive to decreases in bloodflow in the hypertrophic zone. However, since this region had higher initial oxygen partial pressures than the resting-proliferative zone, this decrease in the oxygen tension did not approach the tissue levels normally found in the resting-proliferative zone even at the lowest flow tested. Reduction in oxygen tension and increases in carbon dioxide tension were most effective in diminishing oxygen available to the resting-proliferative compartment if the reduction occurred in blood supply from the articular surface. The hypertrophic cartilage tissue compartment appeared to be much more sensitive to decreases in venous oxygen tension than the resting-proliferative zone.

METHODS

Analysis of Microvasculature of Avian Growth Cartilage

Multiple horizontal sections of avian growth plate cartilage were stained with hematoxylin-eosin. Regions of the growth plate were identified by cell morphology and a vascular channel schematic developed (Figure 1). Vessels were identified as arterioles by the presence of muscle cells in the vessel wall (Figure 2, a & c) . Venules were identified by the presence of blood cells in the lumen and the absence of muscle cells in the vessel wall (Figure 2, b). Sinusoids, which occurred only in the hypertrophic region were identified

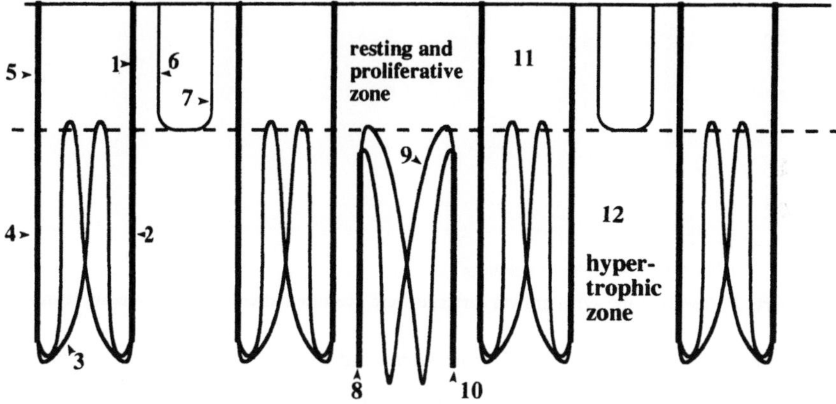

Figure 1 Schematic of avian growth cartilage. Numbers represent microvascular compartments.

by the multichanneled vascular structure and the presence of invading bone cells and bone matrix (Figure 2, d). Arterioles of the resting-proliferative and hypertrophic regions often were accompanied by small venules occupying the same vascular channel as the arteriole (Figure 2, c). These small venules were characterized as accompanying venules. In the resting-proliferative zone, only one accompanying venule was found with the arteriole. However, in the hypertrophic region, as many as three accompanying venules could be identified with an arteriole. Paired arteriolar structures were a feature of the resting-proliferative zone (Figure 2, a). Multiple slices from each region were photographed and elliptical axes of identified vessels were measured. Areas were computed for each type of vascular structure.

The multicompartmental model was designed to represent every vessel type as a

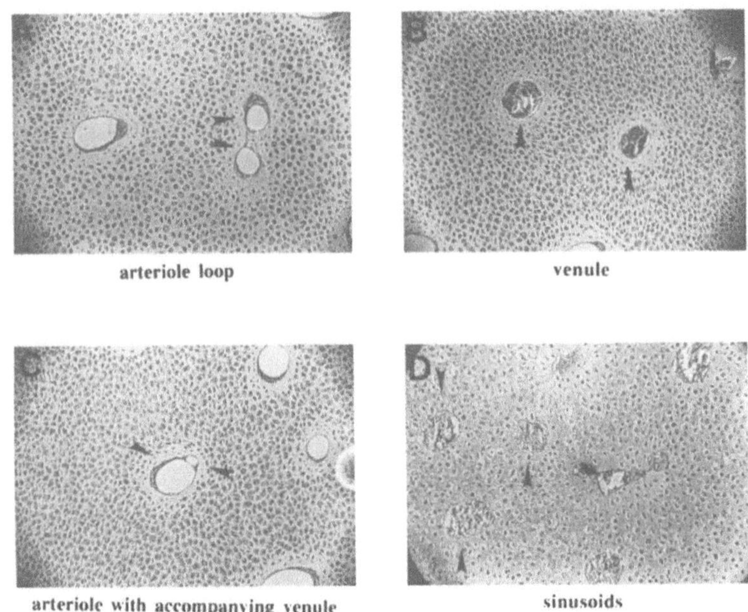

Figure 2 Vessels of the avian growth cartilage were identified after hematoxylin-eosin staining.

separate compartment (Figure 3). Tissue surrounding the vessels was divided into two compartments, resting-proliferative and hypertrophic, which have been shown previously to have different metabolic rates (Silverton et al., 1989a). A radial and axial distribution of pO_2 and pCO_2 was assumed with a unique lumped pO_2 and pCO_2 for each compartment. The multielement compartmental model was defined by appropriate equations (see Ye, this volume) and solved for an O_2-CO_2 coupled, steady state. A computer simulation was utilized to predict the distribution of O_2 partial pressure, CO_2 partial pressure, O_2-Hb binding saturation and pH values for the compartments over a range of oxygen and carbon dioxide concentrations and blood flow rates. Parameters not available directly from experimental data were estimated from the literature. The model was also utilized to estimate pO_2, pCO_2, S and pH if blood flow rate was decreased to 40% and to 10% of its estimated value, or if pO_2 was decreased to 75% of its estimated value while pCO_2 was increased to 110%. These simulations were applied to each of the initial vascular compartments. Thus, the relative influence of changing the arteriole and venous partial pressures was tested in the compartmental model.

RESULTS

In the resting region, arterioles account for 78% of the single vessels, and 32% of these single vessels were accompanied by multiple small venules. In the hypertrophic region, only 14% of the vessels were arterioles, while 67% were sinusoidal structures which

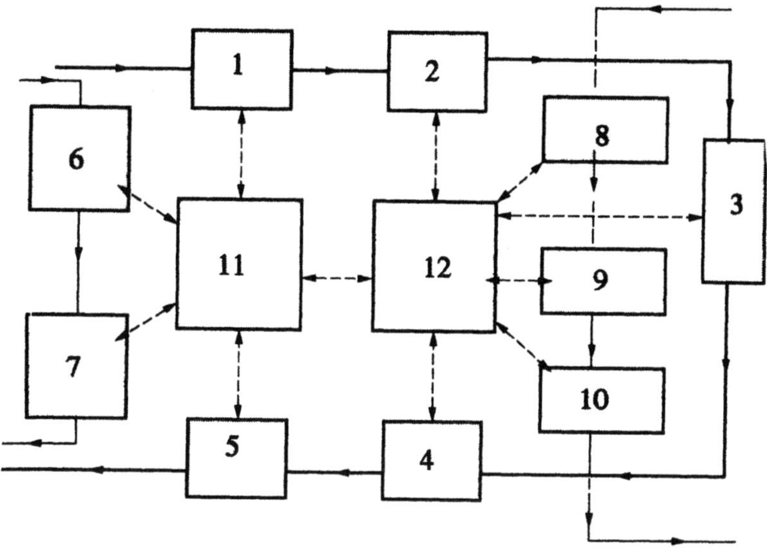

Figure 3 Flowchart for compartmental model. Numbers correspond to Figure 1 compartments.

Table 1. Areas of vessels from representative sections of resting-proliferative and hypertrophic regions

	Resting-Proliferative		Hypertrophic	
	number	area (microns)^2	number	area
arterioles	37	918 ± 151	15	718 ± 185
venules	12	721 ± 208	2	613 ± 433
a. venules	13	48 ± 13	11	97 ± 29
sinusoids		N/A	46	805 ± 119

appeared to be continuations of the bone marrow space. Accompanying 78% of the arterioles present in this zone were multiple small venules. Analysis of the relative .cross-sectional area of these vessel types in each region is shown in Table 1.

Oxygen, and carbon dioxide partial pressures in the growth cartilage were calculated using the compartmental model and employing initial flow rates found in previous studies of the porcine growth cartilage (Silverton et al., 1989). pO_2 and pCO_2 arterial and venous values specific to the avian species were taken from the literature. Figure 4a shows that O_2 partial pressures in the tissue compartments were diminished by not more than 10 mm Hg when flow rate was dropped to 10% of the initial flow. CO_2 partial pressure was only increased by slightly more than one mm Hg with this decrement in flow rate (Figure 4b). Figure 5a and b show the effect on the tissue compartments when the partial pressure of oxygen is decreased to 75% of the estimated value and the partial pressure of carbon dioxide is also increased to 110%. In some simulations, the decrease in arteriolar oxygen and increase in carbon dioxide was applied to the compartments supplying the arteriole loop

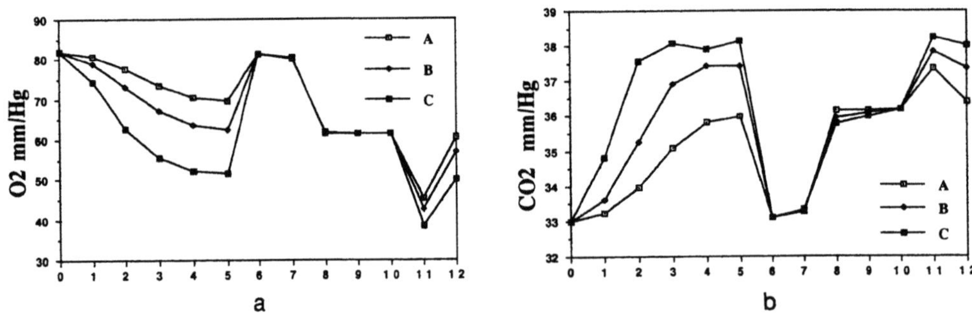

Figure 4 Predicted oxygen (graph a) and carbon dioxide (graph b) for each compartment (x-axis) for normal blood flow (A), 40% flow (B), 10%flow (C).

in the resting-proliferative (Figure 5a & b, line C). In addition, the effect of decreasing the venous oxygen partial pressure was estimated using the compartmental model (Figure 5a & b, line D). Decreasing the arteriolar or venous oxygen pressures to 75% of estimated value did not have a large effect on the resting-proliferative or the hypertrophic regions of the growth cartilage using the compartmental model to predict tissue compartment oxygen and carbon dioxide values.

DISCUSSION

The addition of a compartmental model of the microvasculature to our modeling of the avian growth cartilage has added to our understanding of the oxygen supply of these tissues. Contrary to our original concept of the vascular structure of the avian growth plate, we have found that the vascular channels of the cartilage are heterogeneous. Thus, although some vascular channels may contain both an arteriole and one or several venules, other channels are simple arterioles or venules. In addition, a large percentage of the vessels present in the hypertrophic region of the avian growth plate are sinusoidal in nature and probably represent invaginations of the bone marrow vascular spaces into the cartilage plate.

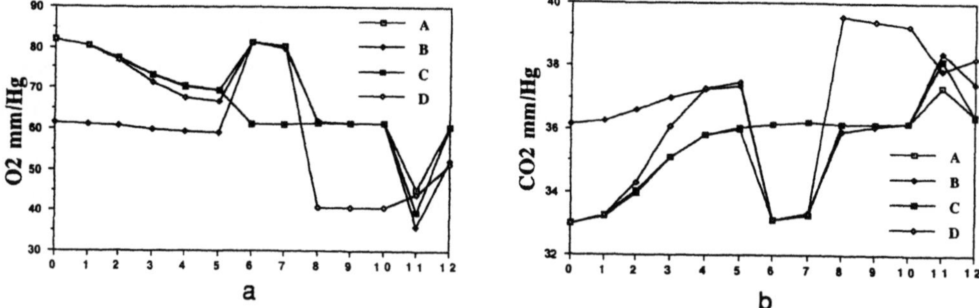

Figure 5 O2 (graph a) and CO2 (graph b) for normal arteriolar partial pressures (A), or decreased oxygen and increased carbon dioxide applied to articular arterioles (B), arteriolar loops (C) or venous supply (D).

The concentrations of oxygen and carbon dioxide in these sinusoidal structures would be expected to be in the range of venous partial pressures rather than arterial tensions. In order to carry out the present modeling of these tissue spaces, we have been forced to account for the relative numbers of arterioles, venules, and sinusoids as well as to relate cross-sectional areas of each to the total microvasculature structure. Thus, our present compartmental model has fitted the experimental data we have gathered on the vascular channels to three oxygen supply systems. The first is a traditional arteriole originating at the articular surface of the bone and penetrating the entire growth plate until the hypertrophic region where it divides into sinusoids and then is drained by venules through the growth plate returning to the articular surface. This first system most closely resembles the arteriole-capillary-venule triad found in other organs. The second oxygen supply structure appears to be an arteriolar loop which originates at the articular surface and loops back to the articular surface at the level of the interface between the resting-proliferative region and the hypertrophic zone. The proportion of these loop structures to the classical triad structure is 1:2.1 in the resting-proliferative region. This construction is not commonly seen in the vascular structure of organs, but when present, as in the kidney, may be employed to maintain a tissue concentration gradient. The rationale for a gradient of oxygen in the cartilage is somewhat compelling as recent research suggests that cartilage aging may be related to oxygen radical production (Michel et al., 1992). The third oxygen supply system originates at the bone marrow interface of the cartilage plate and is composed of sinusoidal structures supplied by the bone sinusoids. The ratio of arterioles to sinusoids is 1:4. We have estimated that half of these sinusoids are supplied by the bone marrow surface alone, without connection to the articular supply. This enormous number of vessels at near venous oxygen tensons may necessitate recalculation of the oxygen concentrations present near hypertrophic chondrocytes. Our original model estimated vessel oxygen partial pressure of 20 mm Hg as was found by microelectrode puncture (Brighton et al., 1971). If two different oxygen partial pressures are present, then modeling oxygen flux between them will be more complex than originally imagined. The result of the triple oxygen supply system of the avian growth cartilage appears to be that decreases in only one part of the system will not cause a precipitous drop in the tissue oxygen level.

REFERENCES

Brighton, C., Heppenstal, R., 1971, Oxygen tension in zones of the epiphyseal plate, the metaphysis and diaphysis, J. Bone Jt. Surg. 53:719.

Davis, T.R.C., Holloway, I., and Pooley, J., 1990, The effect of anaesthesia on the bone blood flow of the rabbit, J. Ortho. Res. 8:479

Haselgrove, J.C., Shapiro, I.M., and Silverton, S.F., 1993, Computer modeling of the oxygen supply and demand of cells of the avian growth cartilage, Amer. J. Phys. 265:C497

Howell, D.S., Pita, J.C., Marquez, J.F., Gatter, R.A., 1969, Demonstration of macromolecular inhibitors of calcification and nucleational factors in fluid from calcifying sites in cartilage, J. Clin. Invest. 48:630

Michel, C., Vincent, F., Duval, C., Poelman, M.C., 1992, Toxic effects and detection of oxygen free radicals on cultured articular chondrocytes generated by menadione, Free Radic. Res. Commun. 17:279.

Silverton, S.F., Wagerle, L.C., Robiolo, M.E., Haselgrove, J.C., and Forster, R.E. II, 1989, Oxygen gradients in two regions of the epiphyseal growth plate, in "Oxygen Transport to Tissue XI," K. Rakusan, G.P. Biro, T.K. Goldstick and Z. Turek, eds. Plenum Publishing, NY, NY.

Silverton, S.F., Matsumoto, H., DeBolt, K., Reginato, A. & Shapiro, I.M., 1989a, Pentose phosphate shunt mechanism by cells of the chick growth cartilage, Bone 10:45.

Silverton, S.F., Pacifici, M., Haselgrove, J.C., Colodny, S.H., Forster, R.E. II, 1990, Two-dimensional model of tissue oxygen gradients in avian growth cartilage, in "Oxygen Transport to Tissue XII," J. Piiper, T.K. Goldstick, M. Meyer, eds. Plenum Publishing, NY, NY.

Ye, G.-F., Moore, T.W., & Jaron, D., 1993, Contributions of oxygen dissociation and convection to the behavior of a compartmental model of oxygen transport model, Microvas. Res. 46:1

A PROGRAM TO CALCULATE MIXED VENOUS OXYGEN TENSION - A GUIDE TO TRANSFUSION?

N. Simon Faithfull, Glenn E. Rhoades, Peter E. Keipert, Andrew S. Ringle* and Ad Trouwborst**

Alliance Pharmaceutical Corp. and *Custom Micro Design, San Diego, CA, USA and **Dept of Anaesthesiology, Amsterdam Medical Center, The Netherlands

INTRODUCTION

Though it is generally accepted that venous blood oxygen tension (PO_2) reflects (but does not measure) PO_2 of the tissue from which it is issuing, it is generally impractical, except under unusual circumstances, to monitor PO_2 in venous blood draining from individual tissues or organs. Hence, the mixed venous PO_2 (PvO_2) is usually taken as an acceptable estimator of the oxygen delivery/consumption ratio in the whole body and is used as a guide to the oxygenation status of the whole body. It would be logical therefore to use PvO_2 as an indication for the need for blood transfusion during surgical procedures and in the trauma situation.

A computer program has been developed to predict PvO_2 under a variety of clinical and experimental conditions. Inputs are grouped into those determining the position of the oxyhemoglobin dissociation curve, such as PO_2, pH, carbon dioxide tension (PCO_2) and temperature, and those determining oxygen transport and delivery, such as cardiac output (CO) and hemoglobin (Hb) concentration. If Hb concentration, arterial and mixed venous blood gas and acid/base parameters are entered, the program will output O_2 delivery and consumption estimates for both red cell contained Hb and for the plasma phase. Alternately, if measured or assumed VO_2 is input, the program can estimate PvO_2.

The program can be used to simulate bleeding during surgery, in which case a bleeding rate must be entered together with the expected cardiac output response to hemodilution; the program assumes that normovolemia is being maintained by volume replenishment. Also calculated is the mass of Hb lost after various volumes of blood loss (or it calculates Hb available for autologous transfusion if preoperative hemodilution is being practiced). By inputting characteristics of plasma phase oxygen carriers, such as Hb solutions or perfluorochemical emulsions, their influence on permitted blood loss and relationship to PvO_2 and Hb concentrations can be determined; this allows calculation of blood loss that may be permitted before transfusion becomes necessary. With this information, techniques can be designed for maximizing the benefit of oxygen carriers in surgical autologous blood strategies. This program has been partially validated using

Oxygen Transport to Tissue XVI
Edited by M.C. Hogan *et al.*, Plenum Press, New York, 1994

retrospective and prospective animal and human data. Additionally animal validation is available using Oxygent™, a concentrated perfluorochemical emulsion containing 90% w/v of perflubron.

THE TRANSFUSION TRIGGER

During the perioperative period, blood transfusions are routinely administered as a "critical" hemoglobin (Hb) concentration or hematocrit is reached. This level has traditionally been at a Hb concentration of 10 g/dL. It was previously thought that allogeneic blood transfusion was a valuable and worthwhile treatment. However, the American College of Physicians has recently published a paper entitled "Practice Strategies for Elective Red Cell Transfusions" in which a physician contemplating giving transfusions is urged to discuss risks and benefits with the patient, anticipate the need for autologous blood and "regard elective transfusion with homologous blood as an outcome to be avoided"[1]. As a result much discussion has taken place as to the lowest acceptable Hb level and the level of a suitable transfusion trigger. Before such a question is answered, it is necessary to first consider the changes that take place during hemodilution as blood is removed and normovolemia is maintained.

As a patient is hemodiluted, either intentionally as part of an autologous blood conservation program, or following surgical bleeding with maintenance of normovolemia, both Hb concentration and arterial O_2 content (CaO_2) decrease. As the red cell concentration falls, a reduction in whole blood viscosity occurs; this, together with the simultaneously occurring increase in venous return, causes a rise in cardiac output (CO) and an improvement in total O_2 transport to the tissues (DO_2). The degree to which this physiological compensation occurs will primarily depend on the response of CO to the reduction in red cell mass. Some authorities have concluded that the relationship between decrease in Hb concentration and CO is linear[2,3] whereas others have maintained that it follows a curvilinear relationship[4]; the degree of curvature found is very minimal, causing many researchers to perform calculations that assume a linear relationship[5].

In man, the extent to which cardiac output increases as Hb concentration decreases varies between 0.25 liters per minute per gm of Hb change[6] to 0.70 L/min/g 1980 [7]). Hence the cardiac output response to hemodilution differs between patients and this will effect the Hb level at which additional oxygen carrying capacity in the blood will be needed. The necessity for transfusion of red blood cells will also vary depending on such factors as vascular tone, which will cause the viscosity contribution to total systemic resistance to vary, and the ability of the myocardium to function at low Hb levels. During moderate hemodilution, myocardial blood flow increases proportionately more than total cardiac output[8,9] and hence, in the absence of significant coronary atherosclerosis, no myocardial ischemia occurs. It has been shown, however, that low postoperative hematocrit (Hct) may be associated with postoperative ischemia in patients with generalized atherosclerosis[10]. Though a number of review articles have attempted to define a critical Hct level[3,4,11,12], most authorities would agree that an empiric automatic transfusion trigger should be avoided and that red cell transfusions should be tailored to the individual patient and be triggered by his or her own response to anemia - indeed, for patients under anesthesia it is recommended that "in the absence of risks, transfusion is not indicated, independent of hemoglobin level"[1].

As arterial blood passes through the tissues, a partial pressure gradient exists between the PO_2 of the blood in the arteriole entering the tissue and the tissue itself. Oxygen is, therefore, released from hemoglobin in the red cells and also from solution in the plasma; the O_2 then diffuses into the tissue. The PO_2 of the blood issuing from the venous end of the capillary cylinder will be a reflection of, but not necessarily equal to, the PO_2 at the distal (venous) end of the tissue through which the capillary passes. Under normal conditions this is essentially the same as that of interstitial fluid in contact with the outside of the capillary[13]. The degree of equilibration between blood and tissue may depend on the speed of passage of blood through the capillary bed and it has been argued that, under conditions of critical oxygen delivery caused by extreme anemia, there may not be time for equilibration of tissue and blood PO_2s[14]; this may lead to higher than expected mixed venous PO_2 (PvO_2)[15]. Nevertheless, in the clinical situation, it is generally accepted that probably "the most reliable single physiological indicator for monitoring the overall balance between oxygen supply and demand is mixed venous oxygen tension"[16]. It might therefore be sensible to use PvO_2 as an indication of the overall adequacy of tissue oxygenation and to use it as a transfusion trigger rather than to use the traditional "10/30 rule" as an indication for red blood cell transfusion.

If PvO_2 is accepted as a reasonable indicator of patient safety, the question arises to what can be considered a "safe" level of this parameter. Though much data exists on critical oxygen delivery levels in animals, there is little to indicate what a critical PvO_2 might be in the clinical situation. The available date indicates that the level is extremely variable. For instance, in patients about to undergo cardiopulmonary bypass, critical PvO_2 varied between about 30 mm Hg and 45 mm Hg[17]; the latter value is well within the range of values found in normal, fit patients. Furthermore, shunting of blood in the tissues will cause elevated levels of PvO_2, such as is found in patients in septic shock, and will result in O_2 supply dependency[18].

A PvO_2 value of 35 mm Hg or more may be considered to indicate that overall tissue oxygen supply is adequate[16], but it must be stressed that this is implicit on the assumption of an intact and functioning vasomotor system. This PvO_2 level is reached at a Hb of about 4 g/dL in patients with good cardiopulmonary function; even lower PvO_2 levels are tolerated in some patients when increased fractional inspired O_2 concentrations (FiO_2s) are employed. In the surgical situation it is necessary to maintain a good margin of safety and it is probably best to pick a PvO_2 transfusion trigger at which the patient is obviously in good condition as far as oxygen dynamics are concerned. In practice, only certain patients will be monitored with a pulmonary artery catheter; thus, PvO_2 will not be available for all patients, leaving the majority to be monitored with the imperfect trigger of Hb concentration.

COMPUTER MODELLING

A computer model has been developed to estimate PvO_2 for a number of input scenarios. The program is thus able to predict Hb levels at which red cell transfusion should occur in any given patient. The program will also run calculations of efficacy of perfluorochemical (PFC) or Hb based O_2 transporting blood substitutes. In a very large Microsoft® Excel spreadsheet, the individual columns represent incremental stages in a progressive hemodilution procedure. A simulated bleeding rate can be set and also a "hemodilution aliquot". For instance if the initial state of a patient is represented by column 1 and the hemodilution aliquot is set to 50 mL, Column 2 will represent the condition of the patient after 50 mL of blood has been removed and replaced with a sufficient amount of a plasma substitute to maintain normovolemia. This process can be repeated as many times as desired.

Prior to starting input of cardiovascular and oxygenation variables, a number of constants are entered such as blood volume, oxygen solubility in plasma and the oxygen content of 1 g of saturated oxyhemoglobin. If the effects of administering a PFC emulsion are to be modelled, oxygen solubility and specific gravity of that PFC are needed, together with its concentration in the emulsion and its circulatory half life at the dose to be administered. Input variables concerned with calculation of CaO_2 include Hb concentration, arterial tension of oxygen (PaO_2) and carbon dioxide ($PaCO_2$), arterial pH (pHa) and body temperature. The position of the oxyhemoglobin dissociation curve is calculated using the Kelman equations[19], which produce a curve that, over the physiological range of O_2 tensions, is indistinguishable from the parent curve proposed by Severinghaus[20]. Inputs for oxygen delivery include cardiac output and it's response to hemodilution in terms of increase in cardiac output for each g/dL reduction in Hb concentration. Arterial and mixed venous saturations (SaO_2 and SvO_2) are calculated from input or assumed blood gas values; whole body oxygen consumption (VO_2) can either be calculated from the program or can be entered as an input. If VO_2 is input, a Solver routine can be used to calculate a PvO_2 that results in the required mixed venous oxygen contents in Hb, plasma and PFC to satisfy the Fick equation[21].

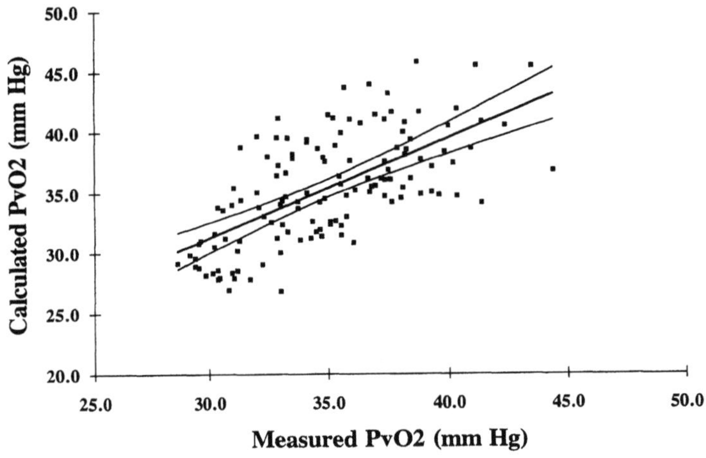

Figure 1. Regression line and 95 percent confidence limits for measured PvO_2 versus PvO_2 calculated by the computer program for 10 patients undergoing normovolemic hemodilution procedures. Data are for both intraoperative and postoperative periods. r = 0.64.

The primary and most important output of the program is PvO_2, which may then be used as a guide to subsequent transfusion practice. Additionally the program will calculate the proportion of total oxygen delivery (DO_2) or VO_2 contributed by O_2 carried in Hb and plasma phases. If an acellular oxygen transporting fluid such as a Hb preparation or a PFC emulsion is present in the plasma phase, it's contribution to DO_2 and VO_2 can also be calculated. The amount of blood that must be removed in order to hemodilute a patient to a certain Hb concentration can be calculated and the amount of Hb present in individual blood "units" removed during the hemodilution procedure are also available.

The program has been validated against a number of studies, both experimental and clinical. Figure 1 is derived from data from 10 patients undergoing surgery under normovolemic hemodilution procedures. The correlation coefficient between measured and calculated PvO_2 was 0.64 - 95 percent confidence limits are shown. Though often presented in scientific literature, the correlation coefficient is unable to indicate the numerical accuracy of comparison of two parameters. Figure 2 shows the difference between measured and calculated PvO_2 plotted against measured PvO_2. The mean calculated value is 0.4 mm Hg below the measured value; 95% confidence limits lie 6.3 mm Hg above and 7.1 mm Hg below the true value at the critical point for PvO_2 of 35 mm Hg.

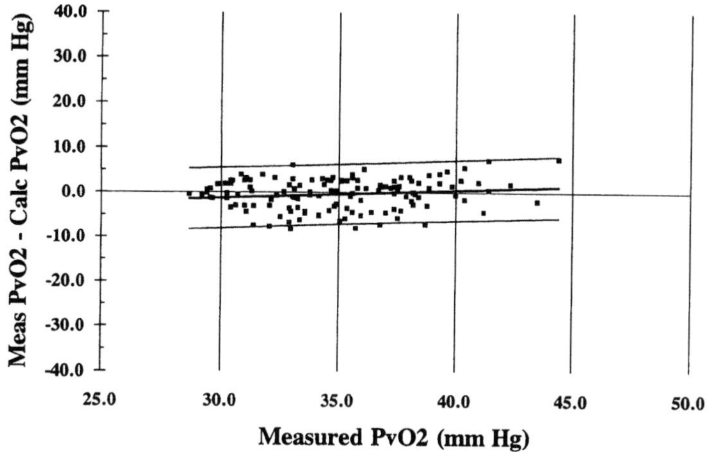

Figure 2. Measured PvO_2 plotted against difference getween measured and calculated PvO_2 for 10 patients undergoing normovolemic hemodilution procedures. Data is for both intraoperative and postoperative periods. 95 percent confidence limits are showm.

Postoperatively, particularly after major surgery, many patients are in a physiologically unstable state and rapid fluctuations may occur in cardiovascular and respiratory parameters. It is to be expected therefore that the PvO_2 prediction program might be more accurate when the patient is in a more stable condition under anesthesia; this is indeed so as shown in Figure 3, which is constructed only from data obtained during the intraoperative period. The correlation coefficient is 0.81 and the mean calculated PvO_2 is 1.1 mm Hg above the measured value (the direction of this error is in the direction of safety, and a physician relying on such a calculated value would tend to intervene to raise PvO_2 earlier than necessary). The spread of error is much less with 95 percent confidence limits lying 5.3 mm Hg above and 2.8 mm Hg below true values at 35 mm Hg (Figure 4).

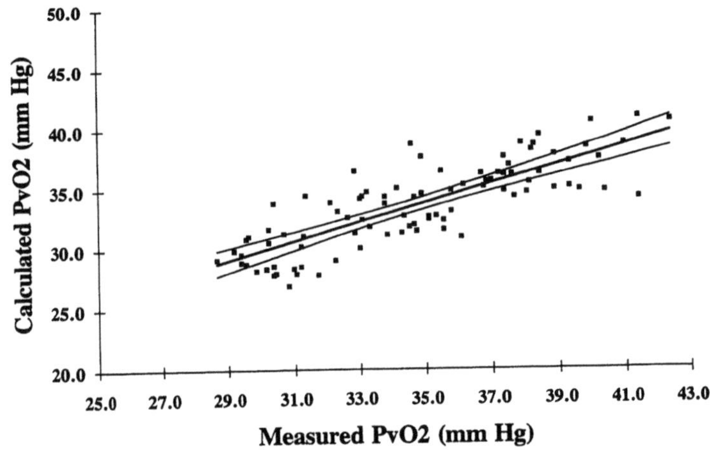

Figure 3. Regression line and 95 percent confidence limits for measured PvO$_2$ calculated by the computer program for 10 patients undergoing normovolemic hemodilution procedures. Data is for only the intraoperative period. r = 0.81.

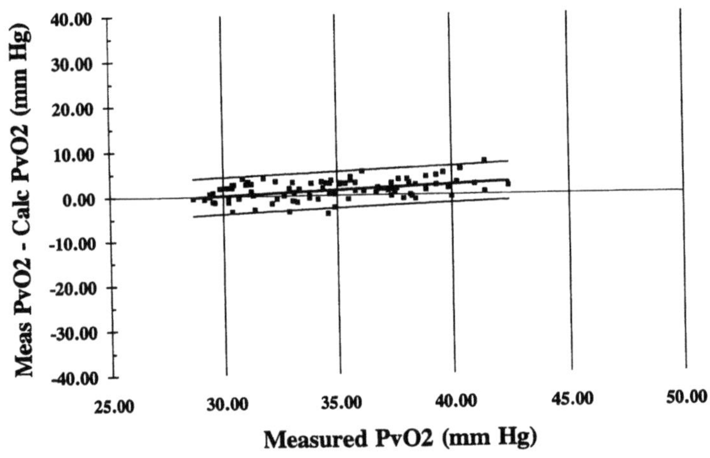

Figure 4. Measired PvO$_2$ plotted against difference between measured and calculated PvO$_2$ for 10 patients undergoing normovolemic hemodilution procedures. Data is for only the intraoperative period. 95 percent confidence limits are shown.

Physiological and clinical studies involving measurement and calculation of oxygenation parameters are usually carried out using cardiac output measurements obtained by thermodilution using a Swan-Ganz catheter. Oxygen delivery and oxygen consumption (VO_2) are then derived from measured or calculated arterial and mixed venous oxygen contents by using the Fick equation. The program described in this paper also uses the Fick principle to calculate PvO_2. If VO_2, which is an important input for the program, is also derived from the Fick principle, it can be argued that mathematical coupling will be occurring and this may lead to a propensity for the program to calculate the "correct" PvO_2. This criticism can be overcome if VO_2 is obtained by direct calculation of oxygen uptake by measurement of inspired and expired oxygen contents.

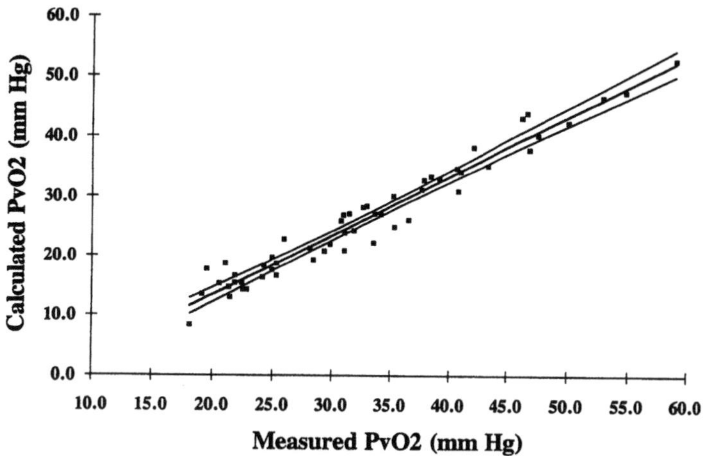

Figure 5. Regression line and 95 percent confidence limits for calculated PvO_2 versus measured PvO_2 for for dogs undergoing progressive hemorrhagic shock. Oxygen consumption was calculated by direct gas measurement. $r = 0.98$.

Figure 5 is a graph of calculated PvO_2 versus measured PvO_2 for a group of dogs involved in a study of critical oxygen extraction coefficients following progressive hemorrhage [22]. In this case, correlation coefficient is 0.98. It should be noted that the regression line is some 5 to 7 mm Hg above the line of identity an this case. This is due to the fact that the Kelman constants used in the calculations are based on human blood with a P50 different from that of dogs. The good correlation obtained would indicate that, contrary to expectations, direct measurement of VO_2 gives better results. This may well be due to the fact that the values are usually averaged over a finite period of time as opposed to being merely a "snapshot" of VO_2. Further validation using clinical data in which VO_2 is derived from gas analysis is indicated.

REFERENCES

1.   American College of Physicians, Practice Strategies for elective Red Blood Cell Transfusion, <u>Annals of Internal Medicine</u> 116.5 (1992): 403-406.

2.   R. C. Fan, R. Y. Z. Chen, G. B. Schuessler, and S. Chien, Effects of hematocrit variations on regional hemodynamics and oxygen transport in the dog, <u>Am J Physiol</u> (1980): H545-H552.

3.   P. G. Robertie, and G. P. Gravlee, Safe Limits of Isolvolemic Hemodilution and Recommendtions for Erythrocyte Transfusion, <u>International Anesthesiology Clinics</u> 28.4 (1990): 197-204.

4.   P. Lundsgaard-Hansen, Hemodilution - New Clothes for an Anemic Emperor, <u>Vox. Sang.</u> 36 (1979): 321-336.

5.   H. Hint, The pharmacology of dextran and the physiological background for the clinical use of Rheomacrodex and Macrodex, <u>Acta Anaesthesiologica Belgica</u> 2 (1968): 119-138.

6.   H. Laks, R. N. Pilon, P. Klovekorn, W. Anderson, J. R. MacCallum, and N. E. O'Connor, Acute Hemodilution: Its effect on hemodynamics and oxygen transport in anesthetized man, <u>Ann Surg</u> 180.1 (1974): 103-109.

7.   D. M. Shah, M. N. Prichard, J. C. Newell, A. M. Karmody, W. A. Scovill, and S. R. Powers, Increased cardiac output and oxygen transport after intraoperative isovolemic hemodilution. A study in patients with peripheral vascular disease., <u>Arch Surg</u> 115 (1980): 597-600.

8.   K. Messmer, L. Sunder-Plassman, F. Jesch, L. Fornandt, E. Sinagowitz, M. Kessler, R. Pfeiffer, E. Horn, J. Hoper, and K. Joachimsmeier, Oxygen Supply to the Tissues during Limited Normovolemic Hemodilution, <u>Res. Exp. Med.</u> 159 (1973): 152-166.

9.   A. S. Geha, Coronary and cardiovascular dynamics and oxygen availability during acute normovolemic anemia, <u>Surgery</u> 80.1 (1976): 47-53.

10.   R. Christopherson, S. Frank, E. Norris, P. Rock, S. Gottlieb, and C. Beattie, (Abstract) Low Postoperative Hematocrit is Associated with Cardiac Ischemia in High-Risk Patients, <u>Anesthesiology</u> 75.3A (1991): A99.

11.   W. Dick, C. Baur, and K. Reiff, Welche Faktoren bestimmen den Kritischen Hamatokrit bei der Indikationsstellung zur Transfusion?, <u>Anaesthesist</u> 41 (1992): 1-14.

12.   H. R. Abel, T. B. Bradley Jr, and H. M. Ranney, "Pathophysiology of the hemoglobinopathies," <u>Clinical Obstetrics and Gynecology</u>, Ed. W. L. Freedman Hoeber Medical Division, Harper and Row, 1969) 15-48.

13.   A. C. Guyton, "Diffusion of oxygen from the capillaries to the interstitial fluid," <u>Textbook of Medical Physiology</u>, Sixth ed. W. B. Saunders Company, 1981) 506.

14.   G. Gutierrez, and J. M. Andry, Increased hemoglobin O_2 affinity does not improve O_2 consumption in hypoxemia, <u>Journal of Applied Phusiology1</u> 66.2 (1989): 837-843.

15.   S. M. Cain, Oxygen delivery and uptake in dogs during anemic and hypoxic hypoxia, (1977): 228-234.

16.   J. V. Snyder, and M. R. Pinsky, <u>Oxygen Transport in the Critically Ill</u>, (Chicago, London: Year Book Medical Publishers, Inc., 1987) 554.

17.   K. Shibutani, T. Komatsu, K. Kubal, V. Sanchala, V. Kumar, and D. V. Bizzarri, Critical level of oxygen delivery in anesthetized man, <u>Crit Care Med</u> 11.8 (1983): 640-643.

18.   Z. Mohsenifar, P. Goldbach, D. P. Tashkin, and D. J. Campisi, Relationship between O_2 Delivery and O_2 Consumption in the Adult Respiratory Distress Syndrome, <u>CHEST</u> 84.3 (1983): 267-271.

19.   G. R. Kelman, Digital computer subroutine for the conversion of oxygen tension into saturation, <u>J Appl Physiol</u> 21.4 (1966): 1375-1376.

20.   J. W. Severinghaus, Blood gas calculator, <u>J. Applied Physiology</u> 21 (1966): 1108-1116.

21.   A. Fick, Ueber die Messung des Blutquantums in den Hertzventrikelen, <u>Würzburg, Physikalisch edizinische Gesellschaft</u> Sitzungsbericht 16 (1870):

22.   N. S. Faithfull, and S. M. Cain, Critical levels of O_2 extraction following hemodilution with dextran or Fluosol-DA, <u>J Crit Care</u> 3 (1988): 14-18.

MICROCIRCULATION AND O_2 EXCHANGE THROUGH THE SKIN SURFACE : A THEORETICAL ANALYSIS

D.W. Lübbers

Max-Planck-Institute for Molecular Physiology
44026 Dortmund, Germany

Introduction

Oxygen is supplied to the upper layers of the human skin not only by blood, but also by surrounding air. Already in 1851 Gerlach measured this O_2 uptake by glueing a horse bladder on the human skin (Gerlach (1851)). The bladder was made gas-tight by varnishing. He found that during a period of 24 hours the O_2 concentration in the bladder decreased from 21.0 % to 19.02 %, whereas at the same time the CO_2 concentration increased from 0 % to 2.5 %. He followed from his experiments that "the cutaneous respiration (i.e. the O_2 uptake from the surrounding air) depends on the amount of blood which perfuses the uppermost capillaries and on its flow velocity. All that increases the amount of blood within the skin increases the cutaneous respiration." To analyse the O_2 supply of the different layers of the skin pO_2 profiles perpendicularly to the skin surface have been measured (Baumgärtl et al. (1987). They reveal that there is a competition between the O_2 supply by blood and that by surrounding air. Starting with the pO_2 of the surrounding air tissue pO_2 first decreases, reaches a minimum and then increases. This demonstrates that the upper part of the skin up to the pO_2 minimum is supplied by the O_2 of the air, i.e. by the O_2 flux through the epidermis, whereas the other parts receive their O_2 from the blood. The oxygen uptake from the air amounts to 80-100 ml $O_2/(m^2 \cdot h)$, i.e. a human being with a skin surface of 1.5 m^2 has an O_2 uptake of 2.0-2.5 ml O_2/min or of about 1 % of its resting O_2 uptake (Fitzgerald (1957). For the total organism it is a small amount , but it can be important for the oxygen supply of the skin. The magnitude of the O_2 flux is influenced by several parameters, mainly by the anatomical structure and the diffusion properties of the skin, by the arterial pO_2, by the blood flow and by the pO_2 at the skin surface. Therefore, for a complete analysis of the oxygen supply of the skin the supply by the blood as well as by the oxygen flux through the skin surface has to be known. We could demonstrate that it is possible to construct a sensor by which the O_2 flux into the skin surface can be monitored. The O_2 flux sensor applies optical O_2 sensors which measure the pO_2 difference across a diffusion test membrane without disturbing the O_2 flux (Lübbers (1992); Holst et al. (1993)).

Oxygen Transport to Tissue XVI
Edited by M.C. Hogan *et al.*, Plenum Press, New York, 1994

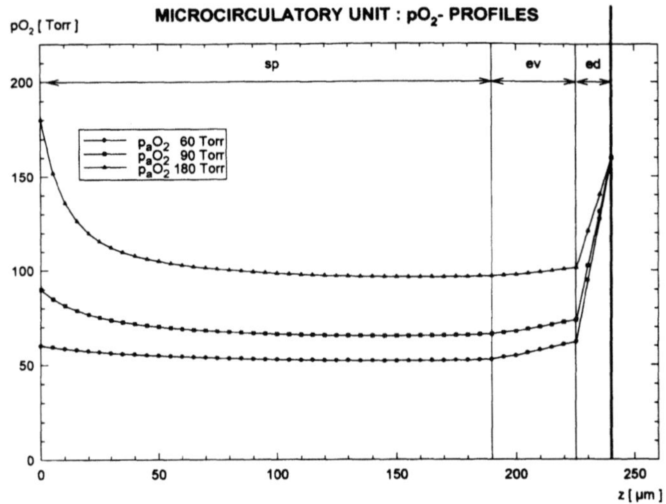

Fig. 1. pO_2 profiles within the microcirculatory unit at different blood flow values ("dry" skin). K (ed)= $4.8 \cdot 10^{-6}$ ml O_2/(cm min·atm), arterial pO_2 (z = 0 µm) = 90 Torr, environmental pO_2 = 160 Torr (air), blood flow (BF) = 1, 10 and 20 ml/(100g·min), skin surface: z= 240 µm. ΔpO_2 (i.e. the O_2 flux into the skin) decreases with increasing blood flow.

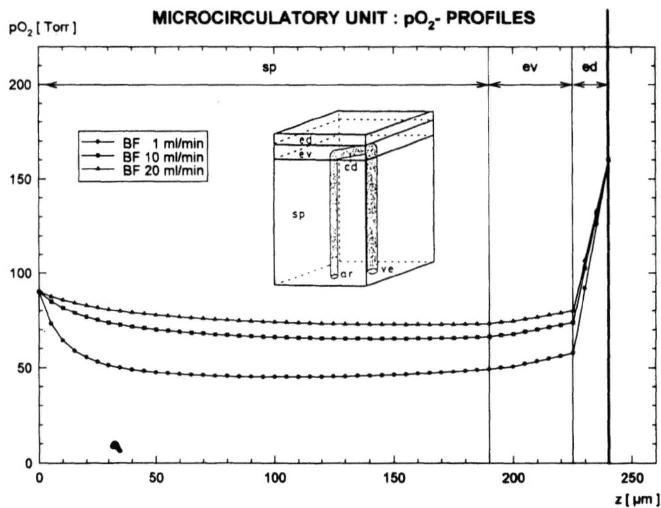

Fig. 2. pO_2 profiles within the microcirculatory unit at different values of arterial pO_2 ("dry" skin). K (ed) as in Fig.1, arterial pO_2 (p_aO_2) = 60 Torr, 90 Torr and 180 Torr, environmental pO_2 = 160 Torr (air), blood flow BF = 10 ml/(100g·min). ΔpO_2 (i.e. the O_2 flux into the skin) decreases with increasing p_aO_2

To analyse the importance and the influence of the different parameters on the oxygen supply of the skin, the O_2 exchange processes of the skin have been theoretically analysed using our capillary loop model by which the oxygen supply of the upper layers of the skin can be simulated (Lübbers and Grossmann (1983)).

Method

Because of the relative constant anatomical structure of epidermis and corium it is assumed that it can be simulated by an assembly of several similar microcirculatory units, mu. Fig. 1 (insert) shows such a microcirculatory unit. It consists out of three layers: the uppermost layer corresponds to the str. corneum (ed = dead layer of the epidermis). The next layer is the viable part of the epidermis (str. basale, str. spinosum and str. granulosum; ev = viable layer) and the third layer contains the capillary loop surrounded by living tissue (sp = str. papillare). Two additional layers can be added to simulate the influence of a sensor. The size of the mu corresponds to mean values taken from literature, e.g. with 51 capillaries/mm^2 a side length of 140 µm is obtained. The mean length of the capillary loop is 190 µm. The mean thickness of the layers are: ed: 15 µm, ev: 25 µm, sp: 200 µm. The model allows to describe the exchange of O_2 and CO_2 in 7 different layers with different O_2 conductivities, K, and O_2 uptakes at different temperatures. Blood is simulated by a hemoglobin-plasma solution. The O_2 dissociation curve can be adapted to different P_{50}, pCO_2 and pH values. The CO_2 dissociation curve is approximated by a straight line. The simulations are calculated for a P_{50} of 26.7 Torr, a pCO_2 of 40 Torr, a pH of 7.4, an O_2 consumption of 0.3 ml O_2/(100g·min) and a temperature of 37° C. Constants given for 37°C are calculated for the actual temperature. The model allows to calculate the distribution of the pO_2 (and the pCO_2) within the mu and the resulting parameters as the local O_2 fluxes under steady state conditions as well as during changes in time. Local O_2 fluxes are used to calculate the balance between O_2 supply and O_2 consumption by which the exactitude of the model calculations can be controlled.

Results

Fig. 1 shows simulated pO_2 profiles within a microcirculatory unit, mu, calculated for different values of blood flow. The pO_2 profiles show the pO_2 changes within the arterial limb towards the skin surface (z = 240 µm). The pO_2 decreases from the arterial inflow (p_aO_2 = 90 Torr, z = 0 µm)), reaches a minimum and then increases to 160 Torr (air). The O_2 flux into the skin, J_s, can be calculated from the ΔpO_2 across the dead layer by the equation $J_s = C \cdot \Delta pO_2$; C is determined by the oxygen conductivity K and the thickness d of the layer ed: $C = K/d$. The lowest pO_2 trace corresponds to a blood flow of BF = 1 ml/(100g·min). With increasing blood flow the ΔpO_2 across the dead layer decreases and consequently the O_2 flux into the skin. At a blood flow of BF = 1 ml/(100g·min) J_s, calculated for the surface of the mu, AMU, amounts to $84.4 \cdot 10^{-10}$ ml O_2/(AMU·min); at BF = 20 ml/(100g·min) J_s distinctly decreases to $66.9 \cdot 10^{-10}$ ml O_2/(AMU·min). The O_2 flux into the skin is also influenced if arterial pO_2 changes: with increasing arterial pO_2 the O_2 flux decreases (Fig. 2).

The large pO_2 decrease in the dead layer shows that its O_2 conductivity influences very much the form of the pO_2 profiles. Therefore, pO_2 profiles using a 10 times increased O_2 conductivity were simulated ("humid" skin). An increase of the O_2 conductivity can be brought about by humidifying the skin surface. Fig. 2 and Fig. 3 are similar simulations, but

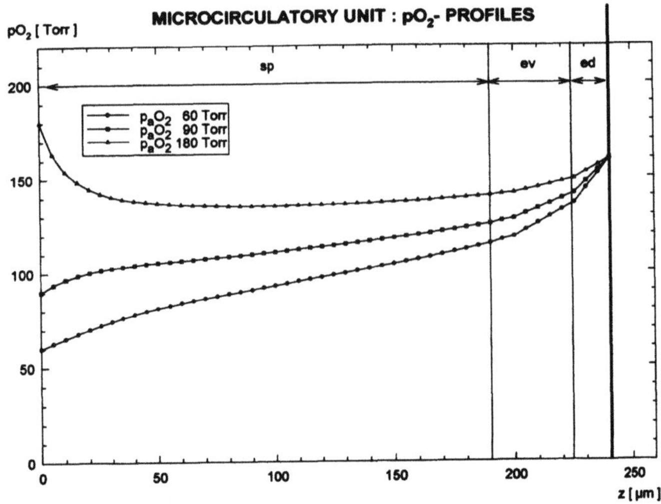

Fig. 3. pO_2 profiles within the microcirculatory unit at different values of arterial pO_2 ("humid" skin). K (ed)= $0.48 \cdot 10^{-6}$ ml O_2/(cm·min·atm), arterial pO_2 (p_aO_2) = 60 Torr, 90 Torr and 180 Torr, environmental pO_2 = 160 Torr (air), blood flow = 10 ml/(100g·min). At p_aO_2 values of 90 and 60 Torrhe O_2 flux into the skin becomes larger than the O_2 consumption of the mu; the pO_2 minimum is consequently shifted to the arterial inflow.

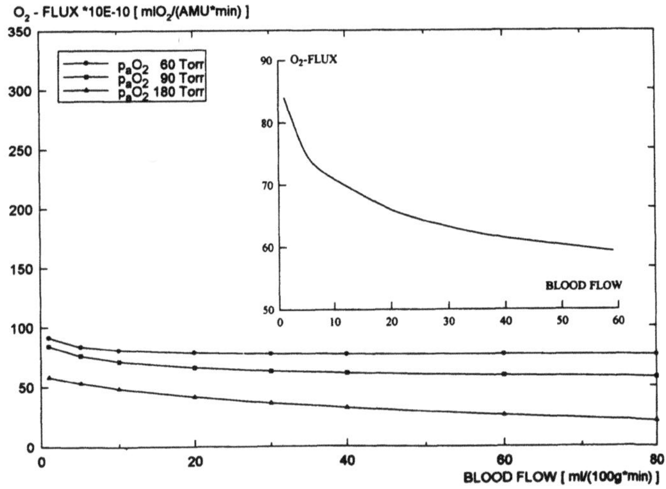

Fig. 4. O_2 flux into the surface of "dry" skin in dependence of blood flow at different arterial pO_2 values. Environmental pO_2 = 160 Torr (air). AMU: area of the surface of the mu. Insert: O_2 flux vs .blood flow (differen scale). p_aO_2 = 90 Torr, p_eO_2 = 160 Torr.

the profiles of Fig. 2, however, the minimum of the other 2 profiles is at the arterial inflow. This demonstrates that under these conditions in "humid" skin arterial blood does not anymore deliver O_2 to the mu, but that the O_2 flux through the skin surface can provide all the oxygen the skin needs and even more.

The behaviour of the O_2 fluxes at "dry" and "humid" skin being in contact with air is shown in Fig. 4 and Fig. 5 the coordinates of which have the same scale. If the O_2 flux through the skin surface is smaller than the oxygen uptake of the microciculatory unit, R_{mu}, the O_2 flux decreases with increasing blood flow; this holds for all the three p_aO_2 values in Fig. 4 and for the p_aO_2 of 180 Torr in Fig. 5. However, at the p_aO_2 values of 60 and 90 Torr in Fig. 5 the O_2 flux becomes larger than R_{mu} and therefore the O_2 flux increases with increasing blood flow. In "humid" skin the effect of blood flow changes is larger than in "dry" skin. The insert of Fig 4 demonstrates that at small blood flow values small changes cause relatively large changes in the O_2 flux. In this range the changes of the O_2 fluxes mirror the changes of blood flow if the other parameter remain constant.

In Table 1 the O_2 flux through the skin surface is compared to the amount of.oxygen wich is needed for the O_2 consumption of the mu; R_{mu} = 130.65·10^{-10} ml O_2/min (= 0.3 ml O_2/(100g·min) is taken as 100%. For example, at a blood flow of 1 ml/(100g·min), a p_aO_2 of 60 Torr and a p_eO_2 of 160 Torr the O_2 uptake through the skin amounts to 124.8% of the O_2 consumption of the mu in"humid", but only to 69.7% in "dry" skin. Table 1 demonstrates the important influence of the O_2 conductivity of the epidermis on the O_2 supply of the skin.

O_2 flux ,J_s,into the surface of the microcirculatory unit of the skin

Table 1. BF = blood flow, "humid" skin: K (ed)= 0.48·10^{-6} ml O_2/(cm·min·atm), "dry" skin: K (ed)= 4.8·10^{-6} ml O_2/(cm min·atm), environmental pO_2 = 160 Torr, 100 % O_2 flux corresponds to a J_s = 130.65·10^{-10} ml O_2/min = R_{mu}.

BF	p_aO_2					
ml/100g·min	60 Torr		90 Torr		180 Torr	
	humid	dry	humid	dry	humid	dry
1 ml	124.8 %	69.7 %	121.9 %	64.6 %	72.6 %	44.7 %
10 ml	149.0 %	61.7 %	116.4 %	54.4 %	68.4 %	37.1 %
20 ml	176.0 %	60.0 %	111.0 %	50.6 %	63.2 %	31.8 %

To measure the O_2 flux into the skin an O_2 flux optode with a diffusion test membrane of polypropylene of a thickness of 12.5μm is placed on the skin surface as shown in Fig. 6. The O_2 flux sensor measures the ΔpO_2 across the polypropylene membrane, i.e. the ΔpO_2 between skin surface pO_2, $p_{ss}O_2$, and environmental pO_2, p_eO_2. If the optode is covered by a gas-impermeable membrane the O_2 flux into the skin becomes zero. Since in this case oxygen is only supplied by blood the pO_2 decreases continuously towards the dead layer and there is no pO_2 difference across the sensor membrane. This pO_2 corresponds to the so called transcutaneous pO_2, $tcpO_2$, that would be measured by an ideal pO_2 electrode (Huch et al. (1981)). The same situation is obtained if the environmental pO_2 equals $tcpO_2$. By increasing p_eO_2 above $tcpO_2$ the ΔpO_2 across the sensor membrane increases showing an increased O_2 flux. Decreasing p_eO_2 below $tcpO_2$ results in an inverse pO_2 difference, i.e. an O_2 flux out of the skin into the surrounding atmosphere. Fig. 6 shows that by applying the O_2 flux sensor the skin surface pO_2 is changed, but that this change can be overcome by manipulating p_eO_2.

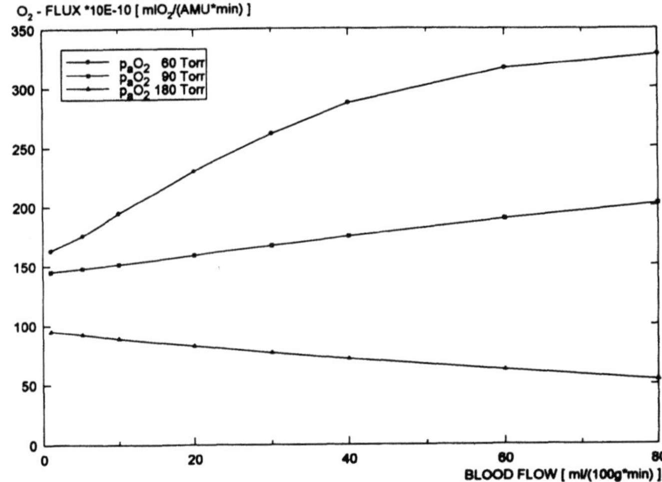

Fig. 5. O_2 flux into the surface of "humid" skin in dependence of blood flow at different arterial pO_2 values. Simulations using the same parameters as in Fig. 4, but for "humid" skin. At p_aO_2 values of 60 Torr and 90 Torr the O_2 flux is larger than R_{mu} and therefore increases with increasing blood flow. Under these conditions the O_2 flux increases the O_2 content of the venous blood.

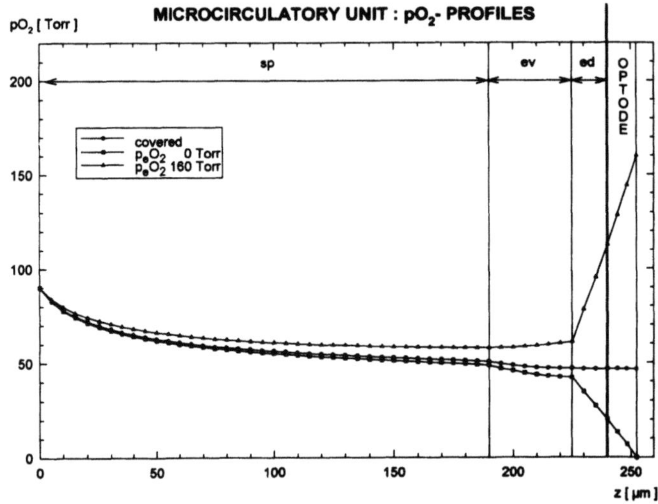

Fig. 6. pO_2 profiles within the microcirculatory unit covered by an optode at different values of the environmental pO_2 ("dry" skin).

Arterial pO_2 =90 Torr, blood flow = 10 ml/(100g·min). O_2 flux and environmental pO_2 (p_eO_2): at 160 Torr (air): O_2 flux J_s = 42.8·10^{-10} ml O_2/(AMU·min), covered (wit a gas-impermeable membrane), at 46.5 Torr: J_s = 0·10^{-10} ml O_2/(AMU·min), at 0 Torr: J_s =-17.4·10^{-10} ml O_2/(AMU·min).

Discussion

If the skin is covered by a gas-impermeable layer the pO_2 distribution within the microcirculatory unit of a given structure depends only on the O_2 supply by the blood and the O_2 demand of the tissue. The O_2 supply by blood is determined by the pO_2, the concentration and the oxygen saturation of hemoglobin as well as by the blood flow. From the arterial inflow the pO_2 constantly decreases along the capillary and reaches its minimum at the skin surface as seen in Fig. 6 (covered). If the skin surface is brought in contact with a pO_2 larger than this $p_{ss}O_2$, oxygen will diffuse through the skin surface into the deeper layers of skin and the minimum of tissue pO_2 will be situated between arterial inflow and skin surface. The O_2 flux into the skin produces a pO_2 gradient across the dead layer the steepness of which is proportional to the O_2 flux into the skin. The steepness of this pO_2 gradient also depends on the O_2 conductivity of the dead layer. Fig. 1 shows the influence of blood flow and Fig. 2 that of p_aO_2 on the ΔpO_2 across the dead layer for "dry" skin, whereas Fig 3 shows a simulation of changes of p_aO_2 for "humid" skin. The corresponding O_2 fluxes are drawn in Fig. 4 and Fig. 5. There are 2 types of the flow dependency of the O_2 flux into the skin: 1.) The O_2 flux is smaller than the O_2 consumption of the microcirculatory unit R_{mu}; then the O_2 flux decreases with increasing blood flow and 2.) The O_2 flux is larger than R_{mu}; then the O_2 flux inreases with increasing blood flow. Under these conditions a "cutaneous respiration" occurs as it was followed from his experiments by Gerlach (1851) since part of the O_2 is taken up by the venous blood. The effect on the O_2 supply of the skin and O_2 uptake by the venous blood is demonstrated in Table 1. For example, at a p_aO_2 of 90 Torr and a blood flow of 20ml/(100g·min) 50 % of O_2 uptake of the"dry" skin is supplied by the skin surface, whereas under the same conditions in "humid" skin the O_2 flux delivers 100 % of the O_2 needed for the R_{mu} and additionally 11% which is used for the oxygenation of venous blood. The analysis of the O_2 supply of the skin demonstrates that many factors determine the amount of oxygen which is taken up through the skin surface. If arterial pO_2, environmental pO_2 and O_2 conductivity of the skin remain constant the O_2 flux is determined by local blood flow. Under these conditions it should be possible to use the optical O_2 flux sensor to monitor the local microcirculation of the skin.

References

Baumgärtl, H., Ehrly, A.M., Saeger-Lorenz, K., and Lübbers, D.W., Initial results of intracutaneous measurements of pO_2 profiles, *in*: Clinical Oxygen Pressure Measurement, Ehrly, A.M., Hauss, J., Huch, R. eds., pp 121-128, Springer Verlag Berlin, Heidelberg, New York, London, Paris, Tokyo (1987).

Fitzgerald, L.R., Cutaneous respiration. *Physiol Rev* 37: 325-336 (1957).

Gerlach, Über das Hautathmen, *Arch. Anat.Physiol.*: 431-479, (1851)

Holst, G.A., Lübbers, D.W., and Voges, E., O_2-flux-optode für medical application, *SPIE Adv. Fluorescence Sensing Technology* 1885: 216-223 (1993).

Huch, R., Huch, A., and Lübbers, D.W., 1981, Transcutaneous pO_2, Georg Thieme Verlag, Stuttgart - New York

Lübbers, D.W. and Grossmann, U., 1983, Gas exchange through the human epidermis as a basis of $tcpO_2$ and $tcpCO_2$ measurements, *in*: Continuous transcutaneous blood gas monitoring, Marcel Dekker, Inc., New York.

Lübbers, D.W., 1992, Transcutaneous measurements of skin O_2 supply and blood gases, *in*: Oxygen transport to tissue XIII, T.K. Goldstick, ed, Plenum Press, New York.

Lübbers, D.W., Fluorescence based chemical sensors. *Adv Biosens* 2: 215-260 (1992).

FRACTAL APPROACHES TO THE MICROCIRCULATION

James B. Bassingthwaighte, Lori Young, Richard P. Beyer, and C. Y. Wang

Center for Bioengineering, University of Washington, Seattle WA 98195
and Department of Mathematics, Michigan State University, East Lansing, MI 48824

That there is marked heterogeneity of regional myocardial blood flows has long been recognized (Yipintsoi et al., *Circ. Res.* 33:573, 1973). The suspicion that much of the heterogeneity might have been an artifact of the microsphere deposition technique has been laid to rest by the studies of Baer et al. (*Am. J. Physiol.* 246:H418, 1984), King et al. (*Circ. Res.* 57:285, 1985), and by Bassingthwaighte et al. (*Circ. Res.* 66:1328, 1990) who showed that the technique had high reproducibility, with error much smaller than the variance in regional flows. The standard deviations of left ventricular myocardial blood flows in 100 mg pieces were 25 to 30% of the mean flow, with ranges from 1/5 of the mean to about twice the mean, while the microsphere variability is only about 8%.

Variability in regional flows is found in all organs. However the variation is NOT random, for there is considerable correlation between flows in neighboring regions. Fractals provide a measure of the variation AND the correlation, simultaneously. The key is self-similarity: the log of the apparent variation increases inversely with the log of the size of the tissue samples in which the flow is measured, meaning that each time the sample size is halved the standard deviation of the flows increases by a constant proportion. The basis is presumably in the branching structure of the vascular tree, with multiple subdivisions of the flow into the ramifying network.

Classic fractals are infinite recursions, repeating the branching forever. Classic fractal correlation gradually tapers off to zero, following this equation for the correlation coefficient:

$$r = 0.5 \left\{ |x + 1|^{2H} - 2x^{2H} + |x - 1|^{2H} \right\},$$

where x is the distance between observed regions of any unit size, independent of the size, and H is the Hurst coefficient, which equals $2 - D$, the fractal dimension, for one-dimensional data and equals $E + 1 - D$ for E-dimensional data. This works for the density of gold nuggets in the Witwatersrand gold field over a 200 mile range, and for flows in the heart and lung over several centimeters. But it does not hold over the full extent of an organ of finite size supplied by one artery. The observation is that is when near neighbors are correlated positively, distant regions are negatively correlated—if flow goes preferentially to one group of near neighbors, the corollary is that distant groups must be relatively deprived, and therefore negatively correlated with the high flow group. Consequently, while the above expression applies only to short distances, the complimentary function applies to near maximum distances within the organ, and to artificial flow networks:

$$r = -1 + 0.5 \left\{ |x + 1|^{2H} - 2x^{2H} + |x - 1|^{2H} \right\}, \quad \text{for} \quad x \quad \text{near} \quad x_{max}, \quad \text{and}$$

$$r = 0, \quad \text{for} \quad x \quad \text{near} \quad x = x_{max}/2.$$

This relationship is being explored in the heart and lung and in artificially constructed flow networks.

Other fractal measures in the heart concern the temporal variability of microvessel flows, the structures of the arterial network, the kinetics of some ionic channels, the spread of Purkinje fiber excitation, and the kinetics of oxygen binding to hemoglobin. In the lung fractals have been used to estimate alveolar surface areas, and branching tracheal and vessel diameter ratios. The fact that image compression ratios of 99% have been achieved by fractal algorithms may mean that image segmentation can be refined using fractal texture characterization. (Supported by NIH grants HL50238, RR1243, and HL19139.)

RECENT ADVANCES IN OXYGEN MEASUREMENTS USING PHOSPHORESCENCE QUENCHING

David F. Wilson and Sergei A. Vinogradov

Department of Biochemistry and Biophysics
School of Medicine
University of Pennsylvania
Philadelphia, PA 19104

PROLOG

Oxygen dependent quenching of phosphorescence quenching is an optical method for measuring oxygen pressure which was first reported in several years ago (see Vanderkooi and Wilson, 1986; Wilson and Vanderkooi, 1987). In the intervening years the method has steadily continued to improve both in the range of application and the precision of the measurements. The goal of the present communication is to review developments which have occurred within the past two years and which should be of significant value to investigators interested in measuring oxygen *in vitro* and *in vivo*. There are five major areas in which significant improvements and/or extensions of the method have occurred: 1. Algorithms have been developed which permit deconvolution of phosphorescence decay curves from probes in the vascular system of tissue into the distribution of decay constants (distribution of oxygen pressures) in the tissue; 2. New phosphorescent probes have been identified which are suitable for measuring oxygen and have both absorption and emission in the near infra-red part of the spectrum (absorptions greater than 620 nm with emission at less than 1,200 nm); 3. Measurements have been made from progressively smaller amounts of tissue, and measurements from areas less than 50 µm in diameter achieved in studies of the microcirculation (see Shonat et al, this procedings). 4. Timing circuits have been developed to allow phosphorescence measurements to be synchronized with rhythmic movements, such as the beating of the heart. This makes it possible, for example, to obtain digital maps of the oxygen distribution in the heart *in vivo* (see Rumsey et al, this proceedings). 5. A new needle type phosphorimeter has been designed with a light guide which can be inserted into tissue for measurements of phosphorescence at defined depths in the tissue. 6. The first images of phosphorescence have been made using confocal techniques (Plant and

Oxygen Transport to Tissue XVI
Edited by M.C. Hogan *et al.*, Plenum Press, New York, 1994

Burns, 1993) and this technique would be expected to provide high spatial resolution maps of oxygen pressure.

BACKGROUND

Phosphorescence is defined as delayed light emission following irradiation of a sample with a flash of light. Historically the name comes from the glow of delayed light light emitted when certain types of rocks were exposed to light and then carried into a darkened room. Many organic molecules also emit phosphorescence when exposed to light. In general the decay constants (lifetimes) of the phosphorescence of the greatest value for oxygen measurements is in the range of μsec to msec, but much longer lifetimes can be observed. Oxygen dependent quenching of phosphorescence makes use of the ability of oxygen molecules to return of the excited triplet state to the ground state with transfer of the energy to oxygen rather than emitting it as light. This quenching process requires collision between oxygen and the excited state molecule, the frequency of which is diffusion limited. Thus the quenching efficiency is related to both the collisional frequency and the probability that energy transfer will occur in any given collision. This relationship is generally referred to as the Stern-Volmer equation which, although it is well known, is presented because it is central to the method:

$$I^o/I \;=\; T^o/T \;=\; 1 \;+\; k_q * T^o * PO_2 \tag{1}$$

The oxygen pressure (PO_2) can thus be determined by measuring either the phosphorescence intensity (I) or phosphorescence lifetime (T). The phosphorescence intensity and/or lifetime is determined at zero oxygen pressure (superscript o) and then at one or more known oxygen pressures such that the value of k_q can be calculated. Once I^o or T^o and k_q are known, phosphorescence can be used to measure in oxygen pressure.

Oxygen dependent quenching of phosphorescence is a method for measuring oxygen which has been rapidly increasing in use, particularly in biological systems (see Vanderkooi et al, 1986; Wilson et al, 1987; 1991; Rumsey et al, 1988; 1990; Shonat et al, 1992a,b; Pawlowski and Wilson, 1991; Ince et al, 1993; Torres Filho and Intaglietta, 1993). Phosphores have been injected into the blood of anesthetized animals and phosphorescence measured either using A. a phosphorimeter which measures in a small area of the tissue surface using light guides to conduct the excitation light to the tissue and phosphorescence from the tissue to the photomultiplier (Pawlowski and Wilson, 1992; Shonat et al, this proceedings; Pastuszko et al, 1993; Tammela et al, 1993) or B. an epifluorescence microscope system which images the tissue phosphorescence and generates digital maps of either phosphorescence intensity (Rumsey et al, 1989; Wilson et al, 1991; Ince et al, 1993) or lifetime (Wilson et al, 1993; Rumsey et al, 1994). The phosphorescence lifetimes are calculated by the best fit to a single exponential and the oxygen pressure calculated for that value of the phosphorescence decay. Although this is a reasonable approximation for high resolution images where the phosphorescence at each pixel is from a very small area of tissue. It is a less correct representation for the light guide system where the total area from which the phosphorescence is collected may be up to several mm in diameter and should contain a representative array of arteriols, capillaries and veins. In the latter case, the best fit single exponential would be expected to yield an oxygen pressure representative of the oxygen pressure in the largest blood volume, that of the veins and adjacent capillaries.

SUMMARY OF NEWER DEVELOPMENTS

Deconvolution of the multi-exponential phosphorescence decay curves

Fit of the phosphorescence decay curves to a single exponential provides very important information about tissue oxygen levels in the venous end of the capillaries and veins. The data, however, contain detailed information on the distribution of oxygen in the tissue. As noted above, the phosphorescence originates from phosphor in the full complement of vessels, arteriols, capillaries and veins. Thus, the observed phosphorescence decay is the sum of a very large number of exponentials with a distribution characteristic of oxygen pressures from arteriolar to the lowest oxygen pressures in the capillaries or veins. The amount of phosphor with each decay constant is determined by the fraction of the sampled blood volume with that oxygen pressure. In the ideal case, it would be desirable to deconvolute the measured decay curve into the distribution of exponentials fully representative of the underlying distribution of blood and oxygen pressures. Dr. Vinogradov suggested the solution (see Vinogradov and Wilson, this proceedings) for multiexponential deconvolution which does not require any *a priori* assumptions concerning the distribution of exponentials. This method can deconvolute a decay curve into a preselected number of exponentials. This provides a histogram of the distribution of exponentials (and therefore of oxygen pressure) at the resolution of the selected number of exponentials. With a reasonably fast microcomputer the number of exponents can be greater than 100, providing a high resolution histogram of the oxygen pressures throughout the vascular system from arteriols to veins.

Phosphorescent oxygen probes absorbing and emitting in the near infra-red

The phosphorescent oxygen probes which have been most successfully used to date have been Palladium complexes of porphyrins. These have many important attributes of great value for oxygen measurements. They have excellent quantum efficiencies for phosphorescence and phosphorescence lifetimes ideally suited for measurements in the physiological range of oxygen pressures. On the other hand, the absorption bands of the phosphorescent porphyrins all lie in the spectral region with extensive absorption by pigments present in tissue. Although their phosphorescence emission occurs in a spectral region with little absorption by tissue pigments, absorption of excitation light limits the depth of tissue in which oxygen can be measured. Although the latter can be used select the depth of measurement in tissue, the depths are limited to less than about 1 mm. Measurements to greater depth could be accomplished if the absorbance were at longer wavelengths, ideally between 630 nm and 750 nm with the emission remaining at wavelengths less than about 1,300 nm. With the longer wavelength absorption both the excitation and emission light would penetrate through at least several cm of tissue, allowing oxygen measurements at substantial depths in the tissue. Preliminary work has permitted identification of several promising phosphors, although it is yet too early to determine which will prove the most useful. The properties of two are given in Table 1.

Table 1. Optical properties of new phosphorescent probes.

Probe name	Absorption maximum (nm)	Emission maximum (nm)
Alfa 1	441 nm 627 nm	784 nm
Alfa 2	440 nm 694 nm	$\approx$ 850 nm

One has an absorption band near 630 nm and emission near 700 nm while the other has an absorption band at 670 nm and emission near 850 nm. In each case, the phosphor binds to albumin and when bound has a phosphorescence lifetime of several hundred microseconds in the absence of oxygen and the quenching constant is suitable for measurements at physiological oxygen pressures. A manuscript with the chemical structures and other properties is in preparation (Vinogradov and Wilson). As this new generation of phosphorescent probes are synthesized in sufficient amounts for general experimental use, they will significantly extend the applicability of this method of oxygen measurement.

A micro-light guide phosphorimeter for local measurements of oxygen pressure at depth in tissue

It is often useful to make oxygen measurements in local regions of tissue which are not readily accessible from the surface. To this purpose, a micro-light guide phosphorimeter is being developed. The light of the flash lamp is conducted to the tissue through a small light guide (< 300 μm in diameter) which can be inserted into the tissue to the desired locality. The excitation light is emitted from the tip of the light guide and spreads through the tissue, the illuminated volume determined by the wavelength of light and the tissue absorbance at that wavelength. Blue light travels only a short distance (< 100 μm) whereas green light travels a longer distance (≈500 μm) etc. The emitted phosphorescence, in contrast, is at wavelengths which are only slightly absorbed by the tissue and are collected at the surface of the tissue. This phosphorescence arises from the small volume of illuminated tissue at the tip of the light guide since this is the only tissue exposed to the pulse of excitation light. Thus the volume of tissue in which oxygen is measured can be varied as desired by the operator, as long as the oxygen probe used has sufficient absorption at the chosen wavelength. This micro-light guide system permits measurements with high spatial resolution at almost any depth in the tissue. Although this approach is invasive, it is no more invasive than are oxygen electrodes designed for the same purpose. Preliminary data indicate the technique provides a very effective method for obtaining oxygen measurements in tissue regions well below the surface.

Phosphorescence measurements in rhythmically moving tissue

One of the very interesting applications of phosphorescence is to measure the oxygen pressure in tissues such as the heart or lung *in vivo*. These tissues move continuously with respiration and/or heart beat. Phosphorescence lifetime measurements using the light guide system are most accurate if the data from several (4 to 20) flashes is averaged to reduce the noise in the digitized decay curve. Rapid tissue movement, such as that of the heart, results in increased noise with a corresponding decrease in the precision with which the decay can be calculated. When imaging phosphorescence it is possible to use flash illumination or short exposure times to obtain single "freeze frame" images of phosphorescence intensity (Rumsey et al, 1989; Ince et al, 1992). Imaging of phosphorescence lifetime in such tissue, and thereby obtaining maps of the oxygen pressure, requires synchronization of the flash with the movement. Any movement between phosphorescence images results in loss of image register and introduces both blurring of the images and error in the calculated phosphorescence lifetimes. Dr. Pawlowski has constructed the appropriate timing circuits for both the light guide phosphorimeter and the phosphorescence imaging system. Rumsey and coworkers (these proceedings) have used phosphorescence imaging synchronized with the electrocardiogram to obtain *in vivo* maps of the oxygen distribution in the hearts of newborn piglets. This technique works only for rhythmically moving objects, however, and for non-rhythmic motion the only solution is still to take the set of images before significant movement occurs or repositioning the images to the same relative position after they are taken.

Application of confocal techniques to increase resolution of phosphorescence measurements

Confocal imaging improves spatial resolution by illuminating a very small area with a relatively high intensity spot of focused light and collecting the emitted light through a pin-hole designed to reduce the amount of light arising from other regions of the sample collected by the detector. The available confocal microscope systems scan the excitation light (a focused laser beam) and are not suited to measuring phosphorescence lifetimes. Phosphorescence is emitted over a time of many microseconds after excitation. Thus with a reasonable rate of scan of the excitation light focal point and the point of emission of phosphorescence do not coincide. Shonat (this proceedings) and Plant and Burns (1993) used confocal techniques to improve spatial resolution of their measurements of phosphorescence lifetime. Plant and Burns (1993) reported increased resolution of images using the ingenious technique of flash illumination and observation through an array of pin holes, simultaneously avoiding the necessity for moving the observation point and obtaining a full image with a single focal plain. This technique attains significant improvement in the depth of focus resolution with relatively simple modification of the imaging apparatus. On the other hand, the increase in resolution was obtained at the expense of total measured light intensity and phosphorescence measurements are often made under conditions for which there is less than optimal light. The approach of Plant and Burns and that "pseudo-confocal" imaging (mathematically removing out of plain light) can reasonable be expected to marked improve the ability to spatial resolve phosphorescence and oxygen pressures.

Supported in part by grants NS-10939 and NS-31465 from the U.S. National Institutes of Health.

REFERENCES

Green, T.J., Wilson, D.F., Vanderkooi, J.M. and DeFeo, S.P., 1989, Phosphorimeters for analysis of decay profiles and real time monitoring of exponential decay and oxygen concentrations. *Analytical Biochem.* 174: 73-79.

Ince, C. Ashruf, J.F., Avontuur, P.A., Spaan, J.A.E., and Bruining, H.A., 1993, Heterogeneity of the hypoxic state in rat-heart is determined at capillary level. *Amer. J. Physiol.* 264: H294-H301.

Ince, C., Ashruf, J.F., Sanderse, E.A., Pierik, E.G., Coremans, J.M., and Bruining, H.A., 1993, *In vivo* NADH and Pd-porphyrin video fluori-phosphorimetry. *Adv. Exptl. Med. Biol.* 317: 267-275.

Lahiri, S., Rumsey, W.L., Wilson, D.F., and Iturriaga, R., 1993, Contribution of *in vivo* microvascular PO_2 in the cat carotid body chemotransduction. *J. Appl. Physiol.* 75(3): 1035-1043.

Pastuszko, A., Saadat-Lajevardi, N., Chen, J., Tammela, O., Wilson, D.F., and Delivoria-Papadopoulos, M., 1993, Effects of graded levels of tissue oxygen pressure on dopamine metabolism in the striatum of newborn piglets. *J. Neurochem.* 60, 161-166.

Pawlowski, M., and Wilson, D.F., 1992, Monitoring of the oxygen pressure in the blood of live animals using the oxygen dependent quenching of phosphorescence. *Adv. Exptl. Med. Biol.* 316: 179-185.

Plant, R.L. and Burns, D.H. 1993, Quantitative, depth-resolved imaging of oxygen concentration by phosphorescence lifetime measurement. *Appl. Spectroscopy,* 47(10): 1594-1599.

Robiolio, M., Rumsey, W.R. and Wilson, D.F., 1989, Oxygen diffusion and mitochondrial respiration in neuroblastoma cells, *Amer. J. Physiol.,* 256: C1207-C1213.

Rumsey, W.L., Robiolio, M. and Wilson, D.F. 1989, Contribution of diffusion to the oxygen dependence of energy metabolism in human neuroblastoma cells. *Adv. Exptl. Med. Biol.,* 248: 829-833.

Rumsey, W.L., Lahiri, S., Iturriaga, R., Mokashi, A., Spergel, D., and Wilson, D.F., 1992, Optical measurements of oxygen and electrical measurements of oxygen chemoreception in the cat carotid body. *Adv. Exptl. Med. Biol.,* 317: 387-395.

Rumsey, W.L., Iturriaga, R., Wilson, D.F., Lahiri, S., and Spergel, D., 1990, Phosphorescence and fluorescence imaging: new tools for the study of carotid body function. *In*: Chemoreceptors and Chemoreceptor Reflexes (H. Acker et al, eds) Plenum Press, New York, pp. 73-79.

Rumsey, W.L., Iturriaga, R., Spergel, D., Lahiri, S., and Wilson, D.F., 1991, Optical measurements of the dependence of chemoreception on oxygen pressure in the cat carotid body. *Amer. J. Physiol.*, 261: C614-C622.

Rumsey, W.L., Schlosser, C., Nuutinen, E.M., Robiolio, M., and Wilson, D.F., 1992, The oxygen dependence of mitochondrial oxidative phosphorylation and its role in regulation of coronary blood flow. *Adv. Exptl. Med. Biol.* 316: 279-284.

Rumsey, W.L., Schlosser, C., Nuutinen, E.M., Robiolio, M. and Wilson, D.F., 1990, Cellular energetics and the oxygen dependence of respiration in cardiac myocytes isolated from adult rat. *J. Biol. Chem.* 265: 15392-15399.

Rumsey, W.L., Vanderkooi, J.M. and Wilson, D.F., 1988, Imaging of phosphorescence: A novel method for measuring the distribution of oxygen in perfused tissue. *Science*, 241: 1649-1651.

Shonat, R.D., Wilson, D.F., Riva, C.E., and Pawlowski, M., 1992a, Oxygen distribution in the retinal and choroidal vessels of the cat as measured by a new phosphorescence imaging method. *Applied Optics* 31, 3711-3718.

Shonat, R.D., Wilson, D.F., Riva, C.E. and Cranstoun, S.D.,1992b, Effect of acute increases in intraocular pressure on intravascular optic nerve head oxygen tension in cats. *Invest. Opthalmology & Visual Sci.*, 33, 3174-3180.

Tammela, O., Pastuszko, A., Lejevardi, N.S., Delivoria-Papadopoulos, M., and Wilson, D.F., 1993, Activity of tyrosine hydroxylase in the striatum of newborn piglets in response to hypocapnic hypoxia. *J. Neurochem.* 60, 1399-1406.

Torres Filho, I.P. and Intaglietta, M., 1993, Microvessel PO_2 measurements by phosphorescence decay method. *Amer. J. Physiol.* 265: H1434-H1438.

Vanderkooi, J.M., Maniara, G., Green, T.J., and Wilson, D.F., 1987, An optical method for measurement of dioxygen concentration based on quenching of phosphorescence. *J. Biol. Chem.* 262: 5476-5482.

Vanderkooi, J.M., and Wilson, D.F., 1986, A new method for measuring oxygen concentration in biological systems. *Adv. Exptl. Med. Biol.* 200: 189-193.

Vanderkooi, J.M., Wright, W.W. and Erecinska, M., 1990, Oxygen gradients in mitochondria examined with delayed luminescence from excited-state triplet probes. *Biochem.*, 29(22): 5332-5338.

Wilson, D.F., 1992, Oxygen dependent quenching of phosphorescence: a perspective. *Adv. Exptl. Med. Biol.*, 317: 195-201.

Wilson, D.F. and Cerniglia, G.J., 1992, Localization of tumors and evaluation of their state of oxygenation by phosphorescence imaging. *Cancer Research*, 52: 3988-3993.

Wilson, D.F., Gomi, S., Pastuszko, A., and Greenberg, J.H., 1993, Microvascular damage in the cortex of cat brain from middle cerebral artery occlusion and reperfusion. *J. Appl. Physiol.*, 74(2), 580-589.

Wilson, D.F., Gomi, S., Pastuszko, A., and Greenberg, J.H., 1992, Oxygenation of the cortex of the brain of cats during occlusion of the middle cerebral artery and reperfusion. *Adv. Exptl. Med. Biol.*, 317: 689-694.

Wilson, D.F., Pastuszko, A., DiGiacomo, J.E., Pawlowski, M., Schneiderman, R., Delivoria-Papadopoulos, M., 1991, Effect of hyperventilation on oxygenation of the brain cortex of newborn piglets. *J. Appl. Physiol.* 70(6): 2691-2696.

Wilson, D.F., Pastuszko, A., Schneiderman, R., DiGiacomo, J.E., Pawlowski, M. and Delivoria-Papadopoulos, M., 1992, Effect of hyperventilation on the oxygenation of the brain cortex of neonates. *Adv. Exptl. Med. Biol.* 316: 341-346.

Wilson, D.F. and Rumsey, W.L., 1991, Factors affecting adaptation of the mitochondrial enzyme covntent to cellular needs. *In:* Response and Adaptation to Hypoxia: Organ to Organelle. (S. Lahiri, N.S. Cherniack and R.S. Fitsgerald, eds.) Oxford Univer. Press, pp. 14-24.

Wilson, D.F., Rumsey, W.L., Green, T.J., and Vanderkooi, J.M., 1988, The oxygen dependence of mitochondrial oxidative phosphorylation measured by a new optical method for measuring oxygen. *J. Biol. Chem.* 263: 2712-2718.

Wilson, D.F., Rumsey, W.L. and Vanderkooi, J.M., 1989, Oxygen distribution in isolated perfused liver observed by phosphorescence imaging. *Adv. Exptl. Med. Biol.*, 248: 109-115.

RECOVERY OF OXYGEN DISTRIBUTIONS IN TISSUE FROM PHOSPHORESCENCE DECAY DATA

Sergei A. Vinogradov and David F. Wilson

Department of Biochemistry and Biophysics
School of Medicine
University of Pennsylvania
Philadelphia, PA 19104

INTRODUCTION

Oxygen dependent quenching of phosphorescence has become one of the most effective and accurate methods for measuring oxygen concentration. Because it is a non invasive optical technique, this method has been successfully used for both in vitro and in vivo studies (see for example Vanderkooi et al., 1991; Wilson et al., 1991). It has great potential as a new clinical tool in the diagnosis and treatment of many diseased states, particularly those leading to altered oxygen pressures in the tissue.

Phosphorescence arises as the excited triplet state molecule returns to the ground state with emission of a photon. Because of rate of decay of the triplet state is proportional to the number of molecules in the triplet state, intensity of phosphorescence after a flash of excitation light follows an exponential law:

$$I(t) = I_0 \cdot \exp(-t / \tau) \tag{1}$$

where τ is phosphorescence lifetime and I_0 - initial light intensity. The characteristic lifetime of the decay, τ, is dependent on the concentration of quencher (molecular oxygen, in biological systems) and this dependence is well described by Stern-Volmer relation:

$$\tau_o / \tau = 1 + k_Q \cdot \tau_o \cdot pO_2 \tag{2}$$

pO_2 is oxygen pressure (concentration). The quenching constant k_Q and phosphorescence lifetime at zero-oxygen concentration τ_o - are characteristics of the probe molecule and they are constant for the given temperature and solvent. Therefore accurate measurements of phosphorescence lifetime (τ) allow direct calculation of the oxygen concentration (pO_2).

Oxygen Transport to Tissue XVI
Edited by M.C. Hogan *et al.*, Plenum Press, New York, 1994

An instrument for these measurements (Pawlowski and Wilson, 1992) uses time-correlated acquisition of phosphorescence intensity. Light, emitted by the sample is conducted to a photo sensor (photomultiplier or photodiode) by a light guide. The detector signal is amplified and digitized to provide a set of numbers (data vector) and subject to numerical analysis to obtain the value of oxygen concentration.

Fitting data with exponential curve (equation 1) automatically assumes homogeneity of the sample, a situation which is practically realized when one measures oxygen in a cuvette with solution of probe at a given oxygen concentration. Then all 'elementary volumes' of the sample are characterized by the same decay constant and their sum is also a single exponential. In this case a linear least square fitting of data gives τ, which can be converted into the corresponding pO_2 according to equation 2.

A function containing several exponential terms (usually no more then 3-4, equation 3) can be used if one assumes that sample has a few discrete 'independent phosphorescence sources' or only few discrete oxygen concentrations. A possible practical realization of this case is solution of several different probes, characterized by different τ_o and k_Q.

$$I = I_1 \cdot \exp(-t / \tau_1) + I_2 \cdot \exp(-t / \tau_2) + ... \qquad (3)$$

Non-linear fitting with varying of both I_i and τ_i leads to its optimal values but the number of components in the probe function still should be chosen in advance. In other words, *a priori* knowledge of discrete decay components, and therefore of the oxygen concentrations, is required. Moreover, choosing wrong number of exponential terms can bring physically meaningless results. Thus two or three exponential decay can be satisfactory fitted with sum of 5-6 exponents, but unfortunately the converse is also true: two or three exponential decay law can conceal a complex set of decays. This property of exponentials is a consequence of their high non-ortogonality and has been discussed previously (Ware et al., 1973).

Oxygenation of tissue *in vivo* characteristically is highly heterogeneous in sense that a wide range of different oxygen concentrations are present. After being exposed to a flash of exciting light, an area of tissue presents, in fact, a sum of very large number of tiny light emitters, each with a phosphorescence lifetime characteristic of the oxygen pressure in its local environment. As a result, acquired data vector represents an integral response of significant tissue volume characterized by broad distribution of oxygen concentrations. Treatment of this signal using single or several discrete exponential functions provides an incomplete description of the physical origin of phenomena. Thus deconvolution techniques which do not introduce bias concerning the lifetimes could provide important information about underlying distribution of oxygen in the tissue.

In the present paper we use an approach based on the Exponential Series Method which can recover the underlying oxygen distributions from data obtained in phosphorescence quenching experiments *in vivo*.

DISCUSSION

If the data $I(t)$ is a convolution of an infinite numbers of exponential decays, characterized by τ's, spaced in the range $[\tau_{min} - \tau_{max}]$ and $g(\tau)$ is signal spectrum (a function describing the distribution), the transformation $g(\tau) \Rightarrow I(t)$ is called a Laplace transform (L) and it is presented by the integral (4) over the exponential kernel.

$$I(t) := \int_0^\infty g(\tau) \cdot \exp\left(-\frac{t}{\tau}\right) d\tau \tag{4}$$

Therefore, recovery of the distribution (finding $g(\tau)$ from measured $I(t)$) includes finding an inverse Laplace transform (L^{-1}) which is numerically strongly ill-conditioned (McWhirter and Pike, 1978). Several algorithms for solution based on integration have been proposed in the literature (Gardner et al., 1959; Provencher, 1976; Provencher, 1976a), however none of them can be solved to satisfactory numerical precision and none provide a method of practical use.

This study utilizes algebraic methodology, namely decomposition of a vector (data vector from the experiment) into the linearly independent basis set of single exponentials.

$$I(t) = \sum_k^N b_k \cdot \exp(-t/\tau_k) \tag{5}$$

We assume that vectors of basis are fixed and only the linear coefficients b_k vary. The obtained set of b_k values plotted versus τ_k will represent the spectrum of the signal. The family of exponentials with lifetimes uniformly spaced in reciprocal τ–space between $1/\tau_{max}$ and $1/\tau_{min}$ was chosen as a basis, however in principal any set of linearly independent functions, such as sets of polynomials or trigonometric functions, can be used instead. When choosing exponentials, on the other hand, certain physical significance can be attributed to the recovered distribution of lifetimes. In fact, equation (2) transforms it directly into the distribution of oxygen concentrations. Because the number of exponentials which one can practically use is finite while the data vector is represented by a continuous sum of decays, this method of decomposition should find the best approximate solution and a criteria of goodness of fit needs to be chosen. The natural choice is χ^2, and therefore weighted least square fitting with the trial function (5) has to be applied. The experimental data was considered to have Poisson noise distribution:

$$\chi^2 = \frac{1}{M} \cdot \sum_i^M \frac{1}{y_i} \cdot (y_i - \sum_k^N b_k \cdot \exp(-t_i/\tau_k))^2 \tag{6}$$

where y_i represents components of data vector and M is a total number of data points. Minimization of χ^2 function provides the best fit of the data vector and simultaneously provides a set of coefficients b_k which represent an approximation of original distribution (see Fig. 2). Formally, the trial function $f(t)$ is linearly dependent upon coefficients b_k (see eq.3) and the optimal set could be found by a non-recursive procedure. However, this method, being essentially curve fitting with free parameters, may include physically impossible negative values of b_k. A non-linear approach, on the other hand, allow coefficients to be constrained to non-negative values when using Marquardt-Levenberg algorithm (James and Ware, 1986).

This method was originally proposed by Ware et al. (1973) for interpretation of data derived from time correlated pulse fluorescence experiments and it was named Exponential Series Method (ESM) (see also James and Ware, 1985). It has been pointed out that similar methodology can be applied to the analysis of fluorescence decays from any heterogeneous system including phosphorescence decays as well. However the poor accuracy of phosphorescence data of the time compared to the those for fluorescence measured by Single Photon Counters (SPC) brought authors to the conclusion that ESM will not be useful for the interpretation of phosphorescence decays. Indeed, high redundancy of exponential basis makes method relatively sensitive to the noise level and it has been extensively tested together with its ability to recover different shapes of artificially simulated and real distributions (Semiartzuk et al., 1990).

The accuracy of time resolved phosphorescence measurements have been significantly improved in the latest years by using lightguide type phosphorimeters designed in this lab. Utilization of fast 12 bit ATD converters in combination with very sensitive photomultipliers allowed increase in the signal-to-noise ratio of acquired data to the extent that it became comparable with SPC accumulated decays. Also, relatively slow changes in the physiological parameters of living samples provide the possibility for long data acquisition: summation and averaging of the decays, and as a result almost noiseless curves can be obtained and subject to numerical analysis. An additional advantage of phosphorescence measurements versus fluorescence ones is relatively long decay times, which allow an excitation pulse and instrument response functions to be filtered out at the stage of data acquisition.

PRACTICAL REALIZATION AND RESULTS

Our numerical procedure follows general Marquardt-Levenberg algorithm (Press et al., 1986), extended to include positive constraints on the parameters values. Software implementation has been written and tested on 486 66 MHz microcomputer. The routine automatically rejects terms which approach values less then 1% of maximal parameter during the search - this run-time reduction of basis set makes the routine much more stable and leads to faster convergence. Generally 100-120 terms were used to construct a trial function, however in principle there is no limitation on a size of the basis set as far as one takes care of matrixes singularity. Usually the procedure converged within 1-2 minutes. In all cases, a flat distribution was used as an initial guess, which is equivalent to having no *a priori* knowledge about the shape of distribution. After the algorithm converged, analysis of residuals and the usual statistical tests have been performed which showed satisfactory results. (As a whole ESM was found to be statistically valid for analysis of multiexponential decays (Ware, 1991).

Our algorithm have been extensively tested on artificially simulated decays with known distributions which were constructed from up to 1000 exponents with random noise of various levels added. Several important cases were considered: i). single exponential decays (Fig. 1a); ii). a few discrete decays (Fig. 1b); iii). bimodal distributions of decays (Fig. 1c); iv). Gaussian distributions with a one separated decay (Fig. 1d). In all cases our algorithm was stable and the distributions were recovered with reasonable accuracy.

Finally the method has been applied to the analysis of experimentally obtained data. The phosphorescence decay collected from the degassed water solution Pd (*meso*-tetracarboxyphenylporphyrin) has been analyzed (Fig. 2a). A decay constant of 698 μsec was obtained with a distribution only 13 μsec at half height. Analysis by fit to a single exponential gave a value of 700 μsec. The phosphorescence decay collected from deoxygenated dimethylformamide solution of one of new phosphorescent probes (see Wilson and Vinogradov in this proceedings) was also analyzed and the narrow distribution obtained had a maximum in a good agreement with single exponential fit.

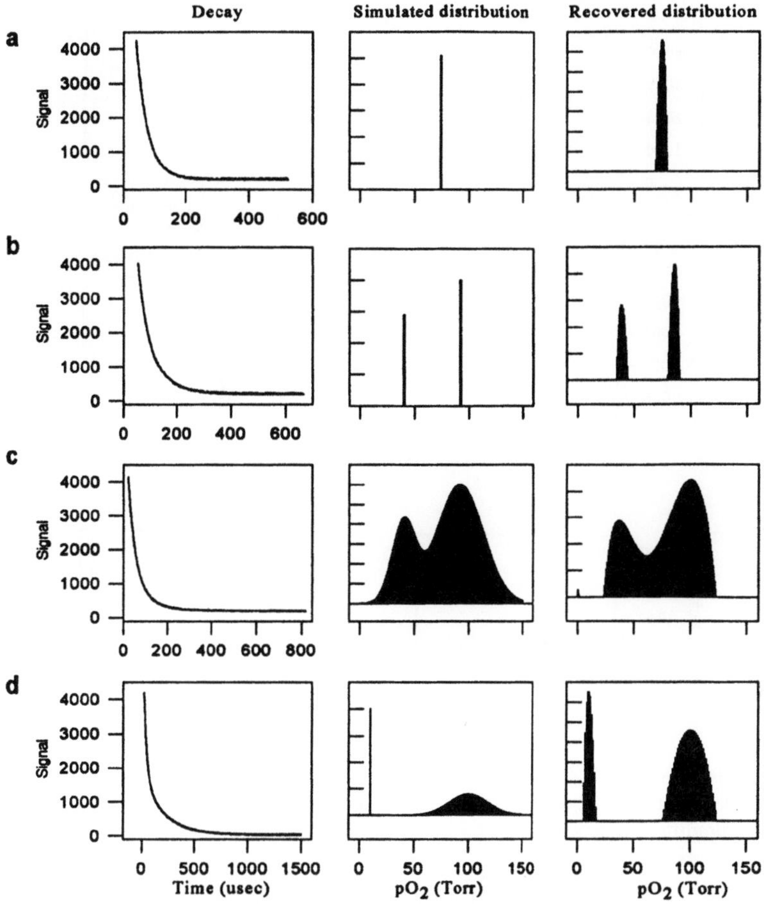

Figure 1. Recovery of underlying distributions from simulated decays. (In all cases random noise was added to the curves). Data points were calculated with 1 µsec interval and with highest level at 4096, consistent with the use of a 1 MHz 12-bit A/D converter. Probe function contained 120 terms over range of 0-150 Torr. Values of k_Q=350 Torr^{-1}sec^{-1} and of τ_0=700 µsec were used for conversion into τ-scale.
(a) Single-exponential at 73.8 Torr. Fitting range 484 data points. A/D delay 40 µsec.
(b) Double-exponential (91.8 Torr and 40.0 Torr). Fitting range 612 data points. A/D delay 54 µsec.
(c) Two Gaussians (positions 91.8 and 40.0 Torr, widths 30 and 15 Torr respectively). Fitting range 552 data points. A/D delay 22 µsec.
(d) Gaussian (position 100.0 Torr, width 40 Torr) with one added single-exponential at 11.0 Torr. Fitting range 900 data points. A/D delay 21 µsec.

Several experimentally obtained decay curves collected from the phosphorescent measurements of brain surface of newborn piglets have been analyzed and the recovered distribution of oxygen is shown in Figure 2c. Analyses of large sets of *in vivo* data are currently in progress and these will establish the physiologically important oxygen distributions in the tissue.

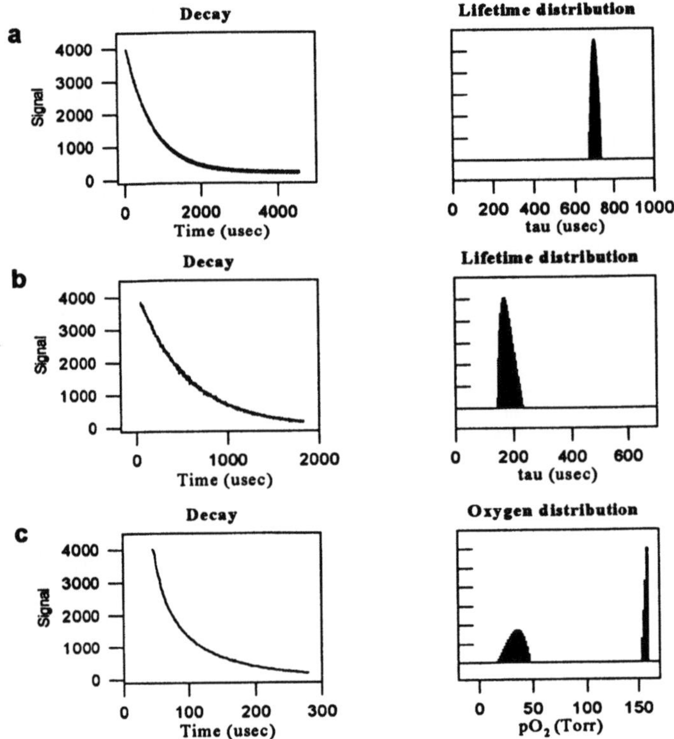

Figure 2. Analysis of real data.

(a) Degassed water solution of Pd (*meso*-tetra-4-carboxyphenylporphyrin) bound to bovine serum albumin. Data colected at sampling frequency 0.5 MHz with 3 averaging cycles. Recovered distribution maximum 698 μsec, independently measured τ_0 of the probe, 700 μsec.

(b) Solution of new organic soluble Pd-porphyrin based probe in degassed Dimethylformamide. Data collected at a sampling frequency of 1.0 MHz with 5 averaging cycles. Recovered distribution maximum 192 μsec, independently measured τ_0 of the probe 189.5 μsec.

(c) Analysis of the phosphorescence decay collected from the surface of the brain of a newborn piglet. Distribution is presented by relatively broad peak at 40 Torr and sharp peak at near 0 Torr. The latter belongs to the flash of the excitation lamp.

CONCLUSION

The ESM approach is recommended for analysis of phosphorescence decay data, particularly that collected from tissue *in vivo*. Software implementation has been written and shown to correctly recover known oxygen distributions from artificially simulated decays with added noise. Initial application has been made to determination of the oxygen distributions in tissue from the phosphorescence decay curves measured *in vivo*.

REFERENCES

Gardner, D.G., Gardner, J.C. and Meinke, W.W., 1959, Method for the analysis of multicomponent exponential decay curves, *J. Chem. Phys.* 31:978

James, D.R. and Ware, W.R., 1985, A fallacy in the interpretation of fluorescence decay parameters, *Chem. Phys. Lett.* 120:455

James, D.R. and Ware, W.R., 1986, Recovery of underlying distributions of lifetimes from fluorescence decay data, *Chem. Phys. Lett.* 126:7

McWhirter, J.C., and Pike, E.R., 1978, On the numerical invertsion of the L*aplace transform and Fredholm integral equations of the first kind.* J.Phys.A.Math.Gen. 11:1729.

Pawlowski, M., and Wilson, D.F. 1992, Monitoring of the oxygen pressure in the blood of live animals using the oxygen dependent quenching of phosphorescence. *Adv. Exptl. Med. Biol.* 316: 179-185.

Provencher, S.W., 1976, A Fourier method for the analysis of exponential decay curves, *Biophys.J.* 16:27

Provencher, S.W., 1976a, An eigenfunction expansion method for the analysis of exponential decay curves, J. Chem. Phys. 64:2772

Press, W.H., Flannery, B.P., Teukolsky, S.A., and Vetterling, W.T. 1986, Numerical Recipes in C. Cambridge: Cambridge University Press.

Semiarczuk, A., Wagner, B.D. and Ware, W.R., 1990, Comparison of the Maximum Entropy Method and Exponential Series Metod for the recovery of distributios of lifetimes from fluorescence lifetime data, *J. Phys. Chem.* 94:1661

Vanderkooi, J.M., Wright, W.W., and Ericinska, M. 1991, Oxygen gradients in mitochondria examined with delayed luminescence from excited-state probes. *Biochemistry,* 29: 5332-5338.

Ware, R., 1991, Recovery of Fluorescence Lifetime Distributions in Heterogenious Systems, *in:* "Photochemistry in Organized and Constrained Media", Ramamurthy, V.,ed., VCH Publishers, New York

Ware, W.R., Doemeny, L.J. and Nemzek, T.L., 1973, Deconvolution of fluorescence and phosphorescence decay curves. A least square method, *J. Chem. Phys.* 77:2038

Wilson, D.F., Pastuszko, A., DiGiacomo, J.E., Pawlowski, M., Schneiderman, R., Delivoria-Papadopoulos, M. (1991) Effect of hyperventilation on oxygenation of the brain cortex of newborn piglets. *J. Appl. Physiol.* 70(6): 2691-2696.

A NEW PHOSPHORIMETER FOR THE MEASUREMENT OF OXYGEN PRESSURES USING PD-PORPHINE PHOSPHORESCENCE

M. Sinaasappel, C. Ince, J.P. van der Sluijs, H.A. Bruining

Department of Surgery, University Hospital Rotterdam
The Netherlands

INTRODUCTION

There is a growing demand for quantitative and rapid determination of low oxygen concentrations in biological samples. The application of oxygen electrodes has so far proven to be the most successful. The use of electrodes, however, is limited by the stability of the electrode surface and diffusion barrier. Furthermore, the electrodes consume oxygen which introduces artifacts at low oxygen pressures.

Measuring the quenching of *fluorescence* decay is a well-known technique for the determination of oxygen concentrations[3]. The major disadvantage of using fluorescence is that the decay times involved are short (ns) which makes the use of lasers and complicated electronics necessary. In recent years a new and promising technique became available. Wilson and co-workers[1,2] described a method based on the oxygen dependent quenching of Pd-porphine *phosphorescence* decay. The use of *phosphorescence* quenching has the advantage of the relatively long decay times (μs). However, for most probes the phosphorescence is only measurable at extreme low oxygen pressures or low temperatures. The phosphorescent probe, Pd-meso-4-tetra-carboxyphenyl porphine[1], which we recently applied in intensity measurements in the isolated saline perfused rat hearts[4], can be used in the physiological range.

In this study we present a new phosphorimeter specially suited for *in vivo* measurement of Pd-porphine phosphorescence decay times, and oxygen. With this new posphorimeter we have re-evaluated the calibration constants, pH and temperature dependency needed to apply Pd-porphine for determination of intravascular oxygen pressures. Finally, we present *in vivo* experiments in the rat intestine and liver.

Theory

Once a Pd-porphine molecule is excited to its first singlet state (S1), it can relax to the first excited triplet state (T1) by inter-system crossing. This first triplet state of Pd-porphine has a relatively long lifetime because the relaxation to the ground state is spin forbidden. When Pd-porphine collides with an oxygen molecule within the lifetime of the triplet state, the energy of this state can be transferred to the oxygen molecule. As a consequence of the energy transfer, the Pd-porphine relaxes to the ground state without emitting a photon. The phosphorescence intensity, therefore, diminishes as a function of the collision frequency. Because the collision frequency with oxygen introduces a second rate process, the decay time is also diminished. The relation between the quenched phosphorescence and the oxygen pressure is described by the Stern-Volmer relation,

$$\frac{P_0}{P} = \frac{\tau_0}{\tau} = 1 + \tau_0 K_q pO_2 \tag{1}$$

where P_0 and τ_0 are the intensity and decay time, respectively, in absence of quenchers and P and τ are the intensity and decay time at a given oxygen pressure pO_2. K_q is the quenching constant which is defined by the Smoluchowski equation as,

$$K_q = \frac{\gamma 4\pi N}{1000}(D_o + D_f)(R_o + R_f) \tag{2}$$

with R_f the collision radius of the chromophore, R_o the collision radius of oxygen, D_f the diffusion coefficient of the chromophore, D_o diffusion coefficient of oxygen, γ the quenching efficiency, and N Avogadro's number.

As can be seen from equation 1 both the intensity and decay time can be used to determine oxygen pressures. Since we want to determine oxygen pressures *in vivo*, measuring decay times is preferable because decay times are independent of the optical properties of tissue and of the dye concentration. Since τ_o and K_q can be determined *in vitro*, measurement of decay times makes it possible to quantify pO_2 *in vivo*. The Smoluchowski equation also shows that K_q dependens on the diffusion of both Pd-porphine and O_2 in the solvent. Because diffusion is temperature dependent, also a temperature-dependency of K_q is to be expected.

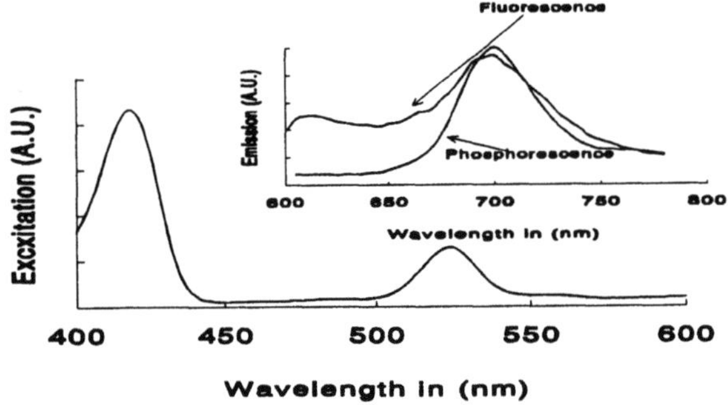

Figure 1. Excitation and emission spectra for Pd-meso-tetra 4-carboxyphenyl porphine bound to albumin. The probe is dissolved as described in Material and Methods. The excitation spectrum is recorded at an emission wavelength of 700nm. The excitation wavelength used for the fluorescence and phosphorescence spectra was 520 nm. For the fluorescence spectrum the Pd-porphine solution is placed in an open vial (pO_2 =157 mmHg). For the phosphorescence spectrum the pO_2 was made zero as described in Materials and Methods. A Hitatchi F4500 spectrophotometer is used in its phosphorescence mode, meaning that the phosphorescence is measured 1 ms after the excitation pulse.

For *in vivo* experiments the dye is bound to albumin so that it is confined to the circulation. In addition, the binding to albumin results in longer decay times at higher pO_2, which is essential for obtaining a measurable signal. In our case the decay time at 150 mmHg increased to $20\mu s$. An other advantage of Pd-porphine is that the phosphorescence emission wavelength is at 695 nm, where the tissue absorption is relatively low.

MATERIALS AND METHODS

The phosphorimeter used for the experiments is shown in Figure 2. The design is based on the instrument described by Pawlowski et al.[2] with several essential modifications.

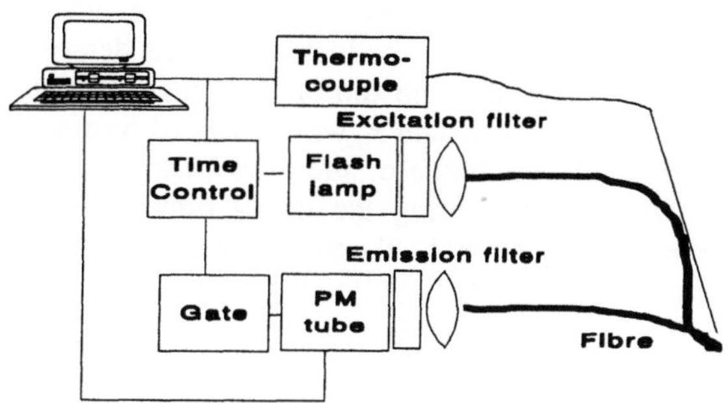

Figure 2. The phosphorimeter.

The light source is an EG&G FX249 with a PS40 power supply with a pulse duration of $3\mu s$. The detector is an R928 photomultiplier (PM) (Hamamatsu) with a C1392-09 gated socket. The current from the photomultiplier is converted to a DC voltage by an amplifier with a response time of less then 500 ns. The signal is converted by a fast 1MHz A/D board (Das-50, MetraByte Corp., Taunton, MA). Finally, the data are processed by a 386DX, 33 MHz personal computer. In front of the flash lamp an interference filter with a transmittance wavelength of 520 ± 20 nm is placed. The excitation wavelength is selected with a long pass filter with a cutoff wavelength at 630 nm. Although there is an interference filter in front of the flashlamp and a long pass filter in front of the photomultiplier, a slight transmittance of excitation light remains. Because the phosphorescence intensity is lower than this stray light, the PM tube is easily saturated. This saturation is inflicted by the build-up of space charge in the tube. The saturation manifests itself by an increase in response time of the tube, which introduces systematic errors in the estimation of the decay time. To ensure that the PM tube does not saturate we introduced a switching device that inverts the voltage drop across the first dynode during the excitation flash. The inversion of the voltage prevents the build up of the space charge and consequent saturation effects. The time control unit takes care of the switching of the gated socket and the triggering of the flash lamp. The excitation light and the phosphorescence light are guided by a randomly mixed bifurcated multi fibre light guide. The calculation of the decay time is performed by taking the natural logarithm of the phosphoresence signal and determining the decay time (τ) and the initial intensity (I_0) by a least square procedure. Because the K_q and τ_0 are temperature-dependent (equation 2), we introduced a thermo-couple (LICOX, Kiel, FRG) for placement on the tissue surface.

The Pd-meso-tetra-4-carboxyphenyl-porphine was dissolved in DMSO (13 mg/ml). This solution was added to 150mM NaCl, 0.5% albumin, 1mM EDTA, and 20mM HEPES to a final concentration of 5.4 μM. This procedure is sufficient to bind the Pd-porphine to albumin[1]. The pH was set with TRISM base. To change the scattering coefficient of the solution without changing the absorption, Intralipid 10%® (Kabi Pharmacia B.V., Woerden, The Netherlands) was used[5].

For the *in vivo* experiments 0.1 ml of 3 mM Pd-porphine-albumin solution was injected i.v. in male Wistar rats weighing around 300 g. The rats were mechanically ventilated with 98% O_2 and 2% enflurane. The blood flow in the superior mesenteric artery was measured using a Transonic (HT106) Medical flow meter in combination with a 1RB probe. The core temperature of the rat was held constant at 39°C by a thermostatic mat. The light guide was positioned on the duodenum or on the liver.

Determination of τ_0 and K_q

To obtain a good signal-to-noise ratio 30 decay traces were averaged. From this averaged trace the decay time was estimated. The goodness of fit was determined by evaluating the residuals[6]. The residuals showed that the time trace became single exponential when a gating time of $6\mu s$ was used. The constant τ_0 was measured by removing the oxygen with 10 mM glucose, 2-5 U/ml glucose oxidase, and 5-50 U/ml catalase[2].

K_q was determined by measuring the decay time in equilibrium with air, the measured values for τ_0, and equation 1. To compensate the pO_2 in water in equilibrium with air for the temperature we used the following equation,

$$pO_2 = (p_B - p_W(T))\ 0.209 \tag{3}$$

where p_B is the barometric pressure and $p_W(T)$ the temperature-dependent solvent vapor pressure.

RESULTS

The pH-dependency of the quenching constant (Figure 3) was determined by estimating the decay time (n=10) and calculating K_q using equation 1 and τ_0 from Figure 5 in the pH range of 5.75-8.25 (T = 37°C).

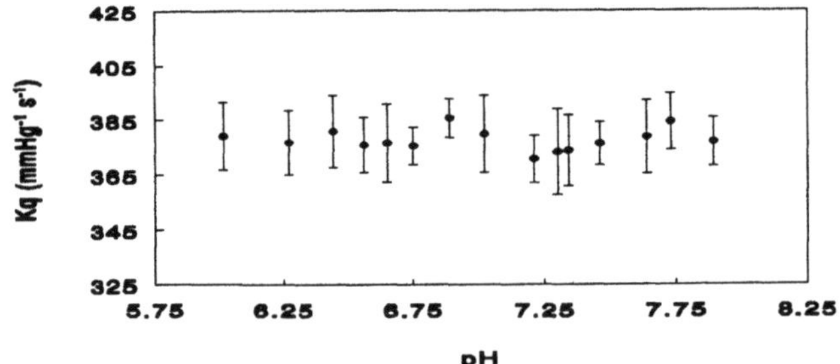

Figure 3. Dependency of the quenching constant to pH at 37 °C and a barometric pressure of 755mmHg. Data represent the mean (± SD) of 10 measurements.

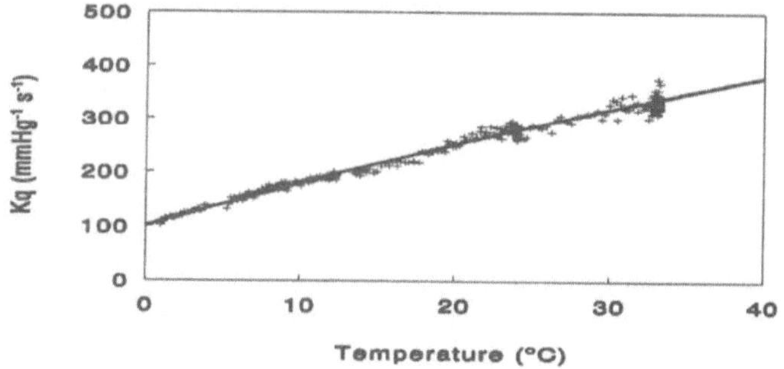

Figure 4. Temperature-dependency of the quenching constant at pH 7.5 and barometric pressure of 760 mmHg.

To measure the temperature-dependency of K_q the open vial, containing the buffered Pd-porphine-albumin complex, was first cooled down to 1 °C and then gradually heated to 35 °C at a rate of 0.5 °C/min. The decay time and temperature were measured every 10 seconds. Figure 4 shows that the K_q increased with increasing temperature, which agreed with the increase of the diffusion coefficient of oxygen with temperature (equation 2). Since K_q is temperature-dependent, correction of K_q is needed for temperature fluctuations in *in vivo* measurements. The relationship between K_q and T was estimated as $K_q = 101 + 8.1T - 0.03T^2$. This relation can be used for correction of pO_2 measurements when the temperature is not constant.

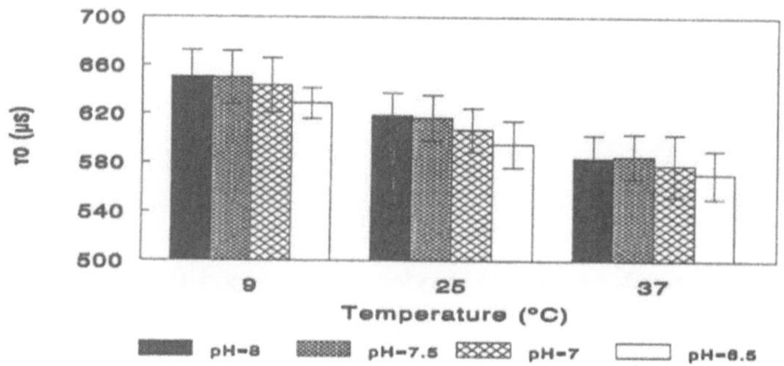

Figure 5. Dependency of τ_0 to pH and temperature. Data represent the mean ($\pm$ SD) of 10 estimations.

Figure 5 shows the estimation of τ_0 at different pH and temperature at zero pO_2. No significant dependency of τ_0 on the pH could be found. Also, the temperature-dependency between 25 °C and 37 °C was not significant.

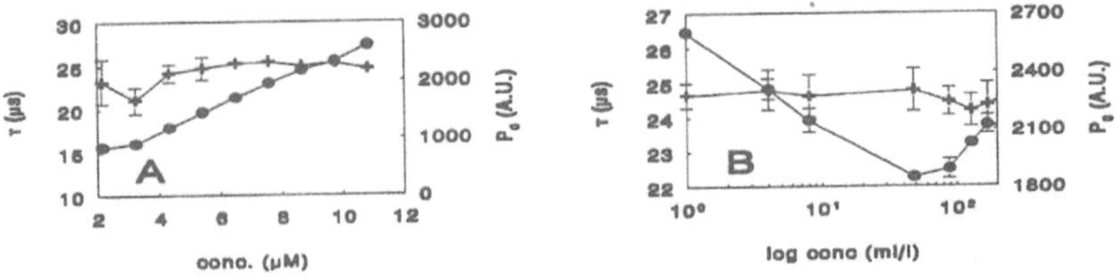

Figure 6. The decay time (+) and I_0(●) as function of: A) concentration Pd-porphine, and B) concentration Intralipid 10%®. Presented are the mean (±SD) of 10 estimations.

To test whether the estimation of the decay time using the new phosphorimeter is indeed independent of the phosphorescence intensity we altered both the emission and excitation light distribution while the decay time was kept constant (constant pO_2 and temperature). In the first experiment the Pd-porphine concentration was increased (Figure 6 A) to increase the emission intensity. In a second experiment the scattering was increased by adding Intralipid® 10% (Figure 6B). Intralipid® 10% is a fatty emulsion with virtually no absorption but a high scattering coefficient. By introducing Intralipid® 10% we can change the scattering without changing the other optical properties of the solution. This makes Intralipid 10% suitable to test the scattering dependency of the measurement[7]. Both experiments indicate that, although the phosphorescence *intensity* is changed several fold, no significant deviations in the decay time can be observed. The oxygen pressure was in equilibrium to air. Figure 6A also shows that the accuracy of the estimation of τ (as can be seen by the error bars) increased with increasing signal.

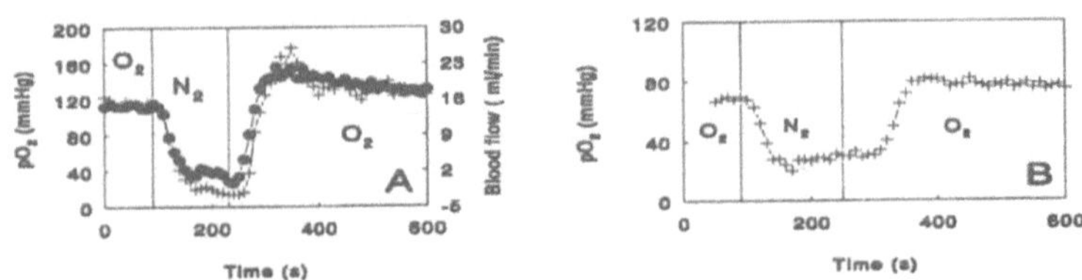

Figure 7. A) The pO_2 (+) in the duodenum and blood flow (●) in the a. mesenterica superior B) The pO_2 in the liver during ventilation with O_2 and N_2.

Figure 7A shows an *in vivo* measurement of the pO_2 and blood flow through the superior mesenteric artery before, during, and after a period of 98% N_2 and 2% enflurane ventilation, Figure 7B shows liver pO_2. The used dye concentration was about 450 nmol Pd-porphine-albumin complex in the rat blood (about 15 ml). Although the rat lay on a warmed mat keeping the core temperature at 39 °C, the temperature on the measured organ was only 30°C which indicates the necessity for temperature compensation. As expected, measured pO_2 values *in vivo* respond well to hypoxic challenge. Moreover, Figure 7A show similar kinetics of both pO_2 and flow reduction during anoxic ventilation.

DISCUSSION

The present paper describes a new phosphorimeter to measure decay times of the oxygen-dependent phosphorescence of Pd-porphine. The main new feature is that the photomultiplier tube is switched off during the excitation flash. Also, a continuous correction for temperature changes is incorporated. With this new phosphorimeter we re-evaluated some of the physical properties of Pd-meso-tetra-4-carboxy-phenyl porphine needed for determination of intravascular pO_2 *in vivo*.

Pd-meso-tetra-4-carboxy-phenyl porphine is a relatively inexpensive probe compared to other Pd-porphine compounds. A disadvantage, however, is its reported pH-dependency[2]. We were not able to confirm this pH-dependency (Figure 3), making its applicability *in vivo* only dependent on temperature changes, a condition also applicable for the more expensive Pd-porphine compounds[2]. For continuous correction of K_q for temperature fluctuations we incorporated a thermocouple and an algorithm to compensate K_q for temperature (Figure 4).

The ideal phosphorimeter shows measured decay times independent of the intensity of the measured phosphorescence. To test whether our phosphorimeter meets to this requirement, we changed the dye concentration and the light distribution of both excitation and emission light. As can be seen from Figure 6A and B the decay time measured by the present phosphorimeter is indeed independent. on the amount of light the PM tube measures.

It is concluded that the presented phosphorimeter provides reliable and reproducible decay times, independent on phosphorescence intensity. The values of K_q and τ_0 presented in this paper should be of use for *in vivo* pO_2 measurements.

Acknowledgements

This study was in part supported by the Netherlands Foundation for Medical Research (grant number 900-519-110).

REFERENCES

1. D.F. Wilson, A. Pastuszko, R. Schneiderman, M. Pawlowski, M. Delivoria-Papadopoulos, Effect of hyperventilation on the oxygenation of the brain cortex in neonates. J Appl Physiol. 70:2691 (1991).
2. M. Pawlowski, D.F. Wilson, Monitoring of the oxygen pressure in the blood of live animals using the oxygen dependent quenching of phosphorescence. Adv Exp Med Biol. (1993) in press.
3. J.R. Lakowicz, 'Principles of Fluorescence Spectroscopy', Plenum Press, . New York,(1983)
4. C. Ince, J.F. Ashruf, J.A.M. Avontuur, P.A. Wieringa, J.A.E. Spaan, H.A. Bruining, Hetrogeneity of the hypoxic state in the rat heart is determined at capillary level. Am J Physiol. 264:H294 (1993).
5. H.J.van Staveren,C.J.M.Moesugo,S.A.Prahl, M.J.C. van Gemert, Light scattering of Intralipid 10% in the wavelength range of 400-1100 nm, Appl Opt. 30:4507 (1991).
6. D.V. O'Connor, W.R. Ware, J.C. Andre, Deconvolution of fluorescence decay curves. A critical comparison of technique, J of Phys. Chem . 38:10 (1979).
7. M. Sinaasappel, H.J.C.M.Sterenborg, Quantification of the hematoporphyrin derivative by fluorescence measurement using dual-wavelength excitation and dual-wavelength detection. Appl Opt. 32:541 (1992).

IMAGING OXYGEN PRESSURE IN TISSUE *IN VIVO* BY PHOSPHORESCENCE DECAY

Marek Pawlowski[1a] and David F. Wilson[2]

[1]Medical Systems Corp., Greenvale, N.Y. 11548
[2]Department of Biochemistry and Biophysics, University of Pennsylvania
Philadelphia, PA 19104

A new imaging system, OxyMap®, has been developed to noninvasively obtain two dimensional maps of the distribution of oxygen concentration in living tissue by imaging the decay of the phosphorescent probes, such as Pd-meso-tetra (4-carboxyphenyl) porphine, in the blood. The phosphorescence of these probes is quenched by oxygen according to the relationship:

$$\frac{T_0}{T} = 1 + k_Q * T_0 * PO_2 \tag{1}$$

Where T and T_0 are the phosphorescence lifetimes in the presence of oxygen at a pressure PO_2 and at an oxygen pressure of zero, respectively. k_Q is the quenching constant that is proportional to the number of collisions that occur between molecular oxygen and the phosphor in the excited state. A flash of excitation light (< 5 μsec at half height) is used to initiate phosphorescence. Images of phosphorescence intensity are captured (at a resolution of 512h x 480v pixels) at different delay times after the flash using a gated, intensified CCD camera. The surface area imaged depends on the optical system used, but standard photographic lens and ring light illumination allow investigation of surface areas up to about 40 mm diameter.

Newly developed, integrated software is designed for use on an i486 microprocessor, operating in 32 bit protected mode, with a TARGA +16/32 frame grabber board. This program controls the timing of the flash lamp and gating of the intensifier of the video camera. In addition it controls the internal circuitry of the camera to permit the sensitivity of the camera to be increased by integration of the light from multiple flashes on the CCD array. Noise reduction can be achieved by averaging and filtering of acquired images. The time needed to collect 7 images with different delay times after the flash is < 6 sec, storing them to disk < 19 sec, and the oxygen pressure maps can be calculated on-line using a 3x3 median filter in less than 47 sec. Off-line processing allows selection of different filters, oxygen

---

[a] To whom the correspondence should be sent.

Oxygen Transport to Tissue XVI
Edited by M.C. Hogan *et al.*, Plenum Press, New York, 1994

"

pressure ranges and other display options for the maps of phosphorescence lifetime and oxygen pressure.

The flash lamp and the camera system can be synchronized to external (TTL standard) signals generated by physiological measuring devices such as those which monitor the heart (ECG), EEG or respiratory function. These monitoring devices can be used to synchronize collection of the images of phosphorescence to precise points in physiological functions that cause rhythmic movement of the area of interest, effectively compensating for any movement during collection of the images.

INTRODUCTION

Currently available instruments based on methods other than phosphorescence quenching do not provide accurate and precise means for oxygen concentration imaging along the investigated area of tissue. Oxygen electrode arrays have been used for multiple point measurements on the tissue surface with a limited spatial resolution, but otherwise oxygen electrodes are limited to invasive measurements along the electrode track during insertion, an invasive technique. In general, oxygen electrodes require calibration before and after insertion and there is a potential for calibration error while in the tissue. Oxygen electrodes have high rates of oxygen consumption at the electrode surface and the measurements are therefore sensitive to any alterations in oxygen diffusion at or near the surface of the cathode.

Oxygen concentration measurement based on the measurement of the decay of the phosphorescent probes is non invasive and as long as the environment of the probe molecules does not change, as is the case for probes bound to albumin in the blood, no calibration errors occur. Thus the values of T_0 and k_Q needed for calculation of the oxygen pressure need only be determined once and then these values can be used for all subsequent measurements providing the experimental conditions are the same. Oxygen consumption by the phosphorescence quenching is very low and does not influence the local oxygen pressure, while the spatial resolution is limited only by the optics of the system, making possible high resolution two dimensional oxygen pressure maps. Spacial resolution of the two dimensional oxygen concentration distribution depends on the optical setup of the camera.

Recent progress in imaging technology, specifically in the quality, sensitivity and gating possibilities of image intensifiers, has made possible collection of very weak luminescence signals for short and precisely measured periods of time. The instrument described here was developed to measure two dimensional oxygen concentration distributions in the surface layer of the tissue. Laboratory prototypes of the instrument and the results achieved have been described in several papers (Shonat et al., 1992, Wilson and Cerniglia, 1992). This paper presents the principles used to develop this instrument, the description of the data acquisition and data processing software, and examples of the applications of the system.

PRINCIPLES OF OPERATION

Phosphorescence is the emission of light from a substance exposed to radiation and persisting as an afterglow after removal of the exciting radiation. This radiation may persist from microseconds to days, Molecules having phosphorescence, such as Palladium complex of porphyrins (see Vanderkooi et al., 1987, Vanderkooi and Berger, 1989), have a triplet state that lies at a lower energy level than that the singlet state and a high rate of conversion between the two states. Return to the ground state from the triplet state requires essentially simultaneous change in the electron spin and photon emission events and thus is "forbidden". The probability of this transition is low and therefore the excited triplet states can be very long lived. For well behaved phosphors the emission of light decays as a single exponential, determined by the probability that any one triplet state molecule will emit a photon. The triplet state to ground state transition can be induced, however, by collision of the excited triplet state molecule with a molecule which can accept energy from the triplet state. The energy transfer provides a mechanism for return of some of the triplet state molecules to the ground state by a nonradiative transition, with a resulting decrease in phosphorescence (quenching). Quenching is a statistical process based on the probability that the triplet state molecule will collide with a suitable energy transfer partner and the probability that energy transfer will occur in any given collision. The effect of a quenching agent is to increase the rate of decay of the triplet state and to shorten the measured phosphorescence lifetime and the amount of light emitted.

In the blood, the primary quenching agent is molecular oxygen, since most other molecules capable of quenching the porphyrin phosphorescence (such as organic free radicals, paramagnetic metal ions, nitric oxide) are generally present at levels far below those of oxygen. Thus, in the blood the phosphorescence of the Pd-porphyrin follows equation (1), and the phosphorescence lifetime provides an accurate measure of oxygen pressure.

The characteristics of the phosphorescence probes can be found in other papers (Wilson et al., 1991, Pawlowski and Wilson, 1992).

Determination of oxygen concentration distribution in the surface layer of the tissue is a two step process:
1. collection of phosphorescence intensity images
2. calculation of the distribution of phosphorescence lifetimes and oxygen concentrations within the investigated area.

Collection of the Phosphorescence Intensity Images

To collect a single phosphorescence intensity image, the system triggers the flash lamp at time zero to excite the porphine molecules in the blood. Then system waits a certain delay time T_d and turns on the CCD camera intensifier to integrate phosphorescence intensity from T_d to infinity, as presented in Figure 1. The default value of the intensifier gating pulse duration is 2.5 msec. This value can be changed depending on the lifetime of the phosphorescence response and on the T_0 of the porphine used. The flash lamp is triggered usually 4 to 8 times at 10 msec intervals with appropriate gating of the intensifier for a given delay time T_d. The CCD array integrates (sums) the collected light from the flash sequence.

The integrated frame is transferred to the frame grabber, digitized and transferred to the computer memory buffer.

The above procedure is repeated, usually for 6 different delay times between the flash of the lamp and the opening of the camera gate. The number of different delay times used should be large enough to allow accurate characterization of the exponential decay.

The set of images of phosphorescence intensity taken with different delays after the flash is accompanied by a so-called background image captured after 2500 µsec delay. The very long delay of the background image assures absence of the phosphorescence response, and permits subtraction of light arising from any non-phosphorescence sources. The background image is captured at the beginning of the measurement procedure and is always subtracted from phosphorescence intensity images.

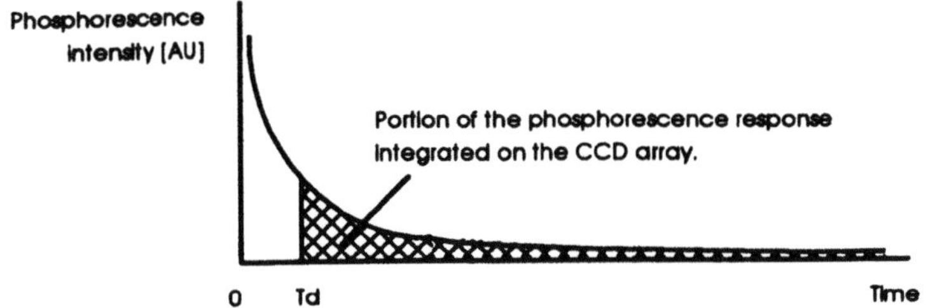

Figure 1. Timing of the phosphorescence intensity image capture.

A typical delay time sequence when acquiring six images, plus background image, for Pd-meso-tetra (4-carboxyphenyl) porphine is: 20 µsec, 40 µsec, 80 µsec, 160 µsec, 300 µsec, 600 µsec, and 2500 µsec. This sequence can be adjusted depending on the expected oxygen concentration level within the investigated area and the T_0 value of the porphine used.

Calculation of the Lifetime and Oxygen Concentration Distributions

Phosphorescence lifetime and oxygen concentration distribution calculations can be carried out with OxyMap system either on-line after completion of the data acquisition, or off-line after the phosphorescence intensity images have been collected and stored on the disk.

Phosphorescence intensity images can be filtered (if needed) using different size (beginning with 3x3 kernel) smoothing median filter, to remove random noise from the images.

For each pixel element with coordinates x, y from each of phosphorescence intensity images a data vector of intensity values is constructed. This data vector forms a plot of integrated intensity versus delay time. Least-squares, single exponential fit is applied to each data vector according to the following equation:

$$\int_{Td}^{\infty} I(T)dT = \int_{Td}^{\infty} I_0 \exp(-T/\tau)dT = I_0\,\tau\,\exp(-T_d/\tau). \tag{2}$$

Three parameters of each single exponential fit are calculated: decay constant, calculated initial value of the decay curve and the correlation coefficient. Upon determination of the decay constant, the oxygen concentration is calculated using the Stern-Volmer relation (1).

Calculated values are stored in data matrices of the same structure as images collected. After completion of the calculation the resulting images can be stored to disk.

DESCRIPTION OF THE INSTRUMENT

Figure 2 presents elements of the OxyMap system. The computer supplied with OxyMap is IBM-PC/AT with 80486/50 MHz processor, 64k static RAM cache memory and 8 Meg dynamic RAM. It has a hard drive of more than 200 MB capacity and an additional tape backup for archival image storage. Approximately 600 images can be stored on one tape. The computer controls the data acquisition and data processing.

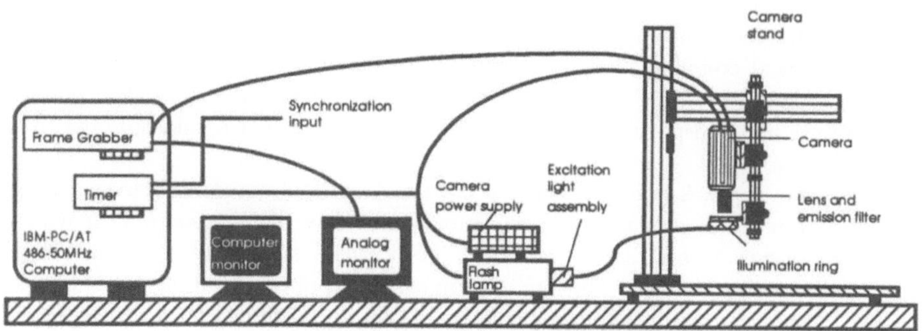

Figure 2. Elements of the OxyMap system.

The counter-timer board is a multi-function and digital expansion board. The board has five independent 16-bit up/down counters, a 1 MHz crystal time base with divider, and separate general purpose 8-bit TTL input/output ports. Fully programmable counter-timer board issues complex sequences of pulses and signals to control the timing of the flash lamp and several functional modules within the camera and the frame grabber.

The frame grabber board is the TARGA+ 16/32 with 1 Meg memory. Due to the characteristics of the camera, TARGA board is working in 8-bit/pixel configuration and 1 Meg memory is configured into 4 frame buffers. The size of the image is 512h x 480v pixels. The mixer of the frame grabber board allows subtraction of a frame stored in a buffer from the incoming one. This enables real time subtraction of a background image.

The ISG-250-R-3, intensified, gated, monochrome CCD video camera (Xybion Electronic Systems Corp., San Diego, CA) is used in OxyMap system. The sensitivity of this camera is better than 10^{-8} lux faceplate illumination. The spectral response of the camera as determined by the use of the Generation 2 (Gen II) single Micro Channel Plate intensifier with S-25 type photocathode covers 390 nm - 920 nm wavelength's range. Photocathode sensitivity to tungsten light at 2854° K is approximately 300 μA/lumen. The luminance gain is typically 18,000. The camera can be gated (turned on or off) in less than 5 nsec and gate widths of down to 20 nsec are possible, providing an electronic shutter between the lens and a solid state CCD array. The camera can be operated in either synchronous or asynchronous mode. The image inhibit signal allows image integration directly on the CCD array.

Phosphorescence emission light is collected by the high quality video zoom lens equipped with the standard C-mount. Parameters of the lens depend on the experimental setup.

A long pass light filter with 630 nm cutoff wavelength at 50% is mounted directly on the camera lens. The filter separates excitation wavelengths, protecting the camera from being saturated by possible reflection of the excitation flash.

The flash lamp (MVS-2601, EG&G, Salem, MA) is a source of high intensity pulsed light. It is used in OxyMap system to deliver excitation radiation to the sample. The flashlamp can be triggered from external source of TTL type signal. The trigger signal is optically separated from the discharge unit and can operate at 1 to 1000 Hz. The flash lifetime is less than 5 μsec at half height with an input energy per flash of 1 J typical.

The light output of the flash lamp is coupled by optical system to the fiber optic bundle to project the light to an area remote from the instrument where the illumination is required. The optical system has been designed to project the image of the flash lamp arc on the surface of the illumination ring light guide through the excitation light filter.

The optical coupling system contains an excitation light filter. It is a typical $\varnothing$ 18 mm bandpass filter (540 ± 35 nm). This filter bandwidth covers the alpha absorption bands of the currently used Pd-porphyrins. Depending on the particular application, the filter can be changed.

The Chiu R-90, 8-point ring light is used in OxyMap system to deliver the light to the area of interest. It provides uniform, interference free illumination over a wide focal distance. The ring's 8-point light sources are the polished ends of optical glass fibers that are mounted in tilting carriers. The illumination ring is equipped with the set of 8 collimating lenses. The ring enables uniform illumination of the area from $\varnothing$ 10 mm to $\varnothing$ 35 mm.

OxyMap SOFTWARE

The phosphorescence lifetime can be calculated from a series of images collected at different delay times after a flash at time zero, with the number of delay times specified to be large enough to characterize the exponential decay. Each image of the size of 512vx480h 8-bit pixels takes approximately 250 kilobyte of memory. Four such images would take the entire memory space accessible in an IBM-PC/AT compatible computer under the DOS operating system with the processor working in so-called real mode. The collection of larger amounts of data can be achieved by using expanded and extended memory mechanisms, but data transfer rates are not fast enough for a real time imaging system.

In order to overcome the above limitations, the OxyMap system software has been developed in C programming language under Intel 386/486 C Code Builder DOS Extender. Under the DOS Extender environment the 80386 or 80486 processors are working in protected mode. Protected mode of operation offers more memory and faster execution, resulting in higher performance. The DOS operating system is designed to run on Intel 86-family processors in 16-bit real mode, which supports a physical address space of 1 megabyte. The Intel 80386 and 80486 processors can operate in 32-bit protected mode, where the physical address space is 4 gigabytes, and where 32-bit operations are significantly faster than their 16-bit counterparts. The flow diagram of the acquisition part of the software is presented in Figure 3. The operation of the software begins with determination of the acquisition protocol.

The largest number of delay times that can be used to determine the decay curve is equal to the number of images that can be collected and stored in the computer memory. Thus, working on the computer with 8 megabytes of memory, 20 images at most can be collected during the data acquisition with the rest of the memory occupied by the program.

The ISG-250 type camera permits integration of several images directly on the CCD array. This feature allows the flash lamp to be fired several times within the same frame exposure but with the intensifier appropriately gated for each flash. The phosphorescence response after each flash falls on the CCD array during the same exposure, resulting in an integration of the light from several flashes into a single image. The resulting image is brighter than that of a single exposure in proportion to the number of flashes integrated. This significantly increases the camera sensitivity and the signal to noise attained in each image.

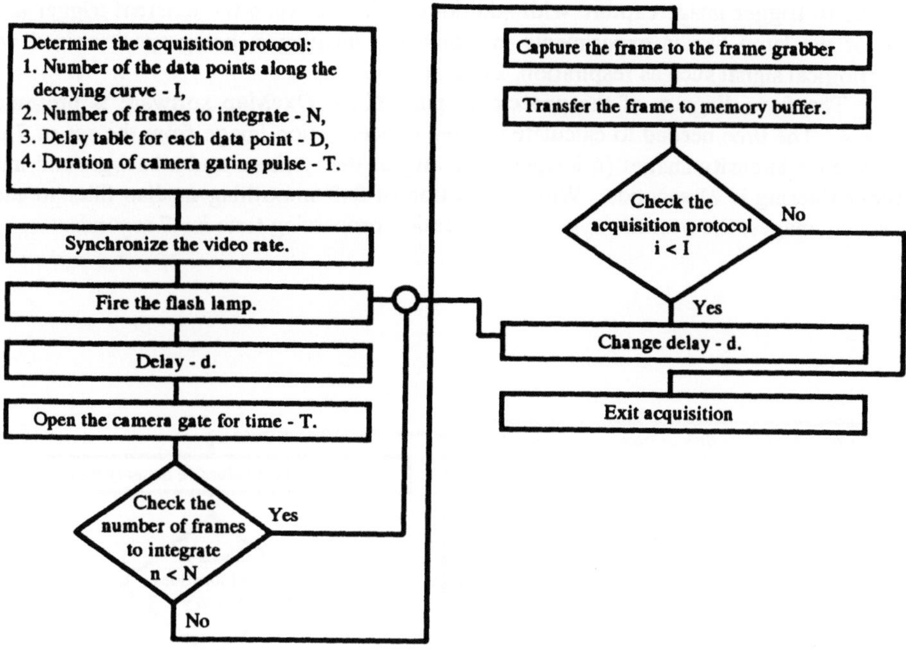

Figure 3. Flow diagram of the acquisition part of the OxyMap software.

The flow diagram in Figure 3 presents the internal measurement loop without external synchronization pulses from, for example, an ECG signal monitor. In this case the camera is gated in the so-called synchronous mode. The flash of the lamp, the opening of the camera gate as well as the frame grabber are synchronized with the video rate of the camera.

After synchronizing pulse, the system waits for the upcoming odd field in the video signal of the camera and for the corresponding vertical drive pulse. Upon detecting the vertical drive pulse, the computer activates the previously programmed counter-timer board. The counter-timer board issues a series of signals that control both the flash lamp and the camera. In the synchronous mode of operation the flash lamp and the camera gate are triggered twice during the 16 ms time period of the odd field with a 5 ms break between the flashes. The break period between the flashes is limited by the decay constant of the camera photocathode response. This feature enables fast data collection without overloading the flash lamp.

After integrating a given number of frames, the image inhibit is released, the frame grabber captures the image released from the camera, and the image is transferred to the memory buffer.

The time needed to collect 7 images (6 images with various decay times and a background image) in synchronous mode of operation (with integration of four frames) is 5.5 seconds.

The acquisition of the above set of data, saving phosphorescence intensity images to disk for off line processing, takes 19 seconds.

In the asynchronous mode of operation, the system waits for the external trigger signal with suspended image release from the camera. The image suspension period cannot be longer than 16 frames because of the noise signal collected on the CCD array. It is also possible to trigger image capture with additional delay between the external trigger and the flash of the lamp like in experiments in which the imaging is being synchronized with a physiological signal such as respiration, ECG or EEG.

The flow diagram of the processing part of the OxyMap software is presented in Figure 4. The time needed to calculate oxygen concentration image from the set of 7 phosphorescence intensity images (6 images with various decay times and a background image), without filtering is 21 seconds. With application of 3x3 smoothing median filter to each of the seven images, the overall oxygen concentration processing time is 47 seconds.

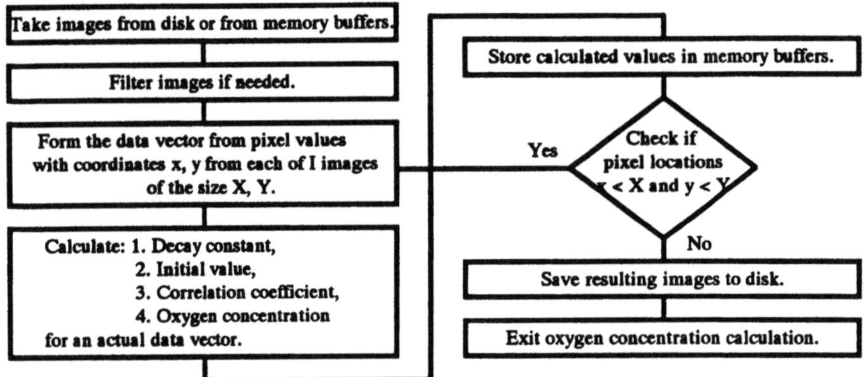

Figure 4. Flow diagram of the image processing part of the OxyMap software. I -- number of phosphorescence intensity images in the measurement sequence. X, Y -- size of the individual image (512hx480v).

OxyMap can acquire, calculate, filter and save oxygen maps every 61 seconds. For rapid acquisition without on-line oxygen calculation, OxyMap can acquire data every 19 seconds.

In the experimental conditions there is usually a mix of different pO_2 values within the investigated area. Thus there might be an error in the measurement associated with fitting a single exponential curve to a multiexponential decay. The computer calculates a fit for each pixel element. Depending on the dimensions of the investigated area the size of the pixel is small. For an image size 3x3 mm the pixel size is about 6 μm and represents a small tissue volume. In this case the error associated with single exponential fit is minimized. This is demonstrated by the fact that the correlation coefficient for the fit at each pixel element is always bigger than 0.95 and usually closer to 0.99.

The OxyMap system offers the following image processing functions:

    Median filters of different sizes for smoothing the images,

    Different pseudo-color palettes in addition to gray scale for image display.

    Arithmetic functions for addition, subtraction, and scaling of the images.

APPLICATIONS

The OxyMap system described here is currently employed to image tissue oxygenation in different experimental animal models. A partial listing includes study of hypoxic-ischemic injury in the adult cat brain (Wilson et al, 1993) and newborn piglets (Wilson et al, 1992), oxygenation and radiation sensitivity in tumor tissue (Wilson and Cerniglia, 1992), blood flow in the cat eye (Shonat et al, 1993), oxygen sensing in the cat carotid body (Lahiri et al, 1993), and oxygen gradients in vascular occlusion in the heart (Rumsey et al, this proceedings).

Acknowledgement: Development of the OxyMap system has been supported in part by SBIR Grant No. R43 NS30265-01 "Quantitative 2D Oxygen Mapping in Tissue".

REFERENCES

Lahiri, S., Rumsey, W.L., Wilson, D.F., and Iturriaga, R., 1993, Contribution of in vivo microvascular PO_2 in the cat carotid body, *J. Appl. Physiol.*, 75:1035.

Pawlowski, M., and Wilson, D.F., 1992, Monitoring of the oxygen pressure in the blood of live animals using the oxygen dependent quenching of phosphorescence, *Adv. Med. Biol.* 316:179.

Rumsey, W.L., Pawlowski, M., Lejavardi, N., and Wilson, D.F., 1993, Oxygen pressure distribution in the heart of newborn piglets in vivo during coronary artery occlusion, *Adv. Med. Biol.* (these proc.).

Shonat, R.D., Wilson, D.F., Riva. C.E., and Pawlowski, M., 1992, Oxygen distribution in the retinal and choroidal vessels of the cat as measured by a new phosphorescence imaging method, *Appl. Optics*, 31:3711.

Shonat, R.D., Wilson, D.F., Riva, C.E., and Cranstoun, S.D., 1992, Effect of acute increases in intraocular pressure on intravascular optic nerve head oxygen tension in cats, *Invest. Ophth. & Vis. Sci.*, 33:3174.

Vanderkooi, J.M., Maniara, G., Green, T.J., and Wilson, D.F., 1987, An optical method for measurement of dioxygen concentration based upon quenching of phosphorescence, *J. Biol. Chem.*, 252:5476.

Vanderkooi, J.M., and Berger, J.W., 1989, Excited triplet states used to study biological macromolecules at room temperature, *Biochim. Biophys., Acta*, 976:1.

Wilson, D.F., Pastuszko, A., DiGiacomo, J.E., Pawlowski, M., Schneidermann, R., and Delivoria-Papado-poulos, M., 1991, Effect of hyperventilation on oxygenation of the brain cortex of newborn piglets, *J. Appl. Physiol.*, 70:2691.

Wilson, D.F., and Cerniglia, G.L., 1992, Localization of tumors and evaluation of their state of oxygenation by phosphorescence imaging, *Cancer Research*, 52:3988.

Wilson, D.F., Gomi, S., Pastuszko, A., and Greenberg, J.H., 1993, Microvascular damage in the cortex of cat brain from middle cerebral artery occlusion and reperfusion, *J. Appl. Physiol.*, 74:580.

MEASUREMENT OF OXYGEN PRESSURE IN THE HEART IN VIVO USING PHOSPHORESCENCE QUENCHING

W. L. Rumsey[4], M. Pawlowski[3], N. Lejavardi[2], and D. F. Wilson[1]

[1]Department of Biochemistry and Biophysics, [2]Department of Physiology;
University of Pennsylvania School of Medicine, Philadelphia, PA 19104;
[3]Medical Systems Corp., Greenvale, NY 11548; and [4]Bristol-Myers
Squibb Pharmaceutical Research Institute, Princeton, NJ 08543

INTRODUCTION

It is generally accepted that a non-invasive method for measuring oxygen pressure in the heart would be a valuable aid to investigators interested in heart and circulatory physiology. We have developed an optical method for imaging the oxygen pressure in the surface layer of tissue, 0.5 to 1 mm in depth, which is based on the oxygen dependent quenching of phosphoresence (Rumsey et al., 1988; 1990; 1991; Wilson et al., 1988; 1991; 1992). From these measurements, two dimensional maps are constructed, thereby providing an accurate analysis of oxygen pressure in tissue in response to various stimuli and experimental conditions. The motion of the heart induced largely by the systolic phase of the cardiac cycle limits the time available for image acquisistion. This necessitates the synchronization of the operation of the flash lamp and imaging of the phosphorescence with the cardiac cycle. In this study we provide preliminary data that show it is possible to obtain maps of oxygen pressure in the epicardium of the heart *in vivo*.

METHODS AND MATERIALS

The measurement of oxygen pressure based on the quenching of phosphorescecne by oxygen has been described in depth previously (see for example, Lahiri et al., 1993; Robiolio et al., 1989; Rumsey et al., 1988; 1990; Vanderkooi et al 1987; Wilson et al., 1988; 1992). In brief, the hearts of anesthetized newborn piglets (3-4 days of age, 2-4 kg) were imaged (OxyMap^TM, Medical Systems Corp., Greenvale, N.Y.) through a Wild Macrozoom microscope with an epifluorescent attachment. The left ventricle was exposed via thoracotomy while maintaining artificial ventilation. Pd-meso-tetra (4-carboxyphenyl) porphine (60 mg in 10 ml of saline containing 60 mg of bovine serum albumin, pH 7.4) was administered to each animal via a cannula placed in the femoral vein. Excitation of the oxygen probe was provided by passing light through an interference filter (center wavelength

537 nm, 45 nm bandwidth at half height). The emitted light, i.e., phosphorescence, was detected with an intensified CCD camera (ISG-250-R-3 Xybion Electronics Systems Corp., San Diego, CA) coupled to the microscope containing a 630 nm cutoff filter . The flashlamp assembly and camera were interfaced with a computer (486/50 MHz, Tri-Star, Chandler, Arizona). A color video monitor (PVM-1341, Sony, San Jose, California) was used for on-line visualization of the lifetime and oxygen pressure distributions from the surface of the heart. A small black suture, about 0.5 mm wide, was positioned on the surface of the ventricle as a reference source for possible motion induced errors.

Heart rate was monitored (MSC-2001, Medical Systems Corp., Greenvale, New York) and the peak of the R wave was used to fire a TTL signal which started the measurement sequence. To synchronize acquisition of data during diastole, a delay time of 160 msec elapsed before the flashlamp was fired. Timing was established using a programmable counter timer board (CTM-05, Keithley-Metrabyte, Taunton, Massachusetts) with resolution of 1 μsec. The camera in combination with the timer board permitted rapid on and off switching of the camera intensifier ($<$ 250 nsec intensifier switching time and gate widths typic",lly 2.5 msec). Images were acquired by a frame grabber (Targa+ 16/32, Truevision, Indianapolis, Indiana) in the form of a full frame. After the lamp flashed, variable delay periods; 20, 40, 80, 160, 300, 600, and 2500 μsec (background image) with the gate width in each case of 2.5 msec, were applied for collection of 7 images.

RESULTS AND DISCUSSION

Figure 1 shows that when the animals were ventilated with room air the microvascular oxygen pressure within the epicardium was, in this case, about 20-30 Torr. It can also be seen that the distribution of epicardial oxygen pressures was heterogenous with individual values varying about a mean value within micro regions. There were, however, no gross regional differences in oxygen pressure.

By lowering the level of inspired oxygen for a brief period, a marked decrease in arterial oxygen pressure resulted and consequently, microvascular PO_2 also declined. It can be seen in Figure 1 that decreasing the ventilated oxygen from 21% (room air) to 14%, microvascular oxygen pressure fell from about 20-30 Torr to about 8-12 Torr. Further reduction of inspired oxygen to 11% oxygen, brought about marked changes in epicardial oxygen pressure, decreasing to about 4-6 Torr. When the animal was returned to normal arterial oxygen pressure to avoid any physiological impairment, microvascular oxygen pressures returned to those obtained during baseline conditions after several minutes.

In general, methods that have been available previously to investigators interested in determining the oxygenation state of myocardial tissue have been limited to either indirect, qualitative methods or those relying on more invasive procedures. For example, it has been possible to obtain measurements of myocardial oxygenation by ultraviolet flash photography for detection of NADH fluorescence (1, 15, 20) arising from hearts perfused in the isolated state with cell free media. NADH fluorescence, however, provides only an indirect estimate of changes in intracellular oxygenation and/or metabolism.   Moreover, NADH fluorescence is subject to interference by changes in absorbance and fluorescence of other endogenous chromophores such as myoglobin, hemoglobin and cytochromes. Recently, a new technique has been developed, based on the electron affinity of nitroheterocyclces, that permits the imaging of hypoxic tisssue in heart using a standard gamma camera (Rumsey et al., 1993; these proceedings). By injecting a [99m]technetium labeled nitroheterocycle into the bloodstream, a bright region emits from hypoxic tissue surrounded by less intense areas of normoxic tissue within the gamma camera image. Although this method may allow for the non-invasive, clinical assessment of myocardial oxygenation, it will not provide direct quantitation of oxygen pressure.

Using oxygen microelectrodes, Schubert and co-workers (1978) measured tissue oxygen pressures in feline hearts perfused in the isolated state with cell-free media gassed with 95%:5% O_2:CO_2. These ex vivo conditions, however, produce abnormally high levels of coronary flow at low levels of cardiac work and do not represent conditions of tissue oxygenation normally found in vivo. Other workers (Gayeski and Honig, 1987) have

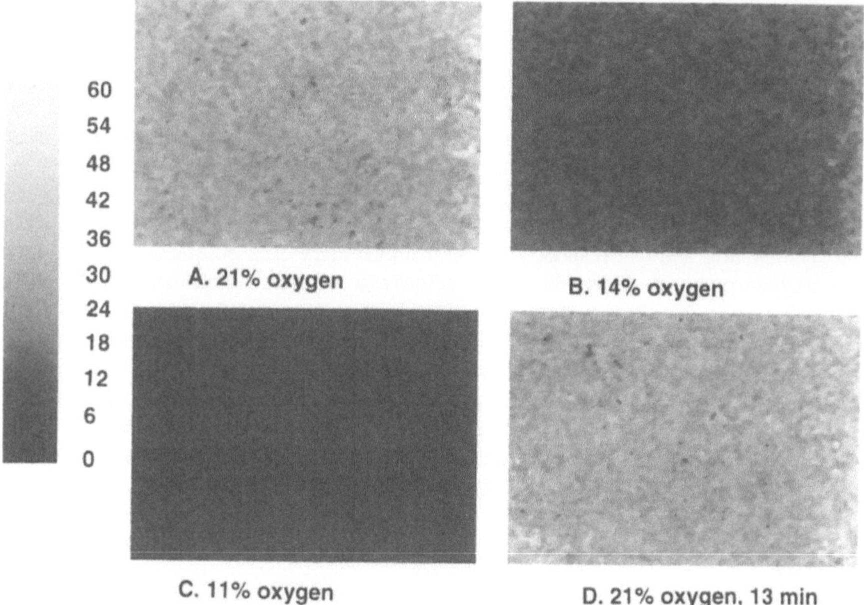

Figure 1. Distribution of oxygen pressures in the epicardial surface of newborn piglet. Two dimensional maps of oxygen pressures in the epicardial surface were constructed from measurements of phosphorescence lifetimes. These values were obtained while ventilating the animal on 21 (A) percent oxygen. The level of inspired oxygen was then lowered stepwise to 14 (B) and 11 (C) percent oxygen. The period of time that the animal was maintained at these low levels of oxygen was about 2 min. Finally the inspired oxygen was increased to normal levels (D). The latter image was aquired 13 min after returning to room air. Units of the scale are in Torr. The magnification 12.5X was equivalent to an area of 4.7 mm X 3.53 mm.

measured the intracellular oxygen pressure of dog and cat heart using myoglobin cryospectroscopy. The level of oxygen in these frozen sections was found to be about 5 Torr and varied by less than 2 Torr in clusters of 10-20 myocytes. Consistent with these findings, Coburn and coworkers (1973) estimated the intracellular oxygen pressures of anesthetized dogs to be between 4 and 6 Torr. These investigators measured carboxymyoglobin concentrations in biopsies of cardiac muscle and related these values to blood carboxyhemoglobin levels after injection of ^{14}CO in order to calculate the tissue oxygen pressure. The invasive nature of these techniques does not allow the investigator to obtain data efficiently from a broad area of tissue. Moreover, it is well known that insertion of oxygen microelectrodes damages cells at the point of impalement.

The utility of measuring oxygen pressure via the method of phosphorescence quenching in tissue has been thoroughly demonstrated in extracardiac tissues (see for example, Rumsey et al., 1988; 1991; Wilson et al., 1991; Wilson and Cerniglia, 1992). This method can be used with either a light sensitive video camera as the photon detector or the video camera can be replaced by a photomuliplier tube coupled to optical fibers. The latter should be capable of transmitting both the excitation light and the emission signal (a system known as "Oxyspot" is commercially available from Medical Systems Inc. Greenvale, NY). Measuring oxygen pressure in tissue using phosphorescence quenching has the distinct advantage that it permits on-line collection of quantitative data (fiber optic system) without disturbing the local environment of the tissue of interest. By imaging the phosphorescence, a broad area of tissue (several millimeters) can be assessed by the investigator. The preliminary data provided in the present report extend our previous observations and demonstrate that this technique can also be applied to the heart.

The work of Coburn and colleagues (1973) and Gayeski and Honig (1987) suggest that the intracellular oxygen pressures in heart are about 5-6 Torr measured in two different animal species, dog and cat, respectively. Our preliminary data obtained in piglets indicate that the oxygen pressure determined in the microvasculature of the heart is about 20 Torr. Similar values were obtained in canine heart by monitoring phosphorescence with the fiber optic system (Rumsey et al., these proceedings). We have shown previously that the oxygen gradient from sarcolemma to mitochondria in cardiac myocytes isolated from adult rat heart is small, about 1 Torr (Rumsey et al., 1990). It has been suggested that large intracellular gradients of oxygen exist in cells (Jones, 1986). Given the results described above, the present findings indicate that the oxygen pressure gradient between the exchange vessels and the mitochondria of cardiac myocytes is primarily extracellular.

Acknowledgments This work was supported by NS-10939, NS-34165, and SBIR-1-R43-NS30265-01.

REFERENCES

Coburn, R.F. Ploegmakers, F., Gondrie, P. and Abboud, R. 1973, Myocardial myoglobin oxygen tension. *Am. J. Physiol.* 224 (4): 870.

Jones, D.P. 1986, Intracellular diffusion gradients of O_2 and ATP. *Am. J. Physiol.* 250: C663.

Gayeski, T.E.J., and Honig, C. R. 1987, Comparison of intracellular PO_2 and conditions for blood-tissue O_2 transport in heart and working red skeletal muscle. *Adv. Exptl. Med.* 215: 309.

Lahiri, S., Rumsey, W.L., Wilson, D.F., and Iturriaga, R. 1993, Contribution of in vivo microvascular PO_2 in the cat carotid body chemotransduction. *J. Appl. Physiol.* 75 (3): 1035.

Robiolio, M., Rumsey, W.L. and Wilson, D.F., 1989, Oxygen diffusion and mitochondrial respiration in neuroblastoma cells. *Am. J. Physiol.* (Cell Physiol. 25) 256: C1207.

Rumsey, W.L., Cyr, J.E., Raju, N., and Narra, R.K., 1993, A novel ^{99m}technetium-labeled nitroheterocycle capable of identification of hypoxia in heart. *Biochem. Biophys. Res. Comm.* 193: 1239.

Rumsey, W.L., Schlosser, C., Nuutinen, E.M., Robiolio, M., and Wilson, D.F., 1990, Cellular energetics and the oxygen dependence of respiration in cardiac myocytes isolated from adult rat. *J. Biol. Chem.* 265 (26): 15392.

Rumsey, W.L., Vanderkooi, J.M., and Wilson, D.F. 1988, Imaging of phosphorescence: A novel method for measuring the oxygen distribution in perfused tissue. *Science* 241, 1649.

Rumsey, W.L., Iturriaga, R., Spergel, D., Lahiri, S., Wilson. D.F. 1991, Optical measurements of the dependence of chemoreception on oxygen pressure in the cat carotid body. *Am. J. Physiol.* 261 (Cell Physiol. 30): C614.

Vanderkooi, J.M., Maniara,G., Green, T.J., and Wilson, D.F. 1987, An optical method for measurement of dioxygen concentration based on the quenching of phosphorescence. *J. Biol. Chem.* 262: 5476.

Wilson, D.F. and Cerniglia, G.J. 1992, Localization of tumors and evaluation of their state of oxygenation by phosphorescence imaging. *Cancer Res.* 52: 3988.

Wilson, D.F., Pastusko, A., DiGiacomo, J.E., Pawloski, M., Schneiderman, R., Delivoria-Papadopoulos, M. 1991, Effect of hyperventilation on oxygenation of the brain cortex in newborn piglets. *J. Appl. Physiol.* 70: 2691.

Wilson, D.F., Rumsey, W.L., Green, T.J., and Vanderkooi, J.M., 1988, The oxygen dependence of mitochondrial oxidative phosphorylation measured by a new optical method for measuring oxygen concentration, *J. Biol. Chem.* 263 (6): 2712.

DETECTING HYPOXIA IN HEART USING PHOSPHORESCENCE QUENCHING AND 99MTECHNETIUM-NITROIMIDAZOLES

W.L. Rumsey, B. Kuczynski and B. Patel

Bristol-Myers Squibb Pharmaceutical Research Institute
Princeton, NJ 08543

INTRODUCTION

The oxygen requirements of cardiac muscle are high and continuous, due largely to the energy expended during cycling of contractile proteins. Any decrease in oxygen flow to the cardiac myocytes places these cells at risk for loss of function or possibly infarction dependent upon the duration of oxygen deprivation. It is well recognized therefore that a non-invasive method for evaluating the oxygenation state of the heart would be a valuable tool for both the research and the clinical setting.

Measurement of oxygen pressure based on the oxygen dependent quenching of the phosphorescence lifetime of selected lumiphors has been used previously for a variety of cell types in vitro, including cardiac myocytes, and in tissues in vivo (Robiolio et al., 1989, Rumsey et al., 1988; 1990; 1991; Wilson et al., 1988a,b; 1991; 1992). For in vivo studies, oxygen measurements were made by acquiring images of phosphorescence lifetimes within a small area of tissue using a light sensitive video camera as the photon detector. These values were then used to construct two dimensional maps of the oxygen pressure distribution in the surface layer (0.5-1 mm depth) at discrete intervals of time. A system is now available ("Oxyspot", Medical Systems, INC, Greenvale, NY) that utilizes a fiber optic bundle interfaced with a photomultiplier tube and that provides continous, on-line measurements of oxygen pressure when positioned on the surface of tissue. At the present time, placement of the fiber optic probe on the myocardial surface necessitates surgical exposure of the heart, a highly invasive maneuver.

In an effort to create a non-invasive procedure for evaluating tissue oxygenation, a 99mTc-labeled nitroimidazole (BMS-181321) has been developed which is preferentially retained in hypoxic cardiac myocytes (Linder et al., 1992; Rumsey et al., 1993). This compound may therefore provide a means for imaging hypoxia in heart using the gamma camera. In the present study, we have developed a model of canine coronary insufficiency utilizing on-line measurements of epicardial oxygen pressure obtained with the "Oxyspot" in order to establish the efficacy of BMS-181321.

Oxygen Transport to Tissue XVI
Edited by M.C. Hogan *et al.*, Plenum Press, New York, 1994

METHODS AND MATERIALS

Mongrel dogs (15-18 kg) were anaesthetized with sodium pentobarbital (50 mg/kg IP, Nembutal, Abbott, Chicago, IL) and instrumented for measurements of aortic and left intraventricular pressures, dp/dt and cardiac electrical activity. Both femoral veins were cannulated for administration of additional anesthesia (when needed), BMS-181321 (50-60 mCi in one ml) and Pd-meso-Tetra (4-carboxyphenyl) porphine (100 mg in saline containing 6% bovine serum albumin). A tracheal cannula was inserted and the animal was respirated at rates sufficient to produce normoxemia (arterial pO_2 = 85-95 Torr) and arterial pH values in the normal range (7.3-7.38, pCO_2 = 37-42 Torr). The heart was exposed via thoracotomy between the 4th and 5th rib. The pericardium was used to cradle the heart in order to minimize motion imparted by filling of the lungs. The left anterior descending artery (LAD) was isolated near its origin and a plastic screw type occluder positioned around the vessel.

For determination of microvascular oxygen pressures (mPO_2) in the epicardium, a fiber optic bundle capable of transmitting light to and from the tissue was placed about 2-3 mm from the surface of the tissue and distal to the LAD occluder. The position of the fiber optic bundle was stabilized using aluminum bars connected to the surgical table and the bundle directed towards an area free of surface conduit vessels. The fiber optic cable was coupled to both a xenon flashlamp assembly and a photomultiplier tube for excitation of the lumiphor (> 510 nm) and detection of the emitted light (> 650 nm), respectively. This system (referred to as "Oxyspot", Medical Systems Inc., Greenvale, NY) was interfaced with a computer (Dell 320LX, Austin, TX) for on-line monitor of phosphorescence lifetimes and oxygen pressures.

Ischemia was induced by tightening the screw clamp on the LAD. The decrease in coronary flow was not monitored directly rather, the lowering of mPO_2 was used as an indicator of ischemia. Coronary flow was reduced sufficient to provide 2-5 Torr of oxygen in the microvasculature. The animal was maintained at this level of mPO_2 for the remainder of the experiment. After the animal was stabilized at this low level of mPO_2, about 8-10 min, BMS-181321 was injected and its cardiac distribution was monitored using the gamma camera (Elscint Apex 409, Hackensack, NJ).

After sacrifice of the animal (Euthanasia-5; Schein, Port Washington, NY), the heart was dissected from the great vessels, removed from the thorax, and rinsed in ice-cold saline. The heart was placed on the camera head and monitored for radioactivity.

Preparation of Radiopharmaceutical

Synthesis of BMS-181321 (Oxo [[3,3,9,9-tetramethyl-1-(2-nitro-1H-imidazol-1-yl)-4,8-diazaundecane-2, 10-dione dioximato] (3-)-N, N', N", N'"]technetium) has been described previously (Linder et al., 1992). In brief, BMS 181321 was prepared by dissolving 2.0 mg of PnAO-1,2-nitroimidazole ligand in 1.5 ml of saline, 0.5 ml of 0.1N $NaHCO_3$ and 0.5 ml of generator eluant (up to 100 mCi $^{99m}TcO_4^-$). The reaction was initiated by addition of 50 μl of a deoxygenated saturated stannous tartrate solution. The reaction was completed within 10 min at room temperature and the compound used immediately after preparation. The radiochemical purity of BMS-181321 was always greater than 90% as determined by High Pressure Liquid Chromatography (HPLC) using a Hamilton PRP-1 column eluted with acetonitrile/0.1M NH_4OAc, pH 4.6.

RESULTS AND DISCUSSION

Figure 1 shows an example of values of phosphorescence lifetime and mPO_2 obtained in canine hearts during resting, baseline conditions and in response to LAD stenosis. The scatter in the data is due to motion induced by contraction and relaxation during the cardiac cycle. Despite this variation, it can be seen that, in this case, resting values of the phosphorescence lifetime and mPO2 are about 100-150 μsec and 20-25 Torr, respectively. These values can be considered typical of 5 animals. By decreasing flow within the LAD microcirculatory bed with the use of the screwclamp, phosphorescence lifetimes increased rapidly after the onset of ischemia, rising to a maximal value of about 600 μsec. Concomitantly, the level of mPO2 decreased to about 4-5 Torr. This value was maintained at this level for the remainder of this recording and was checked intermittently throughout the experimental period.

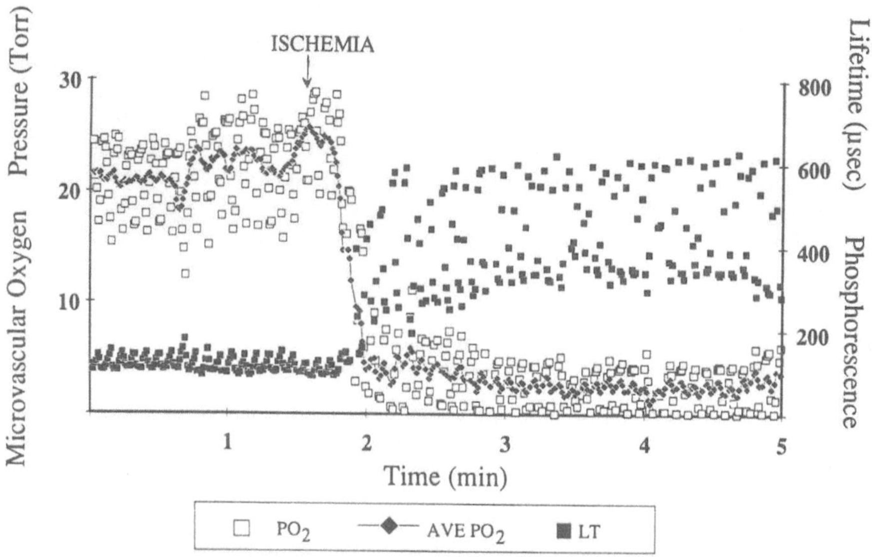

Figure 1. Phosphorescence lifetimes and oxygen pressures in the epicardial surface of canine heart. Phosphorescence lifetimes (LT) and microvascular oxygen pressures are plotted as a function of time. By tightening a screw clamp on the LAD artery at about 1 min and 50 sec of this scan, lifetime values increased and oxygen pressures decreased. For the oxygen pressures, the raw data (PO2) was smoothed (ave PO2) by averaging every seven values, i.e., data points 1-7, 2-8, 3-9, etc.

Immediately after intravenous administration, gamma camera imaging of BMS-181321 resulted in the rapid appearance of radioactivity in the bloodpool of the heart. Sequential images recorded over several minutes, demonstrated an area of greater intensity within the heart, corresponding to the left anterior wall (Figure 2A). The latter was confirmed when the heart was imaged ex vivo (Figure 2B). Oxygen measurements described above indicated that this region of the heart was severely hypoxic. It can also be seen in Figure 2A that radioactivity accumulated within the liver appearing as a bright dome at the base of this image. Despite the association of BMS-181321 with the liver, it was apparent that images of hypoxic tissue could be obtained with the gamma camera in planar mode.

Although the utility of imaging of oxygen pressure via the method of phosphorescence quenching has been thoroughly demonstrated in extracardiac tissues (see for example,

Wilson and Cerniglia, 1992), imaging oxygen pressure in the heart is complicated by its inherent motion. Using a light sensitive video camera as the photon detector requires that the acquisition of the images be synchronized to the electrical impulses of the heart, most especially those of the ventricles since these are the areas of highest interest in   studies of

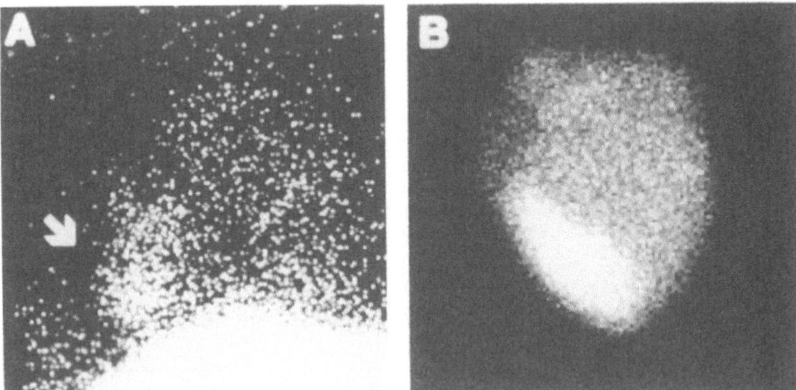

Figure 2A and B. Gamma camera images of BMS-181321. In vivo (A) and ex vivo (B) images were obtained in the planar mode. In (A), the dog was positioned such that the anterior aspect of the animal is to the left side of the image and the inferior aspect at the base. The image was acquired using a left anterior oblique view at 50° at 131 min post-injection. The arrow points to the apical region of the heart. In (B), the heart was removed from the body at the conclusion of the experiment, 182 min post-injection, and images obtained while the heart was positioned atop the head of the gamma camera.

coronary ischemia. In the absence of cardiac gating of the image acquisition with the electrocardiogram, images are blurred and lack adequate resolution for determination of oxygen pressure. On the other hand, replacing the video camera with a photomuliplier tube as the photon detector permits the acquisition of data from the epicardial surface without the absolute requirement of synchronizing cardiac electrical activity with the collection of phosphorescence lifetimes. The present study shows that directional changes in the tissue oxygenation state are easily obtainable using this instrumentation, i.e, "Oxyspot".

In general, studies of myocardial ischemia utilize flow meters placed around either the circumflex artery or the left anterior descending artery to measure changes in coronary flow. Radiolabelled microspheres injected into the circulation can also be used for local distribution of coronary flow but only at defined times within the experiment. Often these methods are coupled to other techniques that provide measures of regional (ultrasonic crystals used for determination of ventricular wall shortening) or global (intraventricular pressure transducer for measuring dp/dt) cardiac function. These types of measurements provide only indirect indicators of oxygen delivery to the parenchymal cells of the left ventricle. By using the technique of phosphorescence quenching described in the present study, direct measurements of oxygen pressure can be made in a continous manner for several minutes as needed throughout the entire experimental period. In this way, the investigator can determine to what extent a 60% reduction in coronary flow effects a loss in microcirculatory oxygen pressure. This technique offers, therefore, a valuable aid in studies of coronary ischemia in the research laboratory.

At present, there are no non-invasive methods for determining the oxygenation state of the heart for use in the clinical setting. Chapman and co-workers (1989) have suggested that radiolabeled nitroheterocycles could be useful for demarcation and imaging of hypoxic tissue using the gamma camera. Previously, these electron affinic compounds were developed for treatment of certain bacterial infections and as radiosensitizers in cancer therapy (Rauth, 1984). Although the mechanism of action for these compounds has not been fully elucidated, it is thought that they are reduced to a radical anion in both normoxic and hypoxic tissue. This radical species reacts with oxygen to produce the superoxide anion and, in so doing, the initial form of the nitroimidazole is regenerated. In hypoxic tissue, however, the nitro radical is believed to be reduced further (a total 4 electron stoichiometry) to yield a hydroxylamine derivative (McClelland et al., 1984) and to covalently bind to intracellular macromolecules (Kedderis et al., 1989; Varghese et al., 1976; Varghese and Whitmore, 1981). According to this hypothesis, a radiolabelled nitroheterocycle would yield a bright region from hypoxic tissue surrounded by less intense areas of normoxic tissue within the gamma camera image. We have chosen to radiolabel a 2-nitroimidazole with [99m]technetium (Linder et al., 1992). This isotope is short-lived (6 hr half-life), is readily available, and has a more suitable photon energy than other isotopes, providing beneficial characteristics for use in humans. The combined properties of nitroheterocycles and [99m]technetium described above, therefore, could provide a means of detecting hypoxic myocardium using a non-invasive technique.

The present data indicate that it is possible to obtain gamma camera images of a hypoxic territory in heart muscle (as measured by phosphorescence quenching) using a novel [99m]technetium labelled nitroimidazole, BMS-181321. Although the oxygen dependence of nitroheterocyle accumulation is not known, the present data suggest that BMS-181321 was retained in severely hypoxic tissue, i.e., less than 5 Torr. Further studies are needed in order to determine the minimum decrease in microvascular oxygen pressure necessary to retain sufficient BMS-181321 for gamma camera image acquisition.

Acknowledgments

The authors are grateful for the skilled technical assistance of Ms Christine Hood.

REFERENCES

Chapman, D., Lee, J., and Meeker, B.E., 1989, Cellular reduction of nitroimidazole drugs: Potential for selective chemotherapy and diagnosis of hypoxic cells, *Int. J. Radiation Oncology Biol. Phys.* 16: 911.

Kedderis, G.L., Argenbright, L.S., and Miwa, G.T., 1989, Covalent interaction of 5-nitroimidazoles with DNA and protein in vitro: Mechanism of reductive activation, *Chem. Res. Toxicol.* 2: 146.

Lahiri, S., Rumsey, W.L., Wilson, D.F., and Iturriaga, R. Control of microcirculatory oxygen pressures in the cat carotid body. *J. Appl. Physiol.* In Press.

Linder, K.E., Cyr, J., Chan, Y.-W., Raju, N., Ramalingam, K., Nowotnik, D.P., and Nunn, A.D., 1992, Chemistry of a Tc-PnAO-nitroimidazole complex that localizes in hypoxic tissue, *J. Nucl. Med.* 33 Suppl (5): 919 (abstr.)

McClelland, R.A., Fuller, J.R., Seaman, N.E., Rauth, A.M., and Battistella, R. 1984, 2-Hydroxylaminoimidazoles-Unstable intermediates in the reduction of 2-nitroimidazoles, *Biochem. Pharmacol.* 33: 303.

Rauth, A.M., 1984, Pharmacology and toxicology of sensitizers: Mechanism studies, *Int. J. Radiation Oncology* 10: 1293.

Robiolio, M., Rumsey, W.L. and Wilson, D.F., 1989. Oxygen diffusion and mitochondrial respiration in neuroblastoma cells. *Am. J. Physiol.* (Cell Physiol. 25) 256: C1207.

Rumsey, W.L., Cyr, J.E., Raju, N., and Narra, R.K., 1993, A novel ^{99m}technetium-labeled nitroheterocycle capable of identification of hypoxia in heart. Biochem. Biophys. Res. Comm. 193: 1239.

Rumsey, W.L., Schlosser, C., Nuutinen, E.M., Robiolio, M., and Wilson, D.F., 1990, Cellular energetics and the oxygen dependence of respiration in cardiac myocytes isolated from adult rat. *J. Biol. Chem.* 265 (26): 15392.

Rumsey, W.L., Vanderkooi, J.M., and Wilson, D.F. 1988, Imaging of phosphorescence: A novel method for measuring the oxygen distribution in perfused tissue. *Science* 241, 1649.

Rumsey, W.L., Iturriaga, R., Spergel, D., Lahiri, S., Wilson. D.F. 1991, Optical measurements of the dependence of chemoreception on oxygen pressure in the cat carotid body. *Am. J. Physiol.* 261 (Cell Physiol. 30), C614.

Varghese, A.J., Gulyus, S., and Mohindra, J.K., 1976, Hypoxia-dependent reduction of 1-(2-nitro-1-imidazoyl)-3-methoxy-2-propanol by chinese hamster ovary cells and KHT tumor cells in vitro and in vivo., *Cancer Res.* 36: 3761.

Varghese, A.J., and Whitmore, G.F., 1981, Cellular and chemical reduction products of misonidazole, *Chem. Biol. Interactions*, 36: 141.

Wilson, D.F. and Cerniglia, G.J. 1992, Localization of tumors and evaluation of their state of oxygenation by phosphorescence imaging. *Cancer Res.* 52: 3988.

Wilson, D.F., Pastusko, A., DiGiacomo, J.E., Pawloski, M., Schneiderman, R., Delivoria-Papadopoulos, M. 1991, Effect of hyperventilation on oxygenation of the brain cortex in newborn piglets. *J. Appl. Physiol.* 70: 2691.

Wilson, D.F., Rumsey, W.L., Green, T.J., and Vanderkooi, J.M., 1988a, The oxygen dependence of mitochondrial oxidative phosphorylation measured by a new optical method for measuring oxygen concentration, *J. Biol. Chem.* 263 (6): 2712.

Wilson, D.F., Rumsey, W.L., Vanderkooi, J.M. 1988b, Oxygen distribution in isolated perfused liver observed by phosphorescence imaging. *Adv. Expt. Med. Biol.* 248: 109.

INTESTINAL ISCHEMIA DURING HYPOXIA AND EXPERIMENTAL SEPSIS AS OBSERVED BY NADH VIDEOFLUORIMETRY AND QUENCHING OF PD-PORPHINE PHOSPHORESCENCE

C. Ince, J.P. van der Sluijs, M. Sinaasappel, J.A.M. Avontuur, J.M.C.C. Coremans, and H.A. Bruining

Department of Surgery, University Hospital, Rotterdam, The Netherlands

INTRODUCTION

The etiology of septic shock, one of the main causes of non-accident death in critical care units today, is as yet ill-understood. The bacterial components thought to be mainly responsible for sepsis syndrome are lipopolysacharides (LPS), also called endotoxins. Presence of these components in the blood stream induces an excessive production of cytokines, oxygen radicals, complement, platelet activating factor, and myocardial depressant factor, which evoke a systemic inflammatory response. Besides their vasoactive action these factors may also cause damage to tissues.

The progress of septic shock is marked by changes in the relationship between oxygen supply and oxygen consumption in various organs. As oxygen supply falls relatively short, compensatory mechanisms redirect blood supply to the heart and the brain, leaving other organs in oxygen debt. The most vulnerable organ systems in this context are the organs of the splanchnic circulation, e.g., the intestines. Compensatory shock occurring as a result of sepsis renders the intestines ischemic and consequently results in degeneration of their epithelial layers. There is evidence to suggest that loss of intestinal viability following ischemia fuels the progress of sepsis syndrome.[1-8] Gastrointestinal (GI) ischemia is thus thought to play a central role in the development and persistence of septic shock. The ability to follow the time course of changes in oxygenation and tissue energetics of the GI tract could provide valuable insight into the etiology of sepsis syndrome.

Under aerobic conditions the oxidative phosphorylation occurring in the mitochondria is the main site for the production of ATP in mammalian cells. Much attention has been directed to the development of non-invasive techniques to measure substrates and intermediates of the oxidative phosphorylation as an indication of the energy state of

Oxygen Transport to Tissue XVI
Edited by M.C. Hogan *et al.*, Plenum Press, New York, 1994

tissue.[9] One technique enables the mapping of the distribution of the energy state of tissue using the fluorescence of reduced nicotinamide adenosine dinucleotide (NADH).[10-12] NADH is a key intermediate in the transfer of reducing equivalents from metabolic substrates in the cytosol to oxygen in the mitochondria. During tissue hypoxia less NADH is oxidized to NAD^+, leading to an increase in the concentration of mitochondrial NADH. When excited with 365nm light NADH, unlike NAD^+, fluoresces at wavelengths centered around 470nm. We have used this property of NADH to develop a videofluorimeter to image NADH fluorescence *in vivo*.[12]

Wilson and co-workers recently introduced a phosphorescent dye to quantitatively determine oxygen pressures in the microvasculature *in vivo*.[13,14] This technique uses the optical properties of Pd-porphine compounds which, when excited at 540nm, phosphoresce at wavelengths larger than 600nm. The time-constant of the decay of the phosphorescence is dependent on pO_2, as described by the Stern-Volmer equation. From this equation it is possible to calculate oxygen pressures.[14,15]

In the present study we describe the application of both NADH videofluorimetry and Pd-porphine phosphorescence quenching to measure tissue energy states as well as pO_2 in the microvasculature of the intestinal tract during hypoxia and experimental sepsis.

MATERIALS AND METHODS

NADH fluorimetry was used to image tissue energy states *in vivo* as shown schematically in Figure 1. Details of the video fluorimeter and image processing can be found elsewhere.[16] Intravascular pO_2 was determined by measuring the quenching of phosphorescence with a phosphorimeter as described elsewhere in this volume.[15] For these

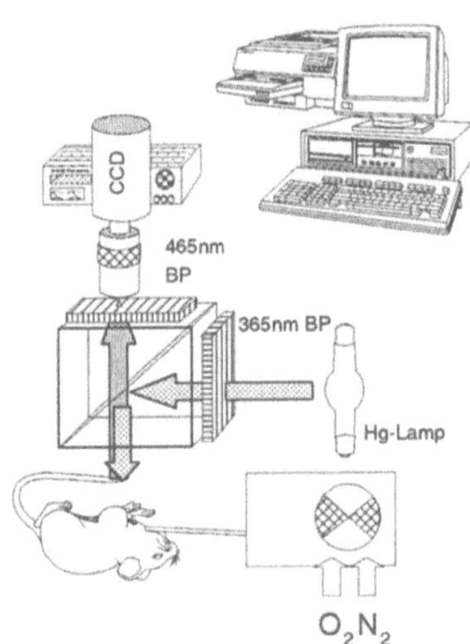

Figure 1. The experimental set-up for NADH video fluorimetric measurements uses a second generation UV sensitive CCD video camera. The fluorimeter provides and detects the needed excitation and emission wavelengths. Images are video recorded and analyzed off-line.

measurements Pd-porphine was bound to albumin to confine the dye to the vasculature,[14] and was administered intravenously. Male Wistar rats weighing 300-350g were mechanically ventilated (49% O_2, 49% N_2O, 2% enflurane), and the intestines were exposed.[17] The mucosa was simultaneously exposed with the serosa by a contravascular longitudinal incision. During the experiments the blood flow in the superior mesenteric artery was measured using a Transonic Medical flow meter (HT106) in combination with a 1RB probe (Transonic Systems, Ithaca, NY). Intestinal ischemia was achieved by ventilating the rats with N_2 instead of O_2. Experimental sepsis was created by intravenous infusion of 3.0mg/kg of body weight of *E. coli* endotoxin (LPS; Sigma, St.Louis, MO), or by i.v. administration of 40μg of muTNF-α.

RESULTS AND DISCUSSION

Exposure of the mucosa of the intestines enables simultaneous imaging of the energy state of the mucosal and serosal tissue. Analysis of the intensity levels of these images showed similar kinetics of mucosal and serosal NADH fluorescence changes during and following an episode of anoxic ventilation (Figure 2^A). These kinetics were characterized by a progressive rise of fluorescence during the entire course of N_2 ventilation. Recovery from anoxic ventilation was associated with an instant decrease of the fluorescence signal,

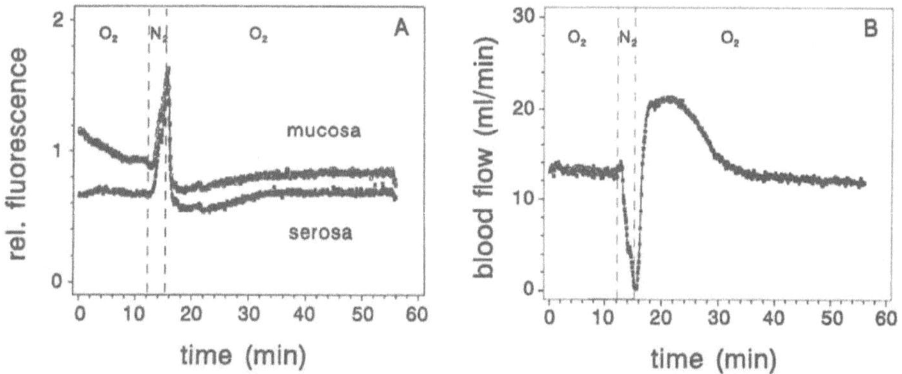

Figure 2. Mucosal and serosal NADH fluorescence of the proximal ileum (A) and blood flow in the superior mesenteric artery (B) before, during, and following a 3 min. episode of N_2 ventilation.

followed by a small undershoot before returning to steady fluorescence levels comparable to the basic fluorescence. It is likely that the increase in fluorescence during the anoxic period is due to the accumulation of NADH within the mitochondria[10] of the epithelial and serosal cells of the intestinal wall. Since hemoglobin molecules absorb the excitation light and the emitted fluorescence light, various amounts of blood in the surface layers of the organs may negatively influence the tissue NADH fluorescence measurements.[18] The undershoot of the fluorescence signal seen in Figure 2^A may thus be caused by an attenuation of the fluorescence signal due to the initial high blood volume in the mesenteric vasculature (changes in blood volume were not compensated for in this study) associated with the reactive hyperemic response to the anoxic period (see Figure 2^B). Furthermore, the fluorescence rise during anoxic ventilation may partially be influenced by the simultaneous decrease in blood volume (indicated by the observed reduction of the blood flow through the superior mesenteric artery to zero; Figure 2^B). Sudden changes in ventilatory oxygen content result in changes in microvascular pO_2 prior to visible changes in blood flow (Figure 3). In these experiments the microvascular pO_2 was estimated using the quenching of phosphorescence of Pd-porphine. The pO_2 was determined in the mucosa of the proximal jejunum before, during, and after an episode of N_2 ventilation. As shown in Figure 3, anoxic ventilation is associated with a sudden fall in pO_2, immediately followed by a drop in the blood flow through the superior mesenteric artery. Recovery from anoxia is associated with a reactive hyperemic flow and with a rapid rise in microvascular pO_2. Both parameters returned to basic levels within approximately 15 minutes.

To study the oxygenation of the intestines during experimental sepsis, LPS was infused in a rat, and the microvascular pO_2 of the serosa of the proximal jejunum as well as the blood flow in the superior mesenteric artery were measured. Figure 4 shows a transient drop in microvascular pO_2 and blood flow following LPS infusion. After recovery a slow decrease in pO_2 and blood flow were observed over a period of two hours. A sudden drop in pO_2, immediately followed by a cessation of the blood flow were associated with the LPS-inflicted death of the rat. In control rats infused with 0.9% NaCl, these parameters did not change during an extended period of time (>6 hours; data not shown).

Since it is thought that the mode of action of endotoxin in inducing shock is mediated by the action of inflammatory agents, we studied some effects of infusion of one of the main mediators that underlie septic shock, namely, TNF-α. Figure 5 shows images of the development of the NADH fluorescence of the serosa of the cecum after infusion of TNF-α. An enhancement in fluorescence is observed over time. Again, the increased fluorescence levels are most probably caused by increased levels of mitochondrial NADH. The blood volume is decreased, indicated by the observed vasoconstriction (Figure 5).

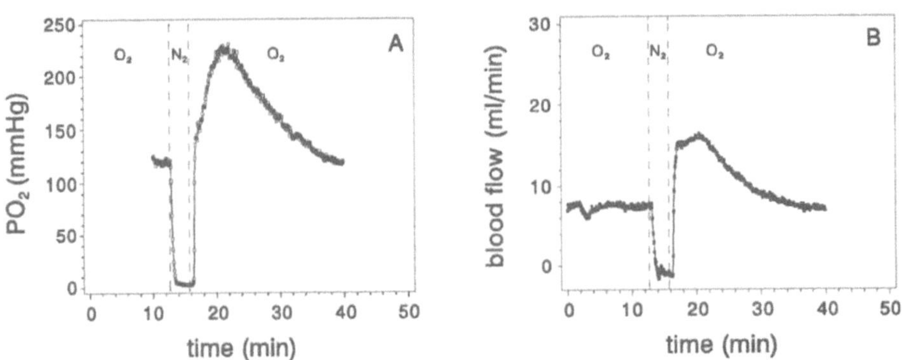

Figure 3. Measurements of the pO_2 of the mucosa of the proximal jejunum using the quenching of Pd-porphine quenching (A), and of the blood flow in the superior mesenteric artery before, during, and after a period of N_2 ventilation.

The absorbance of the short-wavelength excitation light by the blood enables clear visualization of the serosal vascularization, which makes it possible to trace blood vessel diameter changes during the progress of shock.

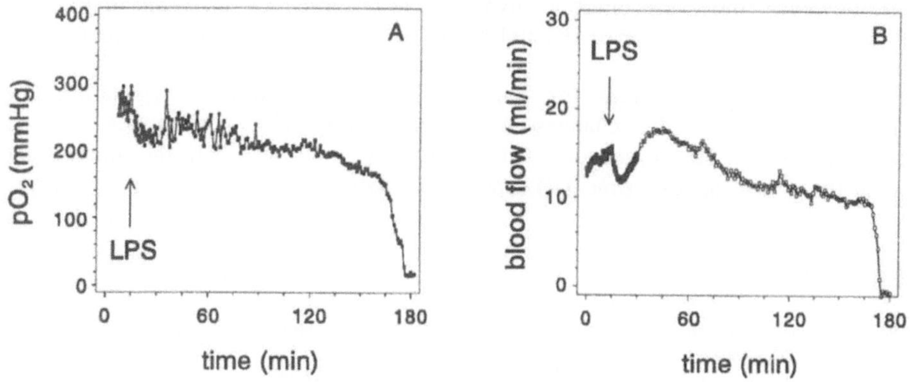

Figure 4. Measurements of the serosal pO$_2$ of the rat jejunum (A) and the blood flow in the a. mesenterica sup. (B) during experimental sepsis induced by a bolus infusion of endotoxin (LPS).

It is indicated that high doses of LPS do not lead to significant levels of intestinal schemia in the first three hours after infusion in the rat. Infusion of a second mediator,

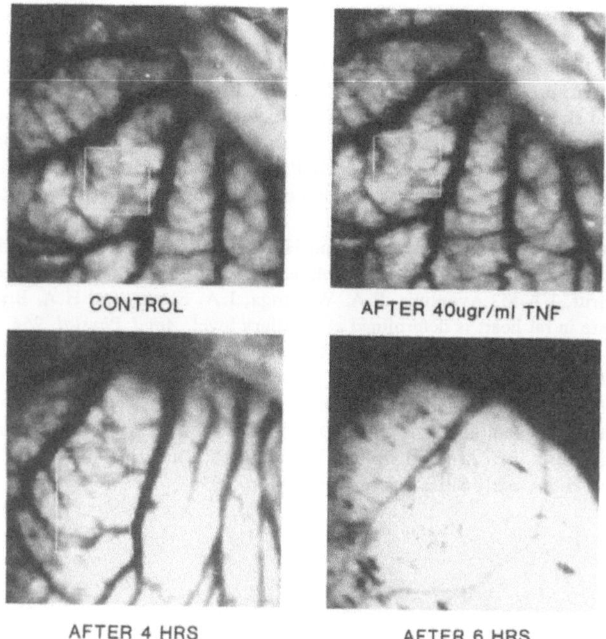

Figure 5. NADH fluorescence images of the serosa of the rat cecum at different times following the infusion of TNF-α.

TNF-α, however, did cause intestinal hypoxia, as observed by increases in fluorescence levels. Therefore, models of experimental sepsis have to be evaluated with great care.

The two optical techniques for the determination of tissue oxygenation showed identical reactions to sudden changes in oxygen supply. These techniques thus can be applied in studies on the oxygenation state of the intestinal wall, both the mucosa and the serosa, in the course of (experimental) sepsis and shock.

REFERENCES

1. J.W. Baker, E.A. Deitch, M. Li, R.D. Berg, and R.D. Specian, Hemorrhagic shock induces bacterial translocation from the gut, *J Trauma.* 28:896 (1988).
2. E.A. Deitch, Bacterial translocation of the gut flora, *J Trauma.* (1990).
3. E.A. Deitch, The role of intestinal barrier failure and bacterial translocation in the development of systemic infection and multiple organ failure, *Arch Surg.* 125:403 (1990).
4. M.P. Fink, K.L. Kaups, H.L. Wang, and H.R. Rothschild, Maintenance of superior mesenteric arterial perfusion prevents increased intestinal mucosal permeability in endotoxic pigs, *Surgery.* 110:154 (1991).
5   M.P. Fink, J.B. Antonsson, H.L. Wang, and H.R. Rothschild, Increased intestinal permeability in endotoxic pigs. Mesenteric hypoperfusion as an etiologic factor, *Arch Surg.* 126:211 (1991).
6. M.P. Fink, Adequacy of gut oxygenation in endotoxemia and sepsis, *Crit Care Med.* (1993).
7. R.J. Goris, I.P. Van Bebber, R.M. Mollen, and J.P. Koopman, Does selective decontamination of the gastrointestinal tract prevent multiple organ failure? An experimental study, *Arch Surg.* 126:561 (1991).
8. M. Papa, Z. Halperin, E. Rubinstein, A. Orenstein, S. Gafin, and R. Adar, The effect of ischemia of the dog's colon on transmural migration of bacteria and endotoxin, *J Surg Res.* 35:264 (1983).
9. R.S. Balaban, Regulation of oxidative phosphorylation in the mammalian cell, *Am J Physiol.* (1990).
10. B. Chance, Pyridin nucleotide as an indicatior of the oxygen requirements for energy-linked functions of mitochondria, *Circ Res.* 38:131 (1976).
11. C. Ince, J.M.C.C. Coremans, and H.A. Bruining, In vivo NADH fluorescence, *Adv Exp Med Biol.* 317:277 (1992).
12. C. Ince and H.A. Bruining, Optical spectroscopy for the measurement of tissue hypoxia, *in:* "Update in intensive care and emergency medicine," J.L. Vincent, ed., Springer Verlag, New York (1991).
13. M. Pawlowski and D.F. Wilson, Monitoring of the oxygen pressure in the blood of live animals using the oxygen dependent quenching of phosphorescence, *Adv Exp Med Biol.* 278 (in press).
14. D.F. Wilson, A. Pastuszko, J.E. DiGiacomo, M. Pawlowski, R. Schneiderman, and M. Delivoria-Papadopoulos, Effect of hyperventilation on oxygenation of the brain cortex of newborn piglets, *J Appl Physiol.* 70:2691 (1991).
15. M. Sinaasappel, C. Ince, J.P. Van der Sluijs, and H.A. Bruining, A new phosphorimeter for the measurement of oxygen pressures using Pd-porphine phosphorescence, (these Proceedings).
16. C. Ince, J.F. Ashruf, J.A.M. Avontuur, P.A. Wieringa, J.A. Spaan, and H.A. Bruining, Heterogeneity of the hypoxic state in rat heart is determined at capillary level, *Am J Physiol.* 264:H294 (1993).
17. C. Ince, J.A.M. Avontuur, M. Sinaasappel, J.M.C.C. Coremans, and H.A. Bruining, 1993, NADH fluorometry and Pd-porphyrin phosphometry of gut and kidney during sepsis, *in:* "Quantitative Spectroscopy in Tissue," M. Kessler and K. Frank, ed., CRC Press, New York (in press).
18. J.M.C.C. Coremans, C. Ince, and H.A. Bruining, NADH fluorimetry and diffuse reflectance spectroscopy on rat heart, *in:* "Medical Optical Tomography: functional images and monitoring," G. Müller, B. Chance *et al.*, eds., SPIE Press, Washington (1993).

^{1}H NMR APPROACH TO OBSERVE TISSUE OXYGENATION WITH THE SIGNALS OF MYOGLOBIN

Thomas Jue, Ulrike Kreutzer, and Youngran Chung

Biological Chemistry Department
University of California Davis
Davis, CA 95616-8635

INTRODUCTION

NMR has opened many new perspectives on metabolic regulation in vivo, with its non-invasive application as a key feature. It can localize metabolite signals from specific tissue and reveal their fluctuation under different physiological conditions, even in humans. The signals of phosphocreatine, inorganic phosphate, ATP, and lactate have helped to detail the cellular response and to illuminate the metabolic regulation (Brown et al, 1982, Koretsky and Williams, 1992).

Despite the advances, techniques to observe oxygen have not kept in step. Although researchers have developed NMR strategies, they are indirect and have uncertain sampling specificity for the intracellular oxygen level. The high energy ^{31}P signals are commonly used to index the cellular oxygen level. As oxygen becomes limiting, the phosphocreatine (PCr) level drops characteristically. In contrast the lactate level rises, reflecting increased NADH. Both the ^{1}H lactate and ^{31}P PCr signals then indirectly index cellular hypoxia. Researchers have also utilized ^{19}F labeled probes, whose relaxation properties are sensitive to the oxygen environment. Matching the experimental results with a standard curve of relaxation rate vs. solution oxygenation leads then to a cellular oxygen concentration (McGovern et al, 1993; Holland et al, 1993). Still others have proposed ^{17}O$_2$ measurements and have adduced H$_2$^{17}O appearance to indicate oxygen metabolism and blood flow (Pekar et al 1991; Fiat and Kang, 1993).

MYOGLOBIN

The ^{1}H NMR signals of cytosolic myoglobin proffer an opportunity to observe precisely the intracellular oxygenation (Jue and Anderson, 1990). Myoglobin is a well characterized, oxygen binding heme protein. It is comprised of approximately 153 amino acids and has 85% α helical structure, separated into A-H segments. The central heme iron coordinates oxygen with a binding affinity about ten times higher than Hb (Antonini and Brunori, 1971). Although key residues are highly conserved, the primary sequence can still vary significantly among different species, leading sometimes to altered oxygen affinity (Dayhoff, 1968). Numerous high resolution crystal and NMR structural studies have established the Mb paradigm (Shin et al, 1993), built upon the first crystallographic analysis (Kendrew et al, 1961).

To measure oxygen level with the NMR signals of myoglobin requires several assumptions: A spectral change must distinguish the ligated and unligated states of Mb. The corresponding spectra must exhibit detectable reporter signals. Signal interferences from other metabolites and proteins are insignificant.

ASSESSING CELLULAR OXYGENATION

Calculating the oxygen level is based directly on the Mb oxygen binding equation:

$$Mb + O_2 \Leftrightarrow MbO_2 ; K = \frac{[MbO_2]}{[Mb][O_2]} ; K = \frac{1}{[O_2]_{50}}$$

If MbO_2 and Mb are measurable with NMR and the K or $[O_2]_{50}$ (partial pressure of oxygen that will half saturate the myoglobin) solution is known, the O_2 calculation is then straightforward, given the underlying assumption that the binding constants for solution and cellular Mb are identical.

$$[O_2] = \frac{[MbO_2][O_2]_{50}}{[Mb]}$$

PROXIMAL HISTIDYL NH SIGNAL AS DEOXYGENATED STATE MARKER

The ^{1}H NMR strategy to detect the reporter signal for deoxygenated Mb is predicated on the electronic structure alteration that accompanies oxygen ligation. Under physiological conditions, the heme Fe is predominantly in the +2 oxidation state. Ligated with oxygen, the heme Fe(II) electrons are paired (S=0), and MbO_2 is diamagnetic. Unligated, the heme Fe(II) electrons are unpaired (S=2), and deoxy Mb is paramagnetic (Weissbluth, 1974). The unpaired electrons in the paramagnetic state can interact with the proton and produce a hyperfine shift in the NMR signal (Jesson, 1973; Horrocks, 1973). In deoxy Mb the unpaired spins interact directly with the histidine F8, proximal histidyl, N_δ H proton (fig 1).

That electron interaction with the proximal histidyl $N_\delta H$ is manifested in a unique ^{1}H NMR signal appearing at ~80 ppm, 25C. The assignment is based on the model study of 2-methyl imidazole axially coordinated to tetraphenylporphyrin (TPP) (Goff and LaMar, 1977) and is amply substantiated with subsequent ^{1}H NMR spectra of deoxy- Mb, which exhibit distinct, exchangeable resonances at ~ 80 ppm (LaMar, 1979). Because of the inequivalence of the α and β subunits of Hb, deoxy Hb A (adult human hemoglobin) yields signals at 76 ppm and 64 ppm, corresponding to the β and α subunits respectively (Ho and Russu, 1981). Corresponding signal for human myoglobin appears at 81 ppm, while sperm whale Mb appears at 79 ppm, 25^{0}C, reflecting the signal's sensitivity to the protein environment (Kreutzer et al, 1993).   Upon oxygenation the signal disappears.

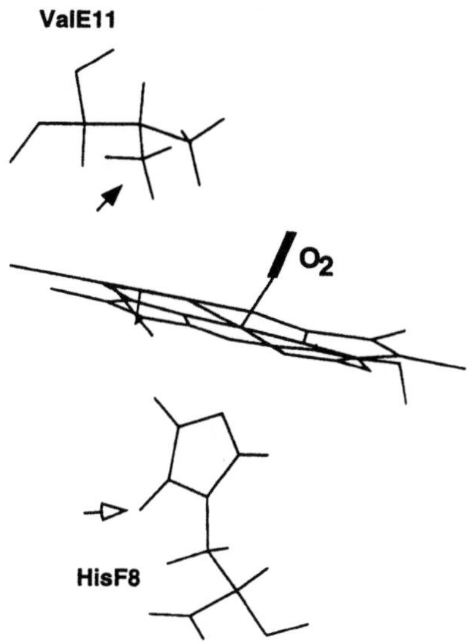

Figure 1. Diagram of the Heme and the Val E11/His F8 Amino Acid Residues:

The Val E11 γ_1 CH$_3$ group is adjacent to the distal side of the heme, whereas the His F8 N$_\delta$H group is on the proximal side. In MbO$_2$, the Val E11 methyl group gives rise to a signal at -2.8 ppm, in deoxy Mb, the proximal histidyl NH proton yields a signal at ~80 ppm.

DETECTION OF PROXIMAL HISTIDYL NH SIGNAL FROM MYOCARDIUM

NMR can indeed detect proximal histidyl NH signal of Mb in myocardium. Fig. 2 shows a bank of spectra from a ~1 gram perfused rat heart under graded ischemic conditions. The top panel tracks the spectral region between 100-60 ppm. Under well oxygenated condition, no signal appears. As the flow rate decreases stepwise, a signal at 80 ppm increases, reaching maximum at 0 ml/min perfusate flow, fig. 2E.   Upon 10 ml/min reflow the signal disappears. The signal's chemical shift and properties correspond directly to the deoxy Mb proximal histidyl NH resonance.   A similar pattern is observed under graded hypoxia conditions (Kreutzer and Jue, 1991).

In these experiments, the Mb signal intensity under 0 ml/min perfusate flow is set to 100%, which then establishes the MbO$_2$ fraction at the intermediate experimental points. With the [O$_2$]$_{50}$ of 1.5 torr, the intracellular oxygen is calculable at each flow rate (Kreutzer and Jue, 1991, Kreutzer et al, 1992). No interfering signals from the cytochromes are observed.

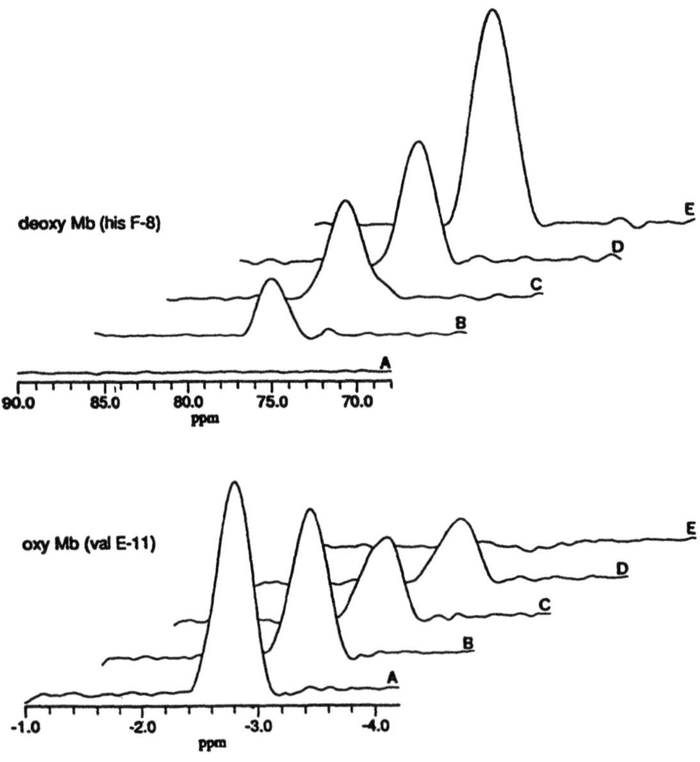

Figure 2. Histidyl NH/Val E11 Signals from Myocardium

^{1}H NMR spectra of MbO_2 and deoxy Mb in myocardium under various ischemic conditions: Hearts were perfused at different flow rate, 11ml/min (A), 3 ml/min (B), 2 ml/min (C), 1 ml/min (D), and 0 ml/min (E). Top trace shows the response of the proximal histidyl NH signal; bottom trace, the Val E11 signal. The Val E11 signal intensity decreases with decreasing flow rate or oxygenation, the proximal histidyl NH signal varies inversely (from Kreutzer et al, 1992).

VAL E11 CH$_3$ SIGNAL AS OXYGENATED STATE MARKER

When Mb is ligated with O2, the Fe is diamagnetic (S=0). However the diamagnetic heme is a conjugated system with an extensive π electron delocalization that creates a ring current and consequently an induced, anisotropic magnetic field. Amino acid residues near the heme experience the local magnetic field and have their resonance positions shifted (Perkins, 1980). The Val E11 γ_1 CH$_3$, positioned on the distal side of the heme experiences a ring current shift to -2.8 ppm. Both model calculations and mutant protein studies support the resonance assignment (Shulman et al, 1970). In the presence of CO, the altered ligand-Fe binding geometry induces a Val E11 shift to -2.40ppm. The Val E11 signal disappears upon deoxygenation. A new set of paramagnetic signals appear in the spectral region (Busse and Jue, in prep).

114

DETECTION OF THE VAL E11 SIGNAL FROM MYOCARDIUM

Fig. 2, bottom panel shows that the Val E11 signal is detectable in perfused rat myocardium. Under well oxygenated conditions, the Val E11 signal is detectable at -2.8 ppm, whereas the proximal histidyl NH signal is not observed, fig 2, top panel. Conversely at 0 ml/min, the proximal histidyl NH signal intensity reaches a zenith, while the Val E11 signal falls to its nadir. A dynamic equilibrium exists between the Val E11 and His F8 signals, such that the one resonance's intensity increase is balanced by the other's decrease under all ischemic conditions (Kreutzer et al, 1992). Upon reoxygenation, the Val E11 signal reappears at the same chemical shift position. Introducing carbon monoxide shifts the Val E11 signal to -2.4 ppm, as noted in NMR protein experiments (Kreutzer et al, 1992; Shulman et al, 1970).

CRITICAL pO_2

The myocardial Mb signals shed light on the critical pO_2, the oxygen level that limits mitochondrial respiration. Fig. 3 shows a correlation between the PCr /ATP ratio and intracellular oxygen level, derived from NMR observations. PCr is readily accessible from the 31P NMR spectra. Above 3 torr of oxygen the PCr level remains stable. However, below the 3 torr threshold, the PCr level drops precipitously. Similar graphs, displaying interaction between intracellular oxygen and oxygen consumption, lactate production, or ATP concentration, all point to a similar value for the critical pO_2 (Kreutzer and Jue, in preparation).

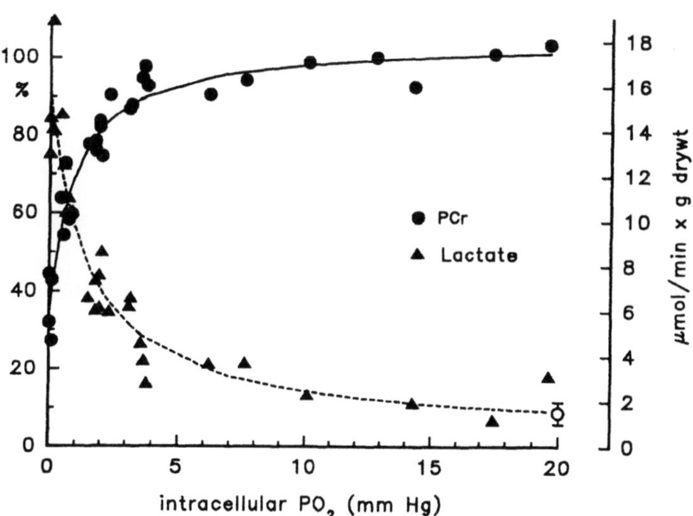

Figure 3.   Graph of PCr/ATP and Lactate vs. O2

The interaction between PCr/ATP and intracellular oxygen level. PCr was measured from the 31P spectra, the oxygen level from the 1H NMR signal of Mb Val E11. The critical pO2 is ~3 torr (from Kreutzer et al , 1992). Left vertical axis refers to [PCr], right vertical axis to [lactate].

CELLULAR ENVIRONMENT OF MYOGLOBIN

The cellular environment does not appear to impose a restriction on the rotational diffusion of Mb. Already tissue Mb produces a proximal histidyl NH line width only slightly greater than the solution state. Such a finding is consistent with the NMR results, suggesting comparable correlation times (Livingston et al, 1983; Kreutzer and Jue, 1991). Indeed field dependent NMR relaxation analyses indicate that the rotational correlation time of Mb in tissue is only 9×10^{-9}s, ~1.5-2.0 times greater than in solution (Wang et al, 1991). Moreover all the cellular Mb appears to be freely diffusive (Kreutzer and Jue, 1991). In contrast the erythrocyte environment is substantially different (Wang et al, 1991). These findings support the hypothesis that Mb may facilitate oxygen diffusion in the cell (Wittenberg et al, 1970, 1985, 1989).

SUMMARY

The myoglobin technique measures oxygen tension in myocytes. It relies on a quantitative measurement of the Val E11 and His F8 signals and an accurate value for the $[O_2]_{50}$ for Mb. Even though the Mb oxygen affinity in the cell is in question, the NMR results still reflect the degree of Mb oxygen saturation. Although magnetic resonance has established a variety of strategies to measure tissue oxygenation, the Mb approach is the most direct and will lead to a better understanding of oxygen's role in regulating cellular activity.

ACKNOWLEDGMENTS

Grants from NIH GM 44916, the American Heart Association 92-221A, and the UCD Hibbard Williams Award have supported the research presented in the article.

REFERENCES

Antonini, E., and Brunori, M., 1971, "Hemoglobin and Myoglobin in their Reactions with Ligands," North Holland, Amsterdam.

Brown, T. R., Kincaid, M., and Ugurbil, K, 1982, NMR chemical shift imaging in three dimensions, *Proc. Natl. Acad. Sci. USA.* 79: 523.

Busse, S and Jue, T., manuscript in preparation.

Dayhoff, M. O. and Eck, R. V., 1968, "Atlas of Protein Sequence and Structure," National Biomedical Research Foundation, Silver Spring.

Fiat, D. and Kang, S., 1993, Determination of the rate of cerebral oxygen consumption and regional cerebral blood flow by non-invasive 17O in vivo nmr spectroscopy and magnetic resonance imaging. part 2. determination of CMRO2 for the rat by 17O NMR, and CMRO2, rcbf and the partition coefficient for the cat by 17O MRI, *Neurological Research.* 15: 7.

Goff, H. and La Mar, G. N., 1977, Spin ferrous porphyrin complexes as models for deoxymyoglobin and -hemoglobin. a proton nuclear magnetic resonance study, *J. Am. Chem. Soc.* 99: 6599.

Ho, C. and Russu, I., 1981, Proton nuclear magnetic resonance investigation of hemoglobins, in: "Methods in Enzymology," E. Antonini, L. Rossi-Bernardi, and E. Chiancone, eds, vol.76, Academic Press, New York.

Holland S. K., Kennan R. P., Schaub M. M., D'Angelo M. J., Gore, J. C., 1993, Imaging oxygen tension in liver and spleen by 19F NMR, *Magn. Reson. Med.* 29:446.

Horrocks, J. DeW., 1973, Analysis of isotropic shifts, in: "NMR of Paramagnetic Molecules," G. N. La Mar, J, DeW. Horrocks, and R. H. Holm, eds, Academic Press, New York.

Jesson, J. P., 1973, The paramagnetic shift, in: "NMR of Paramagnetic Molecules," G. N. La Mar, J, DeW. Horrocks, and R. H. Holm, eds, Academic Press, New York.

Jue, T. and S. Anderson, 1990, ^{1}H observation of tissue myoglobin: An indicator of intracellular oxygenation in vivo, *Magn. Res. Med.* 13:524.

Kendrew, J. C., Watson, H.C., Strandberg, B. E., Dickerson, R. E., Phillips, D. C., and Shore, V. C., 1961, A partial determination by x-ray methods, and its correlation with chemical data, *Nature (London)*. 190:666.

Koretsky, A. P. and Williams, D. S., 1992, Application of localized in vivo nmr to whole organ physiology in animal, *Ann. Rev. Physiol.* 54:799.

Kreutzer, U. and Jue, T., 1991 1H nuclear magnetic resonance deoxymyoglobin signal as indicator of intracellular oxygenation in myocardium, *Am. J. Physiol.* 30:H2091.

Kreutzer, U., Wang, D. S., and Jue, T., 1992, Observing the 1H NMR signal of the myoglobin val E11 in myocardium: an index of cellular oxygenation, *Proc. Natl. Acad. Sci., USA.* 89:4731.

Kreutzer, U., Chung, Y., Butler, D. and Jue, T., 1993, 1H NMR characterization of the human myocardium myoglobin and erythrocyte hemoglobin signals, *Bioch. Biophys. Acta.* 161:33.

Kreutzer, U. and Jue, T., manuscript in preparation.

La Mar, G.N., 1979,. Model compounds as aids in interpreting NMR spectra of hemoproteins, in: "Biological Applications of Magnetic Resonance," R.G. Shulman, ed., Academic Press, New York.

Livingston, D. J., La Mar, G. N. and Brown, W.D., 1983, Myoglobin diffusion in bovine heart muscle, *Science.* 220:71.

McGovern, K. A., Schoeniger, J. S., Wehrle, J. P., Ng, C. E., Glickson, J. D., 1993, Gel-entrapment of perfluorocarbons: a fluorine-19 NMR spectroscopic method for monitoring oxygen concentration in cell perfusion systems, *Magn. Reson. Med.* 29:196.

Pekar, J., Ligeti, L., Ruttner, Z., Lyon, R. C., Sinnwell, T. M., van Gelderen, P., Fiat, D., Moonen, C. T., McLaughlin, A. C., 1991, In vivo measurement of cerebral oxygen consumption and blood flow using 17O magnetic resonance imaging, *Magn. Reson .Med.* 21:313.

Perkins, S. J., 1980, Ring current models for the heme ring in cytochrome c, *J. Magn. Reson.* 38:297.

Shin, H. C., Merutka, G., Waltho, J. P., Wright, P. E., Dyson, H. J., 1993, Peptide models of protein folding initiation sites .2. the G-H turn region of myoglobin acts as a helix stop signal, *Biochemistry.* 32:6348.

Shulman, R. G., Wuthrich, K., Yamane, T., Patel, D. J.,and Blumberg, W. E., 1970, Nuclear magnetic resonance determination of ligand-induced conformational changes in myoglobin, *J. Mol. Biol.* 53:143.

Wang, D. S., Kreutzer, U., and Jue, T., 1991, Separating the intracellular signals of myoglobin and hemoglobin, *Proc. Soc. Magn. Reson. Med.* 301.

Weissbluth, M., 1974, Hemoglobin. cooperativity and electronic properties, in: "Mol. Biol. Biochem. Biophys.," Vol. 15, Springer Verlag, New York.

Wittenberg, J. B., 1970, Myoglobin-Facilitated oxygen diffusion: role of myoglobin in oxygen entry into muscle, *Phys. Rev.* 50:559.

Wittenberg, B. A. and Wittenberg, J. B., 1985, Oxygen pressure gradients in isolated cardiac myocytes, *J. Biol. Chem.* 260: 6548.

Wittenberg, B. A. and Wittenberg, J. B., 1989, Transport of oxygen in muscle, *Ann. Rev. Physiol.* 51:857.

MEASUREMENTS OF pO_2 IN VIVO, INCLUDING HUMAN SUBJECTS, BY ELECTRON PARAMAGNETIC RESONANCE

Harold M. Swartz[1], Goran Bacic[1], Bruce Friedman[3], Fuminori Goda[1], Oleg Grinberg[1], P. Jack Hoopes[2], Jinjie Jiang[1], Ke Jian Liu[1], Toshiaki Nakashima[1,4], Julia O'Hara[2], Tadeusz Walczak[1]

[1]Department of Radiology, [2]Department of Radiation Oncology, [3]Department of Cardiology, Dartmouth Medical School, Hanover, NH 03755
[4]Permanent Affiliation: Third Department of Internal Medicine, Kyoto Prefectural University of Medicine, Kawaramachi-Hirokoji, Kamigyo-ku, Kyoto 602, Japan

INTRODUCTION

The purpose of this paper is to provide an illustrative description of the current state of development of the use of electron paramagnetic resonance (EPR, or completely equivalently, electron spin resonance or ESR) to measure the partial pressure of oxygen (pO_2) in tissues in vivo under physiological conditions. This summary is based on published and unpublished results from our laboratory (1-7) and does not attempt to describe the results of other laboratories which also are working along related lines (8-10). The pertinent features of our technique are illustrated. We also consider the current limitations of the technique and likely developments in the near future. Our evaluation is that: this technique now is suitable for immediate use in small animals; within a short period of time instruments will be available facilitating its use in larger animals; and preliminary studies are imminent in human subjects (7).

The EPR method described in this paper has several features that appear to be advantageous for the measurement of pO_2 in vivo. These features include:

1) sensitivity to a wide range of pO_2 (from <1 Torr to >760 Torr) (1,3,4,7);
2) rapidity of measurements (measurements can be obtained continuously with time resolution of seconds or less) (3,4);
3) measurements can be made non-invasively (after the initial placement of the paramagnetic materials into the tissues of interest) (1,3,4,5,7);
4) measurements are made at well resolved sites (the measurements report on the pO_2 at the site of the paramagnetic material, which can be as small as a 0.2 mm sphere) (3);
5) simultaneous measurements can be made at multiple sites (11);
6) stability in tissues, enabling measurements to be repeated as frequently as desired over periods of at least several months and probably years (1,3,4,7);
7) the oxygen-sensitive paramagnetic materials are very inert biologically and chemically (3,4,12).

Until very recently, however, it appeared that it would be several years before this technique could be applied clinically because of the time required for testing the safety of the particulate paramagnetic materials. We now have discovered a material (India Ink), which already is in extensive use in human subjects (7,13) and which has the desired characteristics for EPR oximetry (7). Using the India Ink in a tattoo we have successfully carried out the

Oxygen Transport to Tissue XVI
Edited by M.C. Hogan *et al.*, Plenum Press, New York, 1994

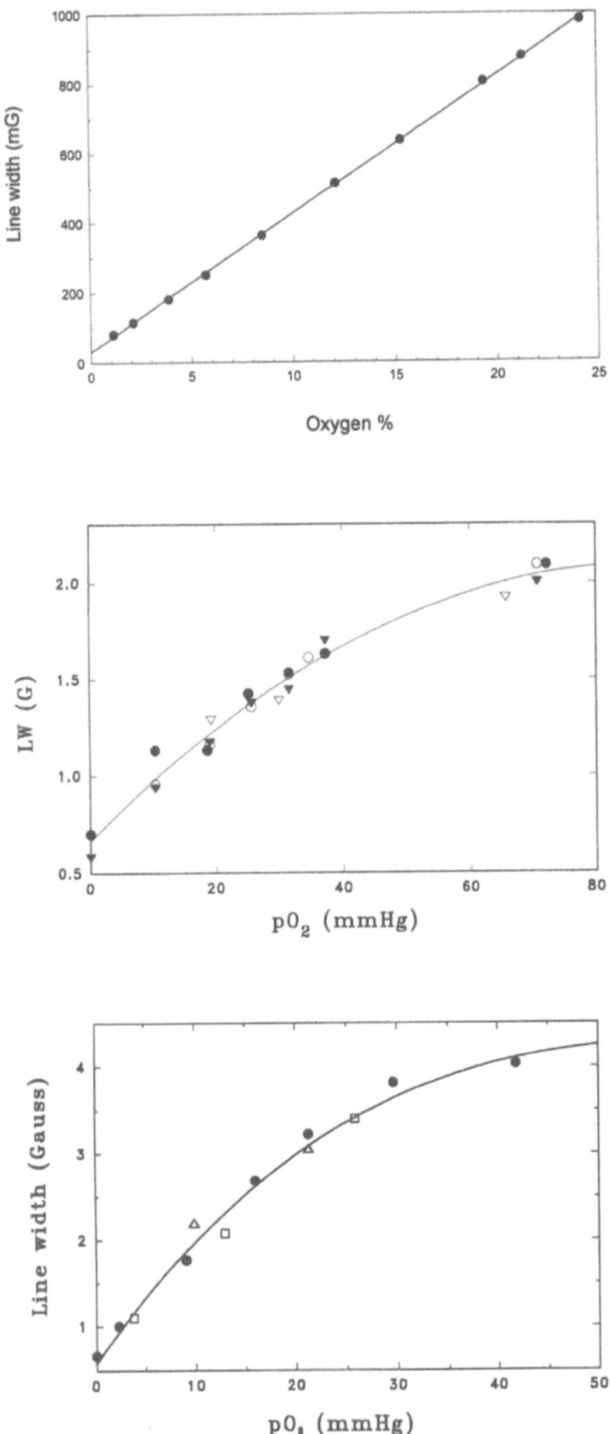

Figure 1 Calibration curves for some paramagnetic materials currently in use to measure pO$_2$ in vivo. a) lithium phthalocyanine, b) fusinite, and c) India ink. Different symbols represents different media, ranging from lipophilic to hydrophilic.

first studies in a human subject. We therefore include the results of several experiments in which India ink is used as the oxygen sensitive paramagnetic material.

PRINCIPLES OF THE METHOD TO MEASURE pO_2 IN TISSUES BY EPR

EPR is a technique that selectively detects species with unpaired electrons (termed paramagnetic species). This provides a very desirable specificity of its response, because there are virtually no paramagnetic species which naturally occur in concentrations that are detectable by the instrumentation used for the measurement of pO_2. It does, however, add the requirement that the oxygen-sensitive paramagnetic species needs to be introduced into the sites where the measurements of pO_2 are to be made.

The technique uses a magnetic field to establish the separate energy levels which will occur in the presence of unpaired electrons, and a resonant frequency to induce transitions which are detected. Although in principle the EPR studies could be carried out at any suitable ratio of the magnetic field and resonant frequency, for practical reasons (principally the need to have maximum sensitivity but to avoid non-resonant absorption of the resonant frequency by the body water) the in vivo studies are likely to be carried out at frequencies between 250-1200 MHz with corresponding magnetic fields of 50-500 Gauss.

The resulting EPR spectra can be affected by the environment in which the paramagnetic material is located. In particular, for some paramagnetic species, the presence of molecular oxygen significantly affects the EPR spectrum (the effects usually are measured by increases in the width of the EPR signals which are proportional to the pO_2, but the effects also may be detected as changes in the intensity of the EPR signal or its lifetime).

The paramagnetic materials used for these experiments have properties which can include: a very sensitive response to pO_2 which is independent of the type of tissue; considerable stability in tissues; lack of response to changes in pH or temperature; and little or no observable toxicity for cells or tissues (3,4,7,12). Some of the most useful paramagnetic substances are particulates with very high densities of unpaired electrons (3,4,7). This enables one to place very small particles in the site(s) of interest which provides a sensitive measure of pO_2 at a well defined site.

The procedure to measure pO_2 in tissues by EPR usually involves the following steps:
1) Place the paramagnetic particle(s) in the region(s) of interest (size up to 0.2 mm diameter); the placement can be confirmed by the use of NMR because of the effect of the paramagnetic materials on NMR images (14);
2) Position the subject between the poles of the magnet (field strength 50-500 gauss);
3) Position the surface detector or loop at or over or around the region of interest (15-17);
4) If multiple sites are to be measured, use a magnetic field gradient to permit differentiation of the spectra from the different sites (11);
5) Tune the EPR spectrometer and obtain spectra either as individual spectra (acquisition time 30-60 seconds) or as continuous readings (by sitting on the peak of the EPR signal);
6) Compare the results to a calibration curve for the paramagnetic substance in a similar environment;
7) Repeat at intervals as desired (probably feasible to do for periods of several years).

RESULTS OF ILLUSTRATIVE STUDIES

The results presented here were obtained with a series of 1.1 GHz instruments built in our laboratory (15-17). These have gradually been modified and improved to meet the special needs of studies in living animals. The most critical technical problems have been how to:
1) achieve maximum sensitivity in the presence of large amounts of materials (fluids, cells, and tissues) which cause nonresonant absorption of the exciting electromagnetic frequency;
2) compensate for the physiological and voluntary motions occurring in live and usually unanesthetized animals;

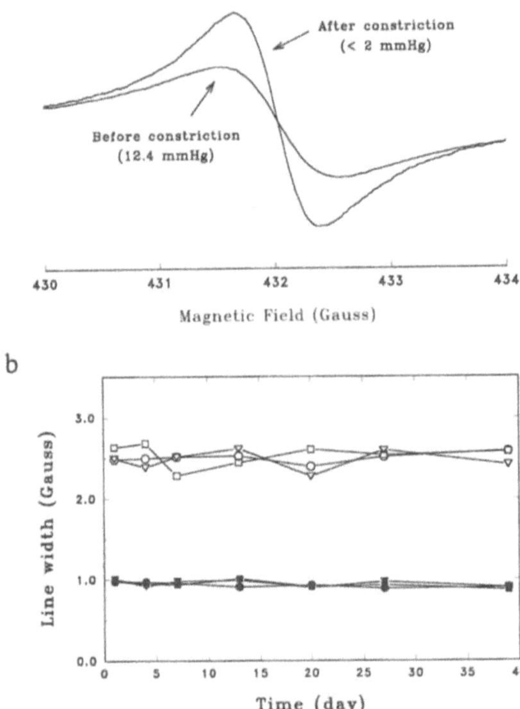

Figure 2. Long term stability and response to pO_2 in skeletal muscle. a) EPR spectrum of fusinite before and after restriction of blood flow, the fusinite was implanted into the muscle 4 months previously; b) India ink; the corresponding pO_2 before (open symbols) and after (solid symbols) the constriction of the blood flow were 14.2 and 1.2 mm Hg, respectively.

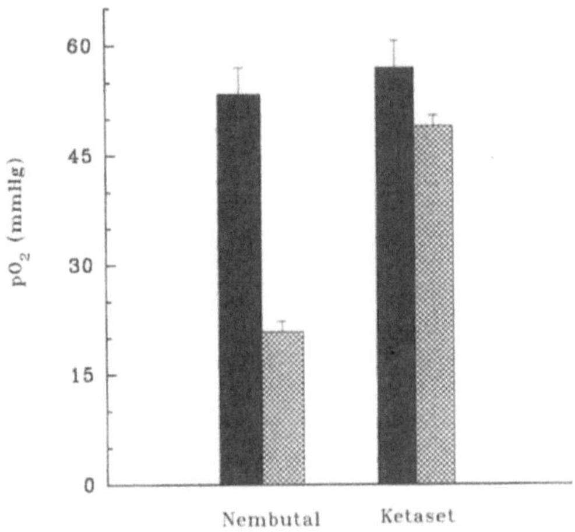

Figure 3. Effects of anesthesia on pO_2 in the rat brain. Measurements were made before (black bar) and after (hatched bar) the i.p. administration of Nembutal and Ketaset. LiPc was inserted into the brain 7 days earlier.

3) process the data to remove effects of motion and other artifacts and maximize the
    amount of information that is used to determine the effects of pO_2 on the EPR
    spectra.

While the technical developments are still in process, as the data that follow indicate,
we already have been able to achieve reasonable resolution of physiologically and
pathophysiologically pertinent pO_2s. Currently a new spectrometer is under construction
which will incorporate the existing developments and add additional features, including the
use of a new magnet which will accommodate specimens with diameters of up to 25 cm.

Calibration And Types Of Use Of Various Paramagnetic Materials

Figure 1 illustrates calibration curves for several of the paramagnetic materials that
are in current use in our laboratory. It appears that different materials will be useful for
different purposes. For example, lithium phthalocyanine has a very narrow line and
therefore is especially good for measuring low values of pO_2 with good spatial and temporal
resolution because of the excellent signal to noise that can be achieved; however, in at least
some tissues, it does not retain its response to oxygen for long periods of time (3). Fusinite
on the other hand, has a somewhat broader line but is very stable in tissues (Figure 2) and
gives very good responses over moderate levels of pO_2 (1-100 Torr) and therefore may be
an excellent probe for long term studies of pO_2 in many tissues (4). In the absence of
oxygen India ink has a linewidth similar to that of fusinite, but the line width broadens much
more rapidly, making it very effective for measuring changes in pO_2 at levels of less than 10
Torr (7).

Stability and Interactions with Cells and Tissues

One of the potential advantages of this technique is the ability to make repeated
measurements of pO_2 without perturbations to the tissue from either repeated invasiveness or
reactions of the tissues to the oxygen sensitive materials. Figure 2 illustrates the stability of
two of the paramagnetic agents in tissue. Table 1 illustrates an apparent lack of toxicity in
cellular systems.

Table 1. Colony forming ability at 9 days (% of control) of anchored CHO cells when
exposed to fusinite or LiPc.

concentration (mg/ml)	size of particles		
	< 1 mm, LiPc, exposed 24 hrs	5 mm, Fusinite, exposed 9 days	10 mm, Fusinite, exposed 9 days
1	--	94	104
5	--	92	104
10	97	96	97
20	99	--	--
30	104	--	--
50	--	--	102

Measurements in Unanesthetized Animals

The ability to make measurements of pO_2 without the use of anesthesia is critical for
many types of studies because of the perturbation of pO_2 by anesthesia. In many cases, even
if the blood pressure and the pO_2 in the circulatory system is kept in the "normal" range by
appropriate procedures during anesthesia, there may be significant perturbations at the tissue
level because physiological controls are likely to be altered. Figure 3 illustrates the potential
for effects from anesthesia and the capability of the EPR technique to make measurements
without anesthesia.

Measurements of Low Values of pO_2 In Vivo (in tumors)

Some of the most interesting and important measurements of pO_2 involve tissues
with a very low pO_2. These includes circumstances where the circulation may be impaired
acutely or chronically, and in tumors. For the latter, the critical values of pO_2 are between 1-

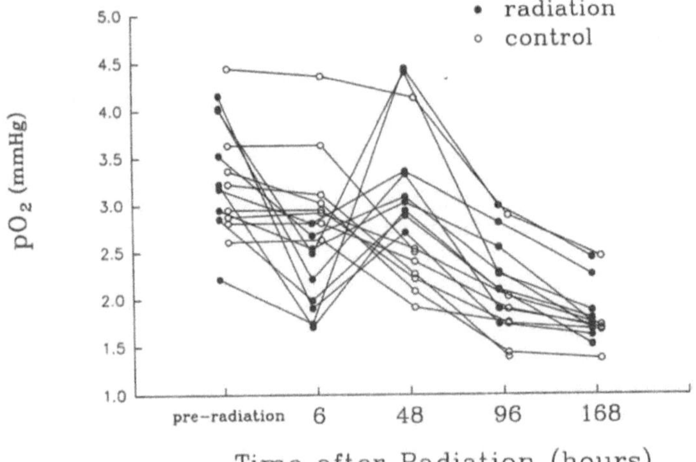

Figure 4. Measurements of low values of pO$_2$ in tumors. Changes of pO$_2$ in MTG-B tumors after a single dose of 20 Gy of ionizing radiation.

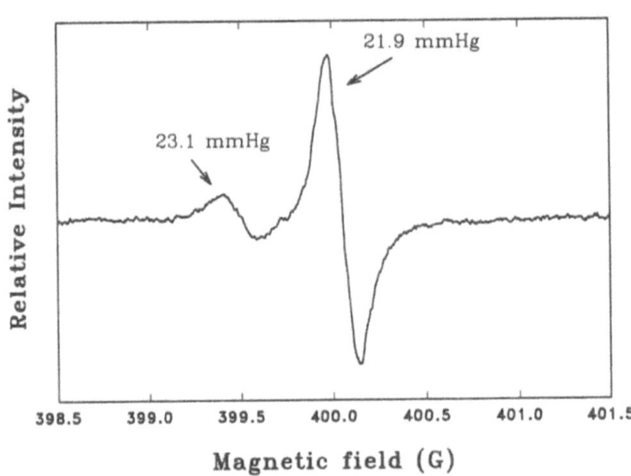

Figure 5. Measurements of pO$_2$ in two sites simultaneously. The EPR spectrum of two lithium phthalocyanine crystals in the left ventricle of a perfused rat heart. The field gradient was about 90 mG/mm.

10 Torr because that is the range in which the response to ionizing radiation and some chemotherapeutic agents changes by a factor of 2-3 (18). It is especially desirable to be able to make these measurements repeatedly and without perturbing the subject or the tumor. Figure 4 illustrates results obtained with the use of India ink in tumors in mice. The method obtained statistically significant differences of less than one Torr in the local pO_2 of the tumors. The studies extended over several days, providing measurements from the same sites in unanesthetized animals.

<u>Simultaneous Measurements of pO_2 in More than One Site</u>

There frequently is a need to obtain measurements of pO_2 from more than one site in order to compare phenomena that are related to local conditions. The EPR method described here provides information on the pO_2 in the tissue immediately in contact with the paramagnetic particles. The ability to obtain measurements of pO_2 at different sites therefore depends on the ability of the technique to obtain spectra that arise solely from each particle. A general discussion of this problem and the solution for a particular paramagnetic agent (lithium phthalocyanine) has been published (6,11). Figure 5 illustrates the use of such a technique. In this situation the technique readily resolved pO_2 at sites that were 5 mm apart; with the use of lithium phthalocyanine and appropriate field gradients, resolution of less than 3 mm should be achieved readily.

<u>Measurements of pO_2 in the Intracellular Compartment In Vivo</u>

The use of oxygen sensitive particles of appropriate dimensions provides an opportunity to achieve localization based on the physical-chemical properties of these particles. An example is the use of particles of India ink which are 2-3μ in diameter and therefore are selectively taken up by the reticulo-endothelial system, especially the Kupfer cells of the liver (7). Figure 6 illustrates the results of experiments in which India ink was administered to a mouse via the tail vein and subsequently more than 95 % of the material was found inside the Kupfer cells. EPR measurements over the liver area therefore provided measurements of the pO_2 within the Kupfer cells; to the best of our knowledge this is the first in vivo measurement of pO_2 in an intracellular compartment.

<u>Oxygen Dependent EPR Measurements in a Human Subject</u>

There are many situations in which the ability to obtain measurements of pO_2 accurately, rapidly, and repeatedly would contribute significantly to clinical care. While the particulate features of many of the paramagnetic materials used with the EPR technique facilitate achievement of these features, they also raise concerns because they may remain indefinitely in tissues. The discovery that India ink may be a good paramagnetic probe for pO_2 therefore was significant because this substance already is in widespread use in humans in vivo and its use is considered quite safe (7,13). An initial study has been carried out to illustrate the potential to obtain oxygen dependent EPR data in human subjects; the results are illustrated in figure 8. These results are considered promising because oxygen dependent changes in the linewidth could be detected even with the very suboptimal circumstances of using a source of India ink that had been in the tissues for several years and which had not been chosen on the basis of its response to pO_2.

DISCUSSION AND CONCLUSIONS

The results illustrated in this article support the conclusion that the measurement of pO_2 in tissues by EPR can be used successfully to obtain measurements that meet the requirements for many studies of physiological and pathophysiological phenomena as described in the introduction.

There appear to be few limitations on uses in small experimental animals such as rodents because of the demonstrated sensitivity, a lack of apparent side effects from the procedures for the measurements, and the depth of measurements is adequate for objects of the size of rodents. The results with the measurements of pO_2 in tumors (Figure 4) indicate that it will be readily feasible to follow process in vivo which result in changes of pO_2 of less than 1 Torr; this is a capability that has not been available by other techniques. The

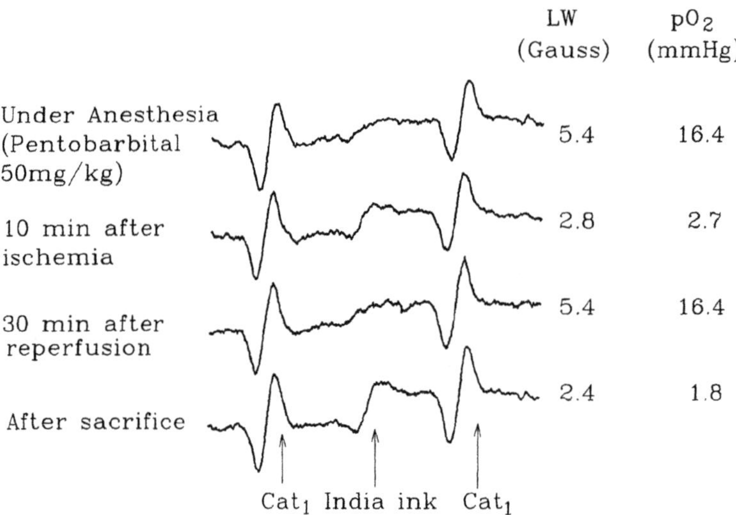

Figure 6. EPR spectra of India ink in the liver of a mouse. 0.04ml of India ink was injected via the tail vein and EPR spectra were obtained under anesthesia (50 mg/Kg of pentobarbital), 10 min after occlusion of the hepatic artery and portal vein, 30 min after reperfusion, and after sacrifice. EPR parameters were: scan range 40 Gauss, scan time 120 seconds, time constant 0.3 second, and modulation 1.6 Gauss. 1 mM of Cat1 was used as a standard sample.

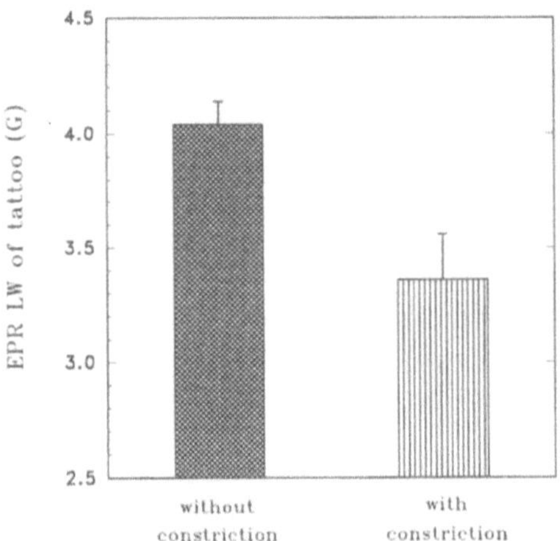

Figure 7. In vivo EPR measurements of pO$_2$ from a human tattoo. Spectra were obtained before and after constriction of the blood flow by means of a rubber tourniquet around the arm. The corresponding values of pO$_2$ were unknown because an oxygen calibration for the material used for the tattoo was not available.

principal experimental limitations in experiments in small animals are likely to be measurements in sites which are near the depth of sensitivity and which also have significant physiological movements. For example, we currently have had to study changes in pO_2 in the myocardial wall with open-chested preparations in order to have adequate signal/noise to measure rapid changes in pO_2. It seems likely, but not certain, that additional technical developments of the current instrumentation will enable us to overcome these limitations.

The use of these methods in larger animals will be quite feasible as long as the site to be measured is within 10 mm of the detector. This includes regions near the surface of the animal; regions accessible via natural or artificial openings (ie. detectors in the form of catheters or needles have been made and are quite sensitive) (17), and regions made accessible through intra-operative studies. Studies that cannot be made by these approaches (eg. non-invasive studies of regions more than 10 mm from the surface) probably will require the development of sufficiently sensitive lower frequency EPR instruments such as those being developed by Halpern (19).

Clinical studies should be feasible in the very near future. It is anticipated that initially these will involve the use of India ink. It may be desirable initially to do the studies with tissues that will be resected soon afterwards, to provide supplemental data on the fate of India ink in tissues. The initial studies are likely to be in tumors, but studies in limbs with potential ischemia are also likely to be carried out very early in the phase of clinical applications.

REFERENCES

1. H. M. Swartz, S. Boyer, P. Gast, J. F. Glockner, H. Hu, K.J. Liu, M. Moussavi, S.W. Norby, T. Walczak, N. Vahidi, M. Wu, and R. B. Clarkson, Measurements of pertinent concentrations of oxygen in vivo, Magn. Reson. Med. 20:333 (1991).

2. J. F. Glockner and H. M. Swartz, In vivo EPR oximetry using two novel probes: fusinite and lithium phthalocyanine, in: "Oxygen Transport to Tissue XIV", W. Erdmann and D.F. Bruley, eds., Plenum Publishing Corp., New York (1993).

3. K.J. Liu, P. Gast, M. Moussavi, S.W. Norby, N. Vahidi, T. Walczak, M. Wu, and H.M. Swartz, Lithium phthalocyanine: a probe for EPR oximetry in viable biological systems. PNAS 90:5438 (1993).

4. N. Vahidi, R.B. Clarkson, K.J. Liu, S.W. Norby, M. Wu, and H.M. Swartz, In vivo and in vitro EPR oximetry with fusinite: a new coal-derived, particulate EPR probe. Magn. Res. Med., in press.

5. H.M. Swartz, S. Boyer, D. Brown, K. Chang, P. Gast, J.F. Glockner, H. Hu, K.J. Liu, M. Moussavi, M. Nilges, S.W. Norby, A. Smirnov, N. Vahidi, T. Walczak, M. Wu, and R.B. Clarkson, The use of EPR for the measurement of the concentration of oxygen in vivo in tissues under physiologically pertinent conditions and concentrations, in: "Oxygen Transport to Tissue XIV", W. Erdmann and D.F. Bruley, eds., Plenum Press, New York (1992).

6. A.I. Smirnov, S.W. Norby, T. Walczak, K.J. Liu, and H.M. Swartz, Physical and instrumental considerations in the use of lithium phthalocyanine for measurements of the concentration of the oxygen, J. Magn. Reson., in press (1993).

7. H.M. Swartz, K.J. Liu, F. Goda, and T. Walczak, India ink: a potential clinically applicable EPR oximetry probe, Magn. Reson. Med. in press.

8. W.K. Subczynski, S. Lukiewicz, and J.S. Hyde, Murine in vivo L-band ESR spin-label oximetry with a loop-gap resonator, Magn. Reson. Med. 3:747 (1986).

9. H.J. Halpern, M. Peric, T.D. Nguyen, D.P. Spencer. Selective isotope labeling of a nitroxide spin label to enhance sensitivity for T_2 oximetry. J. Magn. Reson. 90:40 (1990).

10. J.L. Zweier, S. Thompson-Gorman, and P. Kuppusamy, Measurement of oxygen concentration in the intact beating heart using electron paramagnetic resonance spectroscopy: a technique for measuring oxygen concentration in situ, J. Bioenerg. Biomembr. 23:855 (1991).

11. A.I. Smirnov, S.W. Norby, R.B. Clarkson, T. Walczak and H.M. Swartz, Simultaneous multi-site EPR spectroscopy in vivo, Mag. Reson. Med. 30:213 (1993).

12. M. Wu and H.M. Swartz, Evaluation of the potential cytotoxicity of some paramagnetic materials used in measurements of the concentration of oxygen, Current Topics Biophys. in press.

13. M.B. Fennerty, R.E. Sampliner, L.J. Hixson, H.S., Garewal, Effectiveness of India ink as a long-term colonic mucosal marker, Am. J. Gastroenterol. 87:79 (1992).

14 G. Bacic, K.J. Liu, J.A. O'Hara, R.D. Harris, K. Szybinski, F. Goda, H. Swartz, Oxygen tension in a murine tumor: a combined EPR and MRI study, Magn. Reson. Med., 30:568 (1993).

15. H.M. Swartz and J.F. Glockner, Measurement of oxygen by ESRI and ESRS, in: EPR Imaging and In Vivo EPR, G.R. Eaton, S.S. Eaton, and K. Ohno, eds., CRC Press, Inc., Boca Raton, FL (1991).

16. M.J. Nilges, T. Walczak, and H.M. Swartz, 1 GHz in vivo ESR spectrometer operating with a surface probe, Phys. Med., 5:195 (1989).

17. H.M. Swartz and T. Walczak, In vivo EPR: prospects for the '90's, Phys. Med., in press (1993).

18. J. Chapman, Measurements of tumor hypoxia by invasive and non-invasive procedures: a review of recent clinical studies, Radiother. Oncol. 20:13 (1991).

19. H.J. Halpern, D.P. Spencer, J. van Polen, J.K. Bowman, A.C. Nelson, E.M. Dowey, and B.A. Teicher, Imaging radio-frequency electron-spin-resonance spectrometer with high resolution and sensitivity for in vivo measurements, Rev. Sci. Instrum. 53:1040-1050 (1989).

HALF-LIFE OF PERFLUOROOCTYLBROMIDE IN INNER ORGANS DETERMINED BY FAST 19F-NMR IMAGING

L.J.E. Jäger[1], U. Nöth[2], A. Haase[2], J. Lutz[3]

[1] Institute of Radiology, University of Bonn, D-53127 Bonn, Germany
[2] Inst. of Physics (Dept. of Biophysics)
   University of Würzburg, D-97074 Würzburg
[3] Department of Physiology, University of Würzburg, D-97070 Würzburg

INTRODUCTION

In the group of artificial oxygen carriers perfluorochemicals (PFCs) can serve as contrast agents for 19F-NMR imaging. However, some of the PFCs were accused for a long-term retention by organs of the reticuloendothelial system (RES), especially in liver and spleen (Lutz 1985). Thus the question arose how far PFCs of the second generation would share this disadvantage.
To estimate the concentration of fluorinated substances in different tissues in vivo 19F-nuclear magnetic resonance imaging is a suitable method. In all investigations so far 19F-NMR spectroscopy (Mason et al. 1992, 1993) or 19F-NMR imaging (Ratner et al. 1987, 1988, Freeman et al. 1988, Eidelberg et al. 1988, Meyer et al. 1992) with the spin echo technique were used. The disadvantage of NMR spectroscopy is the missing spatial resolution and that of spin echo imaging is the length of the acquisition time which was reported to be somewhere between 4 min and 30 min.

A systematic determination of the in vivo half-life of PFCs of the second generation by modern means of 19F-NMR imaging was missing so far. Therefore, we examined in this study a highly concentrated emulsion of perfluorooctylbromide (PFOB) by using SNAPSHOT-FLASH 19F-NMR imaging (Haase 1990), a fast gradient echo technique.

METHODS

Male rats of the Wistar strain (240 g - 260 g), kept on a standard diet and water ad libitum were used throughout the experiments. They received a single i.v. dose of either 3 g/kg or 6 g/kg of PFOB in a 90% w/v emulsion (Alliance Pharmaceutical Corp., San Diego). The relative intensity of PFOB in liver and spleen was measured in vivo by 19F-NMR imaging as a function of time, in order to examine the longitudinal course of RES loading by PFOB. The half-life of PFOB in tissue was calculated from the relative intensities by means of the method of least squares on a logarithmic scale.

All experiments were performed on a 4.7 Tesla Bruker BIOSPEC system with a 21 cm horizontal bore. The SNAPSHOT-FLASH technique was applied in combination with STEP-presaturation pulses (Jakob et al. 1992, 1993). A 64 * 128 pixel matrix with 9.2 cm field of view (FoV) was taken. A transaxial slice of 6 mm thickness and an acquisition time of 1.46 s per average was chosen, while the number of averages amounted to 64. During all experiments the surrounding conditions, such as temperature and oxygen supply, remained constant. A phantom of 9% w/v PFOB was used as a control standard for each experiment in order to calculate the relative intensity. This value was obtained in the following way (Nöth et al. 1994):

$$\text{rel. intensity} = \frac{(S^2_{tis} - N^2)^{1/2}}{(S^2_{ref} - N^2)^{1/2}}$$

S_{tis} is the signal intensity of interest, S_{ref} is the signal intensity in the phantom and N is the signal intensity of the noise.

RESULTS

We succeeded in imaging the liver and the spleen within approximately 1.5 min. For a single dose of 3 g PFOB/kg the maximum of uptake in liver and spleen was achieved after 24 hours and 3 hours, respectively. For 6 g PFOB/kg the maxima lay at 36 hours and 12 hours, respectively. The outline of the liver and spleen were precise. A homogeneous signal was obtained from the tissue. In contrast to these findings for a dose of 6g/kg PFOB, the dose of 3g/kg evoked a less homogeneous signal from the tissue and the outline of the liver and spleen were not as precise.

The relative intensity of the 19F-NMR signal as a measure for the extent of PFOB concentration is demonstrated as a function of time. The burdening of cells of the RES with PFOB is dose dependent. The starting point for the calculation of the half-life is given in Figure 1 and 2. The half-life for a dose of 3 g PFOB/kg in tissue amounted to 4.5 days ($r=0.892$) in the liver and to 3.6 days ($r=0.956$) in the spleen, after 6 g PFOB/kg the respective values were 7.4 days ($r=0.980$) in the liver and 8.2 days ($r=0.996$) in the spleen.

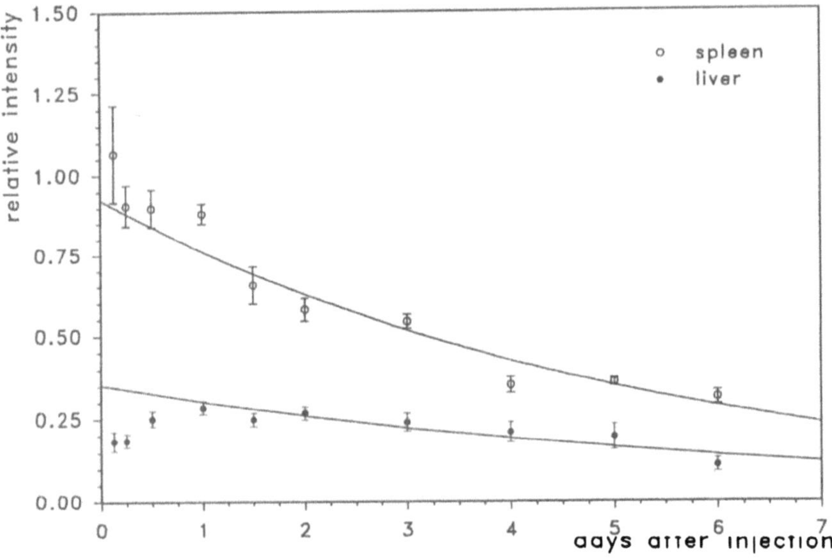

Fig. 1. Longitudinal study over 6 days on the PFOB content in spleen and liver after a dose of 3 g/kg b.wt. in rats. For these curves the regression and half-lifes were calculated, beginning with 1d p.i. Error bars express ± SEM.

130

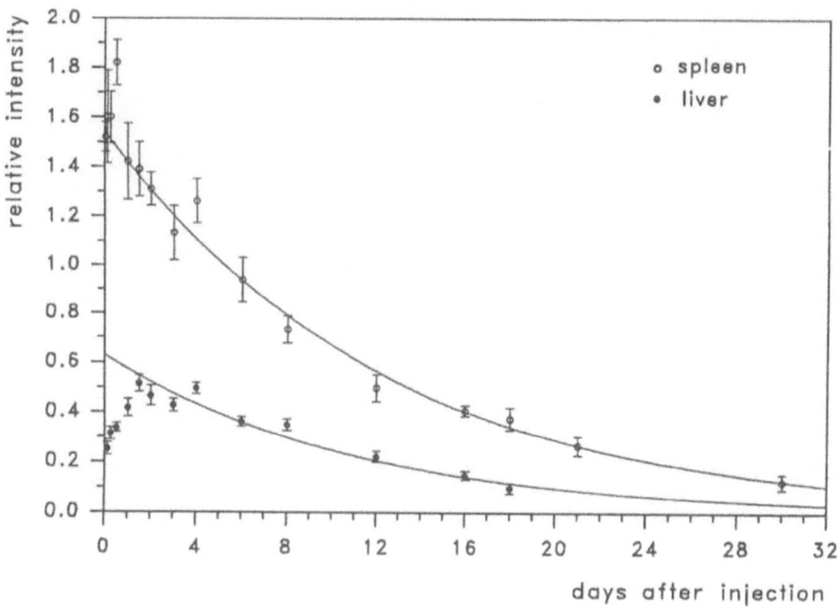

Fig.2. Longitudinal study over 30 days on the PFOB content in spleen and liver after a dose of 6 g/kg b.wt. in rats. Regression and half-life were calculated beginning with 1.5 d p.i. Error bars as in Fig.1.

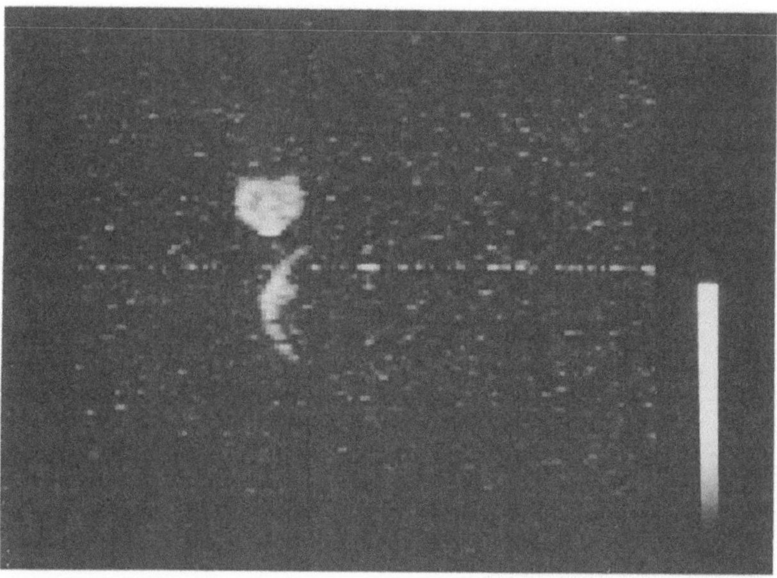

Fig.3. 19F transaxial image of the spleen 3 hours after administration of 3 g PFOB/kg b.wt. On the top of the organ the phantom containing 9 g PFOB/dl can be seen. 64 averages with a total acquisition time of 1 min 33 s.

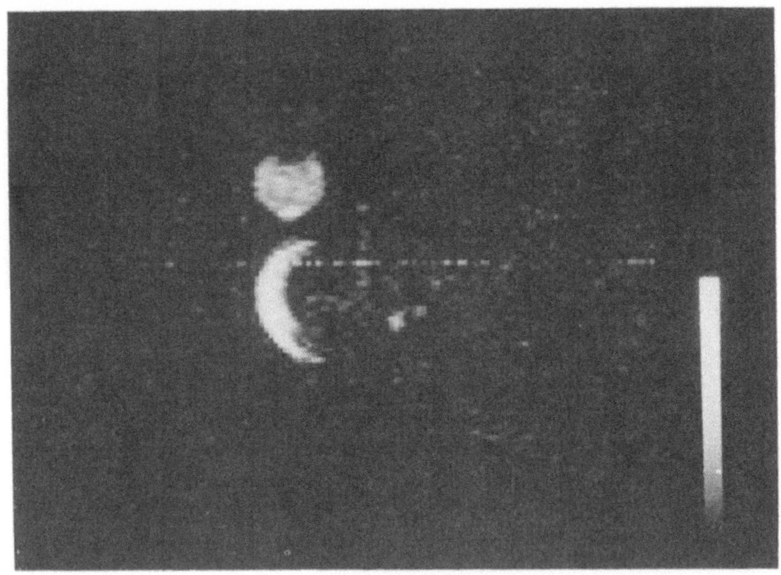

Fig.4. 19F transaxial image of the spleen 12 hours after administration of 6 g PFOB/kg b.wt. 64 averages with a total acquisition time of 1 min 33 s. The horizontal line in the center of the image is a base line artifact.

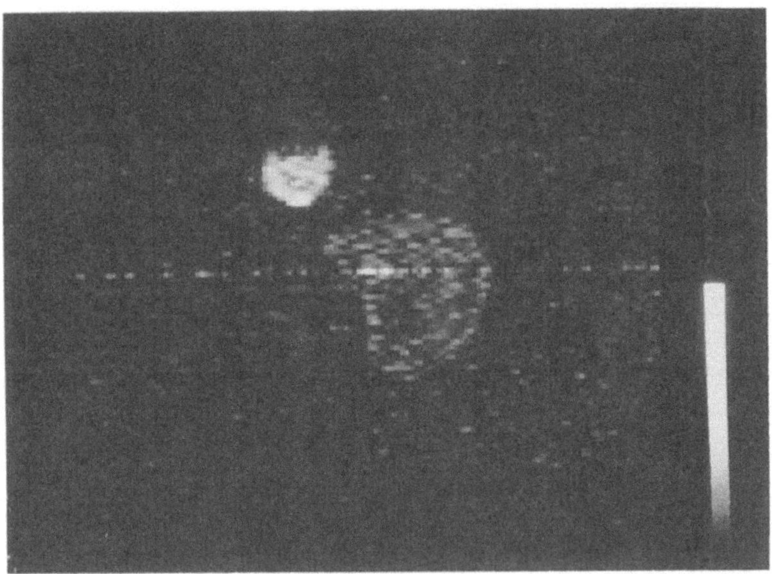

Fig.5. 19F transaxial image of the liver 24 hours after administration of 6 g PFOB/kg b.wt. Conditions like in Fig. 3.

DISCUSSION AND CONCLUSION

The uptake of PFCs by organs of the RES was evaluated up to now by gas chromatography (Yokoyama et al. 1978, Lutz et al. 1980,), histology or CT (Mattrey et al. 1984, Adam et al. 1992) and 19F-NMR experiments (Ratner et al. 1987, Meyer et al. 1992, Nöth et al. 1993, 1994). The great advantage of 19F-NMR is the possibility of longitudinal, noninvasive studies of the content of PFCs within the same animal or object.

By means of this technique the evaluation of half-life of PFCs in select tissues gets a new approach. Comparable determinations generally required the use of new groups of animals for every measurement (Naito and Yokoyama, 1978, Lutz and Metzenauer 1980); only tests of blood levels could be done with the same animal (Lutz and Stark, 1987). The fact of a strong dose dependency of t/2 could be confirmed by our new results.

Our success in demonstrating good 19F-NMR images with a reasonable short total acquisition time argues for the potential of SNAPSHOT-FLASH 19F-NMR imaging.

The relatively short half-life of PFOB in the RES, which is one of the smallest of fluorinated artificial oxygen carriers (Riess and LeBlanc 1982), may indicate the diminished load in organs of the RES and the improvement of modern PFCs (Jäger & Lutz 1993a,b). The feasibility of fast 19F-NMR imaging, even at low dosages (3 g/kg), argues for the use of PFOB as an intravenous contrast agent for 19F-NMR experiments.

REFERENCES

Adam G, Günther RW, Schiffer C, Görtz H, Prescher A, Scherer K, 1992. Computed tomographic enhancement of the liver, liver abscesses, spleen, and major vessel with perfluorooctylbromide emulsion. Invest. Radiol. 27: 698-705

Eidelberg D, Johnson G, Barnes D, Tofts PS, Delpy D, Plummer D, McDonald WI, 1988. 19F NMR imaging of blood oxygenation in the brain. Mag. Res. in Med. 6: 344-352

Freeman DM, Muller HH, Hurd RE, Young SW, (1988). Rapid 19F Magnetic resonance imaging of perfluorooctyl bromide in vivo. Magn. Reson. Imaging 6:61-64

Haase A, 1990. Snapshot FLASH MRI. Application to T1, T2, and chemical-shift imaging. Magn. Reson. Med. 13: 77-89

Jäger LJE, Lutz J, 1993a. Plasma enzyme pattern after administration of different doses of perfluorooctylbromide in rats. Pflügers Arch. 422 [Suppl.] R 117

Jäger LJE, Lutz J, 1993b. Phagocytosis of colloidal carbon after administration of artificial oxygen carriers of first and second generation. Advances in Exp. Med. & Biol., Plenum Press, New York (in press)

Jakob PM, Kotitschke K, Haase A, 1992. STEP pulse: A new presaturation pulse and its application. SMRM 11th Annual Meeting, Book of Abstracts 2: 3910

Jakob PM, Matthaei D, Stodal H, Brinck U, Haase A, 1993. Simultaneous preparation of inversion recovery T_1 and chemical shift selective contrast using snapshot-FLASH-MRI. MAGMA 1: 77-82

Lutz J, 1985. Effect of perfluorochemicals on host defense, especially on the reticuloendothelial system. In: "Perfluorochemical Oxygen Transport". K.K. Tremper ed., Little, Brown & Co. Boston 1985 p.63-93

Lutz J, Metzenauer P, 1980. Effects of potential blood substitutes (perfluorochemicals) on rat liver and spleen. Pflügers Arch. <u>387</u>: 175-181

Lutz J, Stark M, 1987. Half life and changes in the composition of a perfluorochemical emulsion within the vascular system of rats. Pflügers Arch. <u>410</u>: 181-184

Mason RP, Shukla H, Antich PP, 1992. Oxygent^{TM}: a novel probe of tissue oxygen tension. Biomat. Art. Cells, Immob. Biotechn. <u>20</u>, 929-932

Mason RP, Shukla H, Antich PP, 1993. In vivo oxygen tension and temperature: Simultaneous determination using 19F NMR spectroscopy of perfluorocarbon. Magn. Reson. Med. <u>29</u>: 296-302

Mattrey RF, Long DM, Peck WW, Slutsky RA, Higgins CB, 1984. Perfluoroctylbromide as a blood pool contrast agent for liver, spleen, and vascular imaging in computed tomography. J. Comput. Assist. Tomogr. <u>8</u>: 739-744

Meyer KL, Carvlin MJ, Mukherji B, Sloviter HA, Joseph PM, 1992. Fluorinated blood substitute retention in the rat measured by fluorine-19 magnetic resonance imaging. Invest. Radiol. <u>27</u>: 620-627

Naito R, Yokoyama K, 1978. Perfluorochemical blood substitutes. Technical Information Ser. 5, The Green Cross Corp. Osaka Japan

Nöth U, Jäger LJE, Lutz J, Haase A, 1993. Fast 19-F-NMR imaging using PFOB as contrast agent. 10th Ann. Scient. Meeting Europ. Soc. MR in Med. & Biol. Book of Abstracts: 448

Nöth U, Jäger LJE, Lutz J, Haase A, 1994. Fast 19-F-NMR imaging in vivo using FLASH-MRI. Mag. Reson. Imaging <u>12</u> (in press)

Ratner AV, Hurd R, Muller HH, Bradley-Simson B, Pitts W, Shibata D, Sotak C, Young SW, 1987. 19F magnetic resonance imaging of the reticuloendothelial system. Magn. Res. Med. <u>5</u>: 548-554

Ratner AV, Muller HH, Bradley-Simson B, Johnson DE, Hurd RE, Sotak C, Young SW, 1988. Detection of tumors with 19F magnetic resonance imaging. Invest. Radiol. <u>23</u>:361-364

Riess J, LeBlanc M, 1982. Solubility and transport phenomena in perfluorochemicals relevant to blood substitution and other biomedical applications. Pure & Appl. Chem. <u>54</u>:2383-2406

Yokoyama K, Yamanouchi K, Ohyanagi H, Mitsuno T, 1978. Fate of perfluorochemicals in animals after intravenous injection or hemodilution with their emulsions. Chem. Pharm. Bull. <u>26</u>: 956-966

FREQUENCY RESPONSE BY PULSE REDUCTION FOR
THE ANALYSIS OF TRS SPECTRA

K. A. Kang*, D. F. Bruley*, J. M. Londono, and B. Chance

Johnson Foundation, Department of Biochemistry and Biophysics
University of Pennsylvania, Philadelphia, PA 19104
* College of Engineering, University of Maryland Baltimore County
Baltimore, MD 21228

INTRODUCTION

Biological systems can be analyzed by near infrared (NIR) spectroscopy in two ways: (1) A system can be characterized by studying its photon absorption and scattering as a lumped parameter system. A good example of this approach is the measurement of hemoglobin oxygen saturation (or other chromophore concentration) in an organ of the human body under various physiological conditions; (2) Heterogeneities in an organ can be located, for instance imaging of brain or breast tumors. Since the degree of light absorption by various chromophores (oxy/deoxy hemoglobin, cytochrome aa_3, etc.) and that of its scattering by biological scatterers (mitochondria, fat particles, etc.) reflects the physiological state, it is important to obtain accurately the absorption and scattering coefficients of a homogeneous system. At the same time, accurate measurement of the intensity and the phase shift difference at specific positions relative to the surrounding tissue is crucial for imaging.

Frequency response analysis for dynamic systems has been used in the field of engineering system identification and process control for several decades[1,3,4,10,11]. When the input is a sharp pulse and therefore the shape of the pulse is close to the Dirac delta function, the time domain response can be transformed to the frequency domain over a wide range of frequencies by Fourier transform[3]. In the time resolved spectroscopy (TRS) system used here, the FWHM (full width at half maximum) of the input pulse was 150 ps, which is sharp enough to provide the phase and the magnitude information up to 1 GHz. The system parameters that can be obtained from frequency response analysis are the magnitude ratio (MR) and phase shift (ϕ) at various frequencies, the steady state gain (K), the system order (n), and the time constant (τ) of the system. A preliminary study of these parameters in homogeneous and heterogeneous highly scattering media was performed by Kang et al.[8]

In this paper, a method for obtaining absorption and scattering coefficients in infinite homogeneous media is introduced. This method utilizes MR values at two or more distances between the source and the detector (S-D distances, ρ). For a heterogeneous system, MR and ϕ were shown to vary with the position of a localized absorber in a scattering media.

THEORY

Frequency Response Analysis

When an input, x(t), is introduced to a system and an output, y(t), is generated by that input, via the transfer function, G(s), where the transfer function is the ratio of the Laplace transforms of the output [Y(s)] and the input [X(s)]:

$$G(s) = \frac{Y(s)}{X(s)} \tag{1}$$

where 's' is a complex number. Substitution of forcing frequency 'ω' for 's' results in the frequency domain transfer function, G(ω):

$$G(\omega) = \frac{Y(\omega)}{X(\omega)} = \frac{\int_0^{Ty} y(t)e^{-j\omega t}dt}{\int_0^{Tx} x(t)e^{-j\omega t}dt} \tag{2}$$

Theoretically, the lower and upper limits of the integration are $-\infty$ and $+\infty$. For actual data reduction, the integration time limits on the input, x(t), are [0, T_x] and the integration time limits on the output, y(t), are [0, T_y], where T_x and T_y represent the time when the input and output pulse values become zero, respectively.

The two important system parameters that can be obtained from the transfer function are magnitude ratio (MR) and phase shift (ϕ) as functions of forcing frequency. These values can be represented by the following equations:

$$MR\,(\omega) = |G(\omega)| = \sqrt{Re^2(\omega) + Im^2(\omega)} \tag{3}$$

$$\phi(\omega) = \phi\big|_{G(\omega)} = tan^{-1}\left[\frac{Im\,(\omega)}{Re\,(\omega)}\right] = \phi\big|_{Y(\omega)} - \phi\big|_{X(\omega)} \tag{4}$$

where Re(ω) and Im(ω) are the real and imaginary parts of the transfer function, respectively. The normalized input frequency content (NFC) is a very useful indicator for determining if the computed values are meaningful at a given frequency. The NFC of the ideal Dirac delta function is unity throughout the entire frequency domain with zero phase shift. Other shapes have NFC values ranging between 1 and 0.0. As a rule of thumb, reduced data at frequencies where the NFC is less than 0.3 are considered to be inaccurate. Clement and Schnelle[3] illustrated NFC values for various input pulse shapes and it was found that NFC values for TRS pulses are very close to that of a Dirac delta input.

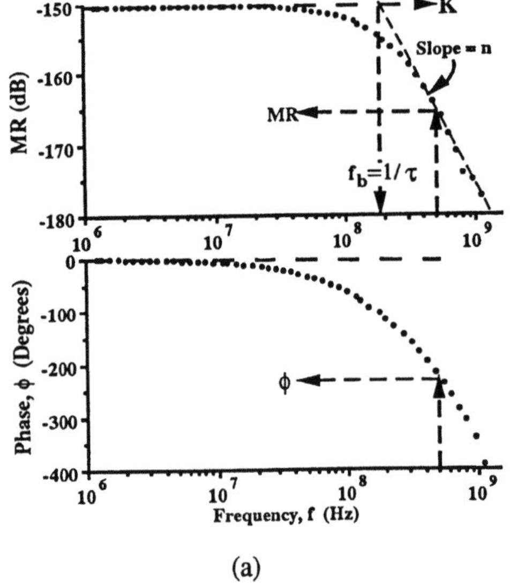

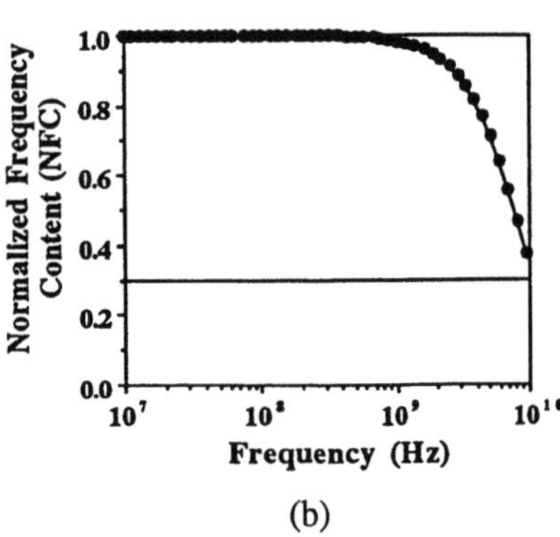

Figure 1. A typical Bode diagram generated from a TRS spectrum obtained from 0.5 % Intralipid solution at a S-D distance of 5 cm. (a) magnitude ratio (MR) and phase shift (ϕ), (b) Normalized frequency content (NFC) at various frequencies.

INSTRUMENTS AND MATERIALS

The following instruments and materials were used to carry out the experiments.
(1) Time resolved spectroscope (TRS): The TRS system was manufactured by Hamamatsu Photonics, KK. (Hamamatsu, Japan). The wavelength used was 780 nm. This TRS is a single photon counting system with a pulse repetition rate of 5 MHz. The detector system was a multi-microchannel plate photomultiplier tube (MCP-PMT)[9] (Fig. 2 (a)].
(2) Optical fibers for the source and the detector: Glass fiber of diameter 200 µm was used for the source (OZ Optics Corp., ONT, Canada) and fiber of diameter 3 mm was used for the detector (Twardy, Daren, CT).
(3) Black India ink (Faber-Castell, Newark, NJ): Black India ink was used to change the photon absorption rate in the medium.
(4) Intralipid™ (Kabi Pharmacia, Clayton, NC): 20 % Intralipid™ was generously donated by Kabi Pharmacia and used as the scattering continuum.
(5) Neutral density filter (Kodak Co., NY, NY): Neutral density gelatin filters of various N.D. numbers were used between the source and the detector for measuring the input TRS spectrum.

METHODS

A 15 L container (20 cm x 40 cm x 26 cm) was used to avoid boundary effects, except for the study of change in scattering induced by varying the concentration of Intralipid™, in which a 5 L container was used. . The source and the detector cables were placed in the middle of the container [Fig. 2 (a)]. To obtain reasonable results in the frequency domain, photon counts at the peak of the TRS spectrum must be higher than 10^3.

To measure the input x(t), the source and the detector were attached to each other and the TRS spectrum was obtained. With this experimental arrangement the laser intensity was too high for the MCP-PMT to handle. To alleviate this problem, neutral density gelatin filters with known N.D. numbers were placed between the source and the detector.

Studies on μ_a and μ_s' of Infinite Media

For experiments designed to determine absorption coefficients, 50 µl portions of 50 % of India ink were added to 15 L of 0.5 % Intralipid™ solution, up to a total added ink volume of 200 µl. TRS spectra were obtained at S-D distances of 3, 4, and 5 cm between the source and the detector, whenever 50 µl of ink solution was added into the media.

For experiments designed to study the effect of the scattering coefficient in a homogeneous medium, the concentrations of Intralipid™ were 1.5 %, 2.0 %, and 2.5 % in a 5 L container. TRS spectra were obtained at S-D distances 3, 4, and 5 cm.

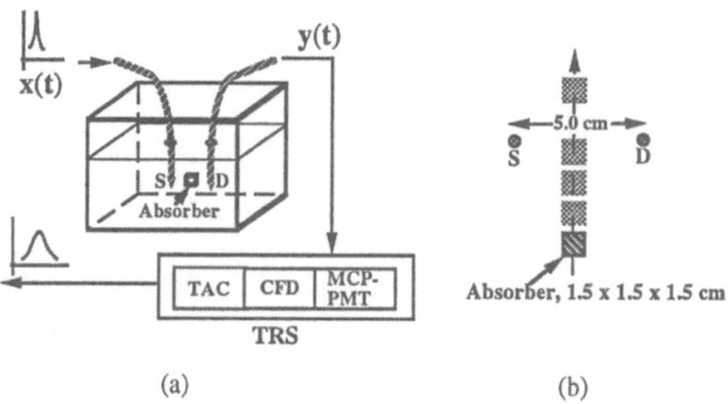

Figure 2. (a) Experimental system. (b) positions of the localized absorber; absorber position is varied along the center line between the source and the detector.

Fig. 1 illustrates a typical Bode diagram, (a) MR (log - log scale) in decibels ; [MR (in dB) = 20 $\log_{10}$ |G(s)|] and (b) ϕ (semi - logarithmic scale), computed from a TRS spectrum of a 0.5 % Intralipid™ solution and (c) NFC for the input source used for this experiment, obtained numerically. The data reduction calculation for this study was performed with a computer code using the Filon's integration method[5], written by Bruley[2].

Determination of Optical Properties of Homogeneous Infinite Media

The intensity (M_{ac}) of the alternating current (AC) and phase shift (ϕ) at a modulation frequency, f (f=$\omega/2\pi$), in a highly scattering infinite medium can be expressed as follows[7]:

$$\frac{M_{ac}}{I_o} = \frac{1}{4\pi D}\frac{1}{\rho}\left[\exp\{-\rho[(\frac{\mu_a}{D})^2 + (\frac{2\pi f}{DC_n})^2]^{1/4}\cos[\frac{1}{2}\tan^{-1}(\frac{2\pi f}{\mu_a C_n})]\}\right] \tag{5}$$

$$\phi = -\rho[(\frac{\mu_a}{D})^2 + (\frac{2\pi f}{DC_n})^2]^{1/4}\sin[\frac{1}{2}\tan^{-1}(\frac{2\pi f}{\mu_a C_n})] \tag{6}$$

where I_o is the initial AC intensity at the source, D is the diffusion coefficient (cm), which is $1/3(\mu_a+\mu_s')$. ρ is the distance between the source and the detector [S-D distance] and μ_a and μ_s' are the absorption and the effective scattering coefficients, respectively. C_n is the speed of light in the medium (cm/s). Since engineers most commonly express MR in units of dB [=20 log (M_{ac})], Eq. (5) can be rewritten for dB units;

$$MR + 20\log(\rho) = -C_1\rho + C_2 \tag{7}$$

where $C_1 = \dfrac{20}{2.303}\bullet\left[(\frac{\mu_a}{D})^2 + (\frac{2\pi f}{DC_n})^2\right]^{1/4}\bullet\cos[\frac{1}{2}\tan^{-1}(\frac{2\pi f}{\mu_a C_n})]$ (a)   (8)

$C_2 = 20\bullet\left[-\log_{10}(4\pi D)\right]$.   (b)

Equation (6) can also be rewritten as:

$$\phi = -C_3\rho \tag{9}$$

where $C_3 = [(\frac{\mu_a}{D})^2 + (\frac{2\pi f}{DC_n})^2]^{1/4}\bullet\sin[\frac{1}{2}\tan^{-1}(\frac{2\pi f}{\mu_a C_n})]$.   (10)

When equation (10) is divided by equation (8) (a), the relationship becomes:

$$C_3 / C_1 = \frac{2.303}{20}\tan[\frac{1}{2}\tan^{-1}(\frac{2\pi f}{\mu_a C_n})] \tag{11}$$

Therefore, $\mu_a = \dfrac{2\pi f}{C_n}\bullet\dfrac{1}{\tan[2\tan^{-1}(\frac{C_3}{C_1}\bullet\frac{20}{2.303})]}$.   (12)

Using equation (10) and the value, μ_a, obtained from equation (12)

$$\frac{1}{D} = 3(\mu_a + \mu_s') = \sqrt{\frac{\{C_3 / \sin[\frac{1}{2}\tan^{-1}(\frac{2\pi f}{\mu_a C_n})]\}^4}{\mu_a^2 + (\frac{2\pi f}{C_n})^2}} \tag{13}$$

and $\mu_s' = \dfrac{1}{3D} - \mu_a$.   (14)

To compare μ_a and μ_s' values computed utilizing MR and ϕ values of the transfer function, TRS spectra were also fit to the analytical solution[6,12] and values of μ_a and μ_s' of the same spectra were obtained.

From the input and output TRS spectra, the magnitude ratio (MR) and phase shift (ϕ) at frequencies with a NFC higher than 0.3 were computed. By curve fitting the TRS spectra to the analytical solution, values of μ_a and μ'_s for the experimental media were also obtained.

Studies on MR and ϕ of heterogeneous media

To study the effect of changes in position of the absorber, the position of a black plexi glass cube (1.5 cm x 1.5 cm x 1.5 cm), was varied along the center line between the source and the detector [Fig. 2 (b)]. The TRS spectrum at each absorber position was obtained at a S-D distance of 5 cm, and MR and ϕ were computed as previously described.

RESULTS AND DISCUSSION

Absorption and Scattering Coefficients of Homogeneous Infinite Media

Fig. 3 (a) and (b) illustrates μ_a and μ'_s computed with MR and ϕ values when the ink concentration in the medium changed linearly without changing the scatterer (Intralipid™) concentration. The amount of 50 % India ink added to 15 L of 0.5 % Intralipid was 0.0, 50, 100, 150, and 200 μL. As can be seen in Fig. 3 (a) μ_a computed with MR and ϕ increases linearly with an increase in the ink concentration. The values for μ'_s remained almost constant at the five different ink concentrations. For comparison, μ_a and μ'_s for the cases above were also obtained by fitting the TRS spectra to the analytical solution [Fig. 3 (c) and

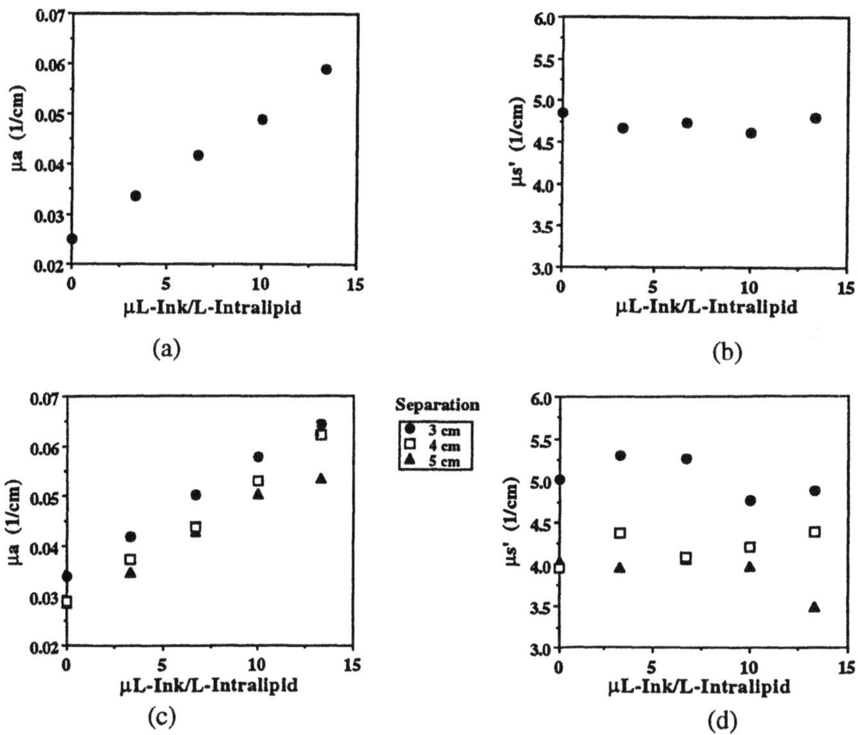

Figure 3. μ_a and μ'_s with changes in the absorber (ink) concentration in a homogeneous media. (a) μ_a and (b) μ'_s obtained using MR values at various frequencies. For μ'_s , the solid and the dotted lines represent the mean and the standard deviation, respectively. (c) μ_a and (b) μ'_s obtained by curve fitting TRS spectra to the analytical solution.

(d)]. The absorption and the scattering coefficients values (μ_a and μ_s') computed using MR and ϕ were in the same range of those obtained by curve fitting. However, the values obtained by curve fitting were less consistent, especially for the scattering coefficients, and strongly depend on S-D distances.

Fig. 4 shows μ_s' and μ_a when the scattering of the media was changed by varying the concentration of the scatterer (Intralipid) at constant absorption. The concentrations of Intralipid employed in the study were 1.5 %, 2.0 %, and 2.5 %. Fig. 4 (a) and (b) are μ_s' and μ_a values computed using MR and ϕ values, respectively. At a constant absorption μ_s' increased linearly as the concentration of Intralipid increased. The absorption coefficient (μ_a) computed at three different Intralipid concentrations showed little variation. Fig. 4 (c) and (d) are μ_s' and μ_a obtained by curve fitting. As in the previous case, μ_s' and μ_a obtained by the fitting were less consistent and also dependent on the S-D distance. The absorption coefficients obtained by curve fitting were slightly greater and the scattering coefficients obtained by curve fitting were less than those obtained using MR and ϕ.

It is difficult to state that one method is more accurate than the other, since there is no device to absolutely measure the optical values of highly scattering media. However, the method using MR and ϕ values seems to be very consistent. If μ_a is greater than 0.5 cm^{-1}, MR changes greatly with a change in the S-D distance, and the S-D distance should be measured very accurately. Fig. 5 shows the relationship of $[MR + 20 \log(\rho)]$ vs. ρ and ϕ vs. ρ at various frequencies at eight different S-D distances. The slopes (C_1 and C_3) were linear with a correlation 1.0 in all cases, which demonstrates the consistency of the computed

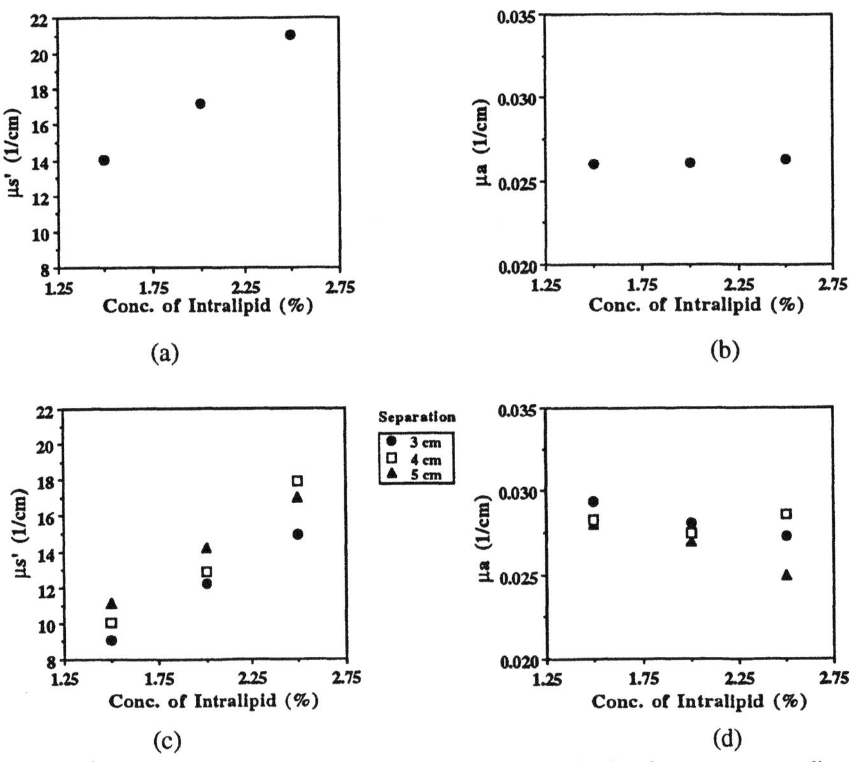

Figure 4. μ_s' and μ_a with changes in the absorber (ink) concentration in a homogeneous medium. (a) μ_s' and (b) μ_a obtained using MR values. For μ_a, the solid and dotted lines represent the mean and the standard deviation, respectively. (c) μ_s' and (d) μ_a obtained by curve fitting TRS spectra to the analytical solution.

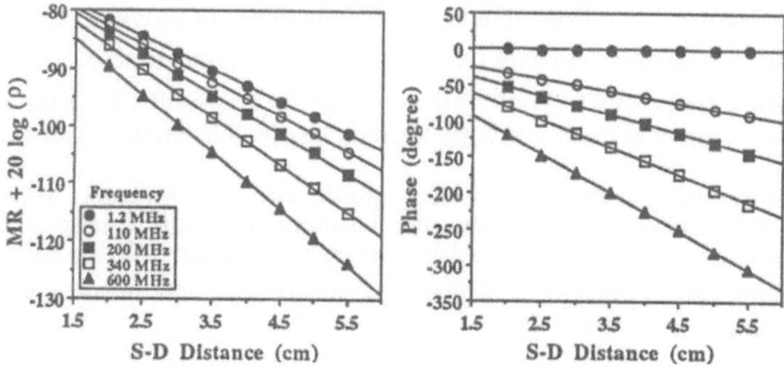

Figure 5. $[MR + 20 \ \log(\rho)]$ vs. ρ and ϕ vs. ρ at various frequencies at eight different S-D distances.

values with varying absorber or scatterer concentrations and the S-D separation. It is found that the variation in computed optical values using MR and ϕ showed less than 2 % difference with 10 times increase in frequency (results not shown).

Heterogeneous System

To study a heterogeneous system, the center of a black cubic absorber (1.5 cm x 1.5 cm x 1.5 cm) was moved along the center line between the source and the detector [Fig. 2 (b)] and TRS spectra were obtained with every 0.5 cm change of the absorber position.

Fig. 6 shows (a) MR and (b) ϕ when the absorber position was changed along the center line between the source and the detector. At all frequencies the MR decreased as the absorber was positioned closer to the center of the source-detector unit. As expected this is due to the absorption of a greater portion of the photon density. However, the change in MR at high frequencies (~ 13 dB) was greater than that at low values (~ 9 dB). This is because, at the middle of the source- detector unit, the photon density of the high modulation frequency is higher than that at the low modulation frequency. This result shows that multi-frequency MR values provide fundamental information on the depth of the absorber.

At all frequencies the phase initially decreased slightly when the absorber was relatively far away from the source-detector. This result shows that the longer pathlengths were reduced by the presence of the absorber. As the absorber was located at the center of the source-detector unit, the phase increase was greater, because photons which take shorter pathlengths were absorbed by the absorber.

CONCLUSIONS

A new approach for analyzing TRS spectra has been applied. The system transfer function of a single TRS spectrum was expressed in terms of the magnitude ratio (MR) and

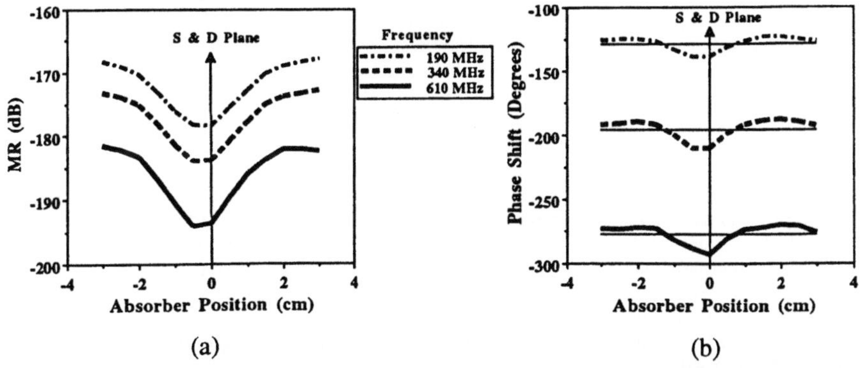

(a)

(b)

Figure 6. (a) MR and (b) ϕ when the absorber position was changed along the center line between the source and the detector, at various modulation frequencies.

phase shift (ϕ) at multiple frequencies. This method has the advantage of providing multiple parameters of a system such as steady state gain (K), system order (n), and time constant (τ) as well as MR and ϕ at multi-frequencies. A preliminary study on K, n, and τ for NIR-TRS spectra were performed by Kang et al.[8] and presently more investigations on the significance of these parameters are in progress.

Since the main chromophore in tissue is hemoglobin and the oxy/deoxy state of hemoglobin can be accurately obtained by the accurate measurement of optical properties, a method for obtaining absorption and scattering coefficients was developed utilizing MR and ϕ values. The absorption and the scattering coefficients calculated using MR and ϕ proved to be mutually independent and consistent with the change in optical properties.

For heterogeneous systems, MR and ϕ were correlated with the position of the absorber. MR at all frequencies decreased when the absorber became closer to the source-detector unit due to the reduction of the more dense portion of photon density by the absorber. When the absorber was located in the longer photon pathlength range, the phase decreased at all frequencies because the mean pathlength decreased. When the absorber was positioned in the short pathlength range, the phase increased due to the reduction of shorter photon paths. When tissue is highly vascular due to tumor growth or the local accumulation of blood due to a pathological condition, the heterogeneity acts as a local absorber. By scanning the area of interest and by studying information on changes of magnitude and phase shifts the local anomaly can be detected. Therefore, this method provides a new strategy for imaging of biological heterogeneities.

This frequency response information from a single TRS spectrum of the system of interest can also be utilized to optimize the design of a phase modulation device.

ACKNOWLEDGMENT

This research was partially supported by NIH grants 2 R01 CA 50766-1, 1 R01 NS27346-04, and 5 R01 HL 44125-4 and by the Pittsburgh Supercomputing Center grant number 1 P41 RR06009 from the NIH National Center for Research Resources. The authors would also like to thank Hammamatsu Optics, KK. for providing the TRS device and their technical support.

REFERENCE

1. Bruley, D.F. and Prados, J.W., "The frequency response analysis of a wetted wall adiabatic humidifier", AIChE Journal, 10(5):612 (1964).
2. Bruley, D.F., Pulse reduction code written for process identification (1974).
3. Clements, W.C., Jr. and Schnelle, K.B., "Pulse testing for dynamic analysis", I & EC Process Design and Development, 2(2):94 (1963).
4. Coughanowr, D.R., in: Process systems analysis and Control, McGraw-Hill, New York (1991).
5. Filon, L.N.G., Proc. of the Royal Soc. Edinburgh, 49:38 (1928).
6. Ishimaru, A. "Diffusion of a pulse in densely distributed scatters", J. Opt. Soc. Am., 68(8):1045 (1978).
7. Kang, K.A., Chance, B., Bruley, D.F, and Londono, J.M., "The application of near field interference of phase modulated spectroscopy for detection of absorbers: Theoretical prediction of phase shift with respect to the location of the absorber by using a probabilistic-numerical technique, the B-W-K method", Proc. SPIE, (Chance B. and Alfano, R.R. eds),1888:340, SPIE, Bellingham, WA (1993).
8. Kang, K.A., Bruley, D.F., Londono, J.M., and Chance, B., "Patho-Physiological System Identification via Frequency Response Analysis of Optical TRS Spectra" (Submitted to Annals of Biomedical Engineering for publication).
9. Koyama, K. and Fatlowitz, D., "Application of MCP-PMTs to time correlated single photon counting and related procedures", Hamamatsu technical information, No. ET-03/OCT (1987).
10. Lewis, C.I., Jr., Bruley, D. F., and Hunt, D.H., "Evaluation of temperature pulse characteristics and pulse testing for thermal dynamic analysis", I&EC Process Design and Development, 6(3):281 (1967).
11. Luyben W.L., Process Modeling Simulation, and Control for Chemical Engineers, 2nd edition, McGraw-Hill, New York (1990).
12. Zhu, J.X., Pine, D.J., and Weitz, D.A., "Internal reflection of diffusive light in random media", Phys. Rev. A., 44:3948 (1991).

AN AUTOMATED SYSTEM FOR THE MEASUREMENT OF THE RESPONSE OF CEREBRAL BLOOD VOLUME AND CEREBRAL BLOOD FLOW TO CHANGES IN ARTERIAL CARBON DIOXIDE TENSION USING NEAR INFRARED SPECTROSCOPY

C.E. Elwell, M. Cope, D. Kirkby, H. Owen-Reece[*], C.E. Cooper[*], E.O.R. Reynolds[*], D.T. Delpy

Departments of Medical Physics and Bioengineering, and Paediatrics[*]
University College London, U.K.

INTRODUCTION

Since it was first described by Jobsis in 1977, near infrared spectroscopy (NIRS) has become widely used as a non invasive technique for monitoring the blood and tissue oxygenation of intact organs and to date has been primarily applied to measurements of cerebral oxygenation and haemodynamics (Brazy et al., 1985, 1986, Ferrari et al., 1986, Hampson et al., 1990). Methods have been developed to make discrete absolute measurements of cerebral blood volume (CBV) (Wyatt et al., 1990) and cerebral blood flow (CBF) (Edwards et al., 1988) by inducing a small change in cerebral oxyhaemoglobin concentration and have largely been applied to neonates (Edwards et al., 1990, Skov et al., 1991). More recently these methods of quantifying cerebral haemodynamics have been applied to adults (Elwell et al., 1992).

NIRS can provide discrete absolute measurements of CBV and also a continuous monitor of changes in CBV. The latter measurement has already been used in neonates to determine the response of CBV to changes in arterial carbon dioxide tension ($PaCO_2$) in preterm and term infants (Pryds et al. 1990, Wyatt et al., 1991). In adults the response of the cerebral circulation to changes in $PaCO_2$ has been investigated using other techniques for example, nitrous oxide clearance (Kety and Schmidt, 1948), positron emission tomography (PET) (Greenberg et al., 1978), and [133] Xenon clearance (Herold et al., 1988).

The measurement of CBV and CBF with NIRS uses oxyhaemoglobin as the tracer which is controlled via the inspired oxygen content (FiO_2) and monitored as arterial oxygen saturation (SaO_2) using a pulse oximeter. Until now, FiO_2 has been controlled manually, using the rotameters of an anaesthetic trolley. The required manipulations are difficult to reproduce manually with any accuracy and also require a certain degree of skill. Multiple measurements are made under the same physiological conditions to estimate a mean and standard deviation and to also allow for some measurements which may have to be discarded, usually due to poor oximetry data. For all these reasons the procedure of making several CBF and CBV measurements can often be a lengthy one. The automated gas blender is designed to overcome many of these problems by providing reproducible waveforms of FiO_2 with a minimum recovery period between measurements.

Another physiological parameter which can be of clinical significance is the responsivity of CBV and CBF to carbon dioxide. $PaCO_2$ can be altered in a number of ways either via respiratory rate and tidal volume or by the inspired carbon dioxide content ($FiCO_2$). It is not always convenient to change the respiratory rate and tidal volume particularly in non ventilated subjects. The automated gas blender design also allows for the CO_2 responsivity of CBV and CBF to be measured by simply increasing $FiCO_2$ which in turn increases $PaCO_2$, and performing the necessary manipulations in FiO_2 while holding $FiCO_2$ constant. Again a sequence of CBV and CBF measurements can be programmed to run at different levels of $FiCO_2$.

The design is initially for use on anaesthetised animals, but it is anticipated that an important use will be on spontaneously breathing patients in whom other methods of varying $PaCO_2$ are most difficult. The development of such a system is a step towards a fully automated NIRS measurement system which will enable CBV and CBF to be determined from preset routines with minimal user input. One could even envisage a constantly varying FiO_2 (with appropriate safety features) about a preset value which could yield "continuous" unattended absolute CBF and CBV measurements every few minutes.

This paper will describe the new, automated, gas blending system and some example data are presented demonstrating the use and validation of the system on an anaesthetised neonatal piglet.

THEORY

Near Infrared Spectroscopy

The details of NIRS have been addressed fully elsewhere (Cope and Delpy 1988, Cope 1991). Briefly the technique depends upon the relative transparency of biological tissue to light in the near infrared region of the spectrum (i.e. between 700 -1000 nm) in order to measure the absorption due to the chromophores oxyhaemoglobin (HbO_2), deoxyhaemoglobin (Hb) and oxidised cytochrome aa_3 ($CytO_2$). A modified Beer-Lambert law which describes optical attenuation in a highly scattering medium can be used to quantify the changes in the concentration of these chromophores (Delpy et al., 1988). This can be expressed as:

$$Attenuation(OD) = \log\frac{I_0}{I} = \alpha cLB + G \tag{1}$$

where OD represents optical densities, I_0 the incident light intensity, I the detected light intensity, α the absorption coefficient of the chromophore ($mM^{-1}.cm^{-1}$), c the concentration of chromophore (mM), L the physical distance between the points where light enters and leaves the tissue (cm), B a "pathlength factor" which takes into account the scattering of light in the tissue and G a factor related to the geometry of the tissue. If measurements are only made of the *changes* in attenuation, then L, B and G can be assumed to remain constant and *changes* in chromophore concentration can be derived from the expression:

$$\delta c = \frac{\delta OD}{\alpha LB} \tag{2}$$

The absorption coefficients of HbO_2 and Hb have been deduced from lysed blood and that of $CytO_2$ in vivo in perfluorocarbon exchange transfused rats (Wray et al., 1988,

Cope et al., 1991). The pathlength factor, B, has been measured by time of flight studies and has a range of 5.93 ± S.D. 0.42 in adult head to 3.85 ± S.D. 0.57 in neonatal head (van der Zee et al., 1992).

In this paper the term Hb_{diff} will be used to represent the difference between the $[HbO_2]$ and $[Hb]$ signals (i.e. $[HbO_2]$ - $[Hb]$) and the term Hb_{sum} to represent the sum of the two signals (i.e. $[HbO_2 + [Hb]]$).

CBV Measurement

The principle of CBV measurements using NIRS has already been discussed in detail with regard to neonatal measurements (Wyatt et al., 1990). The technique involves a slow, small change in arterial oxygen saturation (SaO_2) induced by small changes in FiO_2. If CBF, CBV and oxygen consumption remain constant during the measurement, then the consequent change in cerebral $[HbO_2]$ measured by NIRS, is equivalent to the product of the total cerebral haemoglobin concentration ($[tcHb]$) and the fractional change in arterial saturation (ΔSaO_2):

$$\Delta[HbO_2] \quad - \quad [tcHb] \cdot \Delta SaO_2 \tag{3}$$

In such a manoeuvre of arterial saturation the relationship $\Delta HbO_2 = -\Delta Hb$ is expected. For signal to noise reasons we then rewrite equation (3) as:

$$[tcHb] \ (mM) \quad - \quad \frac{\Delta[Hb_{diff}]}{2 \cdot \Delta SaO_2} \tag{4}$$

It should be noted that the total cerebral haemoglobin concentration derived from this relationship has the units mmol haemoglobin per litre of brain tissue. To convert $[tcHb]$ into cerebral blood volume in $ml.100g^{-1}$ (ml of blood per 100g of brain tissue) the following terms are needed; the molecular weight of haemoglobin (MW_{Hb}), the cerebral tissue density in $g.ml^{-1}$, the total concentration of haemoglobin in the whole blood ($[tHb]$) in $g.100ml^{-1}$, (measured from a venous sample), and the cerebral large to small vessel haematocrit ratio (CLVHR). It must be assumed therefore that all of these parameters remain constant during the course of a measurement or if multiple measurements are to be compared, that they remain constant between measurements. Hence incorporating the above parameters:

$$CBV \ (ml.100g^{-1}) \quad - \quad \frac{(MW_{Hb} \cdot 10^{-3})}{(D_t \cdot 10) \cdot ([tHb] \cdot 10^{-2}) \cdot CLVHR} \cdot \frac{\Delta Hb_{diff}}{2 \cdot \Delta SaO_2} \tag{5}$$

To obtain an absolute quantification of CBV, SaO_2 is altered gradually over a few minutes to produce a slow change in cerebral Hb_{diff} with a period of equilibrium maintained at each level of SaO_2. The resulting Hb_{diff} is then plotted against SaO_2 and the gradient of the regression line is used to calculate CBV.

CBF Measurement

The methodological details of CBF measurements in adults using NIRS have been described previously (Elwell et al., 1992). The technique uses a modification of the Fick principle and oxyhaemoglobin as the intravascular, non diffusible tracer. A sudden increase is induced in SaO_2 and the resulting initial increase in cerebral $[HbO_2]$ represents the accumulation of the tracer in the brain. If measurements are made within the cerebral vascular transit time (T_t), and CBF, CBV and oxygen consumption remain constant during this time, then the accumulation of [HbO_2] is proportional to the arterial inflow of tracer. Flow is then the ratio of the quantity of tracer accumulated to the quantity of tracer introduced over some time t (where $t < T_t$). ΔSaO_2 can be measured using a pulse oximeter on the ear and Δ[HbO_2] is measured by NIRS. The actual quantity of tracer introduced is given by the product of the integral of fractional ΔSaO_2 with respect to time and the total concentration of haemoglobin in the whole blood ([tHb]). CBF is then calculated from the following equation:

$$CBF \ (ml.100g^{-1}.min^{-1}) = \frac{(MW_{Hb} \ 10^{-3})}{(D_t \cdot 10) \cdot ([tHb] \cdot 10^{-2})} \cdot \frac{\Delta Hb_{diff}}{2 \cdot \int_0^t \Delta SaO_2 dt} \qquad (6)$$

It can be seen that some of the same terms used in the CBV calculation are needed here to express CBF in ml.100g^{-1}.min^{-1}. Both measurements require CBV and CBF to remain constant during the calculation period.

To obtain data from which to calculate CBF, the subject's FiO_2 is first slowly reduced to provide a stable baseline at a lowered SaO_2 and then rapidly increased to produce the required sudden rise in SaO_2 (the input of the tracer). It is of prime importance that the rapid change in FiO_2 should take place over one breath to prevent slurring of the input function.

GAS BLENDER REQUIREMENTS

A manual gas blending system based around a modified anaesthetic trolley for measuring CBV and CBF by NIRS in healthy non ventilated adult volunteers has previously been described (Elwell et al., 1992). The design requirements discussed in the introduction led to the following specifications for the automated blender:

Number of gases   :   4   (O_2, CO_2, N_2O, N_2)

Maximum flow rates   : 10 l.min^{-1} (O_2, N_2O, N_2)
                                  :   2 l.min^{-1} (CO_2)

Accuracy   :   1% (O_2, N_2O, N_2)
                                  :   0.2% (CO_2)

Time response   :   < 2 seconds
(5 - 95 %)

The four gases allow nitrous oxide anaesthesia to be used. The maximum flow rates above are adequate for small animal work and for adult human subjects breathing normally. However for adult human subjects breathing carbon dioxide, flow rates of 20 l.min^{-1} are required for which either several 10 l.min^{-1} or higher capacity flow controllers are required. The accuracies reflect the effect those gases have upon cerebral haemodynamics, it being particularly important to hold $FiCO_2$ at a constant value during FiO_2 swings. The fast time response is required as a "step" input response in SaO_2 is necessary for the measurement of CBF.

146

SYSTEM DESCRIPTION

A schematic diagram of the gas blending system is shown in Figure 1. The basic components of the system are voltage controlled mass flow controllers (Type 1259C, MKS, USA) which accurately measure and control the mass flow rate of gases, providing a fast time response and reliable long term stability. Each gas is taken to the blender from a regulated supply (either wall socket or compressed gas cylinders) and passed through a filter. The inlet gas pressure is displayed and a one way valve prevents any return of gas to the supplying circuit. A block diagram of the inlet assembly for a single gas channel is shown in Figure 1(a).

PC based software controls digital to analogue converters which set the voltages to drive the mass flow controllers. The mass flow controllers accept a set point voltage for the required flow rate and compare the required flow with the actual flow measured with an

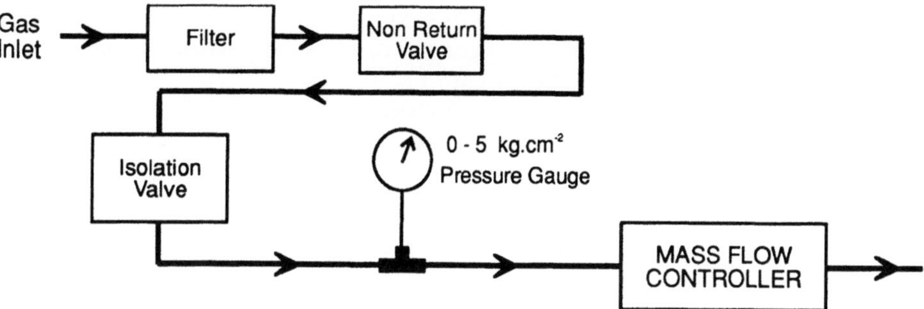

Figure 1 (a) . A schematic of the inlet assembly of the automated gas blender

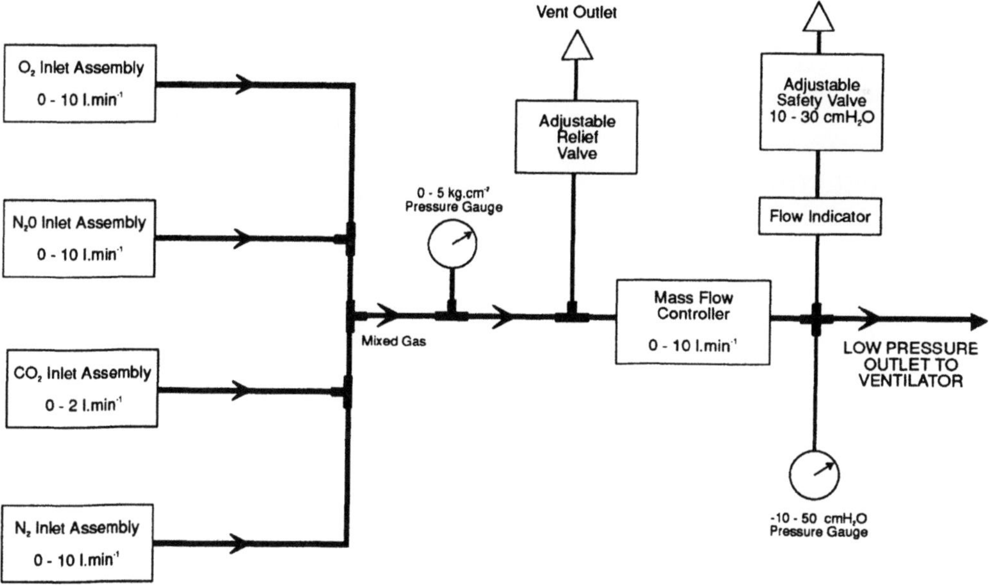

(b) . The gas blender configuration for use on an anaesthetised animal.

internal sensor. A feedback circuit continually modifies the position of a control valve to maintain the flow at the set value. This type of flow controller has the advantage of being independent of the pressure drop across the controller. The mass flow controllers used in this system have a documented time response of < 2 seconds to within 2% of the set flow rate, with a measurement resolution of 0.1% full scale and accuracy of ±0.8% full scale. The flow rate of gas through each controller is displayed on the front panel of the unit along with the total gas flow rate supplied.

In the software the user defines the total flow rate required (in $l.min^{-1}$), followed by the percentage of the three gases (typically O_2, N_2O and CO_2) in the mixture. The balance is always made up with N_2. The mass flow rate of each gas is then set to provide the required mixture and the flow rate and mixture remains unchanged until a new command is sent. The gas mixture can be set to a new value from a single command or a waveform can be defined using a series of commands entered into an ASCII data file. The application of the system for use in humans will require safety checks in the software and hardware controlling the gas mixture.

Any gas can be fed into any controller if the appropriate gas correction factor, which relates mass to volumetric flow, is known. This correction factor is a function of specific heat, density and molecular structure of the gas and can be incorporated into the software to provide accurate flow rates for the chosen gas.

The outlet of the gas blender has a low pressure -10 - 50 cmH_2O gauge. There is also an adjustable safety valve in parallel with the gas outlet which can be set to 'blow off' at any pressure between 10 and 30 cmH_2O.

Figure 1 (b) shows a block diagram of the system configured for use on an anaesthetised animal. The four gases used in the system are O_2, N_2O, CO_2 and N_2 with each gas controlled by a separate mass flow controller. The fifth mass flow controller with a 0 - 10 $l.min^{-1}$ range is therefore used to set the total flow rate of the mixed gas. The vent outlet is used when the sum of the mixed gases exceeds that required by the subject. This situation can be advantageous because the mass flow controllers accuracy is quoted as a percentage of full scale flow rate so that maximum accuracy is reached when larger volumes of gases are mixed, and also because the time response of the system is improved at higher total gas flow rates (see discussion).

Typical gas inlet pressures are 2.5 $kg.cm^{-2}$, with a typical vent pressure of 1.2 $kg.cm^{-2}$ and an outlet pressure of 20 cmH_2O.

SYSTEM PERFORMANCE

The response time of the system was determined by using an end tidal carbon dioxide ($EtCO_2$) monitor (Normocap, Datex, USA) to measure the arrival of CO_2 at the subject's mouthpiece approximately 2 metres from the low pressure outlet of the gas blender. Although the $EtCO_2$ monitor has a good time response (< 600ms) the maximum calibrated range is only 0 - 10%. The CO_2 controller could therefore be tested over the full 0 - 2 $l.min^{-1}$ range and the other controllers tested by measuring dilution of a fixed CO_2 mixture with each gas. Figure 2 shows the 5 - 95% rise time for each of the controllers at a different total flow rate. For flow rates below 20 $l.min^{-1}$ the rise time is longer than that quoted for the mass flow controller alone due to the dead space between the controller outlet and the subject's mouthpiece. Adjustments to improve this rise time, particularly at low flow rates, are suggested in the discussion. The actual arrival time of the input function (i.e. the transit time of the function) is not itself of importance, but more important is the degree of slurring of the function due to response of the mass controllers and additional connecting tubing contributing to dead space of the system.

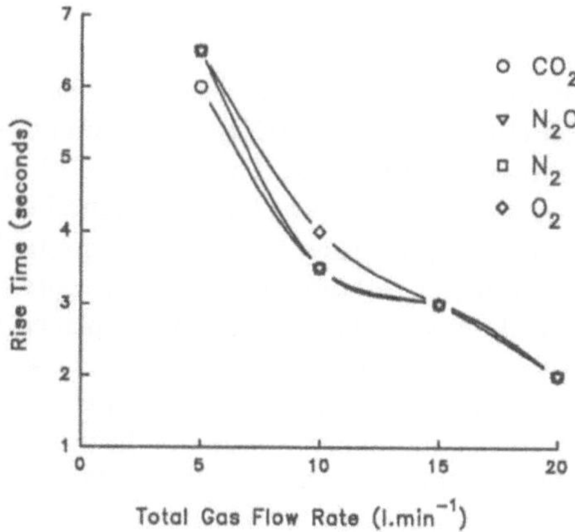

Figure 2 : The 5 - 95% rise time for each of the mass flow controllers measured at different total gas flow rates.

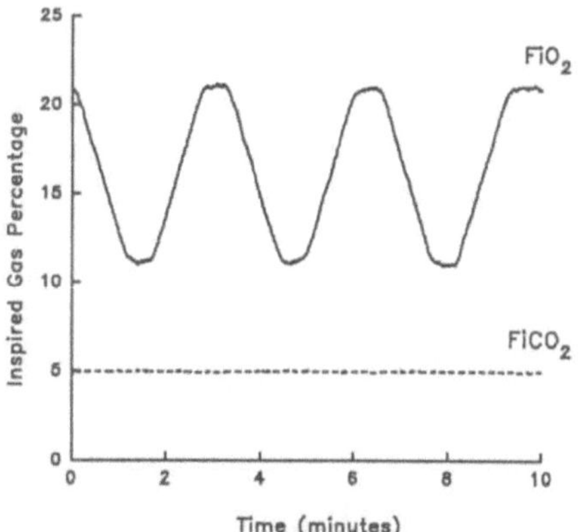

Figure 3 : FiO_2 and $FiCO_2$ measured during a programmed sequence for the automated measurement of CBV.

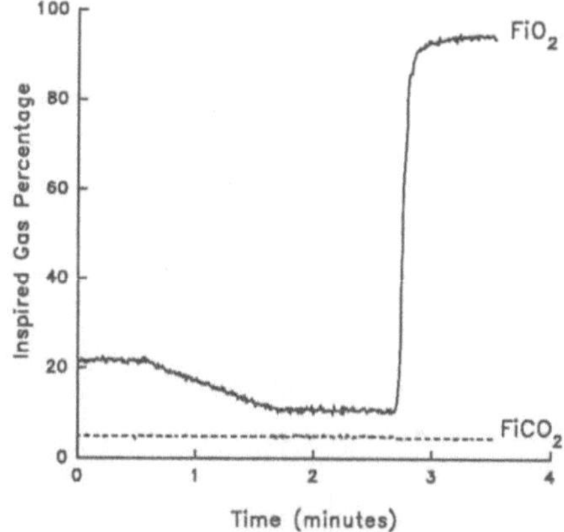

Figure 4 : FiO_2 and $FiCO_2$ measured during a programmed sequence for the automated measurement of CBF.

Figure 3 shows the measured FiO_2 and $FiCO_2$ during programmed automated manipulations necessary for measurement of CBV. FiO_2 was set at 21% initially and then ramped down over 60 seconds to 11%, held at 11% for 30 seconds and then ramped up to 21% over 60 seconds. This pattern was repeated three times in total producing six swings in FiO_2 from which to calculate CBV. During these swings $FiCO_2$ was maintained at 5%.

Figure 4 shows the measured FiO_2 and $FiCO_2$ during the manipulations necessary for a CBF measurement. FiO_2 was reduced over 60 seconds to 11% and then held constant for a further minute to produce a stable baseline. FiO_2 was then instantaneously switched to 95% while $FiCO_2$ was maintained at 5%.

DATA COLLECTION

In a preliminary study the system was used to measure CBV and CBF in an anaesthetised neonatal piglet at two levels of $FiCO_2$. A schematic of the experimental set up used in these studies is shown in Figure 5. The NIRO 500 spectrometer (Hamamatsu Photonics KK, Japan) uses pulsed laser diodes at four wavelengths (779,828,845,and 908 nm) as its light source and a photomultiplier tube for light detection. The near infrared light is carried to and from the spectrometer through flexible fibre optic bundles, truncated at the subject end as small cylindrical optodes. The optodes were placed on either side of the head at a separation of 3.9 cm. Once the optodes were securely fastened the whole area was then wrapped in a light tight cloth to reduce the background light. Data were collected every 0.5 second during CBF measurements and every 5 seconds during CBV measurements.

Arterial haemoglobin saturation and heart rate (HR) were measured using a pulse oximeter (Novametrix 500, USA) via a probe positioned on a skin flap on the right leg. Although it is preferable to measure SaO_2 as close to the brain as possible, it proved very difficult to record reliable and accurate oximetry data on the piglet's ear. The beat to beat facility was used during CBF measurements. The analogue outputs of the oximeter were linked directly to the spectrometer for display and storage alongside the NIRS data.

Anaesthesia was induced with 5% isofluorane, mechanical ventilation was established and the isofluorane level then reduced to 1.5%. The computer controlled gas blender in the configuration shown in Figure 1(b) was used to supply a controlled accurate mixture of O_2, N_2O and CO_2 to the pressure preset, time cycled ventilator (Vickers Model 77, U.K.).

Initially, $FiCO_2$ was set to 0% and at least six CBV measurements were made by slowly changing FiO_2 over several minutes. Three CBF measurements were then made at the same $FiCO_2$. The rapid change in FiO_2 to 100% required for a measurement of CBF was aided by transiently increasing the flow rate at the same time as switching the gas mixture. FiO_2 was then maintained at 21% to produce a stable recording for 10 minutes. $FiCO_2$ was then increased to 5% and no other manoeuvres were performed until the NIRS data reached a new stable baseline. At this point the CBV and CBF measurements were repeated. After the final CBF measurement $FiCO_2$ was returned to 0% and recording continued for a further ten minutes. A venous blood sample was taken from the animal and [tHb] was measured by a Coulter Counter and arterial blood samples were used to measure $PaCO_2$.

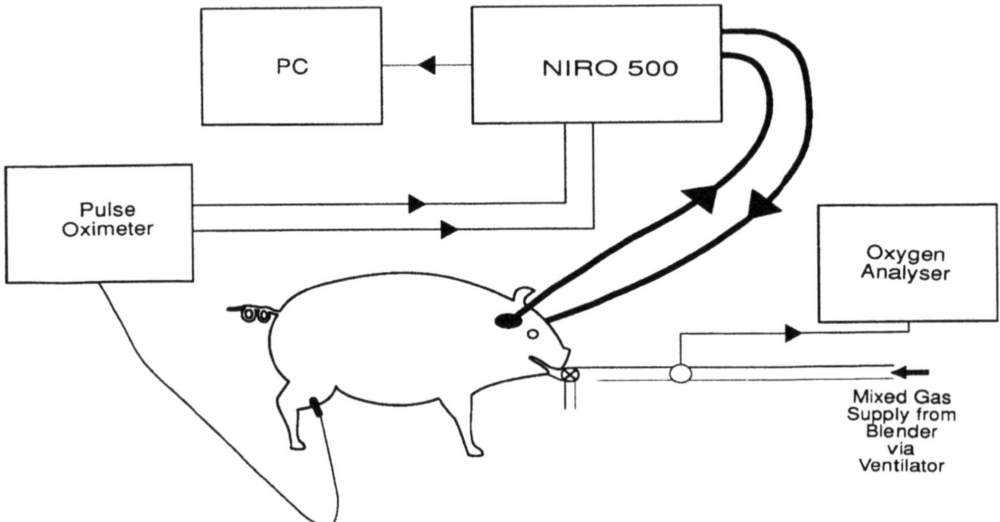

Figure 5 : A schematic of the experimental setup used in a study on a neonatal piglet.

RESULTS

Figure 6 shows the NIRS and SaO_2 data collected during (a) a series of six CBV measurements and (b) a single CBF measurement on the piglet. The change in $FiCO_2$ from 0 - 5% produced a change in $PaCO_2$ from 6.16 - 10.49 kPa. DPF has not yet been measured in neonatal piglet so in all calculations the DPF measured on human neonates was used. A value of 1.05 was taken for the brain tissue density and 0.69 for CLVHR (Lammertsma et al., 1984). The CBV and CBF calculated at the two levels of $PaCO_2$ are shown in Table I with the coefficient of variation (CV) for each group of measurements.

The CBV response to $PaCO_2$ (CBVR) was 2.3 $ml.100g^{-1}.kPa^{-1}$ and the CBF response to $PaCO_2$ (CBFR) was 16.0 $ml.100g^{-1}.min^{-1}.kPa^{-1}$.

To determine the dependence of Hb_{sum} on changes in SaO_2, both Hb_{sum} and Hb_{diff} were regressed against SaO_2 during the CBV calculation period at both levels of $PaCO_2$. Figure 7 shows these regressions during the CBV measurements at the lower $PaCO_2$. The ratio of the gradients of the regression lines is shown in Table I as (ΔHb_{sum} / ΔHb_{diff}).

Table I. CBF and CBV data collected at two levels of $PaCO_2$ with the coefficient of variation (CV) for each group and the (ΔHb_{sum} / ΔHb_{diff}) ratio for the CBV measurements.

$PaCO_2$ (kPa)	CBF ($ml.100g^{-1}.min^{-1}$)	CV	CBV ($ml.100g^{-1}$)	CV	$\Delta Hb_{sum}/\Delta Hb_{diff}$
6.16	65.2 ± 9.7 n = 3	15%	6.0 ± 0.5 n = 19	8%	3.4%
10.49	134.5 ± 13.3 n = 4	10%	15.8 ± 1.5 n = 6	9%	3.7%

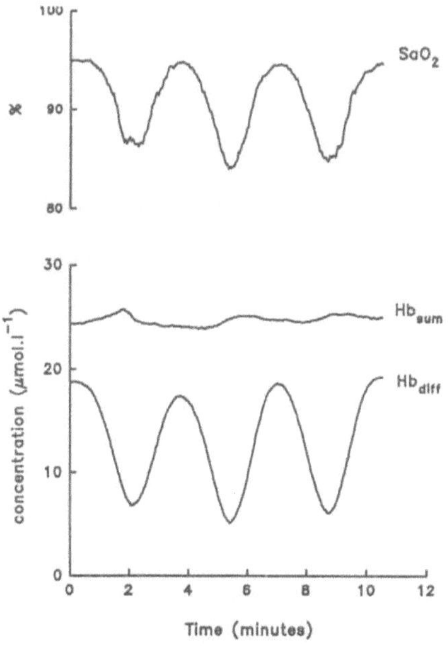

Figure 6(a) : NIRS and SaO_2 data collected on the neonatal piglet during a series of six CBV measurements made at $PaCO_2$ of 6.16 kPa.

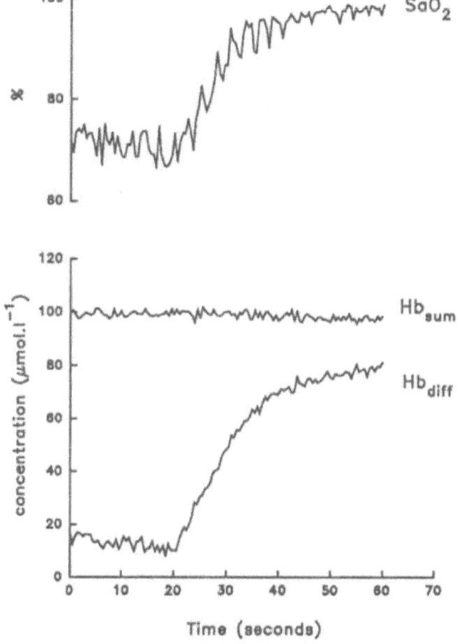

Figure 6(b) : NIRS and SaO_2 data collected on the neonatal piglet during a single CBF measurement made at $PaCO_2$ of 10.49 kPa.

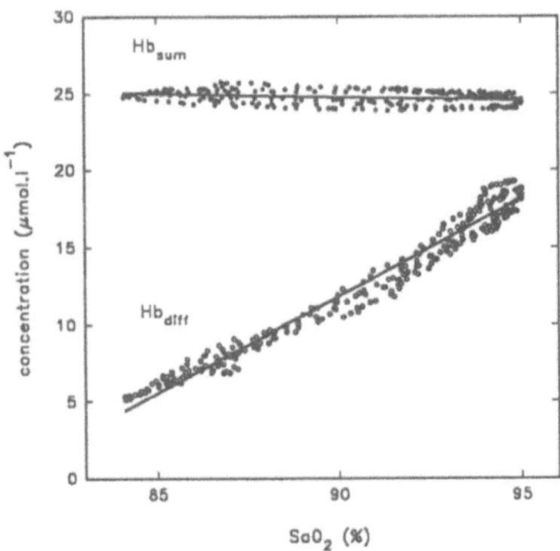

Figure 7 : The regression of Hb$_{sum}$ and Hb$_{diff}$ against SaO$_2$ during the CBV calculation period at PaCO$_2$ of 6.16 kPa.

DISCUSSION

The described system allowed CBV and CBF to be measured repeatedly at different levels of FiCO$_2$ in an anaesthetised piglet. Accurate gas blending was achieved with control over the total flow rate of the gas supplied.

The response time of the system was good at a high total flow rate of 20 l.min^{-1} producing step input functions of FiO$_2$ necessary for CBF measurements. However at lower flow rates the rise times were considerably longer. Although the mass flow controllers are capable of switching mixtures within 2 seconds, the total dead space of the system becomes more significant at lower flow rates producing slurring of step functions. During the time taken for the gas to leave the controller and reach the subject's mouthpiece (i.e. the transit time) there is diffusion and a turbulent mixing of the gas across a concentration boundary established at the point where the gas mixture changes. At lower flow rates the transit time is increased and hence the diffusion of gas is more significant leading to increased slurring of the step function, measured as an increased rise time. These problems of diffusion are also enhanced when the total dead space of the system is increased by connecting a ventilator to the outlet of the blender for anaesthetised subjects. To overcome these problems it is necessary to either maintain high flow rate throughout the circuit or to decrease the dead space. Practically it is more feasible to drive the gases at a high flow rate through the circuit and to place the fifth flow controller external to the main unit just downstream of the subject's mouthpiece. The advantages of a high flow rate system with good response time are then maintained even if a ventilator is used and the dead space between the final flow controller and the subject is kept to a minimum.

The preliminary study on an anaesthetised neonatal piglet demonstrated the use of the system in accurately producing the necessary manipulations in FiO$_2$ and FiCO$_2$. Literature on CBV measurements in neonatal piglets is scarce but the absolute CBF data matches that obtained from measurements made with tracer microspheres where regional values between 47 - 85 ml.100g^{-1}.min^{-1} are quoted (Waglerle et al., 1986). The response of CBV to changes in PaCO$_2$ calculated from absolute CBV measurements are within the range of those quoted in a larger study of CO$_2$ reactivity in neonatal piglets using NIRS (2.3 - 2.9 ml.100g^{-1}.kPa^{-1}) (Takei et al., 1993). The measurements of CBF response to changes in PaCO$_2$ also agree with

152

previously published regional data between 10.5 - 30.8 ml.100g^{-1}.min^{-1}.kPa^{-1} (Helfaer et al., 1991).

The variability in the CBV and CBF measurements, although acceptable is most probably due to unreliable and noisy oximetry data and small changes in CBF during the measurements. The noise on the oximetry data is particularly noticeable on the CBF data which was collected at a high sampling rate of 0.5 second. The NIRS data collected from this animal however was relatively noise free due to minimal movement artefact and lowered optical attenuation compared to a human neonatal or adult head. These data highlight the problems caused by poor quality oximetry data which must be approached if the accuracy and reproducibility of the measurements is to be further improved.

The variation in Hb_{sum} during the CBV measurements shown in Figure 6(a) indicate small changes in CBF over the measurement period, however regression analysis shows that the contamination of the CBV measurements due to these changes is unlikely to be significant. In this study SaO_2 was lowered to 85% during a CBV measurement, compared to 90% in the case of adult volunteer studies. This exaggerated decrease in SaO_2 may enhance the cerebrovascular changes to arterial oxygen tension.

CONCLUSION

An automated gas blending system has been described which provides the ability to accurately control and vary the percentage of at least four different gases in a mixture as well as control the total gas flow rate of the mixture. The system can be used in studies on anaesthetised and non anaesthetised subjects to provide the manipulations in FiO_2 and $FiCO_2$ necessary for measurement of the response of CBF and CBV to changes in $PaCO_2$. Data collected from a preliminary study suggests that a controlled animal model can be used to provide information about the reproducibility of CBV and CBF measurements and the sensitivity of the technique for the measurement of their response to $PaCO_2$.

ACKNOWLEDGEMENTS

The author would like to thank Dr. T. Whitlock and Mr. A. Bishop for their contribution to this study. This work has been supported by grants from Hamamatsu Photonics KK., the M.R.C., the Wellcome Trust, the S.E.R.C., and the Wolfson Foundation. C.E.C. is grateful for a training fellowship from the M.R.C..

REFERENCES

Brazy, J.E., Lewis, D.V., Mitnick, M.H., Jöbsis, F.F., 1985. Noninvasive monitoring of cerebral oxygenation in preterm infants: preliminary observations. *Pediatrics,* 75:217-225

Brazy, J.E., Lewis, D.V., 1986, Changes in cerebral blood volume and cytochrome aa3 during hypertensive peaks in preterm infants. *Pediatrics,* 108:983-987.

Cope, M., Delpy, D.T., 1988. A system for long term measurement of cerebral blood and tissue oxygenation in newborn infants by near infrared transillumination. *Med. Biol. Eng. & Comp.,* 26, 3:289-294.

Cope, M. 1991. The development of a near infrared spectroscopy system and its application for non invasive monitoring of cerebral blood and tissue oxygenation in the newborn infant. *PhD Thesis, University of London.*

Delpy, D.T., Cope, M., van der Zee, P., Arridge, S.R., Wray, S., Wyatt, J.S., 1988. Estimation of optical pathlength through tissue from direct time of flight measurement. *Phys. Med. & Biol.*, 33, 12:1433-1442.

Edwards, A.D., Wyatt, J.S., Richardson, C.E., Delpy, D.T., Cope, M., Reynolds, E.O.R., 1988. Cotside measurement of cerebral blood flow in ill newborn infants by near infrared spectroscopy. *Lancet,* ii:770-771.

Edwards, A.D., Wyatt, J.S., Richardson, C.E., Potter, A., Cope, M., Delpy, D.T., Reynolds, E.O.R., 1990. Effects of indomethacin on cerebral haemodynamics and oxygen delivery investigated by near infrared spectroscopy in very preterm infants. *Lancet* i:1491-1495.

Elwell, C.E., Cope, M., Edwards, A.D., Wyatt, J.S., Delpy, D.T., Reynolds, E.O.R., 1992. Measurement of cerebral blood flow in adult humans using near infrared spectroscopy - methodology and possible errors. *Adv. Exp. Med. & Biol.* 317:235-245.

Ferrari, M., Zanette, E., Giannini, I., Sideri, G., Fieschi, C., Carpi, A., 1986. Effects of carotid compression test on regional cerebral blood volume, haemoglobin oxygen saturation and cytochrome-c-oxidase redox level in cerebrovascular patients. *Adv. Exp. Med. & Biol.*, 200:213-222.

Greenburg, J.H., Alavi, A., Reivich, M., Kuhl, D., Uzzell, B., 1978. Local cerebral blood volume response to carbon dioxide in man. *Circ. Res.*, 43, 2, 324-331

Hampson, N.B., Camporesi, E.M., Stolp, B.W., Moon, R.E., Shook, J.E., Greibel, J.A., Piantadosi, C.A., 1990. Cerebral oxygen availability by NIR spectroscopy during transient hypoxia in humans. *J. Appl. Physiol.* 69:907-913.

Helfaer, M.A., Kirsch, J.R., Haun, S.E., Koehler, R.C., Traystman, R.J., 1991. Age-related cerebrovascular reactivity to CO_2 after cerebral ischaemia in swine. *Am.J. Physiol.* 260:H1482-1488.

Herold, S., Brown, M.N., Frackowiak,R.S.J., Mansfield, A.O., Thomas, D.J., Marshall, J., 1988. Assessment of cerebral haemodynamic reserve: correlation between PET parameters and CO_2 reactivity measured by the intravenous [133] Xenon injection technique. *J Neurol Neurosurg Psychiatry* 51:1045-1050

Jöbsis, F.F. 1977. Noninvasive, infrared monitoring of cerebral and myocardial oxygen sufficiency and circulatory parameters, *Science,* 198:1264-1267.

Kety, S.S., Schmidt, C.F., 1948. The nitrous oxide method for the quantitative determination of cerebral blood flow in man: theory, procedure, and normal values. *J. Clin. Invest.* 27:476-483.

Lammertsma, A.A., Brooks, D.J., Beaney, R.P., Turton, D.R., Kensett, M.J., Heather, J.D., Marshall, J., Jones, T., 1984. In vivo measurement of regional cerebral haematocrit using positron emission tomography. *J. Cereb. Blood Flow Metab.*, 4;317-322.

Pryds, O., Greisen, G., Skov, L., Fris-Hansen, B., 1990. Carbon dioxide related changes in cerebral blood volume and cerebral blood flow in mechanically ventilated preterm neonates. Comparison of near infrared spectrometry and [133]Xenon clearance. *Pediatric Res.* 27:445-449.

Skov, L., Pryds, O., Griesen, G. 1991. Estimating cerebral blood flow in newborn infants: comparison of near infrared spectroscopy and [133]Xenon clearance. *Pediatric Res.* 30:570-573.

Takei, Y., Edwards, A.D., Lorek, A., Peebles, D.M., Belai, A., Delpy, D.T., Reynolds, E.O.R., 1993. Effects of N-ω-nitro-L-arginine methyl ester on the cerebral circulation of newborn piglets quantified in vivo by near infrared spectroscopy. *Pediatric Res.* 34:354-359.

van der Zee, P., Cope, M., Arridge, S.R., Essenpreis, M., Potter, L.A., Edwards, A.D.,

Wyatt, J.S., McCormick, D.C., Roth, S.C., Reynolds, E.O.R., Delpy, D.T., 1992. Experimentally measured optical pathlengths for the adult head, calf and forearm and the head of the newborn infant as a function of interoptode spacing. *Adv. Exp. Med. & Biol.* 316:143-153

Wagerle, C.L., Kumar, S. P., Delivora-Papadopoulos, M., 1986. Effect of sympathetic nerve stimulation on cerebral blood flow in newborn piglets. *Pediatric Res.* 20, 131-135

Wray, S., Cope, M., Delpy, D.T., Wyatt, J.S., Reynolds, E.O.R., 1988. Characterisation of the near infrared absorption spectra of cytochrome aa_3 and haemoglobin for the non invasive monitoring of cerebral oxygenation. *Biochim. Biophys. Acta,* 933:184-192.

Wyatt, J.S., Cope, M., Delpy, D.T., Richardson, C.E., Edwards, A.D., Wray, S.C., Reynolds, E.O.R., 1990. Quantitation of cerebral blood volume in newborn infants by near infrared spectroscopy. *J. Appl. Physiol.* 68, 3:1086-1091.

Wyatt, J.S., Edwards, A.D., Cope, M., Delpy, D.T., McCormick, D.C., Potter, A., Reynolds, E.O.R., 1991. Response of cerebral blood volume to changes in arterial carbon dioxide tension in preterm and term newborn infants. *Pediatric Res.* 29:553-557.

NEAR INFRARED SPECTROSCOPY: *IN SITU* STUDIES OF SKELETAL AND CARDIAC MUSCLE

CA Piantadosi, MD and FG Duhaylongsod, MD*

Departments of Medicine and Surgery*
P.O. Box 3315
Duke University Medical Center
Durham, NC 27710

INTRODUCTION

Assessment of muscle oxygenation during work is an area of considerable research interest owing to questions about the variables regulating the cellular uptake of oxygen (O_2). Cardiac and red skeletal muscles can vary the level of O_2 uptake by 20-fold at steady state in response to sustained work (1). Maximal O_2 uptake has been proposed to be regulated by a number of variables including the flow of O_2, and turnover of NADH, ADP and inorganic phosphate (P_i). A specific rate-limiting step, however, has never been identified experimentally. High rates of oxygen utilization in myoglobin containing muscle are achieved at relatively low sarcoplasmic oxygen tensions. Estimates of muscle PO_2 based upon measurements of myoglobin saturation at near maximal steady state work indicate values in the 2-3 torr range, i.e. myoglobin is about half saturated with oxygen (2).

The PO_2 of working muscle and *in vitro* measurements of high O_2 affinity of cytochrome oxidase suggest that muscle mitochondria can extract O_2 to very low values of PO_2. PO_2 estimates for limiting concentrations of oxygen are given in the range of 0.1 to 0.2 torr (3). Changes in $\dot{V}O_2$, mitochondrial oxidation-reduction state and lactate accumulation are measured when myoglobin is more than 50% deoxygenated. These data, and calculations by others (4), suggest the possibility that maximal O_2 uptake may be limited in some muscles by O_2 availability.

We have used multiwavelength NIR spectroscopy as a tool to investigate the effects of work on the responses in mitochondrial redox state in the dog gracilis near or at $\dot{V}O_2$ max. The NIR method is continuous, nondestructive and sensitive to changes in the relative amounts of $HbO_2 + MbO_2$, $Hb + Mb$ and the oxidized visible copper (Cu_A) of cytochrome c oxidase (5). The technique can be used to study muscle oxidative metabolism in the reflectance mode provided precautions are taken to minimize movement artifacts during muscle contraction. The hypothesis states that a decrease in the oxidation state of the copper moiety helps maintain muscle energy provision at the low PO_2 values encountered at high, sustained work loads.

METHODS

NIR Spectroscopy: O_2 dependent absorption spectra in the NIR region of the spectrum (700-1000 nm) have been defined for Hb, Mb, and the oxidized Cu_A of cytochrome c oxidase (5). On the basis of these spectra, multiwavelength algorithms have been developed and validated to deconvolute NIR signals and provide qualitative estimates of changes in the relative amounts of $HbO_2 + MbO_2$, $Hb + Mb$ and oxidized Cu_A in the tissue. Four wavelength algorithms having a HbO_2 to Cu_A cross talk of less than 5% were used for this study (6).

Muscle Preparations: Two types of dog muscle preparations were studied. The PO_2 dependent behavior of Cu_A in isolated, blood perfused dog heart was studied first. Dog hearts were removed and perfused with blood at a hemoglobin concentration of 10 to 13 g/dl, a temperature of 37°C and a heart rate of 100 bpm. Perfusate PO_2 was varied using premixed gases and values measured using an IL blood gas analyzer (model 1304). Changes in NIR reflectance signals were measured in the cardiac muscle by epicardial reflectance and expressed as fraction of the total labile signal (TLS) achieved with anoxia.

The responses of dog gracilis muscle *in vivo* were studied as reported by Duhaylongsod *et al.* (7). The vascular supply and innervation to the gracilis muscle were isolated in mongrel dogs anesthetized with α chloralose and ventilated mechanically. The skin overlying the muscle was left intact. A catheter was placed in the major vein from the gracilis muscle to allow continuous collection of venous outflow for measurement of blood flow rate. Collected blood was anticoagulated and autotransfused into the dog via the jugular vein. The gracilis nerve was isolated, transected and secured to a platinum electrode. Nerve stimuli were delivered by a Grass stimulator (Model S-88) at 4 volts in trains of 10 stimuli, 1 ms apart and 0.2 ms duration. Five minute stimulations at rates of 2,3,4,5,7,8,10 and 12 per second were used, 30 min apart, three per muscle. Oxygen content of the blood was measured by Co-oximetry (IL model 402) using a canine algorithm.

NIR signals were acquired using separate send and receive fiber bundles in the reflectance mode. The optical fibers (optrodes) were placed in a stereotaxic device with an interoptrode distance of ~ 2.5 cm. Test stimulations of 30 sec were done at 5 Hz to detect motion artifacts. After the final muscle stimulation, the animals were killed with an overdose of KCl and the total labile signal (TLS) determined for the muscle. Changes in the amount of oxidized Cu_A were expressed relative to this TLS.

RESULTS AND DISCUSSION

The results of the dog heart experiments indicated dependence of the oxidized Cu_A signal on the oxygen concentration in the perfusate. At constant $\dot{V}O_2$, this relationship demonstrated a K_m for O_2 of approximately 50 μm on the arterial side.

The Cu$_A$ data from six hearts are graphed as a function of O$_2$ concentration in Figure 1. In Figure 1, a curve for O$_2$ dependent changes in the redox state of cytochrome c in neuroblastoma cells derived from data of Wilson *et al.* (3) is also shown for comparison. Although the Cu$_A$ curve from the heart is shifted to the right relative to *in vitro* curve, this is expected and readily explained by differences in geometry, diffusion gradients and metabolic rate. The NIR data support the concept that the redox state of Cu$_A$ of cytochrome c oxidase varies as a function of physiological O$_2$ supply in beating heart.

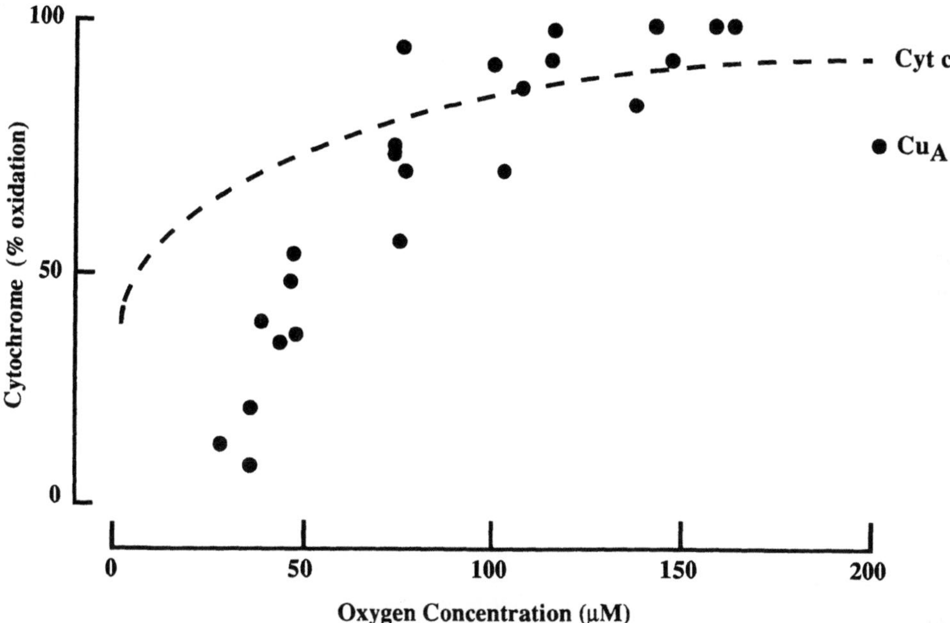

Figure 1. Changes in oxidized Cu$_A$ in dog heart as a function of O$_2$ concentration. Dashed line derived from data of reference 3.

The effects of muscle contraction on the redox state of Cu$_A$ in dog gracilis measured by NIR spectroscopy are shown in Figure 2. The $\dot{V}$ O$_2$ varied as a function of stimulation rate until approximately 8 Hz; when maximum O$_2$ uptake usually was achieved. Notably, the redox state of Cu$_A$ varied directly with both O$_2$ extraction ratio (data not shown) and $\dot{V}$ O$_2$. As $\dot{V}$ O$_2$ approached maximum values, the level of oxidized Cu$_A$ approached the minimal values detected after death. In Figure 2, NIR Cu$_A$ data obtained from the intact gracilis of the dog are shown next to data from isolated liver mitochondria reported by Morgan and Wikstrom (8) for comparison. The redox state of Cu_A of well-coupled liver mitochondria shows reduction responses similar to those of cytochrome c as enzyme turnover rate increases. Morgan and Wikstrom have proposed that Cu_A functions as an electron acceptor between cytochromes c and a. Notably, the Cu_A in the canine gracilis demonstrates behavior analogous to the *in vitro* preparation. Again, the NIR data suggest important quantitative differences in the O_2 diffusion gradients and geometry, but the principle of increased reduction level at increased turnover remains the same and suggests that Cu_A may be a valuable index of oxidative phosphorylation in vivo.

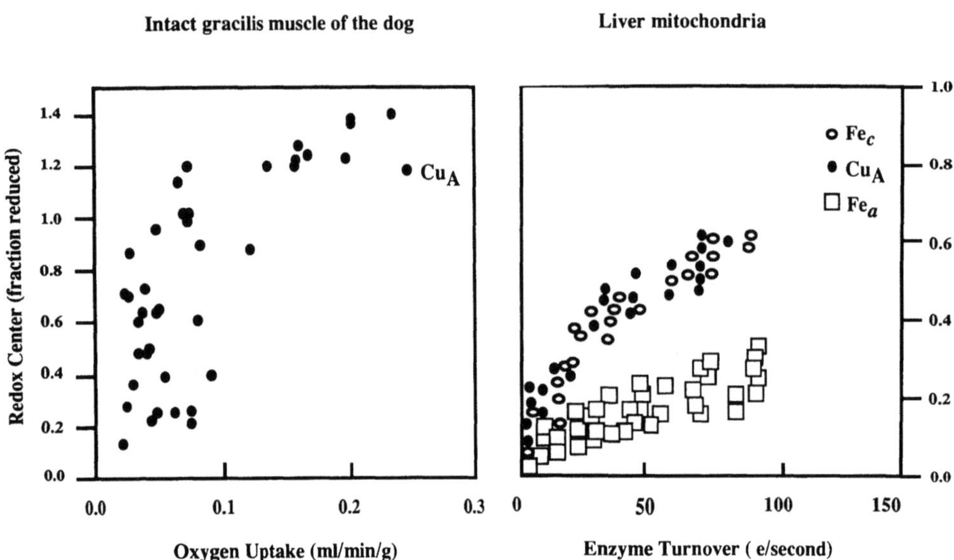

Figure 2. Relationship of redox state to oxygen uptake in dog gracilis. Left panel shows change in oxidized Cu_A by NIR spectroscopy expressed relative to the TLC at death (n=13). Right panel shows redox behavior of Cu_A in well-coupled liver mitochondria from reference 8 for comparison.

REFERENCES

1. Wittenberg B.A., and J.B. Wittenberg. Transport of oxygen in muscle. Annual Rev. Physiol. 51:857(1989).

2. Gayeski T.E.J.,and C.R. Honig. Oxygen gradients from sarcolemma to cell interior in a red muscle at maximal oxygen consumption. Am. J. Physiol. 251:789(1986).

3. Wilson D.F., M. Erecinska, C. Drown, and I.A. Silver. The oxygen dependence of cellular energy metabolism. Arch. Biochem. Biophys. 195(2):485(1979).

4. Roca J., M.C. Hogan, D. Story, D.E. Bebout, P. Haab, R. Gonzales, O. Ueno, and P.D. Wagner. Evidence of tissue diffusion limitation of $\dot{V}O_{2max}$ in normal humans. J. Appl. Physiol. 67:291(1989).

5. Jöbsis F.F. Noninvasive, infrared monitoring of cerebral and myocardial oxygen sufficiency and circulatory parameters. Science Washington, DC, 198:1264(1977).

6. Piantadosi C.A, Absorption spectroscopy for assessment of mitochondrial function *in vivo, in*: "Methods in Toxicology,"Jones, D.P. and L.H. Lash, ed., Academic Press, Inc., Orlando, 107(1993).

7. Duhaylongsod F.G., J.A. Griebel, D.S. Bacon, W.G. Wolfe, and C.A. Piantadosi. Effects of muscle contraction on cytochrome a,a_3 redox state. J. Appl. Physiol. In press, 1993.

8. Morgan J.E., and M. Wikstrom. Steady-state redox behavior of cytochrome *c* oxidase in intact rat liver mitochondria. Biochemistry 30:948(1991).

DIGITAL IMAGING OF THE OXYGENATION STATE
WITHIN AN ISOLATED SINGLE RAT CARDIOMYOCYTE

Eiji Takahashi and Katsuhiko Doi

Department of Physiology
Yamagata University School of Medicine
Yamagata 990-23
Japan

INTRODUCTION

Oxygen molecules released from erythrocytes in the capillary blood traverse plasma, capillary wall, extracellular fluid, cell membrane, cytosol, and mitochondrial membrane, and finally reach the mitochondrial inner membrane where they are utilized for ATP production. This process mainly depends on passive diffusion of the molecule along the oxygen pressure gradient between capillary blood and mitochondria. Therefore, the oxygen pressure at the mitochondrial inner membrane, P_i, can be represented by Eq.1

$$P_i = P_o - R\,\dot{V} \tag{1}$$

where P_o is the capillary blood Po_2 while $\dot{V}$ and R are the oxygen flux (oxygen consumption rate) and the lumped diffusion resistance for oxygen, respectively. Above equation indicates the possibility that mitochondrial oxidative phosphorylation might be limited in the case that either capillary blood Po_2 is lowered or oxygen consumption is increased. However, at least physiologically, it may sound unlikely that this diffusion resistance critically limits oxidative ATP production, since the Km of isolated cytochrome c oxidase for oxygen is considerably low (*i.e.* less than 0.1 Torr) compared to the capillary blood Po_2 (usually well above 20 Torr).

However, above inference based on the biochemical data does not necessarily fit to the result of recent continuous *in vivo* measurement of changes in mitochondrial redox state. Namely, using near infrared (NIR) spectroscopy, Duhaylongsod *et al.* (1993) reported the gradual reduction of cytochrome a,a3 of dog gracilis muscles which was in parallel with increases in the muscular oxygen consumption (induced by electrical stimulation of the muscle). This result may be interpreted as that the increased oxygen flux decreased P_i of Eq.1 to the critical level (<0.5 Torr), suggesting a relatively large R.

We felt that, to clarify the physiological relevance of the oxygen pressure gradient from capillary to mitochondria, it is important to directly measure the changes in oxygen level within a cell. It automatically follows that the method to quantify the oxygen level should provide spatial resolution high enough to resolve changes in the oxygen level around mitochondria (< 1 μm), because steep oxygen pressure gradient might be produced in the vicinity of oxygen consuming mitochondria (Jones, 1986). Furthermore, to determine the accurate location of the oxygen pressure gradient within a cell, isolated single myocyte should be used rather than cell suspension or tissue.

In the present paper, we describe a new spectrophotometric technique which is capable of visualizing oxygenation level within a single isolated cell with a spatial resolution of 0.2 μm/pixel. Using this technique, we demonstrated that changes in the oxygen consumption of a single rat ventricular myocyte considerably affect the oxygen level within the cell.

METHODS

Theory

Oxygen binding of the intracellular pigments such as myoglobin and cytochromes was assessed by way of microspectrophotometry using three wavelengths (406 nm, 420 nm, and 436 nm). Considering the light absorption by intracellular pigments, intensity of the transmitted light at wavelength λ is represented by

$$Y_\lambda = A_\lambda K_\lambda I \tag{2}$$

where A_λ, K_λ, and I represent the light absorption by the pigment, sensitivity of the spectrophotometric system, and intensity of incident light at wavelength λ, respectively. Using a video amplifier capable of adjusting gain and bias, contrast enhancement is carried out as follows

$$Y_\lambda = \alpha (A_\lambda K_\lambda I + \beta) \tag{3}$$

where α and β (<0) denote gain and bias of the video amplifier. After an appropriate calibration, K at respective wavelength can be equalized (*i.e.*, $K_{406}=K_{420}=K_{436}\equiv K$). Then, taking 420 nm as a pivot wavelength, Eq. 3 can be rewritten as follows.

$$Y_{436}^* = \alpha \ (A_{436} \ KI + \beta) \tag{4}$$

$$Y_{420} = \alpha \ (A_{420} \ KI + \beta) \tag{5}$$

$$Y_{406}^* = \alpha \ (A_{406} \ KI + \beta) \tag{6}$$

Finally, to eliminate the unknown parameters such as α, β, I, and K, a new variable Z is defined.

$$Z = \frac{Y_{436}^* - Y_{420}}{Y_{406}^* - Y_{420}} \tag{7}$$

$$= \frac{A_{436} - A_{420}}{A_{406} - A_{420}} \tag{8}$$

Hereafter, we shall show that the actually calculated Z represents the fractional oxygen binding of the intracellular pigments. According to the Beer-Lambert law, light absorption of the cell at wavelength λ can be represented as

$$\log A_\lambda = -(\varepsilon_\lambda^{oxy} C^{oxy} + \varepsilon_\lambda^{deoxy} C^{deoxy} + \varepsilon_0 C_0) L \tag{9}$$

where $\varepsilon_\lambda^{oxy}$ and $\varepsilon_\lambda^{deoxy}$ are extinction coefficients of intracellular pigments in oxygenated/oxidized and deoxygenated/reduced forms, respectively, while C^{oxy} and C^{deoxy} represent concentration of intracellular pigments in oxygenated/oxidized and deoxygenated/reduced forms, respectively. L is the path length (*i.e.* thickness of a cell) and nonspecific light absorption is expressed by the last term of Eq. 9. To eliminate the unknown parameters, the following approximation is applied to Eq. 9,

$$\log(1 - k\Delta) \approx -\Delta \tag{10}$$

where k is a constant (=2.175) and Δ is a small variable. Using this approximation, the newly defined variable Z can be written as follows.

$$Z = \frac{(\varepsilon_{436}^{oxy} - \varepsilon_{420}^{oxy})C^{oxy} + (\varepsilon_{436}^{deoxy} - \varepsilon_{420}^{deoxy})C^{deox}}{(\varepsilon_{406}^{oxy} - \varepsilon_{420}^{oxy})C^{oxy} + (\varepsilon_{406}^{deoxy} - \varepsilon_{420}^{deoxy})C^{deox}} \tag{11}$$

Note that the Z consists only of extinction coefficients of intracellular pigments which are known *a priori* and concentrations of oxygenated/oxidized and deoxygenated/reduced pigments. Finally, fractional oxygen binding of the intracellular pigments (S) can be reported by using the measured Z as follows.

$$S = \frac{C^{oxy}}{C^{oxy} + C^{deoxy}} \tag{12}$$

$$= \left[\frac{1 - \{\varepsilon_{436}^{oxy} - \varepsilon_{420}^{oxy} - Z(\varepsilon_{406}^{oxy} - \varepsilon_{420}^{oxy})\}}{\varepsilon_{436}^{deoxy} - \varepsilon_{420}^{deoxy} - Z(\varepsilon_{406}^{deoxy} - \varepsilon_{420}^{deoxy})} \right]^{-1} \tag{13}$$

Measuring System

Light emitted from a 150 watts short arc metal halide lamp (MSR150E3, Iwasaki Electric Corp.) was introduced, via a liquid light guide, to a rotating plate on which interference filters were mounted. The passing wavelengths of these filters were tuned to 406 nm, 420 nm, and 436 nm (band width 10 nm), respectively, which correspond to the Soret band of myoglobin. The exact position of the filter was controlled by a computer controlled stepper motor. The filtered light was directed to the measuring cuvette on the stage of a microscope, and the transmitted image was converted to electrical signal by a charge coupled device camera (XC-75, Sony) attached to the microscope. The signal was then offset adjusted and amplified by a video amplifier (C2400, Hamamatsu Photonics), digitized (8 bit) at a rate of 30 frames/sec (IQV-50, Hamamatsu Photonics), and stored for subsequent processing. To reduce random noises, image was accumulated for 10 times. Image arithmetics indicated by Eq. 7 were carried out by using a general purpose image processing software (IP Lab, Signal Analytics Corp.) running on a Macintosh IIci computer equipped with a DiimoCache accelerator card (Diimo Technologies Inc.).

Cell Isolation

Single cardiomyocytes were isolated from the ventricles of the male Sprague-Dawley rat weighing 250 - 300 g by the collagenase (Yakult) digestion method, and suspended in the calcium nominally free HEPES buffer solution (150.0 mM NaCl, 3.8 mM KCl, 1.0 mM KH_2PO_4, 1.2 mM $MgSO_4$, 10.0 mM glucose, 10.0 mM HEPES, supplemented with 1.5% bovine serum albumin, pH adjusted to 7.40). Ten μl suspension was placed in the "well" of the measuring cuvette (diameter of the "well", 10 mm). Then the suspension was superfused with a mixed gas of various oxygen concentration. Oxygen partial pressure of the cuvette outflow gas was monitored using a conventional oxygen electrode (17026, Instrumentation Laboratory).

RESULTS

All the experiments were carried out at room temperature (21-23 °C). Figures 1 and 2 exemplify the reconstruction of oxygenation level of a single isolated rat cardiomyocyte. Using the contrast enhancement described above, slight difference of the light absorption characteristics of intracellular pigments could be detected (Fig. 1). Average deflections of the cell image intensity of Y_{406} and Y_{436} from the pivot image (Y_{420}) were 4.0 and 9.8, respectively, for this particular cell. These were <6% of the average light intensity of the pivot cell image. Using these images, a new image representing the oxygenation of the cell was generated according to Eq. 7 (Fig.2). The reconstruction of the oxygen level was conducted with a spatial resolution of 0.2 μm/pixel, when 40x object lens was used. Although the atmospheric Po_2 of this cell was 152 Torr implying that the cell must be fully oxygenated, the reconstructed Z image looked significantly inhomogeneous. This may arise from the different light absorption characteristics of myoglobin and cytochromes.

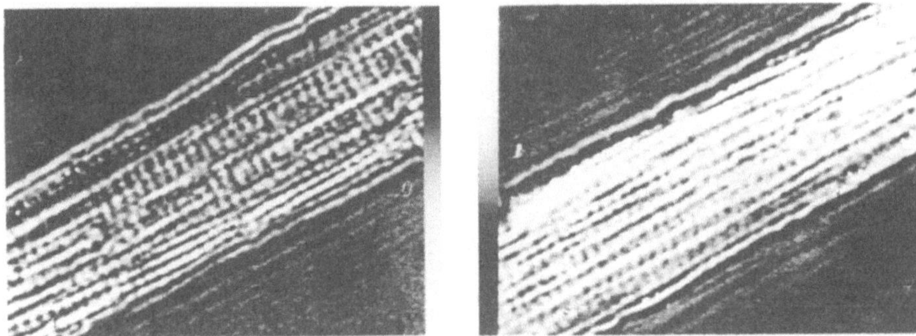

Figure 1. Images demonstrating the difference in light absorption characteristics of an isolated single rat cardiomyocyte. Left and Right panels represent differential images Y_{406}-Y_{420} and Y_{436}-Y_{420}, respectively.

Figure 2. An example of reconstruction of the cellular oxygenation state defined by Eq. 7. The cardiomyocyte was identical to that shown in Fig. 1.

Another way to represent the oxygenation state of the cell is to use a histogram of the Z value (Fig. 3). When Po_2 of the gas superfusing the cell was decreased from 152 Torr to 0.7 Torr, the mode of the Z-histogram shifted from 0.7 to -0.2. Therefore, these Z values would represent the fully oxygenated cell and almost deoxygenated cell, respectively.

We examined whether changes in the oxygen flux into the cell affect the oxygen level of the cell. Oxygen consumption of the cell was altered by 2 mM NaCN or 1 μM cyanide 3-chlorophenylhydrazone (CCCP). One μM CCCP increased the oxygen consumption from 69 ± 18 nM/min/10^6 cells (control, n=7) to 130 ± 46 nM/min/10^6 cells (n=4), whereas 2 mM NaCN completely abolished the oxygen consumption. Table 1 summarizes the mode of the Z-histogram representing the distribution of intracellular oxygenation. The mode Z of untreated cell in normoxia was 0.76 ± 0.07 (n=7). When the cell was made anoxic, mode Z decreased to -0.18 ± 0.09 (n=7). When oxygen consumption was increased by 1 μM CCCP, even a mild hypoxia (Po_2=5.1 Torr, n=7) seemed to completely deoxygenate the cell. In contrast, under severe hypoxia (Po_2=1.2 Torr), NaCN treated cell was significantly well oxygenated (mode Z = -0.02 ± 0.09, n=7) compared to the control cell (mode Z=-0.18 ±0.09).

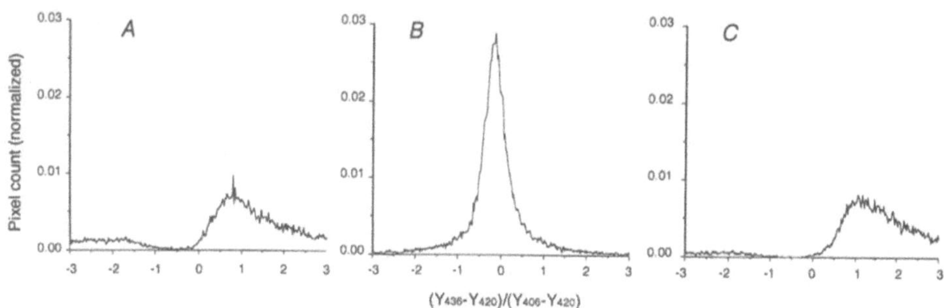

Figure 3. Changes in the Z-value histogram of one cardiomyocyte depending on the oxygenation state of the cell. Plates A, B, and C represent the Z-histograms of normoxic (Po_2=152 Torr), severe hypoxic (Po_2=0.7 Torr), and normoxic (Po_2=152 Torr, after control) cell, respectively.

DISCUSSION

In the present study, a new method for imaging oxygen level of a single cell has been proposed for the purpose of determining whether the intracellular Po_2 gradient, in particular that established near mitochondria, would affect oxidative ATP production. The oxygen pressure gradient may reside either within and outside the cell. To specifically define the exact location of the Po_2 drop *within* a cell, we needed to use single isolated myocytes rather than tissue or cell suspensions. Also, a use of single cell excludes the confounding effect of limitation of oxygen transport caused by topology of capillary network. Thus, by accurately controlling the oxygen tension around the cell, the effects of oxygen consumption and oxygen diffusion on intracellular Po_2 could be defined.

To assess intracellular Po_2 of cell suspension and tissue, spectrophotometry has been used where various intracellular pigments such as myoglobin and cytochromes are used as intrinsic oxygen probes (Vanderkooi *et al.*, 1991). Currently available microspectrophotometric technique, however, would not be readily applicable to isolated single cells because of extremely low light absorption expected for a single myocyte. For example, assuming millimolar extinction coefficient (ε_{mM}) of oxygenated myoglobin at 436 nm being 37 mM^{-1}cm^{-1}, light absorption by oxygenated cytosolic myoglobin (intracellular concentration 0.2 mM) of a single rat ventricular myocyte (thickness 8 μm) would be 1.4% of incident light. When the cell is fully deoxygenated (ε_{mM} = 104 mM^{-1}cm^{-1}), 3.8% of the light would be absorbed in the cell. These theoretical calculations lead to an estimation that light absorption by a single cardiomyocyte would be approximately 0.03% for 1% change in myoglobin oxygen saturation. This is obviously below the limit of detection by commercial spectrophotometric devices.

Video enhanced microscopy (VEM) is a new method which expands the limitation of light microscope defined by the Rayleigh's limit (Shotton, 1988). In VEM, microscopic image is converted to electrical signal using a video camera. Video signal is subsequently amplified to augment the contrast of the image so that the object can be separated from the background. Then, a digital subtraction of the background is conducted on the image, which is an essential procedure in VEM for two reasons; firstly, to confine the object image within dynamic range of the measuring system, and, secondly, to avoid the degradation of signal-to-noise ratio due to mottles (mottle subtraction). However, this contrast enhancement technique would not be readily used in spectrophotometry. Conventional spectrophotometry requires the information regarding absolute intensity (*i.e.*, light intensity measured from the zero level) of transmitted or reflected light to calculate the optical density of the test substance. In contrast, in VEM, the zero intensity level is electrically shifted to out of the range as a result of mottle subtraction. Thus, in VEM, absolute intensity of transmitted light cannot be accurately determined.

The present method, basically using the VEM, does not require the accurate estimation of the zero light intensity level, because the image subtraction procedure between two different wavelengths indicated in Eq. 7 abolishes the unmeasurable bias (β defined in Eq. 3) from the calculation of light absorption. Furthermore, by using an approximation method depicted in Eq. 10, the present method does not require the information regarding the light path length which is quite difficult to estimate in microspectroscopy. This approximation is applicable for $\Delta < 0.1$, if maximum error of 1.5% is allowed. Because Δ's of a single rat cardiomyocyte at the wavelengths used in the present study are <0.02, the use of Eq. 10 would be justified. Unfortunately, the present method cannot be used in a thick tissue because of the limitation imposed by the above assumption.

Using the present new method, faint changes in light absorption of a single cell at different wavelength lights could be successfully imaged (Fig.1). As seen from Fig.3, transition of the extracellular Po_2 from normoxia to severe hypoxia clearly shifted the Z-histogram with little overlap, suggesting that observed changes in the Z value were completely dependent of intracellular Po_2 and that possible interferences from the cell structures such as striation of myofibril, cytoskelton, organella, or nucleus could be ignored. It may be also noted that location of mitochondria appears to be simultaneously identifiable in the Z-image (Fig.2) by virtue of the difference in light absorption characteristics between myoglobin and cytochromes.

Cardiac cells contain relatively large amount of myoglobin. Primary role of this protein is believe to buffer changes in cytosolic Po_2 thereby ensuring constant oxygen supply to mitochondria. Therefore, mitochondrial oxidative ATP production may not be compromised by reduction in oxygen delivery until cytosolic myoglobin is almost completely deoxygenated

Table 1. Summary of the mode of Z histograms.

		Po_2 (Torr)	mode Z
Control	RA	153.2	0.76±0.07
	MH	5.4±0.14	0.50±0.09
	SH	1.3±0.08	-0.18±0.09
1 μM CCCP	RA	153.4	0.84±0.10
	MH	5.1±0.09	-0.19±0.13*
	SH	1.3±0.05	-0.21±0.09
2 mM NaCN	RA	153.2	0.74±0.09
	MH	5.1±0.08	0.58±0.11
	SH	1.2±0.10	-0.02±0.09*

RA, room air; MH, mild hypoxia; SH, severe hypoxia; *, $P<0.05$, comparison with the respective control.

Wittenberg and Wittenberg (1985) reported that, in isolated rat cardiomyocytes, oxygen pressure gradient from cytosol to mitochondrial inner membrane was quite small (<0.2 Torr), and that from extracellular fluid (ECF) to cytosol was negligible in resting cells and <2 Torr in the cell in which respiration was increased to 3.5-folds. Therefore, fluctuations of extracellular Po_2 would not affect oxidative phosphorylation unless extracellular Po_2 is lowered to <2 Torr, where cytosolic myoglobin is entirely deoxygenated.

In contrast, some investigators claim the existence of significant intracellular oxygen gradient by which mitochondrial oxidative phosphorylation might be regulated. Kennedy and Jones (1986) demonstrated in isolated rat cardiomyocytes that oxygenation of myoglobin and oxidation of cytochrome a_3 demonstrated a linear relationship, suggesting considerable oxygen pressure gradient from cytosol to mitochondrial inner membrane. Also, Rumsey *et al.* (1990) suggested that oxygen diffusion induced oxygen concentration gradient between ECF and mitochondria directly affects the mitochondrial respiration. Interestingly, this oxygen concentration difference is a function of mitochondrial respiration.

Present study demonstrated that changes in cellular metabolic rate in fact affect the intracellular oxygen level (Table 1). This seems to suggest the importance of oxygen flux to mitochondria in the regulation of cellular respiration. Diffusion resistance for oxygen molecule stems from unstirred layer in the ECF surrounding the cell, intracellular O_2 diffusivity, and the topological distribution of oxygen sink (*i.e.*, mitochondria), while plasma and mitochondrial membranes seem to offer little resistance. Clustering of mitochondria might result in a steep oxygen gradient around mitochondria (Jones, 1984), and even produce a local anoxic region within the cell when mitochondrial respiration is increased. Benson *et al.* (1980) measured fluorescence of pyrene-1-butyrate in single isolated hepatocyte of the mouse with a spatial resolution of 0.5 μm and found the heterogeneous fluorescence pattern in a single cell. This may imply the heterogeneous distribution of oxygen molecules within a cell. We are hoping that, using the present new microspectroscopic system, intracellular distribution of oxygen molecule and possible microanoxia within the cell could be visualized with a spatial resolution of 0.2 μm.

ACKNOWLEDGEMENT

Part of this study was supported by the research grant provided by the Fukuda Foundation for Medical Technology.

REFERENCES

Benson, D.M., Knopp, J.A., Longmuir, I.S., 1980, Intracellular oxygen measurements of mouse liver cells using quantitative fluorescence video microscopy. *Biochim. Biophys. Acta*, 591:187.

Duhaylongsod, F.G., Griebel, J.A., Bacon, D.S., Wolfe, W.G., and Piantadosi, C.A., 1993, Effects of muscle contraction on cytochrome a,a3 redox state. *J. Appl. Physiol.*, 75:790.

Jones, D.P., 1984, Effect of mitochondrial clustering on O_2 supply in hepatocytes. *Am. J. Physiol.*, 247:C83.

Jones, D.P., 1986, Intracellular diffusion gradients of O_2 and ATP. *Am. J. Physiol.*, 250:C663.

Kennedy, F.G., and Jones, D.P., 1986, Oxygen dependence of mitochondrial function in isolated rat cardiac myocytes. *Am. J. Physiol.* 250:C374.

Rumsey, W.L., Schlosser, C., Nuutinen, E.M., Robiolio, M., and Wilson, D.F., 1990, Cellular energetics and the oxygen dependence of respiration in cardiac myocytes isolated from adult rat. *J. Biol. Chem.* 265:15392.

Shotton, D.M., 1988, Review: Video-enhanced light microscopy and its applications in cell biology. *J. Cell Sci.*, 89:129.

Vanderkooi, J.M., Erecinska, M., and Silver, I.A., 1991, Oxygen in mammalian tissue: methods of measurement and affinities of various reactions. *Am. J. Physiol.*, 260:C1131.

Wittenberg, B.A., and Wittenberg, J.B., 1985, Oxygen pressure gradients in isolated cardiac myocytes. *J. Biol. Chem.* 260:6548.

OPTICAL IMAGING OF HUMAN BREAST CANCER

Shoko Nioka, Mitsuharu Miwa, *Susan Orel, *Mitchell Shnall,
Munetaka Haida, Shiyin Zhao and Britton Chance

Departments of Biochemistry and Biophysics, and *Radiology, University
of Pennsylvania, Philadelphia, PA 19104

INTRODUCTION

Breast cancer has become one of the most serious epidemic diseases of middle aged women since the incidence has increased over the decade dramatically to one eighth or ninth of a woman's life time. Over thirty thousand woman die every year from breast cancer. Mortality can be reduced by early cancer detection. While X-ray mammography has been used to screen breast cancer currently, this technique has difficulty detecting breast cancer in some cases. The reasons are as follows: first, its detection ability has been restricted in some cases, as it is mainly sensitive to calcification, and is hard to distinguish between normal fibrotic tissue and cancer. Secondly for woman under the age of 35 years old, X-ray mammogram does not see through breast tissue with sufficient contrast, since young breast tissue is normally fibrotic. Thirdly X-ray radiation itself is carcinogenic, and frequent usage is not recommended. Recently, Magnetic Resonance Imaging has appeared to be a better technology than X-ray for new mammography (1), but it is more expensive and may not be used as a typical screening test at present. For these reasons, we are still seeking a better imaging technique to satisfy our needs for early breast cancer detection.

Despite the long history of the optical imaging technology, it has not been commonly used for breast cancer detection. Photon diffusion into the tissue makes a shadow of the object too obscure to use for imaging. The shadow imaging with continuous light (diaphanography) has been used in Asia, where average breast size is relatively smaller, but the shadow technique does not allow real imaging. Today, with new technologies, such as the time gated technique (2), long path diffusive photons are eliminated and therefore better images can be obtained. We have studied the ability of the time gated optical technique to human breast cancer.

We have chosen to develop the time gated optical technique for breast cancer detection for the following reasons. First, the breast tissue consists of a relatively low scattering media, namely fat, whose scattering coefficient is as low as 3 cm^{-1}. Second, the breast is located on a surface of the body and accessible to optical instruments. Thirdly, the time gated technique will allow selection of photons which have not diffused extensively (the extensively diffused photons are the ones which create obscure images). We selected 15 % of total photons, which relatively less diffused and arrived earlier than 85% of rest of photons. We emphasize not only possible improvement of sensitivity of detection by this time gated technique, but also high specificity of cancer detection by functional imaging. It creates an oxygen image through hemoglobin oxygenation, a blood volume image and a scattering image, and therefore can be helpful in interpreting the cell function of lesions.

Preliminary studies using time resolved spectroscopy (TRS) for characterizing tumors in the breast has been published elsewhere (2).

SUBJECTS AND METHODS

Thirty five women who had breast tumors detected by X-ray mammogram were subjects for the optical imaging. The age of the women spread from 30 to 84 years old. In them, 15 patients had lesions removed (lumpectomy operation). 20 patients had a single suspicious small tumor shadow in the mammography The tumor sizes are from 3 mm to 1 cm diameter. The post-operative scar tissue contains seroma, fibrotic scar as well as fibro-cystic tissue, which was located initially. The size of the post-operative lesion is 2 cm to 5 cm in diameter. Approximately 30% of the patients have primary fibrocystic disease.

For the optical measurement, the breasts were compressed gently with two surface holders located parallel to each other (Figure 1). One surface holds an incident light guide and the other holds a light guide coupled to a detector, photomultiplier. In between the two light guides, time resolved spectra of the breast are acquired through transmittance mode. Then, the light guides are moved on the two surfaces 5 mm equally to obtain a two dimensional transmittance image. During the acquisition, light guide separation is fixed and constant.

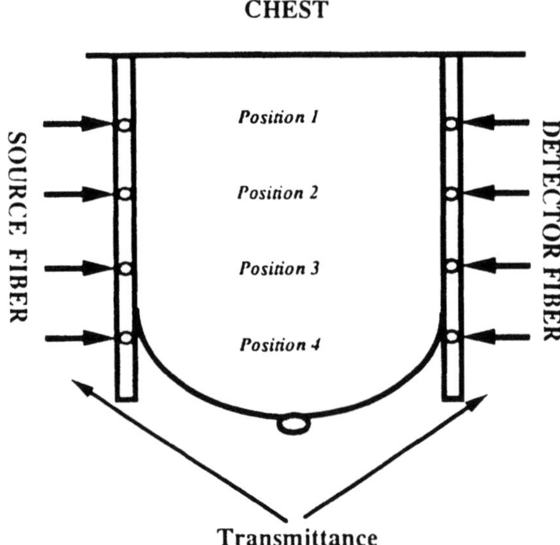

Figure 1. The schema shows a set up for the optical breast imaging. The breast is gentry compressed with two surface holders located parallel to each other. One surface holds an incident light guide and the other holds a light guide coupled to a detector. In between the two light guides, TRS of the breast are acquired through transmittance mode.

A time resolved spectrometer (TRS) consists of two wavelength pulsed lasers, which emit lights at 780 and 830 nm, a photo multiplier (gallium arsenide), a time amplitude converter (TAC), a photon counting system, and data stored in a personal computer through an A/D converter. Time width of the pulsed laser light is 300 picoseconds through the TRS system and is used as an instrument function. This time resolution of the instrument function is good enough for human breast spectra to be used without deconvolution because the tissue spectra gained the half width more than 5 times larger than the instrument function. The two tissue spectra at the two wavelengths are acquired simultaneously by time sharing (delaying one pulse by 10 nanoseconds).

The two tissue spectra with wavelengths at 780 nm and 830 nm are analyzed to obtain absorption and scattering coefficients with a curve fitting program, which in principle, is based upon a diffusion equation using semi-infinite boundary conditions (Patterson and co-workers model) (3). Scattering coefficients are mainly calculated at maximum of photon arrival time in the equation. This requires 15 to 20 % of early arrival photons. The absorption coefficient is mainly calculated by the photon decay curve of the spectra in the equation, requiring 25 to 50 % of the early arrival photons. Hemoglobin saturation with oxygen (SaO_2) is calculated using the two absorption coefficients (4). As a presentation of blood volume, the two absorption coefficients are averaged. Scattering coefficients at two wavelengths are also averaged to yield a scattering image.

Three independent pieces of information are used to process three images from a set of measurements; the oxygen image is based upon hemoglobin saturation with oxygen, blood volume image from absorptions, scattering image from scattering coefficients. Once these three are calculated, they are fed into the image processing program, which translates the location of light guides on the breast surface as a pixel of the image. It therefore creates images of a 5 mm^2 resolution. Usually a 5 x 7 pixel image is acquired.

RESULTS

It is found that the ability to detect breast lesions by the time gated optical method depends upon the size of lesions, the thickness of the breast and contents of breast tissue (fat and fibrous tissue). Usually post-operative lesions can be seen in our optical images, but a small cancer mass of less than 1 cm in diameter is not detectable with 5 to 10 cm thick breasts. Here, we show 3 patients who have post-operative lesions with 4 optical images and MRI.

Case No. 1. The images shown in Figures 2 to 5 are from a 38 year-old patient whose cancer was removed one month previous (lumpectomy), and under pre and post radiation therapy, respectively. A 2 x 1.5 cm cyst is formed in the location, where a cancer mass was before operation (See MRI Figure 2, and 4 corresponding to optical images of Figure 3 and

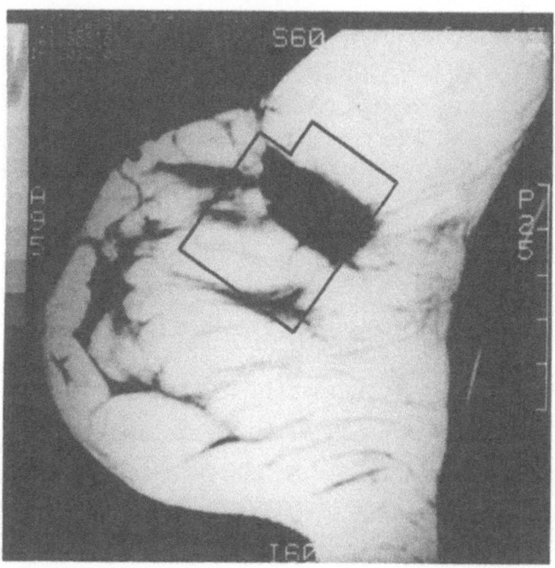

Figure 2. MRI (lateral slice) of case 1, whose cancer was removed (lumpectomy) one month previously. This MRI was taken prior to radiation therapy (pre-radiation). The closed area shows the area that optical imaging was conducted (see Figure 3). Also note that a cyst 2 x 1.5 cm is seen in the location.

173

Figure 5). Light guide separation was 6 cm. The cyst had less oxygen level; O_2 saturation in Hb is 82% in the cyst and 85-87% in the surrounding tissue in pre-radiation (Figure 3-C). After radiation therapy, the size of the cyst became reduced and formed a heterogeneous fibrotic tissue (Figure 4). An overall lower O_2 level was observed in the same area in post radiation than in pre-radiation (Figure 5C as compared to Figure 3C), and similar heterogeneity was observed in the area where the smaller cyst is. The area where the cyst and fibrous scar tissue are located has less blood volume (less absorption (Figure 3-B,5-B)) and more scattering (Figure 3-A, Figure 5-A).

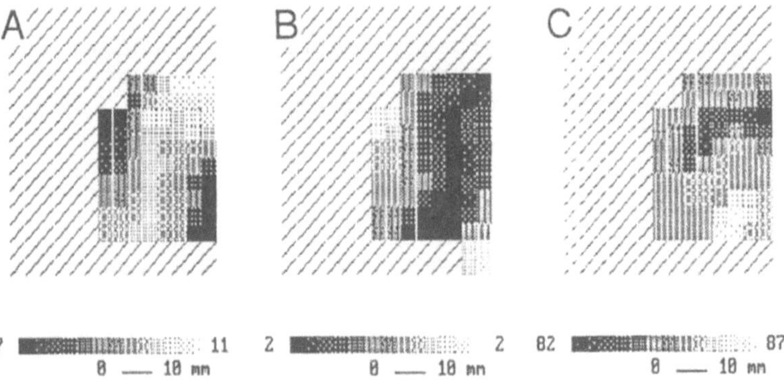

Figure 3. Optical images of case 1, under pre-radiation corresponding to Figure 2. A: scattering image. B: blood volume image C: oxygen image. Note that the cyst has low oxygen concentration than the surrounding tissue.

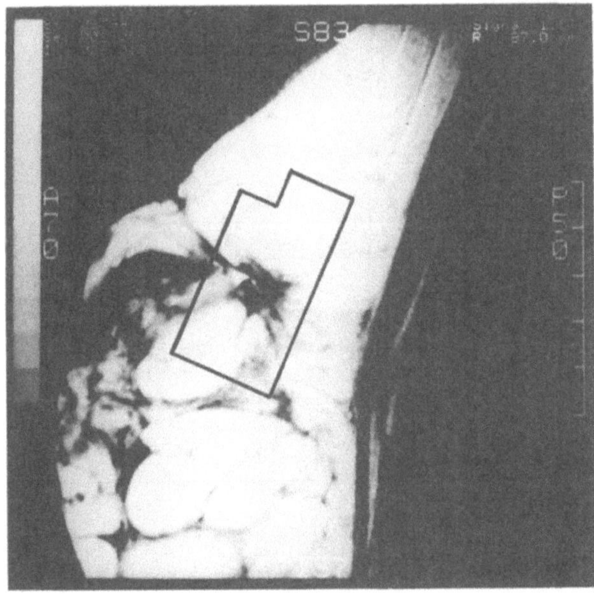

Figure 4. MRI of case 1, after a radiation therapy. The closed area shows the area that optical imaging was conducted (see Figure 5). The size of the cyst became reduced and formed a heterogeneous fibrotic mass due to the irradiation.

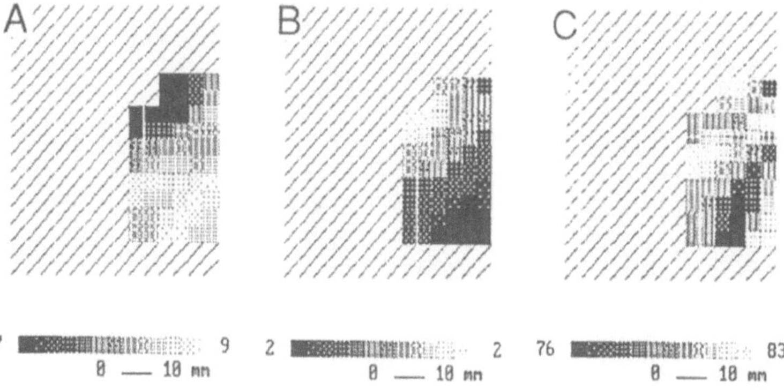

Figure 5. Optical images of case 1, post-radiation therapy, corresponding to Figure 4. A: scattering image. B: blood volume image C: oxygen image. The lesion is more scattering and contains less oxygen than the surrounding tissue.

Case No.2. The MRI and the optical image (Figures 6,7) is from a post lumpectomised cancer patient. She is 48 years old who has small to medium size breasts (light guide separation 4 cm). After lumpectomy, there is a 4.5 x 2.5 cm mass consisting of fibrotic scar tissue and cysts in the middle of her breast (Figure 6). The optical images were taken laterally covering between cysts and fibrotic tissues. The oxygen image shows that a part of the cyst has a low oxygen level indicating a quite heterogeneous profile (Figure 7-C). The oxygen level in the cyst capsule is particularly higher than the surrounding tissue. The blood volume image shows that the cysts have less blood volume (Figure 7-B). In addition, there is heterogeneity in the scattering images in the cysts as well as in the surrounding fibrotic tissue (Figure 7-A) constructing an image of scar tissue.

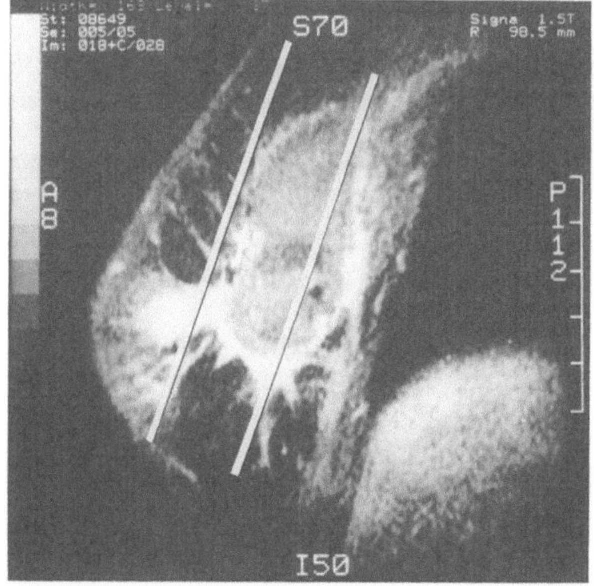

Figure 6. MRI of case 2, who has a large lesion after a lumpectomy. A lesion consists of cysts, fibrous scar tissue, and cysts.

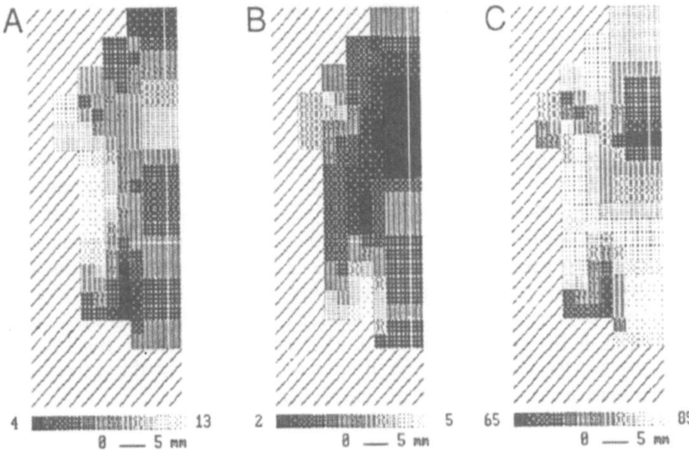

Figure 7. Optical images of case 2. The images were taken from top-to-bottom (cardio-caudal) direction. A transmittance light was shed in the direction of indicated lines and between two lines (Figure 6). The line nearer to the nipple corresponds to the left line of each image. The right side of the images clearly show the lesion. Note that the lesion has a lower oxygen location, which scatters more than surrounding tissue, indicating a heterogeneity.

Case no.3. The patient is 52 years old and weighs 180 lbs. Her tumor in the breast is determined to be a cancer by biopsy. She already has gone through a lumpectomy followed by a radiation therapy (Figure 8).

The optical images (Figure 9) were taken with cardio-caudal direction. In the optical scattering image (Figure 9-A), there are two highly scattered masses, connected to each other by a capsule like structure. This capsule like structure has higher oxygen tension than those in the two masses (Figure 9-C). The two masses differ in blood volume (See Figure 9-B), one near the nipple has more blood volume than the other.

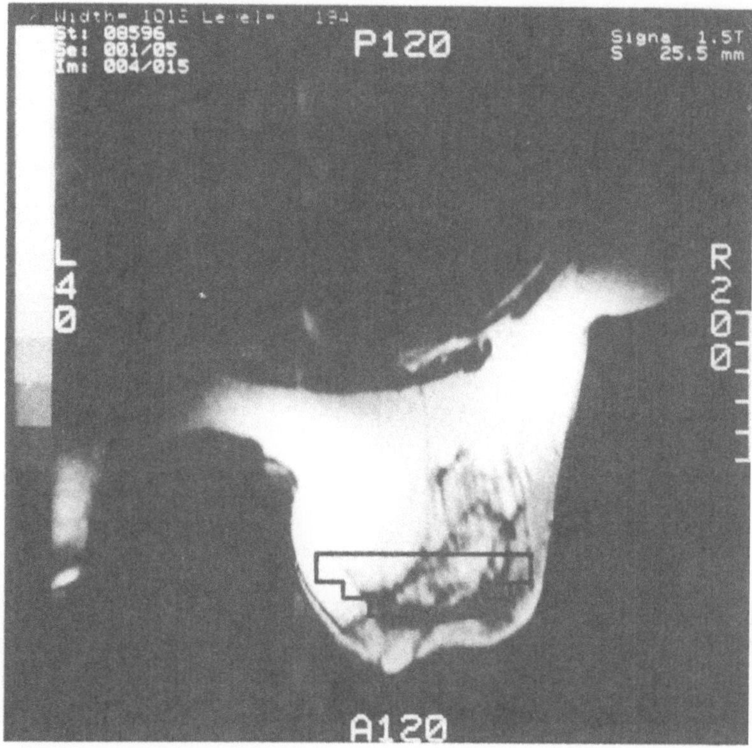

Figure 8. MRI (cardio-caudal direction) of case 3, post-lumpectomy. The closed area shows the area that optical imaging was conducted (see Figure 9).

DISCUSSION

Clearly, we have demonstrated that the time gated optical imaging technique is capable of imaging a relatively large object. However, this technique does not yet clearly detect a small cancer (generally smaller than 1 cm), therefore we did not include those small tumors in this report. It is shown that the time gated imaging system is not yet as precise as X-ray mammography for lesions smaller than 1 cm in size. However it is certainly better than the current diaphanography in principle.

In the optical imaging technique, the main cause responsible for low resolution of image is photon diffusion. For the human breast image, conditions for the continuous wave technique (any technology which uses CW light or current diaphanography) are viable in

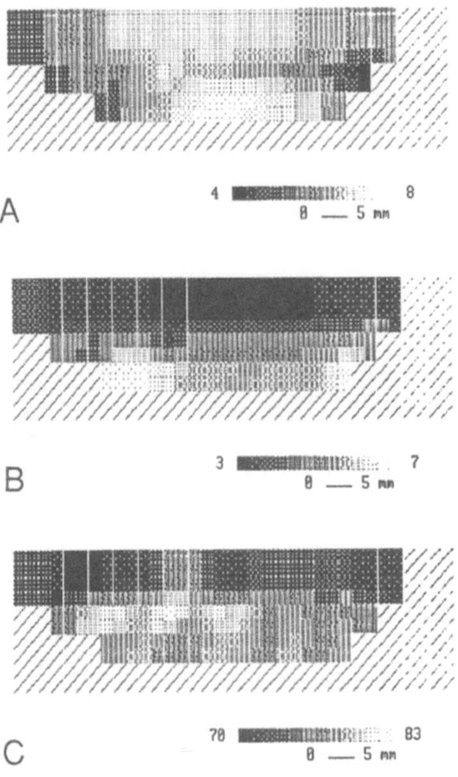

Figure 9. Optical images of case 3, corresponding to Figure 8. A: scattering image. B: blood volume image C: oxygen image. There are two highly scattered masses connected to each other by a capsule like structure. This capsule has higher oxygen tension than in the two masses.

breasts that are small, thin and fatty, and having tumors larger than 1 cm in diameter. Since the CW technique will not satisfy most of needs for early detection of breast cancers, we have developed a time gated photon imaging system to test feasibility for the purpose. Scattering and absorption images use only 15-20% and 40-50% of early arriving photons respectively, which therefore eliminates most of more diffusive photon, that create obscurity. The results show that the scattering image may have more clearer images of lesions than those used absorption (oxygen and blood volume images) because of narrower time windows for scattering images.

It is clear that this time gated system is not adequate to show a high contrast of lesions of less than 1 cm in size from surrounding tissue.The cause of low resolution comes from the many factors. First, although fortunately the breast contains substantial amounts of fat, whose scattering factor is low compared to any other tissue constituents, contrast between fat and fiber and tumor is not always substantial enough. Secondly, differences of oxygen tensions or blood volumes between those tissue components also may not be substantial enough. Thirdly, the technique we use here presents a number in a pixel as a mean value of a long tissue mass across the two light guides, and can not present a small lesion.

These problems can be solved by finding causes of obscurities. A more sophisticated image reconstructing technique such as inverse recovery technique (5), will better adapt photon diffusion characteristics into the image construction procedures. In addition, we will be able to obtain higher resolution images with a contrast agent, such as cardiogreen, which may accumulate in the lesions with permeable capillaries and more circulation. One of the optical techniques, the phased array system using phase modulation device will be expected to give better resolution, and will be suitable for detecting a small object (7). We are in the early stage of developing the optical imaging, and in the future attempts to improve substantially the quality of optical imagings will be pursued.

SUMMARY

Since an increasing number of breast cancers have been reported in recent years, there is a need for improving techniques for early detection of the breast cancer. Here we tested a time gated optical imaging technique as a tool for imaging human breast. Pulsed laser light at wavelengths of 780 and 830 nm are transmitted through human breast tissues and time spectra of the diffused light through the tissue are recorded over nanoseconds. Data from different locations are acquired and used to reconstruct a two dimensional image as a set of spectra in pixel form. The imaging consists of absorption and scattering coefficients, and the absorption coefficients at the two wavelengths are related to oxygen concentration and blood volume. The analysis of these coefficients is based upon the early arrival photons, therefore allowing construction of a better image than those from the current diaphanography. We demonstrate images of breast cancer, cysts created after lumpectomy, and consequences of radiation therapy. Results show that time gated optical imaging can image oxygen concentration in the cancerous and fibrotic breasts. Resolution of the imaging for smaller tumor size needs to be improved.

ACKNOWLEDGMENTS This work was supported in part by NIH grants CA 50766 and USAMRDC DAMD 17-93-C3071.

REFERENCES

1. Orel SG, Schnall MD, Livolsi VA, et al. MR Imaging of suspicious breast lesions with radiologic pathologic correlation. Radiology. in press.
2. Kang, KA, Chance B, Zhao S, Srinivasan, Patterson E, and Troupin T. Breast tumor characterization using near-infra-red spectroscopy. SPIE Proceeding. vol.1888. 1994. in press.
3. Chance B, Leigh JS, Miyake H, Smith D, Nioka S, Greenfield S, Finander M, Kaufman K, Levy W, Young M, Cohen P, Yoshioka H, and Boretsky R. Comparison of time resolved and unresolved measurements of deoxy hemoglobin in the brain. Proc. Nat. Acad. Sci. 85:4971-4975. 1988.
4. Patterson MS, Chance B, and Wilson B. Time resolved reflectance and transmittance for the noninvasive measurement of tissue optical properties. J. Appl. Optics. 28:2331-2336. 1989.
5. Sevick EM, Chance B, Leigh J, Nioka S, Maris M. Quantitation of time- and frequency-resolved optical spectra for the determination of tissue oxygenation. Anal. Biochem. 195:330-351. 1991
6. Schotland J, Haselgrove J, and Leigh JS. Photon hitting density. Applied Optics. 32: 448-453. 1993
7. Chance B, Kang K, He L, Wang J, and Sevick EM. Highly sensitive object location in tissue models with linear in-phase and anti-phase multi-element optical arrays in one and two dimensions. Proc. Natl. Acad. Sci. USA. 90:3423-3427. 1993.

TRANSCUTANEOUS H_2 CLEARANCE - A NEW LEAST-INVASIVE METHOD FOR ASSESSING SKIN BLOOD FLOW

D.K. Harrison[1], R. Abi Raad[2], D.J. Newton[1] and P.T. McCollum[3]

Vascular Laboratory, Directorates of Medical Physics[1] and General Surgery[3], Dundee Teaching Hospitals NHS Trust, Ninewells Hospital and Medical School, Dundee, DD1 9SY, and Department of Biomedical Engineering[2], University of Dundee, Dundee, DD1 4HN, Scotland

INTRODUCTION

Although there is a wide variety of methods available for measuring or, more precisely, assessing skin blood flow (for a review see Swain and Grant, 1989), the only current method available for measuring skin blood volume flow at a local level, i.e. within areas of a few mm^2 and in terms of ml/100g/min, is by use of radioactively labelled freely diffusible indicators (see for example Spence et al., 1985b). Such techniques have proved to be particularly reliable in predicting the outcome of below (versus above) knee amputation in patients with critical lower limb ischaemia (McCollum et al., 1988). However, the technique is invasive (albeit minimally) in that an intracutaneous injection is necessary, it requires the application of a radioactive substance, and above all it is necessary for the patient to remain quite still during the period of recording the clearance, a task which - because of pain - is often difficult for such patients to fulfil.

The aim of this study was thus to investigate the application of a non-invasive hydrogen clearance technique (Harrison and Kessler, 1989b) to measure skin blood flow in humans. The technique has previously been used to measure capillary blood flow in skeletal muscle (Harrison et al., 1990), the beating heart (Harrison et al., 1985) and the brain (Kozniewska et al., 1987), but to date only in animal preparations. It was intended that the method make use of commercially available transcutaneous oxygen ($tcpO_2$) electrodes adapted to render them highly sensitive to hydrogen (Harrison and Kessler, 1989a). The questions to be answered were: could the palladinisation technique render a standard $tcpO_2$ electrode sufficiently sensitive to record H_2 clearances in the skin of normal volunteers (but ultimately of patients), following the inhalation of a small, completely safe, non-explosive concentration of H_2? If so, to what extent would it be necessary to heat the skin in order to obtain reproducible measurements of $tcpH_2$ clearances in normal skin?

The advantage of such a technique would be the possible elimination of the need to administer a radioactive substance in order to measure local skin blood flow. The method would be non-invasive and measurements could be carried out at several sites for a single administration of the indicator.

METHODS

H_2 Clearance Measurements

Two standard transcutaneous pO_2 electrodes (Dräger) were used for the study. They consisted of three 15 µm diameter platinum anodes (wired in parallel), an Ag/AgCl cathode and a thermistor controlled heating element (Spence et al., 1985a). In order to

Oxygen Transport to Tissue XVI
Edited by M.C. Hogan *et al.*, Plenum Press, New York, 1994

increase the sensitivity for H_2, the platinum anodes were coated with palladium 24 h prior to use according to the method of Harrison and Kessler (1989a). In all experiments, the electrodes were polarised at 300 mV for one hour prior to use.

A multi-channel nanoamperemeter (Medical Physics Department, Ninewells Hospital) was used for all experiments. The output from the amplifier was fed to a Cambridge Electronic Design (CED), Cambridge, series 1401 analogue to digital converter. The CED "Chart" and "Sigavg" software packages were used for recording and pre-processing the data and analyses carried out using Lotus 123 and Statgraphics.

After stabilisation of the $tcpH_2$ electrodes at the required temperature for at least 10 minutes the subject was given a gas mixture containing 2% H_2, 20.9% O_2, 77.1% N_2 (BOC Special Gases, London) to breathe until stable $tcpH_2$ values had been reached (or for a maximum period of 20 min). An aviators' mask connected to a 50 l Douglas bag reservoir via two one-way valves (inspiration from Douglas bag, expiration to room) was used initially to administer the gas mixture. Flow from the gas cylinder was regulated to ensure that the reservoir remained inflated. However, later experiments demonstrated that a simple Hudson mask connected directly to the gas cylinder could be used equally effectively. After attainment of stable $tcpH_2$ values, or 20 min whichever was the shorter, breathing of room air was resumed. The subsequent measurements were carried out over a sufficient period for at least 90% of the clearance to be completed. This was a maximum of 20 min following the return to breathing room air. The measured curves were normalised and the semi-logarithmic slope analysis was used for calculation of all $tcpH_2$ blood flow values. Flows were calculated from the semi-logarithmically plotted clearances over the linear part of the curve (Harrison and Kessler, 1989b).

Laser Doppler Measurements

Although the application of a radioactive substance for the measurement of skin blood flow in patients is entirely justifiable on the grounds of its diagnostic efficacy, it was considered that its use in young, healthy volunteers should be avoided if possible. Whilst the laser Doppler technique does not give a quantitative measure of blood flow, it is an extremely useful tool for measuring changes in skin perfusion (Khan et al., 1991). This technique was thus used as the method against which $tcpH_2$ should be compared in normal volunteers. The measurements were carried out at the same skin temperatures as the $tcpH_2$ measurements, at adjacent sites, using two Perimed PF2b instruments.

Experimental Protocol

All procedures were performed according to protocols approved by the Tayside Committee on Medical Ethics after obtaining the informed consent of the subjects. The measurements were carried out after equilibration at a room temperature of 27 ± 1 $^{\circ}C$ for at least 20 min with the subjects lying on a bed in the supine position.

Ten normal subjects within the age range 20-40 years were investigated. The laser Doppler heaters and probes, and $tcpH_2$ electrodes were attached to the volar forearm skin approximately 5 cm apart using double-sided stickers. 1,2-propanediol was used as contact fluid to ensure adequate diffusion of H_2 between the skin surface and the $tcpH_2$ electrode.

The feasibility of using an unheated electrode and an electrode temperature of $37^{\circ}C$ was investigated in early experiments. However, for the main investigation a control clearance was carried out as described above and measured simultaneously at skin temperatures of 40 $^{\circ}C$ and 44 $^{\circ}C$. A second clearance was then carried out during which, two minutes after return to breathing room air, a 4 min suprasystolic cuff occlusion of the brachial artery was performed. The two $tcpH_2$ clearances and laser Doppler values (40 $^{\circ}C$ and 44 $^{\circ}C$) were recorded continuously throughout.

RESULTS

Attempts to measure $tcpH_2$ clearances without heating the electrode or with a heater temperature of $37^{\circ}C$ revealed that although it was possible to measure a clearance in one of the subjects at 37 $^{\circ}C$ (and in none with no heating), it was clear that neither method would be suitable for obtaining reproducible results and were thus not pursued.

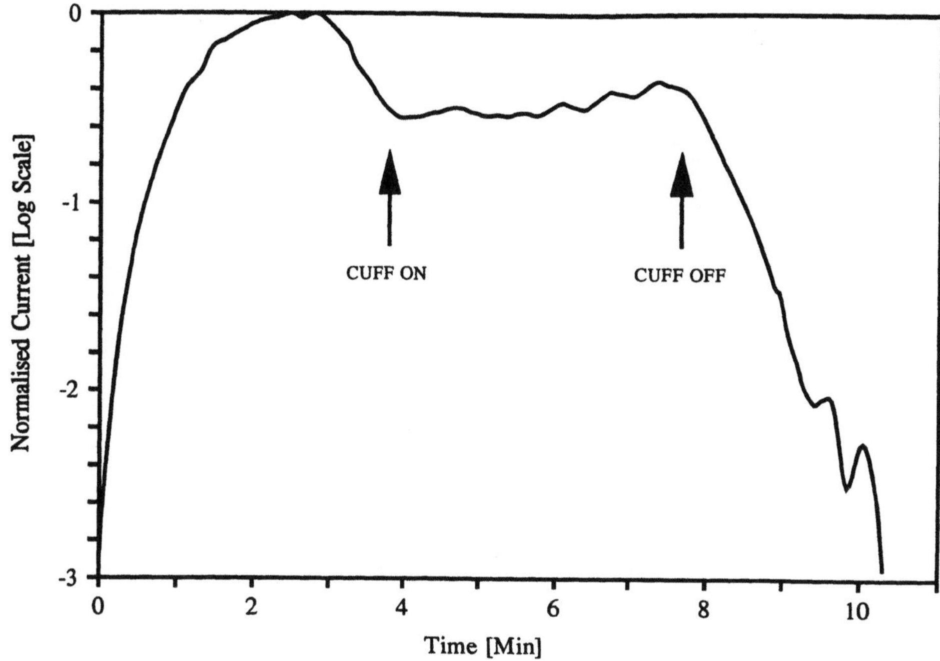

Figure 1. Hydrogen clearance at 40 °C including a 4 min period of tourniquet occlusion.

In the main study, two heater temperatures, 40 °C and 44 °C, were investigated. A clearance in a normal subject is shown in Figure 1 in normalised semi-logarithmic form at 40 °C. It can be seen that a 4 min period of suprasystolic occlusion of the brachial artery caused the clearance to be halted. It can also be seen that, in this case, saturation was achieved after only 3 min of breathing 2% H_2. In fact, the longest period required to saturation was 15 min but it was rarely necessary for the subjects to breath the gas mixture for more than 10 min. Interestingly, $tcpH_2$ increased slightly during the occlusion at 40 °C. The increase in the slope post- compared with pre-occlusion reflects the reactive hyperaemia.

Figures 2 and 3 show the relationships between all normal (ie 1st and 2nd $tcpH_2$ clearance) values and the corresponding LD flux values at 40 °C (Fig. 2) and 44 °C (Fig. 3). Linear correlations between them were significant ($r = 0.61$, $p < .002$ and 0.77, $p < .001$ at 40 °C and 44 °C respectively). Reproducibility between successive normal clearances ($r = 0.88$ at 40 °C and $r = 0.79$ at 44 °C, $p < .001$ in both cases) was high at both temperatures.

Figure 4 shows that there was a highly significant correlation ($r = 0.90$, $p < .001$) at 44 °C between the pre-occlusion and hyperaemic $tcpH_2$ values. The correlation was less significant, however, at 40 °C ($r = 0.57$, $p < 0.01$).

A similar pattern was observed with the LD flux values: at 40 °C there was a slightly lower degree of correlation between the pre-occlusion and hyperaemic values compared with a better correlation at 44 °C. Both correlations were, however, significant ($p < 0.001$). Correlations between 40 °C and 44 °C measurements were not statistically significant ($p > 0.05$).

DISCUSSION

The physical properties and characteristics of palladinised platinum surface hydrogen electrodes have been described in detail elsewhere (Harrison and Kessler, 1989a), and the clearance technique for measurement of capillary blood flow has been described.

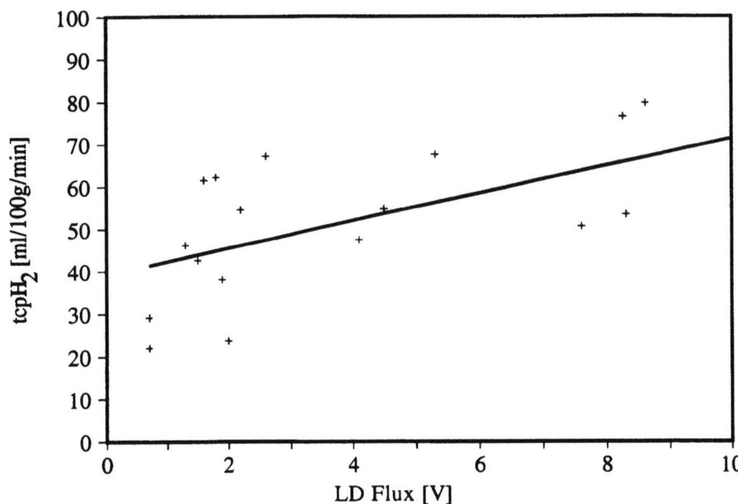

Figure 2. Comparison of blood flow values measured with the tcpH$_2$ technique and laser Doppler flux values at 40 °C.

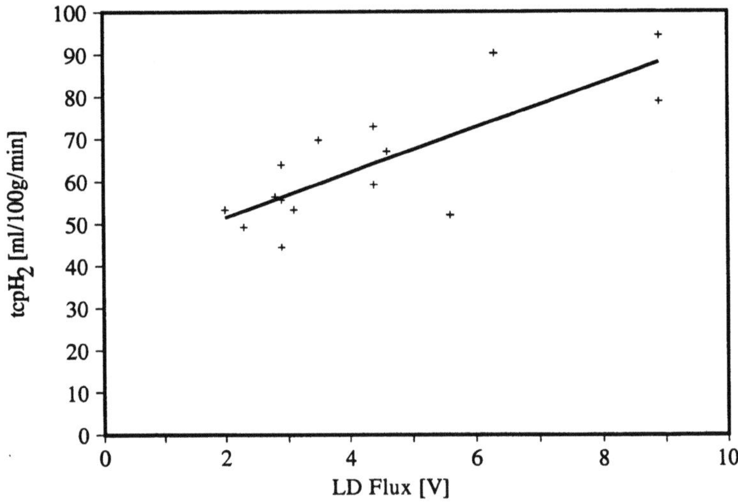

Figure 3. Comparison of blood flow values measured with the tcpH$_2$ technique and laser Doppler flux values at 44 °C.

184

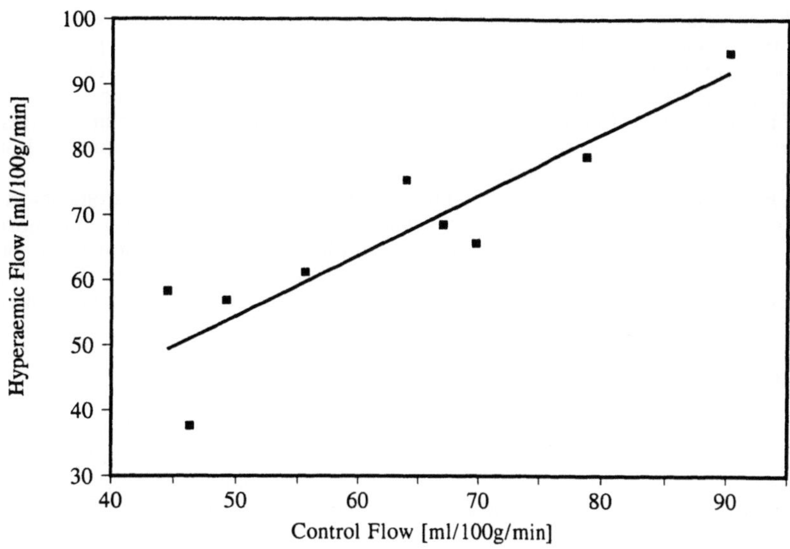

Figure 4. Comparison of post-reactive hyperaemia tcpH$_2$ flow values with pre-occlusion values at 44 ^{O}C.

Use of hydrogen clearance for blood flow measurement in humans has previously been limited to needle electrode techniques (Clark and Bargeron, 1959; Gotoh et al, 1966).

On the other hand, the tcpO$_2$ technique for assessment of peripheral oxygen supply has found a wide range of applications (see, for example Huch et al, 1972; Hauser, 1987). It is clear, however, that tcpO$_2$ values are highly variable and, in particular, no single value can be determined as representing a "critical value" for viability of healing (Hauser, 1987) in critical limb ischaemeia.

The tcpH$_2$ technique offers the potential advantage that H$_2$ is not consumed by the tissue cells but has a similar diffusivity in tissue to O$_2$. It thus fulfils the requirements for a freely diffusible indicator clearance technique.

The experiments have demonstrated that it is possible to record tcpH$_2$ clearance curves, reproducibly, at skin temperatures of both 40 ^{O}C and 44 ^{O}C. Considering the completely different nature of the recorded parameters and the different locations, the tcpH$_2$ results correlated surprisingly well with the laser Doppler measurements at both temperatures.

The clearance was halted during cuff occlusion of the brachial artery. It is thus reasonable to conclude that tcpH$_2$ clearances are a true representation of dermal blood flow. Interestingly, in Figure 1, tcpH$_2$ actually increased slightly during the period of occlusion, most noticeably at 40 ^{O}C. This phenomenon may be indicative of perfusion heterogeneities in the areas surrounding the electrode (Harrison et al., 1993): a neighbouring region of high perfusion may present a positive diffusion gradient for H^2 towards the catchment volume of the tcpH$_2$ electrode. The range of blood flow values reported in this series of experiments for hyperaemic skin is consistent with that reported in the literature by workers using other methods (see Altman and Dittmer, 1974).

The purpose of evaluating the two temperatures was to investigate whether it was possible to record tcpH$_2$ clearances under sub-maximal hyperaemic conditions and how reproducible such measurements would be compared with those at 44 ^{O}C. These results demonstrate that tcpH$_2$ clearances can be measured at both 40 ^{O}C and 44 ^{O}C and that both methods provide reproducible results. However, the poorer correlation against post-occlusion reactive hyperaemic values confirm that 40 ^{O}C does not induce as complete a local hyperaemia as 44 ^{O}C. Both methods provided reproducible values of blood flow but they only correlated well with each other under conditions of reactive hyperaemia, not under normal conditions. The results indicate that both techniques are reproducible, but are simply measurements of different blood flow values - one being still under some control by local and sympathetic mechanisms, the other being maximally hyperaemic.

The advantage offered by the tcpH$_2$ technique is thus the possibility to measure skin blood flow non-invasively, repeatedly and reproducibly under conditions of both supra-maximal and maximal local vasodilation.

The ultimate challenge of extending the investigations to measurements in critically ischaemic skin is showing considerable promise (McCollum et al., 1993), but more work is required to validate the technique for routine application in amputation level assessment. However, it is clear that there is a wide range of areas in both basic physiological and clinical measurement to which this new technique of skin blood flow measurement can be most usefully applied.

ACKNOWLEDGEMENTS

The authors are grateful to Farouk Vawda for his contribution towards early studies which led to the present work and the continuing support of Dr RA Lerski, Director, Department of Medical Physics. David Newton is supported by the Clinical and Biomedical Research Committee of the Scottish Home and Health Department (K/MRS/50/C1948).

REFERENCES

Altman, P.L. and Dittmer, D.S., 1974 "Biology Data Book," 2nd Ed., Vol. 3, FEBS, Bethesda, MD.

Clark, L.C. and Bargeron, L.M., 1959, Detection and direct recording of left-to-right shunts with the hydrogen electrode catheter, *Surgery* 46:797.

Gotoh, F., Meyer, J.S. and Tomita, M., 1966, Hydrogen method for determining cerebral blood flow in man, *Arch. Neurol.* 15:549.

Harrison, D.K., Günther, H., Vogel, H., Ellermann, R., Brunner, M., Höper, J. and Kessler, M., 1985, Oxygen supply and microcirculation of the beating dog heart after haemodilution with Fluosol DA 20, *Adv. Exp. Med. Biol.* 191:445.

Harrison, D.K. and Kessler, M., 1989a, A multiwire hydrogen electrode for in vivo use, *Phys. Med. Biol.*, 34:1397.

Harrison, D.K. and Kessler, M., 1989b, Local hydrogen clearance as a method for the measurement of capillary blood flow, *Phys. Med. Biol.* 34:1413.

Harrison, D.K., Kessler, M., Birkenhake, S. and Knauf, S.K., 1990, Regulation of capillary blood flow and oxygen supply in contracting skeletal muscle in dogs and rabbits, *J. Physiol. (Lond.)* 422:227.

Harrison D.K., Abbot, N.C., Swanson Beck, J. and McCollum, P.T., 1993, A preliminary assessment of laser Doppler imaging for measurement of skin perfusion using the tuberculin reaction in human skin as a model, *Phys. Meas.* 14:241.

Hauser, C.J., 1987, Tissue salvage by mapping of skin surface transcutaneous oxygen tension index, *Arch. Surg.* 122:1128.

Huch, R., Lübbers, D.W. and Huch, A., 1972, Quantitative continuous measurement of partial oxygen pressure on the skin of adults and newborn babies, *Pflüg. Arch. ges. Physiol.* 337:185.

Khan, F., Spence, V.A., Wilso, S.B and Abbott, N.C., 1991, Quantification of sympathetic vascular responses in skin by laser Doppler flowmetry, *Int. J. Microcirc: Clin. Exp.* 10:145.

Kozniewska, E., Weller, L., Höper, J., Harrison, D.K. and Kessler, M., 1987, Cerebrocortical microcirculation in different stages of hypoxic hypoxia, *J. Cerebr. Blood Flow Metab.* 7:464.

McCollum, P.T., Spence, V.A. and Walker, W.F., 1988, Amputation for peripheral vascular disease: the case for level selection, *Br. J. Surg.* 75:1193.

McCollum, P.T., Harrison, D.K., Abi Raad, R., Newton, D. and Holdsworth, R.J., 1993, H$_2$ clearance as a non-invasive technique for measuring skin blood flow in critical ischaemia, *Br. J. Surg.* in press.

Spence, V.A., McCollum, P.T., McGregor I. W., Sherwin, S.J. and Walker, W.F., 1985a, The effect of the transcutaneous electrode on the variability of dermal oxygen tension changes, *Clin. Phys. Physiol. Meas.* 6:139.

Spence, V.A., McCollum, P.T. and Walker, W.F., 1985b, Comparative studies of cutaneous haemodynamics in regions of normal and reduced perfusion, *in:* "Practical Aspects of Skin Blood Flow Measurement," V.A. Spence and C.D. Sheldon, eds., Biological Engineering Society, London.

Swain, I.D. and Grant, L.J., 1989, Methods for measuring skin blood flow, *Phys. Med. Biol.* 34:151.

MEASUREMENT OF CARDIO-RESPIRATORY FUNCTION USING SINGLE FREQUENCY INSPIRATORY GAS CONCENTRATION FORCING SIGNALS

E.M. Williams and C.E.W. Hahn

Nuffield Department of Anaesthetics
University of Oxford
Radcliffe Infirmary
Oxford, OX2 6HE, UK

INTRODUCTION

The possibility of using inspiratory, oscillating inert gas concentration forcing signals to measure the average lung ventilation-perfusion ratio was proposed by Zwart and colleagues (1976, 1978). These authors used halothane as the forcing gas, at sub-anaesthetic concentrations around 0.02% v/v, but this introduced severe measurement problems at these very low concentrations. Recently Hahn *et al* (1993) have described a similar but improved respiratory model which uses nitrous oxide (N_2O) as the forcing gas. The advantages of nitrous oxide over halothane are that it is free from biological toxicity; it is easy to measure N_2O with conventional gas analysis apparatus (especially at the concentrations used by Hahn *et al* (1993), around 2-10% v/v); and because nitrous oxide can be measured continuously in blood, the mathematical model can be extended to include shunt blood flow. In this technique, a three-compartment continuous ventilation mathematical model describes the imposition of a forced oscillating inspired inert gas concentration, PI(t), which oscillates sinusoidally over a range of frequencies (0.6 to 6 min^{-1}), and the subsequent measurement of the attenuation of the end-expired inert gas concentration, PÉ(t), with changing forcing frequency (Figure 1). The plot of PÉ/PI against forcing frequency (the Bode plot) is then used to derive alveolar volume, VA, and pulmonary blood flow, Q̇P. Dead space, VDS, is derived from the sinusoidal mixed expired indicator gas signal. We have further modified this technique by using a single forcing monosinusoid (forcing frequency 0.5 min^{-1}) of two inert gases to measure cardio-respiratory lung function. This binary inert gas oscillation technique uses Argon, which is relatively insoluble in blood, to derive alveolar volume and dead space; and nitrous oxide, which is soluble in blood, to measure pulmonary blood flow.

THEORY

The classical 3-compartment continuous-ventilation model described by Hahn *et al* (1993) is represented as a single ventilated (V̇A), and perfused, (Q̇P), "Ideal" alveolar compartment of volume, VA; a single ventilated but non-perfused dead-space compartment, VDS; and an

Oxygen Transport to Tissue XVI
Edited by M.C. Hogan *et al.*, Plenum Press, New York, 1994

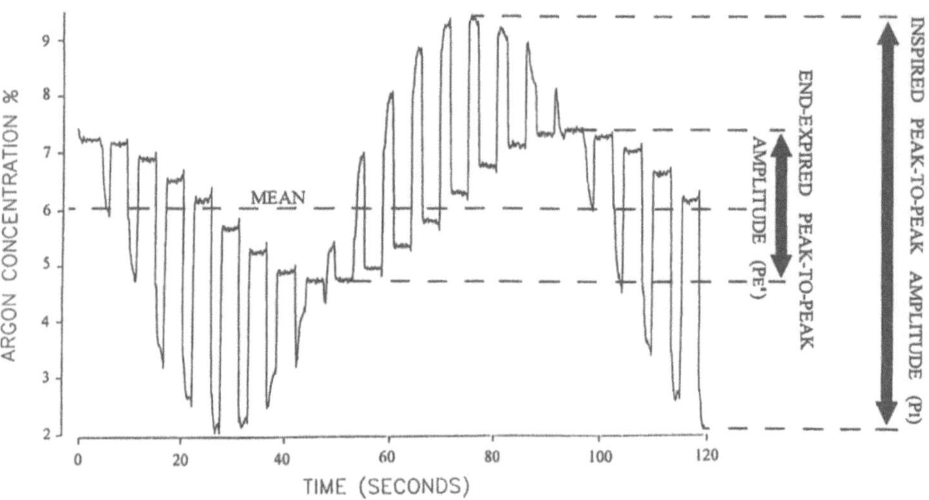

Figure 1a. A recording of an argon signal from a ventilated subject, showing inspired and expired argon concentration (vol %), chopped at the ventilator respiratory rate of 12 breaths min^{-1}. The outer sinusoid envelope (peak-to-peak amplitude PI) traces the breath-by-breath argon forcing sinusoid which is inspired by the subject. The inner sinusoid envelope (peak-to-peak amplitude PÉ) traces the subject's breath-by-breath end-expired argon response to the inspired forcing function. The inspiratory argon forcing function concentration is $6 \pm 4\%$, and the period of the sine wave is 1.5 minutes. The nitrous oxide forcing signal (not shown) looks similar except that it is generated in anti-phase to the argon signal (From Williams *et al.* 1994).

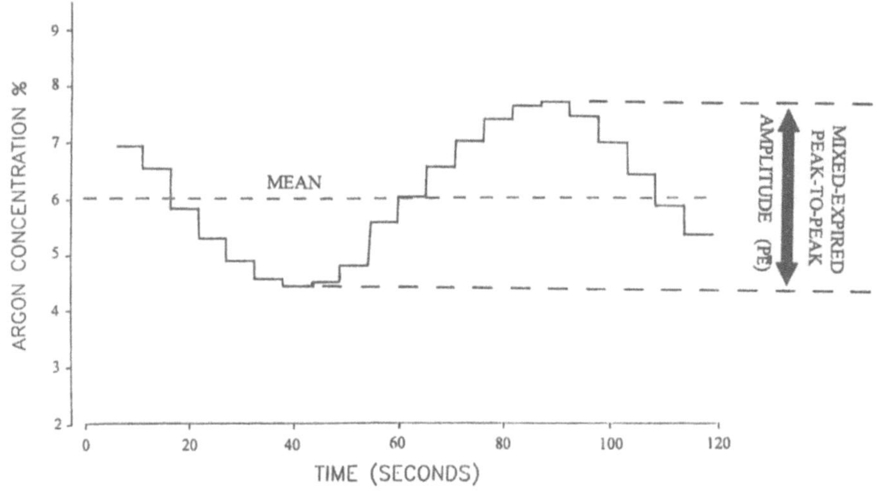

Figure 1b. Illustration of the argon mixed-expired peak-to-peak amplitude, PĒ. This signal is calculated by the computer for each breath by integrating the flow signal with argon end-expired concentration. The mixed-expired signal lies between PI and PÉ, and is used in calculating dead-space (From Williams *et al.* 1994).

188

unventilated but perfused "true shunt" compartment.   When sinusoidally varying concentrations of gases, such as argon and N_2O, are used as the model's input (inspiratory) function then the alveolar and mixed-expired, gas output responses are also sine wave functions. In relation to the amplitude of the input sine wave, the amplitude of the alveolar or end-expired sine wave decrease with increasing forcing frequency in a predictable way.

In the absence of venous recirculation of the inert gas sine waves in mixed-venous blood, and assuming that the mean inspired argon and N_2O partial pressures are in equilibrium throughout the respiratory and cardiovascular systems, then the amplitudes of the time-varying partial pressure of the trace gases can be measured as deviations from their mean values. The absolute values (or modulus) of their amplitudes are written as |P|. The changes in the ratios |PĒ|/|PI| and |PÉ|/|PI| (where |PÉ| and |PĒ| are the amplitudes of the end-expired and mixed-expired sinusoids respectively, and |PI| is the amplitude of the inspired forcing sinusoid), with changes in the frequency of the applied sinusoids, provides all the information required for deriving the cardio-respiratory data in the continuous flow sine wave model.

When a sinusoidal inert gas forcing function is applied to the continuous-ventilation model, the relationship between |PÉ| and |PI|, and forcing frequency, ω, for the inert gas, with a blood-gas solubility, λ, in the absence of recirculation, is given by,

$$\frac{|P\acute{E}|}{|PI|} = \frac{1}{\left(\left(1 + \frac{\lambda \dot{Q}P}{\dot{V}A} \right)^2 + \omega^2 . \tau_L^2 \right)^{\frac{1}{2}}} \tag{1}$$

where $\omega = 2\pi F$ (F = sine wave forcing frequency, min^{-1}) and τ_L is the lung ventilatory time constant, $\tau_L = VA/ \dot{V}A$.

By using a range of forcing periods from 10 seconds to 17 minutes (frequency 6 to 0.06 min^{-1}), a range of |PÉ|/|PI| ratios can be plotted against forcing period, on log-log axes, to form the Bode plot (Figure 2). At different forcing periods, the component parts of equation (1) provide different physiological information as illustrated in Figures 2 and 3. When using a poorly soluble forcing gas such as argon the solubility term, λ, is so low that the term $\lambda\dot{Q}P/\dot{V}A$ approaches zero and equation (1) can be simplified to

$$\frac{|P\acute{E}|}{|PI|} = \frac{1}{\left(1 + \omega^2 . \tau_L^2 \right)^{\frac{1}{2}}} \tag{2}$$

Equation (2) is plotted in Figure 2a, with $\dot{V}A = 2.4$ 1 min^{-1} and VA varied from 0.25 to 4 1. This example approximates to argon (blood/gas solubility coefficient, $\lambda = 0.03$) and Figure 3a shows that the ratio |PÉ|/|PI| decreases (as ω is increased) with increasing frequency (decreasing period) of the forcing sine wave.

Conversely, for a soluble gas like N_2O ($\lambda = 0.47$) both the perfusion and ventilation terms in equation (1) are significant. But at longer forcing periods (>5 minutes) the term $\omega.\tau_L$ becomes negligible so equation (1) approaches

$$\frac{|P\bar{E}|}{|PI|} = \frac{1}{1 + \dfrac{\lambda \dot{Q}P}{\dot{V}A}} \tag{3}$$

which is independent of alveolar volume, $\dot{V}A$, and ω. This is illustrated by the plateau region in Figures 2b and 3, which shows that the ratio $|P\acute{E}|/|PI|$ decreases as $\dot{Q}P$ increases, at constant $\dot{V}A$. However, at sufficiently high ω, the effect of this perfusion/ventilation term diminishes and the volume/ventilation term, $\omega\tau_L$, begins to predominate. Under these conditions equation (1) approaches equation (2), and so even for a soluble tracer gas like nitrous oxide, the N_2O $|P\acute{E}|/|PI|$ ratio becomes, like argon, independent of $\dot{Q}P$. This is illustrated in Figure 3a, which shows equation (1) plotted for both argon and N_2O, with $\dot{Q}P = 2$ l min^{-1}, $\dot{V}A = 2.4$ l min^{-1} and $VA = 0.8$ l (typical physiological values for the ventilated dog) in both instances. For inspired forcing frequencies faster than 1.0 min^{-1} (periods shorter than 1.0 min), the $|P\acute{E}|/|PI|$ ratios for argon and nitrous oxide merge.

Estimation of Airways Dead-Space, VDS

Airways dead-space (VDS) can be derived from the argon signal, at a fixed forcing frequency, using the following equation (Williams *et al* 1994).

$$\frac{\dot{V}DS}{\dot{V}T} = \frac{|P\bar{E}| \cos(\Phi_{\bar{E}}) - |P\acute{E}| \cos(\Phi_{\acute{E}})}{|PI| - |P\acute{E}| \cos(\Phi_{\acute{E}})} = \frac{VDS}{VT} \tag{4}$$

where $|P\bar{E}|$, $|P\acute{E}|$ and $|PI|$ are the argon sine wave amplitudes. Equation (4) takes into account the phase delay between the inspired forcing gas signal and the resulting end-expired ($\Phi_{\acute{E}}$) and mixed-expired ($\Phi_{\bar{E}}$) gas signals. During tidal ventilation, the value of VDS is derived from equation (4) and is then inserted into equation (5) to give alveolar ventilation ($\dot{V}A$),

$$\dot{V}A = BF \, (VT - VDS) \tag{5}$$

where BF is breathing frequency (breaths min^{-1}), and VT, is the tidal volume in litres.

Estimation of Alveolar Volume, VA

Alveolar volume can be derived from the end-expired argon sine wave, once $\dot{V}A$ has been determined from the mixed-expired argon sine wave data, as shown in equations (4) and (5). The alveolar volume can be derived at any given argon forcing period by inverting equation (1), to give

$$VA = \frac{\dot{V}A}{\omega} \left(\frac{1}{\left(\dfrac{|P\bar{E}|}{|PI|} \right)^2} - Z^2 \right)^{\frac{1}{2}} \tag{6}$$

where Z represents the, perfusion/ventilation, term in equation (1),

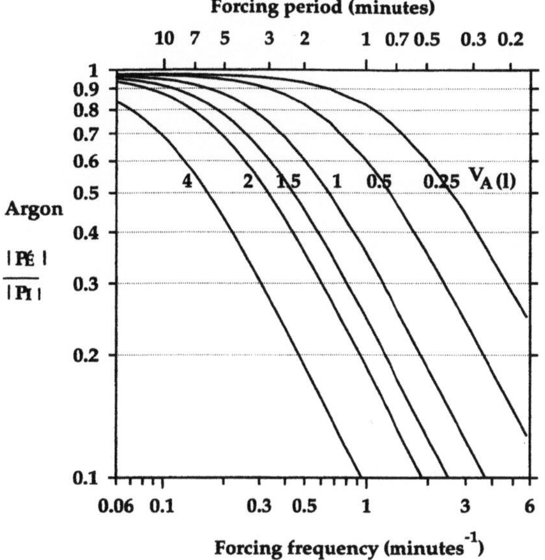

Figure 2a. A theoretical Bode plot of $|P\acute{E}|/|PI|$ against forcing frequency for argon derived using equation (6). In this case alveolar ventilation, $\dot{V}A$, = 2.4 L min^{-1} and VA is varied from 0.25 to 4.0 L.

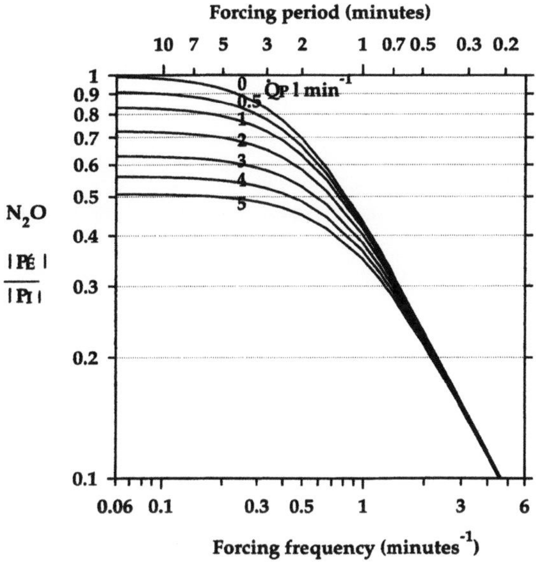

Figure 2b. A theoretical Bode plot for nitrous oxide derived using equation (8), showing $|P\acute{E}|/|PI|$ varying with forcing frequency at a fixed alveolar volume (VA = 0.8 L) and alveolar ventilation ($\dot{V}A$, = 2.4 L min^{-1}) and showing the effect of varying $\dot{Q}P$ between 0.5 and 5 L min^{-1}.

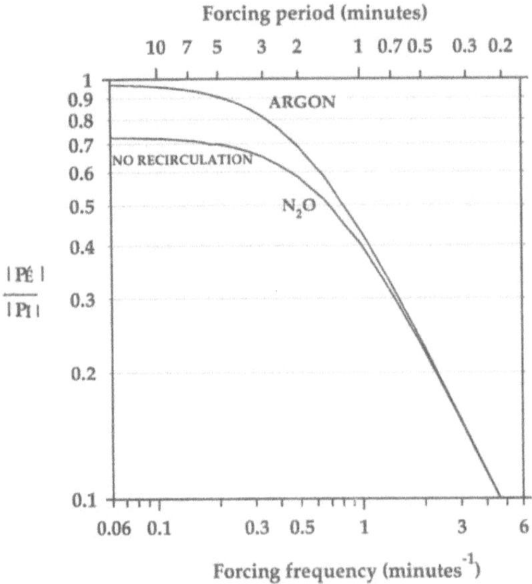

Figure 3a. A theoretical Bode plot of Pé/Pi against forcing period for forcing sinusoids of argon and N_2O. For forcing periods shorter than 0.5 minutes, both argon and N_2O behave like very low solubility inert gases. For forcing periods longer than 0.5 minutes the Pé/Pi relationship for N_2O is governed by pulmonary blood flow, $\dot{Q}_P$, and diverges from the argon Bode plot (From Williams *et al* 1994).

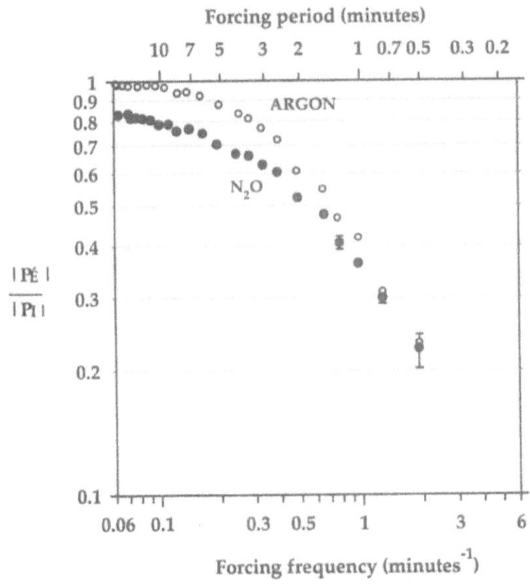

Figure 3b. Experimental values for Pé/Pi for argon and N_2O sinusoids in the dog. The experimental results verify the theoretical model Bode plots shown in Figure 3a above, and also illustrate that N_2O recirculation in the blood begins to take effect for forcing N_2O periods longer than 5 minutes. Calculations for V_A and $\dot{Q}_P$ are made using forcing periods of 2 minutes, to ensure clear separation of the argon and N_2O Bode plots (From Williams *et al* 1994).

$$Z^2 = \left(1 + \frac{\lambda \dot{Q}_P}{\dot{V}_A} \right)^2 \tag{7}$$

and λ (argon) = 0.03. In the absence of regional ventilation/perfusion mismatch, Z^2 in dogs approximates to 1.05. Also, because λ is so low, Z is insensitive to physiological changes in $\dot{Q}_P$ and $\dot{V}_A$.

Estimation of Pulmonary Blood Flow, $\dot{Q}_P$

Pulmonary blood flow can be determined from the N_2O |PÉ|/|PI| data, if the forcing sine wave period is chosen so that the two extremes represented by equations (2) and (3) are avoided. Accordingly, equation (1) is inverted to provide a solution for $\dot{Q}_P$, so that $\dot{Q}_P$ is given by,

$$\dot{Q}_P = \frac{\dot{V}_A}{\lambda} \left(\left(\frac{1}{\left(\frac{|P\acute{E}|}{|PI|} \right)^2} - \omega^2 . \tau_L^2 \right)^{\frac{1}{2}} - 1 \right) \tag{8}$$

Equation (8) can be solved for any forcing frequency, ω, knowing the measured value for |PÉ|/|PI|; and the value for τ_L (derived itself from the measured argon, PI, PÉ and PÉ amplitudes). The solution for $\dot{Q}_P$ from this equation will only make sense if the nitrous oxide forcing frequency lies within the range 0.25 min^{-1} to 1.0 min^{-1} (period 1.0 to 4.0 min). Outside these limits, values for $\dot{Q}_P$ will be prone to error.

MATERIALS AND METHODS

Nine anaesthetized and mechanically ventilated dogs received in their inspired airstream sinusoidal mixtures of argon and nitrous oxide which were generated in anti-phase using a microprocessor-controlled gas mixer, at a forcing frequency of 0.5 min^{-1} or forcing period of 2 minutes (Figure 1). Gas concentration was measured using a respiratory mass spectrometer (VG Mediflex, Fisons, Hastings, UK). Functional residual capacity, FRC, was measured using a multi-breath nitrogen washout technique (Williams *et al* 1994). Alveolar volume was calculated by subtracting the airways dead-space measured using a single breath nitrogen technique. Thermodilution was used to provide a comparable measure of cardiac output ($\dot{Q}_T$). Using a thermodilution computer (Com-2, Baxter Healthcare Corp, St Ana, USA), each estimation of $\dot{Q}_T$ was made by injecting a 10 ml bolus of ice-cold saline (0.9% W/V) into the pulmonary artery via an implanted pulmonary artery catheter. Saline injections were made throughout the respiratory cycle. Pulmonary shunt fraction was measured using an SF_6 elimination technique.

RESULTS

The alveolar volume derived from the argon sine wave response was 0.864 ± 0.138 L (mean ± SD) and was not significantly different (by paired t-test) from alveolar volume measured by a multiple breath nitrogen washout technique, 0.795 ± 0.154 L (Figure 4). The nitrous oxide sine wave derived $\dot{Q}_P$ was 1.76 ± 0.9 L min^{-1} and not significantly different to $\dot{Q}_T$ derived by thermodilution, 1.73 ± 0.59 L min^{-1} (Figure 4). In all these animals, there was only a small pulmonary shunt fraction ($\dot{Q}_S$ = 1.4 ± 0.5 %), and so $\dot{Q}_P$ was effectively equal to $\dot{Q}_T$ in each instance.

CONCLUSIONS

In the healthy canine lung where little pulmonary shunting exists the non-invasive sine wave inspiratory forcing technique provides a good estimate of alveolar volume and cardiac output. The modification of our previous technique allows the non-invasive measurement of cardio-respiratory function using a single 2 minute forcing signal instead of the many forcing signals which are required to construct a full Bode plot. Using a 2 minute forcing period the separation between the argon and nitrous oxide PI/PÉ ratios are within the mathematical model boundaries

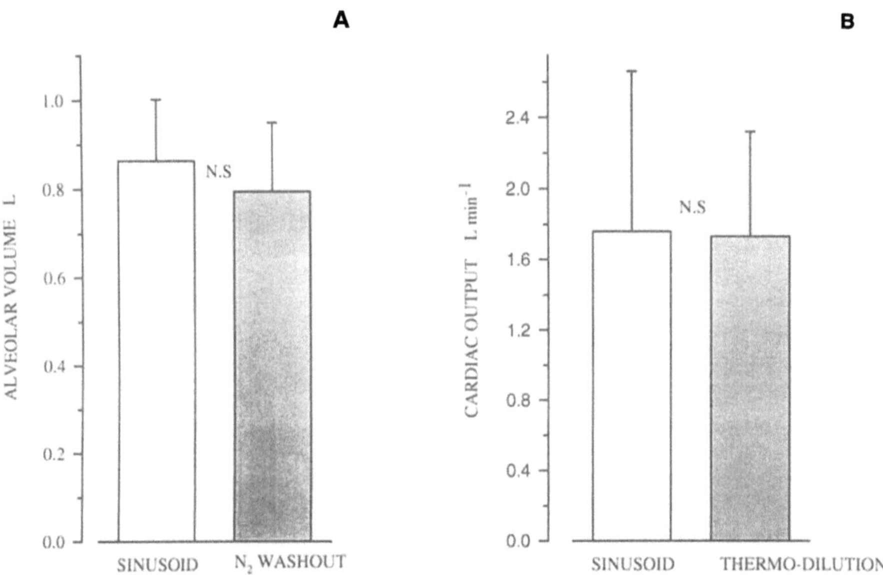

Figure 4. A comparison between A) alveolar volume and B) cardiac output measured by the sine wave and comparator techniques. N.S not significantly different using a Student's paired t-test.

as described above which allows the calculation of VA and Q̇T.   We conclude that this inspiratory forcing technique provides a non-invasive method by which cardio-respiratory function can be comprehensively monitored over short periods and it is hoped that this degree of time resolution will provide additional information on the mechanisms of lung ventilation and perfusion.

Acknowledgements

The authors gratefully acknowledge the support of the Wellcome Trust for this work.

REFERENCES

Hahn, C.E.W., A.M.S. Black, S.A. Barton and I. Scott (1993). Gas exchange in a three-compartment lung model analysed by forcing sinusoids of N_2O. J. Appl. Physiol. **75**, 1863-1876.

Williams, E.M., D.J. Gavaghan, P.A. Oakley, M.C. Sainsbury, L. Xiong, A.M.S. Black and C.E.W. Hahn (1994). Measurement of dead-space in a model lung using an oscillating argon concentration signal. Acta Anaesthesiol. Scand. **38**(2), in press

Williams, E.M., J.B. Aspel, S.M.L. Burrough, W.A. Ryder, M.C. Sainsbury, L. Sutton, L. Xiong, A.M.S. Black and C.E.W. Hahn. Assessment of cardio-respiratory function using oscillating inert gas forcing signals. J. Appl. Physiol. submitted.

Zwart, A., S.C. Seagrave and Van Dieren (1976). Ventilation-perfusion ratio obtained by non-invasive frequency response technique. J. Appl. Physiol. **41**, 419-424.

Zwart, A., J.M. Bogaard, J.R.C. Jansen and A. Verspille (1978). A non-invasive determination of lung perfusion compared with the direct Fick method. Pflügers Arch. **375**, 213-217.

OPTICAL OXYGEN SENSOR USING FLUORESCENCE LIFETIME MEASUREMENT

Shabbir Bambot[1], Raja Holavanahali[2], Joseph R. Lakowicz[3], Gary M. Carter[2] and Govind Rao[1]*

[1]Department of Chemical and Biochemical Engineering and Medical Biotechnology Center of the Maryland Biotechnology Institute, University of Maryland Baltimore County, Baltimore, MD 21228
[2]Department of Electrical Engineering, University of Maryland Baltimore County, Baltimore, MD 21228
[3]Center for Fluorescence Spectroscopy, Department of Biological Chemistry and Medical Biotechnology Center of the Maryland Biotechnology Institute, University of Maryland at Baltimore, Baltimore, MD 21201
*Corresponding author

INTRODUCTION

The industry standard for oxygen measurement, the modified Clark electrode has now been available for more than three decades (Clark, L. C., 1956), and is still being constantly modified and perfected. It is a membrane covered electrode that encloses a platinum cathode, a silver anode and a KCl or Ag/AgCl electrolyte. Oxygen diffusing from the surrounding medium through the membrane gets reduced at the surface of the cathode. The magnitude of the negative bias on the platinum electrode (-0.8 to -1 V) is maintained such that the electrode operates under conditions of membrane controlled diffusion. Since the diffusive flux is a function of the partial pressure of oxygen in the fluid, it is possible to calibrate the electrode current versus oxygen tension. Clark electrodes are calibrated by equilibrating a sample with nitrogen to read zero and followed with either air or oxygen to read 100%. The readings are usually expressed as a percentage of air saturation.

Despite many advances, the Clark electrode suffers from a variety of limitations. These include long term instability, drifts in calibration, flow dependence and susceptibility to electrical interferences when used in bioreactors. Solutions to these problems exist but are generally opposing. For instance, flow dependence may be reduced by employing a thicker membrane but this would occur at the cost of increased response time. As a result, most commercially available systems are design compromises that sacrifice a part of some desirable feature. Inspite of these limitations, the Clark electrode is widely used because it is the best available technique to date. For greater details the reader is directed to the excellent review by Lee and Tsao (1979).

Optical oxygen sensing provides a promising alternative to amperometric methods in solving the problems mentioned above. Using molecular probes whose optical properties change with changing oxygen concentrations, a number of reports in recent years have demonstrated the feasibility of using optical sensing of oxygen (Bacon and Demas, 1987; Carraway et al., 1991; Opitz and Lubbers, 1987; Wolfbeis and Caroline, 1984). It should be noted that an optical measurement technique where oxygen is not consumed will be free of all the drawbacks attributed to electrolyte, electrode and oxygen consumption in the Clark sensor. Additional advantages in using optical methods include small size, no requirement of a reference cell and the ability to do remote sensing using fiber optics.

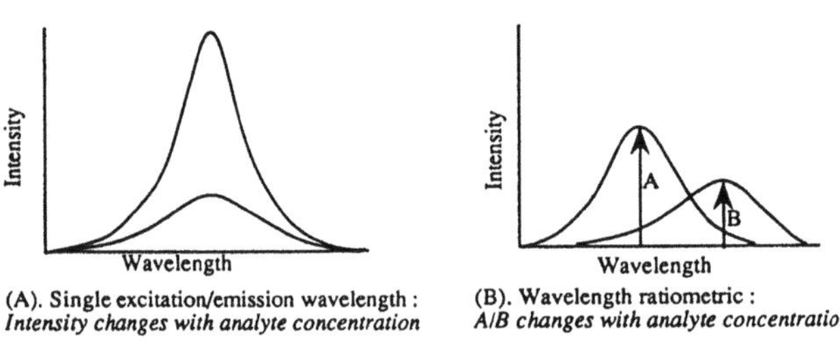

(A). Single excitation/emission wavelength :
Intensity changes with analyte concentration

(B). Wavelength ratiometric :
A/B changes with analyte concentration

Fluorescence intensity based sensing schemes

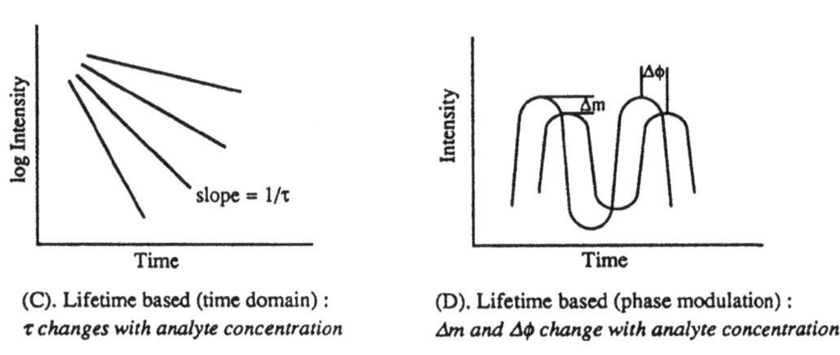

(C). Lifetime based (time domain) :
τ changes with analyte concentration

(D). Lifetime based (phase modulation) :
Δm and Δφ change with analyte concentration

Fluorescence lifetime based sensing schemes

Figure 1. Schemes for fluorescence sensing.

The various methods for optical sensing are illustrated in Figure 1. Intensity methods have not found widespread use due to signal instabilities and the need for frequent recalibration. The fluorescent intensity can vary due to excitation intensity, light losses in fibers, photobleaching or washout of probes, as well as changes in light scattering or absorption characteristics of the sample. Thus, whilst fluorescent intensity methods may yield near perfect results in an optically clean environment, their use in most sensing applications like blood or microbial cultures is severely limited. An alternative to intensity measurements is the wavelength ratiometric method in which (a) the ratio of the intensities at two different emission wavelengths for a single excitation wavelength or (b) a ratio of the intensities obtained at a single emission wavelength for two different excitation wavelengths is plotted as a function of analyte concentration. While wavelength ratiometric methods are promising, they have also not found widespread use, primarily due to the lack of suitable probes and known methods for creating them.

Because of advances in the technology for measuring lifetimes, particularly by the phase-modulation method, lifetime based sensing (Figure 1) offers new opportunities for chemical sensing (Lakowicz, 1983; Lippitsch, et al., 1988; Berndt and Lakowicz, 1992). This is because the lifetimes of fluorescence probes can be sensitive to a variety of factors or chemicals. Moreover lifetime measurements are insensitive to probe concentration, photobleaching, washout and excitation instabilities.

Fluorescence lifetime measurements can be accomplished using either of two methods : (1) time domain or (2) frequency domain fluorometry, often referred to as phase-modulation fluorometry. Of the two, the latter is preferable for sensing applications because of the possibility of constructing low cost instruments to measure even sub-nanosecond lifetimes. In a phase shift fluorometric sensor, the fluorescence lifetime is monitored by measuring the phase shift of the fluorescence radiation relative to the phase of the sinusoidally intensity-modulated excitation light (Lakowicz, 1983).

The application of fluorescent lifetime measurement to oxygen sensing has been particularly effective due to the availability of a variety of of fluorophores that comprise of metal-organic complexes of ruthenium, osmium, iridium and platinum that have strong metal to ligand energy transitions and long lived excited states (upto 5 μs). The long fluorescence lifetime of the probes translates into a low modulation frequency which permits the use of inexpensive electronics in collecting and measuring lifetime related information. Furthermore, the long fluorescence lifetime also means higher quantum yields and lower light intensity requirements allowing the use of low cost, off the shelf blue LED's. All this allows for an inexpensive, compact and rugged setup.

In this report we describe the use of lifetime based sensing in making a workable and autoclavable optical oxygen sensor that is particularly adapted to measuring oxygen levels in bioreactors. We used the oxygen sensitive molecular probe tris [4,7-diphenyl-1,10-phenanthroline] ruthenium (II) complex immobilized in a silicone rubber membrane. The versatility of the experimental setup as well as the methods used makes it readily applicable for measuring other analytes. As novel probes sensitive to other analytes are synthesized, the technique described here could lead to a new generation of sensors.

MATERIALS AND METHODS

Chemicals

The tris [4,7-diphenyl-1,10-phenanthroline] ruthenium (II) complex was synthesized at the Center for Fluorescence Spectroscopy, University of Maryland, Baltimore. All other reagents were purchased from Aldrich or Sigma. The probe embedded silicone membrane was prepared by soaking in a 2.5 mg/ml solution of the ruthenium complex in chloroform for 5 minutes followed by air drying for 5 minutes. The membrane itself was made by spreading a thin layer of GE Silicone II (General Electric Company) on a microscope slide and allowing it to cure overnight. The fluorophore embedded membrane was washed with ethanol to remove surface fluorophore molecules and after drying was laid out on a circular glass window. In the final optrode configuration the membrane was held towards the sample with the glass window facing the excitation beam.

Optics and Electronics

Figure 2 shows a schematic of the experimental setup used to test the optical oxygen sensor. The excitation light from a blue silicon carbide LED (CREE Research Inc., Durham, NC) was coupled into one of the arms of a bifurcated fiber optic bundle (Dolan-Jenner Industries Inc., Woburn, MA) via a 450 nm low wave pass filter (Andover Corporation , Salem, NH). Inspite of the rather low optical output power, the small viewing angle (16^0) of the LED enabled efficient coupling of the output into the fiber optic giving sufficient excitation light at the sensor. The emission was collected from the other arm of the bifurcated bundle. The detector used was a Hamamatsu R928 Photomultiplier tube (PMT) in a Oriel PMT housing (Oriel Corporation, Stratford, CT) with a 8 mm diameter entrance aperture. Stray excitation was prevented from entering by placing a 600nm long wave pass filter in front of the aperture. A focusing lens of 25 mm diameter helped in maximizing fluorescent light collection. The stem of the bundle was housed in a stainless steel tube which held the fluorophore embedded membrane in the excitation path.

The LED was sine wave modulated using a BK Precision Model 3020 sweep/function generator at 76 kHz. The circuit consisted of a 512 Ω resistance placed in series so as to limit the maximum current passing through the diode. The input voltage to the circuit was an AC voltage of 8.3 V (peak to peak) and DC voltage of 8.25 V. This ensured that the LED had a positive bias at all times. This operation of the LED resulted in an AC to DC light output of about 2.45 V AC (peak to peak) and DC voltage of 1.9 V. This is equivalent to 64% of light modulation. Note that light modulation percentage can be easily changed by changing the input voltages.

An EG&G 5206 lock-in analyzer (Princeton, NJ) was used to detect the magnitude and phase of the PMT output. The lock in amplifier affords the flexibility of measuring a pure signal using a narrow (1 Hz) bandwidth amplifier, thus eliminating any out of band noise. Note that once parameters like frequency have been optimized, a simple single frequency phase discriminator can be used to build a significantly cheaper electronic circuit. As shown in Figure 2 the reference electrical signal against which the phase lag of the fluorescence signal can be determined was acquired directly from the signal generator.

Fermentation

The fermentation experiments were conducted using a 2 liter spinner flask equipped with both oxygen sensors, Clark-type and Optical. The Clark electrode was purchased from Ingold (Wilmington, MA). The LB medium contained (in g/l) Tryptone,

10; yeast extract, 5; NaCl, 5. The innocula were prepared using 100 ml of the same medium and grown overnight in a tight flask at 37ºC. The stirring speed was maintained at 600 rpm and the gases were sparged at the rate of about 2 vvm (volume per volume per minute). The fermenter (including media and oxygen sensors) was autoclaved for 20 minutes at 120ºC. The fermentation was carried out at room temperature and 1 ml antifoam 2% (2ml/100 ml water) was added when required.

The data from the two oxygen sensors was obtained on a Macintosh computer through an analog interface (Strawberry Tree Inc. Sunnyvale, CA). The capture rate was one data point per minute over a period of approximately 20 hours.

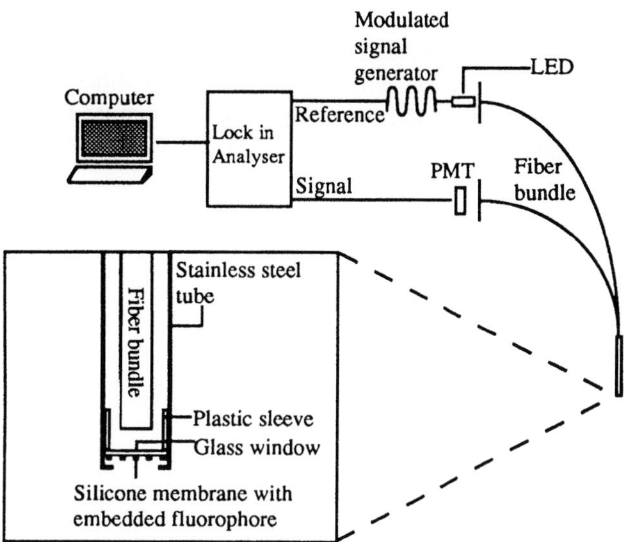

Figure 2. Experimental setup of Optical oxygen sensor. The intensity modulated excitation light enters one arm of a bifurcated fiber bundle. The other arm carries the intensity modulated phase shifted fluorescence to a photomultipler tube. The PMT signal phase is compared with the internal reference from the signal generator using a lock-in amplifier and the phase difference is displayed on a computer.

Sensor Calibration

The sensors were calibrated against known oxygen gas tensions in water. A predetermined mixture of oxygen and nitrogen gas as measured by precision gas "flo-sensors" (model 310, The McMillan Company, Copperas Cove, Texas) was sparged into water contained in a 2 liter spinner flask equipped with both Optical and Clark-type sensors. The flask was adequately vented to the atmosphere. The phase and voltage response of the Optical and Clark-type sensor was recorded and plotted against percent oxygen (Figure 2).

RESULTS AND DISCUSSION

Figure 3, shows the calibration curves for the Clark-type sensor and our Optical sensor in water. While the Clark-type sensor shows a linear calibration the optical sensor shows a non-linear response as predicted by the Stern Volmer type equation (Parker 1968):

$$\tau_0/\tau = 1 + KpO_2 \qquad [1]$$

where τ_0 and τ are the initial and final lifetimes, K is a constant and pO2 is the oxygen tension in the sample.

Equation [1] was used as the calibration for the optical sensor in converting phase angles to percent oxygen. Similarly a linear calibration was used to obtain percent oxygen from volts in the case of the Clark-type electrode. Note that the sensitivity of the optical sensor increases at low oxygen tensions making it superior to the Clark-type electrode in this regime. This is the operating regime for most bioreactor and waste water treatment facilities.

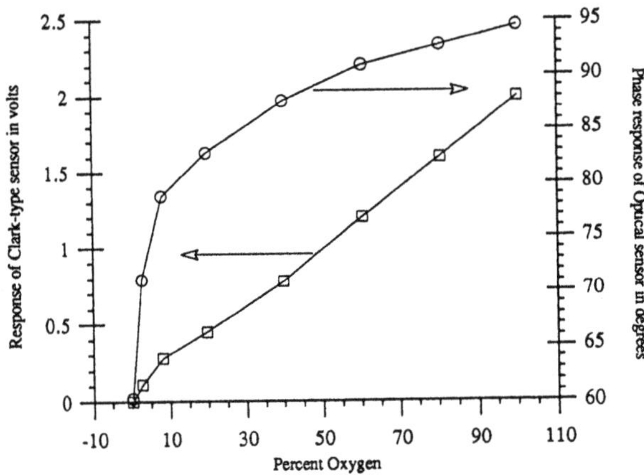

Figure 3. Calibration curves for Optical and Clark-type oxygen electrodes. The phase response of optical sensor and the voltage response of the Clark-type electrode is plotted against percent oxygen in the gas mixture (oxygen and nitrogen) sparged through water contained in a bioreactor. While the Clark-type electrode shows a linear calibration the Optical sensor shows a Stern-Volmer type relationship (see text).

The intensity response of our sensor closely tracked oxygen partial pressure as confirmed by phase readings and the response of a Clark-type electrode placed in the same environment. However the intensity output was found to be extremely sensitive to excitation light intensity, room light and physical perturbation of the system. Figure 4 shows the lifetime (phase) and intensity response of the optical sensor to waving a hand between the emission arm of the fiber optic bundle and the PMT. While the intensity response is severely affected, the phase response is steady and shows no change.

202

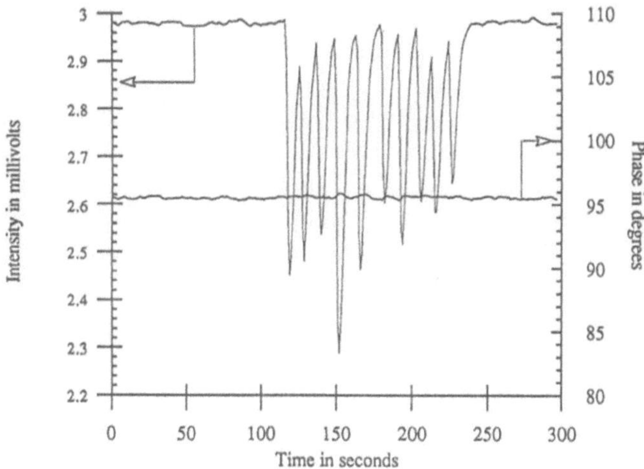

Figure 4. The fluorescent lifetime (phase) response as compared to the fluorescent intensity response upon waving a hand before the input of the photomultiplier tube. While the perturbations do not affect the phase readings the intensity readings show large variations.

Pharmaceutical and biochemical applications require sensors that maintain accuracy and reliability through a requisite number of sterilizations (typically 20). To determine the autoclavability of the optical sensor three different sensor membranes with varying probe concentrations were autoclaved at 120 ºC for 40 minutes. In each case, an initial "setting" of the silicone rubber membrane occurred after the first autoclaving (as evidenced by a change in texture as well as binding with the glass window on which the membrane is placed) that also caused a small change in dynamic range of the sensor (difference in response between 0% oxygen and 100% oxygen). However this was a one-time effect and subsequent autoclaving (upto 20 times) did not result in any measurable change in the dynamic range or in the absolute phase angles measured. This, combined with the fact that there is no long term probe loss from the membrane, allows the sensor to be used in applications which require repeated autoclaving. An additional advantage of the optical sensor is that since it is a non-consumption based system the calibration curve is absolute, requiring only a one time calibration after the first autoclaving.

Performance comparisons with a Clark-type sensor were carried out to determine the applicability of our sensor in monitoring dissolved oxygen (DO) levels in a bioreactor. Figure 5 shows the response profiles of the Optical sensor and Clark-type electrode to DO in an overnight *E. coli* batch fermentation. Both sensors track the variations in oxygen tension, resulting in a similar DO profile. It must be noted that the local fluctuations in the sensor response is not noise but actual fluctuations in DO. In another experiment, a comparison of the response times of the sensors showed that for low oxygen tensions the Optical sensor had a 40% faster response than the Clark-type sensor (data not shown). This is primarily due to the high oxygen permeability of the silicone membrane as well as the non-consumption of oxygen.

CONCLUSIONS

The application of fluorescence lifetime measurements to monitoring dissolved oxygen tensions in bioreactors was investigated in this report. Lifetime-based optical sensors, as mentioned earlier, have been shown to have particular advantages over intensity based optical sensors. In this report, we have demonstrated the advantages of lifetime-based optical methods over Clark-type amperometric sensors. The Optical sensor described is autoclavable, maintenance free, has a fast response, and is particularly useful in the low oxygen regime. Since the sensor does not consume oxygen

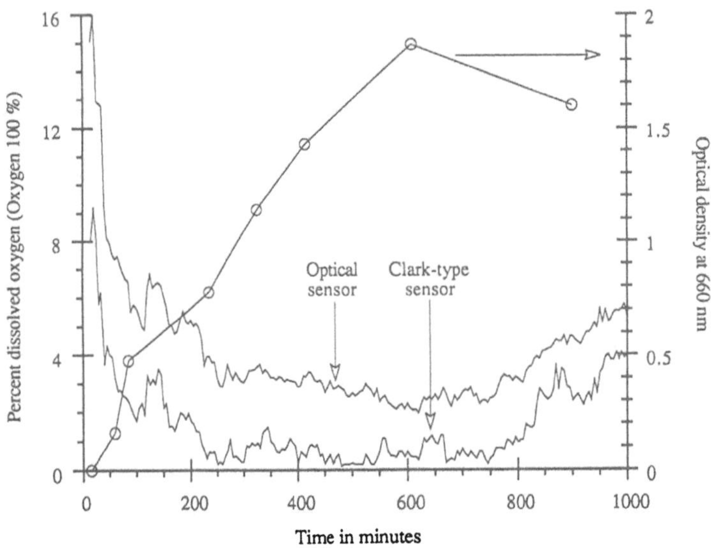

Figure 5 Dissolved oxygen profile during a typical *E. coli* fermentation. Using the calibration obtained from Figure 2, the responses of the two sensors have been converted into percent oxygen and plotted as shown. For clarity the Clark-type electrode response is shown offset by 3%. The optical sensor closely tracks the response of the Clark-type electrode throughout the fermentation.

it is independent of fluid flow rates around it and because it is optical, the sensor is free of electrical interferences. Additional features include, long term stability and a one-time calibration requirement.

The general methodology used above can also be extended to measure other analytes. The field of lifetime based sensing has advanced dramatically during the past year. New lifetime probes have become available. The generic technology described in this article can be directly ported over once effective immobilization methods are developed.

ACKNOWLEDGMENTS

Support from NSF grants BCS9209157 and BCS9157852 with matching funds from Artisan Industries, Inc., Waltham, MA, is acknowledged.

REFERENCES

Bacon, J.R. and Demas, J.N. 1987. Determination of oxygen concentrations by luminescence quenching of a polymer immobilised transition metal complex. *Anal. Chem.* 59: 2780-2785.

Carraway, E.R., Demas, J.N., DeGraff, B.A., and Bacon, J.R., 1991. Photophysics and photochemistry of oxygen sensors based on luminescent transition metal complexes. *Anal. Chem.* 63: 337-342.

Clark, Jr., L.C. 1956. Monitor and control of blood and tissue oxygen tension. *Trans. Am. Soc. Artif. Int.Organs.* 2:41.

Lakowicz, J.R. "Principles of Fluorescence Spectroscopy," Plenum Press, New York (1983).

Lee, Y. H. and Tsao, G. T. 1979. Dissolved Oxygen Electrodes. p. 35-86. *in:* "Advances in Biochemical Engineering". Vol 13. T. K. Ghose, A. Fiechter and N. Blakebrough, eds., Springer Verlag, Berlin.

Lippitsch, M. E. Pusterhofer, J., Leiner, M.J.P., and Wolfbies, O. S. 1988. Fiber optic oxygen sensor with the fluorescence decay time as the information carrier. *Anal. Chim. Acta.* 205: 1-6.

Opitz, N., and Lubbers, D.W., 1987.Theory and development of fluorescence based optochemical oxygen sensors: oxygen optodes. *Int. Anasthesiol. Clin.* 25: 177-197.

Parker, C.A. "Photoluminescence of Solutions," Elsevier, Amsterdam (1968).

Wolfbeis, O.S., and Caroline, F. M. 1984. Long wavelength fluorescence indicators for the determination of oxygen partial pressures. *Anal. Chim. Acta.* 160: 301-304.

TOMOGRAPHIC TIME-OF-FLIGHT OPTICAL IMAGING DEVICE

David A. Benaron, David C. Ho, Stanley Spilman, John P. Van Houten, David K. Stevenson

Medical Imaging and Spectroscopy Section,
Division of Neonatal and Developmental Medicine, Stanford School of Medicine, and the
  Stanford Picosecond Free-Electron Laser (FEL) Center, Hansen Experimental Physics Lab
750 Welch Rd #315, Stanford University, Palo Alto, CA 94305
Phone 415.723-.711, Fax 415.365.0656, E-mail benaron@forsythe.stanford.edu

ABSTRACT

Time-resolved optical imaging has been used to image phantoms, animals, and humans, and offers the potential for the production of functional images of human tissues, such as the oxygenation of brain during stroke. We had previously reported a transmission scanner, and now give an early report on conversion to a rotational tomographic scanner with a non-parallel ray geometry similar to early CAT scanners. Initial scans show that 1) spatial imaging in turbid media using time-of-flight measurements, non-recursive algorithms, and standard tomographic geometry is possible, 2) separation of absorbance and scattering as an image is attainable, a key step in performing spatially-resolved chemometric analysis, 3) imaging of multiple objects buried within scattering material is feasible, demonstrating that equations derived for homogeneous media can be applied in at least some cases to inhomogeneous media such as tissue-like phantoms, and 4) imaging of brain pathology produces recognizable images with sufficient resolution for diagnostic decisions. We conclude that optical tomography is feasible for clinical use and that conversion of the present mechanically scanning device to a clinical scanner should be possible with retention of the current processing algorithms. Such a clinical scanner should ultimately be able to generate images in a few minutes with centimeter resolution at the center of living human brain.

INTRODUCTION

Optical imaging and spectroscopy use light emitted into opaque media such as human tissue to determine interior structure and chemical content, respectively, and have broad application to the field of medicine.[1] Few developments have improved medical diagnostics as much as the ability to noninvasively peer inside the body, and it is expected that newly developing optical imaging techniques will continue this trend. Optical imaging and spectroscopy, key components of optical tomography, center around the simple idea that light passes through the body in small amounts, emerging bearing clues about tissues through which it passed. Rapid progress over the past decade, made possible by the collective output of multiple laboratories and advancements in the opto-electronics field, have brought optical imaging to the brink of clinical usefulness.[1]

Why does medicine need another form of imaging? Many of the recent advances in success of medical care for the intensively ill have come from improvements in our ability to noninvasively monitor ongoing care and to detect threatening changes in body function.[2] However, in order to prevent death and minimize long-term injury, early diagnosis is essential, before such injuries become irreversible. Most imaging techniques remain limited in at least one of four fundamental ways:[2] 1) they are invasive, involving a degree of risk that increases with exposure, 2) they are not portable, while critically ill persons often cannot be moved as such movement may adversely affect their care, 3) they are noncontinuous, and thus useless as monitors for critically ill people who by definition tend to have illnesses that are actively changing processes, and 4) they do not measure tissue function, while tissue injury often occurs due to such functional problems as a lack of oxygen to a portion of the brain. Spectroscopic optical techniques such as niroscopy[3] and regional cerebral spectroscopy[4] have been used to assess hemoglobin oxygen saturation ($HbO_2\%$) in the cerebral circulation, cytochrome oxygenation ($CytO_2$)

in brain, cerebral blood flow (CBF) and blood volume (CBV) in normal and ill infants,[5-8] and to elucidate mechanisms of brain injury from toxic substances such as bilirubin or medications.[9-10] Time- or frequency-re-solved NIRS approaches have allowed our group[11-13] and others[14-22] to generate images from such data. Thus, such measurements hold great potential for noninvasive, regional monitoring of cerebral hypoxia. However, widespread clinical use, while widely anticipated, has been delayed as the collection of the optical data, and conversion of these signals into images, has been problematic.

In our laboratory, we had previously constructed a time of flight optical system operating in the red (780 nm) and near-infrared (850 nm and 905nm), and succeeded using transmission geometries to image both phantoms[23,24] (demonstrating the feasibility of two- and three-dimensional reconstruction) and animals[11,23] (demonstrating the feasibility of imaging internal organs). Such animal images have been similarly produced by another laboratory using a frequency-based approach.[25] In order to assess the feasibility of optical tomography as a clinical tool, we constructed a rotational tomographic device, similar to that used in conventional computer assisted tomography (CAT, or CT scan), and set out to test if such a geometry could be useful in clinical using optical imaging systems.

METHODS

Existing Scanner

We began with an existing translational time-of-flight transmission scanner, reported previously.[11] Briefly, this device consisted of a portable time-of-flight and absorbance (TOFA) system that recorded a tissue transit time for each photon detected. The system was configured around a modified commercial optical time-domain reflectometer rack system (Opto-Electronics TDR-20, Ontario, Canada), and consisted of up to three diode lasers coupled by optical fibers to the base of an automated microscope stage, upon which objects to be imaged were placed. The system had submilimeter average path discrimination, and returned a 256-cell array representing a binned integer photon count over a series of selectable width integration boxcars for time of arrival at the detector fiber. Diode laser beam diameter was 50 µm, temporal half-maximum width was 50-75 ps, peak power was over 1 W with an average power of 100 µW, repetition rate was 1MHz, and triggering was reproducible to within 2 ps. All lasers fire simultaneously, while differences in emitter fiber length temporally separated the signals at the target subject. A 100 µm core diameter detector fiber, located such that linear transmissional photon collection was maximized, collimated the detected light by rejecting light coming more than 12.5 degrees off-axis (half-angle a = 0.22 radians; numerical aperture $N.A.$ = sin a = 0.22), transmitting the collimated detected light

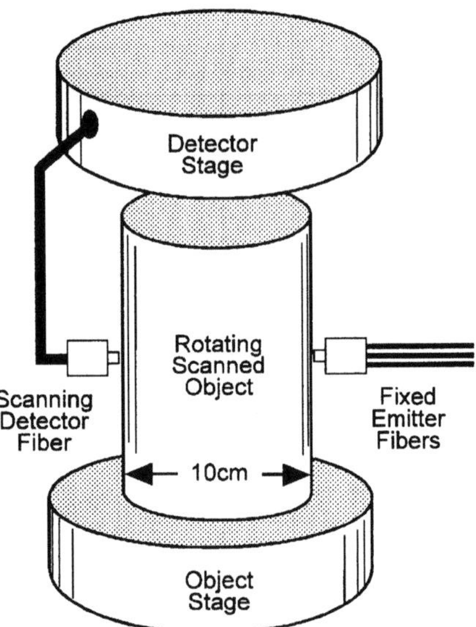

Figure 1. Schematic of the rotational stage. The object to be scanned is placed upon the lower (object) stage, in close proximity to a fixed emitter fiber and a scanning detector fiber. The scanning detector fiber is attached to an upper (detector) rotating stage. Via computer control, nearly any combination of emitter and detector angles can be measured. This set-up has the advantage that it is easily translated into a clinical scanner using fixed emitter and detector fibers.

to a solid-state 75 μm wide over-etched APD configured as a photon counter with 25-50% detection efficiency over the range of wavelengths used.

In order to collect TOFA data, the laser was triggered 256 times as the detector was gated and sampled, once each pulse, over a series of sequential, partially overlapping detection windows, to produce a single 256-bin TOFA curve with 1-bit resolution per channel. This entire process was repeated 128 times to detect up to 32K photons and generate a single 256-bin TOFA curve with 7-bit resolution per channel every 30 ms (3 and 1/2 decades of signal/bin/s). Results were accumulated in a multichannel recorder, and transferred to a PC-based microprocessor system via a GPIB bus (PCL-848A IEEE-488 interface card, PC Labs, Irvine, CA). Minimum transit time increment between adjacent sampling windows can be as small as 2 ps. Mean path length resolution for an unscattered pulse in water is better than 500 μm.[26] Multiple TOFA curves are usually accumulated and binned in each pixel in order to generate data that can be followed for up to 4 to 7 decades of decay (typically for 5 decades) after the laser pulse.

In the translational system, images were generated during a two-dimensional transmission scan by accumulating one binned TOFA curve at each (X,Y) stage location. As the transmission TOFA curve at each location at each location is primarily influenced by the structure of the tissue directly between emitter and detector at that stage location in this early device, only one TOFA curve was needed to produce each transmission image pixel. We had demonstrated that analyzing a constant fraction of the early detected photons using this TOFA device resulted in simplified calculations and successful imaging of phantoms[23,24] and biologic systems.[11,24]

Modifications

We then modified the existing scanner described above by removing the translational stage, and adding two independently rotating stages, both with a common axis of rotation, though separated by 12 cm. The object to be imaged, up to 10 cm in diameter by 15 cm tall, is placed upon the lower stage (object stage), which rotates the object being scanned in front of a fixed-position emitter fiber. The upper stage (detector stage) rotates an arm holding the detector fiber with the tip of the fiber skimming just above the surface of the object being scanned while rotating under computer control around the object (Fig. 1). Both stages are controlled by individual stepper motors (UMD 264-01, Vexta, Oriental Motor, Los Angeles, CA) in a 1:1 drive, and these in turn are connected to custom-built stage driver-controllers. Each stage requires 400 uniform, discrete steps in order to turn one complete revolution, and thus have an angular rotation of 1 gradian per step (although radians are the standard unit of angular measure, we found that the convenience of using one gradian per step facilitated use of faster integer calculations in our processing algorithms, and therefore all rotational angles are presented in these units). Computer control of the stages is provided via connection of the drivers to the RS-232 serial port of the computer. Maximum motor step speed is up to 100K steps per second with 5 arc-minute accuracy; in practice, this speed is limited by the driving software and handshaking to about 10K steps per second due to the speed of the serial port. Software to control the stages and to generate images was written specifically for this system by one of us in compiler Basic (Microsoft Professional Visual Basic for DOS, release 2.0, Redmond, WA), and run as an executable module on a 80386-based 33 MHz PC-compatible system operating under Microsoft Windows 3.1.

As a result of the tomograph configuration which has two independent rotational stages, virtually any emitter-detector angle combination can be measured, provided only that the emitter and detector are each pointed along an axis that passes through the center of the object being scanned, even if the path between the emitter and detector does not. Thus, the tomograph orientation could range from pure "transmission," with emitter and detector opposed 180 degrees, to nearly complete "reflection," with emitter and detector side-by-side. Of course, terms such as transmission and reflection do not apply well to highly scattering media, as nearly all detected photons have greatly scattered by the time of detection. However, the above examples should serve to illustrate how the flexible arrangement of emitter and detector allow testing of multiple different imaging strategies using the same tomographic device. Thus, the object can be illuminated from any angle by rotating the object stage with respect to the fixed emitter fiber, while detection is restricted to about 360 gradians (90% of full circle) due to physical obstruction of detector travel by the emitter fiber and mounting hardware.

In order to zero the system for timing purposes, the emitter and detector are detached from the tomograph, and coupled through a nonscattering mirrored glass wafer of known attenuation and thickness. Next, a series of reference pulses through this known standard are measured, and used by software given the thickness of the glass and its index of refraction, to back calculate a relative time zero that corrects for the length of the fibers and for any optical or electronic delays. In addition, as the reference wafer is a calibrated neutral density attenuation filter, estimation of the unattenuated and unscattered laser intensity can be performed. Once this calibration has been performed, the fibers are reconnected to the tomograph, and scanning can begin.

Scanning

To scan an object, a cylindrical model containing a phantom (or brain sample in a fluid -filled cylindrical holding tank) was mounted upon the lower stage after zeroing of the optical path. The speed of the scan, including rate of stage translation and integration time, as well as the geometry of the scan, including angles measured and step size, was fully programmable. For this series of experiments, a fixed step size of 4 gradians (or 100 sampling points per rotation) was selected for both the object and detector stages. At the start of a scan, the computer requests information regarding the size and estimated refractive index of the sample. Before obtaining the first measurement, the mounted object is rotated until a point on the phantom marked as zero degrees is aligned with the emitter fiber. Next, the detector fiber is positioned as closely as possible to the emitter fiber, generally at an angle of 30 gradians or less. The detector is then scanned through a series of predefined steps until one complete circumferential scan has been obtained. Next, the object is rotated a predefined number of steps, in this case 4 gradians, and the detector sweep is repeated until a second circumferential scan has been obtained. This entire process is repeated, rotating the object slightly more each time, followed by a full circumferential detector sweep each time the object is moved, until the object has been rotated in one full loop, and the point marked zero degrees on the object once again is aligned with the emitter fiber. Thus, the detector makes multiple trips around the object, while the object rotates only once, over the course of one complete scan. In some scans, this raster was reversed, such that the object rotated many times, and the detector rotated only once, with each scan of the object. These two methods are, of course, equivalent.

In order to adjust for drift in system timing, after every n measurements (where n is program-definable and generally falls between 3 and 10), the object and detector are rotated back to a standard position, and a TOFA sample is collected at this standard location. As the standard position does not vary in location during a scan, and is returned to for rescanning on multiple occasions during a full imaging scan, any drift in the calculated measurements at this position must be due to system timing drift in these stable samples, and can be corrected for at time of data processing.

Processing and Imaging

After each TOFA collection, whether the measurement is at an image location or at the reference location, the raw TOFA data, as well as a processed, filtered TOFA curve, intermediate calculations, and identifying information, are stored on disk. Final processing is accomplished by a second computer system (80486/50 or 80586/60), either concurrently with data collection or at a later time, using a program written for this device and reported earlier.[12] Briefly, each time of flight curve was smoothed using known information about the sample, as well as by convolving with a function known to approximate ideal time of flight data, detector response times, and other information. Mean time of flight was calculated, and scattering and absorbance are estimated using equations based upon solutions to the photon transport equations developed by our group and others,[27-34] such as the equations of Sevick or Patterson. Image reconstruction was via simple summation of curvilinear traces based upon physical location of the emitter and detector fibers (see accompanying article on image algorithms[12]). No inverse transforms were employed. Images were etched into array space using a mapping system that was self-modifying based upon a training set of data using known phantoms.[12,35]

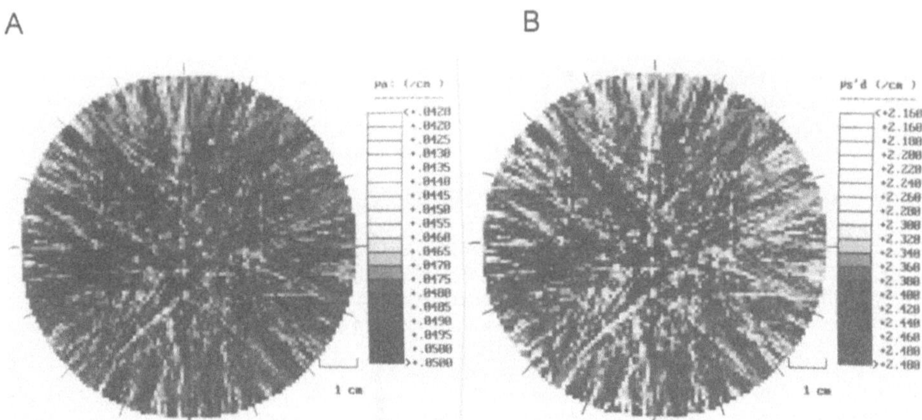

Figure 2. Images of a homogeneous phantom solved for (A) absorbance and (B) scattering. Expansion of image scales reveals fine noise structure shown above, otherwise hidden in relatively homogeneous images. Mean values +/- standard deviation for these images (n=7,845), with measured scattering of 2.34 +/- 0.030 cm^{-1} (actual μ's = 2.50 cm^{-1}) and measured absorbance of 0.048 +/- 0.0006 cm-1 (actual μ_a = 0.046 cm^{-1}).

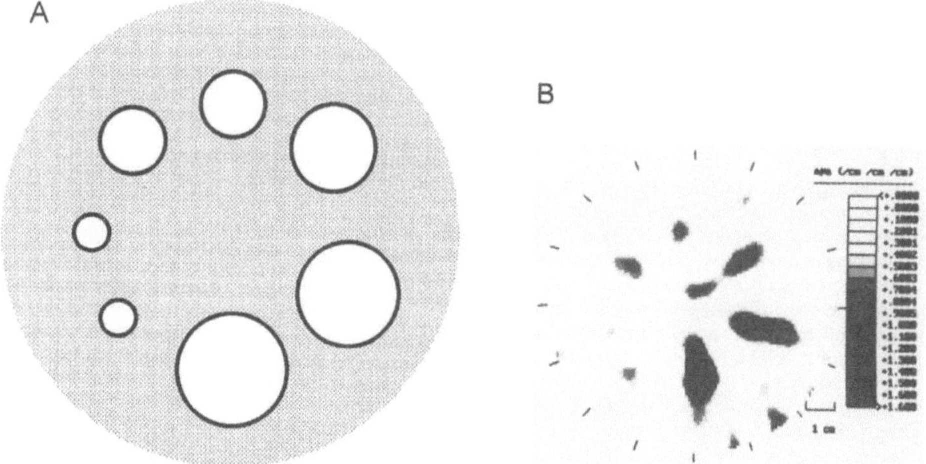

Figure 3. A schematic (A) and image (B) of a phantom with multiple rods. A size-response curve can be made from this image and others (not shown), indicating that the minimum detectable object (MDO) size is about 0.3 cm at 2 cm depth, and about 1.0 cm at 5 cm depth, which is in good agreement with theory.

Experiments

For this series of experiments, three systems were imaged. Each phantom system measured 10 cm in diameter. First, two homogeneous phantoms of known scattering and absorbance were constructed using resin with an absorption coefficient of 0.092 cm^{-1} at 785 nm, 0.046 cm^{-1} at 850 nm, and 0.060 cm^{-1} at 904 nm, and a reduced scattering coefficient of 2.50 cm^{-1} by the addition of TiO$_2$. The homogeneous phantoms were then imaged to test for absorbance and scattering detection, and to measure system noise. Image accuracy, in terms of separation and estimation of absorbance and scattering, was estimated from the absolute values of the pixels in the absorbance/scattering separated images, and image precision was determined from measurements of variation in pixel values. In the second experimental scan, a series of acrylic rods of decreasing radii, placed 3 cm from phantom center, were imaged to test the spatial resolution of the system. The minimum detectable object size (MDO) for clear fluid at a depth of 2 cm in this phantom model was estimated using resolution equations we developed earlier. Last, a sheep brain with collections of fluid and blood was imaged.

RESULTS

Tomograph Operation

Data collection itself required 15 seconds per emitter-detector pair, or about 8 hours for measurements taken every 10 gradians, and 2 days for measurements taken every 4 gradians. These times doubled when a 30 second integration time was used. Once TOFA curves were collected and processed to produce the data files on disk, image processing for an object with nearly 10,000 measurements allowed generation of an image within two minutes. Thus, the actual image generation time for the curvilinear backprojection method is rapid.

Images

Two images at 850 nm of the homogeneous 10 cm phantom are shown on the previous page (Fig. 2). The images of absorbance (Fig. 2A) and scattering (Fig. 2B) have different value ranges, and each appears homogeneous, requiring expansion of the image scale in order to see the fine noise structure of the images. This demonstrates that scattering and absorbance can be separated as an image using time-resolved optical tomography, and that the imaging algorithms are robust to boundary conditions near the edges of the image.

A schematic and an image at 850 nm are shown above of the changes in scattering over space (the second derivative surface) for a series of seven clear rods of varying radii, buried at a constant depth within a 10 cm cylinder (Fig 3). Using the algorithm for detectability that we developed and reported earlier, the resolution of the system can be estimated to be about 0.3 cm at a depth of 2 cm, and about 1.0 cm at 5 cm depth.

Last, a schematic and an optical image of the sheep brain are shown below (Fig. 4). The 1 cm collections of blood and fluid are easily located on a color enhanced image, suggesting that real pathology can be located and visualized using optical tomography.

DISCUSSION

This study demonstrates several key issues central to developing an effective optical tomograph. First, optical imaging can be performed in a true tomographic configuration using a portable system with power levels safe for clinical use and in model systems of comparable scattering, absorbance, size, and shape as tissues in which optical imaging is expected to be helpful. Second, absorbance and scattering can be effectively separated as an image, which is an essential step in chemical analysis using multiwavelength absorbance data and in histological analysis using multiwavelength scattering data. Third, complex biological objects can be resolved. Last, image resolution can be quantified, and appears to be about 3 mm at 2 cm depth, and 10 mm at a 5 cm depth, which is similar to that predicted by several independent theoretical treatments.

The physical layout of our system differs from those used by others.[36-41] For example, Hebden uses multiple parallel rays.[36] Such an approach can have the advantage that if the scattering and absorbance is approximately matched between object and immersion fluid, then the optical path between any two points is a straight line, and conventional analysis fully applies. However, such an index-matching approach is impractical in most cases for the monitoring of living humans (except, as noted above, in the case of breast tumor imaging). In our approach, the detection ring can be of any shape, even irregular, and is closest to CT or EIT scanning in style.[42,43] This difference is important in the transference of our technique into the clinical realm. Our system can easily be converted to a fixed-position fiber-based system that should allow imaging under clinical conditions, and preliminary studies suggest that such an approach can be effective.[44,45]

In the future, our current approach could be improved in multiple ways. First, the speed of the system can be improved over many orders of magnitude. The collection of the full time-of-flight curve with each sampling would result in a 256-fold improvement in speed, increases in light intensity to 100 mW, approaching the current FDA allowance for the illuminance of the body by such systems as endoscopes, would improve collection by 1000-fold, while use of multiple detectors would increase speed by about 32 times in human models. Altogether, we would expect that a 32-point scan (emitters and detectors located at 32 points around a ring) which now requires 5 1/2 hours could take only a few hundred milliseconds. In practice, the scan could be slowed from this maximum speed to allow improvements in image resolution, decreases in costs, and to allow a safety margin in light intensity, such that we expect brain scans to update once a minute Dynamic scans, such as those of the lung, could update once every 30 ms with degraded resolution. Regarding data storage requirements, data storage needs will be overwhelming once ongoing sequential images are collected. For example, if images of the brain are collected every minute, continuous monitoring using the current system would generate terabytes of data a

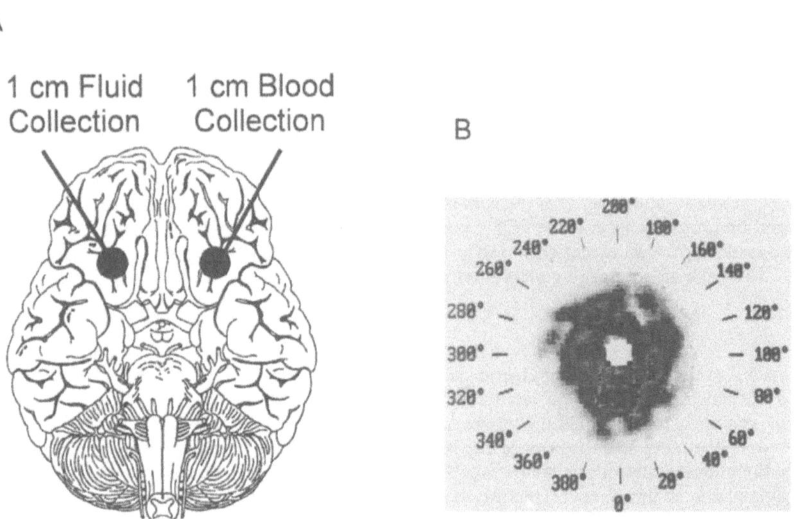

Figure 4. A schematic (A) and an optical image (B) of the sheep brain. The collection of fluid and blood are well visualized in a color-enhanced image, showing that disease in tissue models can be visualized. In this early image, the center of the brain is not imaged due to a limited amount of light passing through this portion of the brain. This problem has since been solved.

day, or more for images collected more quickly (such as images gated to the cardiac or respiratory cycles). Ultimately, this data storage requirement can be shrunk significantly by preprocessing as better algorithms are developed and the raw data no longer needs to be saved. We would expect that a clinical image could be updated continuously while data is being collected, and thus large physical or functional changes could begin to be visualized well before a full optical scan had been repeated. Further, if a bleed or deoxygenation is occurring in one area of the brain, the image algorithms could be trained to rescan and focus upon such an area, allowing faster update of image information in regions in which a change is occurring. A fast clinical system capable of imaging in ten minutes or less is under design.

The use of physiologic conditions for imaging is an oft ignored, but vitally important, factor. The materials imaged as phantoms in reports in the literature range from olives, milk, nondairy creamer, intralipid suspensions, India ink, cold chicken breast, preserved tissues, and others. Further, when actual tissue is used, care is often not taken to ensure that the tissue optical characteristics do not change from those at baseline *in vivo*. For example, dead tissue contains a different volume of hemoglobin than living tissue, and this is particularly true when the tissue has been preserved in formalin. Temperature can also affect tissue optics. For example, breast tissue undergoes dramatic changes with temperature, and this can be envisioned by recalling that the fat of a recently cooked steak is clear, but is highly scattering and opaque when pulled from the refrigerator. Thus, our phantoms were designed to reflect the optical characteristics of intact tissue and fluids, including a background absorbance of the physiological $0.02 - 0.08$ cm^{-1}, and a DPF of 2 - 6 (corresponding to reduced scattering coefficients, μ's, of up to 4 or more). A combination of dyes could be used to nearly exactly reproduce the absorbance of different tissues, as well as of hemoglobin at different saturations with oxygen, and this is under study. As the phantoms are made of a stable resin, this facilitates consistency in testing and retesting, as well as permits exchange of phantoms between laboratories. The use of phantoms of the correct physical size and shape is also often omitted in optical studies. For example, images through breast tissue only millimeters thick may not work in tissue that is as thick as that likely to be encountered in the clinical realm, while imaging software simulations that solve images in "heads" that are a few centimeters across may fail when applied to heads of the correct size. The geometry of the imaging should also be physiologically realistic, and not require transillumination from all sides (unlikely for such tissue as breast or brain, in which the tissue is attached to the body on at least one side), nor require submersion of the tissue in a tank of defined geometry in order to perform imaging. Our models were 10 cm across, corresponding to the transmission diameter of an infant head or the adult female breast, and to the imaging geometries available in the clinical setting. What has been needed is a series of phantoms that are of the correct absorbance and scattering ranges, that have the correct physical dimensions, and realistic imaging geometries, and our phantoms fulfill these conditions. The imaged phantoms had physiologic absorbance and scattering values, were of realistic shape and size. Last, the imaging technique could easily be modified to allow imaging using a clinical imaging harness.

Our study demonstrates that absorbance and scattering images can be separately computed from optical tomographic data using approximations derived from the diffusion approximation for photon transit combined with imaging algorithms. Although the absolute values of our measurements differ from actual values by as much as 6% (and thus the measure is not yet accurate), the variance of the measure is small (suggesting that the technique is precise). Such tightness of variance should allow differences in scattering and absorbance of as small as a few percent to be well identified. Such changes are consistent with those that would be expected, for example, in brain tissue after hypoxic stroke, in which the water content of the tissue cells may change from 88% to as high as 93%. Improved accuracy could be achieved by using more sophisticated equations to estimate absorbance and scattering, or improved calibration techniques. Many such improved estimation equations are possible, for example, and others have used such equations to measure absorbance and scattering in a non-spatial sense, both within homogeneous media and in tissue. Importantly, while these is some distortion in the images made using these non-imaging-based equations, this approach works well in media containing multiple objects of differing absorbance and scattering. This raises the possibility of images processed to segregate the effects of absorbance and scattering, which would be a first step in multiwavelength chemometric analysis of tissue and histological "optical biopsy" for identification and discrimination between different tissue types.

Future studies, already well underway, consist of testing in tissues as well as in animals and humans. Although such images remain to fully analyzed, we are finding good resolution, comparable to the phantom studies. We are in the process of constructing a long-term monitor for humans that we call a "neural net," consisting of a fiber optic web for studying human infants in the intensive care unit, as well as adults with stroke, thus converting our mechanical scanner to an electronically scanned array that is clinically relevant. Lastly, we look forward to improved algorithms allowing better estimation of absorbance and scattering, as well as improved resolution, as our colleagues in physics as well as Radiology continue to explore the field.

ACKNOWLEDGEMENTS

We wish to thank the Office of Naval Research (N-00014-91-C-0170), the Walter Berry Fellowship Fund at Stanford, The NIH (through the PHS General Clinical Research Center Grant M01-RR-00070-30/1 and grant RR-00081), for their partial support for this work.

REFERENCES

1. D.A. Benaron, G. Muller, B. Chance, "A medical perspective at the threshold of clinical optical tomography," In: In: Medical Optical Tomography: functional imaging and monitoring, G. Muller, B. Chance, R. Alfano, et al., eds. (SPIE press, Bellingham WA, 1993), pp 3-9.
2. D.A. Benaron, "Optical imaging reborn with technical advances," Diag. Imag. Jan, 69-76 (1994).
3. F.F. Jobsis, "Noninvasive infrared monitoring of cerebral and myocardial oxygen sufficiency and circulatory parameters," Science 198, 1264-1266 (1977).
4. P.W. McCormick, M.S. Stewart, M.G. Goetting, M. Dujovny, G. Lewis, J.I. Ausman, "Noninvasive cerebral optical spectroscopy for monitoring cerebral oxygen delivery and hemodynamics," Crit. Care Med. 19, 89-97 (1991).
5. J.E. Brazy, "Effects of crying on cerebral blood volume and cytochrome aa_3," J. Pediatrics 112(3), 457-461 (1988).
6. A.D. Edwards, C. Richardson, M. Cope, J.S. Wyatt, D.T. Delpy, E.O.R. Reynolds, "Cotside measurement of cerebral blood flow in ill newborn infants by near-infrared spectroscopy," Lancet ii, 770-771 (1988).
7. D.A. Benaron, C.D. Kurth, J. Steven, L.C. Wagerle, B. Chance, M. Delivoria-Papadopoulos, "Non-invasive estimation of cerebral oxygenation and oxygen consumption using phase-shift spectrophotometry," Proc. IEEE Eng. Med. Biol. Soc. 12(5), 2004-2007 (1990).
8. C.D. Kurth, J.M. Steven, D.A. Benaron, B. Chance, "Near-infrared monitoring of the cerebral circulation," J. Clin. Mon. 9(3), 163-170 (1993).
9. A.D. Edwards, J.S. Wyatt, C. Richardson, A. Potter, M. Cope, D.T. Delpy, E.O.R. Reynolds, "Effects of indomethacin on cerebral haemodynamics in very preterm infants," Lancet 335, 1491-1495 (1990).
10. P. Wu, F.F. Jobsis, "Cerebral changes in cytochrome c as an indicator of bilirubin uncoupling of oxidative phosphorylation," Pediatrics in press (1994).
11. D.A. Benaron, D.K. Stevenson, "Optical Time-of-Flight and Absorbance Imaging in Biologic Media," Science 259, 1463-1466 (1993).
12. D.A. Benaron, D.C. Ho, D.K. Stevenson, "Non-recursive Linear Algorithms for Optical Imaging in Diffusive Media," Adv. Exp. Med. Biol., Oxygen Transport to Tissue XVI, Hogan, ed., in this volume (1994).
13. D.A. Benaron, D.C. Ho, D.K. Stevenson, "Imaging neonatal brain pathology using light," Ped. Research 33(4), 369A (1993).
14. A. Knuttel, J.M. Schmitt, J.R. Knutson, "Spatial localization of absorbing bodies by interfering diffusive photon-density waves," App. Optics 32(4), 381-389 (1993).
15. L.O. Svaasand, B.J. Tromberg, R.C. Haskell RC, T-T Tsay, M.W. Berns, "Tissue characterization and imaging using photon density waves," Opt. Eng. 32(2), 258-266 (1993).
16. L. Wang, P.P. Ho, C. Liu, G. Zhang, R.R. Alfano, "Ballistic 2-D imaging through scattering walls using an ultrafast optical Kerr gate," Science 253, 769-771 (1991).
17. J.B. Fishkin, E. Gratton, "Propagation of photon-density waves in strongly scattering media containing an absorbing semi-infinite plane bounded by a straight edge," J. Opt. Soc. Amer. A 10(1), 127-140 (1993).
18. O. Jarlman, R. Berg, S. Svanberg, "Time-resolved transillumination of the breast," Acta Radiologica 33, 277-279 (1992).
19. J.C. Hebden, R.A. Kruger, "Transillumination imaging performance: spatial resolution simulation studies," Med. Phys. 17, 41-44 (1990).
20. Y. Yamada, Y. Hasekawa, "Simulation of time-resolved optical CT imaging," Proc. SPIE 1431, 73-82 (1991).
21. J.C. Hebden, "Line scan acquisition for time-resolved imaging through scattering media," Opt. Eng. 32(3), 626-633 (1993).
22. B.J. Tromberg, L.O. Svaasand, T-T Tsay, R.C. Haskell, "Properties of photon density waves in multiple-scattering media," App. Opt. 32(4), 607-616 (1993)
23. D.A. Benaron, M.A. Lenox, D.K. Stevenson, "Two-D and three-D images of thick tissue using time-constrained time-of-flight and absorbance (tc-TOFA) spectrophotometry," SPIE 1641, 35-45 (1992).
24. D.A. Benaron, D.C. Ho, B. Rubinsky, M. Shannon, "Imaging (NIRI) and quantitation (NIRS) in tissue using time-resolved spectrophotometry: the impact of statically and dynamically variable optical path lengths," SPIE 1888, 10-21 (1993).
25. E. Gratton, et al. Bioimaging 1, 40-6 (1993).
26. TDR-20 specifications literature, Opto-Electronics, Inc., Ontario, Canada.
27. E.M. Sevick, B. Chance, J. Leigh, S. Nioka, M. Maris, "Quantitation of time- and frequency-resolved optical spectra for the determination of tissue oxygenation," Anal. Biochem. 195, 330-351 (1991).
28. M.S. Patterson, "Time resolved reflectance and transmittance for the non-invasive measurement of tissue optical properties," Appl. Opt. 28, 2331-6 (1989).
29. M. Essenpreis, M. Cope, C.E. Elwell, et al., "Wavelength dependence of the differential pathlength factor and the log slope in time-resolved spectroscopy," Adv. Exp. Med. Biol. 333, 9-20 (1993).
30. S.J. Madsen, B.C. Wilson, M.S. Patterson, et al., "Experimental tests of a simple diffusion model for the estimation of scattering and absorption coefficients of turbid media from time-resolved diffuse reflectance measurements," Appl. Opt. 31(18), 3509-3517 (1992).
31. S.R. Arridge, M. Cope, D.T. Delpy, "The theoretical basis for the determination of optical pathlengths in tissue: temporal and frequency analysis," Phys. Med. Biol. 37(7), 1531-1560 (1992).
32. M. Haida, M. Miwa, A. Shiino, B. Chance, "A method to estimate the ratio of absorption coefficients of two wavelengths using phase modulated near infrared light spectroscopy," Anal. Biochem. 208, 348-351 (1993).
33. B. Chance. N.G. Wang, M. Maris, S. Niok, E.M. Sevick, "Quantitation of tissue optical characteristics and hemoglobin desaturation by time- and frequency- resolved multiwavelength spectrophotometry," In: Adv Exp Med Biol 317, 297-304 (1992).
34. R.F. Bonner, "Model for photon migration in turbid biological Media," J. Opt. Soc. Amer. A 4, 423-432 (1987).
35. D.A. Benaron, D.C. Ho, S. Spilman, J.P. VanHouten, D.K. Stevenson, "Tomographic Time-of-flight Optical Imaging Scanner With Non-parallel Ray Geometry," submitted to Appl. Opt., under review (1994).
36. J.C. Hebden and K.S. Wong. "Time-resolved optical tomography," Appl. Opt. 32, 372-380 (1993).
37. P.C. Jackson, P.H. Stevens, J.H. Smith, D Kear, H. Key, and P.N.T. Wells. "The development of a system for transillumination computed tomography," Br. J. Radiol. 60, 375-380 (1987).
38. J.C. Hebden, R.A. Kruger, and K.S. Wong. "Tomographic imaging using picosecond pulses of light," in Medical Imaging V: Image Physics, R.H. Schnieder, ed. Proc. SPIE 1443, 294-300 (1991).
39. E.M. Sevick, C.L. Burch, B. Chance. "Near-infrared Optical Imaging of Tissue Phantoms with Measurement in the Change in Optical Path Lengths." In: Oxygen Transport to Tissue XV, P. Vaupel, ed. (Plenum Press, NYC, 1994), in press.
40. D.V. Stephens, L. Wang, A.M. Hielscher, S.L. Jacques, F.K. Tittel, "Detection of an object inside a phantom using a spatial filter." Proc. SPIE 2135, in press (1994).
41. J.J. Dolne, K.M. Yoo, R.R. Alfano, "Near-IR fourier space gate and absorption shadowgram images through random scattering media." Proc. SPIE 2135, in press (1994).
42. G.N. Hounsfield, "Computerized transverse axial scanning (tomography), Br. J. Radiol. 46, 1016-1022 (1983).
43. W.A. Kalender, "X-ray computed tomography - state of the art," In: Medical Optical Tomography: functional imaging and monitoring, Muller G, Chance B, Alfano R, et al, eds. (SPIE press, Bellingham WA, 1993), pp 10-27.
44. D.A. Benaron, J.P. van Houten, D.C. Ho, S.D. Spilman, D.K. Stevenson, "Imaging neonatal brain injury using light-based optical tomography," Ped Research, in press (1994).
45. D.A. Benaron, D.C. Ho, S.D. Spilman, D.K. Stevenson, "Imaging of pulmonary pathophysiology using optical tomography in model systems," Ped Research, in press (1994).

NON-RECURSIVE LINEAR ALGORITHMS FOR OPTICAL IMAGING IN DIFFUSIVE MEDIA

David A. Benaron, David C. Ho, Stanley Spilman, John P. Van Houten, David K. Stevenson

Medical Imaging and Spectroscopy Section,
Division of Neonatal and Developmental Medicine, Stanford School of Medicine, and the
  Stanford Picosecond Free-Electron Laser (FEL) Center, Hansen Experimental Physics Lab
750 Welch Rd #315, Stanford University, Palo Alto, CA 94305
Phone 415/723-5711, Fax 415/365-0656, E-mail benaron@forsythe.stanford.edu

ABSTRACT

Optical imaging has been used to image phantoms, animals, and humans. It offers the potential for the production of functional images of tissues, such as oxygenation of brain during stroke. Fast algorithms are needed to allow diagnostically useful images to be generated under realistic conditions, including the likelihood that transmission geometries will not be possible. We proposed a linear algorithm, while less than ideal, may allow rapid reconstruction of images and avoid the pitfalls of recursive, nonlinear solutions. Such techniques may also facilitate the use of varied but physiologic imaging geometries. We found that linear backprojection tomography is feasible for clinical use. Conversion of the present mechanically scanning device to a clinical scanner should be possible with retention of the current processing algorithms. Such a clinical scanner should ultimately be able to generate images in less than one minute with centimeter resolution at the center of living human brain.

INTRODUCTION

Optical imaging and spectroscopy use light emitted into opaque media such as human tissue to determine interior structure and chemical content, and have broad application to medicine.[1] The impetus for such work is reviewed in an accompanying article.[2] We had previously constructed a time-resolved NIR imaging system, and used transmission geometries to image phantoms[3,4] (demonstrating 2-D and 3-D reconstruction) and animals[3,5] (demonstrating imaging of internal organs). Similar images have been made in the frequency-domain.[6]

While transmission images are of reasonable quality, problems arise with clinical application of such approaches. First, a transmission geometry may not be clinically possible. A vanishingly small number of photons traverse the entire adult human head, and this number shrinks further if there is bleeding under the skull or inside the brain. Second, access to the body may be limited, such as due to the need for a sterile field or the need for anesthesia access. Last, many imaging routines are slow to converge, and information such as oxygenation of the brain could come too late if an algorithm takes 20 minutes to converge upon a stable solution. A final barrier to clinical application of optical imaging lies in the difficulty of reconstruction for non-transmission geometries.

In this paper, we describe a fast image algorithm based upon a non-recursive linear model. We hypothesized that light takes a path that can be estimated using simple linear relationships, and that in tissue that is not highly complex such linear approximations will produce a distorted, but usable, image. Further, we propose that in many clinical situations, such images will provide information not otherwise obtainable, and thus are medically useful.

METHODS:

Linear Image Reconstruction Technique.

We proposed a few years ago that simple time-domain backprojection methods could yield approximate (but nevertheless useful) medical images, despite the presence of curvilinear paths taken by photons through tissue. Here we outline our justification for this simplifying approach.

Oxygen Transport to Tissue XVI
Edited by M.C. Hogan *et al.*, Plenum Press, New York, 1994

In an infinite and homogeneous medium, there is a distribution of photon paths between emitter and detector that is convoluted and complex. However, in such a phantom, photons travel pathways that average to a path directly between the emitter and detector. Of course, few if any photons actually travel this direct route. The first arriving photons, those taking the most direct route possible, may have scattered a thousand times between emission and detection. From a statistical viewpoint, however, these early-arriving photons contain the highest information about the material directly between emitter and detector in terms of signal to noise. Each subsequently detected photon is likely to have sampled a larger and larger tissue volume, and thus contains less and less spatial information. Use of the early-arriving photons alone can lead to the construction of powerful imaging techniques based upon standard imaging algorithms. However, such early arriving photons are rare in tissue. Furthermore, when estimating parameters of the medium such as absorbance and oxygenation, a new problem arises. Early-photon filters retain spatial information at the expense of discarding much of the information that would have allowed separation of absorbance and scattering. Conversely, equations that have been derived to allow determination of the absorbance and scattering properties of the medium (whether in the frequency-, spatial-, or time- domain) tend to be based upon simplistic, group behavior rules that apply to ensemble averages of pathways followed by many photons, thus obscuring spatial information. In some cases, these absorbance and scattering measures are based upon photon groups that have sampled different regions of the medium. For example, in the equations of Sevick et al,[7] absorbance is based upon a late decay of photon intensity, thus using photons traversing a large sample volume, while scattering is estimated over a smaller sampling volume.

The use of differing sample volumes for scattering, absorbance, and imaging determinations is not a problem until the medium is no longer homogeneous, such as when there are objects in the medium that differ from background, or when the medium is no longer infinite. Both of these limitations are violated under real conditions. For example, if one looks at the mean optical pathway through a small, cylindrical, homogeneous medium using a least squares fit to the pathways followed by all photons, there will be substantial distortions from linearity in the mean path followed (Fig. 1). No longer are the mean photon pathways linear, except during transmission from one side to a point directly opposite on the phantom (vertical center line, Fig. 1A).

The problem then becomes "How can one use equations predictive of absorbance and scattering, while still retaining the efficient linear imaging algorithms used with the early arriving photons?" In theory, boundary influences can be mitigated by immersing the cylinder in an index matching fluid, but in practice this is often impossible (*i.e.*, most patients strongly object to submerging their head in a fluid bath for extended periods of time, though such an approach can work for the breast). However, some simplifications are possible. If the medium is homogeneous, then the curvature of the mean path is continuous, and may be predictable. If the shape of these paths can be discovered or deduced, then there exists a one-to-one mapping relationship that can be applied to optical measurements allowing correction of this curvilinear distortion of pathway. Mapping of measured optical parameters, such as absorbance or scattering, into an array space using curved paths would allow for rapid imaging. The shape of these curvilinear paths can be estimated using existing software, or may be empirically determined in real samples by using a set of tuning phantoms containing objects of known size

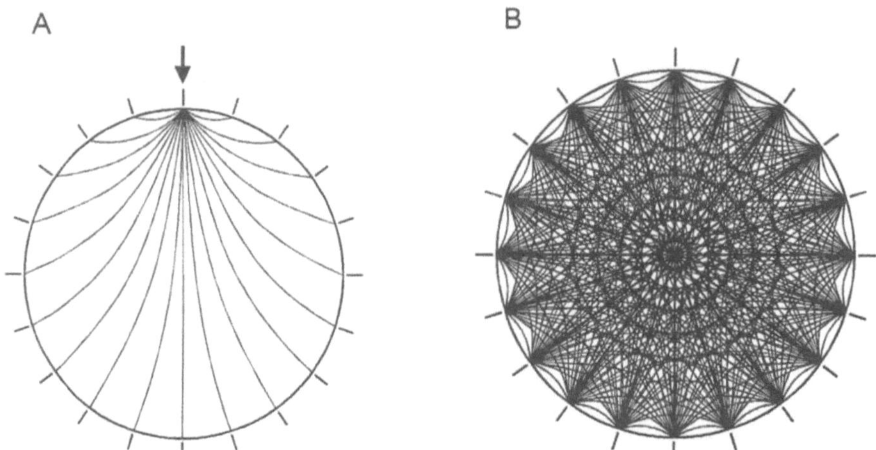

Figure 1. Schematic of mean optical pathways through a homogeneous cylinder, viewed from the top. A) Light from source (arrow) can be detected at various points around the sample. Average path taken by this light is shown as solid lines through cylinder, although in reality there is a diffuse probability-density cloud that describes the paths taken by a group of photons. B) Rotating the photon source around the object and repeating the optical measurements yields a good coverage of the internal structure of the cylinder. In practice, optical density of object center, as well as the difficulty of modeling behavior at the edge and though multiple objects, makes such an approach idealistic. However, as a first approximation, such an approach works well.

and shape, provided the phantoms are similar in optical properties to the tissue under study, and then by tuning the image until for optimal resolution.

Once such a mapping has occurred, errors will arise in the calculation of absorbance and scattering. It is likely that the most accurate determination of absorbance and scattering will be though the optical path that passes through the center of the object, as the mean central pathway in a homogeneous cylinder is linear. This is due in part to the fact that these photons spend a higher proportion of their time away from a boundary, as well as due to the cylindrical symmetry that allows for some cancellation of the boundary effects. We propose that distortion in the scattering and absorbance estimates of the medium at positions that pass other than through the center of the object can be described by a positional offset factor that takes into account emitter and detector position, and other known characteristics of the medium. Provided that such offsets are predictable, then the imaging process reduces to 1) applying curvilinear imaging paths to correct for image distortion, and 2) correcting measured absorbance and scattering values by an offset function. Both the curvilinear distortion and offset function may be determined empirically using known tuning phantoms, and thus introduce no new unknowns.

This linear approach can be taken a step further by including into the model the fact that the mean pathway is not a line, but rather a cloud of photon probability. This cloud can be drawn in array space, rather than using a narrow mean pathway curve, to allow for a more accurate rendering of the image data. Further refinement can include the introduction of any number of serial, compensatory linear operations, many quite simple and thus rapidly implemented and calculated, with each allowing resolution and accuracy to incrementally improve.

In summary, our image reconstruction is via a series of simple, and simplistic, linear operations that have the advantage of high speed at the expense of ideal accuracy. Initially, a good guess is made as to the properties of the medium, though the quality of this guess is ultimately irrelevant as the image itself can be used to tune the offsets and distortion functions. Fine tuning of the curvilinear shape using phantoms with known inclusions, and fine tuning of the scattering and absorbance offsets using phantoms of known optical properties, can increase system accuracy. Once tuning is completed, image reconstruction is via summation of curvilinear traces (or clouds) based upon the known physical location of the emitter and detector fibers and upon the calibration data functions. A curved probability density representing the value of a measured variable is applied into an array space representing real space. If this variable is, for example, absorbance, then absorbance could be estimated using equations such as that presented by Sevick, and is entered in a curved pattern into the array based upon an associated probability of photon transit through that point. No inverse transforms are employed.

Rotational Scanner

All imaging routines were tested using real data collected with an existing optical scanner as described in an accompanying article.[2] Briefly, the scanner consists of a time-of-flight measuring system built around an optical time-domain reflectometer. The scanner has two independently rotating stages, both with a common axis of rotation, though separated by 12 cm. The object to be imaged, up to 10 cm in diameter and 15 cm tall, is placed upon the lower stage (object stage), which rotates the object being scanned in front of a fixed-position emitter fiber. The upper stage (detector stage) rotates an arm holding the detector fiber, with the tip of the fiber skimming just above the surface of the object being scanned, while rotating under computer control around the object.

Virtually any emitter-detector angle combination can be measured. Thus, the tomograph orientation could range from pure "transmission," with emitter and detector opposed 180 degrees, to nearly complete "reflection," with emitter and detector side-by-side. Of course, terms such as transmission and reflection do not apply well to highly scattering media, as nearly all detected photons have greatly scattered by the time of detection, and the above examples are only mentioned to illustrate how the flexible arrangement of emitter and detector allows testing of multiple different imaging strategies using the same tomographic device.

Model System

A series of model phantoms were constructed following the suggestion of the Delpy, Arridge, and Cope that polystyrene resin be used as a base material for phantoms (we used Simmar[tm] unsaturated polyester in monomer, British Petrol, BP chemicals, Advanced Materials Division, Fort Wright KY 41015-0447). This resin has the advantages that it is 1) nonabsorbing and nonscattering in the visible and near-infrared wavelengths, allowing good optical penetration, 2) rugged and nonreactive when polymerized, thus permitting serial measurements on the same phantom over time, as well as sharing among laboratories, 3) tolerant of a wide variety of dopants, thus allowing controlled adjustment of optical characteristics using commercially available dyes and scatterers added to the mixture, and 4) simple to cast and cure, allowing complex models to be manufactured.

We cast models in physiologically relevant sizes, with scattering and absorbance coefficients that approximated real tissue, and scanned these phantoms using geometries that would be feasible for clinical imaging in the future. Such use of relevant phantom characteristics and scanning geometries has been endorsed by Delpy, Chance, our group, and others, and will help standardize comparison between laboratories. Our models consist of styrene monomer ($n = 1.65$, measured using time-of-flight through a nonscattering sample as compared to time through a similar distance in air). We added TiO_2 slurry (TAP plastic, White Color Pigment Concentrate, Dublin CA) for scattering without absorbance, and added a dye (blue resin dye, Environmental Technology Inc., Fields Landing, CA) similar in absorbance to that of hemoglobin, to duplicate tissue optical characteristics. Resin models, once constructed, were hardened using 5 g/L of methyl ethyl ketone peroxide catalyst (Whitco, Argus Div., Marshall Texas), and cast in 10 cm diameter x 12 cm tall polished Kimax[tm] cylinders mounted upon square tempered glass bases. Clear acrylate polymer cubes were used as phantom objects. If absorbing or scattering objects were desired, plastic cubes were painted with flat matte black epoxy paints prior to casting, or could be cast out of resin to meet exact optical specifications. Additionally, tissue samples can be embedded in the models.

Image Reconstruction

After each TOFA collection, the raw TOFA data, as well as a processed, filtered TOFA curve, intermediate calculations, and identifying information, are stored on disk. Final processing is accomplished by a second computer system (80486/50 or 80586/60), either concurrently with data collection or at a later time, using a program written for this device by one of us. Briefly, each time of flight curve was smoothed using known information about the sample, as well as by convolving with a function known to approximate ideal time of flight data, detector response times, and other information. Mean time of flight was calculated, and scattering and absorbance are estimated using equations based upon solutions to the photon transport equations developed by our group and others.[7-14] Image reconstruction was via a parabolic curvilinear backprojection as described above into a spatial array 100 x 100 pixels, using object parameters estimated form the equations of Patterson applied to a best-fit time-of-flight curve

Experiments

Four sets of model systems were constructed and imaged. In each case, the background material was first adjusted to have absorption and scattering coefficients similar to those of living tissues. Prior to the first scan, homogeneous phantoms were used to calibrate the system for estimates of absorbance and scattering.

For the first experimental scan, a series of homogeneous phantoms were imaged to test the effect of the absorbance and scattering offset functions described previously. Next, a cylindrical rod, 2 cm in diameter and made of a material similar in absorbance but less scattering than background material, was buried in a highly scattering background substance at a depth of 2.5 cm from center. The rod phantom was then imaged to test for object localization and image distortion. Image accuracy was estimated using the rod's imaged size, as well as from the amount of distortion from roundness, using a Region of Interest (ROI) software function. The shape of the curvilinear traces was varied to minimize distortion of the position of the rod's center, and also until a sum of squared errors estimation (SSE) of the distortion of the rod from true roundness reached a minimum value. Next, once trained using these known phantoms, a third phantom, consisting of two pairs of 16 mm acrylic cubes (one pair flat black and the other clear) and cast in background material similar to tissue at a mean depth of 2.5 cm from center, was imaged to test for distortion secondary to the presence of multiple objects. Last, a two series

A   B

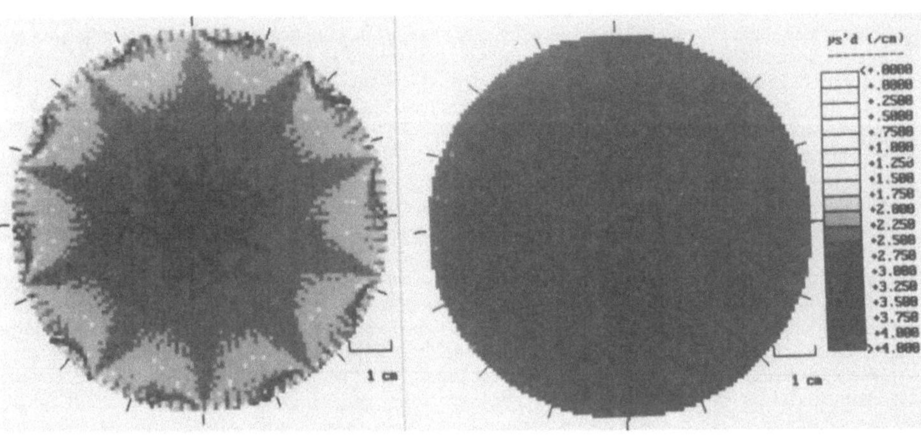

Figure 2. Images of a homogeneous phantom (A) before and (B) after applying offset function. Before offset optimization, there are significant boundary effects (shown analyzed every 40 gradians to emphasize regional distortions). After offset is used, the image is homogeneous, and measured values fall close to actual values.

of four cubes, buried at different depths, were imaged to test for object discrimination and depth resolution accuracy.

RESULTS

Images of estimated scattering at 850 nm for a homogeneous phantom before and after the use of offset correction functions are shown (Fig 2). Without use of an offset, there is considerable depth-dependent distortion (image analyzed every 20 gradians to emphasize these distortions); with an offset used, the image appears homogeneous. Absorbance and scattering images have different value ranges, demonstrating that scattering and absorbance can be separated as images using linear algorithms, and that such methods are robust to boundary conditions. A fine structure of noise in the offset-corrected scattering image can be seen when the scattering scale is expanded further.[2] The total noise in the image was determined by calculating a value histogram for all pixels of the image. In this histogram (n=7,845), the measured scattering was 2.34 +/- 0.030 cm^{-1} (actual μs = 2.50 cm^{-1}), while the measured absorbance was 0.048 +/- 0.0006 cm^{-1} (actual μa = 0.046 cm^{-1}). The error in estimated mean values were 6% for scattering and 2% for absorbance; the standard deviation across the image was 1.2%.

An absorbance image at 850 nm of the 2 cm rod in a 10 cm cylinder is shown (Fig. 3), and was used to calibrate the curvilinear paths. The rod's shape is distorted when paths are curved incorrectly; the correct circular shape and spatial location of the rod are seen in the corrected image, demonstrating that objects can be effectively localized and imaged using linear tomography. The center of the object, calculated from the centroid of iso-value lines on the image at a point where the image value falls 50% between object minimum and background average, is on a vector of -2 gradians and 26 mm from center (actual position was 0 gradians, 25 mm from center). Thus, the imaged center is within about 1 mm of the center of the actual location. The estimated location of the edge of the object, based upon the errors in distance between each pixel in the border in the image filtered image as compared with the actual location of the object edge from center, was 0.1 +/- 0.9 mm (n = 349 pixels for object, n= 71 pixels for border). Derived offset function and rod tuning curves are shown (Fig. 4).

Image of absorbance at 850 nm of the phantoms containing the four cubes are shown (Fig. 5). For the phantom with two transparent and two opaque cubes, the two transparent cubes appear white while the opaque cubes appear black, demonstrating that absorbance (and, in images not shown, scattering as well) can be determined spatially in phantoms containing multiple objects using optical tomography. The centers of the imaged objects are well located, all near the actual angles from center and displaced on average less than 1 mm inward from their actual location. For the phantom containing the four clear cubes at different depths, the depth of the center of each cube is determinable using system software, and is accurate to within a few millimeters.

Once TOFA curves were collected and stored to disk, image processing for a scan with 10,000 measurements allowed generation of an image within two minutes. Thus, the actual image generation time for the curvilinear backprojection method is rapid. Partial angle reconstructions can also be performed, using a restricted range of emitter and detector angles, such as might occur under true clinical conditions (not shown).

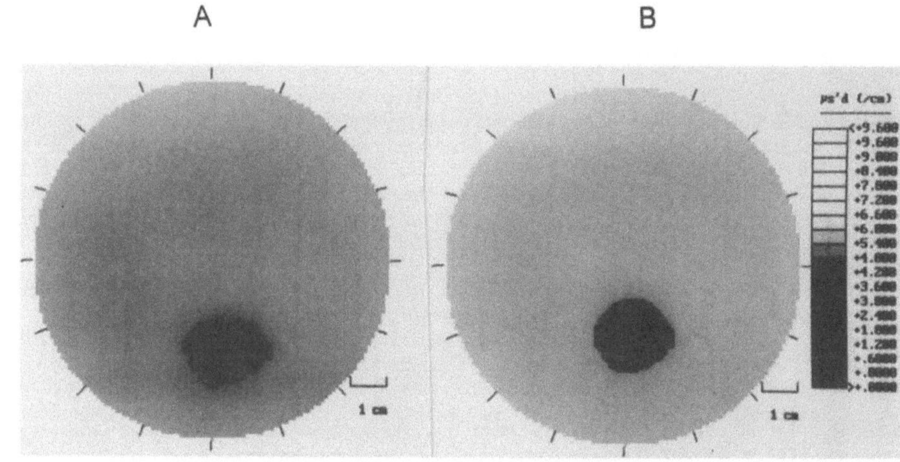

Figure 3. Images of the phantom with single rod (A) before and (B) after correction of curvilinear paths for optimized image shape and location. Before optimization, the shape of the rod is distorted. After optimiziation, image center is located within 1 mm of the actual location, and the mean error from roundness for each pixel is under 1 mm. Both images are after a post-processing filter to better define the edges of the rod.

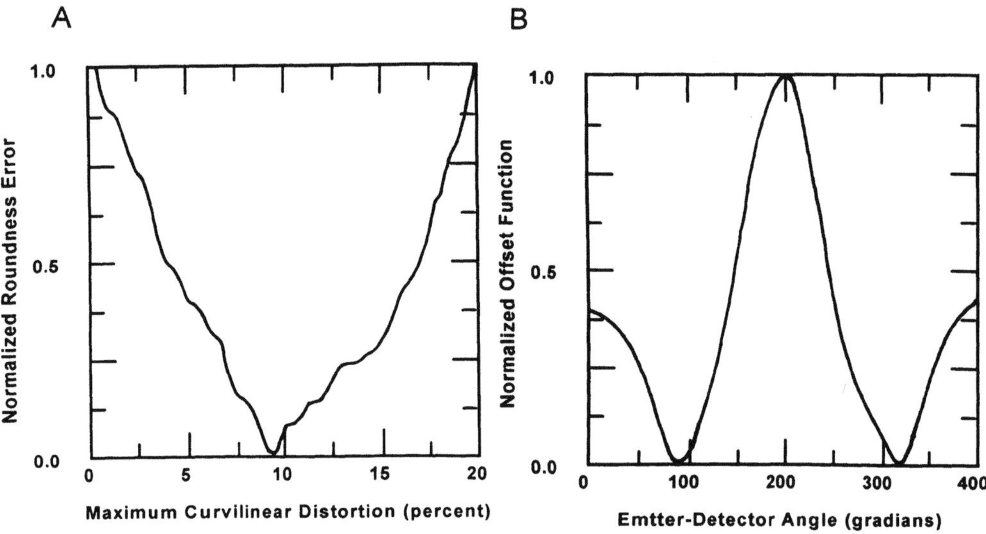

Figure 4. Offset function (A) and curvilinear distortion (B) as calculated from the homogeneous phantom and rod phantom calibrations, shown as normalized fractional distortion of value or path versus detector separation.

DISCUSSION:

This study demonstrates several key issues central to developing an effective optical tomograph. First, optical imaging can be performed in a true tomographic configuration using a fast imaging algorithm that is tolerant of real-world, limited-range data. Second, absorbance and scattering can be effectively separated as an image, which is an essential step in chemical analysis using multiwavelength absorbance data and in histological analysis using multiwavelength scattering data. Last, multiple objects can be effectively imaged, even though the assumptions used in this method are based upon homogeneous media.

Many groups are working upon algorithms to allow optical imaging from tomographic raw data.[15-22] Most are using nonlinear techniques to iteratively determine photon paths. For example, Arridge and Delpy report a recursive procedure in which an image is formed, then a second routine is utilized to estimate what data the calculated image would have produced, which is then compared back to the original data. Based on this comparison, a correction is made and the image is recalculated, and an image is once again produced. The process repeats until the error function falls below a predefined threshold, or until a set number of iterations have taken place.[23,24] Barbour also describes a recursive routine which is self-modifying.[25] However, the utility of such algorithms is currently limited by the enormity of the number of calculations required. On the other hand, a linear algorithm allows reasonable images to be rapidly generated, while retaining absorbance and scattering data.

Linear post-processing algorithms for data and/or images thus may play a large role in optical imaging, and have been underutilized. We show that first-pass linear solutions rapidly generate images, supporting that such techniques are valid. Similarly, Hebden has produced good images uses filtered backprojection. We expect that such images could be substantially improved, either using limited recursion, and look forward to complex solutions from our colleagues to guide us in this pursuit. The rapid convergence of our simple algorithm may allow for fast, though not completely accurate, production of images in vivo. Ultimately, such linear models could be used as a seed image in truly recursive methods, and a better image could be solved once the initial linear image is rapidly determined, if time permits

Once an image is formed, post-processing correction algorithms that are self-refining though non-recursive have allowed further sharpening of such images. For example, the human body has very few density gradients, and thus a smart processor can help eliminate fuzzy edges by applying known information to the image, as demonstrated in this study. For example, the image of the rod was generated by assuming that the rod is solid, and thus that an edge finding post-processing filter can clean the picture up remarkably well. Similarly, the original data is not obtained in a vacuum, as it were, but certain factors are known, such that filter algorithms can be applied to the initial data itself in order to improve the signal. For example, given an emitter-detector spacing which is known, the time of earliest detectable photons is calculable. Also known is the general shape of the time of flight curve, as well as the response times of the laser, detector, etc. All of this information can be used as a filter, convolved with the original data, to increase signal to noise. Applying this filter wisely will give results that allow more accurate imaging. Such an approach, though not for these expressed reasons, has been adopted by Hebden et al., who have been fitting their time of flight curve to a diffusion equation approximation, and find that imaging is improved though an "estimation of early detected photon intensity." It is important to note that, even thought the late-arriving tail of the data contains less spatial information that the leading edge,

220

use of the full time-of-flight curve still allows superior data processing, for example by improving the estimation of the shape of the early portion of the time-of-flight curve in which the signal is typically more noisy than in the later portions. Thus, use of the known shape of the ideal full time-of-flight curve, and use of known factors of the medium, can be used as *a priori* factors to improve images. This suggests, further, that the combination of ultrasound with OI, or CT or MRI with optical imaging, would yield functional information more readily solved onto a structural map of high accuracy obtained via a non-optical method.

While our work has been in the time-domain, our approach would work equally well with other methods. Both time- and space-domain filters have allow estimation of absorbance or scattering in one region of a scattering medium. Use of space-domain continuous wave radiation methods recently developed[26-30] or frequency-resolved measures [31-33] are important steps toward inexpensive regional tissue chemistry and histology determinations, particularly as the cost of instrumentation may be less than for time-resolved approaches.

While such images contain some degree of distortion, we believe that such distortion will be acceptable for certain clinical conditions, such as for monitoring of lung function, imaging of brain hemorrhage, or testing of intestinal oxygenation. Ultimately, the clinical problem addressed will determine whether or not linear algorithms have sufficient resolution for diagnostic accuracy. While imaging of 1 mm breast tumors is difficult,[34] subdural hematomas and areas of cortical stroke can easily be seen using simple probes,[35] and thus should be well visualized using optical tomography. We have had success imaging similar problems in model systems.[36,37]

Last, a method is needed to compare various imaging algorithms. Conventional descriptions of image resolution apply poorly to optical imaging. For example, contrast varies by image type (absorbance, scattering, etc.). We proposed a system of resolution called the minimum detectable object[38] (MDO) in which a series of objects, buried within a defined background material, are imaged. In this model, one parameter varies radially. For example, 1 cm cubes of nonscattering nonabsorbing material could be buried at radially varying depths, such as in Fig. 5B. The resolution factor determined by this model would be the maximum depth at which a similar object could be detected. Alternatively, the objects could all be buried at a depth of 2 cm, while the size of the objects varies, allowing determination of the smallest detectable clear object at a depth of 2 cm. Lastly, the objects could all be of the same size and depth, while the difference from background scattering or absorbance could change, allowing determination of the minimum detectable change in absorbance at a set depth. At some point, standard depths, background scattering and absorbance, and the characteristics of the imaged phantoms will need to be set. This will facilitate comparison between different laboratories. Using this approach, the smallest detectable clear object in this study was 0.3 cm at a depth of 2 cm, and 1.0 cm at a depth of 5 cm, which agrees well with theory.[34,39] The current linear approach can therefore be directly compared to results using other methods on the same data set, allowing objective assessment of progress.

Future studies, already underway, consist of testing in tissues in animals and humans. We are finding good resolution, comparable to the phantom studies. We are constructing a long-term monitor for humans that we call a "neural net," consisting of a fiber optic web for studying infants in the intensive care unit, as well as adults with stroke, thus converting our mechanical scanner to an electronically scanned array that is clinically relevant. Lastly, we look forward to improved algorithms allowing better estimation of absorbance and scattering, as well as improved resolution, as our colleagues in Physics as well as Radiology continue to explore the field.

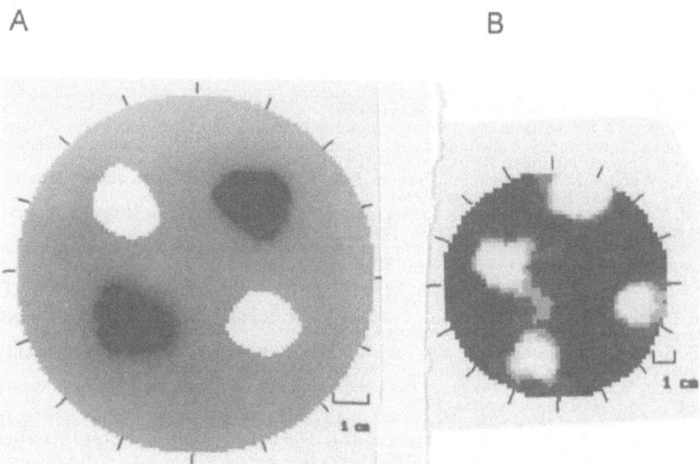

Figure 5. Images of the phantom with four cubes. A) An image of two black and two clear cubes, B) an image of four clear cubes placed at different depths. The location of the centers of each cube is within a few millimeters, and thus is minimally affected by the presence of multiple objects. Both images are after a post-processing filter to better define the edges of the cubes.

ACKNOWLEDGEMENTS

We thank the Office of Naval Research (N-00014-91-C-0170), Walter Berry Fellowship Fund at Stanford, and NIH (PHS General Clinical Research Center M01-RR-00070-30/1 and RR-00081), for partial support.

REFERENCES

1. D.A. Benaron, G. Muller, B. Chance, "A medical perspective at the threshold of clinical optical tomography," In: In: Medical Optical Tomography: functional imaging and monitoring, G. Muller, B. Chance, R. Alfano, et al., eds. (SPIE press, Bellingham WA, 1993), pp 3-9.
2. D.A. Benaron, D.C. Ho, S. Spilman, J.P. vanHouten, D.K. Stevenson, "Tomographic time-of-flight optical imging device," Adv. Exp. Med. Biol., Oxygen Transport to Tissue XVI, Hogan, ed., in this volume (1994).
3. D.A. Benaron, M.A. Lenox, D.K. Stevenson, "Two-D and three-D images of thick tissue using time-constrained time-of-flight and absorbance (tc-TOFA) spectrophotometry," SPIE 1641, 35-45 (1992).
4. D.A. Benaron, D.C. Ho, B. Rubinsky, M. Shannon, "Imaging (NIRI) and quantitation (NIRS) in tissue using time-resolved spectrophotometry: the impact of statically and dynamically variable optical path lengths," SPIE 1888, 10-21 (1993).
5. D.A. Benaron, D.K. Stevenson, "Optical Time-of-Flight and Absorbance Imaging in Biologic Media," Science 259, 1463-1466 (1993).
6. E. Gratton, et al. Bioimaging 1, 40-6 (1993).
7. E.M. Sevick, B. Chance, J. Leigh, S. Nioka, M. Maris, "Quantitation of time- and frequency-resolved optical spectra for the determination of tissue oxygenation," Anal. Biochem. 195, 330-351 (1991).
8. M.S. Patterson, "Time resolved reflectance and transmittance for the non-invasive measurement of tissue optical properties," Appl. Opt. 28, 2331-6 (1989).
9. M. Essenpreis, M. Cope, C.E. Elwell, et al., "Wavelength dependence of the differential pathlength factor and the log slope in time-resolved spectroscopy," Adv. Exp. Med. Biol. 333, 9-20 (1993).
10. S.J. Madsen, B.C. Wilson, M.S. Patterson, et al., "Experimental tests of a simple diffusion model for the estimation of scattering and absorption coefficients of turbid media from time-resolved diffuse reflectance measurements," Appl. Opt. 31(18), 3509-3517 (1992).
11. S.R. Arridge, M. Cope, D.T. Delpy, "The theoretical basis for the determination of optical pathlengths in tissue: temporal and frequency analysis," Phys. Med. Biol. 37(7), 1531-1560 (1992).
12. M. Haida, M. Miwa, A. Shiino, B. Chance, "A method to estimate the ratio of absorption coefficients of two wavelengths using phase modulated near infrared light spectroscopy," Anal. Biochem. 208, 348-351 (1993).
13. B. Chance. N.G. Wang, M. Maris, S. Niok, E.M. Sevick, "Quantitation of tissue optical characteristics and hemoglobin desaturation by time- and frequency- resolved multiwavelength spectrophotometry," In: Adv Exp Med Biol 317, 297-304 (1992).
14. R.F. Bonner, "Model for photon migration in turbid biological Media," J. Opt. Soc. Amer. A 4, 423-432 (1987).
15. J.C. Hebden, R.A. Kruger, "Transillumination imaging performance: spatial resolution simulation studies," Med. Phys. 17, 41-44 (1990).
16. Y. Yamada, Y. Hasekawa, "Simulation of time-resolved optical CT imaging," Proc. SPIE 1431, 73-82 (1991).
17. J.C. Hebden, "Line scan acquisition for time-resolved imaging through scattering media," Opt. Eng. 32(3), 626-633 (1993).
18. B.J. Tromberg, L.O. Svaasand, T-T Tsay, R.C. Haskell, "Properties of photon density waves in multiple-scattering media," App. Opt. 32(4), 607-616 (1993)
19. J.C. Hebden and K.S. Wong. "Time-resolved optical tomography," Appl. Opt. 32, 372-380 (1993).
20. P.C. Jackson, P.H. Stevens, J.H. Smith, D Kear, H. Key, and P.N.T. Wells. "The development of a system for transillumination computed tomography," Br. J. Radiol. 60, 375-380 (1987).
21. J.C. Hebden, R.A. Kruger, and K.S. Wong. "Tomographic imaging using picosecond pulses of light," in Medical Imaging V: Image Physics, R.H. Schnieder, ed. Proc. SPIE 1443, 294-300 (1991).
22. E.M. Sevick, C.L. Burch, B. Chance. "Near-infrared Optical Imaging of Tissue Phantoms with Measurement in the Change in Optical Path Lengths." In: Oxygen Transport to Tissue XV, P. Vaupel, ed. (Plenum Press, NYC, 1994), in press.
23. S.R. Arridge, M. Schwieger, D.T. Delpy. "Iterative reconstruction of near infrared absorption images." Proc. SPIE 1767, 372-383 (1992).
24. S.R. Arridge, "The forward and inverse problems in time-resolved optical imaging." In: Medical Optical Tomgraphy: functional imaging and monitoring, Muller G, Chance B, Alfano R, et al, eds. SPIE press, Bellingham WA (1993). pp 35-64.
25. R.L. Barbour *et al.*," Imaging of diffusing media by a progressive iterative backprojection method using time-domain data," *SPIE* 1641:21-34 (1991).
26. G. Yoon, DN Ghosh Roy, RC Straight, "Coherent backscattering in biological media: measurement and estimation of optical properties." Appl. Opt. 32(4), 580-585 (1993).
27. S. Jacques, A. Gutsche, J. Schwartz, et al., "Video reflectometry to specify optical properties of tissue in vivo." In: Medical Optical Tomography: Functional Imaging and Monitoring, Mller G, Chance B, Alfano R, et al., eds. (SPIE Optical Engineering Press, Bellingham, WA, 1993), pp 211-226.
28. D.V. Stephens, L. Wang, A.M. Hielscher, S.L. Jacques, F.K. Tittel, "Detection of an object inside a phantom using a spatial filter." Proc. SPIE 2135, in press (1994).
29. J.J. Dolne, K.M. Yoo, R.R. Alfano, "Near-IR fourier space gate and absorption shadowgram images through random scattering media." Proc. SPIE 2135, in press (1994).
30. T.J. Farell, M.S. Patterson, B.C. Wilson, "CCD and neural-network-based instrument for the noninbasive determination of tissue optical properties in vivo." Proc. SPIE 2135, in press (1994).
31. A. Knuttel, J.M. Schmitt, J.R. Knutson, "Spatial localization of absorbing bodies by interfering diffusive photon-density waves," App. Optics 32(4), 381-389 (1993).
32. L.O. Svaasand, B.J. Tromberg, R.C. Haskell RC, T-T Tsay, M.W. Berns, "Tissue characterization and imaging using photon density waves," Opt. Eng. 32(2), 258-266 (1993).
33. J.B. Fishkin, E. Gratton, "Propagation of photon-density waves in strongly scattering media containing an absorbing semi-infinite plane bounded by a straight edge," J. Opt. Soc. Amer. A 10(1), 127-140 (1993).
34. A.H. Gandjbakhche, R.J. Nossal, R.F. Bonner, "Theoretical study of resolution limits for time-resolved imaging of human breast." Proc. SPIE 2135, in press (1994).
35. B. Chance, "Early detection of brain ischemia and hemorrhage by optical methods." Proc. SPIE 1641, 162-169 (1992).
36. D.A. Benaron, J.P. van Houten, D.C. Ho, S.D. Spilman, D.K. Stevenson, "Imaging neonatal brain injury using light-based optical tomography," Ped Research, in press (1994).
37. D.A. Benaron, D.C. Ho, S.D. Spilman, D.K. Stevenson, "Imaging of pulmonary pathophysiology using optical tomography in model systems," Ped Research, in press (1994).
38. D.A. Benaron, M. Lenox, M. Goheen, D.K. Stevenson, "Noninvasive measurement and imaging of tissue structure and oxygenation using infrared time-of-flight absorbance (TOFA) spectrophotometry." Proc. IEEE EMB Soc. 14(6), 2402-2404 (1992).
39. J.A. Moon, R. Mahon, M.D. Duncan, J. Reintjes, "Resolution limits for imaging through turbid media with diffuse light." Opt. Lett. 18, 1591-1593 (1993).

RESULTS OF OPTICAL IMAGING OF BRAIN PATHOLOGY. David A. Benaron, MD. Medical Spectroscopy and Imaging Section of the Neonatal and Developmental Medicine Laboratory, and the Hansen High-Energy Physics Laboratory, Stanford University, Stanford, CA 94035.

Optical imaging (OI) and spectroscopy (OS) use light emitted into scattering media such as human tissue to determine interior structure and chemical content, and have broad application to the field of medicine. Existing medical imaging techniques often are limited in suitability as continuous, noninvasive human monitors, and most do not distinguish between normal, impaired, or dead tissues [1]. Visible and near-infrared light will pass through human tissue in small amounts, and this transmitted or reflected light can be used for constructing *in vivo* images. Light is nonionizing at many wavelengths, and known to function well as a medical probe for measuring the *functional* state of tissue. It is believed that such noninvasive optical monitoring techniques potentially offer reduced pain and discomfort associated with medical therapy, improved health outcome, and lowered costs of medical care. Multiple investigators are now studying in vivo optical approaches (including Chance, Kang, Delpy, Arridge, Cope, Hebden, Sevick, Ferrari, Gratton, Alfano, and others). However, due to the high degree of light scattering in tissue, images formed *in vivo* using optical radiation have been poor, and quantitation rendered nonquantitative by long, irregular photon paths.

We developed a portable time-of-flight and absorbance (TOFA) laser system [2]. Using power well within safe FDA-established guidelines, we have produced OI images of objects buried deep within scattering media [2], and successfully generated a whole-body image of an animal, locating major organs and intestinal gas in a rat [1]. We now set out to test if such an apporach would work in humans and on animal models of human diseases. Converting outr TOFA device to a rotational tomographic scanner, we made TOFA measurements from up to 10,000 measurements pairs or emitter/detector locations. Images were reconstructed using a variety of methods, including simple backprojection, linear inverse Radon transforms, and, in the manner of Arridge *et al.*, using a recursive transform/inverse transform algortihm.

Using such an approach, we have recently obtained optical images of sheep brain *in vitro* and human brain *in vivo*, with resolution to better than 1 cm [3]. In test measurements in humans (obtained after informed consent) and tissue samples, optical images were compared to ultra-sonographic results or gross histological analysis, respectively. Results show that superficial hemorrhages are easily detectable, while some forms of deeper bleeding (intraventricular hemor-rhage, for example) are detectable if of sufficient size. Vertricular spaces are located, though not yet well visualized. Lastly, using computer simulations, we have shown that distributions of absorbance, such as those caused by the maldistribution of oxygen, may be imaged using OS in a spatial manner. Our results, combined with earlier results of other laboratories demonstrating the ability to optically measure cerebral blood flow, blood volume, and oxygenation, support a view that light based cerebral imaging is possible, and that such an approach may form the basis of a new type of monitoring system for the study of neurologic disease in neonates.

It is clear that optical imaging and spectroscopy will be a useful clinical tool, and the only question is now when this will happen, and in what capacity. In order to use to improve upon our device, specific features lacking in the TOFA system need to be corrected, including the inability to monitor mid- to far-IR light, and to generate ps pulses with mW average power at variable wavelengths. We are now in the process of constructing such a device, which should be in operation in the latter half of 1993. In addition, clinical correlation between cellular, animal, and human models of OI and OS need to be performed. Such an endeavor will require testing and cross-correlation with outcome measures and existing imaging methodologies, but we expect this to be well underway within three to five years.

REFERENCES
1. Benaron DA, Stevenson DK. *Science* 1993;259:1463.
2. Benaron DA, Lenox M, Stevenson DK. SPIE 1992;1641:35.
3. Benaron DA, Ho DC, Hahn J, Stevenson DK. Ped Research 1993;33(4):369A.

FOREARM BLOOD FLOW MEASUREMENT BY NEAR INFRARED SPECTROSCOPY

Roberto A. De Blasi [1], Marco Ferrari[2,3], Giorgio Conti[1], Annamaria Mega[1] and Alessandro Gasparetto[1]

[1]Istituto di Anestesiologia e Rianimazione, I Universita' di Roma, Policlinico Umberto I, 00161 Rome, Italy; [2]Dipartimento di Scienze e Tecnologie Biomediche, Universita' dell'Aquila, 67100 L'Aquila, Italy; [3]Laboratorio di Biologia Cellulare, Istituto Superiore di Sanità, 00161 Rome, Italy.

INTRODUCTION

Evaluation of muscle contribution to total body flow can be very useful in physiological and clinical conditions. Therefore a simple non-invasive technique for the measurement of muscle blood flow based on near infrared spectroscopy (NIRS) principles could find many applications. The commonly used methods for skeletal muscle flow measurement are complex and difficult to perform repeatedly under different workload conditions. We applied NIRS for the measurement of forearm blood flow (FBF) in volunteers and critically ill patients.

METHODS

Eigth healthy subjects were studied both at rest and after hand exercise during vascular occlusion. FBF was also measured by strain gauge plethysmography. Venous occlusion was realized by a pneumatic cuff rapidly inflated to a pressure of 50 mmHg for 50 s and then released. The reactive hyperemia that follows forearm circulatory arrest and repetitive exercise was also measured.
NIRS measurements were obtained using the NIRO500 (Hamamatsu Photonics, Japan) (1) and a methodology recently described (2). FBF was calculated from NIRS data by evaluating the rate of the oxy-Hb and deoxy-Hb increase ($d[HbO_2]/dt$ and $d[Hb]/dt$) following venous occlusion. FBF was also measured by venous occlusion plethysmography (SPG-16, Meda Sonics, Mountain Wiew, CA).

RESULTS

The mean value for NIRS FBF at rest was 2.0 ± 0.9 and 8.4 ± 3.4 ml·100ml^{-1}·min^{-1} after hand exercise. Sixteen FBF values, determined by NIRS from 8 subjects, were regressed against the corresponding plethysmography measurements. The r value was 0.94 and significant ($P<0.01$). FBF was also measured in four septic patients. Different values were obtained ranging from 0.3 to 7.9 ml·100ml^{-1}·min^{-1}.

CONCLUSIONS

NIRS allows to perform repetitive measurements of FBF in normal and pathological conditions. NIRS determined FBF values are expected to be more specific than plethysmography.

References
1.Cope, M., D.T. Delpy, E.O.R. Reynolds, S. Wray, J.S. Wyatt and P. van der zee. **Adv. in Exp. Med. Biol.** 222: 183-189, 1988.
2.De Blasi, R.A., M. Cope, C. Elwell, F. Safoue and M. Ferrari. **Eur. J. Appl. Physiol.** (in press).

Object localization in brain and breast by a phased array system

B. Chance, K. Kang, L. He[1]

Johnson Research Foundation

D501 Richards Building, Dept. of Biochemistry/ Biophysics, Phila., PA 19104-6089

Last year's ISOTT contained a report on the sensitivity of a phased array system for locating an object of .8 mm in a highly scattering medium at 3 cm within a highly scattering medium. Since then the sensitivity of the method has been exploited and detection of small objects as small as 40 microlitres containing 4 picomoles of absorber (cardiogreen) has been reported[1]. Furthermore, a sharp phase transition and amplitude null has been verified in the human breast. (see Fig. 1)

The automatic localization of objects by electronic scanning methods is described here which is expected to retain the high sensitivity of the static responses of Fig. 1, yet at the same time, allow scanning of the transmitting and receiving position by a number of laser diodes. The system relies for its sensitivity upon the diffusive waves with which the phase coherence is maintained in the near field of the "10 cm waves." Figure 2 shows how a variety of transmitters and receivers can be used to explore the compressed breast for the presence of a hidden absorber, presumably a tissue volume containing hemoglobin concentrations higher than the background due to relatively high metabolic activity of the tumor as compared to the background adipose tissue. A series of transmitters can be connected by a switch to either in-phase or anti-phase operation shown with T1 to T4, phase 0°, and T5 to T8, phase 180°. However, particular pairs of transmitters would be used appropriately to scan the nodal plane through the region in which the object would be expected to be found. Corresponding detectors which would be aligned in the nodal plane of the transmitter system are D1-D4. Thus, any combination of transmitters and receivers would give an undisturbed nodal plane in regions where the object was not present, but a distorted nodal plane where the object was located.

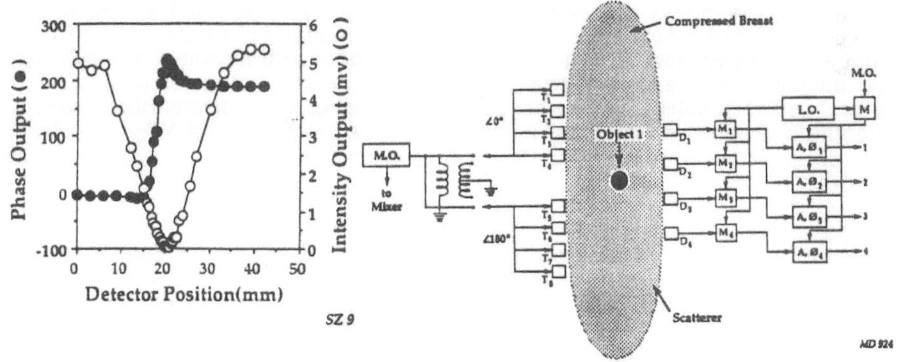

In order to identify the location of the absorber, the nodal plane of the phased array can be perturbed by slightly changing initial amplitude or phase and provide opportunities for fast scanning in two and possibly three dimensions for highly localized single targets. This technology was originally employed as lobe switching radar (as exemplified by SCR584) and was shown to have high precision and excellent signal to noise ratio in automatic tracking. The detection of small optical anomalies in the human breast relies principally on the absorption rather than the scattering coefficient and is largely independent of the modulation frequency.

Mitochondrial redox state as a "Gold Standard" of tissue hypoxia
Britton Chance
Dept. of Biochemistry and Biophysics, University of Pennsylvania School of Medicine, D501
Richards Bldg., Phila., PA 19104-6089

Introduction

The need for a "gold standard" of tissue oxygenation is apparent from the difficulty of ascertaining a "critical level" of tissue sufficent to support biological function. Two measures of mitochondrial capability to produce ATP at low oxygen tensions are available: 1) NMR measurement of PCr/ Pi 2) Optical measurement of the flavoprotein/ pyridine nucleotide redox state of the mitochondrial matrix. The latter is directly related to the membrane potential, and ATP synthesis, and the calcium uptake capabilities.

Localization

The ^{31}PNMR requires large volume elements compared to the capillary/tissue relationship, whilst the redox scan can be measured in three dimensions at <25μ resolution. Myoglobin deoxygenation by both optical and MRS measurements indicates the cytosolic oxygen tension, with the caveat that the location of myoglobin in the pericapillary and the perimitochondrial spaces is unknown and may be non-constant in normoxia and hypoxia. Hemoglobin desaturation localized in the arterial, capillary, and venular bed gives the driving force for oxygen diffusion to the cytoplasm and to mitochondria and maybe quantitated by time resolved optical technology. In addition, exogenous probes of oxygen concentration, pyrene butyric acid, palladium porphyrins etc. are sensitively quenched by O_2, but may leave the blood circulation and thus require distributions.

Speed of Response

The fluorescence method is fast enough to give the kinetics of deoxygenation and reoxygenation consequent to alteration of FIO2 in a variety of organs, particularly in the rat, but also in the mouse and eventually the dog models. The rapid and sensitive in vivo response of NADH redox state to stimulation in vivo afforded further support for the idea that ADP formed from ATP, a primary energy donor and activates ATP synthesis. Nuclear magnetic resonance spectroscopy (MRS) in the late 70's and early 80's put a "lynch pin" to the problem by time resolved measures of phosphocreatine breakdown and inorganic phosphate formation appropriate to large changes of VO_2. The freeze-trapping of the redox state of tissues has a time resolution of <1 sec and can resolve fast changes of metabolic state.

Reversibility

The observed constancy of ATP and the rapid depletion of PCr in muscle function maintains a metabolic homeostasis suggesting that the control of oxidative metabolism in muscle tissue is largely a kinetic one. The lack of thermodynamic control is attributed to the lack of reversal of electron transport in the terminal part of the respiratory chain under aerobic conditions together with the more recent results of LaNoue (pers comm) indicating that the ATPase was essentially irreversible when the ATP concentrations commensurate with that in the mitochondria matrix space.

O_2 gradients

The O_2 gradients from the cytosol to the mitochondrial matrix space can be quantified by time resolved optical spectroscopy of the arterioles, capillary venular hemoglobin saturation, by ^{1}HNMR measurements of myoglobin desaturation in vivo and of the mitochondrial matrix space and by the Fp/Fp + PN redox state by freeze-trap redox scan.   The working myocardium testifies the steepness of the tissue O_2 gradients and the low O_2 concentration at the mitochondria.

DEVELOPMENT OF A CLOSED CIRCUIT SYSTEM FOR PEDIATRIC ANESTHESIA
AND ON-LINE $\dot{V}O_2$ MONITORING

E.P. Eijking, J. Bezstarosti, B. Westerkamp, G. van Dijk, A.E.E. Meursing and W. Erdmann
Depts. of Anesthesiology, Erasmus University, University Hospital Dijkzigt and Sophia
Children's Hospital, Rotterdam, The Netherlands

At previous ISOTT-meetings an automatic feed back controlled closed circuit anesthesia
ventilator system has been described (Physioflex) [1, 2]. In brief: in an internal circuit system
gas is moved unidirectionally by a blower at 70 l/min. This internal circuit (which is connected
to the patient) is separated from the external pressurized circuit by metal membranes in four
membrane chambers. By changing the pressure in the external circuit the metal membranes
move, resulting in ventilation of the lungs. O_2, N_2O and volatile anesthetics are introduced into
the internal system and the concentrations of these gases are feed back controlled. On-line
registration of flow, tidal volume and respiratory pressures, as well as a capnogram, lung compliance and oxygen consumption ($\dot{V}O_2$) are of great value [1]. Nowadays the Physioflex is
accepted for routine anesthesia in adults.

Currently this closed circuit system is being adjusted for use in pediatric anesthesia.
Anesthesia in small children (e.g. newborn babies) is confronted with specific practical pitfalls,
for example, induction of anesthesia by mask and the use of uncuffed endotracheal tubes,
causing large air-leaks. In the development of the Physioflex for use in pediatric anesthesia a
few adjustments have been made:
1.   In small children undergoing general anesthesia with the Physioflex, mask-induction was
complicated by large disturbance of the gas composition in the internal circuit. This was caused
by suction of room-air via the mask due to high flow in the internal circuit (Venturi effect). By
increasing fresh gas flow in the internal circuit (>500 ml/min), this problem could be
overcome.
2.   As a result of accumulation of unwanted gases in the internal circuit (e.g. N_2 wash-out,
and methane absorbed from the gastro-intestinal tract), the system has to be flushed once in a
while. In the "old" system by pressing the flush-key, O_2 and N_2O valves are opened and fresh
gas is delivered into the system with a flow of 2.5 l/min. Due to overfilling of the system the
flush valve is opened ("pop-off") at expiration, with subsequent loss of surplus gas. When
$FiO_2 > 0.8$ is required, it takes 3-4 min to achieve this. In the "new" system, after pressing the
flush-key, fresh gas is delivered to the internal system at 4 l/min with simultaneous regulated
evacuation of surplus gas via the flush valve, which is connected to the vacuum system. During
this so-called "pressure-balanced flush", pressure in the internal system does not change and
very rapid change in the system is achieved (e.g. $FiO_2 > 0.8$ within 1.5 min).

In conclusion: development of a closed circuit system for pediatric anesthesia offers
great improvement for both patient care and clinical (anesthesia) practice. FiO_2 remains exactly
according to set values and can be changed rapidly, whereby O_2 inflow minus fraction of
leakage equals O_2 uptake by the patient. Continuous monitoring of $\dot{V}O_2$ as a vital parameter is
becoming available in small children.

1.   Verkaaik, A.P.K., Erdmann, W., van Dijk, G and Westerkamp, B., 1992, On-line
oxygen uptake measurement ($\dot{V}O_2$): a computer feed-back controlled rebreathing circuit for long
term oxygen uptake registration, Adv. Exp. Med. Biol., 316: 195.
2.   Verkaaik, A.P.K., van Dijk, G., Westerkamp, B. and Erdmann, W., 1992, Gas
exchange in the lung, computer feed back controlled physiological matching of artificial
ventilation, Adv. Exp. Med. Biol., 317: 325.

Gaseous Oxygen Monitoring Using A Membrane Immobilised Phosphorescence Probe

P.M. Gewehr, D.T. Delpy[*].

Dept. de Eletronica, CEFET-PR, Curitiba, Brazil
*Dept. of Medical Physics & Bioengineering, UCL, London, UK.

A range of phosphorescence dyes are now available which have lifetimes at room temperature which are sufficiently long to be measured using relatively inexpensive flashlamp based instrumentation[1,2]. These can be used to monitor oxygen through the quenching effect of this gas, the monitoring of lifetime rather than intensity having the advantage that the measurements are not affected by variations in excitation lamp intensity or alterations in dye concentration. The dye palladium coproporphyrin has been widely used for this purpose for the monitoring of intravascular oxygenation[3].

By immobilising this material in a polymer membrane, it is possible to monitor gaseous oxygen, and potentially to produce an inexpensive but stable sensor. We have examined a range of polymers for this purpose including polyvinyl chloride (PVC), polymethyl Methacrylate (PMMA) and polystyrene. Phosphorescence lifetimes have been measured using a simple flashlamp based system[4]. As expected, the phosphorescence lifetimes vary with the permeability of the membranes, being longest for PVC and shortest in polystyrene. Using a membrane of 25-30μm thickness, a response time (63%) of 2 seconds is achievable using polystyrene.

The Stern-Volmer plots for all of the polymers show a slight non-linearity, this being least for polystyrene and greatest for PVC. This effect can be compensated for by a modification to the Stern-Volmer relationship which assumes either that a percentage of the dye in the membrane remains unquenched (ie. inaccessible to the oxygen), or that the dye exists in regions of differing permeability. If this latter assumption is correct, the use of a more amorphous polymer should improve the linearity of response.

<u>References.</u>

1. Vanderkooi, J.M. et al. 1986. Adv. Exp. Med & Biol., 200, 189-193.
2. Green, T.J. et al. 1988. Anal. Biochem., 174, 73-79.
3. Rumsey, W.L. et al. 1989. Science., 241., 1649-1651.
4. Gewehr, P.M. et al. 1993, Med. & Biol. Eng. & Comp., 31, 2-21.

New Intravital Skeletal Muscle Preparation for Oxygen Transport Investigations

Ravi S Gill[1], Richard F Potter[2], William J Sibbald[2], Christopher G Ellis [1].
Department of Medical Biophysics, Robarts Research Institute [1],
University of Western Ontario
A.C. Burton Vascular Laboratory, Victoria Hospital[2]
London , Ontario Canada.

Introduction

A dual camera intravital video microscopy system has been developed to measure the oxygen saturation of red blood cells flowing through capillaries in vivo (Ellis et al., 1990), and to display and directly evaluate red blood cell dynamics in these same vessels (Ellis et al., 1992). Using this system one can obtain a complete description of the convective transport of oxygen in capillaries in vivo. This new technology has been applied only to the retractor muscle of hamsters (Ellsworth and Pittman, 1990; Stein et al., 1993). The objective of this study was to develop a similar muscle preparation in the rat that could be used to compare oxygen transport findings between species which would be relevant to other ongoing projects in the laboratory studying hyperdynamic sepsis.

Methods

Male Sprague-Dawley rats (weight 150 - 200g) are anaesthetized with pentobarbital 6.5mg per 100g for induction and 3.25mg per 100 g every 45 mins for maintenance. After insertion of hemodynamic monitoring and blood sampling lines (carotid and jugular) the *extensor digitorum longus (EDL)* muscle of the rat hind limb is exposed as described by (Tyml and Budreau, 1990). The distal tendons of the muscle are exposed , ligated and excised. The muscle is then reflected along its length and the dorsal loose connective tissue is excised up to the midpoint where the venous drainage of the muscle empties into the saphenous vein. The animal is then placed prone on the microscope stage with its hind limb extended at the hip, flexed at the knee and the foot supported in a clamp. The EDL is then reflected perpendicular to the limb with the ventral surface on the microscope stage and the dorsal surface viewed by transillumination of the muscle.

The core temperature of the animal is monitored via a rectal probe and maintained at 38°C with a radiant heat lamp. The reflected muscle is kept moist during surgery with warm saline and its temperature is monitored with a fine thermocouple. On the stage the muscle is placed within a well to prevent desiccation and covered with a film of Saran. This is impermeable to oxygen, has appropriate optical properties and allows the unimpeded placement of fine micro PO_2 electrodes (Stein et al., 1993). Transillumination of the muscle is provided by a 100W mercury vapor lamp with a constant light intensity. Observations are made using a Leitz Metallux 3 microscope. The microscope image is simultaneously viewed, at the two wavelengths necessary for O_2 saturation determination, by two high resolution closed circuit video systems. Each system consists of a silicon-intensified-target (SIT) camera (Dage -MTI model 66), video monitor (Panasonic model WV-5410), and videocassette recorder (Panasonic AG-7300). Time code information (Telecom Research) is recorded onto the audio track of each video tape simultaneously.

This preparation has permitted us to obtain high quality video images of microvascular blood flow suitable for analysis of oxygen transport.

VERY LOW FREQUENCY ELECTRON PARAMAGNETIC RESONANCE ALLOWS THE
IMAGING OF OXYGEN CONCENTRATIONS IN DISTINCT PHARMACOLOGIC
COMPARTMENTS DEEP IN LIVING TISSUE

Howard J. Halpern[1], Cheng Yu[1], Miroslav Peric[1], Eugene Barth[1], David Grdina[1] and Beverly A. Teicher[2], [1]University of Chicago, Chicago, IL and [2]Dana Farber Cancer Center, Boston, MA

The use of electron paramagnetic (spin) resonance (EPR) for measurements in living tissue has been limited by the depth of penetration of multi-GHz microwaves which stimulate the transitions detected by the technique. The use of much lower frequency stimulating radiation, 250 MHz, characteristic of the highest frequency human nuclear magnetic resonance imaging (MRI) measurements, allows penetration deep (of order ~10 cm)) in human tissues. Frequency reduction entails a loss in signal-to-noise, although less than is often assumed. Using selectively deuterated spin labels as oximetric spin probes, spectral fitting techniques, and sample size specific resonators, we have regained the signal lost in the frequency reduction. The oximetric spin probes used can be modified at a spectrally irrelevant site to redirect its pharmacologic compartment distribution. Since the detection is a magnetic resonance technique, field gradients allow imaging of the murine tumors used as samples. We present measurments from FSa and NFSa fibrosarcomas grown in the thighs of C3H mice of various sizes above and below 1 cm maximum diameter. We detect significantly lower oxygen concentrations in the larger tumors. This is unlikely to be due to the smaller proportion of normal tissue in the larger tumor samples. The excellent tumor distribution of the spin probe at sub millimeter resolution is documented using the broadening it induces in T_1 weighted ^{1}H MRI. An oxygen sensitive EPR image with a resolution of approximately 3mm using spectral spatial imaging will be presented. These measurements, in small murine tumors, are rigorous tests of the technique, more so than in the larger samples presented by human tissues and tumors.

Work supported by NIH grant CA 90654 and the Balaban Foundation

Local determination of oxygenation and cell vitality in multicellular spheroids by microelectrodes and photometry

G. Holtermann, A. Görlach, N. Opitz, M. Wartenberg, H. Acker

Max-Planck-Institut für molekulare Physiologie, Rheinlanddamm 201, D-4600 Dortmund 1

Multicellular spheroids are nearly spherical aggregates of cultured tumor cells. They develop radial proliferation gradients and are used as models to study radiation as well as drug effects for improving therapeutic strategies. As the knowledge of oxygenation as well as cell vitality in the spheroids is of importance to understand antiproliferative measures methods have to be applied which are able to resolve the three-dimensional architecture of the spheroids without being destructive. Microelectrodes sensitive to oxygen or ph and specifically designed to record extracellularly symmetrical PO_2- and pH-gradients give information about the relationship between oxygen consumption and lactate production in different depths of the spheroids. The quotient delta PO_2/delta pH can be related to radiation sensitivity and growth rate of the tumor spheroids. Combination of measurements with PO_2-sensitive microelectrodes and photometry of the respiratory chain (Cytochromes, NADH, FAD) give information about the critical PO_2 in different depths of the spheroids or regulation of oxygen consumption by glucose. After staining of tumor cells in spheroids with vital/letal fluorescence dyes for characterisation of areas with proliferative or necrotic cells fluorescence intensity measurements in depths up to 200 mm can be obtained by confocal laser microscopy. Symmetrical distribution of FDA (Fluoresceindiacetate) showed a decline of viably stained cells in the centre at a diameter of 40 - 100 mm with an appearance of Lucifer Yellow VS stained cells in the centre, indicating starting necrosis. Using the pH-sensitve fluoroprobe carboxy-SNARF-1 intracellular pH (pH_i) measurements can be obtained from tumor cells in spheroids by confocal laser microscopy. Values of 7.2 recorded under control conditions are comparable with pH_i values measured with pH-sensitive microelectrodes. The use of different techniques to resolve three-dimensional cell activities without disturbing the geometry of viable multicellular spheroids leads to a new approach to assess the efficacy of various therapeutic measures on tumor cells.

FIBER OPTIC PROBE FOR NEAR INFRA-RED MONITORING OF DEEP MUSCLE

Paul Jöbsis
Scripps Institution of Oceanography, La Jolla Ca, 92093-0204, USA

The use of near infra-red spectroscopy to monitor the oxygenation state of tissue *in vivo* with surface mounted probes has been reported (1). However selective monitoring of tissue located deep in the body has required surgery to isolate the tissue and position probes on the tissue surface. To eliminate this requirement a catheter-like fiber optic probe was developed to monitor deep muscle oxygenation.

The fiber optic probe has a six inch stainless steal sheath with a 14 gauge outside diameter. The probe contains 24, 200 micron hard clad silica fibers; 3 light emitting fibers (+ 3 spares), 1 fiber returning light to a near field reference photo detector, and 17 fibers returning light to the signal photo detector. The probe was connected to a Niroscope (Vander corp.) and used to monitor changes in muscle oxygenation during sleep apnea episodes in the iliocostalis lumborum / longissimus dorsi muscle group of Northern Elephant Seal *Mirounga angustirostris* pups during sleep apnea episodes. Examples of probe design and preliminary results will be discussed.

1) Jöbsis, FF Science 198: 1264-7, 1977.

INTRACELLULAR OXYGENATION AND REDOX STATUS IN MUSCLE

Frans F. Jöbsis-VanderVliet[1] and Paul D. Jöbsis[2]
[1]Duke University Medical Center, Durham, NC 27710, USA
[2]Scripps Institution of Oceanography, La Jolla, CA 92037, USA

Near Infra Red (NIR) spectroscopy can be applied non-invasively to organs and tissues *in vivo*. Spectra can be taken through organs as thick as the human head for identification of the absorbing components (1). And it is possible to monitor the reactions of the absorbing molecules to physiological manipulations with a multi-wavelength, differential spectrophotometer designed for the purpose, as was shown for the first time at the 1984 ISOTT meeting (2). The primary variable is oxygen delivery to the tissue or organ and the instrument has therefore been called the NIR Oxygen Sufficiency Monitor or NIROSCOPE (c.f. 3 for a brief review).

Five oxygen-dependent signals are available in the NIR region in which *in vivo* monitoring can be performed (the 700-1300 nm range). First and foremost, the redox state of cytochrome *c* oxidase can be monitored. The enzyme (a.k.a. cytochrome aa_3) catalyzes about 90% of all oxygen usage by the body. Hemoglobin signals, from the oxygenated form (HbO_2) as well as from the de-oxygenated one (Hb), provide separate information on venous and arterially colored blood in the regional circulation. Similarly myolobin, in both its MbO_2 and Mb form, contributes signals concerning the intracellular oxygenation of the muscle cells. Fortunately the hemoglobin and myoglobin absorption spectra are sufficiently similar that they can be monitored with a single algorithm. Unfortunately this similarity has precluded separating their signals on a spectrophotometric basis alone (4).

The purpose of our on-going study is to distinguish between the hemoglobin and myoglobin reactions. In this first report we will show that this can be achieved by the manipulation of physiological and artifical parameters.

1) Jöbsis, FF, Science, 198:1264; 1977; 2) Jöbsis-VanderVliet, FF, Adv Exp Biol Med, 191: 833; 1985; 3) Jöbsis-VanderVliet FF, Fox E and Sugioka K, Internat Anesth Clinics, 25: 209; 1988; 4) Piantadosi CA, Hemstreet TM and Jöbsis-VanderVliet FF, Crit Care Med, 14:698; 1986; Hampson MB, Piantadosi CA, J Appl Phys 64:2449; 1988.

CONCERTED OXYGEN-17/PHOSPHORUS-31 MAGNETIC RESONANCE SPECTROSCOPY:
A NOVEL APPROACH FOR *IN VIVO* CORRELATION OF OXYGEN CONSUMPTION AND
PHOSPHATE METABOLISM

G.D. Mateescu and D. Fercu
Department of Chemistry, Case Western Reserve University, Cleveland, Ohio 44106

After discoveries that initiated the understanding of the structural and functional organization of the cell (A. Claude, C. de Duve and G. Palade, Nobel Prize 1974) and the definitive description of the fine structure of mitochondria (Palade et al., 1948-1952), *in vitro* studies of oxidative phosphorylation (Lehninger et al., 1950; B. Chance et al., 1955, 1985;) yielded valuable information concerning the effects of inhibitors and uncoupling agents (Pietrobon et al., 1987). Recently, a discussion of a paradigm for aging and degenerative diseases (D.C. Wallace, 1992) emphasized the possible relation of mitochondrial DNA mutations with defects in oxidative phosphorylation (OXPHOS). In order to obtain reliable correlations between oxygen consumption and ATP formation it is necessary to avoid, on one side, the uncertainties due to *in vitro* measurements and, on the other, the variations in metabolism due to other factors than mutants or uncoupling agents. We report here preliminary results of the first attempt to make virtually simultaneous *in vivo* ^{17}O and ^{31}P magnetic resonance spectroscopy (MRS) measurements.

Method. Larvae of the *mealworm (Tenebrio molitor)* are introduced into a minirespirator designed to minimize losses of $^{17}O_2$ enriched air and to allow addition of known quantities of chloroform. The minirespirator is placed in the probe of a 4.7 Tesla NMR spectrometer maintained at constant room temperature. ^{17}O and ^{31}P spectra are then taken in immediate succession. Each pair of measurements is repeated at 30 min intervals.

Results and Discussion. The rate of formation of *nascent mitochondrial water* (Mateescu et al., 1991; Pekar et al., 1991) and the evolution of phosphate metabolites during the treatment with chloroform are illustrated in the ^{17}O and ^{31}P figures shown below. The rapid increase of metabolic water and P_i with initial increase of phosphoarginine and decrease of ATP are consistent with an inhibition of the ATPase by chloroform and the enhancement of oxygen consumption. It is thus seen that interleave ^{17}O - ^{31}P MRS constitutes a very promising method for accurate evaluation of the relation between oxygen consumption ant ATP formation as their interdependence is perturbed by external agents. Systematic measurements will be conducted in our laboratory in order to gather statistically significant results. Besides chloroform, which is not a typical OXPHOS uncoupler, we will experiment with classical uncouplers and a series of DNA mutants.

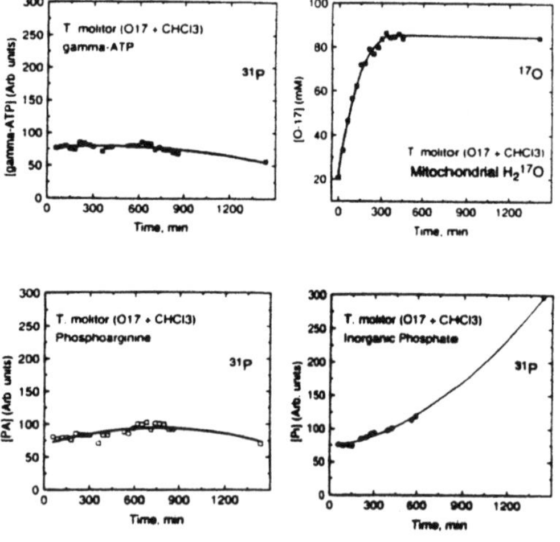

A Preparation for the <u>in vivo</u> Study of Capillary and Fiber Structure-Function Relationships in Rat Diaphragm

<u>David C. Poole</u>, Dept. of Medicine, UCSD, La Jolla, CA. 92093-0623

<u>Introduction.</u> The mammalian diaphragm is adapted superbly to achieve high sustained rates of ATP turnover. Maximal blood flows, capillary volume and oxidative enzyme capacities far exceed those found in other skeletal muscles. Furthermore, decrements in diaphragm function consequent to disease radically impact the physiological capabilities of the individual. The rat costal diaphragm is only 200-500 μm thick depending on the imposed stretch. Thus, structurally the diaphragm lends itself to transillumination and optical examination. A novel preparation for the <u>in vivo</u> examination of the diaphragm microcirculation has recently been described by Boczkowski and colleagues and facilitates examination of vessel diameters and the presence or absence of blood flow (<u>Microvasc. Res.</u> 40, 1990). The purpose of the present report is to detail modifications which permit: a) Avoidance of the Trendelenburg position which can invoke myogenic and sympathetic effects on skeletal muscle microcirculation (Segal and Puri, <u>Med. Sci. Sports Exerc.</u> 25, 1993. b) Avoidance of severe tissue stretching. c) Visualization of muscle sarcomeres and adjacent capillaries. d) Measurement of capillary red blood cell (RBC) velocity.

<u>Methods.</u> Under sodium pentobarbitol anesthesia, Sprague-Dawley rats ($\sim$ 250 g) underwent tracheostomy and laparotomy and iliac arterial cannulation procedures. While being mechanically ventilated, a unilateral thoracotomy was performed in the 5th intercostal space and the diaphragm separated gently from the lungs and mediastinal tissues. The costal diaphragm was exposed by midline and transverse incisions through the abdominal wall and reflecting the liver carefully downward while severing any diaphragm-liver connective tissue attachments. The chest wall was supported and elevated using sutures attached to the overlying skin and secured to a rigid support. The surface of the exposed organs were maintained moist using physiological saline solution. The costal diaphragm was transilluminated via the thorax using a fiber-optic probe (head diameter $\sim$3 mm). The Olympus Videomicroscope (OVM 1000 N) has a high-resolution CCD camera and a 1000x objective (with contact or non-contact head). The probe is small and highly mobile and using clamp support, was maneuvered into position on the inferior (abdominal) aspect of the diaphragm. Images were displayed on a Trinitron high resolution monitor and recorded at 30 frames/s using a JVC (BR-S378U) super-VHS recorder equipped with jog shuttle control.

<u>Preliminary Observations.</u> The optical clarity of the images were excellent and in many instances sarcomeres were clearly visible. Capillary RBC velocity measured using frame-by-frame analysis indicated great heterogeneity of flow. Within the same fields, we observed capillaries in which RBC's were either: a) stationary for 1-3 s, b) moving rapidly i.e., 1000-1600 μm/s, or c) moving at intermediate speeds of 300-1000 μm/s. In addition, there were instances of adjacent capillaries with countercurrent flow.

<u>Conclusions.</u> This preparation facilitates study of the <u>in vivo</u> diaphragm capillary bed. Optical resolution is sufficient to permit measurement of fiber sarcomere length and simultaneously, RBC velocity. Unlike other microcirculation models, the muscle itself undergoes no direct surgical intervention and thus there is little risk of perturbing microvascular responsiveness. Additional proposed modifications include micromanipulator videomicroscope control to increase preparation stability and field coverage.

Supported by: TRDRP 2 KT 0066 and NIHLB HL-17731.

VALIDATION OF NEAR-INFRARED SPECTROSCOPY AS A NON-INVASIVE MONITOR
OF CEREBRAL METABOLISM

EG Whitaker, MD, JR Mault, MD, JS Heinle, MD, AJ Lodge, BA, FF
Jöbsis-VanderVliet, PhD, RM Ungerleider, MD.
Departments of Surgery and Cell Biology, Duke University Medical
Center, Durham, NC 27710 USA.

Photons of the near infra-red range of the spectrum penetrate
scalp and skull relatively easily (1). Near infra-red spectroscopy
(NIRS) can therefore monitor cerebral microcirculation and
mitochondrial oxidative state non-invasively. It utilizes the
differential light absorbances of cytochrome $\underline{c}$ oxidase (a.k.a
cytochrome aa3, Cyt aa3) and of oxy- and de-oxyhemoglobin (2,3).
In this study, we compared Cyt aa3 and oxyhemoglobin (HbO2) signals
to measurements of cerebral oxidative metabolism (CMRO2) and
sagittal sinus venous saturation (ss SO2) at normothermia and
hypothermia. 8 piglets (2-3 kg) were placed on total
cardiopulmonary bypass (CPB) at 100 ml/kg/min. Measurements were
taken on initiation of CPB at 37°C, immediately after cooling to
18°C, after 60 min of hypothermia, and upon rewarming to 37°C and
included cerebral blood flow by ^{133}xenon clearance, arterial and
sagittal sinus blood gases, CMRO2 (mlO2/100 g/min), and continuous
NIRS (Niroscope™, Vander Corp) in units of variation in density.
With cooling, both Cyt aa3 and CMRO2 decreased significantly,
while HbO2 and ss SO2 increased significantly. With rewarming, Cyt
aa3 and CMRO2 increased significantly, while HbO2 and ss SO2
decreased significantly. (All differences significant at p <
0.0001).
In conclusion, non-invasive NIRS accurately monitors changes
in cerebral oxygen delivery and metabolism. As a direct monitor of
aerobic metabolism at the cellular level, further refinement of
this technology may play an important role in the critical care and
operative settings.

(1) Jöbsis FF, Science, 198:1264; 1977. (2) Jöbsis-VanderVliet FF,
Adv Exp Biol Med, 191:833; 1985. (3) Jöbsis-VanderVliet FF,in
Fetal and Neonatal Physiological Measurements, HN LaFeber, JG
Aarnoudse and HW Jongsma, eds., pp. 41-55. Excerpta Medica,
Amsterdam, 1991.

IS RED CELL FLOW HETEROGENEITY A CRITICAL VARIABLE IN THE REGULATION AND LIMITATION OF OXYGEN TRANSPORT TO TISSUE?

Brian R. Duling

Department of Molecular Physiology and Biological Physics
University of Virginia, Charlottesville, Virginia 22908

INTRODUCTION

The vasculature plays more than a purely subservient role in relation to the tissue it supplies - a tissue and its associated vasculature operate in synchrony to modulate the form and function of the organ. Nowhere is the interplay between vascular function and tissue function more apparent than in the supply of O_2 where vascular O_2 supply has the capacity to limit or even to control tissue function. Four closely related, but separable questions have arisen from our efforts to understand close match between O_2 supply and demand. 1.) What is the critical or limiting PO_2 for a tissue? 2.) Is tissue oxygen consumption dependent on flow at PO_2's above those which produce gross tissue hypoxia? 3.) What determines (limits) VO_{2max}? 4.) Does the vasculature control tissue function, or does tissue function control vascular behavior? A closely related issue concerns the location of the tissue oxygen sensor, either that regulating tissue oxygen consumption, or that involved in the control of flow.

Oxygen delivery is usually thought to be a relatively simple diffusive process; and in principle, enzyme kinetics should be easily defined, yet the foregoing questions represent unresolved issues after almost a century of research [1]. Why? Many factors must contribute to our uncertainty, but ultimately it becomes apparent that if one knew the precise PO_2 at two locations in the tissue, many of the puzzling issues would disappear. These two critical locations are the capillary and the locus of whatever oxidative enzyme is involved in the control process (i.e., we must define the source and the sink PO_2's in the diffusive system). Much of the history of the study of these problems can be summarized as a series of efforts of varying success to make estimates of these two parameters, or to define the determinants of the source and sink PO_2's. It is noteworthy that few investigators have simultaneously determined both source and sink PO_2.

Many factors contribute to the complexity of tissue oxygenation; here we will focus on heterogeneity of the distribution of the oxygen carriers, the red blood cells, a major source of our continuing uncertainty regarding the PO_2 at the surface of the capillary. The underlying theme is that any imbalance at the capillary level between oxygen supply

Oxygen Transport to Tissue XVI
Edited by M.C. Hogan *et al.*, Plenum Press, New York, 1994

and oxygen demand will lead to heterogeneities in the distribution of tissue PO_2, and perhaps to heterogeneities of tissue enzyme activity and metabolism. Such heterogeneities, even in the presence of more than adequate total supply of oxygen to the tissue can cloud our understanding of the critical issues in this problem.

Our experiments are based on direct observation of striated muscles in vivo, including the cremaster and the tibialis anterior. Tissues are superfused and the composition of the superfusion solutions can be manipulated to permit study of the regulatory processes including the dimensions of the arterioles, the capillary density, and the red cell flow patterns. Using in vivo preparations, it is possible to change the relation between oxygen supply and oxygen demand by two separate mechanisms. First, the striated muscle can be stimulated directly, either en mass or by using microelectrodes to stimulate single muscle cells. In either case, oxygen consumption and the production of metabolites increase in parallel. Second, the balance between oxygen supply and oxygen demand can be tipped one way or the other by changing the oxygen tension in the superfusion solution covering the tissue. A reduction in superfusion solution PO_2 mimics a contraction-induced elevation in tissue O_2 consumption without the associated increase in tissue metabolism [2].

We have followed two parameters that appear to be critical to the oxygenation of the tissue: lineal density of red cells along the length of the capillary (**n/l**) which is proportional to capillary hematocrit and is used as an index of oxygen capacity of capillary blood, and red cell flow velocity (**vRBC**) as an index of capillary flow. The product of these two is proportional to the capillary red cell flux, which in turn, is a measure of the O_2 flow through the capillary.

Table 1. Measured Flow Heterogeneities

Tissue Sample Size (mm^3)	Method	Coefficient of Variation (SD/mean)	Species-Tissue	Reference
5 x 10^3	Microspheres	-	Cat-Triceps	3
?	Tracer	-	Cat-Gastroc, sartorius	4
?	A-V Transit times	1.07 0.62	Dog-Heart	5
1.5 x 10^2	A-V Transit times	0.36		6
9 x 10^2	Microspheres	0.29	Dog-Heart	7
20 to 2 x 10^2	Tissue [THO]	0.24- 0.59	Dog-EDL	8
1 x 10^2	Microspheres	0.56	Rat-Skeletal muscle	9
2 x 10^{-1}	Histological sections	0.28	Rabbit-Leg muscle	10
1 x 10^{-1}	NADH fluorescence	-	Rat-Heart	11
3 x 10^{-3}	RBC velocity	0.56	Hamster-Cremaster	12

In trying to formulate the relations of heterogeneity to tissue oxygenation, it is important to recognize that flow heterogeneities are neither a pathological nor an occasional event in the circulation, but appear to be a common characteristic of vascular flow as shown in Table I.

Some confusion has been generated by failure to recognize that various investigators use the term "heterogeneity" to refer to quite different things. Table 1 shows a sample of data from the literature obtained from papers reporting "flow heterogeneity". The striking feature is that these flow heterogeneities extend over <u>six orders of magnitude</u> in size, and almost certainly reflect quite different aspects of the relation between supply and demand in the tissue. In this paper I will focus specifically on small-scale heterogeneities at the level of the capillaries, based on the premise that ultimately capillary flow and capillary PO_2 will regulate or limit what the tissue can do and how it functions.

VELOCITY HETEROGENEITY

Flow heterogeneity is estimated in microvascular experiments by measuring red cell velocity. It is assumed that red cell velocity, divided by a conversion factor (1.3), is a measure of capillary blood flow. When measurements of capillary red cell velocity are made under a variety of conditions, the consistent observation is that there is a wide range of red cell velocities within the capillaries. Figure 1 shows histograms of a number of measurements of capillary red cell velocity obtained under several conditions of tissue vascular tone [13]. The degree of vasomotor tone increases as one moves from the from bottom to top panel in the figure. Changes in tone were produced by either use of vasoactive substances, such as oxygen or adenosine, or by stimulating the striated muscle.

Several comments can be made on these data. First, and not surprising, with vasoconstriction, mean capillary velocity falls, presumably in proportion to flow. Second, in each case, there is a substantial range in the observed velocities. Third, the range of velocities observed across the vascular bed increases with vasodilation. The latter observation might be taken as an indication that heterogeneity of flow increases as the vasculature is dilated. Obviously, the range of velocities does increase but this does not mean that the degree of flow dispersion at the bifurcations feeding the capillaries is increased. In order to gain some understanding of the actual flow dispersion involved, one must take into account the absolute value of the flow as well as the heterogeneity. Total dispersion is reported in Table 1 as the standard deviation of the

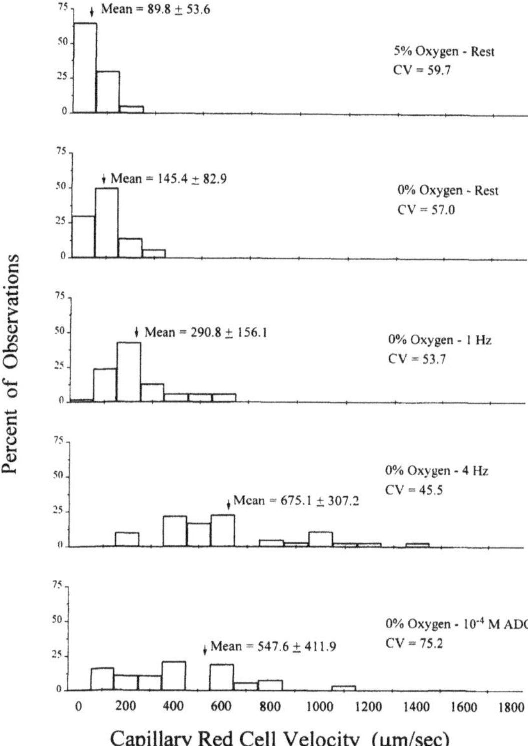

Figure 1. Heterogeneity of capillary flow velocity and the constancy of heterogeneity with changes in vasomotor state. Although mean flow velocity increases with vasodilation, the coefficients of variation change little, indicating a rather constant dispersion at upstream branch points.

velocity measurements. Obviously if a higher flow passes through the capillary bed with the same fractional flow divided among the branches, the range of flows will be greater. There is no well defined theory as to the appropriate means of analyzing such microcirculatory data but in accord with suggestions of Bassingthwaighte and Goresky [5], we have elected to use the <u>coefficient of variation</u> (standard deviation/mean) as an index of flow heterogeneity.

The histograms of capillary red cell velocities in Figure 1 demonstrate quite clearly that, though the range of velocities increases enormously with vasodilation, the coefficient of variation changes little, if at all. These data suggest that the widening dispersion seen with vasodilation reflects an increase in average velocity, with a relatively constant flow fraction at the various branches supplying the capillaries. Others have reported some change in coefficient of variation with vasomotor state [14-17]. However, any changes in dispersion which do occur with vasomotor state appear to be very small compared to the enormous changes in average velocity.

It is interesting to compare the data in Figure 1, with the data in Table 1 where coefficients of variation are seen to be consistently high, regardless of the measurement technique used, or the mass of tissue sampled. This comparison suggests that this degree of heterogeneity is a reflection of the capacity of the vasculature to distribute blood uniformly through the tissues, rather than of a failure in regulation, and emphasizes the importance of incorporating stochastic elements into any model of tissue oxygenation.

A deep appreciation of the significance of these measurements of velocity can only come with corresponding measurement of the nearby tissue oxygen consumption and the mass of tissue perfused by a capillary [18]. If the flow heterogeneities correlate with similar variations in tissue metabolic rate, then the tissue could be perfectly oxygenated, even with the observed flow heterogeneities. It is simply not known whether the variations in red cell velocity seen in Figure 1 correlate with variations either in the mass of tissue perfused or in the oxygen consumption per unit mass. In any case, these data strongly suggest that flow heterogeneities within the microcirculation are so large as to preclude a meaningful analysis of oxygen delivery based solely on any single lumped parameter estimate of capillary blood flow.

Heterogeneities in the microcirculation are not only spatial, as described to this point, but are temporal as well. Moreover, they are related to vascular structure in a complex way. Figure 2 shows the organization of a <u>capillary unit</u> as described by Lund, et al. [19]. The capillaries in striated muscle, originate as small groups of vessels, perfused by a single arteriole, with the flow in those capillaries simultaneously modulated by the contraction of the arteriole. Figure 2 shows sketches made from two different fields of observation in the tibialis anterior of the hamster, and the numbers label capillary units that were identified as groups of capillaries supplied by a single arteriole and drained by a single venule. At times one finds the red cell flow to be continuous and the velocity time independent, and at other times the velocity is cyclical, including being fully shut off on occasion. Figure 2 emphasizes the fact that adjacent units can have quite different behaviors, and thus, not only is the velocity heterogeneous in space, but heterogeneous in time. Similar cyclical variations in velocity and hemotcrit are reported by Ley et al. [20]. Again, there is no good quantitative model of this behavior pattern that allows us to access its impact on tissue oxygenation.

HEMATOCRIT HETEROGENEITY

Blood is two phase system - plasma and red cells - and the two components need not distribute equally within the total volume of the vasculature, nor must the cells distribute

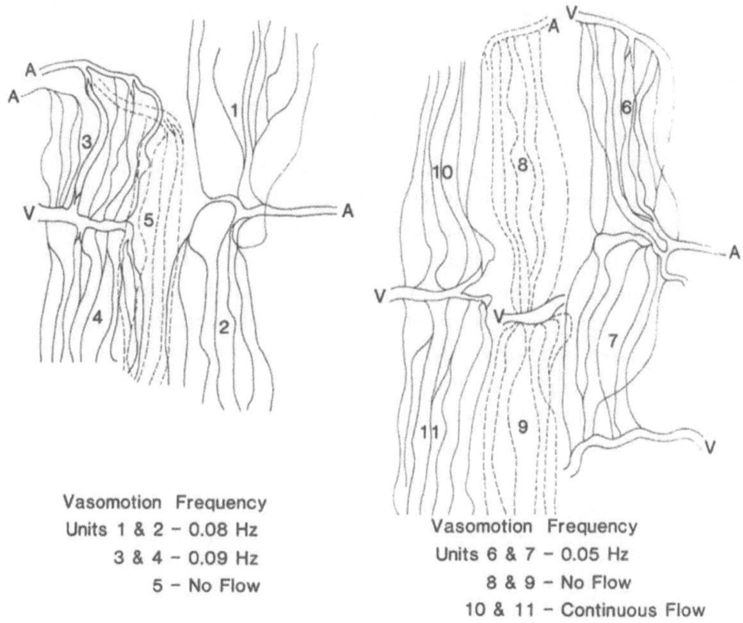

Figure 2. Temporal heterogeneities of red cell velocity in the microcirculation. Capillary units are defined as a group of capillaries fed and controlled by a single arteriole [19]. Some units are closed for long periods of time. Other units show cyclical variations in velocity of all of the perfused capillaries. Frequency of cyclic units is shown.

uniformly at bifurcations [21-24]. As a result, red cells and plasma can and do move with different velocities through the microcirculation; in general, red cells move faster than plasma.

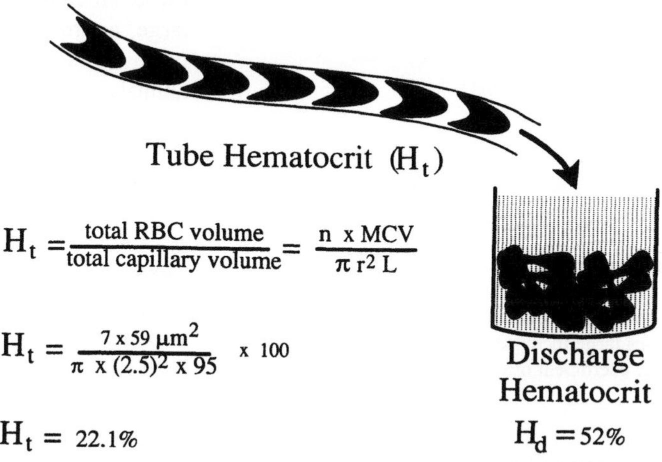

$$H_t = \frac{\text{total RBC volume}}{\text{total capillary volume}} = \frac{n \times MCV}{\pi\, r^2\, L}$$

$$H_t = \frac{7 \times 59\ \mu m^2}{\pi \times (2.5)^2 \times 95} \times 100$$

$$H_t = 22.1\%$$

$$H_d = 52\%$$

Figure 3. Tube and discharge hematocrit in the microcirculation. The hematocrit of the blood contained within a capillary (H_t) may be much lower than the hematocrit of the blood passing through the capillary (H_d). H_t/H_d is proportional to the relative velocity of the red cells and the plasma.

Since plasma and red cells must be conserved across the vascular bed, a more rapid movement of red cells implies a more dilute concentration of red cells within the capillaries, that is, a reduced intravascular hematocrit. This idea is shown schematically in Figure 3, where two different kinds of hematocrit are defined, <u>tube hematocrit</u> (H_t) and <u>discharge hematocrit</u> (H_d). The tube hematocrit is the hematocrit of blood resident at a given instant within the capillary. The discharge hematocrit is the hematocrit of blood which might be collected flowing through and out of the capillary. The figure shows the means by which intracapillary hematocrit is computed, and typical values are given for intracapillary hematocrits that might be observed in striated muscle with in vivo microscopy. The discharge hematocrit is more than double the hematocrit of blood within the capillary. The cause of this lo w intravascular hematocrit has been a subject of much debate and is discussed in detail elsewhere [21-26]. The purpose of this paper is not to raise that issue, rather the intent is to focus attention on the degree of heterogeneity that is observed in intercapillary hematocrit, and presumably, in the capillary oxygen delivery capacity. This heterogeneity is rarely accounted for in models which analyze tissue oxygenation.

Figure 4 shows histograms for capillary hematocrits, obtained from capillaries of animals of different ages [27]. Again, the figure shows enormous heterogeneity in the intracapillary hematocrit, and a striking capacity of the hematocrit to increase with vasodilation in the face of a constant arterial hematocrit. Two points are to be made here, first that capillary hematocrit varies with vasomotor state, and second that the hematocrit varies with the age of the animal. Compare this with Figure 1 and note that with hematocrit as well as velocity, the coefficients of variation are large, approaching 100% in the case of fifty-seven day old animals, and presumably, this variability represents large differences in the oxygen delivery capacity of the capillaries.

It might be argued that although capillary hematocrit is low, it can be assumed to be a constant fraction of the arteriolar hematocrit. However, the situation is much more complex than that, as shown by experiments with hemodilution illustrated in Figure 5.

Under the circumstances tested, the capillary hematocrit does not show a constant relation to the arterial hematocrit as the blood is diluted. In the resting muscle, the

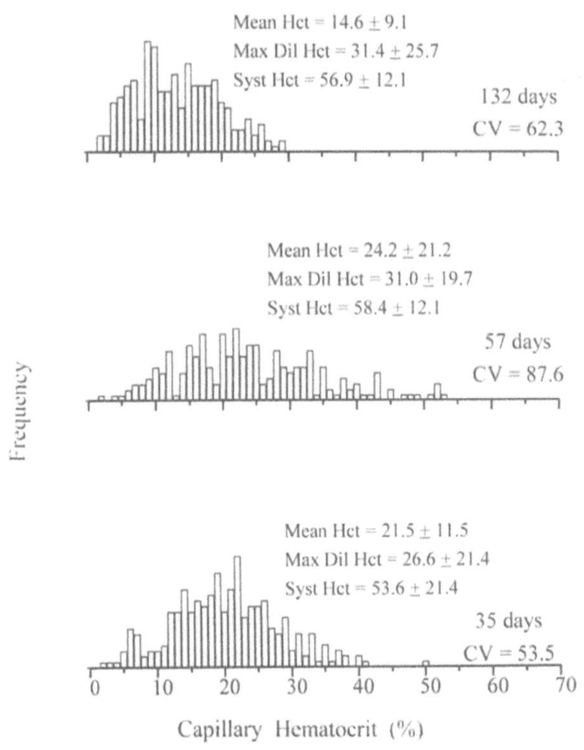

Figure 4. Hematocrit distributions in the microcirculation of animals of various ages [27]. Note the large heterogeneities in hematocrit and the variations with the age of animal studied. Vasodilation causes large increases in average hematocrit with little change in the coefficient of variation.

capillary hematocrit falls less than the arterial blood. Remarkably, the capillary hematocrit in the hyperemic muscle show a greater change than the arterial. The causes for these differences in the two hematocrits remain to be established.

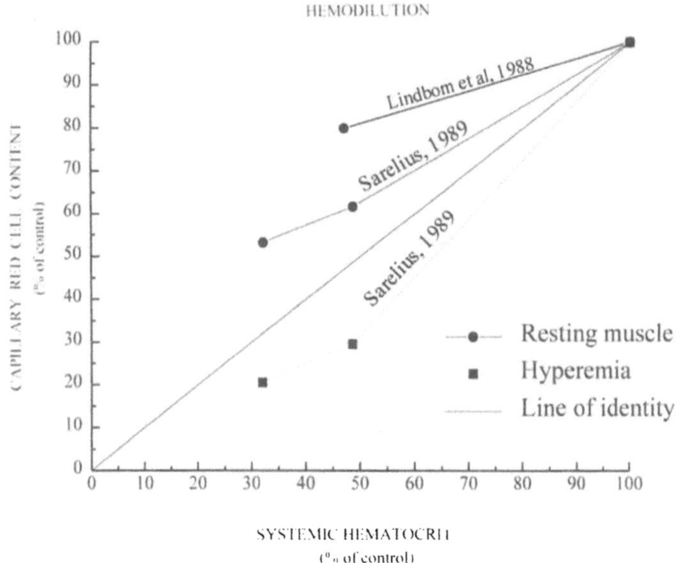

Figure 5. Effects of hemodilution on capillary hematocrit. Note that the capillary hematocrit has been reported to fall more slowly and more rapidly than the systemic hematocrit during hemodilution. Data from Sarelius [28] and Lindbom et al. [29].

POTENTIAL EFFECTS OF RBC FLOW HETEROGENEITIES ON TISSUE PO_2

What is the effect of the heterogeneities in velocity and hematocrit on tissue oxygenation? Put simply, the answer is unknown because the sources of the heterogeneities are unknown. Furthermore, the associated heterogeneities of the mass of tissue perfused and the oxygen consumption per mass of tissue are unknown and at this time are in fact, unmeasurable. If the heterogeneities in capillary oxygen transport parameters were exactly matched to heterogeneities in tissue oxygen consumption, then there would be no effect of these heterogeneities on tissue oxygenation. Alternatively, if the tissue elements perfused by the capillaries sampled in Figures 1 and 3 are homogeneous, the potential effects on oxygenation are great. Based on the assumption of tissue homogeneity, we have made a very simple analysis of what <u>might</u> happen to tissue PO_2 as a result of heterogeneities in velocity. Figure 6 shows an admittedly trivial simulation of the situation at rest and during 4 Hz muscle stimulation based on data comparable to those shown in Figure 1 [30]. While at rest, the velocity heterogeneities would produce a substantial dispersion in tissue PO_2, but the tissue is well oxygenated throughout, and the heterogeneities would not be expected to produce limitations in oxygen consumption. In contrast, the prediction at 4 Hz is that, on average the velocity is adequate to supply the tissue, but with the observed heterogeneities, a fair fraction of the tissue would have its O_2 consumption limited by the supply. Obviously, this is an over-simplified model which ignores many of the critical variables mentioned previously. Its point however, is to once again highlight the need for the incorporation of heterogeneity into more detailed models of microcirculatory O_2 delivery.

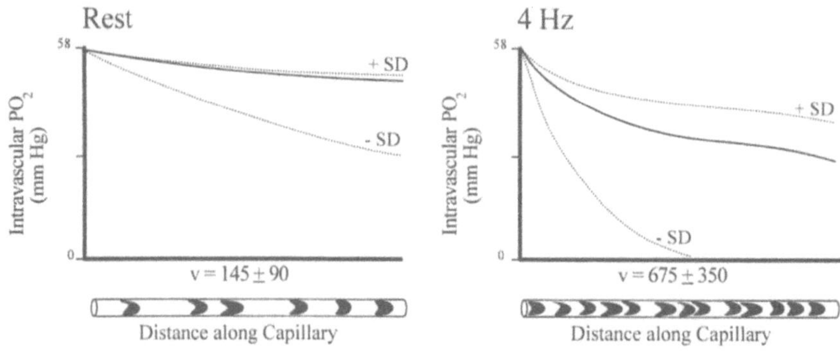

Figure 6. Simulation of the range of capillary PO_2's associated with the type of velocity heterogeneities shown in Fig. 1. Note that the simplified model predicts substantial limitation with muscle work.

It might be argued that the heterogeneities in hematocrit are not critical to tissue oxygenation, since it would be the O_2 flux and not the red cell spacing that is central to the O_2 supply. However, various models suggest that red cell spacing in the capillary may be a critical variable, independent of the capillary red cell flux [31-33]. Figure 7 shows a schematic illustration of the results of one such model. The key thought is that the plasma space between the red cells is predicted to be depleted of O_2 by diffusion into the tissue faster than the O_2 can diffuse out of the red cell to replenish the plasma. At rest this may be of little importance, but as shown in the figure, the PO_2 in the plasma in the contracting tissue is predicted to fall to limiting values before the end of the capillary is reached. In other words, the model predicts that the plasma PO_2 becomes rate limiting. It is noteworthy that two tests of the model in intact tissues have failed to find any evidence that elevation of the plasma O_2 capacity can increase tissue oxygenation delivery[34] (see also Hogan and Wagner, this volume), but no direct measurement has been made on the effect of such a substitution on capillary PO_2.

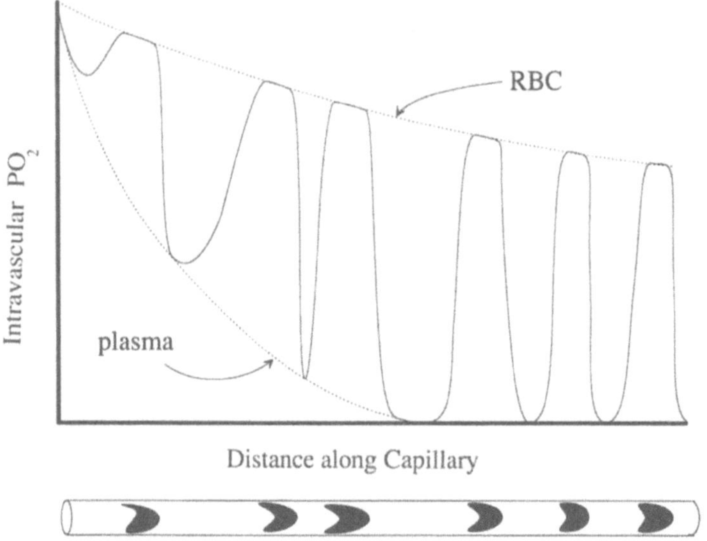

Figure 7. Predicted influence of red cell spacing on intracapillary PO_2. This is a schematic illustration of the effect of red cell spacing on the plasma PO_2 with spacing of red cells commonly observed in vivo. Based on previously published models [31,32].

FUTURE STUDIES

The foregoing comments emphasize the potential importance of heterogeneity of red cell flow and intracapillary red cell spacing in influencing tissue oxygen distribution. The idea is not to place heterogeneity in a preeminent position as the determinant of tissue oxygen delivery. Rather, it is to make the point that our current understanding of heterogeneity suggests that its magnitude is sufficiently great as to mask important tissue events, or to make it impossible to understand other, perhaps ultimately, more interesting aspects of tissue oxygenation, such as limiting enzyme kinetics and oxidative control of tissue function.

Obviously more detailed studies of capillary flow and spatial heterogeneity are warranted, but equally critical is the question of the mass of tissue supplied by a given microvessel, and the local oxygen consumption measured on a scale consistent with the diffusion fields of a capillary unit and individual capillaries. There are no methods available or for that matter even under investigation for the determination of these parameters.

It is interesting to note that one can infer the existence of small-scale heterogeneities in oxygen consumption in a working muscle, based on the fact that motor units are the operative element for normal muscular contraction, and these motor units have a size which is much larger than the size of the capillary unit [35,36]. Thus, it is not unreasonable to expect that velocity heterogeneities would not correspond with metabolic heterogeneities in a working muscle. A new technique for making local measurements of oxygen consumption would be a very important addition to the tools for understanding regulation of tissue oxygenation.

Perhaps the most important missing pieces in the puzzles surrounding the limits to and control of tissue oxygenation are the measurement of PO_2 at the surface of the source for oxygen, i.e. the capillary, and at the sink for oxygen, the mitochondria. Recent developments in phosphorescence and microspectrophotometry suggest that in the near future we may be able to actually make such measurements [37-39].

We also need to measure flow and hematocrit heterogeneity within intact muscles. Preparation of tissues for microcirculatory observation is likely to alter microvascular function, and furthermore, microvascular measurements are made only in thin sheets of tissue, on the superficial vessels, both of which may fail to represent the situation deep within the mass of a large postural muscle. This is likely to be a source of technical challenge for years to come, though the resolution of noninvasive imaging techniques including PET and MRI are approaching the spatial detection necessary.

ACKNOWLEDGEMENTS
Much of the original research reported in this paper was conducted under USPHS grant # HL 12792.

REFERENCES
1. F. Verzar. Influence of oxygen lack on tissue respiration., *J. Physiol. (Lond)*. 45:39 (1912).

2. R.J. Gorczynski and B.R. Duling. Role of oxygen in arteriolar functional vasodilation in hamster striated muscle, *Am. J. Physiol*. 235:H505 (1978).

3. F. Vetterlein, K. Averdunk, and G. Schmidt. Effects of acetylcholine on the regional distribution of microspheres in the skeletal muscle of the cat, *Basic Res. Cardiol*. 72:11 (1977).

4. T.E. Barlow, A.L. Haigh, and D.N. Walder. Evidence for two vascular pathways in skeletal muscle, *Clin. Sci*. 20:367 (1961).

5. J.B. Bassingthwaighte and C.A. Goresky. Modeling in the analysis of solute and water exchange in the microvasculature., 13.*in:* "Handbook of Physiology: The Cardiovascular System IV," Renkin EM and Michel CC, ed., (1984)

6. L. Bass and P. Robinson. Capillary permeability of heterogeneous organs:a parsimonious interpretation of indicator diffusion data, *Clin. Exp. Pharm. &. Physiol.* 9:363 (1982).

7. H. Falsetti, R. Carroll, and M. Marcus. Temporal heterogeneity of myocardial blood flow in anesthetized dogs, *Circulation;* 52:848 (1993).

8. N.F. Paradise, C.R. Swayze, D.H. Shin, and I.J. Fox. Perfusion heterogeneity in skeletal muscle using tritiated water, *Am. J. Physiol.* 220:1107 (1971).

9. M. Laughlin and R. Armstrong. Muscular blood flow distribution patterns as a function of running speed in rats, *Am. J. Physiol.* 243:H296 (1982).

10. E. Renkin, S. Gray, and L. Dodd. Filling of microcirculation in skeletal muscles during timed India ink perfusion, *Am. J. Physiol.* 241:H174 (1981).

11. C. Steenbergen, G. Deleeuw, C. Barlow, B. Chance, and J.R. Williamson. Heterogeneity of the hypoxic state in perfused rat heart, *Circ. Res.* 41:606 (1977).

12. B. Klitzman and P. Johnson. Capillary network geometry and red cell distribution in hamster cremaster muscle, *Am. J. Physiol.* 242:H211 (1982).

13. D.H. Damon and B.R. Duling. Heterogeneity of capillary perfusion in resting and contracting muscle, *Microvasc. Res.* (1984).

14. K. Tyml. Capillary recruitment and heterogeneity of microvascular flow in skeletal muscle before and after contraction, *Microvasc. Res.* 32:84 (1986).

15. K. Tyml. Heterogeneity of microvascular flow in rat skeletal muscle is reduced by contraction and by hemodilution, *Int. J. Microcirc. Clin. Exp.* 10:75 (1991).

16. K. Tyml and C. Budreau. Heterogeneity of microvascular response to ischemia in skeletal muscle, *Int. J. Microcirc. Clin. Exp.* 7:205 (1988).

17. K. Tyml and C. Ellis. Localized heterogeneity of red cell velocity in skeletal muscle at rest and after contraction, *Adv. Exp. Med. Biol.* 248:735 (1989).

18. T. Sweeney and I. Sarelius. Spatial heterogeneity in striated muscle arteriolar tone, cell flow, and capillarity, *Am. J. Physiol.* 259:124 (1990).

19. N. Lund, D.H. Damon, D.N. Damon, and B.R. Duling. Capillary grouping in hamster tibials anterior muscles: flow, *Int. J. Microcirc. Clin. Exp.* 5:359 (1987).

20. K. Ley, L. Lindbom, and K. Arfors. Haematocrit distribution in rabbit tenuissimus muscle, *Acta Physiol. Scand.* 132:373 (1988).

21. G.R. Cokelet. Speculation on a cause of low vessel hematocrits in the microcirculation, *Microcirculation.* 2:1 (1982).

22. B. Fenton, R. Carr, and G. Cokelet. Nonuniform red cell distribution in 20 to 100 μm bifurcations, *Microvasc. Res.* 29:103 (1985).

23. A. Pries, T. Secomb, P. Gaehtgens, and J. Gross. Blood flow in microvascular networks - Experiments and Simulation, *Circ. Res.* 67:826 (1990).

24. A.R. Pries, K. Ley, and P. Gaehtgens. Generalization of the Fahraeus principle for microvessel networks, *Am. J. Physiol.* 251:H1324 (1986).

25. C. Desjardins and B.R. Duling. Microvessel hematocrit: measurement and implications for capillary, *Am. J. Physiol.* 252:H494 (1987).

26. C. Desjardins and B.R. Duling. Heparinase treatment suggests a role for the endothelial cell, *Am. J. Physiol.* 258:H647 (1990).

27. I.H. Sarelius, D.N. Damon, and B.R. Duling. Microvascular adaptations during maturation of striated muscle, *Am. J. Physiol.* 241:H317 (1981).

28. I.H. Sarelius. Microcirculation in striated muscle after acute reduction in systemic hematocrit, *Respir. Physiol.* 78:7 (1989).

29. L. Lindbom, S. Mirhashemi, M. Intaglietta, and K. Arfors. Increase in capillary blood flow and relative haematocrit in rabbit skeletal muscle following acute normovolaemic anaemia, *Acta Physiol. Scand.* 134:503 (1988).

30. B. Duling and D. Damon. An examination of the measurement of flow heterogeneity in striated muscle, *Circ. Res.* 60:1 (1987).

31. W. Federspiel and A. Popel. A theoretical analysis of the effect of the particulate nature of blood on oxygen release in capillaries, *Microvasc. Res.* 32:164 (1986).

32. W.J. Federspiel and I.H. Sarelius. An examination of the contribution of red cell spacing to the uniformity of oxygen flux at the capillary wall, *Microvasc. Res.* 27:273 (1984).

33. B.R. Duling and C. Desjardins. Capillary hematocrit - what does it mean?, *N. I. P. S.* 2:66 (1987).

34. G.P. Biro, P.J. Anderson, S.E. Curtis, and S.M. Cain. Stroma-free hemoglobin: its presence in plasma does not improve oxygen supply to the resting hindlimb vascular bed of hemodiluted dogs, *Can. J. Physiol. Pharmacol.* 69:1656 (1991).

35. M. Ounjian, R.R. Roy, E. Eldred et al. Physiological and developmental implications of motor unit anatomy, *J. Neurobiol.* 22:547 (1991).

36. S. Bodine Fowler, A. Garfinkel, R.R. Roy, and V.R. Edgerton. Spatial distribution of muscle fibers within the territory of a motor unit, *Muscle Nerve.* 13:1133 (1990).

37. C. Ince, J.F. Ashruf, J.A. Avontuur, P.A. Wieringa, J.A. Spaan, and H.A. Bruining. Heterogeneity of the hypoxic state in rat heart is determined at capillary level, *Am. J. Physiol.* 264:H294 (1993).

38. K. Vandenborne, K. McCully, H. Kakihira et al. Metabolic heterogeneity in human calf muscle during maximal exercise, *Proc. Natl. Acad. Sci. U. S. A.* 88:5714 (1991).

39. J.R. Gober, S. Schaefer, S.A. Camacho et al. Epicardial and endocardial localized 31P magnetic resonance spectroscopy: evidence for metabolic heterogeneity during regional ischemia, *Magn. Reson. Med.* 13:204 (1990).

MEASUREMENTS OF HEMOGLOBIN CONCENTRATION AND OXYGEN SATURATION PROFILES IN ARTERIOLES USING INTRAVITAL VIDEOMICROSCOPY AND IMAGE ANALYSIS

Kaushik Parthasarathi and Roland N. Pittman

Departments of Biomedical Engineering and Physiology
Medical College of Virginia
Virginia Commonwealth University
Richmond, Virginia 23298

INTRODUCTION

Oxygen is transported from the blood to the parenchymal cells by passive diffusion across the walls of microvessels. August Krogh (1918/1919) maintained that most, if not all, of the diffused oxygen passed across the walls of the capillaries. This view was modified with the report of Duling and Berne (1970) that significant decreases in PO_2 occurred in vessels of the arteriolar network. These findings were later confirmed with systematic measurements of geometric, hemodynamic and oxygenation variables in four consecutive orders of arterioles of the hamster retractor muscle by Kuo and Pittman (1988, 1990) and Swain and Pittman (1989). These studies were designed to provide estimates of diffusive oxygen flow based on the difference between convective oxygen flows at upstream and downstream sites in unbranched segments of arterioles. Results of a detailed analysis of oxygen mass balance for these arterioles indicated that the observed rate of oxygen diffusion from arterioles was about an order of magnitude higher than expected from standard theoretical considerations (Popel et al., 1990). The possibility that the oxygen permeability of the tissue was much higher than values obtained under in vitro conditions was considered, but this potential explanation could only account for about a factor of two of the discrepancy (Bentley et al., 1993).

Another possible reason for the discrepancy has to do with the photometric measurement procedure for hemoglobin concentration, [Hb], and hemoglobin oxygen saturation, SO_2. In the studies described above, [Hb] and SO_2 were determined from light transmission measured over the centerline of the arterioles. Use of these values to compute convective oxygen flow carried the implicit assumption that both red blood cells and oxygen were uniformly distributed within the arteriolar lumen, an assumption known to be invalid based on work by Ellsworth and Pittman (1986). Pittman and Helber (1992) later showed that neglect of existing nonuniformities in the luminal distributions of red blood cells and oxygen could lead to substantial systematic errors (overestimates, usually) in calculations of convective and diffusive oxygen flow.

Oxygen Transport to Tissue XVI
Edited by M.C. Hogan *et al.*, Plenum Press, New York, 1994

It is reasonable to expect that the degree of nonuniformity in the luminal distribution of red blood cells would be greater just downstream of a bifurcation (i.e., the beginning of an unbranched segment) than further downstream when the cells would be expected to attain a more uniform distribution. Since hemoglobin is contained entirely within the red cells and almost all of the oxygen is bound to the hemoglobin, one would expect similar behavior for the distributions of [Hb] and SO_2. Since the systematic error in calculated convective oxygen flow is larger for larger degrees of nonuniformity, one would expect that the difference in convective flows at upstream and downstream sites, respectively, would overestimate the true difference in convective flow of oxygen and, hence, produce an overestimate of diffusive flow.

Since the diffusive loss of oxygen in arterioles is generally small compared with the convective flow (typically about 1 to 5 %), the difference in systematic error between the upstream and downstream sites needs to be 10 to 50% of the true convective flow in order to account for the observed order of magnitude discrepancy between theory and experiment. In order to maximize the detection of oxygen diffusion from arterioles, upstream and downstream sites were usually separated as widely as possible, with the outcome that the upstream sites were generally close to a bifurcation and thus in a location associated with nonuniform and usually asymmetric distributions of red cells.

In order to achieve more accurate estimates of convective and diffusive flow of oxygen, we developed a video-based system to analyze images of microvessels obtained using intravital microscopy. The analysis yielded full-diameter profiles of [Hb] and SO_2 rather than just single centerline values. The additional information obtained from the new system allows one to observe the longitudinal evolution of the luminal distribution of hemoglobin and oxygen within selected segments of the microvascular network. Profiles in [Hb] and SO_2 were generally quite nonuniform and asymmetric just downstream of bifurcations and achieved more symmetric shapes at distal sites in a given daughter segment. At those segments far enough downstream to have symmetric profiles, a convolution reconstruction procedure was carried out to determine the radial dependence of [Hb] and SO_2.

METHODS

Theoretical

The videodensitometric determination of SO_2 and [Hb] was based upon the methods previously described by Pittman and Duling (1975a, b) for SO_2 and by Lipowsky et al. (1982) for [Hb]. Quantitative measurements of the concentration of any substance by spectrophotometric methods is based on the Lambert-Beer-Bouguer law, which states that the optical density (D)

$$D = \log(I_0/I) = \epsilon cd$$

where I_0 is the incident intensity, ϵ is the millimolar extinction coefficient ($mM^{-1}cm^{-1}$) and is a characterestic of the solute, c is the concentration of the solution (mM) and d is the path length (cm) through the sample. This equation can be applied to whole blood, where a term B for the contribution of light scattering by red blood cells is added to the absorption contribution.

Optical densities are measured at three different wavelengths, viz., 560 nm, 549 nm and 523 nm, where 523 nm and 549 nm are isosbestic wavelengths for oxyhemoglobin and deoxyhemoglobin and 560 nm corresponds to a maximum difference between oxy- and deoxyhemoglobin absorbance (van Assendelft, 1970). For further discussion of the basis for the photometric techniques, refer to Pittman (1986a, b).

Experimental

The system used for data collection is shown schematically in Figure 1. The main component of this system was the dual observation port Zeiss microscope. The object to be viewed was placed on the swivel stage of the microscope and transilluminated by a 75 Watt xenon arc lamp. A built in beam splitter positioned above the objective was used to direct copies of the image toward the two observation ports.

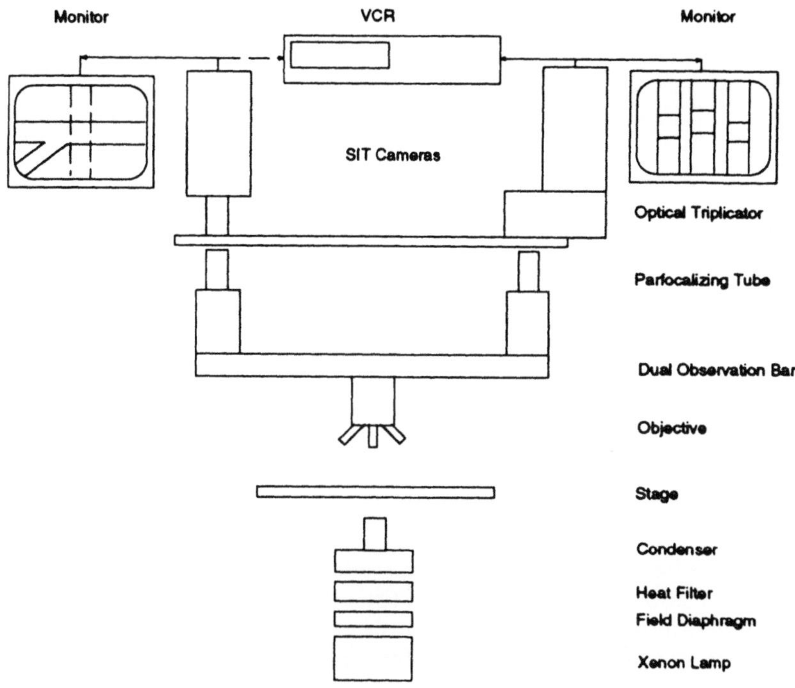

Figure 1. Schematic diagram of the data acquisition system used for recording images of blood vessels.

The image directed toward the right observation port was passed through a parfocalizing tube and then an Optical Triplicator (Mott, 1992). The Optical Triplicator (Figure 2) was designed to produce three replicas of an image. This was achieved using a set of beam splitters. The first beam splitter produced two replicas of the original image. The second beam splitter further duplicated one of the images produced by the first, thus, giving a compete set of three replicas.

The three replicas or subimages contained information at all wavelengths. Information at the desired wavelengths (523 nm, 549 nm and 560 nm) was extracted by a set of bandpass interference filters. The resulting subimages were delimited using rectangular slits and then positioned adjacent to each other by a set of prisms. These modified images were captured by a Silicon Intensified Target (SIT) camera (Dage MTI, model 66).

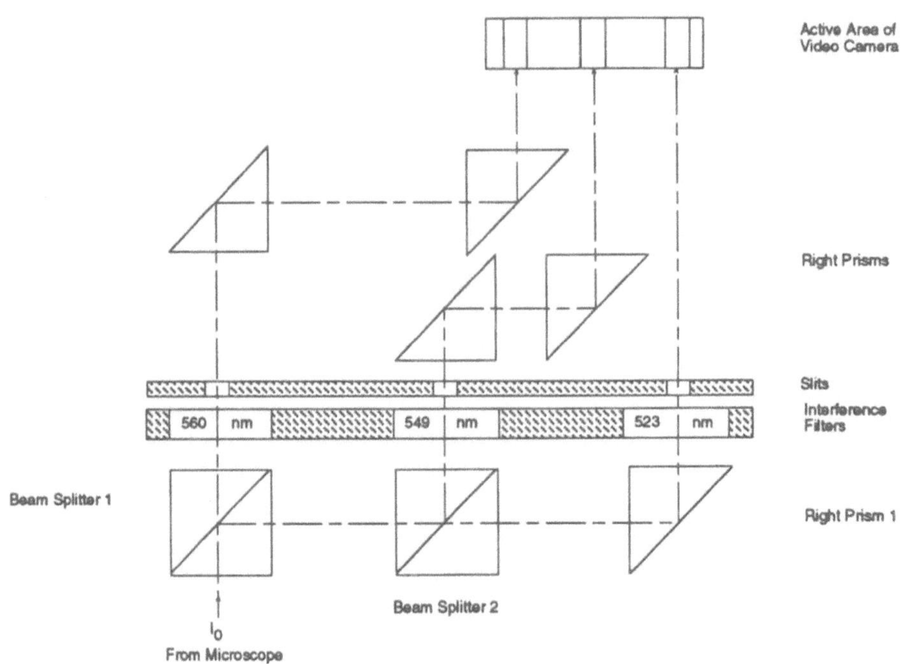

Figure 2. Arrangement of components within the Optical Triplicator.

The composite video output level was controlled manually by varying the gain and the high voltage ("kilovolt") adjustments on the camera. The output of the camera was viewed on a video monitor and simultaneously recorded using a S-VHS video cassette recorder (Panasonic, model AG 7350) for further analysis. The image directed toward the left observation port of the microscope was left unmodified. This image was captured by a second SIT camera and the video output viewed using a second video monitor. The complete image (left port) was used to locate a suitable vessel site for measurements, as the delimited image (right port) offered only a restricted view of a vessel and the muscle itself.

Animal Preparation

Male golden hamsters weighing between 45 and 60 grams and from 20 to 30 days old were used in this study. The animals were initially anesthetized with sodium pentobarbital (6.5 mg per 100 g body weight). A femoral vein was cannulated to allow continuous infusion of pentobarbital at a rate of 0.79 μl/minute. A tracheostomy ensured a clear airway and the animal spontaneously breathed room air. The body temperature was maintained at 37±1 °C. Preparation of the right cheek pouch retractor

252

muscle followed the procedures described by Sullivan and Pittman (1982). The muscle was then covered with a transparent thin plastic film (Saran, Dow Corning) to minimize gas and water vapor exchange between the muscle and the environment.

Experimental Description

The muscle was observed using a 40X objective (Zeiss, UD40, NA 0.65). Vessels with diameters in the range of 20 μm to 50 μm were used in this study. The selected vessel was oriented using the swivel stage, so that the longitudinal dimension of the vessel on the video monitor was horizontal. A freeze frame image of an arteriole as seen on the monitor is shown in Figure 3. The image modified by the Optical Triplicator was recorded on video tape. Recordings at each measurement site were followed by a recording of the black level (i.e., no input light). Subtraction of the black level from the in vivo images referenced them to gray scale "zero," allowing expansion of the image to the full 256 gray scale levels.

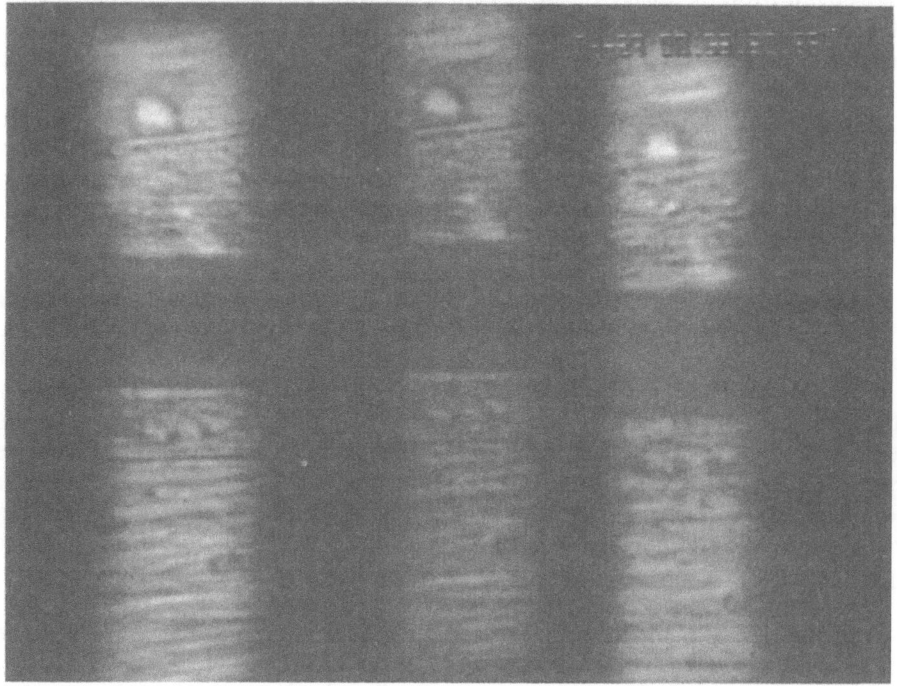

Figure 3. Freeze frame picture of an arteriole as seen on the monitor showing the arrangement of the three sub-images. The extreme left sub-image contains information at 560 nm, the center sub-image information at 549 nm and the extreme right sub-image information at 523 nm.

Analytical

Oxygen saturation and hemoglobin concentration were calculated from recorded light intensity values using the image analysis software Optimas (Bioscan, Inc.). Both custom-defined and user-defined analysis were supported by Optimas, with the user-defined analysis based on a macro compiler, Analytical Language for Images (ALI).

The image analysis operation involved computations on single picture frames. A frame grabber (Imaging Technology, Overlay Frame Grabber VISIONplus-AT) was used to digitize the images. The frame grabber was capable of storing one 640 x 480 bits image, providing 8-bits for image intensity data and 4-bits for non-destructive overlay information. The capture and display functions of the frame grabber were controlled by Optimas.

To locate corresponding points (pixels) in all three sub-images, the relative horizontal and vertical translations of the three sub-images were required. The image of a stage micrometer with a square grid pattern etched on it was recorded and using a pair of points in the grid, the relative translations among the three subimages were calculated.

Optical Density (D) depends on the values of incident light intensity (I_0) and the transmitted light intensity. Values of transmitted light intensity were obtained by direct measurement from the image of the vessel. Following Pittman and Duling (1975b), the light intensity transmitted through nearby avascular tissue was used as a close approximation of I_0. Pixel locations in the avascular region that contained fluctuations in light intensity were eliminated from consideration while estimating I_0, since it was probable that these tissue regions contained out-of-focus microvessels whose interference could lead to inaccuracies in the estimate of I_0. The spatial heterogeneity of the tissue introduced an ambiguity in the selection of I_0. This was overcome by using ratios of the light intensity from any two subimages. The ratios were termed K-factors and were defined previously by Pittman and Duling (1975b).

Image replication in the Optical Triplicator reduced the light intensity levels in all three subimages. Moreover, use of neutral density filters within the Optical Triplicator to equalize the light intensity levels of the three subimages further reduced the light intensity reaching the video camera. To compensate for the low input signal to the video camera, its gain was increased, resulting in higher output signal levels and lower signal to noise ratios. Improvement of the signal to noise ratios was achieved by temporally averaging several frames of the same image. The number of averaged frames varied from six to eleven, depending on the gain employed for the video camera. Oxygen saturation and hemoglobin concentration were calculated using this processed image of averaged frames.

Profiles of oxygen saturation and hemoglobin concentration were obtained by determining their values vertically across the image of the vessel. Thus, from any given vessel image, several profiles were generated. Standard profile smoothing techniques were employed and these were followed by intraprofile averaging and interprofile averaging. In intraprofile averaging, starting from the midpoint of the profile, sets of five consecutive values in a vertical column were averaged and a new profile created using these mean values. The sets did not overlap (i.e., there were no values common to adjacent sets) and effectively the new profile contained one-fifth the number of values as the original. In interprofile averaging, sets of five profiles (longitudinally adjacent columns) were averaged together to produce a single profile. This reduced the effective number of profiles by five. This averaging sharply reduced the standard deviation of values in the averaged profile. Averages using more pixels per set for intraprofile averaging and more profiles per set for interprofile averaging did not significantly lower the standard deviation of values any further.

Individual values of oxygen saturation and hemoglobin concentration within these profiles are the result of averaging oxygen saturation and hemoglobin concentration along the vertical paths within a vertical cross section of the vessel. The profiles are thus the result of averaging the radial distributions of oxygen saturation and hemoglobin concentration along the vertical cross section of the vessel. To reconstruct the radial distributions from these profiles (i.e., a circular cross section was assumed), we used the convolution method developed by Ramachandran and Lakshminarayanan (1971), modified by Shepp and Logan (1974) and later used by Mancuso et al. (1990). This reconstruction procedure requires profiles obtained by projections in several directions. Using the current experimental setup, a projection in only one direction was possible. We circumvented this constraint by using only laterally symmetric profiles. We assumed that if the distribution of oxygen saturation and hemoglobin concentration were axisymmetric, then the resulting profiles would be laterally symmetric. Even though the converse need not necessarily be true, we assumed that laterally symmetric profiles were the result of axially symmetric distributions. As profiles derived from axially symmetric distributions using projections in different directions are identical, we substituted the laterally symmetric profile obtained from one projection for profiles from multiple projections and computed the radial distributions of oxygen saturation and hemoglobin concentration.

RESULTS

Determination of the extinction coefficients used in the estimation of hemoglobin concentration and oxygen saturation profiles was done using 300 μm path length rectangular glass Microslides (Vitro Dynamics , Rockaway, NJ) filled with hemoglobin solutions. Fully oxygenated hemoglobin solutions (SO_2 = 100%) were used to determine the optical density of oxyhemoglobin for the 560, 549 and 523 nm filters. Hemoglobin solutions that were completely deoxygenated with sodium dithionite were used in similar determinations for deoxyhemoglobin. A rectangular Microslide filled with distilled water served as the reference (I_0).

Light intensity values measured laterally across the diameter were used to calculate diametric profiles of hemoglobin concentration and oxygen saturation. The light intensity is given in eight-bit gray scale units (i.e., full scale = 256 gray scale units). Portions of profiles closer than about 7 μm from the two walls were excluded due to the unpredictable nature of the profiles close to the wall caused by optical refraction effects associated with the curved wall (Ellsworth and Pittman, 1986).

Using the profiles, we analyzed the oxygen saturation and hemoglobin concentration distributions near a bifurcation. We found that the profiles of hemoglobin concentration and oxygen saturation downstream of a bifurcation depended on several factors, including the nature of the corresponding profiles in the parent vessel, orientation of the daughter vessels with respect to the parent and diameters of the parent and daughter vessels. Generally, the profiles at sites just downstream of a bifurcation were highly asymmetric.

To explore the possibility that these asymmetric profiles might develop into more symmetric profiles further downstream, profiles were obtained at several sites along single unbranched arteriolar segments. The results obtained from one such arteriolar segment are shown in Figure 4. It was found that the length of the vessel segment had to be several vessel diameters for the profiles to become symmetric. In many cases within the set of vessels observed in this study, the evolution of asymmetric profiles occurred at distances greater than fifteen vessel diameters. However, it was found that this was not a rigid criterion because the evolution of the profiles, particularly oxygen saturation, would be expected to depend on several other factors (e.g., red cell velocity, hematocrit, vessel diameter and the presence of nearby vessels that could act as sources or sinks for oxygen) in addition to vessel segment length.

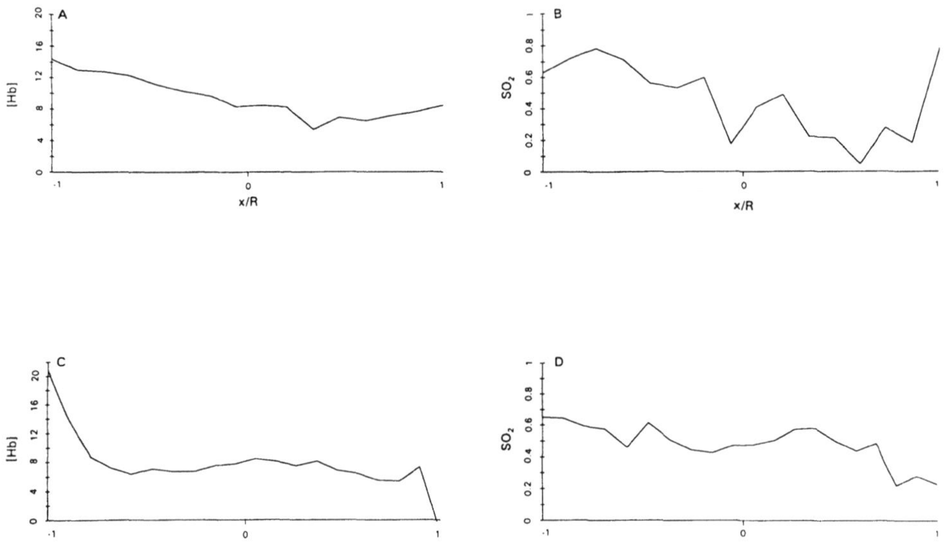

Figure 4. Profiles of hemoglobin concentration (mM) and fractional oxygen saturation determined at two widely separated sites along an unbranched arteriolar segment are shown in panels A-D to demonstrate the evolution of asymmetric profiles near bifurcations into laterally symmetric profiles further downstream. The diameter of the vessel used in these measurements was approximately 30 μm. The horizontal axis in each panel, x/R, represents the location of the profile point (x) normalized to the radius (R) of the arteriole. Panels A and B show profiles of [Hb] and SO_2, respectively, determined at a site 400 μm downstream from a bifurcation. Panels C and D show profiles of [Hb] and SO_2, respectively, determined at a site 950 μm downstream from the same bifurcation.

Laterally symmetric profiles were identified to compute the radial dependencies of hemoglobin concentration, H(r), and oxygen saturation, S(r), using the convolution reconstruction method. The radial dependencies of hemoglobin concentration and oxygen saturation for a representative vessel with laterally symmetric profiles are shown in Figure 5. These preliminary results demonstrate that the luminal distributions of red blood cells (hemoglobin) and oxygen can be determined by this procedure.

256

DISCUSSION

In order to obtain more accurate estimates of diffusive oxygen flow in single microvessels, we have developed a method based on measurements of diametric profiles of oxygen saturation and hemoglobin concentration. As discussed previously, the existing methods based on centerline measurements fail to take into account nonuniform distributions of red cells and oxygen, factors that are capable of having a marked influence on the calculations of diffusive oxygen transport. The use of radial distributions of oxygen saturation and hemoglobin concentration computed from their profiles should give a more reliable estimate of convective and, hence, diffusive oxygen flow.

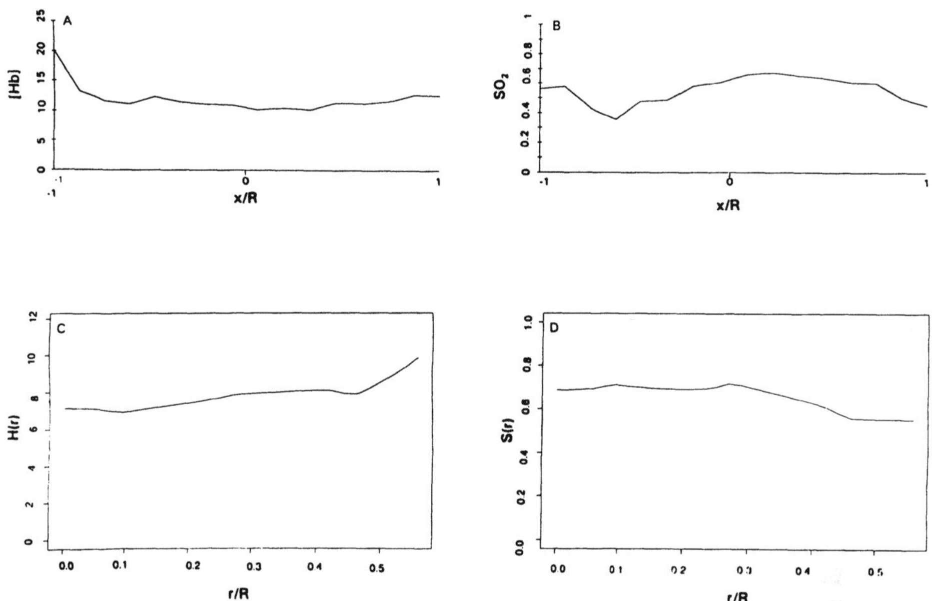

Figure 5. Radial dependencies of hemoglobin concentration and oxygen saturation, respectively, in panels C and D were derived from the corresponding profiles displayed in panels A and B. The diameter of the arteriole at this site was 38 μm; the horizontal axes in panels A and B represent the normalized lateral position in the profile as in Fig. 4; and the horizontal axes in panels C and D represent the radial distance (r) from the center of the arteriole normalized to the radius (R). The vertical axes in the four panels have the same meaning as those in Fig. 4.

Hemoglobin concentration and oxygen saturation have been measured, for the first time, using profiles of light intensity data obtained with intravital video microscopy. Previous methods have combined separate single point measurements made at several locations across the cross-section of an arteriole to generate a profile (Ellsworth and Pittman, 1986). The current method simplifies the measurement process while simultaneously increasing measurement accuracy. Also, since the measurements of light intensity were done using video pixels, spatial variability of oxygen saturation and hemoglobin concentration over distances less than 2 μm could be detected.

A possible explanation for the large discrepancy between observed and predicted oxygen diffusion from the arterioles is based, in part, on the expectation that the distributions of red cells and/or oxygen is more non-uniform just downstream of a bifurcation (the typical upstream measurement site in an unbranched segment) than at more downstream locations. Thus, we obtained profiles of hemoglobin concentration and oxygen saturation at several sites along unbranched segments of arterioles, starting at a site very close to the upstream bifurcation. Flow is divided between the two daughter branches of a bifurcation in a way that depends on characteristics of the bifurcation, such as the relative sizes of the daughter branches with respect to the parent vessel and their branching angle. Assuming that the distributions of hemoglobin and oxygen were axially symmetric just upstream of the bifurcation, the distributions in the daughter branches close to the bifurcation could be highly asymmetric. We would expect that the hemoglobin and oxygen should redistribute as they proceed further downstream and eventually their distributions should become symmetric and possibly uniform. However, this redistribution requires vessel segments with lengths equal to several vessel diameters, greater than fifteen in some cases. Such long vessels are seldom found in the retractor muscle and most other vascular networks, indicating that the luminal distributions might not become symmetric by the end of the vessel segment. The resulting non-uniform distribution now at the entrance to the following bifurcation region would give rise to daughter branches with more complex hemoglobin and oxygen distributions.

Using this method we have attempted to follow the evolution of the asymmetric profiles near bifurcations to more symmetric profiles further downstream. Of the limited sample considered in this study, the characteristic distance needed to attain a symmetric profile was found to be about 15 to 20 vessel diameters. As one moves downstream along an arteriole starting at a point close to a bifurcation, the luminal distribution of hemoglobin, which is confined to the red blood cells, tends to become more symmetric. Once symmetric, the hemoglobin profiles and, hence, luminal distributions tended to remain that way until the next downstream bifurcation was reached, in the absence of other disturbing factors. However, the oxygen saturation profiles often became more complex due to the continual diffusion of oxygen across the walls of the arteriole. The presence of nearby microvessels could lead to diffusive interactions between the arteriole and other vessels and thus complicate interpretation of the observed profiles.

The asymmetric profiles typically evolved into laterally symmetric profiles beyond a critical distance. Assuming the laterally symmetric profile to be the result of an axially symmetric distribution, the radial dependencies of hemoglobin concentration, H(r), and oxygen saturation, S(r), were calculated. The radial dependence of hemoglobin concentration was computed directly by using the profile values as input data for the convolution reconstruction algorithm. However, since oxygen is bound to hemoglobin and in all computations weighted by hemoglobin concentration, the values of S(r) were computed by first computing the radial dependence of the product [Hb]•SO_2. The radial dependence of the product, HS(r) is reduced to the radial dependence of oxygen saturation, by dividing it by H(r). The radial dependencies thus calculated could be combined with the velocity distribution to compute more accurate estimates of convective and diffusive oxygen flow.

In conclusion, we have determined oxygen saturation and hemoglobin concentration in arterioles by using full diameter profiles of transmitted light intensity. Within the limited data set we have obtained, we found that the profiles at sites close to a bifurcation were generally asymmetric and remained so at sites several vessel diameters downstream from the bifurcation. Thus, one cannot assume that profiles in hemoglobin concentration and oxygen saturation are uniform and calculations of diffusive oxygen flow based on such assumptions could contain substantial errors. Further work is required to gain a better understanding of the evolution of luminal red cell and oxygen distributions that is needed for accurate estimates of oxygen transport.

ACKNOWLEDGEMENTS

The authors gratefully acknowledge the technical assistance of Mei Dong, Robert Helber, Li Wang and Lei Zheng. This research was supported by grant HL18292 from the National Heart, Lung and Blood Insitute.

REFERENCES

Bentley, T.B., Meng, H. and Pittman, R.N., 1993, Temperature dependence of oxygen diffusion and consumption in mammalian striated muscle, *Amer. J. Physiol.* 264:H1825-H1830.

Duling, B.R., and Berne, R.M., 1970, Longitudinal gradients in periarteriolar oxygen tension: a possible mechanism for the participation of oxygen in the local regulation of blood flow, *Circ. Res.* 27:669-677.

Ellsworth, M.L., and Pittman, R.N., 1986, Evaluation of photometric methods for quantifying convective mass transport in microvessels, *Amer. J. Physiol.* 251:H869-H879.

Krogh, A., 1918/1919, The number and distribution of capillaries in muscles with calculations of the oxygen pressure head necessary for supplying the tissue, *J. Physiol. (London)* 52:409-415.

Kuo, L., and Pittman, R.N., 1988, Effect of hemodilution on oxygen transport in arteriolar networks of hamster striated muscle, *Amer. J. Physiol.* 254:H331-H339.

Kuo, L., and Pittman, R.N., 1990, Influence of hemoconcentration on arteriolar oxygen transport in hamster striated muscle, *Amer. J. Physiol.* 259:H1694-H1702.

Lipowsky, H.H., Usami, S., Chien, S., and Pittman, R.N., 1982, Hematocrit determination in small bore tubes by differential spectrophotometry, *Microvasc. Res.* 24:42-55.

Mancuso, T., Diller, T.E., Ellsworth, M.L., and Pittman, R.N., 1990, Intravascular transport of oxygen in microvessels, *in*: "Biofluid Mechanics * 3," D.J. Schneck and C.L. Lucas, ed., New York Univ., New York, 149-158.

Mott, E.A., 1992, Microvascular oxygen transport: Development of an Optical Triplicator, M.S. Thesis, Virginia Polytechnic Institute and State University.

Pittman, R.N., 1986a, In vivo photometric analysis of hemoglobin, *Ann. Biomed. Eng.* 14:119-137.

Pittman, R.N., 1986b, Microvessel blood oxygen measurement techniques, *in* "Microvascular Technology," C.H. Baker and L. Nastuk, ed., Academic Press, New York, 367-389.

Pittman, R.N., 1991, Oxygen transport in the striated muscle microcirculation: Reconciliation of theory and experiment, *in*: "Proc. IEEE Southeastcon '91," R.R. Mielke, ed., IEEE, New Jersey, 149-152.

Pittman, R.N., and Duling, B.R., 1975a, A new method for the measurement of percent oxyhemoglobin, *J. Appl. Physiol.* 38:315-320.

Pittman, R.N. and Duling, B.R., 1975b, Measurement of percent oxyhemoglobin in the microvasculature, *J. Appl. Physiol.* 38:321-327.

Pittman, R.N. and Helber, R.W., 1992, Effects of nonuniform distributions of hemoglobin ([Hb]) and oxygen saturation (SO$_2$) on the photometric estimation of oxygen in microvessels, *FASEB J.* 6: A2082.

Popel, A.S., and Gross, J.F., 1979, Analysis of oxygen diffusion from arteriolar networks, *Amer. J. Physiol.* 237:H681-H689.

Popel, A.S., Pittman, R.N., and Ellsworth, M.L., 1989, The rate of oxygen loss from the arterioles is an order of magnitude higher than expected, *Amer. J. Physiol.* 256:H921-H924.

Ramachandran, G.N., and Lakshminarayanan, A.V., 1971, Three-dimensional reconstruction from radiographs and electron micrographs: Application of convolutions instead of Fourier transforms, *Proc. Nat. Acad. Sci. USA* 68:2236-2240.

Shepp, L.A., and Logan, B.F., 1974, The Fourier reconstruction of a head section, *IEEE Trans. Nuc. Sci.* NS-21:21-33.

Sullivan, S.M., and Pittman, R.N., 1982, Hamster retractor muscle: A new preparation for intravital microscopy, *Microvasc. Res.* 23:329-335.

Swain, D.P., and Pittman, R.N., 1989, Oxygen exchange in the microcirculation of hamster retractor muscle, *Amer. J. Physiol.* 256:H247-H255.

Van Assendelft, O.W., 1970, "Spectrophotometry of Hemoglobin Derivatives," Royal Van Gorcum Ltd., Assen, The Netherlands.

THE INFLUENCE OF HYPOXIC HYPOXIA ON CORONARY BLOOD FLOW HETEROGENEITY

Robert W. Baer

Department of Physiology
Kirksville College of Osteopathic Medicine
Kirksville, MO 63501-1497 U.S.A.

INTRODUCTION

The delivery of oxygen and other nutrients to myocardial tissue is a two-step process involving first, convective delivery to the microcirculation and then, diffusive delivery across the capillary wall and interstitium into the tissue cells. The diffusive part of this process is important for substances like oxygen which are normally highly extracted. The amount of oxygen extracted is a function of capillary transit time (Honig et al., 1992). Capillary transit time, in turn, is directly related to capillary flow and inversely related to capillary path length. Heterogeneities in capillary transit time lead to heterogeneities in the extraction process. There is direct evidence that coronary blood flow shows heterogeneity (Marcus et al., 1977; King et al., 1985) and indirect evidence for heterogeneity of capillary transit times (Rose and Goresky, 1976; Rose et al., 1980). This can cause diffusion heterogeneity for a substance like oxygen which exhibits partially flow-limited diffusion.

During hypoxic hypoxia the convective portion of the oxygen delivery process is aided by hypoxic vasodilation which elevates mean coronary blood flow (Powers and Powell, 1973). But what of the diffusive part of the delivery process? If changes in vasomotor tone were uniform, then hypoxic vasodilation might not significantly alter the distribution of coronary capillary transit times. However, Bourdeau-Martini et al. (1974) observed that hypoxic hypoxia increased the uniformity of capillary spacing on the surface vessels of rat right ventricles. Intercapillary spacing also decreased. More recently, Wolpers et al. (1990) have presented evidence from indicator dilution experiments with inert gases to suggest that hypoxic vasodilation is associated with decreased transit time heterogeneity in dogs. The purpose of the present study was to directly investigate the effect of hypoxic vasodilation on coronary blood heterogeneity using radiolabeled microspheres. Specifically, we asked whether observed changes were related to the hypoxic vasodilation *per se*, or whether they were secondary to the increased convective flow. We examined hypoxia with constant perfusion pressure and elevated coronary flow, and hypoxia with reduced pressure and constant coronary flow.

Oxygen Transport to Tissue XVI
Edited by M.C. Hogan *et al.*, Plenum Press, New York, 1994

METHODS

Animal Preparation

Experiments were performed on 6 mongrel dogs of either sex weighing 20.8 ± 0.8 Kg. Pentobarbital, sodium (25-35 mg/Kg, i.v., supplemented as necessary) was used for anesthesia. Ventilation was maintained with a constant volume respirator (Harvard Apparatus) attached to a rebreathing circuit containing a sodalime canister for absorption of CO_2. During instrumentation and under control conditions, 100 % O_2 was supplied at a rate of about 700 ml/min. This produced an $F_IO_2 > 0.90$ (Beckman OM11 Oxygen analyzer).

The heart was exposed through a left lateral thoracotomy at the 4th intercostal space. Extracorporeal perfusion of the left coronary circulation was established as follows. The left main coronary artery was dissected free of surrounding tissue, and a silk suture was passed around it at its junction with the aorta. The tip of a metal cannula was inserted into the left main coronary artery via the left subclavian artery, and was secured in place with the silk suture. The cannula was attached to a pressure controlled reservoir which was refilled with blood from the left femoral artery using a servo-controlled pump. The perfusion line between the reservoir and the cannula included an electromagnetic flow transducer (Carolina Medical Electronics), a microsphere injection port, and a downstream mixing chamber with a magnetic stir bar. A constant heart rate was maintained by injecting the bundle of His with formaldehyde and electrically pacing the right ventricle.

Characterization of Flow Heterogeneity

Coronary flow heterogeneity was assessed by injecting microspheres radiolabeled with ^{153}Gd, ^{57}Co, ^{51}Cr, ^{113}Sn, ^{85}Sr, or ^{46}Sc. Microspheres were placed in injection vials (VWR Scientific), vortexed, and ultrasonicated prior to injection (Heymann et al., 1977). Approximately 5×10^5 microspheres labeled with each tracer were injected into the coronary cannula over 20 s using heparinized blood as the injection vehicle. At the end of the experiment, the perfusion territory of the cannula was marked by perfusion fixing with a saline solution containing formaldehyde and Monastral Blue dye. In one of the seven completed experiments, a substantial portion of the left ventricular freewall and septum did not stain blue. Therefore, this experiment was not included in further analysis. In all other experiments, the left ventricular freewall, the intraventricular septum, and portions of the right ventricle and atria were included in the cannula perfusion territory. The atria and right ventricular freewall were removed and not used for the analysis. The left ventricle (freewall plus septum) was divided into 288 pieces by first dividing it into 6 ring-shaped slices from base to apex, dividing each slice into 8 radial sectors, and dividing each radial sector into 6 layers from endocardium to epicardium. Left ventricular piece weight averaged 0.33 ± 0.03 g. Pieces were counted on a gamma counter (Searle 1185) attached to a multichannel pulse-height analyzer (E G & G Ortec), and individual nuclide activities in each piece were determined using a matrix inversion approach (Baer et al., 1984).

Experimental Protocol

Six measurements of coronary flow heterogeneity were made during each experiment. Concurrent with each measurement, the following hemodynamic data were recorded: left ventricular pressure, aortic pressure, coronary artery pressure, and coronary flow. In addition, blood samples were taken from the arterial reservoir and the coronary sinus for measurement of blood gases (Instrumentation Laboratories, IL213), hemoglobin levels, and O_2 saturations (Radiometer, OSM2).

The animals were stabilized at a coronary artery pressure of 100 mmHg with coronary autoregulation left intact. This was used as the control condition. A microsphere tracer was injected into the coronary flow line, followed 5 min later by a second tracer under the same conditions.

Hypoxic hypoxia was induced by changing the gas supply to the respiratory rebreathing system from 100% O_2 to a gas mixture of O_2 and N_2. (Gas flows approximately: 3.5 L/min N_2; 250 ml/min O_2). The arrangement of the system was such that there was a lag in the onset of hypoxic hypoxia due to the dead space of the respiratory recirculation system, the delay introduced by the extracorporeal coronary perfusion circuit, and possibly body stores of oxygen. To accelerate the induction of hypoxia, we flushed the system with 100% N_2 during the initial phases of induction. The F_1O_2 was monitored continually as it gradually approached 7%. This induction generally took 10-15 min. After a lag caused by the factors cited above, coronary flow rose as hypoxic hypoxia was induced, and gradually plateaued. Once this flow plateau had been reached, an additional 5 min period was allowed for stabilization. A third measurement of flow heterogeneity was then made, with a fourth measurement following 5 min later.

To differentiate between flow increases due to hypoxic vasodilation *per se*, and those caused secondarily by the attending increase in coronary flow, we then lowered coronary flow to control levels by reducing coronary perfusion pressure while maintaining the reduced F_1O_2. Under these conditions, we made a fifth, and 5 min later a sixth, measurement of flow heterogeneity.

Data Analysis

The deposition density of each microsphere tracer in each piece of the myocardium was presumed to be proportional to the flow to those pieces. Coronary flow heterogeneity could therefore be quantified in terms of the heterogeneity of microsphere deposition densities. Because flow variation was of greater interest than absolute flow, we constructed normalized deposition density functions which, by construction, had an area equal to unity and a mean relative deposition density equal to unity (Bassingthwaighte, 1982). The height of the density function at any given deposition density is a function of the fractional weight of the heart that all shared a similar blood flow. If the heart were homogeneously perfused, the density function would appear as a spike with a relative weight of one at the mean deposition of one.

Mean deposition density for each nuclide was calculated as total activity in all left ventricular pieces divided by the total left ventricular weight (g). The relative deposition density (d_i) of an individual piece was calculated as the activity per gram of that piece divided by the mean deposition density of the left ventricle. The probability density function was constructed by dividing relative deposition densities into 62 bins having a class width (Δd_i) of 0.1 (10% of the mean). The lowest bin included all pieces with deposition densities below 0.5, and the highest bin included all pieces with deposition densities above 6.0. In practice, very few pieces had deposition densities above 3.0. The height of the density function (w_i) for each deposition density bin is calculated by summing the fractional weights, ($w_i/\Sigma w_i$), of all left ventricle pieces meeting the bin's deposition density criterion and dividing this sum by the class width, Δd_i, to produce a curve with unit area.

To characterize flow heterogeneity by a single number, we calculated the relative dispersion of deposition densities for each radionuclide. The observed relative dispersion (RD_{obs}) is calculated as the standard deviation divided by the mean of the density function. This is also known as the coefficient of variation. The RD_{obs} represents a combination of dispersion due to true spatial differences ($RD_{spatial}$) in the mean flow of across pieces plus local dispersion (RD_{local}) with individual pieces about the piece's mean. Local dispersion will occur because of true temporal fluctuations in a piece's flow about its mean (twinkle), and because the measurement of local flow is subject to methodological errors. To examine the impact of true spatial dispersion, we made the assumption, used by King et al. (1985), that the observed variance equals the sum of spatial and local variance. Therefore,

$$RD^2_{spatial} \quad = \quad RD^2_{observed} \quad - \quad RD^2_{local} \tag{1}$$

Local variation was evaluated for each condition by calculating the average deviation of two sequential injections made at 5 min intervals. The average of the 288 left ventricular pieces was used in calculating true spatial relative dispersion from observed relative dispersion.

The magnitude of observed relative dispersion depends on how finely the left ventricle is sectioned. The relationship between piece size and relative dispersion appears to be fractal in nature. The expression given by Bassingthwaighte (1988) for this relationship is:

$$RD(m) \quad = \quad RD_{1\,g} \cdot m^{1-D} \tag{2}$$

where m is the average mass of left ventricular pieces for a given cutting scheme; D is the fractal dimension; $RD_{1\,g}$ is the observed relative dispersion for standardized pieces weighing 1 g; and RD(m) is the observed relative dispersion for pieces of mass size, m. To analyze the effect of hypoxia on this fractal relationship, we arithmetically reassembled the left ventricles by summing counts and weights of adjoining pieces. Our reassembly schema allowed us to calculate relative dispersion for each heart when sectioned into 288, 144, 72, 36, 12, 6, and 2 pieces. A plot of log (RD) against log (m) was generally linear for 6 to 288 piece "cutting" schemes. From the slope and intercept of this relationship, we were able to calculate the fractal dimension, D, and the RD of standardized 1g pieces.

Statistical Analysis

Data are expressed as mean ± standard deviation unless otherwise indicated. Analysis of variance was used to make statistical comparisons across conditions. When desirable, individual between group comparisons were made using Newman-Kuels multiple range test. A $p < 0.05$ was accepted as statistically significant. Standard linear regression on the logarithmic values was used in determining fractal dimensions and standardized relative dispersion.

RESULTS

Hemodynamic data are shown in Table 1. By design, coronary flow rose with the induction of hypoxic hypoxia, but coronary artery pressure did not change. Hypoxic vasodilation was associated with a fall in coronary vascular resistance. To return coronary flow to control levels, it was necessary to lower coronary perfusion pressure to 42 ± 13 mmHg. Coronary vascular resistance did not differ significantly from the previous hypoxia condition following this manipulation. Left ventricular end-diastolic pressure did not change significantly with hypoxia. Aortic systolic pressure did not fall with hypoxia alone, but it fell significantly when coronary flow was restored to the control level. Aortic diastolic pressure dropped with the induction of hypoxia, and fell further when coronary flow was lowered to the control level. Because of the electrical pacing, heart rate did not change significantly from its control value of 84 ± 10 bpm.

Table 1. Hemodynamic Data

Variable	Control	Hypoxia	Hypoxia -- Flow Match	Significance by ANOVA, p<
Coronary flow (ml/min/100g)	109 ± 43	258 ± 78	110 ± 4	0.001
Coronary Artery Pressure (mmHg)	99 ± 1	99 ± 4	42 ± 13	0.001
Coronary Vascular Resistance (mmHg/ml/min/100g)	1.02 ± 0.34	0.40 ± 0.11	0.40 ± 0.06	0.001
Left Ventricular End-diastolic Pressure (mmHg)	4 ± 4	4 ± 6	9 ± 11	n.s.
Aortic Systolic Pressure (mmHg)	99 ± 21	94 ± 31	71 ± 32	0.05
Aortic Diastolic Pressure (mmHg)	55 ± 15	36 ± 19	22 ± 17	0.001
Heart Rate (beats/min)	84 ± 10	88 ± 11	84 ± 10	n.s.

Values represent mean $\pm$ standard deviation. Significance is for ANOVA. Results of Newman-Kuels Test for individual mean differences are discussed in text. n.s. = not significant, $p > 0.05$.

Table 2. Blood Gas Values

Variable	Control	Hypoxia	Hypoxia -- Flow Match	Significance by ANOVA, p<
P_aO_2 (mmHg)	112.7 ± 79.7	27.0 ± 5.9	31.9 ± 4.4	0.001
$P_{cs}O_2$ (mmHg)	36.8 ± 6.0	17.6 ± 4.8	12.5 ± 5.0	0.05
$\%S_aO_2$	95.8 ± 3.9	$34.6 \pm .5$	40.9 ± 17.4	0.001
$\%S_{cs}O_2$	61.7 ± 8.4	17.5 ± 7.0	8.1 ± 3.4	0.001
C_aO_2 (ml/dl)	18.5 ± 1.4	6.8 ± 1.2	7.5 ± 2.1	0.001
$C_{cs}O_2$ (ml/dl)	11.9 ± 1.0	3.4 ± 0.9	1.5 ± 0.6	0.001
pH	7.40 ± 0.08	7.3 ± 0.09	7.18 ± 0.11	0.001
P_aCO_2 (mmHg)	27.5 ± 5.3	25.1 ± 6.0	22.8 ± 6.6	n.s.
MVO_2 (ml/min/100g)	6.9 ± 2.2	8.7 ± 2.4	6.0 ± 2.1	0.05

Values represent mean $\pm$ standard deviation. Significance is for ANOVA. Results of Newman-Kuels Test for individual mean differences are discussed in text. n.s. = not significant, $p > 0.05$.

Blood gas values are shown in Table 2. As expected, hypoxic hypoxia was associated with a significant fall in P_aO_2, $\%S_aO_2$, and C_aO_2. The corresponding coronary sinus values also fell. Because we failed to compensate completely during the induction of hypoxia, a slight acidosis accompanied the onset of hypoxia. P_aCO_2 did not change significantly. Myocardial oxygen consumption rose slightly during hypoxia and fell again when flow was reduced to its control value.

The normalized probability density function curves (constructed from 1728 left ventricular pieces in 6 dogs) are shown in Figure 1. Under all conditions, the curves are slightly right skewed, but the mode shifts to the right with the induction of hypoxia. When coronary flow was returned to control levels, the mode became more peaked and was fairly close to the mean deposition density of one. Note that duplicate measurements under each condition produce remarkably similar distributions of deposition densities. Individual plots for the 288 pieces of each dog all reflect this same pattern.

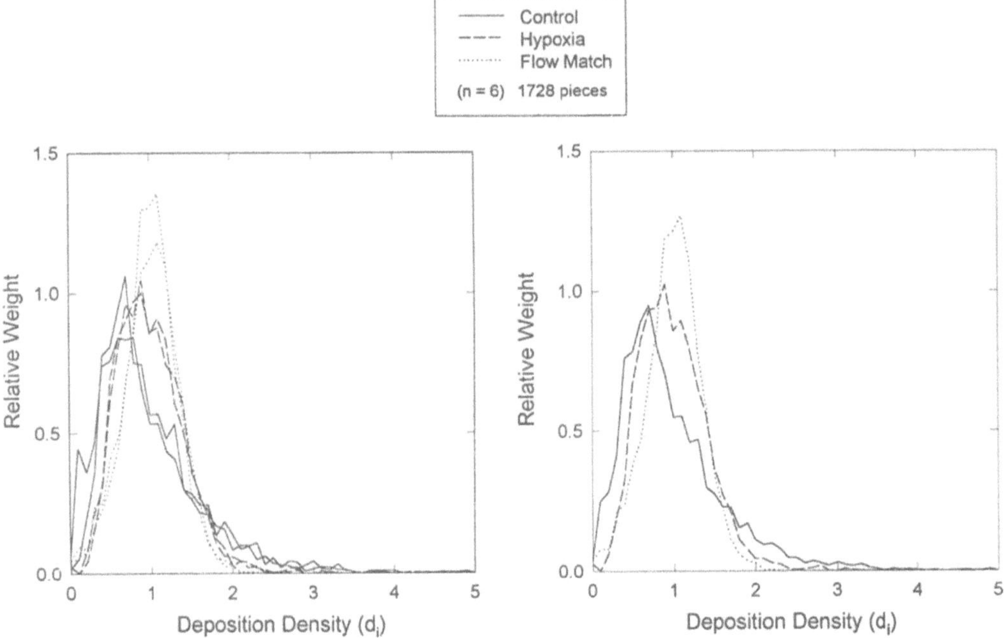

Figure 1. Normalized probability density plots for microsphere deposition density in the dog. Distributions were constructed from 1728 left ventricular pieces taken from six dogs. Left panel shows data for each of the six individual microsphere injections under the 3 experimental conditions. Right panel shows the same data as pooled microsphere pairs (3456 density measurements) for the 3 conditions.

Observed relative dispersion averaged 27.3 ± 10.4 % under control conditions and fell significantly to 17.0 ± 3.2 % with the hypoxia (Figure 2). Matching coronary flow to control levels produced a slight further drop in relative dispersion to 14.1 ± 2.0 %, but this was not significant by the Newman-Kuels multiple range test. Local relative dispersion averaged 15.8 ± 12.3 % under control conditions and slightly, but not significantly, less with hypoxia, 9.0 ± 3.0 % when coronary flow was elevated and 9.4 ± 1.4 % when flow was returned to control levels. True spatial relative dispersion averaged 21.2 ± 3.9 % under control conditions and decreased significantly (p < 0.001) to 14.3 ± 2.2 % and 10.0 ± 3.9 % under the two hypoxic conditions.

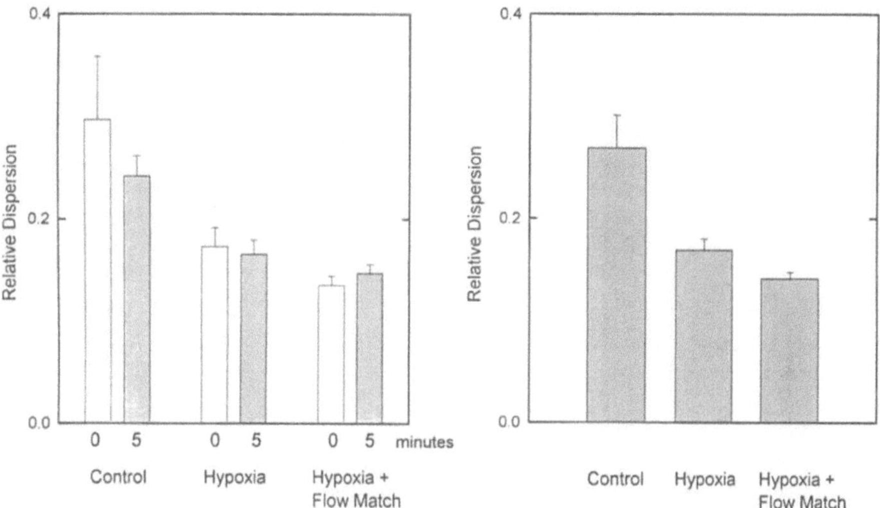

Figure 2. Average (n=6) relative dispersion of microsphere deposition densities in dog left ventricles sectioned into 288 pieces and weighing 0.33 ± 0.03 g. Left panel shows data for each of the six individual microsphere injections under the 3 experimental conditions. Right panel shows the same data as pooled microsphere pairs for the 3 conditions.

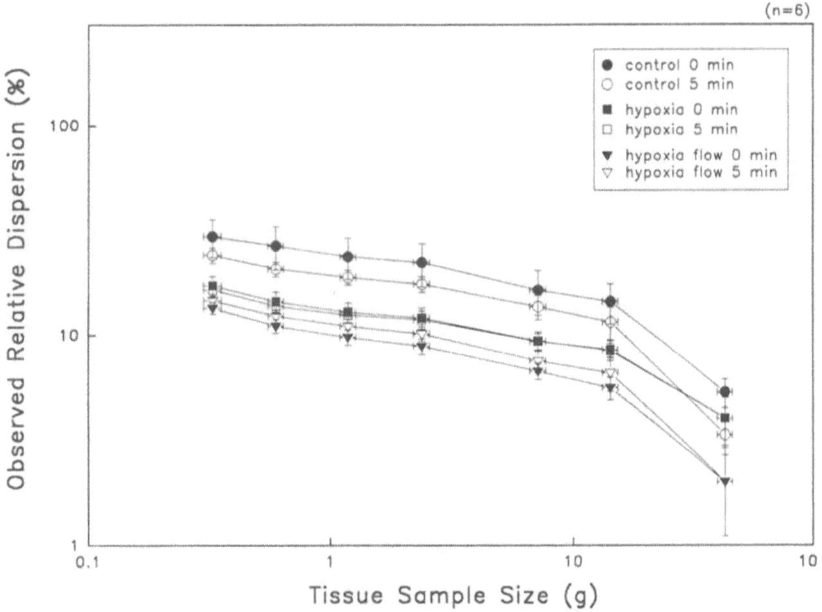

Figure 3. Fractal nature of left ventricular coronary flow. Reconstruction of the ventricles into neighboring pieces of similar size shows that the relative dispersion of microsphere deposition densities is predictably related to piece size over a range of sizes.

Fractal analysis resulted in a linear relationship between log RD and log m for pieces of less than 10 g (Figure 3). Pieces of ~10 g are the result of nearest-neighbor reconstructions of the hearts into 6 spatially-defined pieces of similar size. The fractal dimension, D, for the linear portion of these ventricular reconstructions was 1.19 ± 0.08 and did not change significantly with hypoxia whether or not flow was returned to control levels. The $RD_{1 g}$ was 21.9 ± 9.1 % under control conditions and fell significantly ($p < 0.001$) to 13.4 ± 3.4 % and 10.9 ± 1.9 %, respectively, under hypoxia without and hypoxia with flow matching.

DISCUSSION

The principal finding of this study is that hypoxic vasodilation results in a more uniform distribution of coronary flow. This decreased flow heterogeneity is independent of the increase in coronary flow normally associated with hypoxic vasodilation. Our finding is based on the assumption that microspheres distributed themselves in proportion to flow. We took precautions to assure that the microspheres were disaggregated, well suspended, and uniformly mixed. However, due to the particulate nature and density of microspheres we can not exclude the possibility of preferential streaming. Bassingthwaighte et al. (1987) investigated this issue by comparing distribution of microspheres to that of 2-iododesmethylimipramine. They concluded that although there was a small systematic bias for microspheres to enter high flow regions they were generally adequate for measuring regional flows.

The reduced flow heterogeneity during hypoxia is associated with a probability density function that has become more symmetrical about its mean. The flow distribution throughout left ventricle, like the capillary density of the surface capillaries of the right ventricle (Bourdeau-Martini et al., 1974), becomes more uniform with the onset of hypoxic hypoxia. By itself, the reduced flow heterogeneity that we observed can explain the reduced heterogeneity of capillary transit times during hypoxia observed by Wolpers et al. (1990) with indicator dilution curves from inert gases. We still need further information to know whether or not the capillary path length was also altered in that previous study.

Under control conditions, we observed a relative dispersion in dog left ventricles of 27.3%. This is generally comparable to that found by other investigators in a number of species. For dog left ventricles, values of 21.4% (Marcus et al., 1977), 21.7% (Sestier et al., 1978), and 41% (Kuikka et al., 1986) have been reported. King et al. (1985) observed a value of 28% in conscious baboons. In rabbits, values of 30% (Gorman et al., 1989) and 33% (Conway and Weiss, 1985) have been observed. Bassingthwaighte et al. (1989) have demonstrated that observed relative dispersion is a function of piece size, and that the relationship is fractal in nature. Fractal analysis of the present data yielded a fractal dimension, D, of 1.19. This is similar to the values obtained by Bassingthwaighte's group and others (Overholser et al., 1991). Interestingly, fractal dimension does not change with hypoxic vasodilation and must therefore be related to some more intrinsic property of the ventricle, perhaps large vessel branching pattern (van Beek et al., 1989). The stability of fractal dimension with hypoxic vasodilation supports its proposed role as a generalized system descriptor that can be used to correct for size differences in different studies coming from various laboratories. Fractal correction of our data to standard 1 g pieces yielded an average relative dispersion of 21% which is identical to the value observed by Marcus et al. (1977) for 1.08 g pieces of dog left ventricle.

The observed relative dispersion appears to be a spatially stable pattern in any individual animal. However, there are apparently temporal fluctuations about the mean in individual pieces (King et al., 1985). The magnitude of such fluctuations also appears to be somewhat higher in pieces with elevated flow (King and Bassingthwaighte, 1989). Based on measurements made at 5 min intervals, we obtained a measure of local relative dispersion of between 9.0 and 15.8 %. If we assume, as King et al. (1985) did, that the variances contributing to observed variance are independent and additive, we obtain an estimate of true spatial relative dispersion. When we make this calculation, we still arrive at the conclusion that hypoxic vasodilation *per se* leads to a reduced blood flow heterogeneity.

Conway and Weiss (1985) also observed decreased relative dispersion with hypoxic hypoxia, but they attributed this decrease to the increase in coronary flow. The present study shows that the hypoxia-induced decrease in relative dispersion is related to the vasodilation, not the flow increase. Polansky and Weiss (1993) have recently proposed the general hypothesis that relative dispersion is inversely related to mean coronary flow rate. Hypoxic vasodilation would appear to be at least one exception to this hypothesis. Our data are more consistent with the suggestion by Rose et al. (1977) that transit time heterogeneity is proportional to coronary vascular resistance. At present, it would seem prudent to evaluate flow heterogeneity under each physiological condition of interest until we can develop a more complete understanding of the factors which influence it.

In summary, it appears that hypoxic vasodilation leads to improved nutrient delivery by affecting both steps in the delivery process. It not only increases overall convective delivery, but it also improves the uniformity of flow distribution and thus optimizes the overall diffusion to the myocardium.

ACKNOWLEDGEMENTS

Support for this study was provided by grants from the American Heart Association, National Center and the Missouri Affiliate. The author gratefully acknowledges the superb technical assistance of Carla Connelly Workman and Donna J. Buckley.

REFERENCES

Baer, R.W., Payne, B.D., Verrier, E.D., Vlahakes, G.J., Molodowitch, D., Uhlig, P.N., and Hoffman, J.I.E. (1984) Increased number of myocardial blood flow measurements with radionuclide-labeled microspheres, *Am. J. Physiol.* 246 (*Heart Circ. Physiol.* 15): H418-H434.

Bassingthwaighte, J.B. (1982) Calculation of transorgan transport functions for a multipath capillary-tissue system, *Natl. Tech. Info. Serv.* PB82-229535.

Bassingthwaighte, J.B., Malone, M.A., Moffett, T.C., King, R.B., Little, S.E., Link, J.M., and Krohn, K.A. (1987) Validity of microsphere depositions for regional myocardial flows, *Am. J. Physiol.* 253 (*Heart Circ. Physiol.* 22): H184-H193.

Bassingthwaighte, J.B. (1988) Physiological heterogeneity: Fractals link determinism and randomness in structures and functions, *News in Physiol. Sci.* 3: 5-10.

Bassingthwaighte, J.B., King, R.B., and Roger, S.A. (1989) Fractal nature of regional myocardial blood flow heterogeneity, *Circ. Res.* 65: 578-590.

Bourdeau-Martini, J., Odoroff, C.L., and Honig, C.R. (1974) Dual effect of oxygen on magnitude and uniformity of coronary intercapillary distance, *Am. J. Physiol.* 226: 800-810.

Conway, R.S., and Weiss, H.R. (1985) Dependence of spatial heterogeneity of myocardial blood flow on mean blood flow rate in the rabbit heart, *Cardiovasc. Res.* 19: 160-168.

Gorman, M.W., Wangler, R.D., and Sparks, H.V. (1989) Distribution of perfusate flow during vasodilation in isolated guinea pig heart, *Am. J. Physiol.* 256 (*Heart Circ.Physiol.* 25): H297-H301.

Heymann, M.A., Payne, B.D., Hoffman, J.I.E., and Rudolph, A.M. (1977) Blood flow measurements with radionuclide-labeled particles, *Prog. in Cardiovasc. Dis.* 20: 55-78.

Honig, C.R., Connett, R.J., and Gayeski, T.E.J. (1992) Oxygen transport and its interaction with metabolism; a systems view of aerobic capacity, *Med. Sci. Sports Exerc.* 24: 47-53.

King, R.B., Bassingthwaighte, J.B., Hales, J.R.S., and Rowell, L.B. (1985) Stability of heterogeneity of myocardial blood flow in normal awake baboons, *Circ. Res.* 57: 285-295.

King, R.B. and Bassingthwaighte, J.B. (1989) Temporal fluctuations in regional myocardial flows, *Pflugers Arch.* 413: 336-342.

Kuikka, J., Levin, M., and Bassingthwaighte, J.B. (1986) Multiple tracer dilution estimates of D- and 2-deoxy-D-glucose uptake by the heart, *Am. J. Physiol.* 250 (*Heart Circ. Physiol.* 19): H29-H42.

Marcus, M.L., Kerber, R.E., Erhardt, J.C., Falsetti, H.L., Davis, D.M., and Abboud, F.M. (1977) Spatial and temporal heterogeneity of left ventricular perfusion in awake dogs, *Am. Heart J.* 94: 748-754.

Overholser, K.A., Bhatte, M.J., and Laughlin, M.H. (1991) Modeling the effect of flow heterogeneity on coronary permeability-surface area, *J. Appl. Physiol.* 71: 758-769.

Polansky, L, and Weiss, H.R. (1993) Effect of flow reduction on coronary blood flow heterogeneity, *Proc. Soc. Exp. Biol. Med.* 202: 97-102.

Powers, E.R., and Powell, W.J. Jr. (1973) Effect of arterial hypoxia on myocardial oxygen consumption, *Circ. Res.* 33: 749-756 .

Rose, C.P., and Goresky, C.A. (1976) Vasomotor control of capillary transit time heterogeneity in the canine coronary circulation, *Circ. Res.* 39(4):541-554.

Rose, C.P., Goresky, C.A., Belanger, P, and Chen, M.J. (1980) Effect of vasodilation and flow rate on capillary permeability surface product and interstitial space size in the coronary circulation: A frequency domain technique for modeling multiple dilution data with Laguerre functions, *Circ. Res.* 47: 312-328.

Sestier, F.J., Mildenberger, R.R., and Klassen, G.A. (1978) Role of autoregulation in spatial and temporal perfusion heterogeneity of canine myocardium, *Am. J. Physiol.* 235 (*Heart Circ Physiol 4*): H64-H71.

Wolpers, H.G., Hoeft, A., Korb, H., Lichtlen, P.R., and Hellige, G. (1990) Heterogeneity of myocardial blood flow under normal conditions and its dependence on arterial PO_2, *Am. J. Physiol.* 258 (*Heart Circ. Physiol.* 27): H549-H555.

van Beek, J.H.G.M., Roger, S.A., and Bassingthwaighte, J.B. (1989) Regional myocardial flow heterogeneity explained with fractal networks, *Am. J. Physiol.* 257 (Heart Circ. Physiol. 26): H1670-H1680.

OXYGEN DELIVERY AND INTENTIONAL HEMODILUTION

Stephen M. Cain

Department of Physiology and Biophysics, University of
Alabama at Birmingham, Birmingham, AL 35294-0005, U.S.A.

INTRODUCTION

Normovolemic hemodilution can progress to a classic form of hypoxia
called anemic anoxia in the terminology first used by Barcroft (1920). It differs
from anoxic anoxia, or hypoxic hypoxia in more modern terminology, in that
arterial PO_2 can be quite normal but the arterial O_2 concentration is lower than
normal. The question naturally arises, therefore, of why one would intentionally hemodilute in a clinical setting. The reason becomes clearer when
consideration is given to the desirability of sparing as much as possible the use
of allogeneic blood because of its additional risk factors. Furthermore,
hemodilution within specified limits may actually be desirable by virtue of
increased blood fluidity and consequently better tissue oxygenation under
some circumstances. There is even evidence that tissue PO_2 in various organ
systems may actually increase with mild hemodilution (Messmer et al., 1973).
Given the desirability of preoperative autologous blood donation and/or
volume expansion with cell-free diluents such as dextran, albumin, and
hetastarch, it is worthwhile to ask what the lower limits of hemodilution might
be. Such limits are set to some extent by the ability of the body to compensate
for the decrease in oxygen carrying capacity. The nature of those compensatory actions, their effectiveness, and the net effect upon tissue oxygenation in
health and disease states are the topics that will be discussed.

COMPENSATORY MECHANISMS IN HEMODILUTION – SYSTEMIC

In experiments that compared progressively more severe anemic and
hypoxic hypoxias in separate groups of anesthetized dogs, O_2 uptake was
preserved until a "critical" O_2 delivery was reached and that was the same for
both types of hypoxia (Cain, 1977). In other words, as long as the total
amount of O_2 presented to the tissues per unit time was above that critical
value, systemic O_2 uptake and, presumably thereby, tissue oxygenation were at
levels sufficient to maintain essential functions. Because systemic O_2 delivery
is the product of arterial O_2 concentration and cardiac output, it is obvious that
a principal compensation for reduced hemoglobin concentration in blood,
which defines anemic hypoxia, is to increase cardiac output.

An immediate aid to increase cardiac output with hemodilution is the
consequent lowering of blood viscosity as formed elements are removed from
the blood. Figure 1 was generated from data of Fan et al. (1980) who measured change in viscosity as a function of blood hematocrit. Although the

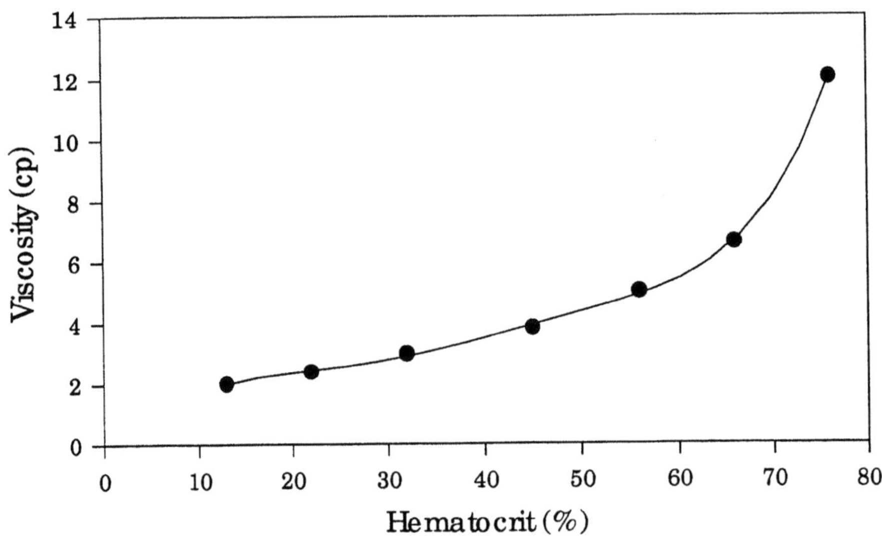

Fig. 1   Viscosity of blood as a function of hematocrit

relationship becomes very steep at hematocrit values above and relatively flat below the normal average of 45%, a further reduction to 20% will still halve the viscosity. The significance of that can best be appreciated by the fact that vascular resistance is the product of viscosity and hindrance. The latter term reflects the contribution of vessel geometry. Accordingly, if no other compensatory action were to take place in reaction to lowering hematocrit to 20%, the resultant decrease in vascular resistance would increase cardiac output. If the increase were exactly in proportion to the decrease in O_2 content of arterial blood, then oxygen delivery would be the same as that before hemodilution. Because this is not the case, the concept of an optimal hematocrit arises. As Murray et al. (1963) described, systemic O_2 delivery is maximal in the normal range of hematocrit from 35% to 45%. In later studies, Sunder-Plassmann et al. (1971) stated that the optimal hematocrit with respect to O_2 delivery to the tissues was actually closer to 30%. At hematocrits greater than that, the increased viscosity of blood increases vascular resistance and the consequent decrease in cardiac output is more than the gain in O_2 carrying capacity. Similarly, below that range, the increase in cardiac output with decreased blood viscosity and vascular resistance is insufficient to compensate for the decreased carrying capacity. Nevertheless, alterations in cardiac output with normovolemic hemodilution will at least partially compensate with respect to systemic O_2 delivery.

In addition to this passive compensatory mechanism of lowered resistance and increased cardiac output, more reactive mechanisms are called into play. Although anemia is not a stimulus to the carotid body if arterial PO_2 is normal, it will stimulate the aortic body with many of the same responses as a result (Hatcher, et al., 1978). In addition to the sympathetic augmentation of cardiac output, this also raises the possibility of some preferential distribution of blood flow amongst organ systems. This was examined in anesthetized dogs by Fan et al. (1980) who found that the only organ system that actually decreased flow with hemodilution was the spleen. Liver, gut,, and kidney all maintained blood flow as hematocrit was lowered but did not share in the increased cardiac output that was occurring at the same time. Heart and brain, however, increased flow as they either maintained or decreased vascular resistance in contrast to the other organ systems in which resistance was increased with lowered hematocrits. Preferential

distribution of increased cardiac output by means of heightened vasoconstrictor tone is another compensatory mechanism of some importance to the
preservation of essential functions in the whole organism. The importance of
vigorous sympathetic constrictor tone to increase and to optimize O_2 extraction
in the periphery can best be appreciated by the results obtained when α–adrenergic receptors were blocked with phenoxybenzamine in anesthetized dogs
made hypoxic by ventilation with low O_2 gas mixtures. The ability of the
body to increase O_2 extraction to compensate a decrease in O_2 delivery was
very significantly compromised in the treated group (Cain, 1978).

Skeletal muscle was not included in the study of Fan et al. (1980). As
an organ system that comprises 40% or more of body weight and which
receives as much as 25% of cardiac output even at rest, it needed to be
included in this picture. Cain and Chapler (1978) found that resting skeletal
muscle did not redirect its blood flow even when hemodilution was carried
past the point of limiting systemic O_2 uptake. The role of sympathetic
vasoconstriction in the responses of both whole body and skeletal muscle to
anemic hypoxia were further examined in studies by Chapler and Cain (1982)
who used phenoxybenzamine to block α–adrenergic receptors. They found
that hemodilution reduced vascular resistance, as expected, but that no further
change was noted after α–adrenergic blockade. The most notable response to
the addition of an α–blocking agent after hemodilution was a decrease in
cardiac output. It required volume expansion on the order of 12 ml/kg to
restore cardiac output to the pre-block value. The loss of this compensatory
increase in cardiac output in response to hemodilution in the presence of
α–block was sufficient to reduce O_2 uptake significantly even though it had
been previously maintained at an hematocrit of 13%. These results pointed to
the important role of sympathetic activation in anemic hypoxia; their essential
action was not on the resistance vessels but rather to increase venous return by
means of increased tone in capacitance vessels. That this did occur was very
nicely demonstrated by Chapler et al. (1981) who measured the hindlimb
weight and venous pressures in dogs before and after hemodilution to
hematocrit of 14%. They saw a decrease in limb weight and an increase in
venous pressure which were good indications of the constriction of venous
capacitance vessels with the stimulus of hemodilution.

To summarize the effects of normovolemic hemodilution at the whole
body and regional levels:

     1. The principal compensation for decreased hemoglobin
concentration in blood is to increase cardiac output.

     2. Although heart and brain are better perfused than
other organ systems, there is relatively little regulation of peripheral distribution of blood flow.

     3. Any loss of venous tone or decrease in venous return
will immediately detract from the ability of the body to cope with
hemodilution.

COMPENSATIONS TO HEMODILUTION IN THE MICROCIRCULATION

Passive compensatory responses to intentional hemodilution are present
not only at the organ system level but also at the level of the microcirculation.
The principal protection offered at the microcirculation devolves from the
observation that capillary or "tube" hematocrit is normally less than the
systemic hematocrit measured in large conduit vessels. This was first noted by
Fahraeus (1929) who reported that the hematocrit of blood in a system of tubes
decreased with decreasing tube diameter even though the hematocrit of the
blood in the reservoir which served as the source was unchanged. The
principal reason for this was the disparate rates of movement of plasma and
red blood cells through capillary size tubes. The biorheology that explains this
phenomenon is complicated but a simple explanation can give the gist of the
reason. One factor is that red blood cells are more concentrated at the axis of
flow. At the same time, a plasma layer becomes more stationary because of

the greater resistance to its movement by virtue of the frictional forces encountered at the lining of the vessel. At any moment in time, therefore, more plasma is found in the microcirculation than red blood cells even though the total number of red cells and the volume of plasma entering and exiting the microcirculation remains the same. Another way to state this is that a relative state of hemodilution always is present in the capillary bed. Of importance to the topic at hand, systemic hemodilution has a lesser effect in the capillary so that red cell density there for the purpose of gas exchange is also less affected.

The relative lack of change in capillary hematocrit as systemic hematocrit was lowered was illustrated by Intaglietta (1989). Direct in vivo measurements of capillary hematocrits in mesentery, muscle, and skin showed little to no decline as systemic hematocrit was halved from normal. In practical terms, this means that down to hematocrits as low as 20%, there was little decrease in capillary density of red cells so that O_2 delivery would thus be little affected as well. What little decrease was seen in the tissue beds that were examined was compensated by the increase in flow velocity as hematocrit was reduced. Consequently, red blood cell flux held constant as systemic hematocrit was halved from its normal value. The practical significance of these observations was well illustrated by studies of Messmer et al. (1973) who measured tissue PO_2 in several organ systems of anesthetized dogs with platinum multiwire surface electrodes as the animals' blood was exchanged for dextran. They found that mean tissue PO_2 in skeletal muscle, liver, pancreas, small intestine, and kidney either did not decrease or actually increased as hematocrit was progressively and acutely lowered from 42% to 19%. They stressed, however, that these results depended upon a vigorous cardiac output response and the absence of hypovolemia and/or hypotension.

Finally, one other favorable factor may be operating to preserve tissue oxygenation with intentional hemodilution. As pointed out by Duling and Berne (1973), there can be a loss of O_2 from arterioles by precapillary diffusion either to surrounding tissue or to parallel veins in close proximity to the arterioles. This O_2 "leak" or diffusive shunt is fixed by the mean PO_2 gradient that exists between the vessel and the surrounding tissue or parallel vein. As a function of transit time, therefore, an increase in red cell velocity will result in less loss of O_2 from any particular red cell before it reaches the capillary to participate in the more useful route of O_2 transfer to tissues. One would expect that losses in PO_2 by precapillary diffusion would be decreased or minimized with intentional hemodilution so that tissue oxygenation would benefit by this mechanism as well.

To summarize the effects of intentional hemodilution at the level of the microcirculation and upon tissue oxygenation:

1. Capillary hematocrit is normally about one-half the systemic hematocrit found in large vessels.

2. Capillary hematocrit stays nearly constant until systemic hematocrit is decreased below 20%.

3. Capillary red blood cell flux increases with hemodilution so that tissue oxygen delivery is actually greatest at systemic hematocrit of 30% and tissue PO_2 either rises or does not fall until a hematocrit of 20% is reached.

4. Precapillary shunting of PO_2 is decreased with increased red blood cell velocity that ensues with hemodilution so that tissue oxygenation is further preserved.

INTENTIONAL HEMODILUTION IN EXPERIMENTAL ANIMALS DURING RESUSCITATION FROM ENDOTOXIC SHOCK

The conclusion reached on the basis of information obtained in normal anesthetized animals was that tissue oxygenation did not suffer when systemic hematocrit was lowered to as low as 20%. Clearly, intentional hemodilution for purposes such as preoperative autologous blood donation or for volume expansion would not be contraindicated if the heart and vascular systems were normally capable in patients. The next question is whether, in instances where the cardiovascular system may not be normally capable, hemodilution would still be acceptable down to the same limits of hematocrit. Sepsis or endotoxemia provides one such example in that both the heart and the microcirculation may be dysfunctional as a result of the many reactions caused by endotoxin (Cain, 1986). In addition to pathological manifestations of endotoxin that depress myocardial contractility and grossly disturb the microcirculation, hemodilution itself may increase diffusional resistance to O_2 movement from the interior of the capillary to the interior of cells using O_2. Once the spacing between red blood cells in the capillary does increase, the increased diffusion distance through plasma could impede the movement of O_2 from out of the red cell because of the low O_2 solubility of plasma (Homer, et al., 1981; Gutierrez, 1986).

With these factors in mind, Cain et al. (in press) tested the relative efficacy of infusing dextran alone or with the addition of perflubron, a brominated perfluorocarbon compound, to increase the O_2 solubility of the plasma phase of blood in the resuscitation of anesthetized dogs from endotoxic shock. The protocol can be seen in Figure 2 which illustrates events in the whole body. After a control measurement, 2 mg/kg of *E. coli* lipopolysaccharide (LPS, endotoxin) was slowly infused intravenously over a period of one hour. A classic picture of endotoxic shock developed in which mean arterial blood pressure fell to ~60 mmHg and cardiac output decreased even though vascular resistance was falling at the same time. Although whole body O_2 uptake did not fall significantly, total O_2 delivery approached the critical limit in normal dogs (Cain, 1977) as O_2 extraction increased to near maximum. With the development of circulatory shock arterial lactate also increased. The events noted in hemodynamic and metabolic measurements in these dogs infused with endotoxin replicated those observed in human septic patients.

At the end of the endotoxin infusion period in Figure 2 resuscitation was instituted by a continuous infusion of dextran (MW 70,000) at the rate of 0.5 ml/kg/min. In half of the animals, perflubron was added to the infusate so that they received a total of 10 ml/kg over the 2-hr resuscitation period. The others had the same amount of vehicle in which the perflubron emulsion was made. Because no red blood cells were added to the infusate, the hematocrit decreased over the same period to ~19% on the average. Cardiac output responded almost immediately to the volume expansion and continued to increase in both groups until it was more than doubled above the control value by the end of the resuscitation period. O_2 delivery increased back to the control level by the end of the first hour but then became level as the cardiac output increases only balanced the loss of O_2 carrying capacity with the progressive hemodilution. Although O_2 uptake had not changed significantly during the onset of endotoxic shock, with resuscitation the increased O_2 demand was noted by the significant increase in uptake. What was very evident in these results was the fact that even the endotoxic dog was able to compensate for hemodilution to half normal hematocrit in that O_2 delivery was increased back to the control level and O_2 uptake was actually increased. The addition of perflubron to increase O_2 solubility in the plasma phase had no important effect. At the whole body level, therefore, no detectable diffusion barrier existed between the red blood cell interior and the capillary wall with hemodilution to hematocrits of ~20%.

In the same experiments shown in Figure 2, similar measurements were made in the regional circulation to the left hindlimb skeletal muscle. Those results may be seen in Figure 3. In general, the results paralleled those seen in the whole body. Although O_2 delivery was increased by volume resuscitation, it was not returned to the control level. The perflubron may have exerted some beneficial effect on O_2 transfer in that some significant increases were seen in the limb muscle O_2 uptake and O_2 eatraction was generally higher.

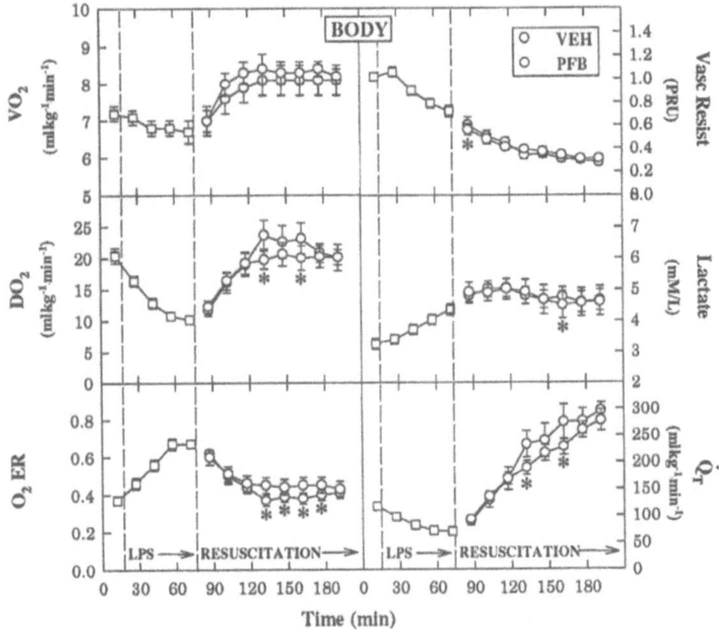

Figure 2. Whole body values (mean±SE) for O_2 uptake (VO_2), O_2 delivery (DO_2), O_2 extraction ratio (O_2ER), vascular resistance, arterial lactate concentration, and cardiac output (Q_T). Asterisks denote $p<0.05$; filled circles are the perflubron (PFB) group; unfilled circles are the vehicle (VEH) group (from Cain, et al., in press, with permission).

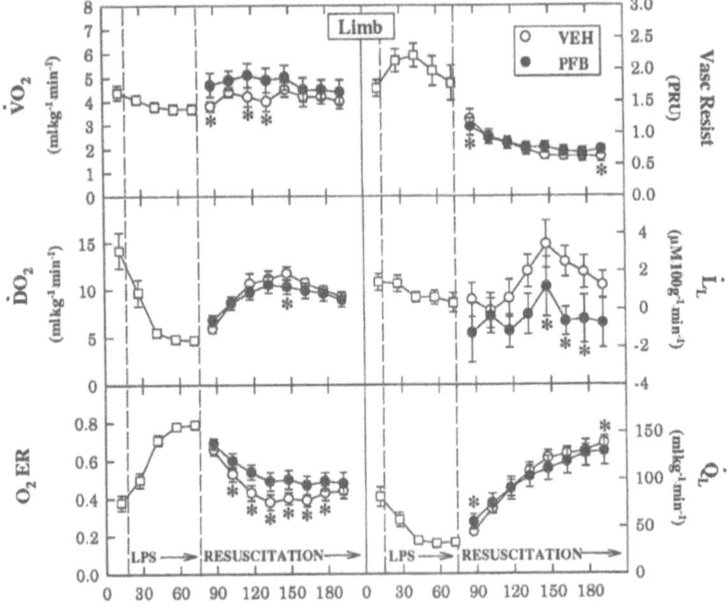

Figure 3. Limb values (mean±SE) for O_2 uptake (VO_2), O_2 delivery (DO_2), O_2 extraction ratio (O_2ER), vascular resistance, lactate flux (L_L), and blood flow (Q_L). Asterisks denote $p<0.05$ (from Cain, et al., in press, with permission).

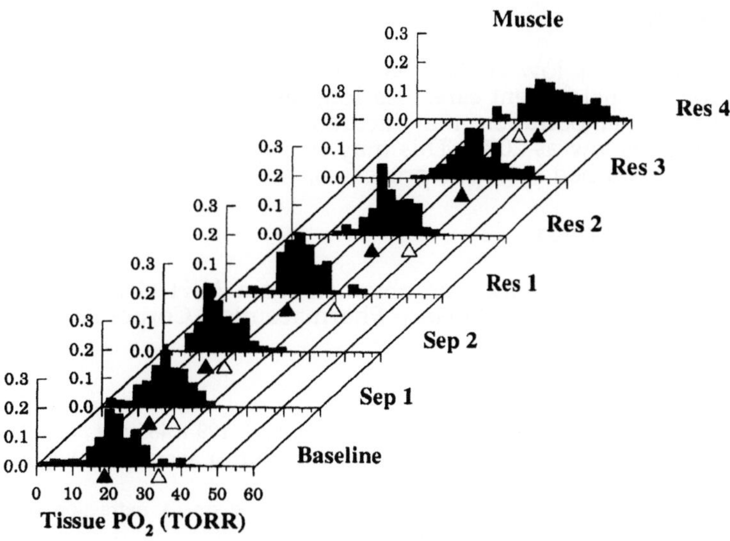

Figure 4. Histograms of tissue PO$_2$ distributions in skeletal muscle obtained before endotoxin infusion, at 30 and 60 min of endotoxin infusion (LPS1, LPS2), at 30, 60, 90, and 120 min after starting volume resuscitation with dextran (RES1, RES2, RES3, RES4). Filled triangles indicate values for mean tissue PO$_2$; unfilled triangle indicate mean venous PO$_2$ (adapted from Vallet, et al., in press, with permission).

Were the results obtained with addition of perflubron indicative of some minor impediment to diffusion of O$_2$ into skeletal muscle with hemodilution in the endotoxic dog? Other experimental results indicated that it was not sufficient to cause any tissue hypoxia. In experiments that followed the same protocol just described except for the use of perflubron, Vallet et al. (in press) measured tissue PO$_2$ at the surface of skeletal muscle using the multiwire electrode described by Lund et al. (1980). Their results, shown in figure 4, display the typical normal frequency distribution of tissue PO$_2$ in resting skeletal muscle with an average value of ~25 torr and the higher mixed venous PO$_2$ of 40 torr. With the induction of endotoxic shock, the distribution was shifted leftward to lower values but maintained the same Gaussian shape. Within 30 min of beginning resuscitation with a continuous infusion of dextran (0.5 ml/kg/min), average tissue PO$_2$ was restored close to the pre-endotoxin value and continued to increase to well above that norm as muscle blood flow continued to increase. The final histogram was obtained when the hematocrit was 17.0±1.6%. Clearly, no major impediment to tissue oxygenation existed in skeletal muscle of endotoxic and hemodiluted dogs. Because of the success in maintaining tissue oxygenation with volume expansion and hemodilution with dextran, it is worthwhile to speculate that the thinning of blood may actually have been beneficial to that end. Although the experiments have not yet been done, resuscitation with similar volumes of whole blood after induction of endotoxic shock may well be less successful in maintaining or raising tissue PO$_2$ levels because of microcirculatory dysfunction and increased tendency to sludging at higher hematocrits.

277

CONCLUSION

Hematocrits as low as 30% have long been deemed acceptable or even desirable for optimal patient care. As long as cardiac output can be increased and sustained, hemodilution may actually benefit tissue oxygenation even down to levels as low as 20%. This may be especially true if the microcirculation has been disturbed by septic processes.

ACKNOWLEDGEMENT

Support for this work was obtained from NIH Grant #RO1 HL 26927 and from funds supplied by Alliance Pharmaceutical Corporation.

REFERENCES

Barcroft, J., 1920, Presidential address on anoxaemia. *Lancet* 199 ii:485-489.

Cain, S. M., 1977, Oxygen delivery and uptake in dogs during anemic and hypoxic hypoxia. *J. Appl. Physiol.* 42:228-234.

Cain, S. M., 1978, Effects of time and vasoconstrictor tone on O_2 extraction during hypoxic hypoxia. *J. Appl. Physiol.* 45:219-224.

Cain, S. M., 1986, Assessment of tissue oxygenation. *Crit. Care Clin.* 2:537-550.

Cain, S. M. and Chapler, C. K., 1978, O_2 extraction by hindlimb versus whole dog during anemic hypoxia. *J. Appl. Physiol.* 45:966-970.

Cain, S. M., Curtis, S. E., Vallet, B., and Bradley, W. E., in press, Resuscitation of dogs from endotoxic shock by continuous dextran infusion with and without perflubron added. *Adv. Exp. Med. Biol.*

Chapler, C. K. and Cain, S. M., 1982, Effects of α-adrenergic blockade during acute anemia. *J. Appl. Physiol.* 52:16-20.

Chapler, C. K., Stainsby, W. N., and Lillie, M. A., 1981, Peripheral vascular responses during acute anemia. *Can. J. Physiol. Pharmacol.* 59:102-107.

Duling, B. R. and Berne, R. M., 1973, Longitudinal gradients of periarteriolar oxygen tension. *Circ. Res.* 27:669-678.

Fahraeus, R., 1929, The suspension stability of blood. *Physiol. Rev.* 9:214-274.

Fan, F. C., Chen, R. Y. Z., Schuessler, G. B., and Chien, S., 1980, Effects of hematocrit variations on regional hemodynamics and oxygen transport in the dog. *Am. J. Physiol.* 238:H545-H552.

Gutierrez, G., 1986, The rate of oxygen release and its effect on capillary PO_2 tension: a mathematical analysis. *Resp. Physiol.* 63:79-96.

Hatcher, J. D., Chiu, L. K., and Jennings, D. B., 1978, Anemia as a stimulus to aortic and carotid chemoreceptors in the cat. *J. Appl. Physiol.* 44:696-702.

Homer, L. D., Weathersby, P.K., and Kiesow, L. A., 1981, Oxygen gradients between red blood cells in the microcirculation. *Microvasc. Res.* 22:308-323.

Intaglietta, M., 1989, Microcirculatory effects of hemodilution: background and analysis. In: Tuma, R. F., White, J. V., and Messmer, K. (eds.) *The Role of Hemodilution in Optimal Patient Care.* W. Zuckschwerdt Verlag München, pp 21-40.

Lund, N., Ödman, S., and Lewis, D. H., 1980, Skeletal muscle oxygen pressure fields in rats. *Acta Anaesth. Scand.* 24:155-160.

Messmer, K., Sunder-Plassmann, L., Jesch, F., Gornandt, L., Sinagowitz, E., and Kessler, M., 1973, Oxygen supply to the tissues during limited normovolemic hemodilution, *Res. Exp. Med.* 159-166.

Murray, J. F., Gold, P., and Johnson, B. L., Jr., 1963, The circulatory effects of hematocrit variations in normovolemic and hypervolemic dogs. *J. Clin. Invest.* 42:1150-1159.

Sunder-Plassmann, L., Klövekorn, W. P., Holper, K., Hase, U., and Messmer, K., 1971, The physiological significance of acutely induced hemodilution. *Proc. 6th Eur. Conf. Microcirculation, Aalborg,* Karger, Basel, pp 23-28.

Vallet, B., Lund, N., Curtis, S. E., Kelly, D., and Cain, S. M., in press, Gut and muscle tissue PO_2 in endotoxemic dogs during shock and resuscitation. *J. Appl. Physiol.*

EFFECTS OF ISOVOLEMIC HEMODILUTION ON MICROCIRCULATORY PARAMETERS AND SKELETAL MUSCLE OXYGENATION DURING ANAESTHESIA

C.G.H.M. Kooiman[1], A.J. van der Kleij[2], Ch. P. Henny[1], D.A. Dongelmans[1] and [1], M. Günderoth

Departments of [1]Anaesthesia and [2]Surgery, Academic Medical Centre of the University of Amsterdam, Meibergdreef 9, 1105 AZ Amsterdam, The Netherlands. [3]Eppendorf, Hamburg

INTRODUCTION

Isovolemic hemodilution shortly before starting a major surgical procedure has been developed to reduce the risks of homologous bloodtransfusions[1,2].

During isovolemic hemodilution the hemoglobin content decreases and systemic hemodynamic parameters change: systemic vascular resistance decreases proportionally to a decrease in bloodviscosity and cardiac output increases compensating for the lowered oxygen transport capacity of the blood[3,4].

In clinical practice bloodgas analysis gives data about O_2 availability. It offers no information, however, on peripheral tissue oxygenation. This is the main reason for reluctace to introduce hemodilution as a routine clinical procedure. During hemodilution tissue pO_2 measurements can be used to monitor peripheral tissue perfusion[5-7]. Heinrich *et al.*[8] have shown that isovolemic hemodilution in unanaesthetised healthy volunteers leads to an improved muscular perfusion as reflected by a right shift of pO_2 histograms.

The aim of this study was to evaluate the effect of isovolemic hemodilution using gelatine solutions (Gelofusine) on skeletal muscle oxygenation measured with a needle pO_2 electrode in patients undergoing elective surgery.

MATERIALS AND METHODS

Ten patients (7 men, 3 women) aged 28-69 years (mean age 56.5 years) scheduled for elective surgery and eligible for intentional isovolemic haemodilution (HD) participated in this study. Inclusion criteria: age >18 years and a haemoglobin content >7 mmol/l (11.4 g/dl). Exclusion criteria: patients with chronic compensated or decompensa-

Oxygen Transport to Tissue XVI
Edited by M.C. Hogan *et al.*, Plenum Press, New York, 1994

ted organ dysfunction, such as liver disease, renal insufficiency, severe heart failure (NYHA 3-4), pulmonary insufficiency, use of β-blockers, drugs affecting hemostasis, known bleeding disorders, acute infections or septic conditions, pregnancy and history of drug abuse. After informed consent was obtained isovolemic HD was performed according to the normogram of Zetterström and Wiklund to obtain a haematocrit (Hct) ranging between 0.25 and 0.30[9]. Gelofusine was given in a proportion of 1:1.5. During HD heart rate (HR), arterial blood pressure (ABP), ECG and respiratory parameters (FiO_2, tidal volume, frequency) were monitored. All patients were treated to the anaesthesiologist's policy: respiratory parameters could be changed and fluid infusion with cristalloids were continued during HD.

Skeletal muscle pO_2 was measured polarographically as described by Fleckenstein[10] in the deltoid muscle using a needle pO_2 electrode Histograph 6650 (Eppendorf-Netheler-Hinz GmbH) immediately before and after HD. Results of the skeletal muscle pO_2 measurements were presented in a pO_2 histogram (n=100).

Simultaniously with the pO_2 measurements bloodsamples were collected. For determination of haematological parameters blood was drawn into a 1.5 ml vacutainer containing 0.04 ml EDTA. Haemoglobin (Hb) and Hct counts were performed on a H1 Technicon (Technicon instruments corporation, Tarrytown, NY). Counts for arterial and venous bloodgas parameters were performed on a ABL 4 or ABL 300 (Acid-Base-Laboratory; Radiometer, Copenhague). Blood was drawn into a 10 ml vacutainer containing 1 ml sodium-EDTA for measurement of viscosity and erythrocyte deformability. Blood viscosity was measured at 37°C with the Contraves LS 30 viscometer (Contraves, Basel, Switzerland) at three shear rates; high shear (HS): 70/s, medium shear (MS): 0.5/s and low shear (LS): 0.05/s, representing approximately the shear rates in the capillaries, arteries and veins respectively. All measurements were performed in duplicate. Erythrocyte deformability was measured by means of the LORD (Laser Optical Rotational Deformability meter) at shear stresses of 0.95, 3, 9.49, 30 and 53,35 Pa.

Results of this study are expressed as means $\pm$ SD. Statistical comparisons of the control and the haemodilution parameters were made using the Student's t-test. Statistical significance was assigned to be p=0.05, when p<0.08 the change was assigned a tendency.

RESULTS

In two patients haemodilution was performed in two steps because of probable inoperability. The second step was initiated when patho-anatomical examination showed no malignant tumor cells. During haemodilution ABP, HR and ECG remained within normal limits in all patients.

Skeletal muscle pO_2 measurements. Study of individual pO_2 histograms revealed three types of alterations: a shift to the left indicating hypoxia, a shift to the right indicating a higher oxygenation and a less heterogeneous distribution indicating altered perfusion. Two patients showed no change in the pO_2 histograms. Pooling of patient pO_2 histograms except three, revealed that mean skeletal muscle pO_2 was 33.0 $\pm$ 14.6 and 37.0 $\pm$ 15.5, p50 was 30.7 $\pm$ 6.3 and 35.0 $\pm$ 11.3 before and after HD respectively. These increases were statistically not significant. Three patients were excluded from evaluation. One patient incidentally received epidural infusion of marcaine 5% during the HD period. Another patient had a bleeding during the second pO_2 measurement. In the third patient the HD period lasted 2 hours and 19 minutes, so other factors than HD influencing O_2 supply could not be excluded. The histograms of the last two patients showed an extreme left shift.

Laboratory Parameters. In all patients except one, Hct was significantly decreased to a range between 0.25 and 0.30 (table 1). Mixed venous bloodgas analysis are not available from two patients. Arterial pH significantly decreased from 7.49 $\pm$ 0.02 before HD to 7.45 $\pm$ 0.04 after HD and mixed venous pH significantly decreased from 7.45 $\pm$ 0.02 before HD to 7.40 $\pm$ 0.04 after HD (table 2).
Statistical analysis of arterial oxygen pressure (P_aO_2) on 5 patients in whom respiratory parameters remained unchanged during the HD period showed a tendency to increase

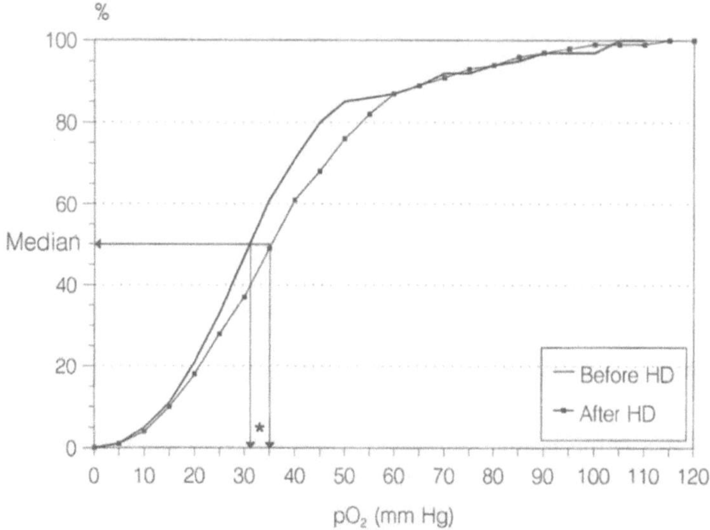

Figure 1. Non significant right shift of the cumulative pO_2 distribution curve induced by HD (mean Hct from 0.40 to 0.26). * right shift of the p50 from 30.7 $\pm$ 6.3 to 35.0 $\pm$ 11.3.

from 163.4 $\pm$ 50.3 to 192.0 $\pm$ 72.5 before and after HD respectively (table 3).

Haemorheology. Hct significantly decreased from 0.38 $\pm$ 0.04 to 0.26 $\pm$ 0.02. Rheological parameters demonstrated a significant decrease in blood viscosity at all shear rates (LS: from 31.9 $\pm$ 17.9 to 8.2 $\pm$ 2.8, MS: from 16.9 $\pm$ 4.5 to 6.7 $\pm$ 1.7, HS: from 4.6 $\pm$ 0.49 to 2.97 $\pm$ 0.37 before and after HD respectively). The parameters indicative for the microcirculation (the ratio of Hct to HS Viscosity) did not change significantly (before HD mean 0.085 $\pm$ 0.006, after HD mean 0.087 $\pm$ 0.007, p=0.5). Measurement of the elongation index (EI) of the RBC's revealed no significant change in erythrocyt deformability.

Table 1. Hct reduction and pO$_2$ histogram shifts, mean ± SD

% Hct Reduction	Hct before HD	Hct after HD	duration HD	P$_m$O$_2$ before HD	P$_m$O$_2$ after HD	R-shift L-shift reduced heterogeneity
41%	0.46	0.27	2.19h	26.5 *	11.4 *	L-shift
40%	0.35	0.21	2.47h	28.3 *	12.7 *	L-shift
38%	0.42	0.26	41	34.9	29.8	no change
38%	0.42	0.26	41	34.8	50.8	R-shift
35%	0.41	0.27	54	30.2	23.7	L-shift
34%	0.38	0.25	31	31.1	32.7	no change
27%	0.37	0.27	47	30.5	37.5	RH
26%	0.34	0.25	25	54.3	41.6	L-shift + RH
20%	0.35	0.28	43	34.2	54.9	R-shift + RH
20%	0.35	0.28	55	36.4 *	91.4 *	R-shift + RH
mean ± SD	0.38 0.04	0.26 0.02		33.0 14.6 #	37.0 15.5 #	

* excluded patient data, # mean patient data, except excluded patient data (*).

Table 2. Bloodgas analysis before and after haemodilution, mean ± SD

pH$_a$ before	after	P$_a$O$_2$ before	after	P$_a$CO$_2$ before	after	pH$_v$ before	after	P$_v$O$_2$ before	after	P$_v$CO$_2$ before	after
7.51	7.42	223	188	34	40						
7.46	7.49	127	137	35	32	7.45	7.44	48	38	36	36
7.50	7.45	116	131	30	30	7.46	7.36	41	35	35	34
7.44	7.48	422	198	36	30						
7.48	7.41	300	148	34	35	7.44	7.36	45	37	38	40
7.45	7.41	159	174	36	37	7.45	7.37	39	41	39	43
7.51	7.38	564	130	31	40	7.41	7.36	40	41	39	45
7.53	7.48	195	239	29	32	7.49	7.44	53	59	34	35
7.51	7.50	171	209	32	30	7.48	7.45	41	38	35	34
7.49	7.45	244	309	35	37	7.48	7.38	41	50	36	41
mean 7.49	7.45	252.1	186.3	33.2	34.3	7.45	7.40	43.5	42.4	36.5	38.5
± SD 0.03	0.04	142.2	56.5	2.5	4.0	0.03	0.04	4.84	8.10	1.9	4.3

Table 3. Relationship between FiO_2 and P_aO_2 and mean $\pm$ SD

FiO_2	Tidal Volume	P_aO_2 before HD	P_aO_2 after HD
33	600	127	137
31	600	116	131
33	550	159	174
40	700	171	209
60	550	244	309
mean $\pm$ SD		163.4 $\pm$ 50.3	192.0 $\pm$ 72.5

DISCUSSION

In all but one patients haemodilution resulted in a decrease of the Hct to a range of 0.25 to 0.30. Statistical analysis of seven pO_2 histograms, revealed a non-significant increase of the p50 and the mean skeletal muscle pO_2. Evaluation of individual pO_2 histograms revealed three types of alterations. All patients were treated according to the individual regime of the anaesthesiologist. The P_aO_2 values in table 3 reflect the liberal use of oxygen during induction of anaesthesia.

Kuo and Sarelius described an increase in P_aO_2 after HD[12,13]. This phenomenon was also observed in this study (table 3). Other authors have found a positive correlation between tissue oxygen pressure and an increase in FiO_2[14,15]. In this study one patient incidentally received epidural analgesia (marcaine 5%, between L2-L3) during the HD procedure. The pO_2 histogram extremely shifted to the right. This may be attributed to the combined effect of HD, a rise in FiO_2 (before HD 35, after HD 40) and a vasoactive and redistributive effects of marcaine. It may also be hypothesized that the pO_2 measurement was within the vicinity of an artery; P_mO_2 was higher than P_vO_2.

Several studies reported that RBC distribution in the capillaries of skeletal muscle becomes more uniform due to increased vasomotion[17,18]. Moreover, vasomotion also increases when FiO_2 is reduced[16]. We observed one patient with an obvious shift to the right and a less heterogeneous distribution after HD and a FiO_2 reduction from 100 to 50%. This phenomenon may be attributed to both HD and increased vasomotion.

Under normal conditions oxygen consumption is independent of oxygen delivery. When oxygen supply is reduced oxygen extraction ratio (OER) increases[19]. Two patients with a Hct reduction of 40% showed a left shift of the pO_2 histogram, indicating skeletal muscle hypoxia (table 1). In the first patient the pO_2 measurement was performed during haemorrhagic hypovolemia due to a surgical bleeding (Hct 0.21), in the second patient during hypovolemic hypovolemia (urine output 50 ml/hour). These left shifts were accompanied by an increased (calculated) OER of 93% and 70%, respectively, reflecting reduced oxygen supply.

The Hct is one of the major rheologic determinants of blood viscosity. In the microcirculation oxygen delivery is a function of the ratio of Hct to HS viscosity and vascular hindrance[20]. In this study blood viscosity was significantly reduced by HD. The Hct to HS viscosity ratio did not change. Erythrocyte deformability was not affected by the use of Gelofusine for HD. HD as performed in this study seems not to affect oxygen delivery in the microcirculation.

It is concluded that isovolemic haemodilution using Gelofusine in patients undergoing elective surgery does not impair skeletal muscle oxygenation. It is a safe method for reduction of homologous blood transfusions. Moreover, skeletal muscle pO_2 measurement during the HD procedure provides information about the microcirculation which can not be derived from macrocirculatory parameters (e.g. early hypovolemia).

REFERENCES

1. J.P. Waymack, Ph. Miskell and S. Gonce, Alterations in host defence associated with inhalation anesthesia and blood transfusion. *Anest Analg* 69: 163 (1989)

2. P.A. Schriemer, D.E. Lonecker and P.D. Mints, The possible immunosuppressive effects of perioperative blood transfusion in cancer patients. *Anesthesiology* 68: 422-428 (1988)

3. K. Messmer, D.H. Lewis, L. Sunder-Plasman, W.P. Klövekorn, N. Mendler and K. Holper, Acute normovolemic hemodilution. *Europ Surg Res* 4: 55-70 (1972)

4. K. Messmer and L. Frey, Das Prinzip der Hämodilution. *Chirurg* 62: 769-774 (1991)

5. A. Gosain, J. Rabkin, J.P. Reymond, J.A. Jensen, T.K. Hunt and R.A. Upton, Tissue oxygen tension and other indicators of blood loss or organ perfusion during graded hemorrhage. *Surgery* 109: 523-532 (1991)

6. F. Gottrup, R. Firmin, J. Rabkin, B.J. Halliday and T.K. Hunt, Directly measured tissue oxygen tension and arterial oxygen tension assess tissue perfusion. *Crit Care Med.* 15: 1030-1036 (1987)

7. T.K. Hunt, J. Rabkin, J.A. Jensen, K. Jonsson, K. von Smitten and W.H. Goodson III, Tissue oxymetry: an interim report. *World J Surg* 1987; 11: 126-132 (1987)

8. R. Heinrich, M. Gunderoth-Palmowski and N. Machac, Effects of hemodilution with middle molecular HES solution on muscle tissue pO_2 in healthy volunteers *in*: Clinical Oxygen Measurement II, A.M. Ehrly, ed., Blackwell Ueberreuter Wissenschaft Berlin 1990

9. H. Zetterström and L. Wiklund, A new nomogram facilitating adequate haemodilution. *Acta Anaesthesiol Scand* 20: 300-304 (1986)

10. W. Fleckenstein, R. Heinrich, Th. Kersting, H. Schomerus, Ch. Weiss, A new method for bedside recording of tissue pO2 histograms. *Verh Dtsch Ges Inn Mede* 90, 439 (1984)

11. L. Kuo, R.N. Pittman, Effect of hemodilution on oxygen transport in arteriolar networks of hamster striated muscle, *Am J Physiol* 254, H331-H339 (1988)

12. I.H. Sarelius. Microcirculation in striated muscle after acute reduction in systemic hematocrit, *Resp Physiol* 78, 7-17 (1989)

13. N. Chang, W.H. Goodson III, F. Gottrup, T.K. Hunt, Direct measurement of wound and tissue oxygen tension in postoperative patients, *Ann Surg* 197, 470-478 (1983)

14. K. Jonsson, J.A. Jensen, W.H. Goodson III, J.M. West, T.K. Hunt, Assessment of perfusion in postoperative patients using tissue oxygen measurements, *Br J Surg* 74, 263-267 (1987)

15. F. Gottrup, R. Firmin, N. Chang, W.H. Goodson III, T.K. Hunt, Continuous direct tissue oxygen tension measurement by a new method using an implantable silastic tonometer and oxygen polarography, *Am J Surg* 146, 399-403 (1983)

16. E. Bouskela and W. Grampp, Spontaeous vasomotion in hamster cheek pouch arterioles in varying experimental conditions, *Am J Physiol* 262, H478-H485 (1992)

17. S. Mirhashemi, K. Messmer, K.E. Arfors and M. Intaglietta, Microcirculatory effects of normovolemic hemodilution in skeletal muscle. *Int J Microcirc Clin Exp* 6: 359-369 (1987)

18. K. Tyml, heterogeneity of microvascular flow in rat skeletal muscle is reduced by contraction and by hemodilution. *Int J Microcirc Clin Exp* 10;75-86 (1991)

19. W.J. Federspiel, IH. Sarelius, An examination of the contribution of red cell spacing to the uniformity of oxygen flux at the capillary wall. *Microvasc Res* 27, 273-285 (1984)

20. S. Chien. Physiological and pathophysiological significance of hemorheology, *in* Clinical Hemorheology, S. Chien, J. Dormandy, E. Ernst and A. Matrai ed., Martinus Nijhoff Publishers, Dordrecht (1987).

THE ROLE OF ENDOTHELIUM-DERIVED RELAXING FACTOR (EDRF) IN THE WHOLE BODY AND HINDLIMB VASCULAR RESPONSES DURING HYPOXIC HYPOXIA

C.E. King[1], S.E. Curtis[2], M.J. Winn[3]*, J.D. Mewburn[1], S.M. Cain[4], and C.K. Chapler[1]

Department of Physiology[1], Queen's University, Kingston, Ontario, Canada, K7L 3N6 and Departments of Pediatrics[2], Pharmacology[3], and Physiology and Biophysics[4], University of Alabama at Birmingham, Birmingham, AL, USA, 35294

INTRODUCTION

We and others have postulated that the most efficient oxygen extraction from a diminished oxygen supply is achieved when locally-generated vasodilators promote blood flow to areas of greater O_2 demand while others remain under vasoconstrictor tone (Cain and Chapler, 1980; Granger and Shepherd, 1979). Previous experiments in our laboratory have shown that administration of N^ω-nitro-L-arginine methyl ester (L-NAME), a nitric oxide synthase (NOS) inhibitor, doubles both total and hindlimb peripheral resistance in the anesthetized dog. The current study was undertaken to test the hypothesis that the greater levels of whole body and hindlimb resistance during NOS inhibition would limit the ability of the animal to effectively utilize the reduced oxygen supply available during severe hypoxic hypoxia (HH). In particular, we wished to assess the functional significance of endothelium derived relaxing factor on oxygen utilization and vascular resistance during severe hypoxic hypoxia in the whole body and skeletal muscle.

METHODS

Mongrel dogs (n=7), ranging in weight from 16.8 to 27.7 kg, were anesthetized with sodium pentobarbital (32 mg/kg, i.v.), paralyzed with succinylcholine chloride (20 mg/kg, i.m. and 0.1 mg/min, i.v.) and ventilated to maintain arterial PCO_2 between 30-35 mmHg. Additional anesthetic (65 mg) was given every hour after administration of the paralytic agent.

A Swan-Ganz catheter was advanced into the pulmonary artery via the right jugular vein and used to obtain body temperature measurements, mixed venous blood samples

---

*deceased July 29, 1993

and thermodilution cardiac output values. A catheter was placed in the left brachial artery for measurement of arterial blood pressure and withdrawal of arterial blood samples. The right cephalic vein was cannulated for i.v. infusion of succinylcholine chloride.

Both the venous outflow and arterial inflow of the left hindlimb were isolated. Ties were passed through the upper thigh on either side of the femur, crossed, and then tightened in order to direct all venous outflow from the hindlimb through the femoral vein. A tourniquet around the ankle excluded paw flow from the hindlimb circulation. The arterial flow to the hindlimb was isolated to the femoral artery by placing ties around the circumflex and external and internal iliac arteries at the junction of the abdominal aorta. In some cases, other arteries which arose from the abdominal aorta which appeared to supply the left hindlimb were also ligated. Heparin (1,000 units/kg) was given upon completion of all surgical interventions.

A two-channel cannula was inserted in the left femoral vein. One channel contained an electromagnetic flow probe (Narco Bio-Systems) for measurement of hindlimb blood flow, while the second channel was used to obtained a zero recording from the flow probe without occluding venous outflow. Venous outflow from the hindlimb was returned to the animal via a reservoir connected to a cannula in the right femoral vein. The proximal end of another two-channel cannula was inserted into the right femoral artery while the distal end was inserted caudally into the isolated left femoral artery. One channel served as an auto-perfusion line while the second channel was connected to a Masterflex peristaltic pump to control blood flow. Our goal was to maintain hindlimb blood flow within 10% of the control autoperfused value throughout the experiment. A third branch of the cannula was used to monitor perfusion pressure of the left femoral artery. Arterial isolation was tested by occlusion of blood flow to the hindlimb and was considered successful when venous outflow was less than 1 ml/min and perfusion pressure was < 10-20 mmHg. Once surgical procedures were completed, blood flow to the hindlimb was changed from auto-perfusion to pump-perfusion.

The experimental protocol consisted of 2 major sections. The first was a control section which was subdivided into a 30 minute period of normoxia followed by a 30 minute period of hypoxic hypoxia. In the second section, the animals were first given the L-arginine analog, N^{ω}-nitro-L-arginine methyl ester (L-NAME)(20 mg/kg, i.v.) and then data were obtained, as above, during a 30 minute normoxic period and a 30 minute hypoxic period. The animals were ventilated with 9% O_2 in N_2 to produce hypoxic hypoxia. During the last five minutes of each period, thermodilution cardiac output determinations were made and arterial, mixed venous, and muscle venous blood samples were taken. Hindlimb blood flow, arterial pressure, heart rate, hindlimb perfusion pressure, and body temperature were monitored continuously.

Some preliminary experiments (n=7) were also performed in which anesthetized, paralyzed, and ventilated mongrel dogs were volume expanded and/or given 8-bromo-cyclic guanosine monophosphate (8-bromo-cGMP) during normoxia following L-NAME. Control cardiac output and total peripheral resistance were determined and then L-NAME (20 mg/kg, i.v.) was given and measurements were obtained again at 30 minutes post-L-NAME. One of three treatments was then performed; two dogs received a volume expansion of 10 ml/kg dextran (6% in Tyrode's solution), two received a volume expansion of 20 ml/kg dextran, and three received 8-bromo-cGMP (2 mg/kg, i.v.). Measurements were made at 30 min following each of the above treatments.

All blood samples were analyzed for PO_2, PCO_2, and pH using a BMS-MKII Radiometer blood gas analyzer. Oxygen concentration was measured using an IL-482 Co-oximeter and the values were corrected for O_2 solubility using the value 0.003 ml O_2/mmHg/dl. Whole body O_2 uptake ($\dot{V}O_2$) was determined using open circuit spirometry; on-line measurements of ventilation and expired gas concentrations provided a minute by minute determination of $\dot{V}O_2$. Hindlimb O_2 uptake was calculated using the Fick principle. Oxygen extraction, O_2 delivery, and vascular resistance were calculated for the whole body and hindlimb using standard equations.

The data are presented as the means ± standard error. Differences between experimental periods were tested using a repeated measures analysis of variance and Duncan's multiple means test. Significance was accepted at $p < 0.05$.

RESULTS

The values for cardiac output, mean arterial pressure (MAP) and total peripheral resistance (TPR) are shown in Figure 1. Cardiac output increased from 114 ± 10 to $172 \pm 21\ \text{ml} \cdot \text{kg}^{-1} \cdot \text{min}^{-1}$ ($p < 0.01$) during HH ($PaO_2 = 20 \pm 3$ mmHg) in the control period.

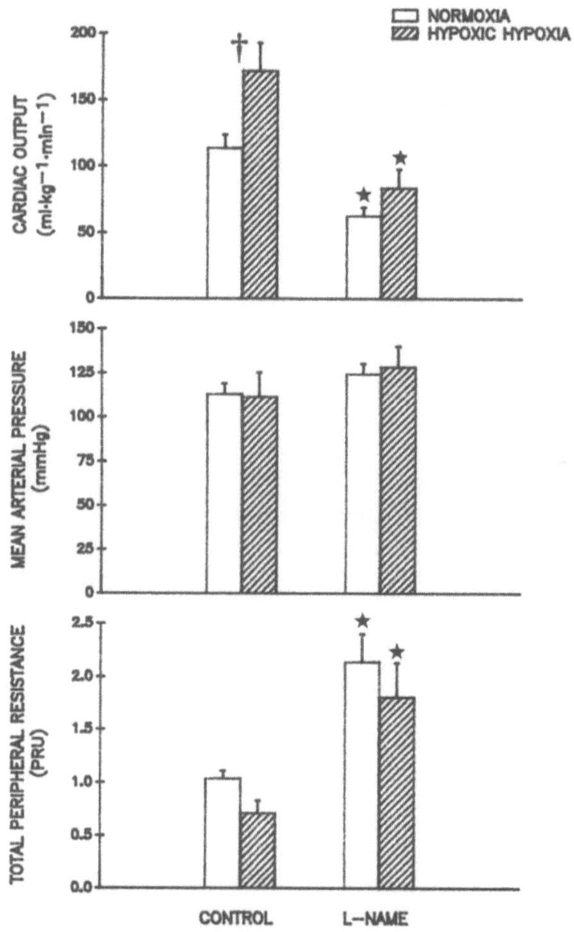

Fig. 1 The mean values ± SE for cardiac output, mean arterial pressure and total peripheral resistance. (†) indicates significant difference ($p < 0.05$) between normoxia and hypoxia during either the CONTROL period or following L-NAME. (∗) indicates significant difference ($p < 0.05$) between the normoxic values OR between the hypoxic values in the CONTROL and L-NAME sections.

Following administration of L-NAME, cardiac output fell to 63 ± 7 ml·kg⁻¹·min⁻¹ ($p < 0.01$) and failed to increase significantly during hypoxia ($PaO_2 = 24 \pm 2$ mmHg). The values for MAP and TPR did not change during hypoxia as compared to normoxia during either control or after L-NAME. The values for TPR were greater during L-NAME as compared to control ($p < 0.05$) during both normoxia and hypoxia respectively.

Hindlimb blood flow was maintained within 10% of the control normoxic values throughout the experiment (Figure 2). Before L-NAME, both hindlimb resistance and perfusion pressure decreased ($p < 0.05$) during hypoxia as compared to normoxia. After administration of L-NAME, no differences in the normoxic or hypoxic values for hindlimb resistance or perfusion pressure were observed as compared to the control period.

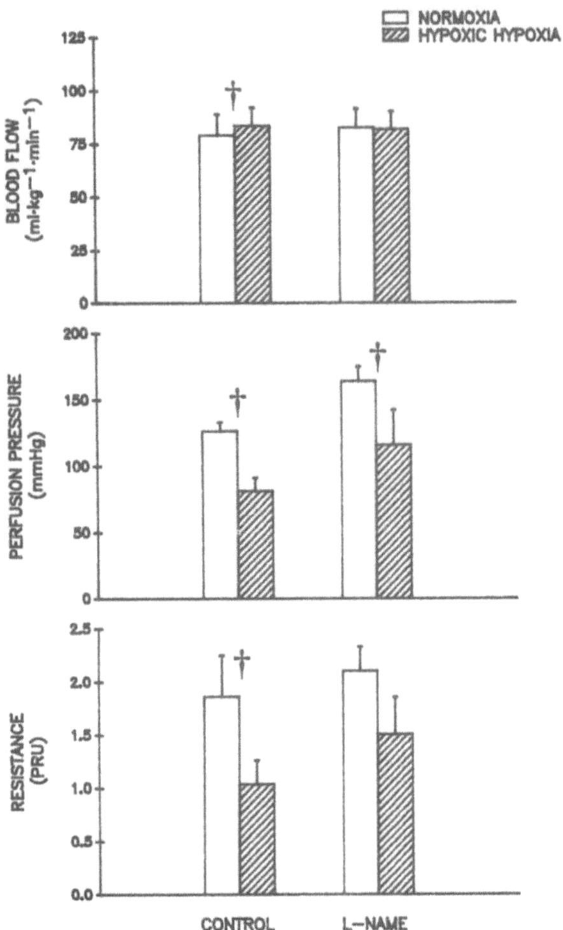

Fig. 2   The mean values ± SE for hindlimb blood flow, perfusion pressure and resistance.   Symbols as indicated in Fig. 1.

The values for whole body and hindlimb O_2 uptake, delivery, and extraction ratio are shown in Figures 3 and 4 respectively. Whole body and hindlimb O_2 uptake decreased significantly during hypoxia in both the control and L-NAME sections, however, the respective normoxic and hypoxic values for $\dot{V}O_2$ were not different between the two sections. During hypoxia, the values for O_2 delivery were significantly reduced but were not different between control and L-NAME. The substantial reduction in cardiac output

288

after L-NAME resulted in a reduction (p<0.05) in whole body O_2 delivery during normoxia; no such reduction in O_2 delivery was seen in the hindlimb which was kept at constant flow. In the whole body, the O_2 extraction ratio increased from a mean value of 0.24 to 0.53 during hypoxia in the control section (p<0.05) and again increased to that level following administration of L-NAME during normoxia. The whole body O_2 extraction ratio increased further to 0.71±0.03 during hypoxia after L-NAME (p<0.05). The normoxic and hypoxic values for hindlimb O_2 extraction ratio were not different between the control and L-NAME sections respectively, however, the hindlimb O_2 extraction ratio increased during hypoxia in each case (p<0.05).

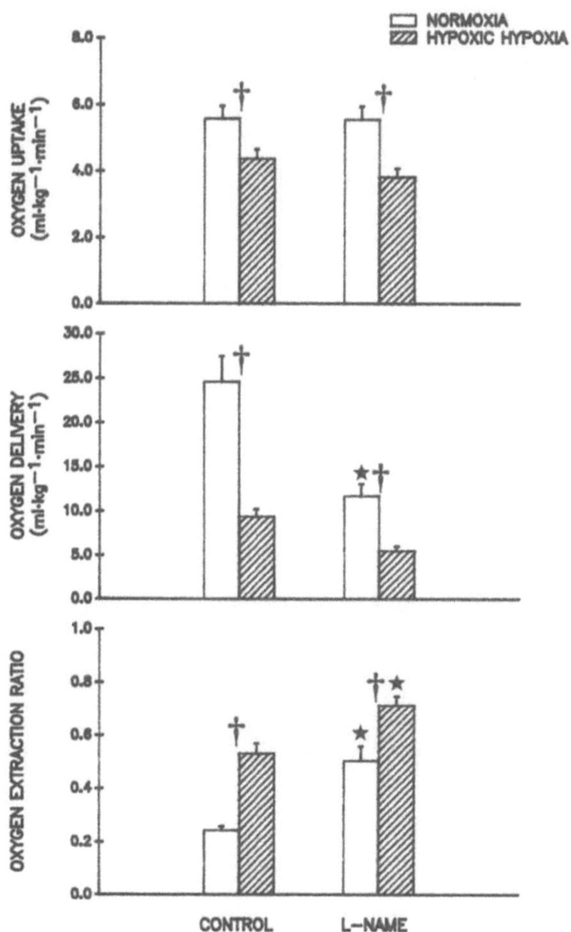

Fig. 3   The mean values ± SE for whole body O_2 uptake, O_2 delivery, and O_2 extraction ratio. Symbols as indicated in Fig. 1.

In the preliminary set of experiments, cardiac output decreased to 54% and total peripheral resistance increased to 197% of the control level after L-NAME (p<0.05) (Figure 5). Volume expansion (10-20 ml/kg dextran) returned cardiac output to 76% and 89% of the control value respectively and total peripheral resistance returned to 190% and 140% of the control value. Cardiac output did not change following administration of 8-bromo-cGMP, despite a 33% decrease in total peripheral resistance.

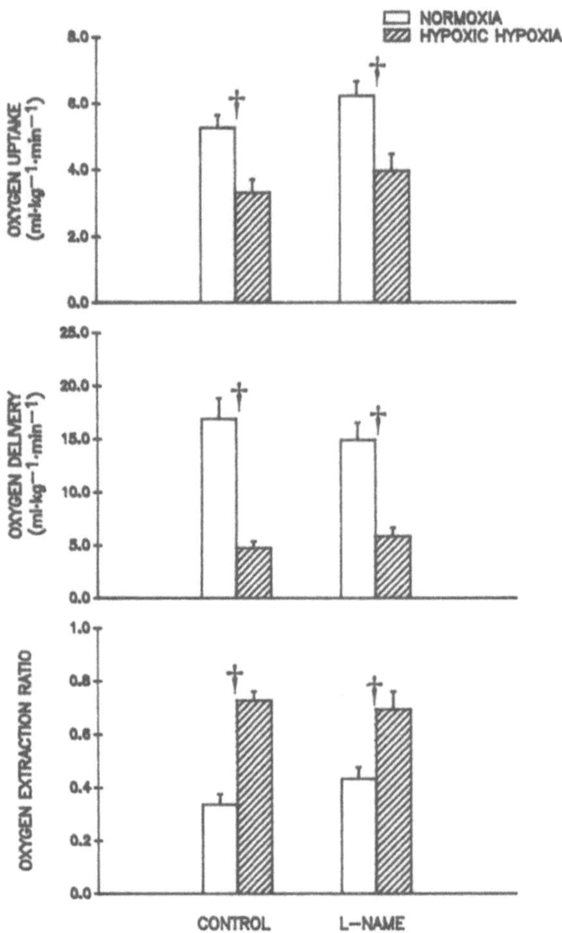

Fig. 4   The mean values ± SE for hindlimb O_2 uptake, O_2 delivery, and O_2 extraction ratio. Symbols as indicated in Fig. 1.

DISCUSSION

The decrease in cardiac output and the increase in total peripheral resistance observed following administration of L-NAME were expected. Similar findings have been reported by others in anesthetized cats (Bower and Law, 1993; Loeb and Longnecker, 1992), anaesthetized dogs (Kilbourn et al., 1990; Klabunde et al., 1991) and conscious dogs (Elsner et al., 1992). The most prominent result from our study was that the compensatory increase in cardiac output normally observed during severe hypoxic hypoxia (Cain & Chapler, 1979; King & Cain, 1986; King et al., 1985; Kubes et al., 1989) was abrogated following administration of the nitric oxide synthase (NOS) inhibitor L-NAME. The mechanism(s) responsible for this outcome remains to be fully appreciated although there is no obvious reason to suspect that it differs from that which underlies the reduction in cardiac output observed following NOS inhibition during normoxia.

290

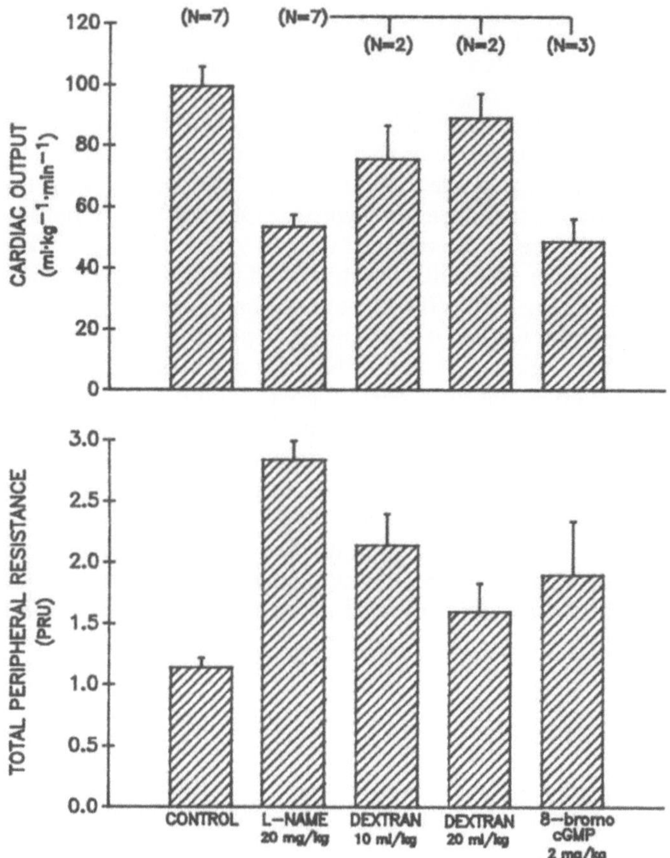

Fig. 5   The mean values ± SE for cardiac output and total peripheral resistance in preliminary experiments with volume expansion and 8-bromo-cGMP.

An increase in both mean circulatory filling pressure and venous resistance have been shown to occur following NOS inhibition with L-NAME in anesthetized cats (Bower and Law, 1993) and with L-NMMA in the awake rat (Glick et al., 1993). The latter would be expected to decrease cardiac output while the former would have the opposite effect. On the basis of their findings, Bower and Law (1993) concluded that the relatively modest increase in mean circulatory filling pressure which they observed following administration of L-NAME, was more than offset by a substantial increase in venous resistance with the result that the cardiac output was reduced. Our preliminary finding that volume expansion (10-20 ml/kg dextran) restored cardiac output to the pre-treatment control level in normoxic L-NAME treated dogs tended to support the above conclusion. Presumably the volume expansion resulted in an increase in mean circulatory filling pressure which lead to an increase in venous return. Interpretation of our data was complicated by the decrease in vascular resistance which resulted from the reduction in blood viscosity following hemodilution. However, administration of 8 bromo-cGMP decreased resistance without an accompanying increase in cardiac output.

If whole body oxygen uptake was taken as the reference point, the failure of cardiac output to rise during hypoxic hypoxia following L-NAME appeared to have little or no physiological consequence in the present experiments. Whole body oxygen extraction increased during hypoxic hypoxia following L-NAME to a level that resulted in oxygen uptake not being different from that observed during hypoxia in the control period.

However, the higher value for oxygen extraction during hypoxic hypoxia in the L-NAME treated animals (Figure 3) demonstrated that little, if any, safety margin remained to offset further reductions in oxygen delivery.

We previously observed that NOS inhibition did not compromise oxygen extraction in canine hindlimb skeletal muscle during local hypoxic hypoxia (Vallet et al., 1993). In those experiments, no centrally mediated increase in vasoconstrictor tone was expected since only the hindlimb vasculature was hypoxic. Our working hypothesis in the current study was that efficient use of the reduced oxygen supply during hypoxic hypoxia would depend upon a balance of centrally mediated vasoconstrictor activity and local vasodilatory factors (Cain, 1978). If this were the case, the expectation was that skeletal muscle oxygen extraction would be less during hypoxic hypoxia after administration of L-NAME. Our observation was that L-NAME abolished the decrease in hindlimb muscle resistance observed during hypoxic hypoxia establishing that the balance between the vasoconstrictor and vasodilator factors had indeed been altered, however, this had no effect on the utilization of the limited oxygen supply available to skeletal muscle during hypoxia. This finding did not differ from our earlier result observed in skeletal muscle during local hypoxia (Vallet et al., 1993).

In conclusion, NOS inhibition abrograted the rise in cardiac output and the decrease in skeletal muscle vascular resistance normally observed during severe hypoxic hypoxia, but did not compromise the utilization of oxygen in either the whole body or skeletal muscle.

ACKNOWLEDGEMENTS

This work was supported by the Medical Research Council of Canada and the National Institutes of Health (Grant No. HL-26927).

REFERENCES

Bower, E.A., and A.C.K. Law, 1993, The effects of N^{ω}-nitro-L-arginine methyl ester, sodium nitroprusside and noradrenaline on venous return in the anesthetized cat, Br. J. Pharmacol., 108:933.

Cain, S.M., 1978, O_2 extraction by hindlimb versus whole dog during anemic hypoxia, J. Appl. Physiol., 45:966.

Cain, S.M., and C.K. Chapler, 1979, Oxygen extraction by canine hindlimb during hypoxic hypoxia, J. Appl. Physiol., 46:1023.

Cain, S.M., and C.K. Chapler, 1980, O_2 extraction by canine hindlimb during α-adrenergic blockade and hypoxic hypoxia, J. Appl. Physiol., 48:630.

Elsner, E., A. Müntze, E.P. Kromer, and G.A. J. Riegger, 1992, Inhibition of synthesis of endothelium-derived nitric oxide in conscious dogs; hemodynamics, renal and hormonal effects, Am. J. Hypertension, 5:288.

Glick, M.R., J.S. Gehman, and J.A. Gascho, 1993, Endothelium-derived nitric oxide reduces baseline venous tone in awake instrumented rats, Am. J. Physiol., 265:H47.

Granger, H.J., and A.P. Shepherd, 1979, Dynamics and control of the microcirculation, in: "Advances in Biomedical Engineering," J. Brown, ed., Academic Press, N.Y.

Kilbourn, R.G., A. Jubran, S.S. Gross, O.W. Griffith, R. Levi, J. Adams, and R.F. Lodato, 1990, Reversal of endotoxin-mediated shock by N^{G}-methyl-L-arginine, an inhibitor of nitric oxide synthesis, Biochem. Biophys. Res. Comm., 172:1132.

King, C.E., and S.M. Cain, 1986, Adrenergic and local control of O_2 uptake during and after severe hypoxia, J. Appl. Physiol., 61:1920.

King, C.E., S.M. Cain, and C.K. Chapler, 1985, Peripheral vascular responses to hypoxic hypoxia after aortic denervation, Can. J. Physiol. Pharmacol., 63:1197.

Klabunde, R.E., R.C. Ritger, M.C. Helgren, 1991, Cardiovascular actions of endothelium-derived relaxing factor (nitric oxide) formation/release in anesthetized dogs, Eur. J. Pharmacol., 199:51.

Kubes, P., S.M. Cain, and C.K. Chapler, 1989, Neural regulation of canine skeletal muscle blood flow during hypoxic hypoxia, Am. J. Physiol., 257:H1581.

Loeb, A.L., and D.E. Longnecker, 1992, Inhibition of endothelium-derived relaxing factor-dependent circulatory control in intact rats, Am. J. Physiol., 262:H1494.

Vallet, B., S.E. Curtis, M.E. Winn, C.E. King, C.K. Chapler, and S.M. Cain, 1993, Hypoxic vasodilation does not require nitric oxide (EDRF-NO) synthesis, The FASEB J., 7:A761.

CRITICAL OXYGEN EXTRACTION IN DOG HINDLIMB AFTER INHIBITION OF NITRIC OXIDE SYNTHASE AND CYCLOOXYGENASE SYSTEMS

*Mark J. Winn[1], Benoît Vallet[2], Scott E. Curtis[3],
Christopher K. Chapler[4], Cheryl E. King[4], and Stephen M. Cain[2]

Departments of Pharmacology[1], Physiology and Biophysics[2],
Pediatrics[3], University of Alabama at Birmingham,
Birmingham, AL 35294, U.S.A. and Department of Physiology[4],
Queen's University, Kingston, Ont. K7L 3N6, Canada

INTRODUCTION

When anesthetized dogs were given the α–adrenergic blocking agent phenoxybenzamine and then made hypoxic, their ability to extract oxygen from a supply that was limiting to oxygen uptake was significantly less than in unblocked animals (Cain, 1978). This was evident by a lower slope of the line relating O_2 uptake to oxygen delivery as O_2 uptake became linearly dependent upon O_2 delivery. The reason for the lesser efficiency in extracting oxygen by the α–blocked animals was postulated to depend upon the loss of vasoconstrictor tone. The hypothesis that was offered stated that a vigorous constrictor tone was necessary in hypoxia so that blood flow in excess to need would not occur in any organ system or tissue. The constrictor tone in areas where O_2 demand exceeded O_2 supply would then be offset by the production of vasodilator metabolites in proportion to the imbalance of supply and demand. In this manner, blood flow and O_2 delivery would be matched to local O_2 need so that O_2 would not be shunted through areas that were overperfused relative to their O_2 uptake.

Although this appeared to be true for the whole body, presumably at the level of organ systems, it did not hold for an homogenous tissue group, skeletal muscle of the canine hindlimb. Cain and Chapler (1980) showed that even though whole body O_2 extraction was less in α–blocked dogs made hypoxic, the hindlimb extracted O_2 as well as in a group which did not receive phenoxybenzamine. Similarly, β–adrenergic blockade with propranolol did not alter O_2 extraction by limb muscles during severe hypoxia (Cain and Chapler, 1979). It appeared that blockade of adrenergic control systems in skeletal muscle did not affect its O_2 extraction capability.

Recently, however, other non-adrenergic vascular controls mediated by endothelium have been shown to be important in the vascular responses to hypoxia. Pohl and Busse (1989) found that hypoxia stimulated release of endothelium derived relaxing factor (EDRF) which is considered by most to be nitric oxide (NO). Others have shown that EDRF produced by activity of NO synthase is responsible for a continuous state of vasodilator tone even in

---

*Deceased July 29, 1993

Oxygen Transport to Tissue XVI
Edited by M.C. Hogan *et al.*, Plenum Press, New York, 1994

resting, normoxic skeletal muscle (White, et al., 1993). Hypoxia has also been shown to elicit increased release of prostaglandins such as prostacyclin that could contribute to increased vasodilator tone (Busse, et al., 1984; Messina, et al., 1992).

Our question for these studies was whether endothelial mediation of increased vasodilator tone in hypoxia was essential to optimize O_2 extraction. A corollary of that question was whether increased vasoconstrictor tone caused by withdrawal of endothelial derived vasodilator tone would increase O_2 extraction during severe hypoxia in skeletal muscle. To avoid sympathetic activation by chemoreceptors or baroreceptors, we restricted hypoxia to the left hindlimb skeletal muscle by controlling blood flow to the area in the presence and absence of appropriate inhibitors of NO synthase and prostaglandin production.

METHODS

Dogs (n = 16) were anesthetized (30 mg/kg pentobarbital sodium iv), paralyzed (30 mg succinylcholine chloride im + 0.1 mg/min iv), and pump-ventilated to maintain arterial PCO_2 at 30 to 35 torr. Catheters were placed in carotid and pulmonary arteries and in the right femoral vein. Arterial inflow to the left hindlimb was isolated to the femoral artery by ligating the internal and external iliac and deep circumflex arteries at their origin on the abdominal aorta. Perfusion was maintained from the contralateral femoral artery. An occlusive roller pump was interposed in the circuit to control blood flow to the left hindlimb muscles. Perfusion pressure was measured by a pressure transducer at a t-connector placed in the femoral artery catheter. Initially, flow was set to 100 ml/min per kg of estimated limb muscle weight. Venous outflow was restricted to the left femoral vein by two tourniquets placed at the groin level and flow from the paw was excluded by a third tourniquet at the ankle. The venous outflow was measured by a cannulating-type flow probe and then drained into the right femoral vein through a reservoir suspended above it. Vascular isolation was confirmed by cessation of flow when the perfusion pump was stopped briefly and reactive hyperemia thereafter confirmed that the vasculature remained normally reactive. This was checked before and after each experiment.

Mean systemic arterial pressure was measured at the carotid artery as well as arterial blood gas tensions, pH, and O_2 content. The latter quantity was also measured in mixed venous blood from the pulmonary artery. Whole body O_2 uptake was measured from expired gas analyses and cardiac output was calculated by the Fick Equation. For the hindlimb, the arteriovenous O_2 difference was multiplied by the limb blood flow to obtain O_2 uptake which was normalized to the muscle weight determined after each experiment. O_2 delivery was taken as the product of blood flow and arterial O_2 content. Total vascular resistance was the ratio of mean systemic arterial pressure and cardiac output and limb resistance was the ratio of perfusion pressure and limb blood flow.

The dogs were divided into two groups of 8. A pretreatment period of 30 min was followed by an additional 30-min observation period after N^G-nitro-L-arginine methyl ester (L-NAME, 20 mg/kg body weight) and indo-methacin (3 mg/kg) were given as a bolus. This was the experimental group (EXPT) but the control group (CTRL), which received no drugs, were observed for a similar period. At that time, the hindlimb was made progressively ischemic by lowering flow from ~100 ml/kg/min to ~5 ml/kg/min in 10 steps at 5-min intervals. Earlier experiments had shown that a new steady state was established in approximately 3 min after such flow changes. Samples were taken and measurements made at the end of each period.

Critical O_2 delivery to the limb muscle was determined for each experiment by finding the point of intersection of two linear regression lines fitted to the "plateau" region and the supply-dependent slope of O_2 uptake when graphed against O_2 delivery. The critical O_2 extraction ratio was the ratio of O_2 uptake at the critical point to the critical O_2 delivery. Differences between groups were tested by Student's t-test with $p<0.05$ accepted as significant.

RESULTS

Inhibition of the NO synthase and cyclooxygenase systems with L-NAME and indomethacin had prompt and well marked effects on systemic hemodynamics. These are shown in Table 1. Before treatment, there were no significant differences between the two groups in any systemic measurement.

Table 1: Systemic responses to L-NAME and indomethacin.

	Before	After
Mean art. press (mmHg)	124±16	172±7*
Cardiac output (ml/kg'min)	119±51	76±18*
Vasc. resist. (PRU·kg)	1.25±0.50	2.38±0.50*
O_2 uptake (ml/kg'min)	5.8±0.7	6.1±0.7
O_2 extract.	0.33±0.15	0.51±0.11*

* $p < 0.05$

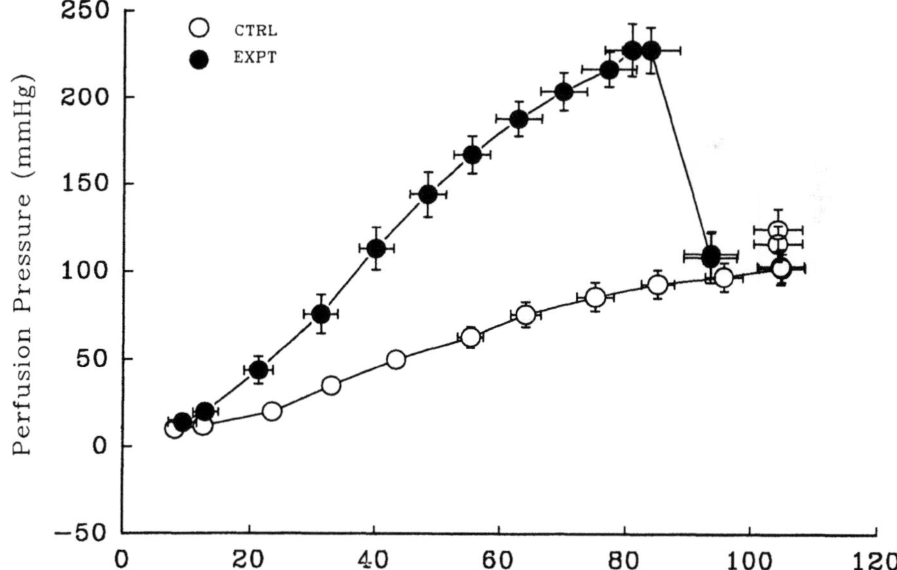

Fig. 1: Perfusion pressure to hindlimb muscles as flow was progressively reduced (mean ± SE).

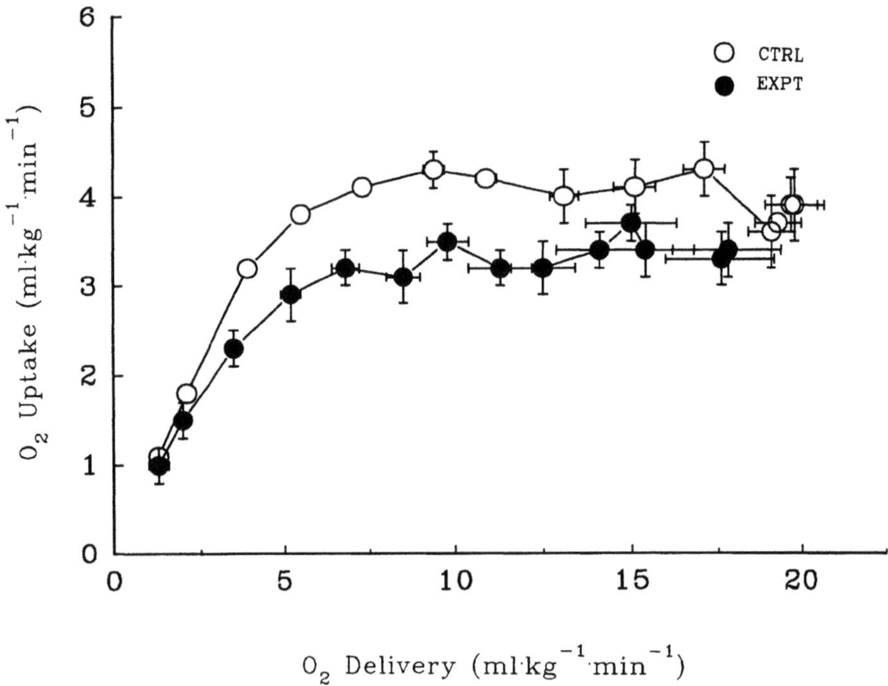

Fig. 2: O_2 uptake as a function of O_2 delivery to hindlimb muscles at each step of flow reduction (mean ± SE).

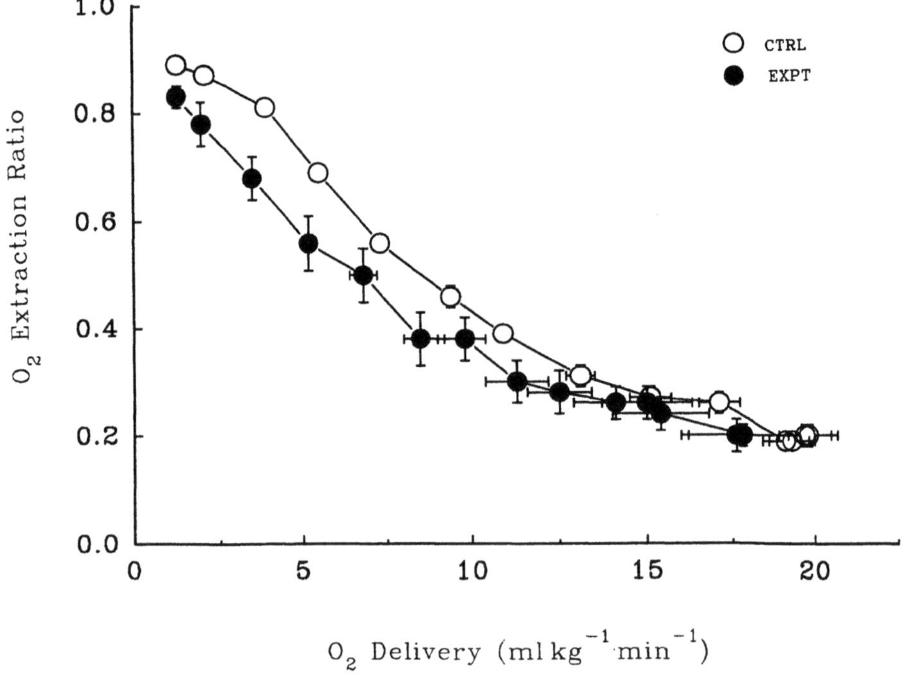

Fig 3: O_2 extraction ratio in hindlimb as O_2 delivery was progressively reduced (means ± SE).

Similarly, the relationship between perfusion pressure and blood flow in the hindlimb muscles, as illustrated in Fig. 1, showed no differences between the two groups prior to giving the blocking agents. The increase in perfusion pressure occurred after L-NAME and indomethacin and represented an increase in vascular resistance of almost three-fold from 1.18 ± 0.13 to 3.02 ± 0.26 resistance units. That difference in resistance was maintained throughout the entire range of flows but remained parallel down to the lowest flow.

In Figure 2, averages of limb O_2 uptake are shown with respect to O_2 delivery for each flow step. Although the two groups started at the same level, the limb O_2 uptake in the control group increased slightly during the experiment whereas that in the experimental group tended to drift down. Based on the results obtained in individual dogs by the dual regression method, the O_2 uptake at the critical delivery level was 4.4 ± 0.3 ml/kg/min in the control group but was 3.1 ± 0.3 ml/kg/min in the experimental group at their critical O_2 delivery, which was significantly less (p<0.05). The critical O_2 delivery values did not differ significantly between the groups, 5.4 ± 0.4 vs. 5.1 ± 0.6 ml/kg/min.

Although critical O_2 deliveries did not differ, the O_2 uptake at that point was significantly less in the treated group. Hence, there was a significant difference in the O_2 extraction ratio at the critical point. This is evident in Fig. 3 in which a lesser O_2 extraction ratio was seen at each level of O_2 delivery. The actual values of O_2 extraction ratio at the critical O_2 delivery were 0.81 ± 0.02 in the control group and 0.66 ± 0.07 in the experimental group (p<0.05).

DISCUSSION

The basic premise underlying this study was that the efficient use of a diminished O_2 supply between and within organ systems depended upon an optimal balance of vasoconstrictor and vasodilator tone. As O_2 delivery was progressively decreased, O_2 extraction was raised to maintain O_2 uptake. To accomplish this, a vigorous vasoconstrictor tone would act to decrease blood supply to tissues and areas that had a low O_2 demand whereas those with a high demand would have sufficient vasodilator tone to offset vasoconstriction and insure blood flow appropriate to their need. The source of that dilator tone has been assumed to be directly related to the imbalance between O_2 supply and demand. In this manner, O_2 shunting would be minimized through areas with a high rate of perfusion relative to their O_2 uptake. Any factor that altered the natural balance between constrictor and dilator influences would, therefore, be expected to lower O_2 extraction at a given rate of O_2 delivery.

This premise was supported by earlier results that showed that the intact anesthetized dog was less able to extract O_2 from a deficient supply after α-adrenergic blockade with phenoxybenzamine (Cain, 1978). Subsequently, it was shown that a similar lessening of O_2 extraction ability was not seen in the hindlimb muscles of dogs although the whole body effect was the same in hypoxia after α-block with phenoxybenzamine (Cain and Chapler, 1980) and that β-adrenergic blockade with propranolol also did not interfere with O_2 extraction in hypoxic limb muscles (Cain and Chapler, 1979). The lack of effect of α-block on limb muscles, however, could have been attributed to the incomplete blockade achieved by phenoxybenzamine. Kubes et al. (1992) showed that as much as 76% of constrictor tone in resting skeletal muscle was due to α_2-adrenergic receptors which were minimally blocked by phenoxybenzamine. Sufficient constrictor tone would thus have still been present to promote O_2 extraction. Blockade of β-receptors, conversely, would only have heightened constrictor tone which did not interfere with O_2 extraction in hypoxia.

Another test of the basic premise was essayed by Cain and Bredle (1990) who infused dopexamine, a potent dopaminergic and β_2-adrenergic agonist, into anesthetized dogs while progressively lowering blood flow to the hind-limb muscles. The drug caused marked vasodilation which was mainly attributable to its β–agonist properties since dopaminergic stimulation had been shown to have little effect on skeletal muscle in the dog (Jackson, et al., 1982). O_2 extraction at the critical O_2 delivery was significantly less than in a control group and venous PO_2 at that point was significantly higher. This indicated an increase in functional peripheral shunting through the muscle that was in accord with results obtained by Yonekawa et al. (1981) who saw a decrease in tissue PO_2 of canine gracilis muscle as they infused isoproterenol. This occurred in spite of an increase in O_2 delivery to the muscle and was the result of the increased shunting that they calculated as a result of the unopposed β–vasodilation. In these instances, an overbalance toward increased dilator tone definitely affected O_2 extraction and tissue oxygenation adversely.

If a loss of constrictor tone and/or an increase in dilator tone can decrease the ability of skeletal muscle to extract O_2 during hypoxia, then would an increase in constrictor tone possibly increase that ability? King and Cain (1986) used methoxamine to overdrive α_1–adrenergic receptors while anesthetized dogs were ventilated with an hypoxic gas mixture. Both whole body and limb muscle O_2 extraction were as high, but no higher, than in an hypoxic control group. Although hypoxia was not imposed in their experiments, Cain and Chapler (1981) observed limb vascular resistance to be more than doubled while blood flow was decreased by 30% when norepinephrine was infused intraarterially into canine hindlimb. O_2 uptake was able to be maintained by increased O_2 extraction. This provided another indication that increased vasoconstrictor tone, if it did not promote greater O_2 extraction, at least did not interfere with it.

With these facts in mind, the findings of the present study can now be put into perspective. The efficacy of the drug doses used to block the NO synthase and cyclooxygenase systems was clear in both the whole body and limb hemodynamic effects. In the case of the limb, the increase in perfusion pressure at the control flow rate indicated a marked vasoconctriction with nearly 3-fold increased vascular resistance. As the limb muscles were made progressively ischemic in the absence of these endothelial sources of vasodilator tone, local hypoxia was insufficient to direct blood flow according to regional O_2 demand. This was evident because the critical O_2 extraction ratio was significantly reduced compared to the unblocked control group. Clearly, increasing vasoconstrictor tone by withdrawal of endothelial vasodilator sources did not benefit O_2 extraction in hypoxia. Equally clear was the fact that EDRF-NO and eicosanoid mediated relaxant control were essential to optimize O_2 extraction.

Which of the two endothelial systems might have been of greater importance in hypoxia? Separate blockade of each system in a protocol similar to the one used here would provide a direct answer to that question. Although those experiments remain to be done there is some pertinent information that points toward the answer. Vallet et al. (in press) examined the role of EDRF-NO in hypoxic vasodilation using both ischemic and hypoxic hypoxia in the limb muscles of separate groups of dogs. The vasoconstriction they saw with inhibition of only the NO-synthase system was less than we observed with the combined block in these experiments. Furthermore, they found that hypoxic vasodilation occurred in the absence of EDRF-NO activity and that O_2 extraction was as complete as in the unblocked groups. This information leads us to suggest that eicosanoid activity may be the key factor to promote O_2 extraction when O_2 supply becomes limiting to O_2 uptake. Experiments with indomethacin block of cyclooxygenase with other systems left functional remain to be done.

ACKNOWLEDGEMENTS

Funds for these studies were provided by NIH HL26927 (SMC) and HD 28831 (SEC), AHA (MJW), and MRC (CEK and CKC) grants. The expert technical assistance of W. E. Bradley and G. Clyde-Davis is also gratefully acknowledged.

REFERENCES

Busse, R. Förstermann, U., Matsuda, H., and Pohl, U. (1984) The role of prostaglandins in the endothelium-mediated vasodilatory response to hypoxia. *Pflügers Arch.* 401:77-83.

Cain, S. M. (1978) Effects of time and vasoconstrictor tone on O_2 extraction during hypoxic hypoxia. *J. Appl. Physiol.* 45:219-278.

Cain, S. M. and Bredle, D. L. (1990) Actions of a dopaminergic and β_2-adrenergic agonist on O_2 extraction by canine skeletal muscle. *Adv. Exp. Med. Biol.* 277:569-575.

Cain, S.M. and Chapler, C. K. (1979) Oxygen extraction by canine hindlimb during hypoxic hypoxia. *J. Appl. Physiol.* 46:1023-1028.

Cain, S. M. and Chapler, C. K. (1980) O_2 extraction by canine hindlimb during α-adrenergic blockade and hypoxic hypoxia. *J. Appl. Physiol.* 48:630-635.

Cain, S. M. and Chapler, C. K. (1981) Effects of norepinephrine and α-block on O_2 uptake and blood flow in dog hindlimb. *J. Appl. Physiol.* 51:1245-1250.

Jackson, L. K., Key, B. M., and Cain, S. M. (1982) Total and hindlimb O_2 uptake and blood flow in hypoxic dogs given dopamine. *Crit. Care Med.* 10:327-331.

King, C. E. and Cain, S. M. (1986) Adrenergic and local control of O_2 uptake during and after severe hypoxia. *J. Appl. Physiol.* 61:1920-1927.

Kubes, P., Melinyshyn, M., Nesbitt, K., Cain, S. M., and Chapler, C.K. (1992) Participation of α_2-adrenergic receptors in neural vascular control of canine skeletal muscle. *Am. J. Physiol.* 262 (*Heart Circ. Physiol.* 31):H1705-H1710.

Messina, E. J., Sun, D., Koller, A., Wolin, M. S., and Kaley, G. (1992) Role of endothelium-derived prostaglandins in hypoxia-elicited arteriolar dilation in rat skeletal muscle. *Circ. Res.* 71:790-796.

Pohl, U. and Busse, R. (1989) Hypoxia stimulates the release of endothelium-derived relaxant factor (EDRF). *Am. J. Physiol.* 256 (*Heart Circ. Physiol.* 25):H1595-H1600.

Vallet, B., Curtis, S. E., Winn, M. J., King, C. E., Chapler, C. K., and Cain, S. M. (in press) Hypoxic vasodilation does not require nitric oxide (EDRF/NO) synthesis. *J. Appl. Physiol.*

White, D. G., Drew, G. M., Gurden, J. M., Penny, D. M., Roach, A. G., and Watts, I. S. (1993) The effect of N^G-nitro-L-arginine methyl ester upon basal blood flow and endothelium-dependent vasodilation in the dog hindlimb. *Br. J. Pharmacol.* 108:763-768.

Yonekawa, H., Berk, J. L., Neuman, M. R., and Liu, C. C. (1981) Tissue hypoxia and increased physiological tissue shunt caused by beta-adrenergic stimulation. *Eur. Surg. Res.* 13:325-338.

INTRAVITREAL PERFLUOROCARBON AND OXYGEN DELIVERY IN
INDUCED RETINAL ISCHAEMIA

S.J. Cringle, D-Y Yu, V.A.Alder and E-N Su

Lions Eye Institute, University of Western Australia
Perth, Western Australia

INTRODUCTION

The original interest in liquid perfluorocarbons in medical research stemmed from their very high oxygen carrying capacity (Faithfull, 1992). Increased oxygen delivery to the retina has already been demonstrated following intravascular administration of perfluorocarbon emulsions (Braun et al, 1992: Thoreson and Purple, 1987). Pure liquid perfluorocarbons have also been used in vitreal surgery, where their low viscosity, high specific gravity, and optical clarity are attractive properties in the treatment of retinal detachment (Chang, 1987). The feasibility of vitreous exchange with liquid perfluorocarbons therefore raises the possibility of exploiting the oxygen carrying properties of such materials to relieve retinal hypoxia in ischaemic disorders of the retina.

The present study employed an acute model of retinal ischaemia in the cat, and tested the ability of a highly oxygenated intravitreal perfluorocarbon (perflubron, Alliance Pharmaceutical Corp.) to sustain retinal function. Three different methods of producing ischaemia were employed.

- A sudden increase in intraocular pressure (IOP) to above systolic blood pressure in order to produce total ischaemia of the retinal and choroidal circulations.
- Graded increases in IOP to reduce the net perfusion pressure to both the retinal and choroidal circulations.
- An intraocular probe pushing on the vessels at the optic disk to selectively occlude the retinal vasculature.

Retinal function was monitored by recording the b-wave amplitude of the light adapted electroretinogramme (ERG) at critical stages in each procedure.

METHODS

A total of 9 adult cats were used. In each case the left eye was chosen for perflubron treatment and the right eye used as the control. All surgical procedures were performed under general anaesthesia. The animals were induced with a 2 ml loading dose of Saffan (alphaxalone 9 mg/ml, aphadalone acetate 3 mg/ml) and then maintained on an infusion at 10 ml/hour of a diluted solution of 10 ml Saffan in 40 ml saline.

Vitrectomy

The left eye was immobilised by suturing the conjunctiva adjacent to the limbus to a fixation eye ring. The pupil was then dilated with neosynephrine (10%), and homatropine (2%). Standard pars plana vitrectomy was performed under microscope observation (Zeiss OPMI 1FC) by an experienced ophthalmologist, using an Ocutome II vitrectomy system. A thorough vitreous removal and balanced salt solution (BSS) replacement was achieved in each case. Once the surgery was complete the wound was closed and subconjunctival antibiotic and steroid administered and the animal allowed to recover. Aseptic conditions were maintained throughout.

Acute Experiment

Using the same anaesthetic regime as in the initial vitrectomy surgery the animals were prepared for the acute phase of the experiment two weeks later. Artificial ventilation, continuous monitoring of systemic blood pressure, and periodic blood gas analysis were employed (Alder et al, 1990). The vitrectomised eye was once again stabilised by an eye ring sutured to the conjunctiva at the limbus.

Electroretinography

A Nicolet Compact 4 evoked potential suite was used for all electroretinography. Full field, Ganzfeld stimulation was employed. Specially manufactured plano-concave contact lens electrodes were used to pick-up the ERG and allow microscope observation of the fundus. Pupils of both eyes were dilated with neosynephrine (10%), and homatropine (2%). Simultaneous ERG recording from both eyes was used and the data stored to disk. The stimulus and recording conditions chosen were: Flash stimulus 1.5 log units (Nicolet calibration standard), background (level 2) 78 $\mu W/cm^2$, (both white light), 8 averages for each measurement with 0.9 seconds between flashes. The bandwidth of the amplifier was 1 Hz - 1 KHz. These stimulus and background conditions were chosen to enable recording of light adapted ERG's without having to wait for adaptation when switching from microscope observation of the fundus during the retinal blocking probe experiments. The amplitude of the b wave of the ERG was taken as the indicator of retinal function.

Perflubron Instillation

A specially adapted double barreled needle was used to deliver the perflubron and remove the saline at a similar rate so that IOP changes during the exchange were minimized. The perflubron had been equilibrated with oxygen and warmed to body temperature prior to instillation. The entry hole was closed as soon as the needle was removed by tightening preinserted sutures. The same entry hole was then used in those experiments in which the retinal blocking probe was employed. Particular care was taken to ensure that there was no leakage from this wound even during induced episodes of high intraocular pressure.

Intraocular Pressure (IOP) Manipulation

A means of producing simultaneous increases in IOP in both treated and control eyes was developed. A high precision gas regulator was used to pressurise a sealed bottle of Krebs solution. This bottle was connected to the anterior chamber of both eyes *via* a 25 gauge needle in each eye. A second needle was positioned in the anterior chamber of each eye to measure the resultant IOP *via* small volume pressure transducers. Systemic blood pressure in a femoral artery and rectal temperature were also continuously monitored. The IOP in each eye was recorded on a strip chart recorder, along with the systemic blood pressure and the trigger pulses from the Nicolet system. The IOP was either raised in a stepwise fashion to produce graded ischaemia, or in a single step to a pressure above systolic blood pressure to produce total retinal and choroidal ischaemia.

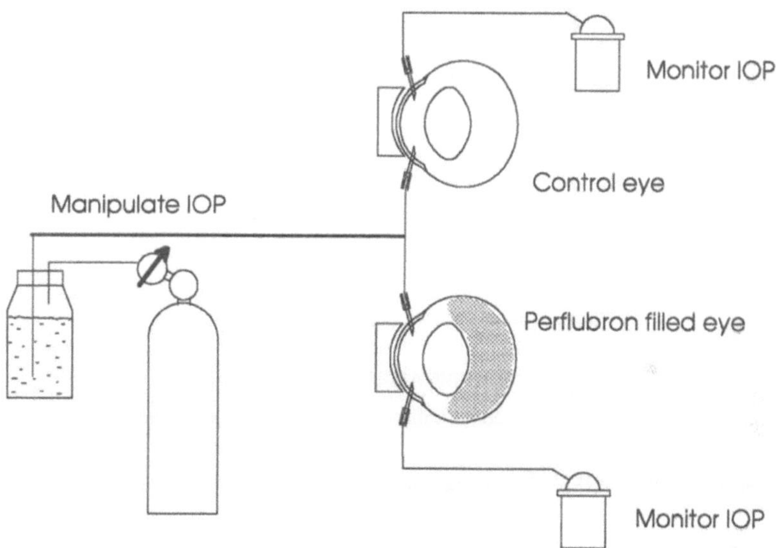

Figure 1. Schematic of simultaneous IOP manipulation in the control and perflubron filled eyes

Retinal Blocking Probe

We manufactured a retinal blocking probe which consisted of a 1.2 mm glass sphere on the end of a 200 µm diameter tungsten wire in a specially shaped hypodermic carrier. This device was inserted into the eye through the pars plana incision and manipulated on a concentric arc system to place the glass ball immediately in front of the optic disk, and at normal incidence. When the tungsten wire was advanced the glass ball could be seen pressing on the centre of the disk and the stasis of blood in the larger retinal vessels verified.

RESULTS

Perflubron Instillation

The interface between the perflubron and the saline was clearly visible and with careful techniques a very full exchange could be achieved. The high specific gravity and low viscosity of the perflubron made the instillation considerably easier than experienced with other intraocular tamponades such as silicone oil. Perflubron instillation was found to have no deleterious effects on the ERG.

Intravitreal Perflubron Oxygen Tension

In four experiments a sample of perflubron was aspirated from the eye and the oxygen tension measured in a blood gas analyser (Corning 166). The combined data is plotted in Figure 2 as a function of time in the eye and provides a reasonable estimate for the period for which the perflubron remains highly charged with oxygen. The oxygen tension of the perflubron has reduced to half the original instillation value in approximately 45 minutes. Even after 5 hours the perflubron has an oxygen tension greater than seen in control eyes.

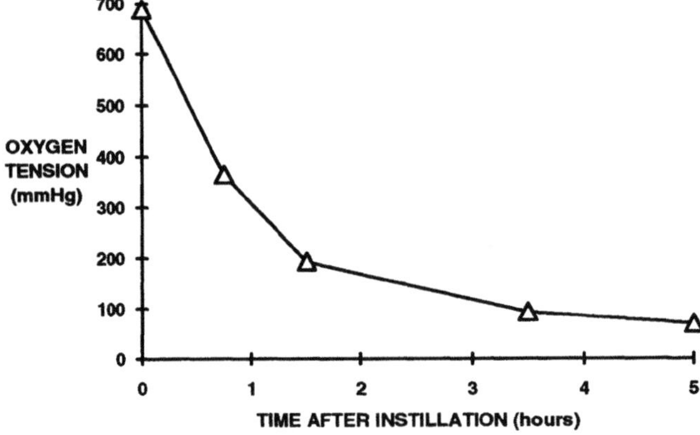

Figure 2. Oxygen tension of aspirated perflubron from four eyes at different times after instillation.

Intraocular Pressure (IOP) Manipulations

Total Ischaemia. Manipulation of the IOP was found to be a very convenient means of creating ocular ischaemia. IOP is normally 15-20 mmHg. If this is raised then the net perfusion pressure (mean blood pressure - IOP) across the eye is reduced. Raising the IOP to sufficiently high levels (above systolic blood pressure), produces unequivocal total ischaemia of both the retinal and choroidal circulations. This is effectively the ultimate test of the ability of the intravitreal perflubron to support the entire thickness of retina.

In the following example total ischaemia was induced in both treated and control eyes 30 minutes after perflubron instillation and the time course of the b wave suppression followed. The b wave amplitude of the treated eye fell more slowly than in the control eye but in both eyes it was effectively extinguished within two minutes of the onset of ischaemia. Equivalent results were also found in two other animals tested 10 minutes and 1 hour after perflubron instillation.

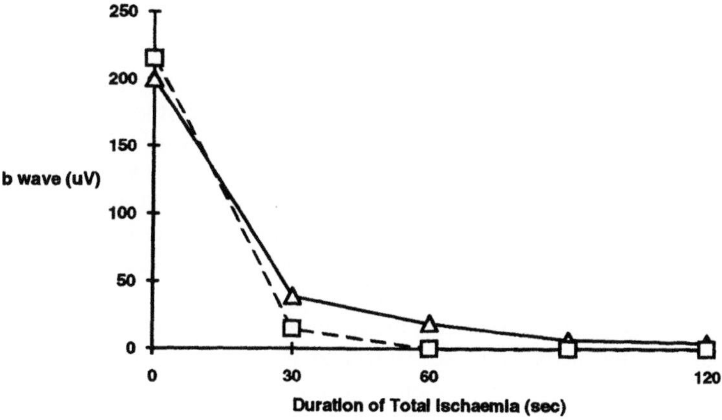

Figure 3. Example of ERG suppression in treated (triangles) and control (squares) eyes following total retinal ischaemia 30 minutes after perflubron instillation.

Graded Ischaemia. More moderate increases in IOP were employed to determine the relative resistance of the treated and control eyes to the same level of graded ischaemia. The precise control and monitoring of IOP allowed highly reproducible reductions in net perfusion pressure to be produced. The perflubron treated eyes had slightly higher b wave amplitudes at low perfusion pressures in each case. Normalizing the b wave amplitude to that present at normal IOP in each eye allowed the data from all experiments to be combined. B wave amplitudes at perfusion pressure ranges of 20-40 mmHg were averaged together and so on for pressures of 40-60, 60-80 etc. The averaged data shows no significant difference between the treated (n=4) and control eyes (n=6) as a function of perfusion pressure.

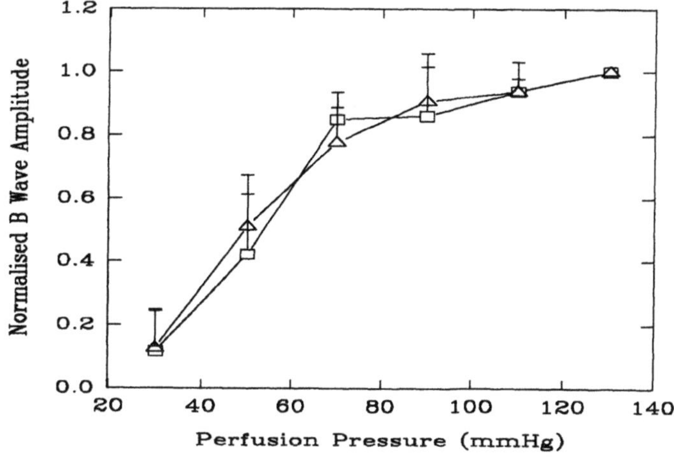

Figure 4. Mean and standard deviation of normalized b wave amplitude as a function of perfusion pressure in treated (triangles) and control (squares) eyes.

Retinal Blocking Probe

The positioning and observation system allowed the glass ended blocking probe to be positioned at normal incidence just over the optic disk. Once the probe was advanced we could clearly see the blood in the major retinal vessels stop flowing, and the vessels of the optic disk become devoid of blood in the region behind the glass ball. We found however, that there was a tendency for the retinal blood flow to recover even though the probe was still visibly pushing on the disk. This was also reflected in b wave recovery and occurred in normal as well as treated eyes. Attempts to create a more severe occlusion by pushing the probe harder onto the optic disk resulted in further suppression of the b wave but caused severe damage to the retinal vessels and the tissue around the disk. This indicated that we could not rely on the blocking probe technique to produce a reliable and reproducible degree of retinal ischaemia. It is our opinion that comparison of b wave suppression by successive use of the probe in the same eye before and after perflubron instillation is unreliable given the inability to guarantee complete occlusion or the same level of ischaemia in the two situations. It was clear however that even in the perflubron treated eye a marked degree of suppression was produced once the blocking probe was applied. In the example shown in Figure 5, a 50% suppression of the b wave amplitude was produced one minute after pushing the probe onto the optic disk in a treated eye 30 minutes after perflubron instillation.

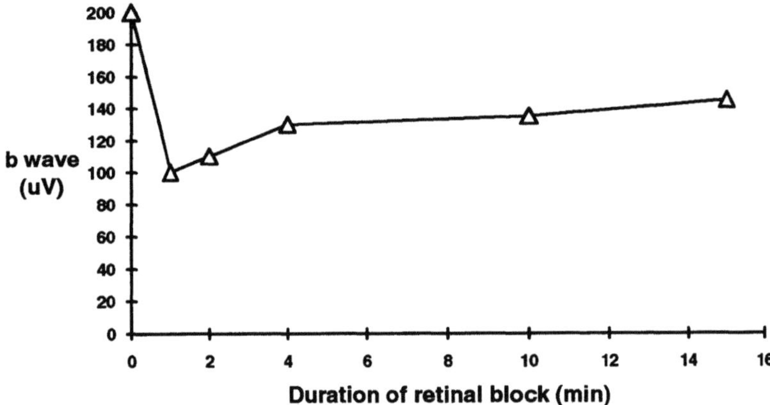

Figure 5. b wave amplitude as a function of time after attempting to selectively block the retinal circulation with a probe pushing on the vessels at the optic disk.

DISCUSSION

The physical properties of liquid perflubron allowed a very efficient instillation into the vitreous cavity to be achieved and no short term adverse effects on retinal function were noted. The oxygen content of the perflubron remained very high in physiological terms for several hours after instillation (Fig. 2), a finding which is supported by earlier work with a different perfluorocarbon (Berkowitz et al, 1991). This situation would appear to optimize the potential for the perflubron to act as an oxygen source and sustain the retina during periods of induced retinal ischaemia. Despite this situation, total ischaemia of the retinal and choroidal circulations, induced by a sudden rise in IOP to levels above systemic blood pressure resulted in the ERG being extinguished in both treated and control eyes (Fig. 3). The decay of ERG amplitude was slightly delayed in the perflubron treated eyes but the effect was marginal. Experiments in which a graded increase in IOP was produced also showed no significant resilience of the perflubron treated eyes when compared to the control eyes (Fig. 4). Reductions in the net perfusion pressure induced by raising the IOP affect both the retinal and choroidal circulations, and are therefore not a particularly good model for ischaemic disorders of the retinal circulation alone. We attempted to produce selective and reversible occlusion of the retinal circulation by means of a mechanical blocking probe pushing on the vessels at the optic disk. We found this to be unreliable, in that the same degree of ischaemia could not be reproduced and some degree of spontaneous recovery was evident even though the probe continued to push on the vessels at the optic disk. However, the observation of significant ERG suppression in the treated eye, once again suggests that the perflubron was not able to maintain retinal function following at least partial occlusion of the retinal circulation.

The question therefore is why is the ERG not sustained? The most likely explanation is that oxygen supply is not the most critical factor, and that some other ischaemic mechanism is involved. Alternatively, it may be that as the oxygen is consumed by the ischaemic tissue a large oxygen gradient may develop in the stationary perflubron layer adjacent to the retinal surface, and the flux of oxygen may not be enough to sustain the retina under ischaemic conditions. We planned to address this point by utilizing oxygen sensitive microelectrodes to look for such a gradient. However, a preliminary bench test demonstrated that such electrodes would not operate in a pure perflubron environment, presumably due to the very high electrical resistivity of such materials (one of the early industrial uses of liquid perfluorocarbons was as transformer coolants). Figure 6 illustrates the bench experiment, along with the measured oxygen gradients. As the oxygen microelectrode was driven across the layer of saline above the highly oxygenated reservoir of perflubron the oxygen gradients in the saline reflect the oxygen source from the perflubron. However, once the electrode reached the perflubron layer at depth of about 3500 µm, the measured current fell, even though the oxygen tension was known to be high. It was clear from this bench experiment that intraocular measurements of oxygen gradients in the perflubron treated eyes was not appropriate.

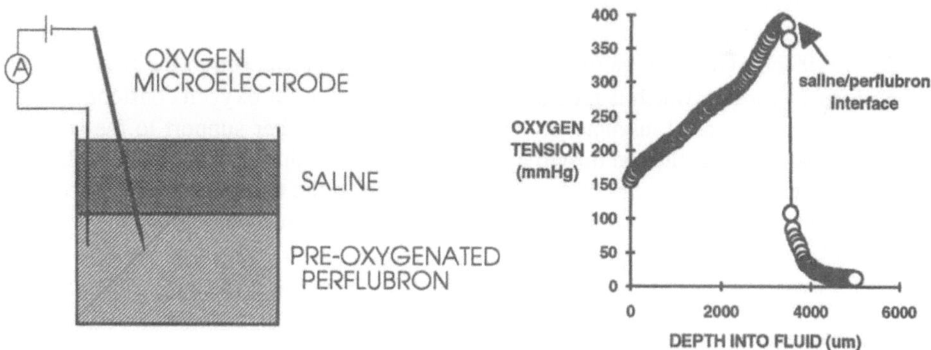

Figure 6. Measurement of oxygen gradients across the layer of saline illustrate the oxygen flux from the perflubron. Penetration of the perflubron itself leads to artefactually low currents due to the high electrical resistivity of the perflubron

It is interesting, given the electrical insulating properties of the perflubron, that the ERG amplitude was not suppressed by the presence of the perflubron. The question of ERG suppression due to electrical insulation of intravitreal materials has been the subject of continuing debate in the study of the toxic effects of intravitreal silicone oil (Doslak, 1987). Some authors have attributed reduced ERG amplitudes to the insulation effect of intravitreal silicone oil rather than impaired retinal function due to retinal toxicity. The implication from the present study in which the vitreous cavity was almost completely full of perflubron in the treated eye is that the insulation effects are negligible and that a normal corneal ERG is not dependent on current pathways through the bulk of the vitreous.

The question of the long term survival of ischaemic retina is not addressed in this study. The inability of the perflubron to sustain the ERG does not mean that the hypoxic component of the ischaemia is not reduced, or indeed eliminated. We do not know what effect such treatment may have on the ability of the retina to recover from extensive periods of ischaemia. Improved longer term recovery from ischaemic insult has been demonstrated in histological studies following vitreal perfusion with oxygen and glucose rich solutions (Blair et al, 1989). On the other hand there is a body of opinion that relief of the hypoxic component of an ischaemic insult may be more harmful to tissue recovery than no treatment at all (Hall, 1993).

The hypothesis that hypoxia is not the most important factor in acute retinal ischaemia is not in agreement with some earlier studies. Although it would appear that work of Flower and Patz (1971) and Ben Nun et al (1988) demonstrate conclusively that enhanced oxygen delivery to the inner retina either from the choroid or the vitreous can restore the dark adapted ERG, these studies depend on the complete abolition of the retinal circulation. Even a small residual flow could confound the results. We were not able to produce reliable occlusion of the entire retinal vasculature even when pushing the probe hard enough to create extensive damage to the retinal vasculature.

There is other evidence that hypoxia is not the dominant effect in induced ischaemia. In studies of ERG changes with raised IOP, Yancey and Linsenmeier (1988) also found that the b wave was suppressed. Alder et al (1990) have previously demonstrated that oxygen tension at the retinal surface is well maintained at perfusion pressures as low as 40 mmHg. So the b wave suppression seems unlikely to have been due to hypoxia. Yancey and Linsenmeier also employed a short period of hyperoxic ventilation, which failed to restore the b wave, once again suggesting that hypoxia was not the key element.

In our experimental situation we can obtain a good estimate of the rate of oxygen delivery from the perflubron. The volume of intraocular perflubron was typically 2ml and it's oxygen tension decreased by 323 mmHg in the first 45 minutes (Fig 2.). The solubility of oxygen in neat perflubron is 53 ml of oxygen per 100 ml at 760 mmHg and 37°C (Faithfull, 1992). The average oxygen delivery from the perflubron in the first 45 minutes is therefore 10 μl/min, which is greater than the estimated total oxygen consumption of the dark adapted cat retina (Linsenmeier, 1986). This lends further support to the suggestion that the retina in the perflubron treated eye was not hypoxic during the induced ischaemia.

The present study provides clear evidence that whilst the conditions for a vitreal source of oxygen have been optimized, this is not sufficient to sustain retinal function in the presence of acute retinal ischaemia.

REFERENCES

Alder VA, and Cringle SJ (1989). Intraretinal and preretinal PO2 response to acutely raised intraocular pressure in cats. American Journal of Physiology. 256(Heart & Circulatory Physiology 25):H1627-H1634.

Ben-Nun J, Alder VA, Cringle SJ, and Constable IJ (1988). A new method of oxygen supply to acute ischemic retina. Investigative Ophthalmology and Visual Science 29:298-304.

Berkowitz BA, Wilson CA, and Hatchell DL (1991). Oxygen kinetics in the vitreous substitute perfluorotributylamine: a ^{19}F NMR study in vivo. Investigative Ophthalmology and Visual Science. 32:2382-2387.

Blair NP, Baker DS, Rhode P, and Solomon M (1989). Vitreoperfusion: A new aproach to ocular ischaemia. Archives of Ophthalmology 107:417.

Braun RD, Linsenmeier RA, and Goldstick TK (1992). New perfluorocarbon emulsion improves tissue oxygenation in cat retina. Journal of Applied Physiology 72(5):1960-1968.

Chang S (1987). Low viscosity liquid fluorochemicals in vitreous surgery. American Journal of Ophthalmology 1987;103:38-43.

Doslak MJ (1987). Theoretical analysis of the effects of intraocular silicone oil on the ERG. Investigative Ophthalmology and Visual Science ARVO Suppl. 28:p115.

Faithfull NS (1992). Artificial oxygen carrying blood substitutes. In, Oxygen Transport to Tissue XIV, Edited by W. Erdmann and D.F. Bruley, Plenum Press, New York 1992, pp55-72.

Faithfull NS (1992). Second generation fluorocarbons. In, Oxygen Transport to Tissue XIV, Edited by W. Erdmann and D.F. Bruley, Plenum Press, New York 1992, pp441-452.

Flower RW and Patz A (1971). The effect of hyperbaric oxygenation on retinal ischaemia. Investigative Ophthalmology 10:605-616.

Hall ED (1993). Cerebral ischaemia, free radicals and antioxidant protection. Biomedical Society Transactions 21:334-339.

Linsenmeier RA (1986). Effects of light and darkness on oxygen distribution and consumption in the cat retina. Journal of General Physiology 88:521-542.

Thoreson WB and Purple RL (1987). Effect of an artificial oxygen carrier on the isolated arterially perfused cat's eye. Investigative Ophthalmology and Visual Science. ARVO abstract 28:408.

Yancey CM, and Linsenmeier RA (1988). The electroretinogram and choroidal PO2 in the cat during elevated intraocular pressure. Investigative Ophthalmology and Visual Science 29:700-707.

WHOLE ANIMAL, LUNG AND MUSCLE HEMODYNAMICS AND FUNCTION ARE MAINTAINED DURING AND AFTER OXYGENT™ HT INFUSION

T.E. Gayeski[1], R.J. Connett[1], W.A. Voter[1], J.L. Frierson[1], P.E. Keipert[2], N.S. Faithfull[2]

[1]University of Rochester, Rochester, NY
[2]Alliance Pharmaceutical Corp, San Diego, CA

INTRODUCTION

Oxygent™ HT is a second generation concentrated, stable fluorocarbon emulsion based on perflubron (perfluorooctyl bromide [PFOB], Alliance Pharmaceutical Corp.). One purpose for the development of Oxygent HT is to increase oxygen delivery to the tissues through increased plasma oxygen solubility. We studied the hemodynamic and the oxygen transport characteristics of the pentobarbital anesthetized dog in the presence of low doses of Oxygent HT.

These investigations were designed to examine effects on the whole animal, the pulmonary system and skeletal muscle. Animals were studied under conditions of both normoxia and hyperoxia. To examine the influence of the drug under conditions of high and low oxygen extraction from blood, both resting and exercising gracilis muscles were studied. Our principal finding is that whole animal and organ function is maintained during and after Oxygent HT infusion.

METHODS

Hound type mongrel dogs weighing ~20 kg were anesthetized with pentobarbital. We induced anesthesia with intravenous pentobarbital (35 mg/kg) and maintained anesthesia with intermittent bolus injection of pentobarbital (approximately 130 mg/hr). Animals were mechanically ventilated with either 100% oxygen or room air. All animals had their electrocardiograms and core temperatures recorded and were invasively monitored using a pulmonary artery and systemic arterial catheters.

The gracilis muscle was surgically, neurally and vascularly isolated. The details of this gracilis preparation are given elsewhere [1]. After isolation, the muscle was attached to a force transducer so that both resting and exercising muscle tension could be recorded. The venous effluent of the preparation could be diverted into a graduated cylinder containing mineral oil so that blood flow and venous blood gases from the muscle could be measured. The gracilis muscle was observed under both resting and working (~67% of maximal oxygen consumption) conditions.

To provide adequate hydration of the animals, normal saline was infused at a rate of 6 cc/kg/min. Oxygent HT was infused at rates of ~0.2 cc/kg-min to total doses of 1 cc/kg of dog weight (0.9 g perflubron/kg). No vasoactive or inotropic drugs were administered at any time.

The cardiorespiratory effects of Oxygent infusion on the whole animal were monitored by using systemic and pulmonary artery pressures (systolic, diastolic and mean pressures) and right and left ventricular filling pressures (right atrial (RA) and mean pulmonary capillary wedge (PCW) pressures). We measured thermal dilution cardiac outputs using 10 cc injectates of ice-cooled normal saline. Finally we obtained blood for determination of PO_2, PCO_2, pH, hemoglobin saturation, hemoglobin content, methemoglobin content, carbonmonoxyhemoglobin content, lactate concentrations and fluorocrits from the systemic artery, pulmonary artery and gracilis muscle vein. Determinations were made before infusion of Oxygent HT and at 5, 30 and 55 minutes after infusion was complete.

RESULTS

The data and any observed trends are summarized in the text. Representative data from individual animals is presented in Tables 1 and 2.

WHOLE ANIMAL

Hemoglobin content was normal in all animals and was unchanged after infusion of Oxygent HT. The fluorocrit (Fct) is defined as the percent of perfluorocarbon in the blood. In these experiments it was determined by centrifuging blood in a hematocrit tubes. At 5 minutes after infusion of 1 cc/kg of Oxygent HT Fct was ~1% and reached a constant value of ~0.5% by 30 minutes after infusion.

Systemic arterial pressure was maintained during and after infusion of Oxygent HT. In 11 of the 12 dogs ventricular filling pressures were initially normal (~ 1-3 torr for both RA and PCW pressures) and remained so after infusion of Oxygent HT. Because these filling pressures were low, as expected in these normal hearts, small changes in pressure could have large effects on ventricular function. What was observed was a trend towards a small decrease in PCW pressure after infusion of Oxygent HT with little change in the subsequent 55 minutes. This decrease was accompanied by a small increase in heart rate (~ 10%), a decrease in cardiac output (~25%) and an increase in systemic vascular resistance (~25%). We conclude that the effects of Oxygent HT on central hemodynamics are small and that normal physiologic mechanisms for maintaining blood pressure remain intact.

To test if there was any primary defect in cardiac function we infused normal saline (300 cc) into an animal until PCW pressure returned to the preinfusion value. The changes in cardiovascular parameters observed with infusion returned to the preinfusion status, suggesting that there was no primary cardiac effect due to Oxygent HT. These findings are typified in the data in Table 1.

The cardiac function of one dog was found to be grossly abnormal from the onset. After anesthesia was induced but before any other interventions were undertaken, left ventricular filling pressure was elevated (~ 24 torr) and cardiac function was modestly depressed

Table 1. Whole Animal Hemodynamics

Time after end of Oxygent infusion

1 ml/kg Oxygent		Control	5 min	30 min	55 min	73 min*
P	beats/min	131	135	130	128	120
BP	Systolic	162	180	175	170	165
	Diastolic	113	140	142	138	130
	Mean	140	160	158	150	145
PAP	Systolic	26	26	26	22	35
	Diastolic	12	9	7	11	10
	Mean	18	17	15	16	18
PCWP		7	5	4	4	8
RAP		2	3	2	2	1
CO	Liters/min	4.7	3.7	2.6	2.1	4.2
SV	cc/beat	36	27	20	16	35
LVSW	g-m/beat	66	58	43	33	66
RVSW	g-m/beat	8	5	4	3	8
SVR	PRU	29	43	60	71	34
PVR	PRU	2	3	4	6	2

* values after 350 ml of normal saline

Table 1. Representative example of hemodynamics of dog following 1 ml/kg infusion of Oxygent. Times indicate minutes following completion of Oxygent infusion. All pressures are given in torr. Pulse (P) and blood pressure (BP) were not affected by infusion of Oxygent. However, pulmonary capillary wedge pressure (PCWP), stroke volume (SV) and cardiac output (CO) decreased prior to infusion of normal saline. BP was maintained by an increase in systemic vascular resistance (SVR) through 55 minutes post-infusion. At 55 minutes 350 cc's of normal saline was infused over approximately 10 minutes and hemodynamic values were recorded at 73 minutes. Note that hemodynamic values returned to control values, particularly left and right ventricular stroke work (LVSW and RVSW respectively). Right atrial pressure (RAP) was essentially unchanged throughout. Pulmonary vascular resistance (PVR) increased slightly but remained within normal limits.

(See table 2). All transducers were checked and recalibrated to make certain these values were not measurement error. In this animal there was little change in central hemodynamics during and after infusion of Oxygent HT. This finding is consistent with Oxygent HT having no measurable effect on cardiac function even when left ventricular function is below normal.

PULMONARY SYSTEM

After the infusion of Oxygent HT, a small increase in arterial PO_2 was found while PCO_2 and hemoglobin content remained unchanged. The increase in PO_2 was <5 torr in normoxia and no greater than 10 torr during 100% oxygen ventilation. There was no change in pulmonary artery blood gas values during the protocol. Hence, there was either no change or perhaps a slight improvement in gas transport properties of the lung after infusion of Oxygent HT.

Table 2. Whole Animal Hemodynamics

1 ml/kg Oxygent		Control	5 min	Time after end of Oxygent infusion	
				30 min	55 min
P	beats/min	140	155	151	147
BP	Systolic	149	176	165	156
	Diastolic	133	150	143	137
	Mean	140	163	155	145
PAP	Systolic	39	38	51	43
	Diastolic	28	28	35	27
	Mean	33	34	42	34
PCWP		27	24	32	28
RAP		5	3	5	3
CO	Liters/min	1.54	2.41	1.92	1.99
SV	cc/beat	11	16	13	14
LVSW	g-m/beat	17	30	22	22
RVSW	g-m/beat	4	7	7	6
SVR	PRU	88	66	78	71
PVR	PRU	4	4	5	3

Table 2. Representative example of hemodynamics of dog following 1 ml/kg infusion of Oxygent. Times indicate minutes following completion of Oxygent infusion. P, BP PAP, RAP were not affected by infusion of Oxygent. In this animal that had spontaneously high PCWP and decreased ventricular performance, left ventricular function was at least sustained, if not improved, by Oxygent HT infusion.

There was no increase in pulmonary vascular resistance and a slight fall in pulmonary artery pressures following Oxygent HT infusion. The change in pressure was anticipated given the decrease in cardiac output and no change in pulmonary vascular resistance.

GRACILIS MUSCLE

In response to the decrease in cardiac output, muscle vascular resistance would be expected to increase to maintain systemic blood pressure. Consequently, blood flow in a resting muscle would be expected to decrease. This autoregulation was observed in all of our muscles, both innervated and denervated.

Under resting conditions, there was an increase in extraction of oxygen as muscle blood flow decreased. This increased extraction compensated for the decreased flow so that muscle oxygen consumption was maintained (See Table 3). The presence of hyperoxia had no effect on either the measured or derived parameters in either the control or post-infusion states.

The entire muscle mass is ~50% of the animal's weight. If all of the muscle beds responded as the gracilis muscle did in this protocol, then the change in muscle vascular resistance is

Table 3. Muscle Synopsis

1 ml/kg Oxygent	Control	Exercise Control	Recovery	Time after end of Oxygent infusion			
				5 min	30 min	55 min	Exercise
Flow (ml/100g.min)	8.2	71	6.1	5.5	4.1	5.5	82
Resistance (PRU)	3.9	1.6	18.4	19.1	27.3	21.8	1.3
Extraction (ml/100ml)	1.7	13.7	2.6	3.0	3.1	2.7	14.2
O2 offered (ml/100g.min)	1.5	12.8	1.1	1.0	0.8	1.0	15.4
VO2 (ml/100g.min)	0.14	9.8	0.16	0.17	0.13	0.15	11.6

Table 3. There were no discernible trends in muscle blood flow, O_2 offered (CaO_2 x muscle blood flow), and O_2 consumed (VO_2) amongst muscles. The muscle could vasoconstrict and hence muscle resistance varied at rest (5, 30 and 55 min points) increased. Note that it was also able to vasodilate in response to stimulation. While there was an increase in VO_2 in this muscle, there was no increase in VO_2 at rest or during exercise when all muscles are included for these animals with a normal hematocrit.

sufficient to account for the change in systemic vascular resistance observed. This increased resistance had no measurable effect on muscle oxygen consumption of this oxidative muscle.

In this preparation, the gracilis muscle is a small fraction of the dog's muscle mass. Hence, exercising this isolated muscle does not influence systemic vascular resistance or systemic blood pressure significantly. Consequently, muscle blood flow during exercise should be less effected by the central hemodynamic status than resting values. This was observed in our muscles. In both normoxia and hyperoxia, muscle blood flow, oxygen extraction and oxygen consumption were unchanged after infusion (See Table 3).

LACTATE AND PYRUVATE CONCENTRATIONS

Neither lactate nor pyruvate concentrations or their ratios showed measurable change in systemic arterial, pulmonary arterial and gracilis venous blood after infusion of Oxygent HT. Hence, there was no increase in production by the lung or gracilis muscle, assuming removal of lactate and pyruvate were constant - by inference whole body, implying that tissue values and oxygen availability were unchanged.

DISCUSSION

When first generation fluorocarbons were administered to critically ill patients, there were reports of pulmonary [2] and cardiac [3] complications. There were also reports of vascular effects from several laboratories using animal models. In particular, vascular integrity in skeletal muscle was reported to be adversely affected [4,5]. Despite the absence of clear documentation of this problem with these earlier fluorocarbons, questions about the effect of fluorocarbons on vascular integrity have persisted at least in the minds of animal investigators. The results of this study demonstrated that Oxygent HT has no adverse effect on whole animal oxygen uptake or vascular resistance, pulmonary gas exchange or vascular resistance, and muscle function or organ vascular resistance.

The ability of the integrated cardiovascular system to respond to changes in filling pressures remained intact. In the presence of adequate hydration, Oxygent HT was accompanied by small decreases in ventricular filling pressure and cardiac output. Modest increases in systemic vascular resistance were invoked to maintain blood pressure. If all muscle vasculature responded as the gracilis muscle vascular system did, the increase in skeletal muscle resistance was sufficient to account for the entire change in systemic vascular resistance. In the clinical setting that would include the potential of continued blood loss, use of Oxygent HT would not be expected to interfere with the normal homeostatic mechanisms necessary to compensate for this insult.

At the low dose tested and in the presence of normal hemoglobin contents the small fluorocrits achieved in this study would be predicted to have, at most, a modest effect on tissue oxygenation. However, in the presence of anemia due to blood loss the oxygen delivery capacity of Oxygent HT can supplement significantly the oxygen offered to the tissue through oxygen bound to hemoglobin. Under conditions of anemia (hemoglobin concentration of ~8.5 g/dl) Hogan et al (6) have shown that exercising muscle oxygen consumption is increased by ~12% when 6 cc/kg of Oxygent HT was administered in the presence of hyperoxia compared to muscles without this supplementation of oxygen offered. This increase in oxygen consumption may have been due to the increased oxygen offered to the tissues because of the increased dissolved oxygen present in Oxygent HT or more efficient delivery of oxygen at the cellular level. Studies to clarify the relationship between the oxygen offered through the convective channels and the myocyte PO_2 are in progress.

In summary, whole animal hemodynamics and oxygen consumption are well maintained in the presence of Oxygent HT. Both pulmonary and muscle functions are unchanged as well. Unlike its first generation counterparts, infusion of Oxygent HT preserves the organ function while increasing oxygen solubility in blood.

References

1. T.E.J. Gayeski, R.J. Connett, and C.R. Honig, O_2 transport in the rest-work transition illustrates new functions for myoglobin, Am J Physiol 248:H914-H921 (1985).

2. A.M. Police, K. Waxman, and G. Tominaga, Pulmonary complications after Fluosol administration to patients with life-threatening blood loss, Critical Care Medicine 13:96-98 (1985).

3. K.K. Tremper, G.M. Vercellotti, and D.E. Hammerschmidt, Hemodynamic profile of adverse clinical reactions to Fluosol-DA 20%, Critical Care Medicine 12:428-431 (1984).

4. N.S. Faithfull and S.M. Cain, Critical oxygen delivery levels during shock following normoxic and hyperoxic haemodilution with fluorocarbons or dextran, AEMB 215:79-87 (1987).

5. N.S. Faithfull, C.E. King, and S.M. Cain, Peripheral vascular responses to fluorocarbon administration, Microvascular Research 33:183-193 (1987).

6. M.C. Hogan, D.C. Willford, P.E. Keipert, N.S. Faithfull, and P.D. Wagner, Increased plasma O_2 solubility improves O_2 uptake of in situ dog muscle working maximally, J. Appl. Physiol. 73:2470-2475 (1992).

THE INTERACTION OF ACTEOSIDE WITH MITOCHONDRIAL LIPID PEROXIDATION AS AN ISCHEMIA/REPERFUSION INJURY MODEL

Ning Pan and Hitoshi Hori *

Department of Biological Science and Technology
Faculty of Engineering, The University of Tokushima
Tokushima, 770 Japan

INTRODUCTION

Since most of the O_2 utilized by mammals takes place in mitochondria, the organelle would be particularly subject to free radical-induced changes.   Attempts to decrease the rate of production of these changes with antioxidants may be thwarted because of the highly selective permeability of the inner membrane and/or adverse effects of antioxidants on mitochondrial function (Darley-Usmar et al. 1987; Horrum et al. 1987; Szabados et al. 1989).   Recently, formation of oxygen free radicals was reported to be involved in the development of reperfusion injury (Gauduel and Duvelleroy 1984; Powell and Tortolani 1992).   One of the sources of oxygen free radicals is suggested to be the mitochondrial electron transport system (Gonzalez-Flecha et al. 1993; Powell and Tortolani 1992). Acteoside is a phenyl-propanoid glycoside widely distributed in plants, especially in tonic drugs (Birkofer et al. 1968; Miyase et al. 1982; Kobayashi et al. 1984; Imakura et al. 1985; Takeda et al. 1985).   We present the anti-peroxidative effects of acteoside on the lipid peroxidation in isolated rat liver mitochondria.

MATERIALS AND METHODS

Acteoside was isolated, as one of the major components, from *Phacellathus tubiflorus* Sieb. et Zucc. (Pan et al. 1993).   Caffeic acid and ferulic acid were obtained from Tokyo Chemical Industry Co. Ltd.   These chemical structures were shown in Fig.1. DPPH ($\alpha,\alpha,$-

---

*Address correspondence to this author.

diphenyl-β–picrylhydrazyl), α-tocopherol, and Iron(II) sulfate heptahydrate were obtained from WAKO Pure Chemical Industries, Ltd.

Mitochondria were prepared from male Wistar rats (250 - 300g) by homogenization followed by differential centrifugation in ice-cold medium (pH 7.4) (Myers and Slater 1957; Cain and Skilleter 1987). The mitochondrial protein content was determined by the

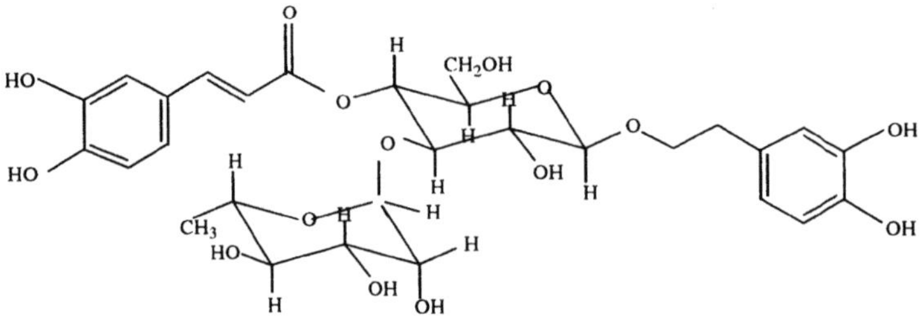

acteoside

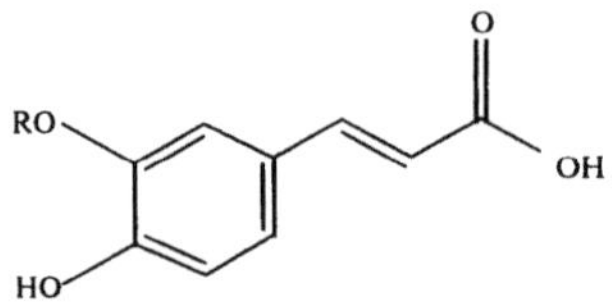

$R = H$ caffeic acid
$R = CH_3$ ferulic acid

Fig. 1. The structures of Acteoside, Caffeic Acid, and Ferulic Acid

biuret method using bovine serum albumin as a standard (Gornall et al. 1949). Lipid peroxidation in mitochondria was measured by monitoring oxygen consumption with an oxygen electrode and assuming an oxygen concentration of 516 nmol O/ml in the experimental medium at 25°C. The Inhibition (%) on mitochondrial lipid peroxidation was calculated by using Eq. 1 developed by us. The interaction of acteoside and other

$$\text{Inhibition\%} = \left(1 - \frac{R_p t}{t_{inh}} k \right) 100\% \tag{Eq.1}$$

$$\text{control coefficient } k = \frac{t_{inh_0}}{R_{p_0} t_0}$$

R_p = Rate of lipid peroxidation (nmol O/min)
R_{p_0} = Rate of lipid peroxidation of control (nmol O/min)
t = total-time (min) t_{inh} = induction time (min)
t_0 = total-time of control (min) t_{inh_0} = induction time of control (min)

compounds with free radical was measured by using stable free radical DPPH at 517nm according the method of Blois (1958). Respiration rate of mitochondria was measured using the oxygen electrode.

RESULTS AND DISCUSSION

Acteoside were initially examined for its ability to inhibit iron-induced lipid per-oxidation. The caffeic acid and ferulic acid were tested as compounds comparable to

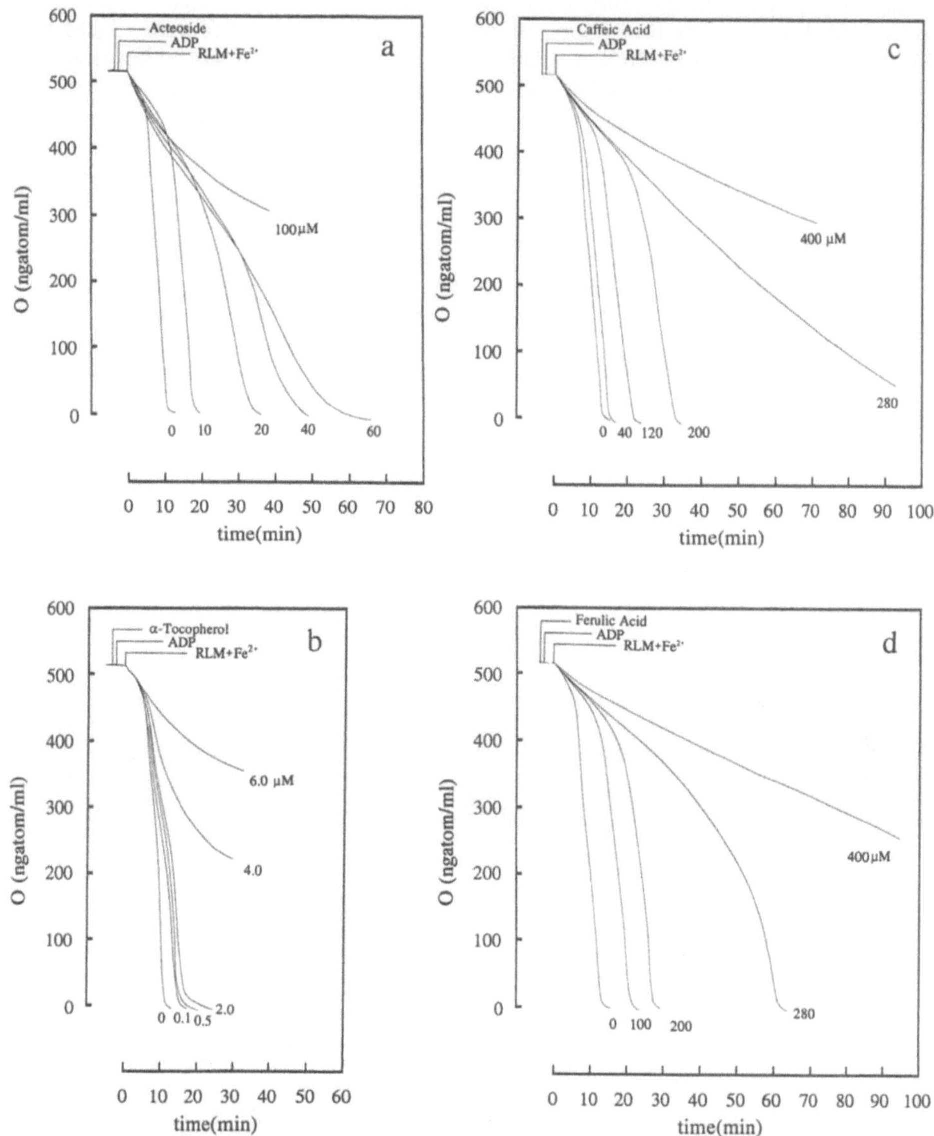

Fig. 2. Effect of acteoside, caffeic acid, ferulic acid, and α-tocopherol on O_2 consumption due to iron-depen-dent lipid peroxidation in rat liver mitochondria (the typical experimental results). a) acteoside, b) a-toco-pherol, c) caffeic acid, d) ferulic acid. To 2.53 ml of the incubation medium (175 mM KCl, 10 mM Tris-HCl, pH 7.4, 25°C), 1 mM ADP, 100mM Fe^{2+} (FeSO4), and mitochondria (0.7 mg protein /ml) were added.

a phenyl propanoic acid moiety in acteoside. α-Tocopherol was selected as a standard antioxidant. As showing in Fig. 2, iron-dependent lipid peroxidation in mitochondria showed first slow oxygen consumption (initiation) and followed rapid oxygen consumption (propagation) in the absence of inhibitor. Acteoside, caffeic acid, ferulic acid, and α-tocopherol showed concentration-dependent inhibition on oxygen consumption due to iron-dependent lipid peroxidation in isolated rat liver mitochondria in different profile. Acteoside inhibited both initiation (evaluated by the induction time: t_{inh} in Eq. 1) and propagation (evaluated by the rate of lipid peroxidation: R_p in Eq. 1) of the lipid peroxidation, α-tocopherol almost inhibit only the latter (Fig. 2. a, b). Caffeic acid and ferulic acid (caffeic acid 3-methyl ether) mainly affect the initiation (Fig. 2. c, d). The order of inhibitory potency of the compounds was: α-tocopherol > acteoside > caffeic acid > ferulic acid (Fig. 3). The IC50 values of acteoside and α-tocopherol were 15 μM and 2.7 μM. The IC50 values of caffeic acid and ferulic acid were not obtained because of their abnormal dose-dependent curves, probably due to their acidity. The effect of

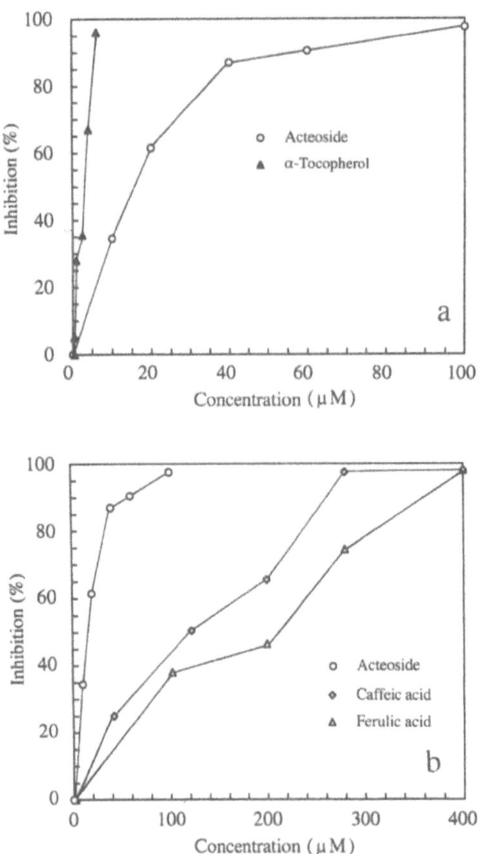

Fig. 3. Inhibition on iron-dependent lipid peroxidation in liver mitochondria. a) acteoside and α-tocopherol.. b) acteoside, caffeic acid , and ferulic acid. The calculation method is described in the Materials and Methods (Eq. 1).

acteoside on the induction time of ADP-Fe^{2+} induced lipid peroxidation shows that acteoside inhibited the formation of the active complex which acts as the initiator.

For a better understanding of these compounds' behavior in the iron-dependent lipid peroxidation in mitochondria, we determined the interaction of the compounds with a stable free radical DPPH. The results (Fig. 4) showed that the radical-scavenging abilities of these compounds were graded as follows: acteoside > α-tocopherol, caffeic acid > ferulic acid. In the reaction of α-tocopherol with DPPH, the molecular ratio is one for two (Blois 1958). Thus, acteoside, caffeic acid, and ferulic acid interacts with DPPH in the ratio of 4 : 1, 2 : 1, and 1 : 1 respectively. Acteoside showed strong radical scavenging ability.

These results indicated that the strength of the inhibitory effect of the test compounds on mitochondrial lipid peroxidation was dependent on the type of phenolic groups and their

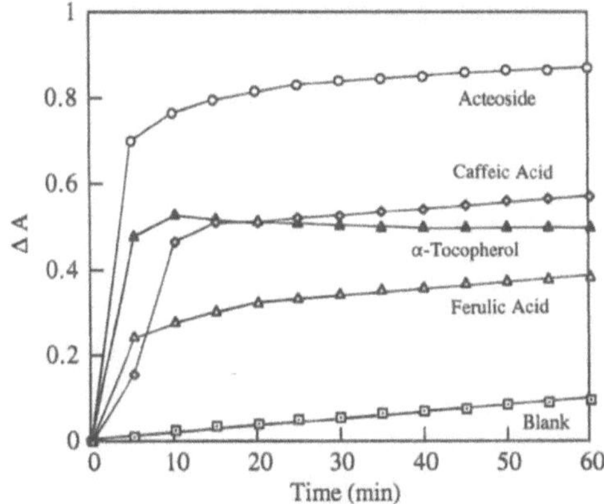

Fig.4. The interaction of acteoside, caffeic acid, ferulic acid, and α-tocopherol with DPPH. To 2.5 ml of 100μM DPPH solution, containing 20 mM 2-(N-Morpholino) ethanesulfonic acid (MES), pH 5.5, 5 μl final concentration of 20 μM test compound was added. The change in optical absorbance at 517 nm was measured every 5 min.

number in each molecule. Recently Okuda et al. (1992) have also been reported a similar results from *in situ* ESR measurements of tanins. We suggest that scavenging lipid peroxyl radicals represents a primary mechanism of the inhibition. Although this mechanism operate on molecular level well, the other factors on organelle, organ, and body level, such as the high resistance of mitochondrial membranes to permeability changes (Darley-Usmar et al. 1987), the metabolic stability and other biological activities of compounds, might affect the radical-scavenging action.

Mitochondria are one of the main sources of oxygen free radicals in the liver under physiological conditions (Boveris and Cadenas 1973; Chance et al. 1979). Recently,

Gonzalez-Flecha et al. (1993) reported that mitochondria were the first oxyradical source to undergo marked changes, starting after reperfusion following short periods of ischemia and leading to increased rates of superoxide anion and hydrogen peroxide productions. Mitochondrial uncouplers were reported to reduce the rate of mitochondrial hydrogen peroxide production to negligible levels (Boveris and Cadenas 1973). Acteoside, as a proton-donor to active oxygen species and lipid free radicals, might show protective effects against reperfusion injury. Further studies, both *in vitro* and *in vivo*, on acteoside and its related compounds as antioxidants with inhibitory effect against the mitochondrial electron transport system, might play an important role on the research on tonic drugs.

ACKNOWLEDGMENTS

This work was supported in part by a Grant-in-Aid for Scientific Research (No. 05671872 to H.H.) from The Ministry of Education, Science and Culture, Japan. The authors are grateful to Dr. H. Tarada, Dr. Y. Takeda, and Mr. K. Kogure of The University of Tokushima, for valuable discussions concerning mitochondria, acteoside, and iron-dependent lipid peroxidation.

REFERENCES

Birkofer, L., Kaiser, C. and Thomas, U., 1968, Sugar esters. IV. acteoside and neoacetoside, sugar esters from *Syringa vulgaris*, *Z. Naturforsch.* 23B:1051.

Blois, M.S., 1958, Antioxidant determination by the use of a stable free radical, *Nature* 181:1199.

Boveris, A. and Cadenas, E., 1973, The mitochondrial production of hydrogen peroxide, *Biochem. J.* 134:707.

Cain, K. and Skilleter, D.N., 1987, Preparation and use of mitochondria in toxicological research, in: "Biochemical Toxicology: a practical approach," K. Snell, and B. Mullock, ed., IRL Press Ltd., washington DC.

Chance, B., Sies, H. and Boveris, A., 1979, Hydroperoxide metabolism in mammalian tissues, *Physiol. Rev.* 59:527.

Darley-Usmar, V.M., Rickwood, D. and Wilson, M.T., 1987. Mitohondria: a practical approach. IRL Press Ltd., Washington DC.

Gauduel, Y. and Duvelleroy, M.A., 1984, Role of oxygen radicals in cardiac injury due to reoxygenation, *J. Mol. Cell. Cardiol* 16:459.

Gonzalez-Flecha, B., Cutrin, J.C. and Boveris, A., 1993, Time course and mechanism of oxidative stress and tissue damage in rat liver subjected to invivo ischemia-reperfusion, *J. Clin. Invest.* 91:456.

Gornall, A.G., Bardawill, C.J. and and David, M.M., 1949, Determination of serum proteins by means of the biuret reaction, *J. Biol. Chem.* 177:751.

Horrum, M.A., Harman, D. and Tobin, R.B., 1987, Free radical theory of aging: effects of antioxidants on mitochondrial function, *Age* 10:58.

Imakura, Y., Kobayashi, S. and Mima, A., 1985, Bitter phenyl propanoid glycosides from *Campsis Chinensis*, *Phytochemistry* 24:139.

Kobayashi, H., Karasawa, H., Miyase, T. and Fukushima, S., 1984, Studies on the constituents of Cistanchis herba. III. Isolation and structures of new phenyl-propanoid glycosides, cistanosides A and B, *Chem. Pharm. Bull.* 32:3009.

Miyase, T., Koizumi, A., Ueno, A., Noro, T., Kuroyanagi, M., Fukushima, S., Akiyama, Y. and Yakemoto, T., 1982, Studies on the acyl glycosides from *Leucoseptrum japonicum* (MIQ.) Kitamura *et* Murata, *Chem. Pharm. Bull.* 30:2732.

Myers, D.K. and Slater, E.C., 1957, The enzymic hydrolysis of adenosine triphosphate by liver mitochondria I. Activities at different pH values, *Biochem. J.* 67:558.

Okuda, T., Yoshida, T. and Hatano, T., 1992, Antioxidant effects of tannins and related polyphenols, in: "Phenolic Compounds in Food and Their Effects on Health II: Antioxidants and Cancer Prevention," Huang, M.-T., Ho, C.-T. and Lee, C.Y., ed., American Chemical Society, Washington, DC.

Pan, N., Yokoyama, H., Xu, G.J. and Hori, H., 1993, The constituent of *Phacellathus tubiflorus*: acteoside and its biological activity. *Proceeding 2 of 113th Annual meeting of Pharmaceutical Society of Japan* : 217.

Powell, S.R. and Tortolani, A.J., 1992, Current research review: recent advances in the role of reactive oxygen intermediates in ischemic injury, *J. Surg. Res.* 53:417.

Szabados, G., Tretter, L. and Horvath, I., 1989, Lipid peroxidation in liver and Ehrlich ascites cell mitochondria, *Free Rad. Res. Commun.* 7:161.

Takeda, Y., Fujita, T., Satoh, T. and Kakegawa, H., 1985, On the glycosidic constituents of *Stachys sieboldi* MIQ. and their effects on hyarulonidase activity, *Yakugaku Zasshi* 105:955.

ISCHEMIA REPERFUSION DAMAGE IN THE GUT AND ITS TREATMENT WITH DRUGS OF THE AMINOSALICYLIC ACID GROUP

Joachim Lutz and Albert Josef Augustin

Dept. of Physiology, University of Würzburg, Röntgenring 9, 97070 Würzburg, Germany
Dept. of Clinical Ophthalmology, University of Bonn, S.-Freud-Str. 25, 53127 Bonn, Germany

INTRODUCTION

Lipid peroxides play an important role in inflammatory diseases and ischemia reperfusion damage of the gut. Thus, antioxidants and oxygen free radical scavengers gain increasing significance in clinical therapy. Furthermore, some predominantly empirically approved drugs have also scavenging effects. This may be the case for sulfasalazine, a sulfapyridine - aminosalicylate compound. In addition to the antibiotic action of sulfapyridine, 5-amino salicyclic acid (5-ASA) is supposed to act as a scavenger. However, because of the appearance of side effects, some modifications of the molecule were produced. We tested the effects of different azo-linked drugs on an improved model (Augustin et al., 1988-1992, Lutz et al. 1989-1992) for producing reperfusion damage in rat intestine.

METHODS

28 male Wistar rats of 250-350 g body weight were randomly divided into seven groups of four animals. One group served as untreated control (basic). The other six underwent reversible occlusion of the superior mesenteric artery for 90 min, followed by a reperfusion period of 150 min. Five of these operated groups received either 5-ASA intravenously, mesalazine (MES) , benzalazine (BEN), olsalazine (OLS) or sulfasalazine (SUL), mostly perorally two hours before operation at a dose of 1 g/kg b.wt. each. One operated group served as a second control (MAO). MES and SUL were given in enema form to be comparable and to prevent resorption in the upper gastrointestinal tract. Only 5-ASA was tested by an i.v. route. After the reperfusion period the animals were killed under ether anesthesia by exsanguination and the tissue of the intestine was homogenized in its entirety. With this material, several kinds of tests were performed. The content of thiobarbituric acid reactive substances (TBARS) was determined according to a modification of the method of Ohkawa et al. (1979). HPLC was used to determine the content of malondialdehyde-like substances (MLS) in a part of the batches, according to Esterbauer et al. (1990).
As a further test for oxidative stress, the content of glutathione in the reduced (GSH) and oxidized (GSSG) form was determined according to the method of Griffith (1980). The tissue level of myeloperoxidase (MPO) was determined according to Krawisz et al. (1984).
All tests were performed in duplicates. Statistical analysis was performed with Student's two-tailed t-test with an extension according to Behrens-Fisher if variances were unequal. Values given in text or figures are expressed as means ± SEM.

RESULTS

At the end of the reperfusion period, all tests revealed dramatic and highly significant changes in the operated control group as compared to the basic levels of the totally untreated group. The highest increase appeared for TBARS (**Fig.1**). 5-ASA i.v. turned out to be slightly effective, however, caused by quick inactivation by acetylation, levels at the side of damages seem to have been not high enough. MES was more effective, though significantly ($p < 0.05$) less than the other three drugs used. Closest to the best effect of SUL came OLS: both agents differed only insignificantly from each other. Since a controversy exists concerning the capacity of the TBA reaction to evaluate real products of lipid peroxidation (Janero and Burghardt 1988, Ceconi et al. 1991, Fantini et al. (1992), a comparison of results from 16 unselected values of TBARS and HPLC determinations of malondialdehyde-like substances (MLS) was undertaken. The correlation between both methods was found to follow the equation

$$TBARS = 2.24 \; MLS - 174.61 \quad \text{with } r = 0.9099$$

Thus the test for TBARS as a screening method for products of lipid peroxidation fully attained the intended aim. **Fig.2** depicts a large depression of GSH following MAO; the loss was best compensated by BEN, OLS and SUL, the former two not significantly different from the latter. **Fig.3** shows the increase of oxidized GSSG, best compensated by OLS ($p < 0.001$). For the ratio of GSSG/GSH (**Fig.4**), the increase was significantly depressed only by BEN ($p < 0.05$), OLS and SUL ($p < 0.01$). **Fig.5** demonstrates that all azo-linked drugs used depressed the increase of MPO, an indirect measure of neutrophil immigration into gut tissue ($p < 0.01$). OLS contributed most prominently to the depression of MPO, though statistically not significantly different from the other two agents.

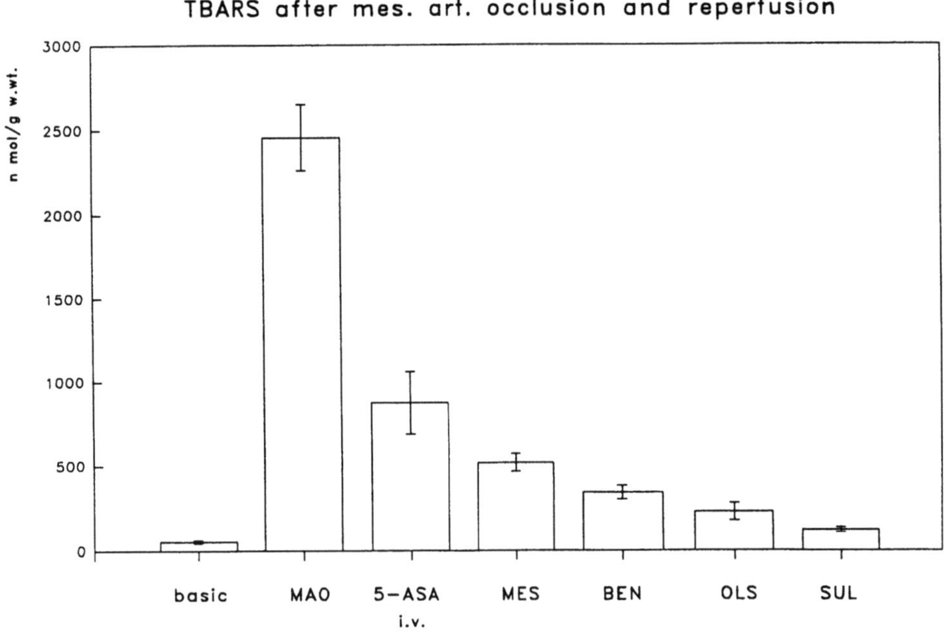

Figure 1. TBARS levels increased dramatically ($p < 0.001$) after MAO and reperfusion. This increase could be depressed very significantly by all azo-linked agents ($p < 0.001$), best by SUL, followed by OLS, though not significantly different from SUL.

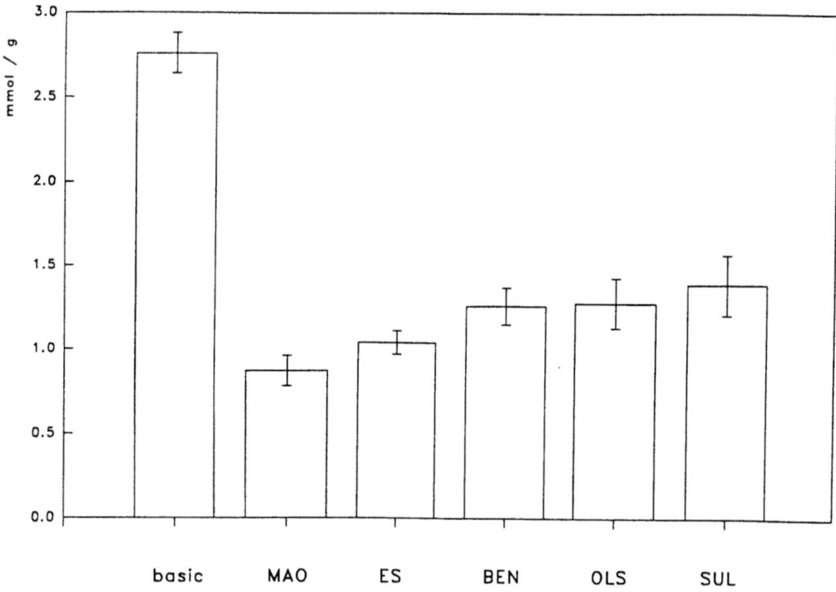

Figure 2. The content of GSH in gut tissue of the tested groups. After MAO and reperfusion the tissue level was found to be depressed to 31.5 ± 3.9 %. This depression was counteracted by the tested drugs, significantly by the last three agents ($p < 0.05$).

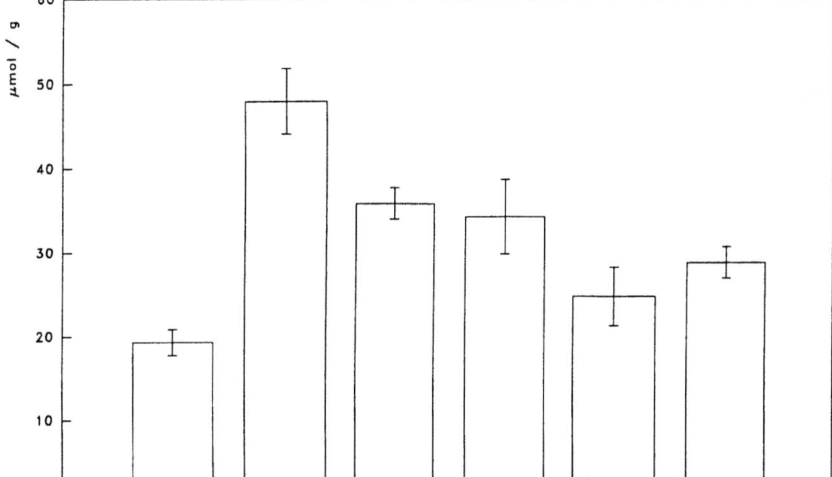

Figure 3. Tissue levels of GSSG in intestinal tissue increased after MAO and reperfusion 2.48 ± 0.19 fold ($p < 0.001$). This elevation could be depressed by MES and BEN with $p < 0.05$, by OLS and SUL with $p < 0.01$.

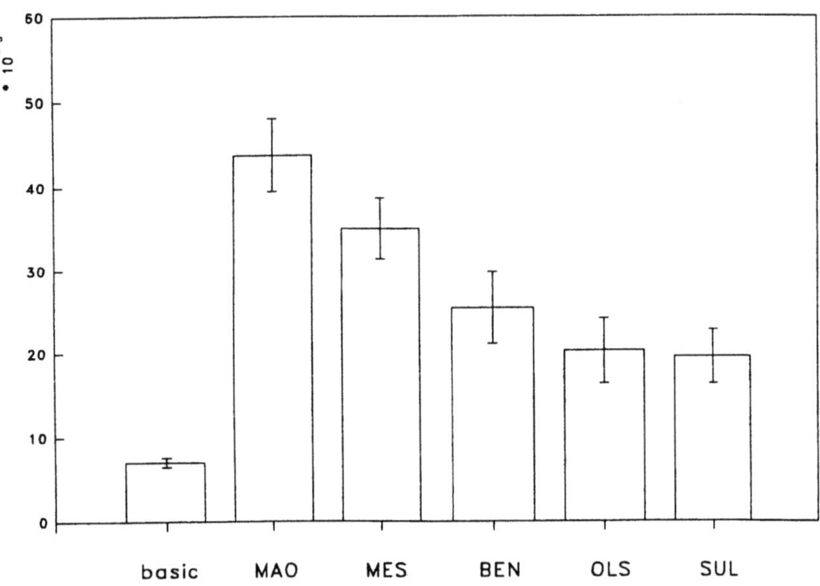

Figure 4. The ratio of GSSG/GSH increased 7.3 ± 0.7 fold after MAO and reperfusion (p<0.001). This effect was depressed insignificantly by MES, however with p<0.05 by the other three agents.

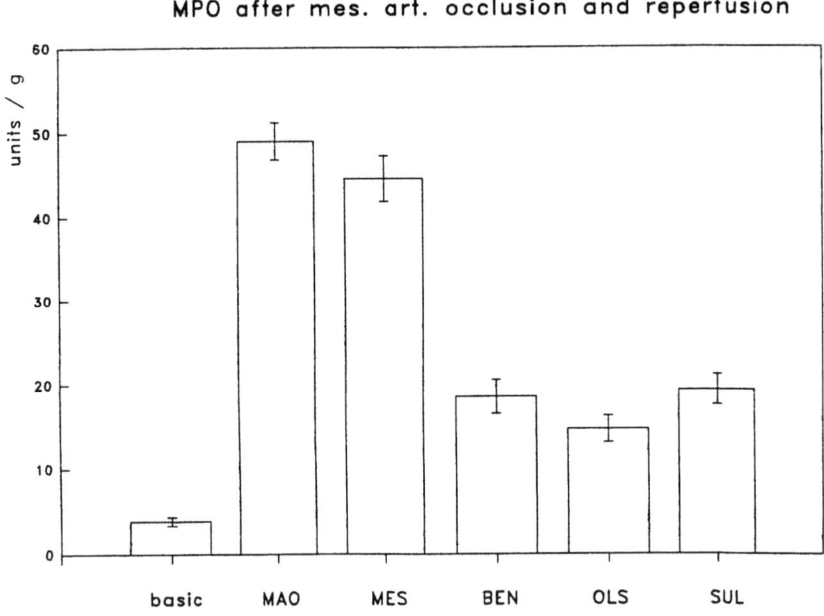

Figure 5. The content of myeloperoxidase increased after MAO and reperfusion 12.7 ± 0.6 fold (p<0.001). Whereas administration of MES resulted in only an insignificant depression, BEN and OLS reached or undercut the value of SUL; all three agents significantly (p<001) lower than the MAO control.

DISCUSSION AND CONCLUSION

Recently Allgayer et al. (1991) demonstrated in a model experiment that in contrast to 5-ASA the compound 4ASA did not inhibit radical production, though such an effect theoretically should be expected by both agents. Thus their results cast doubts on the concept that the chief effect of 5-ASA is produced by radical scavenging of reactive oxygen products.
Kvietys and Grisham (1991) have relativated this objection by referring to multiple common actions of both agents, e.g. in case of scavenging the porphyrin cation radical of myeloperoxidase and further by an equal efficacy in scavenging OH radicals, generated by the interaction between iron and ascorbate. In this way the results of Allgayer et al. seem to be limited to their special methodological approach.

In our experiments 5-ASA, when given in a sustained-release formulation as MES, was effective in depressing the radical damage caused by MAO, though not to such a degree like the other applied drugs. This may be because an enema had to be given perorally, though an enema on this way was very effective in the case of SUL. Intravascular half life of 5-ASA is described to amount dose dependent to 0.6 - 1.4 hours (Klotz et al. 1985), apparently too short to let the i.v. applied drug reach sufficient levels during ischemia and reperfusion in our model. Therefore the other tests were done without this kind of administration.

SUL, which was first introduced in therapy by Svartz already in 1942, was found still to be the most effective agent in our experiments. 5-ASA is considered to bear the anti-inflammatory moiety of the molecule (Khan et al. 1977, Ahnfelt-Ronne and Nielsen 1987), but there remain additive effects. Its augmented efficacy can be attributed to the antibiotic action of the sulfapyridine group. Since SUL is limited in its use by several adverse reactions, further agents have been synthesized by coupling 5-ASA to other molecules.

BEN was first applied by Bartalsky in 1982. It is made by linking the 5-ASA molecule to p-aminobenzoic acid, thus contains an diazo bond which itself may support free radical scavenging (Niel et al. 1987). Furthermore it is slowly resorbed until bacterial reductive degradation occurs and thus reaches the lower intestinal tract in effective doses.

OLS (Willoughby et al. 1982) consists of two azo-linked 5-ASA molecules; its protective action came nearest to the effect of SUL. Here again the Azo coupling may exert the above mentioned amplifying effect. As a further action of OLS an inhibition of xanthine oxidase is described (Carlin et al. 1985).

Altogether the products with a 5-ASA group turned out to be very effective not only in suppressing the increase of TBARS or MLS, but also in reducing the generation of GSSG and in diminishing the decrease of GSH. This pronounces the free radical scavenging effect of the related drugs.

The increase of MPO after MAO is an important indication to neutrophil immigration into tissue. It occurred by a factor of more than ten in the operated animals. This reaction could scarcely be suppressed by MES, as also was found in former experiments (Augustin et al. 1992). Here very high levels of the respective agent in tissue seem to be necessary which could not be reached in all cases. The most effective substance was OLS, though not significantly different from BEN and SUL in a statistical manner. OLS as a dimer of 5-ASA might have released the highest local effective concentration of the tested agents.

REFERENCES

Ahnfelt-Ronne I, Nielsen OH, 1987. The anti-inflammatory moiety of sulfasalazine, 5-aminosalicylic acid, is a radical scavenger. Agents Actions 21, 191-194

Allgayer H, Gugler R, Böhme P, Schmidt M, 1991. Is radical scavenging necessary in the treatment of inflammatory bowel disease? Gastroenterology 100, 581-582

Augustin AJ, Lutz J, 1991. Intestinal, hepatic and renal production of thiobarbituric acid reactive substances and myeloperoxidase activity after temporary aortic occlusion and reperfusion. Life Sciences 49, 961-968

Augustin AJ, Goldstein RK, Milz, J, Lutz J, 1992. Influence of anti-inflammatory drugs and free radical scavengers on intestinal ischemia induced oxidative tissue damage. Advanc. Exp. Med. Biol. 316, 239-251

Bartalsky A, 1982. Salicylazobenzoic acid in ulcerative colitis. Lancet 1982, 1, 960

Carlin G, Djursäter R, Smedegard G, Gerdin G, 1985. Effect of anti-inflammatory drugs on xanthine oxidase induced depolymerisation of hyaluronic acid. Agents Actions 16, 377-384

Ceconi C, Cargnoni A, Pasini E, Condorelli S, Curello S, Ferrari R, 1992. Evaluation of phospholipid peroxidation as malondialdehyde during myocardial ischemia and reperfusion injury. Am. J. Physiol. 260, H 1057-1061

Esterbauer H, Cheeseman KK, 1990. Determination of aldehydic lipid peroxidation products, malondialdehyde and 4-hydroxynonenal. In: Methods in Enzymology. Free Radicals and Metal Ions in Human Disease. (Eds. L.Packer and A.N. Glarer) New York. Academic 190, 407-421

Fantini GA, Yoshioka T, 1992. Use and limitations of thiobarbituric acid reaction to detect lipid peroxidation. Am. J. Physiol. 263, H981-H982

Griffith, OW, 1980. Determination of glutathione and glutathione disulfide using glutathione reductase and 2-vinylpyridine. Anal. Biochem. 106, 207-212

Janero DR, Burghardt B, 1988. Analysis of cardiac membrane phospholipid peroxidation kinetics as malondialdehyde: nonspecificity of thiobarbituric acid-reactivity. Lipids 23, 452-458

Khan KA, Piris J, Truelove SC, 1977. An experiment to determine the active therapeutic moiety of sulfasalazine. Lancet 1977 II, 892-895

Klotz U, Maier KE, Fischer C, Bauer KH, 1985. A new slow-release form of 5-aminosalicylic acid for the oral treatment of inflammatory bowel disease. Biopharmaceutic and clinical pharmacokinetic characteristics. Arzneimittel-Forsch. 35, 636-639

Krawisz JE, Sharon P, Stenson WF, 1984. Quantitative assay for acute intestinal inflammation based on myeloperoxidase activity. Gastroent. 87, 1344-1350

Kvietys P, Griesham MB, 1991. Is radical scavenging necessary in the treatment of inflammatory bowel disease? Reply. Gastroenterology 100, 582

Lutz J, Augustin AJ, Friedrich E, 1990. Severity of free oxygen radical effects after ischemia and reperfusion in the intestinal tissue and the influence of different drugs. Advanc. Exp. Med. Biol. 277, 683-690

Lutz J, Augustin AJ, Purucker E, Milz J, 1992. Combination of treatment with perfluorochemicals and free radical scavengers. Biomat Art Cells Immob. Biotech. 20, 951-958

Niel TM, Winterbourn CC, Vissers MCM, 1987. Inhibition of degranulation and superoxide production by sulfasalazine. Biochem. Pharmac. 36, 2765-2768

Ohkawa H, Ohishi N, Yagi K, 1979. Assay for lipid peroxides in animal tissues by thiobarbituric acid reaction. Anal. Biochem. 95, 351-358 (1979)

Svartz N, 1942. Salazopyrin: A new sulfanilamide preparation. Acta Med. Scand. 60, 577-598

Willoughby CP, Aronson JR, Agback H, Bodin NO, Truelove SC, 1982. Distribution and metabolism in healthy volunteers of disodium azodisalicylate, a potent therapeutic agent for ulcerative colitis. Gut 23, 1081-1087

GASTRIC INTRAMUCOSAL ACIDOSIS DURING WEANING FROM MECHANICAL VENTILATION

Z. Mohsenifar

Division of Pulmonary Medicine
Cedars-Sinai Medical Center
UCLA School of Medicine
8700 Beverly Boulevard
Los Angeles, California 90048

INTRODUCTION

Hussain and Roussos[1] showed that during endotoxic shock, respiratory muscle blood flow in spontaneously breathing dogs increases significantly from baseline values, and in the animals who are mechanically ventilated, respiratory muscle perfusion decreases to control values. Similarly, blood flow in organs like brain, gut, and skeletal muscles, in the mechanically ventilated group was significantly higher than the spontaneously breathing group. Therefore, during mechanical ventilation, a fraction of the cardiac output used by the respiratory muscles can be made available to the other organs and, conversely, during spontaneous ventilation, organs such as skeletal muscles, brain and gut fall under the disadvantage of losing out to respiratory muscles. The magnitude of increase in respiratory blood flow in the spontaneously breathing group was from 51 ml/min to 101 ml/min at 60 minutes of shock. Mohsenifar et al.[2] tested the hypothesis that gastric intramucosal pH can be used as a trend indicator of gut and/or splanchnic ischemia in patients with chronic obstructive lung disease, respiratory failure, during weaning from mechanical ventilation. This chapter will summarize the data on the role of gastric intramucosal pH as an indicator of weaning success or failure in these patients.

Fidian-Green[3,4] reported that during multiple organ failure, one could measure gastric intraluminal CO_2 and calculate gastric intramural pH using the Henderson-Hasselbalch equation. He felt that intramural pH could be used as a predictive indicator, since the splanchnic circulation is affected during hemodynamic stress due to redistribution of blood flow secondary to endogenous vasoconstrictors. At times, this inadequate regional tissue perfusion and oxygenation may occur despite an apparently adequate whole body systemic oxygen transport. In a different study, Fidian-Green et al.[5] demonstrated that in patients who developed evidence of ischemic colitis, pH_I can be used as an indicator of sigmoid ischemia during the intraoperative period. The issue of weaning from mechanical ventilation is a timely one; it is important to predict success or failure of weaning. With ideal predictors, one conceivably can save many patient days in the intensive care units which is very important in this day and age of cost-savings. On the other hand, predicting weaning failure is, again, important. At times, patients go through a very traumatic experience of being reintubated. Despite at least one major recent publication regarding the use of traditional weaning parameters[6], little has entered the critical care literature concerning good or poor prognosticators of weaning from mechanical ventilation. The issue of regional tissue oxygenation as measured by gastric pHI has never been addressed as a predictor for weaning from mechanical ventilation. Yet, weaning is a form of exercise during which respiratory muscles undergo severe stress, particularly in patients with chronic obstructive lung disease. At the same time, the cardiovascular system is responsible to provide the metabolites to a system which may have a rise in its VO_2 by as high as 25-30%[7,8]. As shown in Figure 1, patients with lung disease and adequate oxygen transport will have a decrease in diaphragmatic blood flow when the patient is on full assisted mechanical ventilation. This patient, when undergoing weaning trials, will demonstrate an increase in diaphragmatic blood flow to support 25 to 100% increase in oxygen consumption that is needed to deal with the excess work of breathing. Patients with adequate oxygen delivery will, perhaps, deal with this increase in demand with no difficulty. However, if the patient has inadequate oxygen delivery or borderline oxygen delivery, one will have to divert the blood flow from the splanchnic bed to meet the goals of the respiratory muscles (Figure 2). Over the years, this issue has been completely ignored, and there has been no attention paid to oxygen transport in patients who undergo weaning trials. Mohsenifar et al.[2] studied 29 patients who were intubated due to chronic obstructive lung disease, muscular weakness, and/or pneumonia. During the stages of weaning, they measured conventional weaning parameters and blood gases. In addition, they sampled the gastric juice and measured the gastric CO^2 and derived gastric intramucosal pH

TABLE 1[+]

VARIABLES ON ASSISTED MECHANICAL VENTILATION

	SUCCESSFULLY WEANED (n=18)	FAILED WEANING (n=11)	DIFFERENCE (95% CI)	P-VALUE
Age	69 ± 13	61 ± 20	8 (-10 to 12)	.561
Days on mechanical ventilation	6.2 ± 8	7 ± 4	.8 (-2 to 6)	.589
FIO2 (%)	35 ± 8	40 ± 7	5 (-6 to 7)	.426
Respiratory (resp/min)	16 ± 4	17 ± 6	1 (-2 to 5)	.483
Tidal volume (ml)	626 ± 105	608 ± 120	18(-105 to 69)	.660
Systolic blood pressure (mmHg)	128 ± 24	140 ± 22	12 (-7 to 31)	.211
Heart rate	96 ± 18	95 ± 17	1 (-15 to 13)	.891
Arterial pCO_2 (mmHg)	39 ± 7	40 ± 14	1 (-7 to 9)	.788
Arterial PO_2 (mmHg)	95 ± 21	93 ± 29	2 (-14 to 16)	.679
Gastric pCO_2 (mmHg)	37 ± 12	49 ± 23	12(-1.3 to 25)	.075
Gastric intramural pH	$7.45 \pm .13$	$7.36 \pm .20$	.09(-.24 to .01)	.073

Mean $\pm$ 1 SD
CI = Confidence interval; FIO_2 = Inspired oxygen tension

Table 1 demonstrates characteristics of the patients on assisted mechanical ventilation in both groups.

[+]From Annals of Internal Medicine, by permission[2]

TABLE 2[+]

SUCCESSFULLY WEANED (n = 18)

	DURING MECHANICAL VENTILATION	MEAN DURING WEANING	DIFFERENCE (95% CI)	P-VALUE
Respiratory rate	16 ± 4	22 ± 6	6 (2 to 10)	.002
Tidal volume (ml)	626 ± 105	494 ± 16	-132 (-191 to -73)	.002
Systolic blood pressure (mmHg)	128 ± 24	128 ± 26	.1 (-7 to 7)	.988
Heart rate	96 ± 18	97 ± 20	1 (-4 to 6)	.723
Arterial pCO_2 (mmHg)	39 ± 7	41 ± 10	2 (-.2 to 4)	.082
Arterial PO2 (mmHg)	95 ± 21	93 ± 18	-2 (-5 to 7)	.823
Arterial pH	7.48 ± .05	7.46±.05	-.02 (-.03 to .02)	.280
Gastric pCO_2 (mmHg)	37 ± 12	36 ± 12	-1 (-4 to 2)	.434
Gastric intramural pH	7.45 ± .13	7.46±.12	.01 (-.01 to .03)	.290

Mean ± 1 SD
CI = Confidence intervals; MV = Mechanical ventilation

Table 2 demonstrates changes in physiologic variables during weaning from successful group.

[+]From Annals of Internal Medicine, by permission[2]

via Henderson-Hasselbalch equation. They calculated gastric pH_I through the Henderson-Hasselbalch equation, which is:

$$6.1 + \log \text{bicarbonate/gastric } PCO_2 \times .03$$

where bicarbonate is the bicarbonate concentration obtained from the blood gases and gastric PCO_2 is measured directly. The hypothesis of this measurement is that the gastric juice CO_2 is the same as tissue CO_2, and as ischemia develops in the splanchnic region, hydrogen protons are generated and react with the tissue bicarbonate which results into excessive amounts of CO_2. The CO_2 will diffuse through transgastric membrane inside the gastric juice which is sampled. This method has been validated by Fidian-Green et al. and Cunningham et al.[9,10] In this study, there were 18 patients who were successfully weaned and 11 patients who failed weaning. As shown in Table 1, the characteristics of patients on assisted mechanical ventilation were very similar to that of patients who were weaning failures. As shown in Table 2, patients who were successfully weaned had a slight increase in their respiratory rate and reduction in tidal volume during weaning trials. The weaning trials entailed pressure support of 7-8 cmH_2O for twenty minutes. Arterial blood gases did not change during weaning trials in patients who were successfully weaned. Also, gastric PCO_2 and gastric intramucosal pH did not change. However, as shown in Table 2, in patients who failed weaning, tidal volume decreased significantly, and respiratory rate increased. As in the previous group, there were no changes in heart rate, arterial PCO_2, and PO_2; however, gastric PCO_2 increased from 49 to 111, and gastric intramucosal pH dropped from 7.36 to 7.09. Figure 3 demonstrates the individual data for gastric intramucosal pH in patients who are successfully weaned and in those patients who were weaning failures. As shown in Table 2 and Figure 3, a drop in gastric pH_I of more than .1 classifies 9 out of 11 unsuccessful cases and correctly identifies all 18 cases of successfully weaned group. Similarly, a rise in gastric PCO_2 of 10 mm or greater identifies 9 out of 11 unsuccessful cases and all 18 successful cases. The data has shown that a value of pH_I of greater than 7.30 or a change of less than .09 has 100% positive predictive value with confidence interval of 81-100 and a negative predictive value of 100% with confidence interval of 72-100. Many current studies investigating gastric intramucosal pH use gastric tonometry which requires up to 60 minutes of equilibration between gastric juice and CO_2 of the tonometer. Recently, Sun et al.[11] measured gastric intramural PCO_2 and pH in a model of anaphylactic shock. They demonstrated that the gastric PCO_2 response time was 52 seconds. Although in the ICU setting, gastric tonometry is a reasonable way of looking at overall splanchnic ischemia, detecting real time changes in gastric

TABLE 3[+]

FAILED WEANING (n = 11)

	DURING MECHANICAL VENTILATION	MEAN DURING WEANING	DIFFERENCE (95% CI)	P-VALUE
Respiratory	17 ± 6	28 ± 12	11 (4 to 18)	.006
Tidal volume	608 ± 120	431 ± 158	-177 (-284 to -70)	.004
Systolic blood pressure (mmHg)	140 ± 22	139 ± 31	-1 (-15 to 12)	.819
Heart rate	95 ± 17	98 ± 22	3 (-3 to 9)	.326
Arterial pCO_2 (mmHg)	40 ± 14	42 ± 12	2 (-.8 to 4)	.177
Arterial $PO2$ (mmHg)	93 ± 29	91 ± 27	-2 (-8 to 9)	.729
Arterial pH	7.49 ± .05	7.46±.03	-.03 (-.05 to .02)	.280
Gastric pCO_2 (mmHg)	49 ± 23	111 ± 62	62 (22 to 102)	.006
Gastric intramural pH	7.36 ± .20	7.09±.23	-.27 (-.42 to -.12)	.003

Mean ± 1 SD
CI = Confidence intervals; MV = Mechanical ventilation

Table 3 demonstrates changes in physiologic variables during weaning for the failed group

[+]From Annals of Internal Medicine, by permission[2]

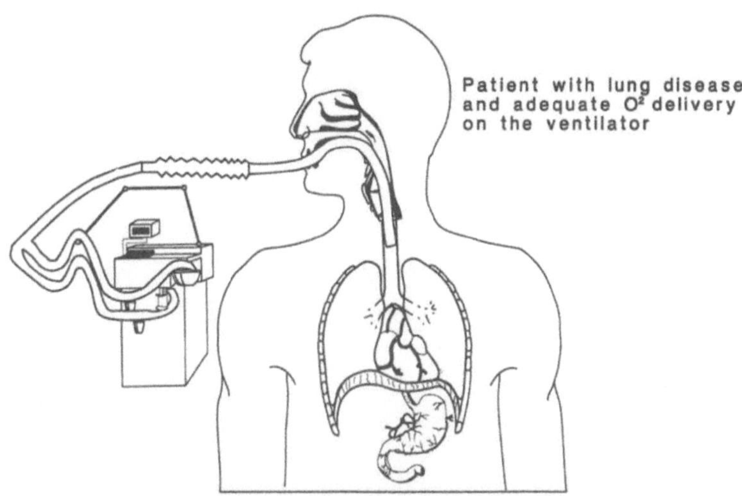

Figure 1 shows schematic demonstration of diaphragmatic and gastric blood flow in a patient on mechanical ventilation

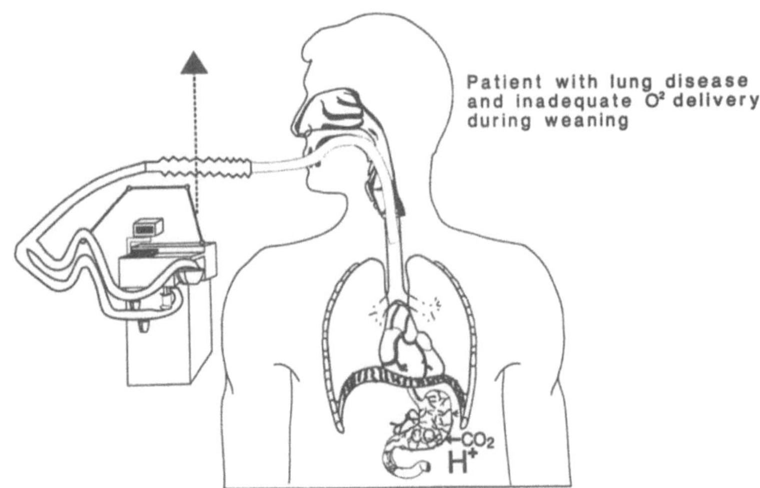

Figure 2 demonstrates a patient with lung disease and inadequate cardiac reserve who has, during weaning trial, demonstrated an increased blood flow to diaphragmatic muscle and a decrease in blood flow to splanchnic ischemia with resultant proton generation and CO_2 diffusion from splanchnic region to the gastric juice.

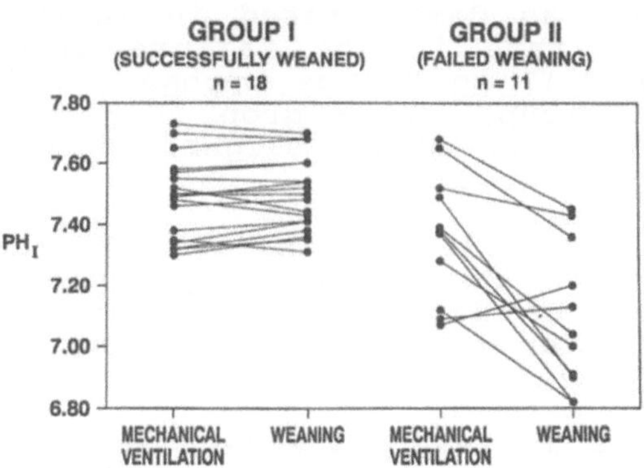

Figure 3 shows individual pH_I in patients who were successfully weaned and in those who were weaning failures.

CO_2 is essential in critically ill patients. One additional factor is that all of these patients have to be on H2-receptor blockers in order to eliminate the proton secretion by the gastric mucosa.[12]

One of the main differences between the two groups of patients during weaning is that patients who failed weaning had a significant drop in their gastric intramucosal pH, and their initial intramucosal pH was lower. This phenomenon can be interpreted as indicating tissue hypoxemia. It is unlikely that the weaning increases metabolic demands of the stomach and, most likely, it implies a reduction in the gastric blood flow. The possible reasons for the fall in gastric blood flow could be the following:

1. Weaning requires an increase in respiratory muscle blood flow 25-50%, but this cannot be the only factor since it should be seen in both successful and failure groups.

2. The failure group may have needed a higher respiratory blood flow. In this case, they should have shown some evidence of increased work of breathing. Although not significant, as demonstrated in Table 2, the rise in respiratory rate in the failure group was slightly higher, from 17 to 28, as opposed to 16 to 22 in the successful group, and a drop in tidal volume in the failure group was, again, slightly more significant than the successful group.

3. The group that failed had inadequate cardiac function and could not meet the increased O_2 demand requirements by increasing cardiac output. This was primarily responsible for both the respiratory pump failure and gastric hypoxemia.

Our hypotheses is that inadequate ability to increase O_2 transport and cardiac output to support the excessive work during weaning trials was primarily responsible for the redistribution of blood flow and splanchnic ischemia. Further evidence to support this is that three patients who failed weaning trials were put back on assisted mechanical ventilation, and gastric PCO_2 levels were repeated; the repeated numbers were similar to the numbers originally obtained on assisted mechanical ventilation.

In summary, gastric intramucosal pH measurement is an important indicator of tissue ischemia and hypoxemia and can be used as a sensitive means in predicting weaning failure or success, particularly in patients who have inadequate cardiac reserve.

REFERENCES

1. S.N.A. Hussain and C. Roussos, Distribution of respiratory muscle and organ blood flow during endotoxic shock in dogs, JAP 59:1802 (1985).
2. Z. Mohsenifar, A. Hay, J. Hay, M. Lewis, S.K. Koerner, Gastric intramural pH as a predictor of success or failure in weaning patients from mechanical ventilation, Annals of Internal Medicine (in press).
3. R.G. Fiddian-Green, Hypotension, splanchnic hypoxia and arterial acidosis in ICU patients, Circ Shock 21:326 (1987).
4. R.G. Fiddian-Green, Studies in splanchnic ischemia and multiple organ failures, in: "Splanchic Ischemia and Multiple Organ Failure", A. Martson, G.B. Bulkley, R.G. Fiddian-Green, U. Haglund, editors, Edward Arnold, London/C.V. Mosby Company, St. Louis, Missouri (1989).
5. R.G. Fiddian-Green, P.M. Amelin, J.B. Herrmann, E. Arous, B.S. Cutler, M. Schiedler, H.B. Wheeler, S. Baker, Prediction of the development of sigmoid ischemia on the day of aortic operations, Arch Surg 121:654 (1986).
6. K.L. Yang, M.J. Tobin, A prospective study of indexes predicting the outcome of trials of weaning from mechanical ventilation, NEJM 324:1445 (1991).
7. R.D. Hubmayr, L.M. Loosbrock, D.J. Gillespie, J.R. Rodarte, Oxygen uptake during weaning from mechanical ventilation, Chest 94:1148 (1988).
8. S. Field, S.M. Kelly, P.T. Macklem. The oxygen cost of breathing in patients with cardiorespiratory disease, ARRD, 126:9 (1982).
9. R.G. Fiddian-Green, E. McGough, G. Pittenger, E. Rothman, Predictive value of intramural pH and other risk factors for massive bleeding from stress ulceration, Gastroenterology 85:613 (1983).
10. J.A. Cunningham, C.D. Cousar, J.H. Jaffin, J.W. Harmon. Extraluminal and intraluminal pCO_2 levels in the ischemic intestines of rats, Curr Surg 44:229 (1987).
11. S. Sun, M.H. Weil, W. Tang, R.J. Gazmuri, V. Desai, Gastric intramural pCO_2 as an indicator of organ ischemia during anaphylactic shock, Clin Research 39(3):708A (1991).
12. S.O. Heard, C.M. Helsmoortel, J.C. Kent, A. Shahnarian, M.P. Fink, Gastric tonometry in healthy volunteers: Effect of ranitidine on calculated intramural pH, Crit Care Med 19(2):271 (1991).

ALTERATIONS IN ERYTHROCYTE DEFORMABILITY UNDER HYPOXIA: IMPLICATIONS FOR IMPAIRED OXYGEN TRANSPORT

Joseph M. Rifkind and Omoefe O. Abugo

National Institutes of Health
National Institute on Aging
Gerontology Research Center
Baltimore, MD 21224

INTRODUCTION

Oxygen delivery to the tissues requires that the erythrocyte be highly deformable in order to pass through the narrow pores of the capillary bed. It has been shown that oxidative damage produces crosslinking of hemoglobin with membrane proteins (Shaklai et al., 1987; Snyder et al., 1985; Reinhart et al., 1986) resulting in a more rigid, less deformable membrane. Hypoxia as a source for oxidative damage and thereby a possible loss in erythrocyte deformability is based on recent studies which indicated enhanced autoxidation and superoxide formation (Rifkind et al., 1989, 1991) within erythrocytes at intermediate oxygen pressures.

In this paper we investigate the effect of hypoxic incubation on erythrocytes. It is demonstrated that the oxidative stress at intermediate oxygen pressures produces membrane damage associated with enhanced lysis and membrane protein crosslinking. The membrane damage is also shown to result in decreased deformability.

MATERIALS AND METHODS

Blood was obtained from BLSA program participants at the Gerontology Research Center. Blood was washed twice in PBS buffer pH 7.4 containing 0.1 mM EDTA to remove buffy coat and plasma.

Deoxygenation of Red Cells

5% hct red blood cell suspensions were deoxygenated in a Labconco controlled atmosphere glove box (Labconco Corp., Kansas City, MO) to desired O_2 pressure. Orbisphere pO_2 electrode was used to monitor pO_2 in the glove box. At each desired pressure, red cells were gently rocked for one hour to equilibrate with the glove box pO_2. Red cells were then transferred to specially designed spectroscopic cells and sealed with ground glass tops. Five samples were prepared at each of four oxygen pressures - 159 mmHg, 15.7 mmHg, 6.4 mmHg and 0.2 mmHg. Immediately after the deoxygenation, measurements were made (day zero). Remaining samples were incubated at room temperature, and further measurements made on days one, two, three and six. One set of spectroscopic cells was opened to the atmospheric oxygen on each day for determinations of lysis, hemoglobin oxidation, deformability and crosslinking of membrane proteins.

Lysis and Hb Oxidation

Red cell suspensions were taken out of spectroscopic cells, and spectra determined in 1 mm path length spectroscopic cells using the Perkin-Elmer Lambda 6 UV-Visible spectrophotometer equipped with a scattering accessory. The spectrum of the red cell suspension from 490 to 640 nm was compared to that of the supernatant obtained after centrifugation at 3000 rpm for 6 minutes. The quantities of hemoglobin present in the red cell suspension and supernatant were determined by multi-component fitting of acquired spectra. Percent lysis was calculated as the ratio of hemoglobin species in the supernatant compared to that in the suspension. Oxidation was the percent methemoglobin in the supernatant (lysate) determined by the multi-component fitting of spectra.

Deformability Measurements

The cell transit analyzer, CTA (ABX, Montpellier, France), was used for deformability measurements. From incubated cells, 0.1% hct red cell suspensions were prepared in PBS buffer pH 7.4 containing 0.1 mM EDTA. The cell transit times were then determined as previously described (Koutsouris et al., 1988), using a 5 μ polycarbonate filter containing 30 pores, at 5 cm H_2O pressure. The transit time is determined by the duration of the drop in electric conductance across the filter when a red cell is in one of the pores. For each measurement, at least 2,000 cells were analyzed to obtain a mean cell transit time (MCTT). From the distribution of the transit times it was possible to determine the percent of red cells with relatively longer transit times. It was also possible to determine FR, the filtration rate (i.e. the number of cells passing through the pore per second). The filter was always cleaned by sonication between measurements. For these measurements, the same filter was used to eliminate inter-filter variability (Koutsouris et al., 1989).

Membrane Protein Crosslinking

Red cells were lysed in 30 volumes of 5 mM sodium phosphate buffer pH 8.0 at the end of each day's measurements, and membranes separated from lysate. White ghosts were prepared by washing membranes twice in 30 volumes of 5 mM sodium phosphate buffer pH 8.0. White ghosts were then stored at -20°C until all measurements were complete. Membrane proteins at the different pressures for the different days were then resolved using SDS-PAGE electrophoresis, on a pre-cast 4-12% tris-glycine gradient gel (Novex, San Diego, CA). All buffers used for the electrophoresis were non-reducing so that crosslinking of proteins could be observed. Gels were stained using the Coomassie blue stain. Gels were then scanned with the model 1650 scanning densitometer (Bio-Rad, Hercules, CA) so as to obtain densitograms for the SDS-PAGE profiles of each sample.

RESULTS AND DISCUSSION

Lysis and Oxidation

Under sterile conditions it was found that six days of incubation was necessary for lysis to reach 5% of the hemoglobin. At this time, a comparison was made of the percent lysis for samples incubated at different partial pressures of oxygen (Table 1). Intermediate pressures of 15.7 mmHg and 6.4 mmHg were found to undergo greater lysis, with the highest levels of lysis produced at 15.7 mmHg.

Table 1 LYSIS AND OXIDATION ASSOCIATED WITH HYPOXIC INCUBATION FOR SIX DAYS

Oxygen Pressure (mmHg)	Lysis (%)	Oxidation of Hemoglobin in Hemolysate (%)
159	5.6	13.4
15.7	7.6	39.9
6.4	6.1	21.4
0.2	5.5	7.2

Within the cells, oxidation of hemoglobin is minimized by the enzymatic processes which reduce oxidized hemoglobin. However, in the hemolysate it was possible to quantitate the oxidative processes by measuring hemoglobin oxidation. Also shown in Table 1 are the results for hemoglobin oxidation which indicate a correspondence between the lysis and hemolysate oxidation, with the highest level of oxidation also at 15.7 mmHg.   These results confirm the previously reported finding that hemoglobin autoxidation and thereby superoxide production is enhanced at intermediate oxygen pressures. A relationship between these oxidative processes and membrane damage is implicated by the coincident increase in lysis and increase in oxidation (Table 2).

Membrane Crosslinking

It has been shown that oxidative damage to the membrane can result in crosslinking reactions between membrane proteins and with cytoplasmic proteins (Shaklai et al., 1987; Snyder et al., 1985; Reinhart et al., 1986). Figure 1 shows densitometric scans of SDS-PAGE for the erythrocyte membrane proteins obtained from erythrocytes incubated at 15.7 mmHg of O_2 for various times. In this figure it can be seen that a band of slightly higher molecular weight than spectrin band 1 is gradually formed during incubation. By day six this band is clearly resolved.

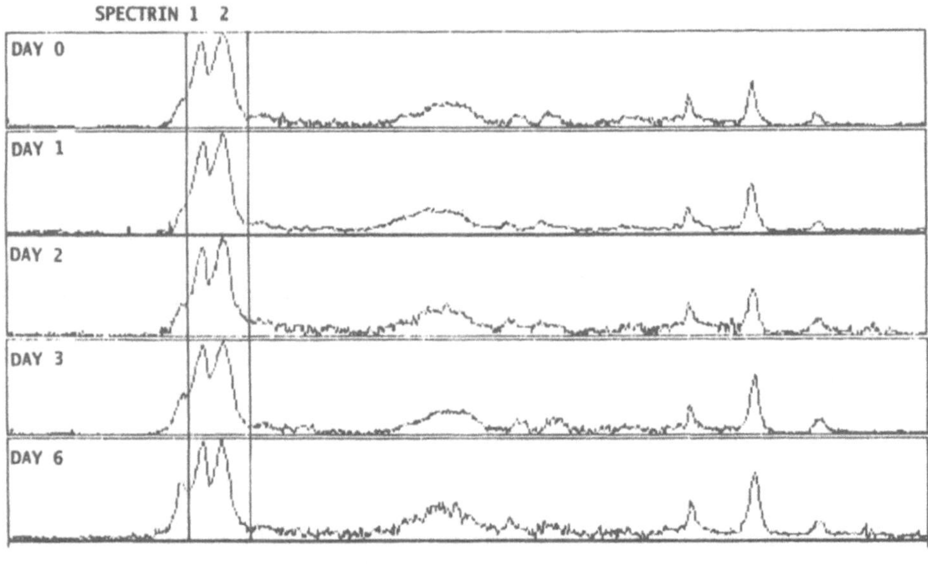

Figure 1. Densitometric scans of the SDS-PAGE profile of erythrocyte membranes obtained from erythrocytes incubated at 15.7 mmHg for various periods of time. The vertical lines indicate the regions used for integration to determine relative amounts of crosslinked material (See Table 2).

The high molecular weight crosslinked material was quantitated by integrating the dye absorbance from the densitometric scans for material with molecular weight greater than spectrin relative to all the absorbance through the region containing spectrin band 1 and band 2 (Figure 1). Table 2 shows the quantitation of this high molecular weight crosslinked material after incubation at various oxygen pressures for both day 3 and day 6. The highest level of crosslinked material is again found at 15.7 mmHg, with the levels decreasing at both higher and lower partial pressures of oxygen. The same trend found for crosslinking, lysis and oxidation suggests a relationship between crosslinking and oxidative processes in the cell.

The dominant band observed in Figure 1 corresponds to the 255,000 molecular weight band found when hemoglobin is crosslinked to the α-band of spectrin (Fortier et al., 1988; Snyder et al., 1985; Sauberman et al., 1983). This crosslinked material has been shown to be produced by the addition of H_2O_2, during cellular aging and in a number of abnormal situations (Sauberman et al., 1983). Furthermore, a relationship has been established between this membrane crosslinking and red cell membrane rigidity (Fortier et al., 1988).

Table 2 HIGH MOLECULAR WEIGHT CROSSLINKED MATERIAL
AFTER HYPOXIC INCUBATION

Oxygen Pressure (mmHg)	3 Day Incubation (%)	6 Day Incubation (%)
159	16.5	19.4
15.7	18.1	22.8
6.4	17.9	16.1
0.2	15.0	15.2

Erythrocyte Deformability

In order to determine the effect of hypoxic incubation on deformability , a cell transit analyzer, which measures the time required for each cell to pass through a 5 μ pore, was used. From these measurements three separate measures of deformability are provided. 1) The mean cell transit time (MCTT) is the mean for the whole population of cells measured (usually 2,000 cells); 2) Percent slow cells is the percent of the cells with a transit time greater than 2.1 msec at a pressure of 5 cm. The slow cells may actually control the flow through the capillary bed by limiting the flow of the more deformable cells. 3) A third measure of deformability is the filtration rate (FR). When results are compared at the same hematocrit, the FR is a measure of the relative number of cells which successfully enter and pass through the pores. Figure 2 shows that all three measures of deformability indicate impaired deformability at the intermediate oxygen pressure of 15.7 mmHg. Thus, the MCTT and the percent slow cells were highest while there is a decrease in the relative number of cells which successfully pass through the pores.

Similar conclusions have also been suggested on the basis of the ratio of the discoid volume to that of a swollen volume. This ratio provides a measure of the excess surface area required to deform and pass through the capillary bed. Here again it was found that the cells after hypoxic incubation are more spherical with less excess surface area (Rifkind et al., 1992).

The loss in deformability under hypoxic conditions can be attributed to the enhanced autoxidation (Levy et al., 1988) and the resultant membrane damage (Rifkind et al., 1991) leading

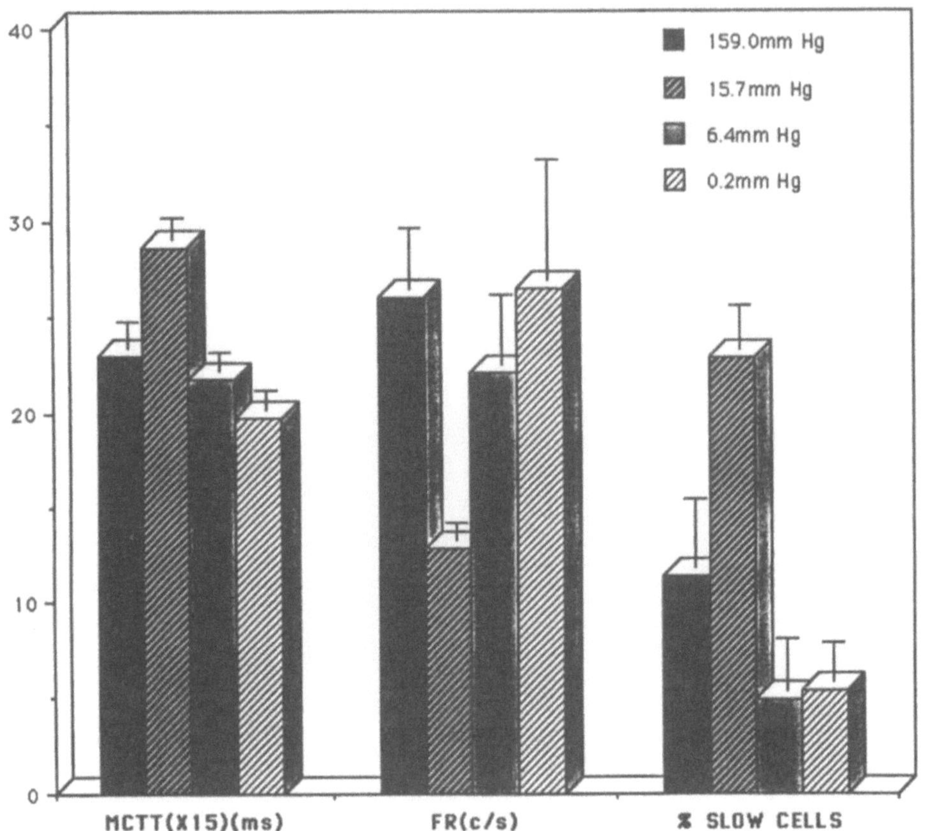

Figure 2. The effect of incubation for three days at various partial pressures of oxygen on three different measures of deformability.

to crosslinking between hemoglobin and membrane proteins (Reinhart et al., 1986; Fortier et al., 1988) which increase the membrane rigidity and thereby lead to a decrease in deformability.

Our studies involve in vitro incubation for extended periods of time; however, the changes observed are analogous to those found during aging of rats (Rifkind et al., 1992; Abugo et al., submitted). In those studies it was found that in old rats the deformability decreases and the lysis, which reflects oxidative processes, increases. These results suggest the possibility that hypoxic stress during aging may contribute to these erythrocyte changes. The possible functional impairment associated with decreases in deformability produced by hypoxia is currently being investigated.

REFERENCES

1. N. Shaklai, B. Frayman, N. Fortier, and M. Snyder, Crosslinking of isolated cytoskeletal proteins with hemoglobin: a possible damage inflicted to the red cell membrane. *Biochim. Biophys. Acta.* 915:406 (1987).
2. L.M. Snyder, N.L. Fortier, J. Trainor, J. Jacobs, L. Leb, B. Lubin, D. Chiu, S. Shohet, and N. Mohandas, Effect of hydrogen peroxide exposure on normal human erythrocyte deformability, morphology, surface characteristics, and spectrin-hemoglobin cross-linking. *J. Clin. Invest.* 76:1971 (1985).
3. W.H. Reinhart, L.A. Sung, and S. Chien, Quantiative relationship between Heinz body formation and red blood cell deformability. *Blood* 68:1376 (1986).
4. N. Fortier, L.M. Snyder, F. Garver, C. Kiefer, J. McKenney, and N. Mohandas, The relationship between in vivo generated hemoglobin skeletal protein complex and increased red cell membrane rigidity. *Blood* 71:1427 (1988).
5. J. Rifkind, L. Zhang, J.M. Heim, and A. Levy, The Role of Hemoglobin in Generating Oxyradicals, *in*: "Oxygen Radicals in Biology and Medicine," M.G. Simic, K.A. Taylor, J.F. Ward, and C. von Sonntag, eds., Plenum Publishing Corporation, New York (1989).
6. J.M. Rifkind, L. Zhang, A. Levy, and P.T. Manoharan, The hypoxic stress on erythrocytes associated with superoxide formation. *Free Rad. Res. Comms.* 12:645 (1991).
7. D. Koutsouris, R. Guillet, C. Lelievre, M.T. Guillemin, P. Bertholom, Y. Beuzard and M. Boynard, Determination of erythrocyte transit times through micropores. I. Basic operational principles. *Biorheology* 25:763 (1988).
8. D. Koutsouris, R. Guillet, R.B. Wenby, ănd H.J. Meiselman, Determination of erythrocyte transit times through micropores. II. Influence of experimental and physiochemical factors. *Biorheology* 26:881 (1989).
9. L.M. Snyder, F. Garver, S.C. Liu, L. Leb, J. Trainor, and N.L. Fortier, Demonstration of haemoglobin associated with isolted, purified spectrin from senescent human red cells. *Brit. J. Haematol.* 61:415 (1985).
10. N. Sauberman, N.L. Fortier, W. Joshi, J. Piotrowski, and L.M. Snyder, Spectrin haemoglobin crosslinkages associated with *in vitro* oxidant hypersensitivity in pathologic and artificially dehydrated red cells. *Brit. J. Haematol.* 54:15 (1983).
11. J.M. Rifkind, O. Abugo, and J. Heim, A possible relationship between decreased filterability of erythrocytes in old subjects and oxyradical damage. Abstracts, 45th Annual Meeting of the Gerontological Society of America, Washington, D.C., November 1992.
12. A. Levy, L. Zhang, and J.M. Rifkind, Hemoglobin: A Source of Superoxide Radical under Hypoxic Conditions, *in*: "Oxy-Radicals in Molecular Biology and Pathology," Alan R. Liss, New York (1988).
13. O. Abugo, J. Heim, and J.M. Rifkind, Effect of aging on the deformability of rat red blood cells. Submitted for publication.

THE HALDANE EFFECT UNDER DIFFERENT ACID-BASE CONDITIONS IN PREMATURE AND ADULT HUMANS

H Kalhoff*, F. Werkmeister, H. Kiwull-Schöne, L. Diekmann*, F. Manz**, and P. Kiwull

Department of Physiology, Ruhr-University, 44780 Bochum,
Pediatric Clinic*, 44137 Dortmund, and
Research Institute of Child Nutrition**, 44225 Dortmund, Germany

INTRODUCTION

The Haldane effect (HE) is characterized by the binding of hydrogen ions and CO_2 accompanying deoxygenation of hemoglobin (Hb) due to a negative heterotropic allosteric ligand interaction (Bauer, 1974; Siggaard-Andersen, 1974; Baumann et al., 1987). The binding sites of the oxygen-linked hydrogen ions have been mapped for adult hemoglobin (Perutz, 1970; Kilmartin, 1972), being in part responsible for the pH-differences between oxygenated and deoxygenated blood. The question arises whether structural differences of Hb, either among different mammalian species or during developmental life, may influence the quantity of the Haldane effect, at least within the limits of accuracy achieved by blood-gas and acid-base analysis in both basic research and clinical practice.

Comparative studies in rabbits, cats and dogs did not reveal significant differences in HE-induced pH-differences over a wide range of respiratory and metabolic acid-base conditions (Kiwull-Schöne et al., 1992). There are, however, striking differences between Haldane effect data of these species and those for adult human blood reported in the literature (v.Mengden et al., 1969; Siggaard-Andersen, 1974). Therefore, these human data may not be adequate for acid-base and blood-gas analysis in basic research on animal physiology.

Likewise, it is unclear whether the ontogenetic differences between the structures of adult and fetal Hb may result in Haldane effect variations of practical importance. In clinical practice, knowledge of the HE in terms of pH-difference between Hb and HbO_2 is important for indirectly estimating the base excess (BE) value from simultaneous measurements of pH and P_{CO_2} in samples with incomplete oxygen saturation.

The objective of our study was to find out, if HE-data valid for human adult blood are suitable for blood gas analysis in human prematures as well, in spite of their high percentage of fetal hemoglobin (HbF). Although HE-data for adult human blood are extensively available in the literature, due to their great variability, we performed own measurements in an adult control group.

Oxygen Transport to Tissue XVI
Edited by M.C. Hogan *et al.*, Plenum Press, New York, 1994

METHODS

Blood-gas and acid-base analysis was performed in a control group of 15 healthy adult volunteers (4 male, 11 female) and in an experimental group of 10 prematures (8 male, 2 female), their average percentage of HbF being about 80%. The mean gestational age ($\pm$ SEM) was 30.6 $\pm$0.3 weeks and birth weight 1354 $\pm$98 g, the mean postconceptional age was 33.6 $\pm$0.5 weeks and the actual weight 1569 $\pm$63 g. The premature infants were nurtured with human milk or standard premature formula (Prematil®, Milupa). Patients suffering from diseases involving renal or hepatic insufficiency and those with severe anemia or hypoxemia were excluded.

In the adult control group, a capillary sample was taken from the finger tip (50-100μl) using a heparinized glass capillary (D 941-10-100, Radiometer, Copenhagen). Shortly after this, venous blood was taken from a cubital vein (1.5-2.0 ml) under nearly anaerobic conditions using heparinized syringes not permeable for CO_2 (Qs90 syringes, Radiometer, Copenhagen; 120 IU Heparin/syringe). In the premature experimental group, after obtaining written parental consent, the venous blood samples were taken first, shortly before the capillary samples. All samples were either analyzed instantly or kept on ice for less than 2 hours before measurements.

From capillary samples, the total hemoglobin concentration (g/dl) was automatically determined together with partial concentrations of different Hb-derivates (COHb, MetHb, Hb, HbO_2). In the prematures, the proportion of HbF was additionally determined after full oxygenation. From the same sample, routine blood-gas and acid-base analysis was carried out (ABL 520, Radiometer, Copenhagen). The venous samples were investigated for pH, Pco_2 and bicarbonate concentrations by the equilibration technique (Astrup and Schrøder, 1956; Siggaard-Andersen and Engel, 1960), using a microtonometer/pH-meter unit (BMS2 MK2, Radiometer, Copenhagen) and 4 precision gas mixtures containing either 4% or 8% CO_2 in oxygen or nitrogen, respectively.

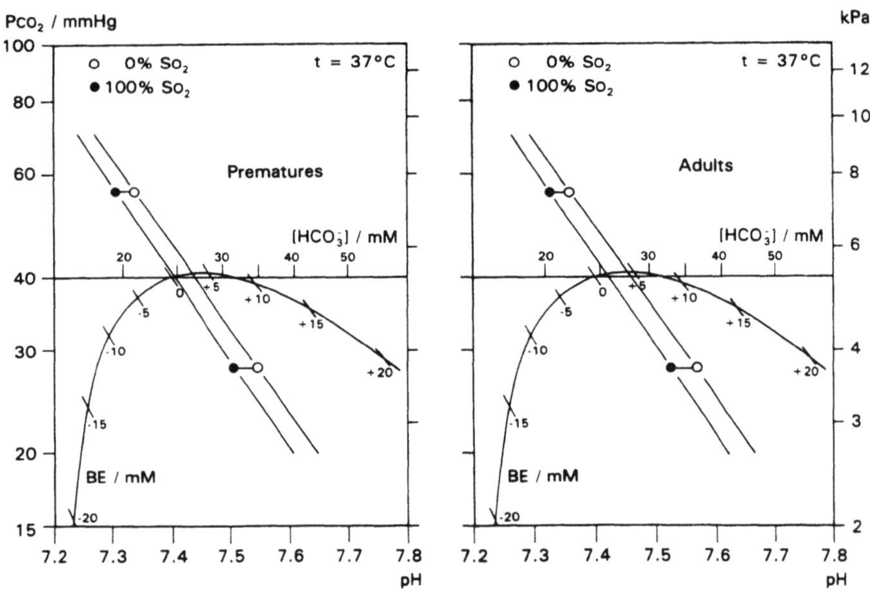

Figure 1. The Haldane effect (HE) in human blood under normal metabolic acid-base conditions. Each point represents the mean of 4 repeated pH-determinations in 24 and 15 blood samples of prematures and adults, respectively. Evaluation of the lgPco₂/pH-lines for slopes, for HCO₃⁻ concentrations at a Pco₂ of 40 mmHg, and for pH-shifts by deoxygenation, see Table 1.

The accuracy of the gas-mixtures was tested by tonometry of the 1:4 Sørensen phosphate buffer. Occasionally, plasma lactate concentrations were determined photometrically (Hitachi) by the enzymatic UV-method (Boehringer Test-Combination Lactate). To exclude systematic errors due to time-dependent anaerobic lactic acid accumulation, equilibration in nitrogen was kept short, with high and low Pco_2 in random sequence.

RESULTS

The Haldane effect under normal metabolic acid-base conditions

The $lgPco_2$/pH-diagrams in Fig. 1 show the average pH-values obtained by equilibration of the blood samples with 4% and 8% CO_2 in either O_2 or N_2. Under each condition, pH was determined 4-fold, so that any mean value in Fig. 1 (and Table 1) represents $(4\cdot15)$ 60 and $(4\cdot24)$ 96 pH determinations in 15 and 24 blood samples of adults and prematures, respectively. The HE was quantified by the pH-difference (ΔpH) of oxygenated and deoxygenated blood. Evaluation of these differences is shown by Table 1. It can be seen that the HE-induced ΔpH is inversely related to Pco_2. Due to the linear relationship between $lgPco_2$ and pH, ΔpH can be read for any level of Pco_2. Thus, under near standard conditions (Pco_2 = 40 mmHg, $[HCO_3^-st]$ ~ 24-25.5 mM), the HE-induced ΔpH was 0.036 $\pm$0.001 at a mean [Hb] of 14.0 $\pm$0.3 g/dl in the adults, being not significantly different from 0.038 $\pm$0.002 at a mean [Hb] of 13.8 $\pm$0.6 g/dl in the prematures. When normalized to [Hb] = 15 g/dl, these values reached 0.039 $\pm$0.001 and 0.041 $\pm$0.001, respectively. In the prematures the percentage of HbF was 78.4 $\pm$2.2 %.

Table 1. The Haldane effect in humans with normal metabolic acid-base status.

Variable	Condition	Prematures (n = 24)	Adults (n = 15)
$\Delta pH_{(deox-ox)}$	4% CO_2	0.042 $\pm$ 0.002	0.041 $\pm$ 0.001
	8% CO_2	0.034 $\pm$ 0.002	0.031 $\pm$ 0.001
$[HCO_3^-]$ mM	100% So_2	24.4 $\pm$ 0.3	25.6 $\pm$ 0.4
	0% So_2	26.6 $\pm$ 0.3	27.9 $\pm$ 0.4
$\beta = \Delta lg\, Pco_2/\Delta pH$	100% So_2	-1.51 $\pm$ 0.02	-1.55 $\pm$ 0.03
	0% So_2	-1.45 $\pm$ 0.02	-1.47 $\pm$ 0.02

Means $\pm$SEM of HE-induced pH-differences between deoxygenated and oxygenated blood at different levels of Pco_2, as well as bicarbonate concentrations $[HCO_3^-]$ at Pco_2=40 mmHg (5.3 kPa) and CO_2-buffering capacities (β) for complete and zero O_2-saturation. Data derived from measurements presented by Fig. 1.

The Haldane effect under different respiratory acid-base conditions

For both groups, prematures and adults, Fig. 2 shows an inverse linear correlation between the HE-induced ΔpH and the logarithm of the HCO_3^--concentration $[HCO_3^-]$, calculated for different levels of Pco_2 in a rather normal range of metabolic acid-base status (see Fig. 1). The average ΔpH $\pm$SEM at $[HCO_3^-]$ = 24 mM turned out to be quite similar in both groups, with 0.042 $\pm$ 0.001 in the adults and 0.041 $\pm$0.001 in the prematures. These values were significantly higher than 0.034 $\pm$0.001 as correspondingly to be calculated from the data of v.Mengden et al. (1969), Table 2.

Since the Haldane effect decreases with a rise in Pco_2 but increases with a rise in fixed acids (Siggaard-Andersen, 1974), the inverse relationship between ΔpH and $lg[HCO_3^-]$ has

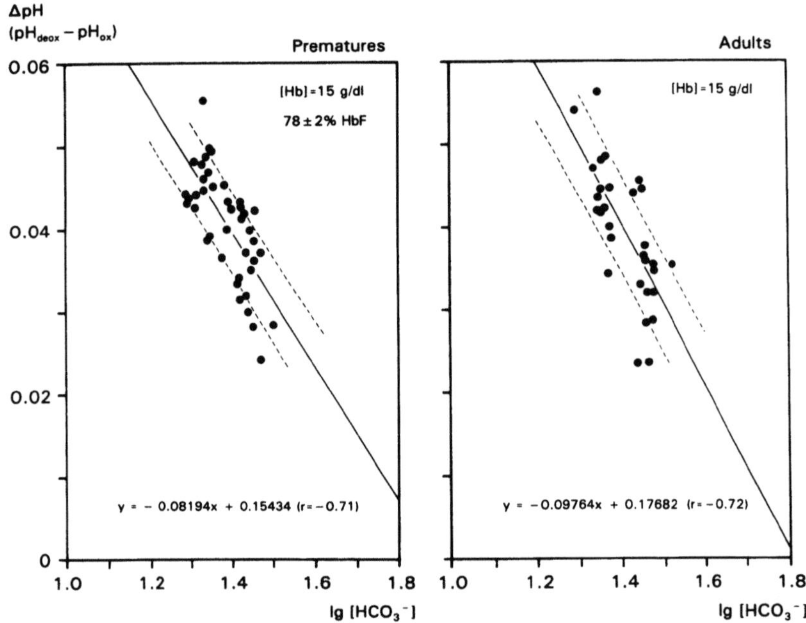

Figure 2. Comparison of the Haldane effect (HE) in premature and adult humans. Linear regression analysis was performed of the pH-difference (ΔpH) between deoxygenated and oxygenated blood as a function of lg[HCO_3^-]. ΔpH-values were initially normalized to 15 g/dl [Hb].

been found to be linear over a wide range of both respiratory and metabolic acid-base deviations in human adults (v. Mengden et al., 1969) as well as in rabbits, cats and dogs (Kiwull-Schöne et al., 1992). According to ΔpH = b·lg[HCO_3^-]+a, linear regression analysis revealed that both slopes (b) and intercepts (a) did not significantly differ between adults and prematures in our own measurements. If, however, the data of v.Mengden et al.(1969) were taken into account, their intercepts were significantly lower (Table 2). Including the actual Hb-concentration, a more general expression (Equation I)

Table 2. Regression analysis of the HE-induced ΔpH as function of lg[HCO_3^-].

	Underlying data	Slope $b \pm s_b$	Intercept $a \pm s_a$	$A \pm s_A$	$B \pm s_B$	r	n
(i)	Human adults, own measurements	-0.09764 ±0.01799	0.17682* ±0.02539	153.6* ±28.3	1.811 ±0.0423	-0.72	30
(ii)	Human prematures, own measurements	-0.08194 ±0.01286	0.15434* ±0.01786	183.1* ±28.7	1.884 ±0.367	-0.71	42
(iii)	Human adults, v.Mengden et al.(1969)	-0.06549 ±0.00558	0.12457 ±0.00791	229.1 ±19.5	1.902 ±0.202	-0.96	15

Regression characteristics, standard deviations (s) and correlation coefficients (r) of the linear relationship y = bx + a (y = ΔpH, x = lg[HCO_3^-]), calculated from n pairs of data, whereby ΔpH was normalized to [Hb] = 15 g/dl. Values for A and B were obtained as A = [Hb]/b and B = b/a.
*Significant difference compared to (iii); unpaired t-test, $P_D < 0.05$

can be achieved, characterized by the parameters B=b/a (i.e. slope/intercept) and A=[Hb]/a (Hb concentration/intercept). For simplicity, our HE-induced ΔpH values in Fig. 2 were normalized to [Hb] = 15g/dl before being subjected to regression analysis. The resulting values for A and B given by Table 2 are not significantly different between our groups of adult and premature humans. Again, the data of v. Mengden et al.(1969) differ from ours in that their value for A is significantly greater.

Correction procedures

Blood-gas and acid-base analysis under conditions of incomplete oxygenation needs consideration of the Haldane effect. If the HE-induced pH-difference between deoxygenated and oxygenated blood is neglected, several errors may influence the resulting P_{CO_2} or standard $[HCO_3^-]$, depending on whether the "indirect" equilibration technique with pH-measurement (Astrup and Schrøder, 1956) or direct measurement of P_{CO_2} and pH is performed.

Two customary procedures to estimate the HE-induced pH-difference between fully deoxygenated and fully oxygenated blood were proposed:

(I) $\qquad \Delta pH = (B - \lg[HCO_3^-]) \cdot [Hb]/A \qquad$ (v.Mengden et al., 1969)

(II) $\qquad \Delta pH \sim \Delta[BE] = 0.3 \cdot [Hb] \qquad$ (Siggaard-Andersen and Engel, 1960)

These procedures were tested for their goodness of prediction, compared to direct measurement. The test results are shown by Table 3 under standard conditions (P_{CO_2} = 40 mmHg, HCO_3^-st = 24 mM). HE-correction (II) for the base excess (BE) and actual [Hb] would mean in case of 14 g/dl [Hb] a Δ[BE] of 4.2 mmol/l, corresponding to a HE-dependent ΔpH of about 0.059. Comparatively, our measured ΔpH values reach only about 65% of this obviously overestimated HE in both adults and prematures.
The other proposed HE-correction (I), considering bicarbonate concentrations and actual [Hb], appears to be the optimal approach, if the presently determined values for A and B are used. However, this correction does reach only about 85% of the measured values when using A and B of v.Mengden et al.(1969).

Table 3. HE-induced pH-differences under standard blood-gas and acid-base conditions.

Procedure		ΔpH_{40}	
		Prematures	Adults
Direct measurement		0.038 ±0.002	0.036 ±0.001
Estimation with Equation I	(i)	0.039 ±0.002	0.037 ±0.001
	(ii)	0.038 ±0.002	0.037 ±0.001
	(iii)	0.031 ±0.002	0.031 ±0.001
Estimation with Equation III		0.058	0.059

ΔpH_{40} means the HE-induced pH-difference between deoxygenated and oxygenated blood at a $PaCO_2$ of 40 mmHg and a $[HCO_3^-st]$ of 24 mM, at actual Hb-concentrations.
(i)-(iii) calculated with the different values for A and B shown by Table 2.

DISCUSSION

The Haldane effect in adult humans

Under standard conditions, we found HE-induced pH-changes of 0.042 at 24 mM $[HCO_3^-]$ and 15 g/dl [Hb] in human adult blood. The more commonly used proton Haldane factor ($f_H = \Delta H^+/\Delta HbO_2$) can be calculated by multiplying ΔpH with the factor 0.148, including Donnan distribution and pH-ratio between plasma and red cells (Siggaard-Andersen, 1974), thus yielding a value of -0.28.

In the literature, higher numerical proton Haldane factors between -0.31 and -0.40 have been determined by "differential" titration (Siggaard-Andersen, 1974, Zwart et al., 1984). CO_2 Haldane factors ($f_H = \Delta CO_2/\Delta HbO_2$) determined by manometric methods were reported to be -0.34 under standard in vitro conditions (Müller et al., 1988) and -0.28 for ambient hypocapnic CO_2 partial pressures in vivo (Loeppky et al., 1983). The great variability in literature for f_H values even under normal acid-base conditions may be due to different definitions of the Haldane factor, to different techniques and to possible variations of 2,3-DPG concentration (Müller et al., 1988; Siggaard-Andersen, 1974).

From the data of v.Mengden et al.(1969) HE-induced ΔpH-values as low as 0.034 could be derived for standard conditions, but more recently values between 0.039 and 0.044 were reported (Müller et al., 1988; Eiring et al., 1988), being in close agreement with our present results.

Changing respiratory and metabolic acid-base conditions are expected to take considerable influence on the Haldane effect, being characterized by the increase in CO_2 content and decrease in H^+ activity of the blood during Hb-deoxygenation. These components referring to both oxylabile carbamate and HCO_3^- formation were extensively studied. A most complex interactive dependency on pH and P_{CO_2} has been found, with predominance of oxylabile HCO_3^- formation at pH below 7.5 and of oxylabile carbamate in the more alkaline range (for reviews see Bauer, 1974; Baumann et al., 1987; Klocke, 1987).

The overall HE-induced change in plasma pH was found to decrease with a rise in P_{CO_2} but to increase with a rise in fixed acids over a wide range of respiratory and metabolic acid-base conditions, in adult humans (v. Mengden et al., 1969) as well as in some laboratory animals (Kiwull-Schöne et al., 1992). Thus, the empirically established inverse linear relationship between ΔpH and $lg[HCO_3^-]$, including two parameters A and B, is a plausible and useful tool to estimate the Haldane effect in the physiological and pathophysiological acid-base range. However, the linear regression analysis of the present data yielded quantitatively different results from that reported originally by v.Mengden et al. (1969). Our values for the parameters A and B are much lower (Table 2). Interestingly, rather supportive values for A around 165 and for B around 1.85 can be calculated from more recent literature, referring to changes in P_{CO_2} (Müller et al., 1988) and to changes in both P_{CO_2} and fixed acids (Böning et al., 1993).

The Haldane effect in premature humans

In the human fetus, adult hemoglobin appears at around eight weeks of gestation, but not before 30 weeks there is an acceleration in the "switch" from the fetal (HbF) to the adult (HbA) type of hemoglobin synthesis (Bard, 1973), the exact nature of the switch not yet being identified. Thus, in our premature group, at a mean gestational age of 31 weeks, the average percentage of HbF was still as high as 80%. Nevertheless, we did not detect any measurable difference in the HE-induced ΔpH between prematures and adult humans, neither under standard conditions nor over the physiological range of bicarbonate changes. Moreover, the present results quantitatively agree with those we reported earlier for rabbits, cats and dogs (Kiwull-Schöne et al., 1992). This would agree with the finding that the amino acids ß146 His and ß94 Asp, responsible for binding oxygen-linked hydrogen ions, were found to be invariant in different types of mammalian Hb (Baumann et al., 1987). In much the same way as the proton Haldane effect, the fixed-acid Bohr effect is not distinguishable in rabbits and cats (Kiwull-Schöne et al., 1987) as well as in fetal and adult hemoglobin (Bauer and Schröder, 1969).

On the other hand, differences in the quaternary and tertiary structure of the hemoglobin molecule should be considered to influence the magnitude of the combined CO_2 and proton Haldane effect. The increased oxygen-affinity of fetal blood could be attributed to structural alterations of the binding site for 2,3-DPG, resulting in a poorer affinity for this phosphate molecule (Bauer et al., 1969; Bauer, 1974). Due to less competition, the carbamate formation during deoxygenation is therefore greater in fetal than in adult blood (Bauer and Schröder, 1969). We did not determine the carbamate portion directly, but there are literature data showing that even dramatic differences in this component are not reflected in measurements of plasma pH (Böning et al., 1993).

Consequently, at least the measurable HE-induced ΔpH in whole blood samples seems to depend neither on ontogenetic nor on phylogenetic variations of the Hb molecule, so that empirical Haldane effect data, once determined, can be used equally for adult and fetal human blood as well as for blood of different mammalian animal species.

Correction procedures

The exact knowledge of the HE is of importance when blood gas analysis is done in blood samples with incomplete oxygen saturation. The HE-dependent increase of proton buffer capacity during deoxygenation should not wrongly be attributed to the CO_2/HCO_3^- buffer system. This is important for "indirect" estimation of P_{CO_2} from two-gas equilibration and pH-measurement, as well as in clinical praxis, for secondary calculation of standard $[HCO_3^-]$ from "direct" measurement of pH and P_{CO_2}.

By the "indirect" equilibration method, serving as reference method for determination of metabolic acid-base status and CO_2 buffering capacity, linear lgP_{CO_2}/pH-lines are established to read the P_{CO_2} of a given blood sample from its actual pH (Astrup and Schrøder, 1956). As these buffer lines normally rely on complete oxygen saturation, the pH-increase due to deoxygenation in hypoxic blood samples would lead to an underestimation of the actual P_{CO_2}. On the other hand, values obtained by the "direct" method in

Table 4. Validity of different HE-corrections for estimation of P_{CO_2} or $[HCO_3^-st]$.

Procedure		P_{CO_2} [mmHg]		$[HCO_3^-st]$ mM	
		Prematures	Adults	Prematures	Adults
True value		40.0	40.0	24.0	24.0
Value without HE-correction		35.4	35.2	26.2	26.1
Correction with Equation I	(i)	40.4	40.1	23.9	23.9
	(ii)	40.3	40.1	24.0	23.9
	(iii)	39.4	39.2	24.4	24.3
Correction with Equation II		43.8	43.6	22.9	22.8

Examples are standard values of 40 mmHg P_{CO_2} and 24 mM $[HCO_3^-st]$ to be approached from "indirect" or "direct" blood gas analysis, respectively, in fully deoxygenated blood.
(i)-(iii) use of Equation I with the different values for A and B shown by Table 2.

hypoxic blood samples would be too alkaline in relation to the CO_2/HCO_3^- buffer, this leading to an overestimation of standard bicarbonate (or base excess) values.

Taken different estimates of the HE-induced ΔpH into account, as given by Table 3, related errors in determination of P_{CO_2} and $[HCO_3^-st]$ are shown by Table 4, being more or less properly corrected. Given standard reference values of 40 mmHg for P_{CO_2} and of 24 mM for $[HCO_3^-st]$ are best approached by the bicarbonate related HE-correction (Equation I), particularly when using appropriate values for A and B. The base excess related HE-correction (Equation II), which is quite commonly used, would lead to considerable overestimation of P_{CO_2} and underestimation of $[HCO_3^-st]$, at least in the range of normal acid-base conditions. For concomitant respiratory and metabolic acid-base changes, the latter method appears even less suitable, since the influence of P_{CO_2} in addition to that of BE on the Haldane effect is not taken into account (Werkmeister, 1992).

Considering the special risk of prematures to develop metabolic acid-base disturbances during their first weeks of life (Kalhoff et al., 1993), further optimization and higher accuracy of acid-base analysis during concomitant blood-gas disorders is of primary interest.

SUMMARY

The Haldane effect (HE) was investigated in human adults and prematures under normal metabolic acid-base conditions but at different levels of P_{CO_2}. Venous blood samples were equilibrated with low and high P_{CO_2} in either O_2 or N_2. The change in plasma pH of oxygenated blood by deoxygenation did not differ between both groups. Thus, ontogenetic differences of human hemoglobin structure do not influence the net proton Haldane effect measured in terms of whole blood pH-changes. Since the present data quantitatively agree with those we reported earlier for rabbits, cats and dogs (Kiwull-Schöne et al., 1992), phylogenetic differences in hemoglobin structure of these mammalian species do not either seem to play a role in this respect.

The influence of the Haldane effect on plasma pH has to be considered in blood-gas and acid-base analysis of samples with incomplete oxygenation. This is important for the indirect determination of P_{CO_2} through pH by the equilibration method (Astrup and Schrøder, 1956), serving as reference method for determination of metabolic acid-base status and CO_2 buffering capacity. Likewise, HE-correction is important for indirect estimation of metabolic acid-base status (BE and HCO_3^-st) from clinical routine P_{CO_2}- and pH-measurement.

In spite of the vaste amount of literature on the Haldane effect in human blood, quantitative data for practical purpose are less available and still equivocal. By the present study, a strong inverse linear correlation between the HE-induced ΔpH and $lg[HCO_3^-]$ could be shown over a wide range of acid-base changes. The determined parameters A and B were checked for accuracy to approach the true value when estimating either P_{CO_2} or standard $[HCO_3^-]$ indirectly. Thus, a highly valuable empirical $[HCO_3^-]$-related pH-correction of the Haldane effect is established to improve blood-gas and acid-base analysis in hypoxic blood.

Acknowledgement. The presented results are part of a Diploma-Thesis by Frauke Werkmeister, Ruhr-University Bochum, 1992. The expert technical assistance of Sabine Adler and Claudia Bräuer is gratefully acknowledged.

REFERENCES

Astrup P. and Schrøder, S., 1956, A simple electrometric technique for the determination of carbon dioxide tension in blood and plasma, total content of carbon dioxide in plasma, and bicarbonate content in "separated" plasma at a fixed carbon dioxide tension (40 mmHg), Scand. J. Clin. Lab. Invest. 8:33

Bard, H., 1973, Postnatal fetal and adult hemoglobin synthesis in early preterm newborn infants, J. Clin. Invest. 52:1789

Bauer, C., 1974, On the respiratory function of haemoglobin, Rev. Physiol. Biochem. Pharmacol. 70:1

Bauer, C. and Schröder, E., 1972, Carbamino compounds of haemoglobin in human adult and foetal blood, J. Physiol. 227:457

Bauer, C., Ludwig, M., Ludwig, I. and Bartels, H., 1969, Factors governing the O_2-affinity of human adult and foetal blood, Respir. Physiol. 7:271

Baumann, R., Bartels, H. and Bauer, C., 1987, Blood oxygen transport, in: "Handbook of Physiology", Sect. 3, Vol. IV, L.E. Farhi, S.M. Tenney, eds., Am. Physiol. Soc., Washington, D.C.

Böning, D., Schünemann, H.J., Maassen, N. and Busse, M.W., 1993, Reduction of oxylabile CO_2 in human blood by lactate, J. Appl. Physiol. 74:710

Eiring, P., Grote, J. and Rouwen, D., 1988, Influence of hemoglobin oxygen saturation on CO_2 and hydrogen ion binding in normal human blood, Pflügers Arch. 411/1:R 58

Kalhoff, H. Manz, F., Diekmann, L., Kunz, C., Stock, G.J. and Weisser, F., 1993, Decreased growth rate of low-birth-weight infants with prolonged maximum renal acid stimulation, Acta Paediatr. 82:522

Kilmartin, J.V., Fogg, J. Luzzana, M. and Rossi-Bernardi, L., 1973, Role of the α-amino groups of the α- and β-chains of human hemoglobin in oxygen-linked binding of carbon dioxide, J. Biol. Chem. 248:7039

Kiwull-Schöne, H., Gärtner, B. and Kiwull, P., 1987, The effects of CO_2 and fixed acid on the O_2-Hb affinity of rabbit and cat blood, Pflügers Arch. 408:451

Kiwull-Schöne, H., Werkmeister, F. and Kiwull, P., 1992, The Haldane effect under different acid-base conditions, Adv. Exp. Med. Biol. 316:11

Klocke, R.A., 1987, Carbon dioxide transport, in: "Handbook of Physiology", Sect 3, Vol. IV, L.E. Farhi, S.M. Tenney, eds., Am. Physiol. Soc., Washington D.C.

Loeppky, J.A., Luft, U.C. and Fletcher, E.R., 1983, Quantitative description of whole blood CO_2 dissociation curve and Haldane effect, Respir. Physiol. 51:167

v.Mengden, H.J., Schultehinrichs, D. and Thews, G., 1969, Dependence of plasma pH on oxygen satura tion, Respir. Physiol. 6:151

Müller, R., Grote, J. and Steinhausen, F., 1988, Die CO_2-Bindungskurve des normalen menschlichen Blutes und ihre Beeinflussung durch die Oxygenation des Hämoglobins, in: Funktionsanalyse biologischer Systeme, Gustav Fischer Verlag, Stuttgart, New York 18:61

Perutz, M.F., 1970, Stereochemistry of cooperative effects in haemoglobin, Nature 228:726

Siggaard-Andersen, O., 1974, "The Acid-Base Status of the Blood", Munksgaard, Copenhagen

Siggaard-Andersen, O. and Engel, K., 1960, A new acid base nomogram, Scand. J. Clin. Lab. Invest. 12:177

Werkmeister, F., 1992, Der Haldane-Effekt bei respiratorischen und metabolischen Säure-Basen-Veränderungen. Eine vergleichende Untersuchung bei verschiedenen Säugetierspezies einschließlich des Menschen, Diploma-Thesis, Faculty of Biology, Ruhr-University Bochum, Germany

Zwart, A., Kwant, G., Oeseburg, B. and Zijlstra, W.G., 1984, Human whole-blood oxygen affinity: effect of temperature, J. Appl. Physiol. 57:429

THE PROTON BOHR FACTOR OF NATIVE AND CROSSLINKER TREATED HEMOGLOBINS
- ITS POSSIBLE SIGNIFICANCE FOR THE EFFICACY OF HEMOGLOBIN BASED
ARTIFICIAL OXYGEN CARRIERS

W.K.R. Barnikol

Institut für Physiologie und Pathophysiologie
Johannes Gutenberg-Universität
D-55099 Mainz
BR Deutschland

INTRODUCTION

The classical (alkaline) Bohr effect is the shift of the oxygen (O_2) hemoglobin (Hb) binding curve of whole blood by carbon dioxide (CO_2). The effect has two components, which may be assessed separately: (1) Direct interaction of CO_2 with amino groups of hemoglobin (CO_2-Bohr effect) and (2) indirect influence via change of pH (proton Bohr effect). The second can be quantitatively characterized by the so called proton Bohr factor (PBF): $\partial \log P50 / \partial pH$ (P50: O_2 partial pressure at half saturation of hemoglobin). Generally PBF depends on all parameters affecting O_2-Hb binding, e.g. pH, PCO_2, temperature.

Presumably the alkaline proton Bohr effect is the basis for an important self regulating compensatory mechanism to supply specifically tissues with oxygen suffering from O_2 deficit. These tissues switch from aerobic to anaerobic metabolism producing locally lactic acid and decreasing also the pH of blood down to about 7.1, which is a lower limit for the organism; so the O_2-Hb-binding curve shifts locally to the right and facilitates O_2 release from hemoglobin within the capillaries to the tissues: The higher the absolute value of the proton Bohr factor at this pH is the better this mechanism works. So, PBF-values in the "acid" range, at said pH of 7.1, are of interest. Several hemoglobins, suitable as basis for artificial oxygen carriers, have the maximal alkaline proton Bohr factor at pH about 7.1. The value of normal human blood is about -0.5 (temperature/T 37 °C, CO_2 partial pressure/PCO_2: 40 mmHg). Freshly prepared red blood cells have, according to Kister et al.[1], at

PCO_2 = 0 and at pH 7.1 a PBF of -0.53 as related to extracellular pH; it is known that CO_2 decreases the Bohr effect[2].

So it is convenient to define for a practical purpose the proton Bohr factor at pH 7.1 ($PBF_{7.1}$) besides P50 and n50 (see legend table 1), as an additional characteristic quantity for an artificial oxygen carrier. Fluorocarbons, of course, have neither a P50 value nor a cooperativity and nor a Bohr effect.

Hemoglobins, dissolved in plasma, are used as artificial carriers because of their favourable binding properties, also with regard to the Bohr effect. But native hemoglobin is not suitable as the kidneys clear it quickly. So crosslinking is necessary, and, besides O_2 cooperativity, the P50 must be adjusted properly to a normal value.

As the modified hemoglobins used as artificial oxygen carriers are dissolved in an extracellular milieu - also within liposomes -, the proton Bohr effect has to be investigated under these conditions. Therefore we have studied the oxygen binding properties, including the Bohr effect, within a simulated (ional) milieu of plasma investigating first native human, bovine and porcine hemoglobin. Then we have allowed to react human and bovine hemoglobin with two crosslinkers: (1) 2,5-diiso-thiocyanatobenzenesulfonate (DIBS) and (2) 4,4'-diisothiocyanatostilbene-2,2'-disulfonate (DIDS); these crosslinkers react with the aminogroups of the hemoglobins. Thereby we have chosen certain reaction conditions in order to get products which were suitable for use as artificial oxygen carriers regarding their P50- and n50-values. Reaction in highly concentrated Hb solution results in soluble hyperpolymer hemoglobins, reaction in diluted state gives only monomer products[3]. Those hyperpolymer hemoglobins have molecular weights of some 10^6 g/mol as evidenced by light scattering investigations[6]. The concept behind this is to develop an artificial oxygen carrier with negligible oncotic pressure for use as a blood additive in contrast to a substitute[20].

MATERIALS AND METHODS

Human blood was taken from the cubital vein of young healthy donors, bovine and porcine blood we got from a slaughter-house. The plasma of the blood samples was removed by washing the red blood cells with Biku-solution (:125 mM NaCl, 4.5 mM KCl, 20 mM $NaHCO_3$; 1 M = 1 mol/l). Hemolysis of the cells was achieved by cryolysis. The resulting solution had a Hb content of about 300 g/l and a pH of 8.0; for monomer preparations the cryolysate was diluted to about 10 g/l with Biku-solution, for hyperpolymer preparations the solution remained undiluted. Deoxygenation was achieved by overflow of the stirred solution with nitrogen. The volume of the reaction mixtures were 300 µl, the reaction temperature was 4 °C, and the time about 20 h. The crosslinkers were added in ten-fold molar excess related to Hb (66000 g/mol); in case of reaction at high concentration, the amount of crosslinker and reaction time were optimized to get a maximum yield of soluble polymers (about 90%). DIDS and DIBS were dissolved in dimethylsulfoxide (about 1 M) before addition. The reactions were terminated with lysine in 30-fold molar excess related to the amount of the crosslinker. The hyperpolymers were separated from low molecular weight material by ultrafiltration with a cut off at 300000 g/mol in Biku- solution. The overall yield was about 40%. The monomers were ultrafiltrated into Biku-solution using a 10000 g/mol filter before measuring oxygen binding properties. Met-Hb contents of the samples was lower than 10%.

Gelchromatography (size exclusion chromatography) of the hyperpolymers was done with Sephacryl 200, 300 and 400 HR, analysis of covalent binding between $\alpha\beta$-dimers in monomer preparation was done with the aid of Sephadex G-100 SF in 1 M $MgCl_2$, gels are purchased from Pharmacia, D-Freiburg.

The oxygen binding properties were assessed with the aid of a thin layer method which is described in detail elsewhere[4]: the layers of hemoglobin solutions were equilibrated with different gas mixtures via a teflon membrane. The gas mixtures were analyzed according to Scholander[5]. Conditions of measurements: T: 37 °C, PCO_2 = 40 mmHg, Hb concentrations (cHb) about 20 g/l. The error of the method is approximately 10% in P50.

RESULTS

Table I comprises the results of the measurements: It gives the above defined proton Bohr factor of the native hemoglobin from man, ox, pig and that of reaction products prepared with crosslinkers as mentioned at pH = 7.1 and PCO_2 = 40 mmHg ($PBF_{7.1}$).

Table 1 demonstrates, that the (absolute) value of the proton Bohr factor of human Hb, dissolved in an extracellular milieu, is substantially smaller than in whole blood: 0.31 versus 0.5. The PBF-values of human and bovine hemoglobin are very similar, whereas the hemoglobin from pig exhibits a smaller proton Bohr effect.

The n50 values in table 1 show, that the homotropic O_2-cooperativities of the three hemoglobin are scarcely modified by change of the solvent milieu. But in all cases the P50 values are decreased considerably, so that the native hemoglobins are not suitable to be used as extracellular artificial oxygen carriers.

After reaction of human and bovine hemoglobin with the crosslinkers used here, once again the proton Bohr factor is substantially lowered to one third. All the reaction products would be suitable as artificial oxygen carriers in regard to P50 values and fairly suitable as to n50 value. No principal differences of the effect is seen in this respect neither comparing monomer and hyperpolymer products nor comparing DIBS and DIDS. Human hemoglobin, reacted with DIBS, even results in a proton Bohr factor near zero.

Table 1. Oxygen binding properties of human whole blood, native and chemically modified hemoglobins with DIBS and DIDS

hemoglobin	state of hemoglobin	crosslinker	$PBF_{7.1}$	P50 (mmHg)	n50
human	whole blood	—	-0.5	26	2.6
human	native	—	-0.31	15	2.6
bovine	native	—	-0.34	22	2.6
porcine	native	—	-0.19	16	2.5
human	monomer	DIBS	-0.002	21	1.6
human	monomer	DIDS	-0.10	28	1.5
bovine	monomer	DIDS	-0.13	59	1.3
bovine	hyper-polymer	DIBS	-0.11	25	1.3
bovine	hyper-polymer	DIDS	-0.06	31	1.3

Remarks: $PBF_{7.1}$: Proton-Bohrfactor at pH = 7.1, P50 and n50 refer to pH = 7.4; all three quantities are measured at 37 °C and with PCO_2 = 40 mmHg in 125 mM NaCl, 4.5 mM KCl, 20 mM $NaHCO_3$; reaction of the hemoglobins with the crosslinkers in deoxy state, n50 is an average value between 0.4 and 0.6 of O_2 saturation. DIBS: 2,5- diisothiocyanatobenzenesulfonate; DIDS: 4,4'-diisothiocyanatostilbene-2,2'-disulfonate.

Size exclusion chromatography done with Sephadex G-100 SF with 1 M $MgCl_2$ as eluent indicates that the monomer products are covalently linked between the two αβ-dimers of the Hb molecule at a degree of 85%. In case of human hemoglobin reacted with DIDS this was formerly found by Kavanaough et al.[7]. Size exclusion chromatography done with the Sephacryls reveals a broard molecular weight distribution of the hyperpolymer hemoglobins, but almost no (> 3%) polymer hemoglobins in monomer preparation are found.

DISCUSSION

Values of proton Bohr factors from human hemoglobin under similar (extracellular) conditions may be evaluated also from data of other investigators. So Antonini et al.[8] come to -0.41 as $PBF_{7.1}$ (cHb: 5-10 g/dl; T: 40 °C; PCO_2:0). From measurements of Kilmartin et al.[9] a value of -0.30 follows (cHb: 130 g/l; T: 37 °C; PCO_2: 40 mmHg; in 100 mM KCl).

Duhm[10] finds a proton Bohr factor around pH = 7.4 of -0.25 (cHb: 260 g/l; PCO_2: 40 mmHg; T: 37 °C; in 130 mM KCl, 18 mM $NaHCO_3$, 2.5 mM $MgCl_2$). Bucci et al.[11] have found a $PBF_{7.1}$ value of -0.32 (cHb: 1 g/l; T: 37 °C; PCO_2: 36 mmHg; in 100 mM Tris/HCl); Fronticelli et al.[13] give a value of -0.24 (cHb: 80-100 g/l; T: 37 °C; PCO_2: 0; in 150 mM Tris/HCl).

From the same paper with the same conditions one gets -0.45 as a $PBF_{7.1}$ value for bovine hemoglobin; increasing PCO_2 to 36 mmHg in 100 mM Tris/HCl brings $PBF_{7.1}$ "down" to -0.28[11]. Ilan et al.[12] have found $PBF_{7.1}$ to be -0.37 (cHb: 2 g/l; PCO_2: 0; T: 37 °C; 150 mM Tris/HCl).

Kim et al.[21] have done oxygen binding measurements with porcine hemoglobin. Their proton Bohr factor around 7.3 is -0.11 (cHb: 230 g/l; PCO_2: 40 mmHg, T: 37 °C; in 150 mM KCl; 18 mM $NaHCO_3$, 5 mM $MgCl_2$). At pH 7.1 the proton Bohr effect is expected to be greater.

All proton Bohr factors from other investigators, mentioned here, are also substantially lower than that of whole blood, at least as concerns human hemoglobin.

Regarding O_2 delivery preferably into tissues suffering from O_2 deficit the CO_2 Bohr effect seems to be minor importance mainly for two reasons: (1) hypercapnic areas in tissues are unlikely because of the high diffusivity of CO_2 and because of a quickly regulating mechanism of the lung for CO_2, (2) CO_2 Bohr factors ($\partial log/\partial PCO_2$) at pH 7.1 in the ional milieu mentioned above, as far as known, are low. For native human, bovine, and swine hemoglobin the values are 0.28, 0.06, and 0.26 respectively. As the CO_2 Bohr effect is mediated by the amino groups and as these are blocked by the crosslinkers modified hemoglobins, if at all, they exhibit a very low CO_2 Bohr effect. The CO_2 Bohr factor of hyperpolymer bovine hemoglobin, crosslinked with DIBS and DIDS, for instantance is almost zero.

DIDS and DIBS were already used by other investigators for modification of hemoglobin. Fuhrmann et al.[15] as well as Kavanough et al.[7] have let react human hemoglobin with DIDS in deoxy state and have found a decrease of both the O_2 affinity and the O_2 cooperativity; both is confirmed in table 1. DIBS, using inositol hexaphosphate as a co-reagent, was applied by Manning et al.[16] to crosslink the subunits of human hemoglobin; they got several reaction products.

Also in this study the reaction of the hemoglobin with crosslinkers in dilute solution may have resulted into a heterogeneity of monomer reaction products, which were not separated; in that case the oxygen binding data reported here, including the proton Bohr factors, may be average values.

Brouwer et al.[14] have measured the proton Bohr effect of human hemoglobin reacted with benzene-penta-carboxylate coupled with the aid of carbodiimide. Taking the $PBF_{7.1}$-values from their measurements also a considerable change the proton Bohr effect is seen comparing native hemoglobin with the product: from -0.74 to -0.27.

Ilan et al.[12] have bovine hemoglobin allowed to react with divinylsulfone; from their results also follows a change of proton Bohr effect ($PBF_{7.1}$) of native, monomer and polymer hemoglobin from -0.37 to -0.15 and -0.10, respectively, indicating that the polymerization decreases the proton Bohr effect.

An interesting molecular design seems to come from Fronticelli, Bucci et al.[11,17,18,19] by pseudocrosslinking hemoglobins with mono (3,5-dibromo-salicyl)-fumarate (FMDA) in order to maintain better a high P50 value and high homotropic O_2-cooperativity. The authors have used human, bovine and porcine hemoglobin. The reaction products unlike native hemoglobin do not decay during sedimentation, and in rats their half life time is increased fourfold. But also FMDA diminishes the proton Bohr effect: In case of human Hb $PBF_{7.1}$ changes from -0.32 to -0.11, in case of bovine Hb from -0.28 to -0.20.

The same investigators have also used benzene-1,3,5-tricarboxylate (BTC), activated with 3,5-dibromo-salicyl, as crosslinking agent. In this case indeed a slight increase of the proton Bohr effect is found: from -0.28 to -0.34. But, by far, the value of whole is not achieved. In case of human hemoglobin the proton Bohr effect again changes from -0.32 to -0.23 (all expressed as $PBF_{7.1}$ values).

In all examples, mentioned above, the crosslinker increases the P50 values. So, as a rule, decrease of proton Bohr effect and decrease of oxygen affinity go together - with the said exception of bovine Hb reacted with BTC.

The decrease of the proton Bohr effect presumably is due to a loss of binding sites for oxylabile protons caused by the crosslinker.

As regard to the development of an artificial oxygen carrier, which is able to delivers specifically oxygen to tissues suffering from O_2 deficit, it would be very desirable to find out modified hemoglobins with the same or even higher proton Bohr effect than whole blood has.

ACKNOWLEDGEMENTS AND REMARKS

I thank Beate Krumm and Simone Prätorius for skilful technical assistance and I thank H. Pötzschke for helpful discussions. Part of the results presented here belong to the theses of Ulrike Klein and Stephanie Heimann.

SUMMARY

Especially the (alkaline) proton Bohr effect seems to provide an important self regulating mechanism of the organism to deliver specifically oxygen into tissues suffering from O_2 deficit. In this way these tissues switch from aerobic to anaerobic metablism, get lactacid, thereby shifting oxygen hemoglobin binding curve to the right and thus facilitating the oxygen release. The higher the absolute value of the proton Bohr factor (: $\partial logP50/\partial pH$) is the better this mechanism works. To get one characteristic number the proton Bohr factor at pH 7.1 is taken. This pH in blood is about a lower limit for organism and human blood has at this pH its maximum proton Bohr factor which is about -0.5. When designing a hemoglobin based artificial oxygen carrier such a high or even a higher

proton Bohr factor should be aimed at. But bringing human hemoglobin into
extracellular milieu decreases the said proton Bohr effect down to -0.31;
about the same values have bovine and porcine hemoglobin under these
conditions. Before native hemoglobin can be used as an artificial oxygen
carrier outside the red blood cells, they must be crosslinked; otherwise
they are cleared quickly by the kidneys. Reaction of human and bovine
hemoglobin with the crosslinkers DIBS (2,5- diisothiocyanatobenzenesulfonate) and DIDS (4,4'-diisothiocyanatostilbene-2,2'-disulfonate) decreases the proton Bohr effect once again substantially down to about
-0.1 irrespective of the degree of polymerization (monomer and hyperpolymer). Proton Bohr factors of reaction products from various hemoglobins
and different crosslinkers evaluated from measurements of other investigators largely confirm the findings of this study. It is speculated that
by the reaction of the hemoglobins with crosslinker oxygen dependent
proton binding sites get lost.

REFERENCES

1. J. Kister, M.C. Marden, B. Bohn, Cl. Poyart, Functional properties of
    hemoglobin in human red cells: II. Determination of the Bohr
    effect, Respir. Physiol. 73:363 (1988).
2. L. Garby, M. Robert, B. Zaar, Proton- and carbamino-linked oxygen
    affinity of normal human blood, Acta Physiol. Scand. 84:482
    (1972).
3. W.K.R. Barnikol, Influence of the polymerizatin step alone on oxygen
    affinity and cooperativity during production of hyperpolymers from
    native hemoglobin with crosslinkers, Biomater., Artif. Cells,
    Artif. Organs, 1993, in press.
4. W.K.R. Barnikol, W. Döhring, W. Wahler, Eine verbesserte Modifikation
    der Mikromethode nach NIESEL und THEWS (1961) zur Messung von O_2-
    Hb-Bindungskurven in Vollblut und konzentrierten Lösungen, Respi-
    ration 36:86 (1978).
5. P.F. Scholander, Analyzer for accurate estimation of respiratory
    gases in one-half cubic centimeter samples, J. Biol. Chem. 167:235
    (1947).
6. K.-H. Massenkeil, R.G. Kirste, H. Pötzschke, W.K.R. Barnikol, Her-
    stellung und polymeranalytische Charakterisierung fraktionierter,
    mit Glutardialdehyd vernetzter und mit Natriumcyanoborhydrid
    reduktiv stabilisierter Hyperpolymerer aus Rinder- und Human-
    Hämoglobin, Biol. Chem. Hoppe-Seyler 373:798 (1992).
7. M.P. Kavanaugh, D.T.-B. Shih, R.T. Jones, Affinity labeling of hemo-
    globin with 4,4'-diisothiocyanostilbene-2,2'-disulfonate: covalent
    crosslinking in the 2,3-diphosphoglycerate binding site, Biochem-
    istry 27:1804 (1988).
8. E. Antonini, J. Wyman, M. Brunori, Cl. Fronticell, E. Bucci, A.
    Rossi-Fanelli, Studies on the relations between molecular and
    functional properties of hemoglobin: V. The influence of tempera-
    ture on the Bohr effect in human and in horse hemoglobin, J. Biol.
    Chem. 240:1096 (1964).
9. J.V. Kilmartin, L. Rossi-Bernadi, Interaction of hemoglobin with
    hydrogen ions, carbon dioxide, and organic phophates, Physiol.
    Rev. 53:836 (1973).

10. J. Duhm, 2,3-Diphosphoglycerate metabolism of erythrocytes and oxygen transport function of blood, In: Erythrocytes, thrombocytes, leukocytes - recent advances in membrane and metabolic research; E. Gerlach, K. Moser, E. Deutsch, W. Wilmanns (eds.); Georg Thieme Publisher, Stuttgart, 1973, 149-157.

11. E Bucci, C. Fronticelli, A. Razynsha, V. Militello, R. Koehler, B. Urbaitis, Hemoglobin tetramers stabilized with polyaspirins, Biomater., Artif. Cells & Immob. Biotechn. 20:243 (1992).

12. E. Ilan, P.G. Morton, Th.M.S. Chang, The anaerobic reaction of bovine hemoglobin with divinylsulfone: structural changes and functional consequenses, Biochim. Biophys. Acta 1163:257 (1993).

13. C. Fronticelli, E. Bucci, Ch. Orth, Solvent regulation of oxygen affinity in hemoglobin, sensitivity of bovine hemoglobin to chloride ions, J. Biol. Chem. 259:10841 (1984).

14. M. Brouwer, R. Cashon, J. Bonaventura, Carbodiimide-mediated coupling of benzenepentacarboxylate to human hemoglobin: structural and functional consequences, Biomat. Artfif. Cells and Immob. Biotechn. 20:323 (1992).

15. G.F. Fuhrmann, C. Creutzfeldt, K. Rudolphi, H. Fasold, The anion-transport inhibitor H_2DIDS cross-links hemoglobin interdimerically and enhances oxygen unloading, Biochim. Biophys. Acta 946:25 (1988).

16. L.R. Manning, Sh. Morgan, R.C. Beaves, B.T. Chait, J.M. Manning, J.R. Hess, M. Cross, D.L. Currell, M.A. Marini, R.M. Winslow, Preparation, properties, and plasmaretention of human hemoglobin derivatives: comparison of uncrosslinked carboxymethylated hemoglobin with crosslinked tetrameric hemoglobin, Proc. Natl. Acad. Sci. USA 88:3329 (1991).

17. E. Bucci, A. Razynska, B. Urbaitis, C. Fronticelli, Pseudocross-linking of human hemoglobin with mono-(3,5-dibromosalicyl)fumarate, J. Biol. Chem. 264:6191 (1989).

18. C. Fronticelli, E. Bucci, A. Razynska, J. Sznajder, B. Urbaitis, Z. Gryczynshi, Bovine hemoglobin pseudo-crosslinked with mono-(3,6-dibromosalicyl)fumarate, Eur. J. Biochem. 193:331 (1990).

19. B. Urbaitis, Y.S. Lu, C. Fronticelli, E. Bucci, Renal Excretion of pseudo-crosslinked human, porcine and bovine hemoglobins, Biochem. Biophys. Acta 1156:50 (1992).

20. H. Pötzschke, St. Guth, W.K.R. Barnikol, Divinyl sulfone-crosslinked hyperpolymeric human hemoglobin as an artificial oxygen carrier in anesthetized spontaneously breathing rats, Adv. Biol. Med. (1993), in press.

21. H.D. Kim, J. Duhm, Postnatal decrease in the oxygen affinity of pig blood induced by red cell 2,3-DPG, Am. J. Physiol. 226:1001 (1974).

METHODS FOR MEASURING LEVELS OF O_2 IN TISSUES

Peter D. Wagner[1] and Peter Scheid[2]

[1]Division of Physiology, 0623A
University of California, San Diego
9500 Gilman Drive
La Jolla, CA 92093-0623

[2]Institut für Physiologie
Ruhr-Universität Bochum
D-44780 Bochum, Germany

INTRODUCTION

Of two symposia programmed for the San Diego meeting, one dealt with techniques for assessing oxygenation directly or indirectly in the tissues.

Table I indicates the participants in that symposium and the techniques about which they spoke.

Table I. Symposium participants.

Topic	Speaker
Frozen myoglobin spectroscopy	T. Gayeski, U. Rochester
Microelectrodes	H. Acker, Max-Planck-Inst., Dortmund
Phosphorescence quenching	D. Wilson, U. Pennsylvania
Magnetic Resonance Spectroscopy	T. Jue, U. California, Davis
Near-infrared Spectroscopy	C. Piantadosi, Duke University
NADH Fluorescence	B. Chance, U. Pennsylvania
Electron Spin Resonance	H. Swartz, Dartmouth College

The basic objectives of the symposium were to learn more about the advantages and limitations of each technique, to better understand the principles of each method, to assess the information content of each and to get some feel for the cost and technical difficulty involved.

A full hour was reserved for discussion of these methods, and this discussion resulted in a direct comparison of the key features of each, item by item. The end product is shown in Table II which is hopefully self-explanatory. This table was completed by the symposium co-chairs and subsequently edited by all of the speakers listed in Table I.

All involved hope it is of use to the members of ISOTT as they strive to understand O_2 transport and utilization within tissues.

<h3 style="text-align:center">TABLE II, part 1. Methods for measuring levels of O_2 in tissues</h3>

	Frozen Myoglobin Spectroscopy	Phosphorescence Quenching	Magnetic Resonance Mb Spectroscopy
Variable Actually Measured	Mb and Hb saturation	Phosphorescence decay time	Proton NMR
Time Resolution	single one-time measurement	1 milli sec	minutes
Spatial Resolution	light microscopy 50μm field	light microscopy 50 μm field is possible	~ ½ gm tissue
Specificity	Mb, Hb only - According to location in muscle/blood vessels	virtually complete - some pH interference	Mb, Hb saturation (distinguishable from each other)
Degree of animal	completely	ess. none	none
Invasiveness man	cannot be done	not evaluated	none
Technical Skills Required	major	modest to major	major
Expense of Equipment, US $	150K ? not available anymore	20K - single site 70K - P_{O_2} MAP	300K to millions
Special Advantages	Spatial Resolution - 50 μm	fast and continuous very specific for O_2	coupling to ^{31}P and other MRS spectra feasible in man
Special Disadvantages	invasive single point in time	plasma P_{O_2} only at present	low temporal and spatial resolution. Limited to muscle (Mb)
Sensitivity	0.2 Torr at low P_{O_2} < full Mb saturation	Good: 0.1 Torr	0.2 Torr at low P_{O_2} < full Mb saturation
Signal Stability	very stable	very stable	very stable
Toxicity	Requires removal of organ	none known possible light damage	none

TABLE II, part 2. Methods for measuring levels of O_2 in tissues

	NIR Spectroscopy		NADH Surface Fluorescence
	Continuous Light	**Frequency and Time Domain**	
Variable Actually Measured	Mb, Hb & Cyt a saturation and blood volume	Saturation of Hb	Mitochondrial NADH
Time Resolution	seconds	seconds	20 milli sec
Spatial Resolution	scale of cm	scale of mm	15 μm
Specificity	high: Hb, Mb low: Cu_A of Cytochrome a, a_3	hi: Hb/Mb has imaging capability	hi: but interference from blood
Degree of animal **Invasiveness man**	none none	none none	organ surface intraoperative
Technical Skills Required	can be major for Cyt. a detection	modest for spectroscopy high for imaging	modest
Expense of Equipment, US $	75K commercial 28-36K home-build	20K spectroscopy 60-100K imaging	20K - scanner 30K - on-line video
Special Advantages	continuous, non-invasive	continuous, non-invasive quantitative saturation (measures pathlength)	specificity localization
Special Disadvantages	semi-quantitative; separation of Mb & Hb not possible. Separation of Hb/Hb & Cu_A difficult. Does not measure optical pathlength	pathlength calibrator is desirable	invasive; Hb interferes. indirect - does not measure O_2
Sensitivity	μ molar	μ molar	μ molar levels (n moles of NADH)
Signal Stability	0.3-1.0%	$\pm$ 3% saturation	1% full scale
Toxicity	none at FDA limits	none at FDA limits	minimal

<h2 style="text-align:center;">TABLE II, part 3. Methods for measuring levels of O_2 in tissues</h2>

	Electron Spin Resonance	P_{O_2} Needle Electrodes	P_{O_2} Surface Electrodes
Variable Actually Measured	P_{O_2}	P_{O_2}	P_{O_2}
Time Resolution	seconds	seconds to minutes depends on membrane thickness	seconds to minutes depends on membrane thickness
Spatial Resolution	500 μm	1 μm	15 μm
Specificity	high for P_{O_2}	high for P_{O_2}	high for P_{O_2}
Degree of animal	ess. none	invasive	non invasive
Invasiveness man	none for depths up to 1 or 5 cm	invasive	non invasive
Technical Skills Required	modest	high	modest
Expense of Equipment, US $	50K	1K	20K
Special Advantages	rapid repeatable over months	high spatial resolution	generates P_{O_2} histogram
Special Disadvantages	sensitivity for deep organs	invasive, single point in time	surface only
Sensitivity	0.1 Torr	1 Torr	1 Torr
Signal Stability	very stable	stable	stable
Toxicity	probably none but requires study	none	none

PARTIAL SUBSTITUTION OF RED BLOOD CELLS WITH FREE HEMOGLOBIN SOLUTION DOES NOT IMPROVE MAXIMAL O_2 UPTAKE OF WORKING IN SITU DOG MUSCLE

M.C. Hogan, S.S. Kurdak, R.S. Richardson, and P.D. Wagner

Department of Medicine, University of California, San Diego, La Jolla, CA 92093

INTRODUCTION

It has been proposed (Gayeski et al., 1987; Honig and Gayeski, 1993) that the O_2 carrier-free region between the erythrocyte and the myocyte sarcolemma, which would include the plasma space, may serve as the major source of resistance to the diffusion of oxygen. Federspiel and Popel (1986) have suggested, using theoretical modeling, that the flux of O_2 from a red blood cell through the capillary wall is limited to the near vicinity of the red blood cell, and that O_2 diffusion contributes negligibly to overall O_2 flux out of the capillary in the plasma spaces between erythrocytes.

For a given O_2 delivery (flow X arterial O_2 content), the amount of O_2 that can be extracted and used by the working tissue, according to Fick's law of diffusion, is determined by the diffusing capacity of the muscle (DmO_2) and the resulting O_2 partial pressure gradient from the red blood cell to the mitochondria. The DmO_2 is a lumped conductance parameter that takes into account all of the variables that determine the resistance to O_2 diffusion from the red blood cell to the mitochondria (ie. O_2 offloading kinetics from hemoglobin, membrane resistances, diffusional distances, O_2 solubilities, capillary surface area, etc.). We have demonstrated previously (Hogan et al., 1991) that when [Hb] was decreased (by whole blood hemodilution) in blood perfusing maximally working muscle, while keeping muscle blood flow constant, the estimated DmO_2 decreased considerably. This in part might have been a result of an increase in the red blood cell spacing causing less of the capillary surface area to be available at any instant for O_2 diffusion. To examine this possibility, we recently demonstrated (Hogan et al., 1992) that when the O_2 solubility of the plasma was improved twofold by a blood substitute (perfluorocarbon emulsion), the maximal O_2 uptake ($\dot{V}O_2$) was increased by only the amount that the convective O_2 delivery was improved, with no change in the estimated DmO_2. This therefore did not support a significant role for low plasma O_2 solubility as being a principal factor in O_2 diffusional resistance in working tissue, although a change in the estimated DmO_2 may have been obscured by the relatively small increase in plasma O_2 solubility.

It was the purpose of the present study to further test the hypothesis that the inter-erythrocyte space is a major resistance factor for O_2 diffusion by partially replacing erythrocytes with a cell-free hemoglobin solution. This strategy allows a large increase in the O_2 carrying capacity of the plasma while maintaining O_2 delivery to the working muscle the same as whole blood with similar [Hb]. This should result in an increase in the capillary surface area available for O_2 diffusion and a more uniform distribution of Hb along the capillary. If in fact the inter-erythrocyte spacing is a critical determinant of the DmO_2, this strategy should improve the oxygen diffusion capacity of O_2 supply-limited working muscle so that a higher percentage of the oxygen delivered to the tissue could be extracted, thereby achieving a higher oxygen uptake.

METHODS

The left gastrocnemius from four dogs was isolated as described previously (Hogan et al., 1991). All experimental periods were conducted using pump perfusion of blood from the contralateral artery so that muscle blood flow could be controlled.

Each muscle was stimulated to contract isometrically (tetanic) at 0.75 Hz for 5 min, which results in a sub-maximal $\dot{V}O_2$ (approximately 80%) under normal O_2 delivery conditions. The experimental protocol consisted of 4 separate 5-min contraction periods for the isolated muscle, separated by 20-min rest periods. During each 5-min contraction period, the first 3 min was with pump controlled perfusion (approximately 130 mm Hg) with the animal's own blood at natural [Hb]. The last 2 min of the 5-min work period consisted of the muscle being perfused at a slightly lower perfusion pressure (to insure O_2 supply-limited conditions) with a quick-switch to either: 1) blood of the animal diluted with saline to hemoglobin concentration = 8.8 g/100 ml; or 2) a mixture consisting of 33% normal dog blood and 67% cell-free hemoglobin solution to a final total perfusate concentration of 8.7 g hemoglobin/100 ml (so 1/3 of Hb was contained within red blood cells and 2/3 was cell-free in the plasma). Each muscle underwent 2 contraction periods with each of the 2 treatments (total of 4 work bouts) in which the treatment used within a work bout was reversed from the preceding work bout.

The cell-free hemoglobin was a cross-linked hemoglobin between α subunits at lysine 99 residues, which results in a stable cross-linked tetramer possessing nearly normal hemoglobin characteristics and P_{50}, but is slightly less viscous and was obtained from Dr. Robert Winslow from the Letterman Army Institute.

Arterial blood samples were drawn from the arterial line just before entering the muscle and venous samples were obtained from the left popliteal vein as close to the gastrocnemius as possible. A numerical integration technique was used to calculate an overall value for muscle O_2 diffusing capacity (DmO_2) and mean capillary PO_2 using Fick's law of diffusion and a simple model of capillary gas exchange, as outlined previously (Hogan et al., 1990). Two way analysis of variance was used for the statistical analysis and the 0.05 level of significance was used.

RESULTS

Muscle O_2 delivery and $\dot{V}O_2$ were significantly less, despite unchanged electrical stimulation parameters, during the 2-min treatment periods (low [Hb] conditions) compared to the prior 3-min with normal perfusion and [Hb]--indicating oxygen supply-limited conditions during the 2-min treatment periods. Table 1 presents the important measurements made during this study.

Table 1. Measurements (means ± SEM, n = 8) of blood parameters and principal variables related to O_2 transport and gas exchange during the whole blood and 2/3 cell-free Hb conditions.

	whole blood	2/3 cell-free Hb
[Hb], g/100 ml	8.8 ± 0.3	8.7 ± 0.2
Hct, %	27.3 ± 0.5	8.5 ± 0.3*
P_{50}, Torr	34.0 ± 0.7	32.9 ± 0.3
PaO_2, Torr	115 ± 9	126 ± 9
$PaCO_2$, Torr	33 ± 2	30 ± 2
pHa	7.33 ± 0.03	7.40 ± 0.02
MVR, mmHg·ml⁻¹·min⁻¹·100 g⁻¹	1.6 ± 0.2	1.2 ± 0.2*
CaO_2, ml·100 ml⁻¹	11.3 ± 0.4	10.9 ± 0.4
Muscle blood flow, ml·min⁻¹·100 g⁻¹	83 ± 11	84 ± 11
O_2 delivery, ml·min⁻¹·100 g⁻¹	9.6 ± 1.7	9.4 ± 1.5
$\dot{V}O_2$, ml·min⁻¹·100 g⁻¹	6.5 ± 1.3	6.0 ± 1.4
PvO_2, Torr	31 ± 3	29 ± 3
O_2 extraction, %	66 ± 5	61 ± 6
DmO_2, ml·100 g⁻¹·min⁻¹·Torr⁻¹	0.14 ± 0.03	0.15 ± 0.04

MVR, muscle vascular resistance; DmO_2, calculated muscle O_2 diffusing capacity. * indicates significantly different (p < 0.05) from other condition.

The total concentration of Hb was not different between the two conditions, even though the manner in which Hb was distributed (free vs. within erythrocytes) was by design quite different. As muscle blood flow was deliberately kept the same between the two treatments, and arterial [O_2] was the same also, there was no significant difference in the O_2 delivery to the muscle. Because of the lowered [Hb] during both the 100% red blood cell or partial cell-free Hb perfusion, O_2 delivery and $\dot{V}O_2$ was reduced below that seen under the control conditions, so that the muscle was certainly O_2 supply-limited. The principal observation was that $\dot{V}O_2$ was not significantly different between the two different conditions. The PO_2 of the muscle effluent venous blood (PvO$_2$), the O$_2$ extraction ratio (V̇O$_2$/O$_2$ delivery), and DmO$_2$ were also not significantly different between the two conditions.

DISCUSSION

The important result of this study was that at similar muscle blood flow, total Hb concentration, arterial O$_2$ content, and thus O$_2$ delivery, there was no difference in V̇O$_2$ in O$_2$ supply-limited working dog muscle between a condition in which O$_2$ was carried almost entirely by Hb contained within erythrocytes and a condition in which 67% of the O$_2$ was carried by cell-free Hb in solution. This suggests that a more uniform O$_2$ distribution throughout the capillary does not improve the overall coductance for O$_2$ transfer from the capillary to the mitochondria.

We recently reported (Hogan et al., 1991) that reductions in [Hb] resulted in a decrease in the estimated DmO$_2$ at maximal V̇O$_2$ in isolated, in situ working muscle perfused with whole blood. In support of the notion that the DmO$_2$ is reduced with lower [Hb], we have also compared the amount of O$_2$ extracted by maximally working muscle from similar O$_2$ deliveries that have been reduced by either lowering [Hb] (Hogan et al., 1991), inducing hypoxemia (Hogan et al., 1990), or causing ischemic blood flow reduction (Hogan et al., 1993). Muscle O$_2$ extraction and maximal O$_2$ uptake are significantly less, at the same O$_2$ delivery, when [Hb] is reduced then for conditions of either hypoxemia or ischemia (approximately 70% vs 80% vs 90%, respectively). Honig and Gayeski (1993) suggested that the increased resistance to O$_2$ diffusion between erythrocyte and sarcolemma seen in anemia during submaximal work was because at any instant O$_2$ was available from fewer red blood cells. However, other possible factors that might decrease O$_2$ conductance capacity (or increase diffusional resistance) with reductions in [Hb] could be: 1) changes in the distribution of capillary erythrocyte flow, 2) the increased plasma space between erythrocytes, or 3) O$_2$ off-loading kinetics from Hb being sufficiently slow as to be rate-limiting at high flow rates (rapid capillary transit times for erythrocytes).

It has been demonstrated (Piiper and Haab, 1991) that a heterogeneous distribution of blood flow throughout the muscle can influence the interpretation of a calculated O$_2$ diffusing capacity. In the present study, we can not exclude the possibility that there was a difference in muscle blood flow distribution between the two conditions of this study. In fact, there was significantly less muscular vascular resistance with the partial cell-free Hb solution, probably resulting from this solution having a slightly lower viscosity because of the fewer erythrocytes. However, with fewer erythrocytes present and 2/3 of the Hb free in plasma, it would be expected that there would be a more uniform distribution of O$_2$ throughout the working tissue during the partial cell-free Hb condition. This would make the muscle more homogeneous in terms of O$_2$ distribution, which should have improved the ability of the muscle to extract O$_2$ if red cell heterogeneity were indeed a problem. We attempted to keep as many hemodynamic variables similar between the two treatments in the present study (blood flow, acid-base characteristics, total Hb, etc.) to minimize differences in blood flow distribution between the two conditions.

It has been suggested (Federspiel and Popel, 1986; Groebe, 1990) that the plasma space between erythrocytes in the capillary contributes little to O$_2$ flux into the cell, so that O$_2$ diffusion into the cell is virtually limited to the area subjacent to the erythrocyte. This would mean that as the erythrocyte spacing becomes even greater as red cell concentration is reduced, there would be more capillary surface area exposed to plasma and therefore unavailable for O$_2$ diffusion, and the DmO$_2$ would be reduced.

It was suggested by Homer et al. (1981) that the erythrocyte-to-plasma PO$_2$ gradient becomes greater during anemia, so that there is less PO$_2$ equilibrium along the length of the capillary, and that this might be alleviated by increasing the O$_2$ carrying capacity of the plasma. In a previous experiment (Hogan et al., 1992) to examine the role of plasma spacing between erythrocytes as being an important determinant of the low DmO$_2$ that we found during anemia, we increased the O$_2$ solubility of the plasma compartment (from 0.003 to 0.005 ml O$_2$· 100 ml blood^{-1}·Torr^{-1}) by using a perfluorocarbon emulsion and examined whether the subsequent increase in the plasma O$_2$ diffusion coefficient and plasma O$_2$ content would contribute to a higher DmO$_2$. At the same muscle blood flow, the perfluorocarbon emulsion carried more O$_2$ to the working tissue and therefore the O$_2$ delivery was greater. Under the conditions of that study, the increase in V̇O$_2$max that we measured during perfluorocarbon perfusion was strictly related to the increase in convective O$_2$ delivery, with no increase in the muscle O$_2$ diffusing capacity. However, it was possible that the increase in plasma O$_2$ solubility achieved in that study was not of the magnitude necessary to measure small changes in DmO$_2$. In that study, the increase in O$_2$ carried by plasma was from 1.5% of the total (control) to 2.5% with fluorocarbon.

We had expected a higher O$_2$ extraction and DmO$_2$ would result from adding cell-free Hb to the plasma if the increase in inter-erythrocyte spacing that occurs during anemia was the cause of the reduced DmO_2 seen during low [Hb] conditions (Hogan et al., 1991). The results of the present study are in agreement with our previous report (Hogan et al., 1992). These results are also consistent with a previous report by Biro et al. (1991). Our studies therefore do not directly support a role of the inter-erythrocyte spacing as being the principal cause of the reduced DmO_2 seen during anemia.

However it should be noted that in the present study, red cell concentration (hematocrit $= 8.5 \pm 0.3$) was less in the mostly cell-free Hb solution compared to the whole blood condition (hematocrit $= 27.3 \pm 0.5$), even though total [Hb] was kept constant. Therefore, red cell spacing in the microcirculation was likely greater in this condition than when all Hb was contained within red cells, and it is possible that a reduced DmO_2 resulting from this fact was offset by increasing the capillary surface area in contact with O_2 (from the added Hb solution). However, we argue that a more uniform distribution of Hb between red blood cells and plasma at the same total [Hb] should improve O_2 conductance if the normally heterogeneous distribution of Hb between plasma and red blood cells limits maximal O_2 efflux in the first place.

At the high flow rates measured during maximal work intensities, it is possible that the capillary erythrocyte transit time becomes a critical factor in determining how much O_2 can be unloaded from the Hb. We suggest that because both the calculated DmO_2 and the measured O_2 extraction were not different between the two conditions of the present study, it remains to be shown that the diminished DmO_2 during reduced [Hb] perfusion that was reported previously (Hogan et al., 1991) was not the result of slow O_2 off-loading kinetics.

In conclusion, under the conditions of this study, at constant O_2 delivery and when 2/3 of the O_2 transported to the working muscle was carried by cell-free hemoglobin in the plasma, there was no change in maximal oxygen uptake or oxygen extraction and no improvement in the calculated diffusional conductance of O_2 into the muscle (compared to perfusion with whole blood at the same [Hb]). These results suggest that the reduced O_2 conductance capacity (DmO_2) that is seen when [Hb] is lowered is not likely the result of Hb distribution heterogeneity between red blood cell and plasma.

This research was supported by National Institutes of Health Grant HL 17731 and AR 40155.

REFERENCES

Biro, G.P., P.J. Anderson, S.E. Curtis, and S.M. Cain (1991). Stroma-free hemoglobin: its presence in plasma does not improve oxygen supply to the resting hindlimb vascular bed of hemodiluted dogs. <u>Can. J. Physiol. Pharmacol.</u> 69: 1656-1662.

Federspiel, W.J., and A.S. Popel (1986). A theoretical analysis of the effect of the particulate nature of blood on oxygen release in capillaries. <u>Microvas. Res.</u> 32: 164-189.

Gayeski, T.E.J., R.J. Connett, and C.R. Honig (1987). Minimum intracellular PO_2 for maximum cytochrome turnover in red muscle in situ. <u>Am. J. Physiol.</u> 252 (<u>Heart Circ. Physiol.</u> 21): H906-H915.

Groebe, K (1990). A versatile model of steady state O_2 supply to tissue. Application to skeletal muscle. <u>Biophys. J.</u> 57: 485-498.

Honig, C.R., and T.E.J. Gayeski (1993). Resistance to O_2 diffusion in anemic red muscle: roles of flux density and cell PO_2. <u>Am. J. Physiol.</u> 265 (<u>Heart Circ. Physiol.</u> 34): H868-H875.

Hogan, M.C., D.E. Bebout, P.D. Wagner, and J.B. West (1990). Maximal O_2 uptake of in situ dog muscle during acute hypoxemia with constant perfusion. <u>J. Appl. Physiol.</u> 69: 570-576.

Hogan, M.C., D.E. Bebout, and P.D. Wagner (1991). Effect of hemoglobin concentration on maximal O_2 uptake in canine gastrocnemius muscle in situ. <u>J. Appl Physiol.</u> 70: 1105-1112.

Hogan, M.C., D.C. Willford, P.E. Keipert, N.S. Faithfull, and P.D. Wagner (1992). Increased plasma O_2 solubility improves O_2 uptake of in situ dog muscle working maximally. <u>J. Appl. Physiol.</u> 73: 2470-2475.

Hogan, M.C., D.E. Bebout, and P.D. Wagner (1993). Effect of blood flow reduction on maximal O_2 uptake in canine gastrocnemius muscle in situ. <u>J. Appl. Physiol.</u> 74: 1742-1747.

Homer, L.D., P.K. Weathersby, and L.A. Kiesow (1981). Oxygen gradients between red blood cells in the microcirculation. <u>Microvasc. Res.</u> 22: 308-323.

Piiper, J., and P. Haab (1991). Oxygen supply and uptake in tissue models with unequal distribution of blood flow and shunt. <u>Respir. Physiol.</u>, 84: 261-271.

<u>DEPENDANCE OF OXYGEN DELIVERY ON HEMATOCRIT</u>

K. Messmer, Institute for Surgical Research, Klinikum Grosshadern, University of Munich, Germany

Oxygen delivery (DO_2) is derived from arterial oxygen content and cardiac output. DO_2 does, however, not directly correlate with changes of hematocrit since the latter affects cardiac output by virtue of its influence on blood viscosity and hence viscous resistance to flow. At normal pulmonary function linear increments of hematocrit increase arterial oxygen content, nevertheless, DO_2 tends to fall during hemoconcentration and polycythemia as result of reduced venous return and stroke volume respectively. Despite low cardiac output, low flow velocity and due to prolonged transit of the red cells through the capillary network tissue oxygenation can be preserved. On the other hand, reduction of oxygen content by hemodilution (low hematocrit while maintaining normal blood volume) similarly preserves adequacy of tissue oxygenation, however, by opposite mechanisms, namely increase of cardiac output and organ blood flow on account of reduced blood viscosity, lowered total peripheral resistance and augmented venous return. These beneficial effects of dilutional anemia are enhanced by augmented cardiac contractility and increased venomotor tone. Compensating mechanisms in acute dilutional anemia are the rise of cardiac output, enhancement of blood oxygen extraction and maintenance of adequate mixed venous PO_2. The net effect on oxygen tissue delivery depends further on physiological and interventional factors, e.g. redistribution of cardiac coutput, angioarchitectonic organ differences, effect of diluents on corpuscular and plasma components (RBC fluidity and plasma viscosity) and local adjustment of the cross-sectional area of the microvascular network. As a rule regional organ flow is redistributed in favor of heart and brain, however, different responses to hemodilution have been reported depending upon species, awakeness or anaesthesia. Oxygen flux to most organs is maintained during isovolemic hemodilution until hematocrit reaches 9% while oxygen consumption remains unchanged. These more recent findings explain the complicationfree survival of extreme hemodilution (7% hct) of dogs. In moderate and more so in severe ischemia the reduction of microvascular red cell passage time elicited by hemodilution may be beneficial to tissue oxygenation. Recent studies on RBC-flow distribution in microcirculatory networks show that hemodilution is able to increase tissue oxygenation in regions endangered by underperfusion. Normal cardiac function and absence of silent coronary ischemia are the most important prerequisits to implement intentional hemodilution as alternative to homologous blood transfusion or therapeutic strategy in ischemic diseases.

Selected readings

Brückner UB, Messmer K: Blood rheology and systemic oxygen transport. Biorheology 1990; 27: 903-12

Intaglietta M: Microcirculatory effects of hemodilution: background and analysis. In: Tuma RF, White JF, Messmer K (eds) The role of hemodilution in optimal patient care. Zuckschwerdt, München, 1989, pp 21-41

Menger MD, Sack FU, Barker JH, Feifel G, Messmer K: Quantitative analysis of microcirculatory disorders after prolonged ischemia in skeletal muscle: Therapeutic effects of prophylactic isovolemic hemodilution. Res Exp Med 188: 151-165, 1988

Messmer K, Sunder-Plassmann L, Klövekorn WP, Holper K: Circulatory significance of hemodilution: rheological changes and limitations. In: Harders H (ed) Advances in microcirculation, vol. 4, Karger, Basel, 1972, pp 1-77

Messmer K: Acute preoperative hemodilution: physiological basis and clinical application. In: Tuma RF, White JF, Messmer K (eds) The role of hemodilution in optimal patient care. Zuckschwerdt, München, 1989, pp 54-73

Pries AR, Fritzsche A, Ley, Gaehtgens P: Redistribution of red blood cell flow in microcirculatory networks by hemodilution. Circ Res 1992; 70: 1113-1121

Van Woerkens ECSM, Trouwborst A, Van Lanschot JJB: Profound hemodilution: What is the critical level of hemodilution at which oxygen delivery-dependent oxygen consumption starts in an anesthetized human? Anest Analg 1992; 75: 818-21

Van Woerkens ECSM, Trouwborst A, Duncker DJGM, Koning MMG, Boomsma F, Verdouw PD: Catecholamines and regional hemodynamics during isovolemic hemodilution in anesthetized pigs. J Appl Physiol 1992; 72: 760-769

MODERATE ANEMIA DOES NOT DECREASE SUBCUTANEOUS TISSUE OXYGEN TENSION IN RABBITS

Harriet W. Hopf*, Dax Swanson and Thomas K. Hunt
Departments of Anesthesia* and Surgery
University of California, San Francisco, CA, USA

Introduction: Numerous animal studies have shown that normovolemic anemia does not decrease wound tensile strength until hematocrit falls below 17%. Although anemia does decrease arterial oxygen content, decreased blood viscosity may allow increased tissue perfusion and thus normal tissue oxygenation. Prior studies of anemia and healing have not evaluated oxygen delivery to the wound. Since the rate of collagen deposition in wounds is proportional to local oxygen tension, the lack of a detrimental effect of anemia on wound healing may be due to increased tissue perfusion. In the present study, subcutaneous tissue oxygen ($PsqO_2$) was measured in normovolemic anemic rabbits to evaluate the effect of anemia on tissue perfusion.

Methods: Anemia was induced in 3-4 kg New Zealand white rabbits (n=7) by phlebotomy every other day until a stable hematocrit of 30% was achieved. Control (n=5) rabbits were phlebotomized and immediately retransfused every other day for the same period of time. Once a stable hematocrit was achieved, rabbits were anesthetized with 1% halothane in air and a 15 cm Luer-hubbed, Silastic tube was implanted in the dorsal subcutaneous tissue. $PsqO_2$ was then measured continuously by a fluorescent oxygen sensor (InnerSpace Medical, Irvine, CA) which was inserted into the tube and flushed with hypoxic saline. $PsqO_2$ was recorded at an FiO_2 of 0.21 and then 1.0, after an equilibration period of at least 25 minutes at each FiO_2. Finally, a fluid bolus of 15 cc/kg was infused over 15 minutes, and the maximum $PsqO_2$ was recorded at an FiO_2 of 1.0. Results were compared using Student's t-test.

Results: $PsqO_2$ (mean±SD mmHg) was no different in control vs anemic rabbits at FiO_2 0.21 (63±10 vs 57±18, p=0.5), FiO_2 1.0 (99±11 vs 146±80, p=0.22), or after the fluid bolus (135±17 vs 191±81, p=0.17).

Conclusions: Because wound tissue oxygen delivery is diffusion limited, perfusion rather than hemoglobin determines oxygen delivery. Moderate anemia does not decrease wound oxygen delivery because perfusion increases markedly as red cell mass decreases. At high FiO_2, in fact, increased dissolved oxygen may even increase wound oxygen delivery.

Supported by NIH GM 27345 and InnerSpace Medical

MECHANISM OF TISSUE OXYGEN TRANSPORT ENHANCEMENT BY FLUOROCARBON BLOOD SUBSTITUTES AT NORMAL HEMATOCRIT

Thomas K. Goldstick, Robert A. Linsenmeier, Lyle F. Mockros, and Christopher M. Waters, Departments of Chemical and Biomedical Engineering, Northwestern University, Evanston, IL 60208-3120 USA

Fluorocarbon based emulsions have long been suggested as artificial blood substitutes. In previous experimental studies with *Fluosol DA®* (Green Cross Corp., Osaka, Japan) these emulsions have been found to enhance oxygen transport to the cat retina following exchange transfusion in cats breathing 100% oxygen. These cats had severely reduced hematocrits. Nevertheless enhancement occurred even though the arterial blood carried much less oxygen than before the exchange. In these studies the reduced blood viscosity contributed to a significant increase in blood flow. In more recent studies with a concentrated emulsion, *Oxygent™ HT* (Alliance Pharmaceutical Corp., San Diego, CA), similar enhancements have been observed at normal hematocrit even with extremely small amounts of added emulsion. In these studies there was only a negligible increase in arterial blood oxygen content and no apparent change in either blood viscosity or flow. Although the early results with exchange transfusion could be explained by increased blood flow, the more recent results are quite puzzling. Why did the tissue PO_2 increase with no change in blood PO_2, content or flow even with less than one volume percent of the fluorocarbon added to the blood?

The mechanism we propose involves an increased overall mass coefficient between blood and tissue. It seems apparent that the major oxygen transport resistance resides in the blood plasma either between the red cells in capillaries, and/or in the annulus between the red cells and the blood vessel wall in both capillaries and arterioles. The particular fluorocarbon tested most recently is perflubron (perfluorooctylbromide). This perfluorocarbon dissolves 24 times more oxygen than plasma. In addition, a near wall excess of particles smaller than red cells has been observed by Eugene Eckstein and his coworkers. This near wall excess would amplify the effect of even a small amount of perflubron and could play a major role in enhancing oxygen transport to tissue from arterioles. This phenomenon is also currently being studied in highly oxygen permeable (microporous polypropylene), 200 μm tubes in an extracorporeal membrane oxygenator. Inside each tube the mean velocity of the blood is 3 cm/s. Blood passes through each tube for 4 s. The results indicate a dramatic increase in the rate of blood oxygenation with the addition of even small amounts of the perflubron emulsion.

Based on the proposed mechanism, enhancement of oxygen transport by the perflubron emulsion will undoubtedly be clinically useful both in treating tissue hypoxia in patients as well as in improving the efficiency of extracorporeal membrane oxygenators.

EFFECTS OF PERFLUBRON EMULSION (AN ENHANCER OF OXYGEN CARRIAGE IN PLASMA) ON OXYGEN TRANSPORT IN THE ANEMIC ANESTHETIZED DOG

Emily C. Johnson,[1] B. Kipp Erickson,[1] Eric K. Birks,[1] Peter E. Keipert,[2] N. Simon Faithfull,[2] Peter D. Wagner.[1] [1]Department of Medicine, University of California, San Diego, La Jolla, CA 92093-0623 USA, [2]Alliance Pharmaceutical Corp., San Diego, CA 92121 USA

Oxygent™ HT, an oxygen transporting emulsion of perflubron (perfluorooctyl bromide [PFOB], Alliance Pharmaceutical Corp.), raises blood oxygen solubility, but its effects on hemodynamic and gas exchange parameters at moderate and high doses are incompletely documented. The purposes of this study were to examine consequences of increasing doses of Oxygent HT on 1) pulmonary gas exchange; 2) pulmonary and systemic hemodynamics; and 3) mixed venous P_{O_2} ($P\bar{v}_{O_2}$) and saturation. We studied intact, anesthetized dogs breathing 100% oxygen, in which Hct had first been reduced to 24-26% by exchange with 5% albumin (ALB). Dogs were infused intravenously with either Oxygent HT (n = 6) or 5% ALB (n = 5) in incremental doses of 3 ml/kg, up to 12 ml/kg. Arterial and venous blood gases, pH, base excess, and mean arterial pressure (MAP) were determined from femoral arterial and venous blood. O_2 content was determined by the LexO$_2$Con since O_2 saturation cannot be measured spectrophotometrically in the presence of Oxygent HT. Pulmonary artery mean and wedge pressures were measured using a Swan-Ganz catheter in the pulmonary artery; pulmonary (PVR) and systemic vascular (SVR) resistances were calculated. The multiple inert gas method was used to determine lung $\dot{V}_A/\dot{Q}$ relationships. Expired gases were collected and analyzed by mass spectrometry, and $\dot{Q}$ determined by Fick principle.

Results:

1. Four incremental doses of Oxygent HT (3 ml/kg each; 12 ml/kg total) increased blood O_2 solubility by a total of 0.002 ml/100ml/Torr, causing an increase in arterial O_2 concentration of 1.28 ml/100ml.

2. Pa_{O_2} remained unchanged at about 650 Torr, Pa_{CO_2} did not rise, and pulmonary $\dot{V}_A/\dot{Q}$ relationships, $\dot{V}_{O_2}$ and $\dot{V}_{CO_2}$ also remained unchanged as Oxygent HT was administered.

3. $\dot{Q}$ rose by 21% after 3 ml/kg Oxygent HT, but then fell with increasing doses (-18% from baseline after 12 ml/kg), while both pulmonary and systemic mean arterial pressures increased progressively, as did airway pressure.

4. $P\bar{v}_{O_2}$ rose from 66 to 73 (3 ml/kg) and to 77 Torr (6 ml/kg), then fell to 72 Torr (12 ml/kg). This was predicted quantitatively from corresponding changes in $\dot{Q}$ and O_2 solubility. When $P\bar{v}_{O_2}$ changes were predicted specifically ignoring the rise in O_2 solubility afforded by Oxygent HT, $P\bar{v}_{O_2}$ was considerably lower, and mixed venous O_2 saturation would have been about 10% lower after 12 ml/kg.

Conclusions

We conclude that Oxygent HT in this model:
1. Causes no measurable pulmonary gas exchange defect in doses as high as 12 ml/kg;
2. Leads to progressively higher PVR and SVR, and fall in $\dot{Q}$ when the dose exceeds 3-6 ml/kg, possibly due to increased blood viscosity;
3. Augments $P\bar{v}_{O_2}$ as expected from the increase in plasma O_2 solubility given the values of other pertinent variables.

(Supported by Alliance Pharmaceutical Corp.)

DETERMINANTS OF RED CELL MOTION IN THE MICROCIRCULATION

Shu Chien, Institute for Biomedical Engineering, Departments of AMES-Bioengineering and Medicine, and Center for Molecular Genetics, University of California San Diego, La Jolla, CA 92093-0412, U.S.A.

In order for red blood cells (RBCs) to perform their function of oxygen transport and exchange, they must be able to flow through capillaries with diameters smaller than the major diameter of the RBC. Thus, the motion of RBCs in traversing the microcirculation necessitates their change of shape in response to hemodynamic forces. The remarkable RBC deformability is an important factor in keeping the viscous resistance of the normal blood at a low level and facilitating blood flow through the microcirculation.

The viscosity of the blood in vessels larger than ~300 μm is essentially independent of the vessel diameter, but it decreases progressively with smaller vessel diameters to reach a minimum viscosity value at a diameter of about 7-10 μm (the Fahraeus-Lindqvist effect). One of the major reasons for this reduction in blood viscosity in small vessels is the concomitant decrease in hematocrit. Further reductions in vessel diameter lead to a rise in blood viscosity (inversion of the Fahraeus-Lindqvist effect) because of the necessity for greater RBC deformation in these narrow channels.

The three major determinants of RBC deformability are the relatively low viscosity of the interior hemoglobin-rich fluid, the excess surface area for the cell volume, and the cell membrane flexibility. RBC membrane is composed of a lipid bilayer and integral proteins which are connected to an underlying network of cytoskeletal proteins. With microrheological techniques, the viscoelastic properties of RBC membrane can be determined in normal conditions, following manipulations of membrane composition, and in disease states. Recent biochemical and molecular biological studies have elucidated the molecular organization of the RBC membrane. The major RBC membrane proteins have been cloned and sequenced, leading to the emergence of a picture for the molecular basis of membrane structure and function. Molecular biological studies performed in conjunction with rheological investigations have generated new insights into the molecular basis of the rheological characteristics of the RBC membrane. These findings have provided the input for theoretical modeling of the micromechanical behavior of RBCs during deformability tests and flow through narrow vessels. Such interdisciplinary studies have contributed to our understanding of the molecular basis of the dynamic behavior of RBCs in health and disease.

When the deformability of the red blood cell is reduced due to abnormalities in any of the determinants mentioned above, the resulting decrease in blood flow would tend to reduce the rate of oxygen delivery. However, there is usually a concomitant decrease in hematocrit to improve the blood flow rate and minimize the reduction in oxygen delivery. An example of this is sickle cell anemia, where the effect of the sickle cell rigidity on blood flow and oxygen delivery is partially offset by the anemia.

Ischemia, Reperfusion, and White Blood Cell Function in the Microcirculation

G. W. Schmid-Schönbein, M. Suematsu, J. Barroso-Aranda, Chavez-Chavez RH, Yee TT, DeLano FA, and Zweifach BW.

Institute for Biomedical Engineering, University of California, San Diego, La Jolla, California, 92093-0412, USA

Even under normal physiological conditions, the presence of leukocytes in microvessels influences local flow field and perfusion. Entry of leukocytes into capillaries may temporarily interrupt red cell motion. In capillaries, leukocytes move with a lower velocity than red cells, which leads to accumulation of erythrocytes upstream of the leukocyte and a plasma region downstream. Leukocyte adhesion to the post-capillary endothelium may also reduce microvascular perfusion. Consequently leukocytes may cause elevation of whole organ resistance.

Under conditions of ischemia or shock, the microvascular motion of leukocytes is further impaired. There are two mechanisms: Reduction of the perfusion pressure and decreased degree of deformation of leukocytes in microvessels, as in the case of acute local ischemia or hemorrhagic shock; activation of blood cells associated with elevated membrane adhesion, by expression of adhesion glycoproteins, as well as pseudopod formation, as in the case of endotoxic shock or inflammatory reactions.

One of the initial vascular responses after induction of ischemia is the microvascular entrapment of circulating neutrophil and monocytes in capillaries and adhesion to postcapillary venules. In rat skeletal muscle microvessels, with low spontaneous activation of circulating neutrophils, adhesion to the endothelium during early ischemia (< 1hr) is weak and neutrophils have only a mild influence on tissue cytotoxicity. During restoration of central blood pressure and reperfusion, the majority of leukocytes are rapidly washed out of the microvessels. However, the number of capillary plugging and venular adherent neutrophils increased again in time to levels above the ischemic period after about 30-45 minutes. Prior to this period (within about 15 min. of reperfusion) explosive cell damage is observed, starting along parallel capillaries which are well perfused. Glycolytic myocytes formed the majority of damaged cells, with significant less or undetectable damage to oxidative fibers. These results suggest, that in reperfusion the major initial damage to the tissue is mediated by endothelial cells, possibly via a xanthine oxidase mechanism, which is followed by neutrophil entrapment and enhancement of the organ injury. Thus, these results suggest that in experimental ischemia and reperfusion both neutrophil dependent and endothelium dependent cytotoxicity leads to parenchymal cell injury.

Supported by NIH grants HL 17682 (Dr. John Ross SCOR Director) and HL 43026 (Dr. Shu Chien PPG Director).

Relaxing and Contracting Factors in the Microcirculation
Robert L. Engler, M.D.
VAMC and UCSD, San Diego, CA

The five main determinants of coronary artery blood flow are: metabolic regulation, coronary perfusion pressure (auto-regulation), systolic compression, autonomic nervous system, and circulating substances and/or endothelial factors. Recent information on endothelial factors which regulate coronary blood flow are of interest to this meeting because of important interactions with molecular oxygen and with hemoglobin when used as a blood substitute for oxygen transport. Circulating substances/endothelial factors which alter coronary blood flow include catacholamines, prostacyclin, endothelial derived relaxant factor (EDRF, nitric oxide, NO), endothelial derived contracting factor, thromboxane A_2, endothelin, angiotensin-2, vasopressin, and neuropeptide Y. Nitric oxide which activates guanylate cyclase directly is released by endothelial cells in response to shear as well as certain pharmacologic stimuli including: thrombin, acetylcholine bradykinin, endotoxin, F-MET-LEU-PHE, leukotrienes, platelet activation factor, and some other non physiologic stimuli. Shear is an important initiator of EDRF release because of its role in auto-regulation. Through shear related NO release, other factors which cause large vessel coronary dilation, microvascular dilation or an increase in coronary perfusion pressure will result in increased vasorelaxation and a positive feedback cycle for coronary vasodilation. Thus, nitric oxide will counteract the intrinsic auto regulatory system which keeps blood flow constant when metabolic need is constant during changes in coronary perfusion pressure. Nitric oxide may also act as a feedback link between metabolically induced microvascular dilation and large vessel dilation, the signal being transmitted by increased shear. Thus metabolic microvascular dilation can act indirectly to counteract sympathetic macrovascular, α-adrenergic constriction during exercise or stress. The importance of these observations to oxygen transport is that superoxide directly antagonizes the nitric oxide signal by chemical interaction. Increases in tissue pO_2 or situations where superoxide radical formation is enhanced will diminish nitric oxide signaling. Of particular interest is the observation that atherosclerotic and hypercholesterolemic vessels in experimental animals and man showed diminished nitric oxide production in response to stimulation, thus decreasing an important factor for large vessel coronary dilation in coronary artery disease patients. These patients may be more sensitive to the effects of superoxide or free Hb. Hemoglobin when applied as a blood substitute or when liberated during hemolysis combines directly with nitric oxide to form methemoglobin and nitrate, which is ineffective at increasing guanylate cyclase activity and subsequent vasodilation. Normally hemoglobin confined to red blood cells has little effect on nitric oxide signaling because of the extremely short half life of nitric oxide and the fact that the signaling pathway does not physically allow for red cell interference. When hemoglobin is free in the plasma or diffuses into the interstitial space perhaps between the endothelial cell and target vascular smooth muscle, it may have profound effects on vascular tone. Thus, infusion of hemoglobin in clinical trials causes a significant, sustained increase in peripheral vascular resistance. In addition nitric oxide has other important physiologic roles including anti-inflammatory activity, inhibiting platelet aggregation, and functioning as a neurotransmitter. The extent to which hemoglobin or superoxide physiologically interfere with these functions of nitric oxide are important areas for future investigation.

ASSESSMENT OF ORGAN FUNCTION IN MULTIPLE SYSTEM ORGAN FAILURE

Guillermo Gutierrez M.D., Ph.D. Pulmonary and Critical Care Medicine Division. U. of Texas
Health Science Center . Houston, TX 77030

Regional tissue hypoxia is a possible cause of multiple system organ failure (MSOF) in patients
with sepsis or with the adult respiratory distress syndrome (ARDS). However, a consensus is
emerging among researchers and clinicians that systemic measures of oxygen transport and
consumption are inadequate indexes of regional tissue hypoxia. This appears to be true also for
other systemic parameters of tissue oxygenation, such as arterial lactate concentration and mixed
venous PO_2. Arterial lactate concentration may be misleading, since it may be normal in spite of
regional tissue hypoxia. Conversely, elevated arterial lactate concentrations may occur without
peripheral tissue hypoxia. The reason for this discrepancy lies in the complicated kinetics of
whole body lactate utilization, since some organs consume and others produce lactate. In ARDS
the lung may be a major source of lactate production, perhaps the result of increased metabolic
activity of inflammatory cells. Increases in lung lactate production have been correlated with the
severity of ARDS, as characterized by the Lung Injury Score.

The utility of measuring aerobic metabolic activity to monitor organ dysfunction remains to be
established. These metabolic parameters may provide an early monitoring of tissue hypoxia and
perhaps help prevent the onset of MSOF. Among these measures are the degradation products of
adenosine monophosphate. Arterial concentrations of uric acid, hypoxanthine, and inosine,
measured with high pressure liquid chromatography, have been correlated with the severity of
critical illness, although individual organ production of these metabolites is difficult to measure.

Sepsis and hypoxia are characterized by major changes in the distribution of the cardiac output,
to the various organs, resulting in profound alterations in regional tissue perfusion. Blood flow
maldistribution in the critically ill may be the consequence of a uniform vascular response to any
type of systemic stress, be it endotoxemia, hypoxia, or fear, in the form of the "fight or flight"
reaction. The combination of neurological vascular control, as well as local increases in
capillarity, combine to maintain, or even augment, blood flow to organs involved in locomotion,
such as skeletal muscle, the heart, the adrenal glands, and the brain. Therefore, organs lacking
the capacity to autoregulate blood flow tend to receive less than their share of perfusion. Among
these organs are the kidneys and the gut, which exhibit a countercurrent microvascular
arrangement where O_2 diffuses from arteriole to venule, creating a longitudinal PO_2 gradient.
Parameters that reflect changes in aerobic metabolism in these organs may be useful in
monitoring early organ hypoxia. Tonometrically measured intestinal or gastric mucosal PCO_2
appears to be a useful clinical measurement. Increases in mucosal PCO_2 occur as hydrogen ions
produced by the hydrolysis of anaerobically generated ATP are buffered by extracellular
bicarbonate. Tonometric measures of PCO_2, when combined with measures of arterial
bicarbonate in the Henderson-Hasselbalch equation, provide an estimate of the gut mucosal pH
(pHi). Low pHi values (< 7.32) have been associated with greater ICU mortality and a
prospective study, in which resuscitation of critically ill patients was guided by changes in pHi,
showed improved survival in patients admitted to the intensive care unit with pHi ≥ 7.35.

CHANGE IN OXYGEN AFFINITY IN PRENATAL AND POSTNATAL YOUNG OF A
VIVIPAROUS AUSTRALIAN ELAPID SNAKE (*Pseudechis porphyriacus*)
*Robert A.B. Holland, Sandra L. Butler, and Susan J. Calvert. School of Physiology and
Pharmacology, University of New South Wales, Kensington, Sydney, New South Wales,
2033, Australia*

Of the six species of snake in the genus *Pseudechis*, only óne, the Red-bellied Black
Snake, *(Pseudechis porphyriacus)* is viviparous. The snakes mate in early October and give
birth in early February, the gravid period extending over part of spring and summer. The
young are born enclosed in membranes from which they emerge usually within 24 hours.
Adult snakes are about 1.2 metres long and newborn snakes about 18 cm long.

In preliminary work we found that the maternal-fetal P_{50} difference was due in part
to the fetal oxygen equilibrium curve (OEC) being to the left of the non-gravid adult OEC
but also due to the maternal OEC being to the right of the non gravid adult OEC.

This work describes the OEC of 88 developing snakes at different stage from 34
days before birth to 32 days after birth.

Adult snakes were captured in the wild in early October soon after mating, and were
held until several weeks after they had given birth. Embryos were surgically removed
(halothane anaesthesia) and blood taken from the anaesthetized embryos. Post natal young
were killed by decapitation, and blood taken.

The OECs were measured on a modified Hemoscan (thin film method) the standard
PCO_2 being 21 Torr and standard temperature 30°C. P_{50} was read from the curves and
Hill plots made of the OEC data points. ATP in the red cells was measured by the UV
method with phosphoglycerate kinase, following the change from NADH to NAD at 340
nm (Sigma kit).

The mean P_{50} of red bellied black snakes before birth was 26 Torr. After birth the
mean P_{50} was 31.5 Torr at 0-1 days, 33 Torr at 2-3 days and 36 Torr at 4-10 days. By 4
weeks after birth, all snakes had a P_{50} in the range 40-45 Torr, our normal range for non-
gravid adults.

The Hill plots were bent and n_H was calculated for lower and upper ranges (around
25% and 75% saturation). In the lower part n_H was usually between 1.8 and 3. The value
of n_H in the upper part was 4.2 (mean) and it was 5 or greater in 10 snakes before birth or
within 15 days after birth.

No met Hb was found to be present. Previous isoelectric focusing of haemoglobin
from snakes of this species before and just after birth showed only one haemoglobin, which
had a pI similar to that of adult Hb.

The red cell concentration of ATP, a co-factor known to decrease the O_2 affinity of
haemoglobin, was found to increase during the period over which the young snakes were
observed postnatally.

ACTIVE AMINO ACID TRANSPORT AND OXYGEN METABOLISM IN THE PERFUSED HUMAN PLACENTAL
LOBULE

[1]D.J. Maguire, [1,2]S.M. Marshall, [2]R.H.Mortimer and [2]G.R. Cannell.

[1]Faculty of Science and Technology, Griffith University, Australia Q4111,
[1]Conjoint Endocrine Laboratory, Royal Brisbane Hospital, Australia Q4029.

Amino acids exhibit predominantly maternofetal transport due to specific channels within
the human placenta. Our investigations have established that a change in oxygen delivery was
correlated with changes in glucose consumption, lactate production, and the clearance of [14C]-
leucine and a reference marker (3H-Water) in the perfused lobule.   Lobules were subjected to
oxygenated and hypoxic conditions under varying fetal perfusate flow rates (1,3,6 ml/min).
Substrates were dosed in the maternal compartment of an established model (Cannell et al 1988).

Clearance of [3H]-water increased with corresponding fetal perfusate flows both in the
oxygenated and hypoxic series.   Clearance reached maximins 0.261 (+/- 0.182) under oxygenated
conditions and 0.191 (+/- 0.129) under hypoxic conditions.   Clearance of 14 C-leucine exhibited a
similar response.   Clearances were 0.051 (+/- 0.046)- 0.145 (+/-0.05) for oxygenated and 0.041 (+/-
0.024)-0.105 (+/-0.133) for hypoxic series.

Lactate production increased as fetal perfusate flow rates decreased for the oxygenated
(43.54 +/-15.8 - 31.31+/-5.98) and hypoxic (56.09+/-34.05 - 36.9=10.8) series respectively.

Glucose consumption appeared to plateau at a fetal perfusate flow rate of 3 ml/min under
oxygenated and hypoxic conditions.

Cannell, G.R., Kluck, R.M., Hamilton, S.E., Mortimer, R.H., Hooper, W.D., and Dickinson, R.G.(1988)
Clin Exp Pharm Physiol 15, 837-844.

PLACENTAL PROPRANOLOL METABOLISM IN NORMOXIA AND HYPOXIA

[1]G.R. Cannell, [1,2]A.J. Fletcher, [1]R.H. Mortimer and [2]D.J. Maguire

[1]Conjoint Endocrine Laboratory, Royal Brisbane Hospital, Herston, Australia Q4029
[2]Faculty of Science and Technology, Griffith University, Nathan, Australia Q4111

Propranolol is a non-selective ß-blocker which has been used to treat hypertension during pregnancy. In humans and dogs, the metabolism of propranolol is microsomally mediated, and stereoselective (von Bahr et al, 1982). The disposition of this drug has also been studied in the isolated perfused liver, the major organ of detoxification (Jones et al, 1984). In that investigation, anpoxia was induced in the perfusion system and it was reported that metabolism of propranolol was markedly reduced despite normal uptake of the drug. We report an investigation into the effect of reduced oxygen supply upon placental metabolism of propranolol.

Term placentas were obtained following either vaginal birth or caesarian section from patients with no significant history of illness or drug use. The foetal and maternal circulations of a placental lobule were perfused as previously described (Cannell et al, 1988). The perfusate consisted of M199 tissue culture medium, heparin (25 IU/ml), gentamicin (100 mg/l), glucose (2 g/l), sodium bicarbonate (2.9 g/l), and dextran (29 g/l), foetal and 7.5 g/l, maternal) with the pH maintained between 7.35 and 7.45. Both the maternal and foetal circulations were recirculated for 6 hours (Miller et al., 1985; Schneider et al.,1985) with flow rates of 25 and 3 ml/min respectively. The system was dosed with racemic unlabelled propranolol and L-[4-^{3}H]-propranolol. Perfusion samples were extracted with Sep-pak C18 solid phase cartridges and analysed by HPLC using a radiochemical detector and by GC-MS. The perfusion was initally established under normal (oxygenated) mode. Hypoxia was then induced by equilibrating both the foetal and maternal perfusates with N_2/CO_2. The perfusion chamber was open to the atmosphere.

Under these conditions, a pO2 of approximately 100 mm Hg was maintained throughout the placenta. By comparison, in the standard oxygenated perfusion system, a distinct oxygen gradient exists between maternal and foetal circulations and the lowest foetal perfusate oxygen levels are significantly above 100 mm Hg.

Propranolol equilibrated between the maternal and foetal circulations within one hour. In 6 hour perfusate samples, two compounds were detected, one being unchanged propranolol and the other being desaminoisopropylpropranolol formed by partial degradation of the side chain of the parent compound. The ratio of propranolol to its major metabolite was essentially the same in oxygenated and deoxygenated experiments. However, a ratio which was significantly different from oxygenated perfusion experiments was found in a smoking cohort.

Cannell, G.R., Kluck, R.M., Hamilton, S.E., Mortimer, R.H., Hooper, W.D., and Dickinson, R.G.(1988) Clin Exp Pharm Physiol 15, 837-844.
Jones, D.B., Mihaly, G.W., Smallwood, R.A., Webster, L.K., Morgan, D.J. and Madsen N.P. (1984) Hepatology 4, 461-466.
von Bahr, C., Hermansson, J. and Lind, M. (1982) J. Pharm. Exp. Therap. 222, 458-462.

CAPNOGRAPHIC CURVE AND CARDIAC OUTPUT MEASUREMENT IN CRITICALLY   ILL PATIENTS

Renzo L. Zatelli, Departement of Anesthesiology and Intensive Care,
S.Anna Hospital, Ferrara, Italy

Capnography is a non invasive, easy to use and routine monitoring in Intensive Care Unit (I.C.U.).Changes of Ventilation/Perfusion ratio (V_A/Q) can modify the shape of the capnographic curve very drammatically and the onset of broncospasm decreases the slope of second portion of capnogram increasing the angle α between PQ and QR lines. If a correlation between angle α and V_A/Q exists [Kalenda], it could allow to compute cardiac output (C.O.) in a non invasive way.

METHODS : In I.C.U. patients capnographic curves were recorded by a side-stream capnograph (Oscar Datex) and analyzed by a computer program to obtain: end tidal CO_2 ($ETCO_2$), mean expiratory CO_2 ($mECO_2$), angle α, fractional expiratory time to reach 50% of $ETCO_2$ value (FET_{50}) and PQ line slope. External ventilation (Siemens Servoventilator C), $ETCO_2$ and $mECO_2$ were recorded in order to compute alveolar ventilation (V_A). C.O. was measured by thermodilution method (Edwards Baxter REF.1) through a Swan-Ganz catheter on a base of three measurements and computed by angle α and V_A [C.O. = V_A / (a + b*α)]. Measured and computed C.O. were compared by paired T test and correlation was tested. P < 0.05 was accepted as significant.

RESULTS: Computed C.O. are smaller than measured C.O. (mean : $8.65 \pm$ s.d.1.99 vs. 8.81 ± 1.96, P = 0.768) and their correlation is weak: r = 0.589 (P = 0.034). No correlation there is between V_A/Q and angle α , r = 0.18.

CONCLUSIONS: In studied critically hill patients capnographic curve pattern (mainly angle α and F.E.T.$_{50}$) was not able to quantify C.O. values in clinically useful way.

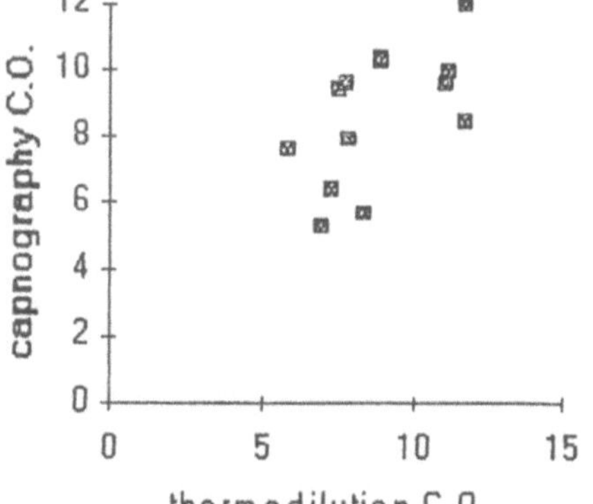

REFERENCE::

Kalenda Z: Mastering Infrared Capnography, Kerckebosch BV - Zeist - The Netherlands 1989 p 101.

EFFECTS OF ENERGY DEMAND IN ISCHEMIC AND IN HYPOXEMIC ISOLATED RAT HEARTS

Michele Samaja, Stefania Casalini, Sonia Allibardi and Antonio Corno*.

Dept. of Biomedical Science & Technology
Scientific Institute San Raffaele, University of Milan,
*Hospital San Donato, Milan, Italy[1]

INTRODUCTION

While myocardial ischemia is characterized by low coronary flow rate (CFR) at high arterial PO_2 (P_aO_2), during hypoxemia P_aO_2 is low at high CFR. Although both situations may potentially lead to dysoxia, defined as a condition with unbalanced O_2 supply/demand ratio (Connett *et al.*, 1990), the washout of membrane-diffusible catabolites such as lactate is depressed during ischemia but not during hypoxemia because of the different CFR's that determine different washouts of intracellular lactate. It is therefore tempting to speculate that, when ischemia and hypoxemia are matched for the O_2 supply, hypoxemia becomes equivalent to an ischemic condition with enhanced washout of lactate. This hypothesis provides a good opportunity to evaluate the role of O_2 and lactate in dysoxic contractile systems.

The isolated Langendorff-perfused rat heart is particularly suitable to study the effects of hypoxemia and ischemia because CFR and P_aO_2 can be regulated by a pump and a membrane oxygenator to yield selected O_2 supplies:

$$O_2 \text{ supply} = CFR \times P_aO_2 \times \alpha \qquad (1)$$

where α, which represents the O_2 solubility coefficient, remains constant in aqueous buffers. Furthermore, isolated hearts are accessible to several physiological and metabolic measurements. Finally, use of blood in the perfusing medium was avoided at the expense of dealing with unphysiological conditions, but with clear advantages of accuracy and precision in measuring the O_2 content. The purposes of this study were: 1) Determining whether hypoxemia and ischemia at the same O_2 supply elicit the same responses, and hence if O_2 is critical regulator of myocardial function and metabolism in dysoxia; 2) Defining the effect of increased energy demand (electrical stimulation) in hypoxemic and ischemic hearts to assess if these hearts have a reserve of energy.

---

[1]Address correspondence to: Michele Samaja, Dept. of Biomedical Science and Technology, via Olgettina 60, I-20132 Milano, ITALY.

MATERIALS AND METHODS

General

We perfused isolated rat hearts with oxygenated buffer by a Langendorff technique monitoring myocardial function, O_2 uptake (VO_2) and lactate production rate (J_{Lac}). Hearts were stabilized for 20 min at CFR=15 ml/min and P_aO_2=670 mmHg. The volume of the intraventricular balloon was set to yield end-diastolic pressure (EDP)=7.0±0.5 mmHg and was kept constant throughout. Under these conditions, the O_2 supply was (see eq.1):

$$\text{Baseline: } 15 \times 670 \times \alpha = 14.1 \ \mu\text{moles } O_2/\text{min} \qquad (2)$$

where α is 1.4×10^{-6} moles/L/mmHg (Roughton and Severinghaus, 1973). At t=0 min, the O_2 supply was shortened reducing either CFR (ischemic group, n=6) or P_aO_2 (hypoxemic group, n=6) to 10% of the baseline. The O_2 supplies under these conditions were, respectively:

$$\text{Ischemia: } 1.5 \times 670 \times \alpha = 1.41 \ \mu\text{moles } O_2/\text{min} \qquad (3)$$

$$\text{Hypoxemia: } 15 \times 67 \times \alpha = 1.41 \ \mu\text{moles } O_2/\text{min} \qquad (4)$$

Hearts were initially allowed to adjust their heart rate (HR), but at t=20 min HR was set to 300 min^{-1} for 10 min in all groups. Measurements were taken at the end of the various phases at stable myocardial function.

Apparatus

The perfusing buffer (115.6 mM NaCl, 4.7 mM KCl, 1.2 mM KH_2PO_4, 0.5 mM EDTA, 1.2 mM Na_2SO_4, 28.5 mM $NaHCO_3$, 2.5 mM $CaCl_2$, 1.2 mM $MgCl_2$, 16.6 mM glucose, pH 7.4 at 37°C) was equilibrated in Sylastic membrane oxygenators (Dideco, Italy) at 37°C with gases containing either 94/6/0 or 0/6/94 $O_2/CO_2/N_2$ to yield P_aO_2=670 or 67 mmHg at constant P_aCO_2 (43 mmHg). A roller pump delivered the buffer at either 15 or 1.5 ml/min to a filter (8 μm pore size, 47 mm diameter, Nuclepore Corp., Pleasanton, CA), a preheater and the aortic cannula.

Hearts from male Sprague Dawley rats (250-280 g), anesthetized by i.p. heparinized sodium thiopental (10 mg/100 g b.w.), were mounted on the system and immersed in the buffer kept at 37°C. The venous return was collected by the pulmonary artery and a saline-filled Latex balloon was introduced into the left ventricle. A square wave stimulator (Harvard, South Natick, MA) with 5 ms pulse duration and 10 V pulse amplitude was connected to electrodes placed on the aortic cannula and on the apex of the ventricle.

Measurements, calculations and statistics

A pressure transducer (Harvard Apparatus mod.52-9966, Natick, MA) connected to the balloon provided EDP and the developed pressure (LVDP) by means of a dedicated LabVIEW®2 system (National Instruments, Austin, TX) running on Macintosh Quadra 700 computer. The venous return was analyzed for PO_2 (YSI mod.5300 Oxygen Monitor, Yellow Springs Inc., OH) and lactate (Sigma Diagnostic, St.Louis, MO). Data are expressed as mean±SEM. The Student's t-test for unpaired and paired observations was used to compare hypoxemic and ischemic hearts and to evaluate the effects of pacing, respectively. The significance level was set to p=0.05 (two-tailed).

RESULTS

HR decreased in ischemic (p=0.001) but not in hypoxemic hearts (Fig.1). Hypoxemic hearts underwent diastolic contracture (p<0.0005) but their LVDP was greater than in ischemic hearts (p=0.001). Pacing did not affect EDP and decreased LVDP (p=0.01) in hypoxemic hearts.

Although P_vO_2 was higher in ischemic than hypoxemic hearts (Fig.2, p<0.0005), VO_2 was the same in both groups. Since P_vO_2 was not changed by pacing, VO_2 remained constant. Venous [lactate] was not affected by pacing in hypoxemic hearts, but increased in ischemic hearts (Fig.3, p=0.002). The lactate production rate (J_{Lac}) was calculated from venous [lactate] and CFR, and was higher in hypoxemic than in ischemic hearts (p<0.0005). Pacing, however, increased J_{Lac} in ischemic hearts only (p=0.002).

The myocardial contractile work was expressed as LVDP×HR (Fig.4) and was lower in ischemic than in hypoxemic hearts (p<0.0005). Pacing increased LVDP×HR in ischemic hearts (p=0.005) only without influencing hypoxemic hearts. The turnover of ATP (J_{ATP}) was calculated using steady-state stoichiometry $6.42 \times VO_2 + 1.25 \times J_{Lac}$ (Paul, 1980) assuming no mitochondrial uncoupling. As for LVDP×HR, J_{ATP} was higher in hypoxemic hearts (p<0.0005) and pacing increased J_{ATP} in ischemic hearts only (p=0.001) without affecting hypoxemic hearts.

DISCUSSION

Myocardial function was more depressed during ischemia than during hypoxemia. However, ischemic hearts upgraded their performance when stimulated, while this ability was blunted in hypoxemic hearts. Despite different P_vO_2 in the two groups, VO_2 was the same and was unaffected by pacing. Both venous [lactate] and J_{Lac} increased in ischemic hearts upon pacing but was constant in hypoxemic hearts.

Thus, when matched for the O_2 supply, hypoxemia and ischemia elicited different responses in our model consistently with recently reported data (Stainsby *et al.*, 1990; Dodd *et al.*, 1993). In another study (Hogan *et al.*, 1992), however, working dog gastrocnemius muscles *in situ* were exposed to either hypoxemia or ischemia applying an O_2 reduction of ~50%, but no bioenergetic differences were observed at three levels of stimulation. The discrepancy between these results is likely due to major procedural differences and to the more severe energy imbalance in our hearts.

One the main differences between hypoxemia and ischemia is the washout of diffusible catabolites. Thus, it may be inferred that this feature, and not the O_2 supply *per se*, determined the observed differences between ischemic and hypoxemic hearts. Since VO_2 was the same in both groups and was not varied by pacing, O_2 was not an adequate reserve of energy in this model. The different P_vO_2 values in ischemic and hypoxemic hearts reflect the low perfusion pressure in ischemic hearts secondary to their lower CFR. Such condition may have diminished the number of open capillaries per unit tissue volume increasing the artero-venous shunt and decreasing the tissue ability to extract O_2 (Hogan *et al.*, 1993).

The greater performance of hypoxemic hearts with respect to matched ischemic hearts, and the ability of ischemic hearts to increase their performance when stimulated appear linked to the anaerobic glycolysis capacity. Both venous [lactate] and J_{Lac} increased upon stimulation in ischemic hearts, but remained constant in hypoxemic hearts. Further, J_{Lac} was higher in hypoxemic than in ischemic hearts, irrespectively of pacing, suggesting that during hypoxemia anaerobic glycolysis was working at or near maximum, irrespectively of the actual demand of energy. This observation is consistent with the

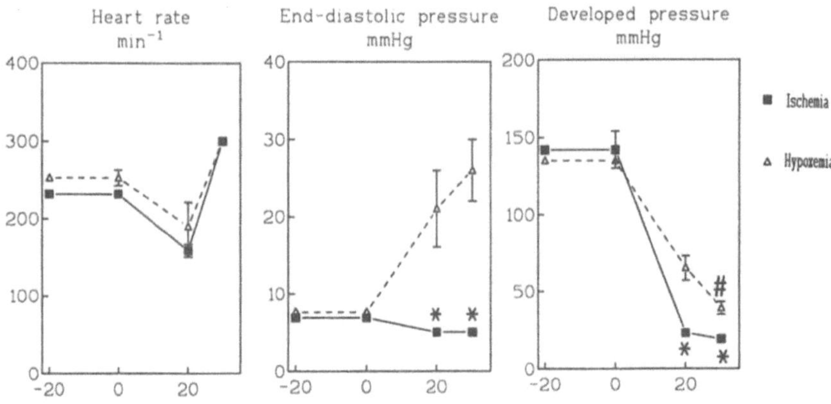

Figure 1. Heart rate, end-diastolic pressure and developed pressure in ischemic (■) and hypoxemic (Δ) hearts (n=6 for each) during baseline (t=0 min), dysoxia (t=20 min) and in dysoxic paced hearts (t=30 min). *=significant difference between hypoxemic and ischemic hearts, unpaired Student's t-test; #=significant difference between paced and spontaneous hearts, paired Student's t-test.

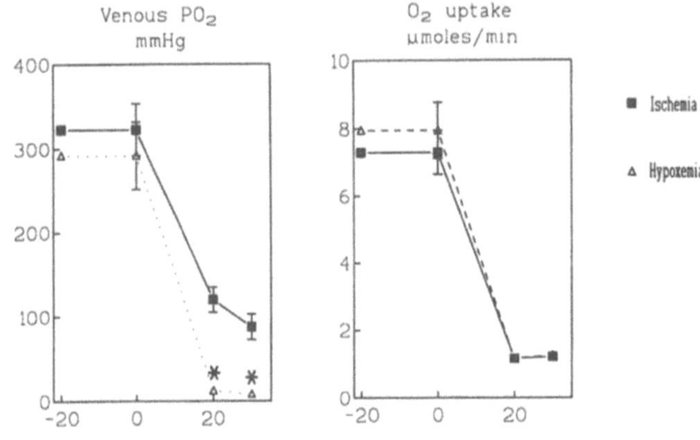

Figure 2. Metabolism of O_2 (venous PO_2 and O_2 uptake) during baseline (t=0 min), dysoxia (t=20 min) and in dysoxic paced hearts (t=30 min). Other information in Fig.1.

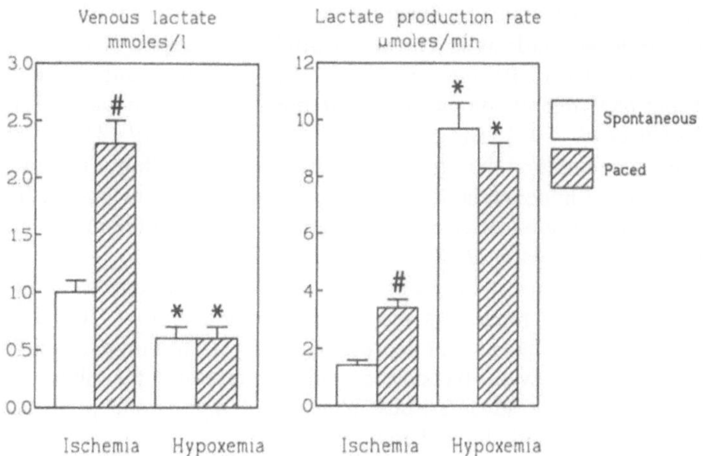

Figure 3. Metabolism of lactate (venous [lactate] and lactate production rate in ischemic and hypoxemic hearts, with and without pacing. Other information in Fig.1.

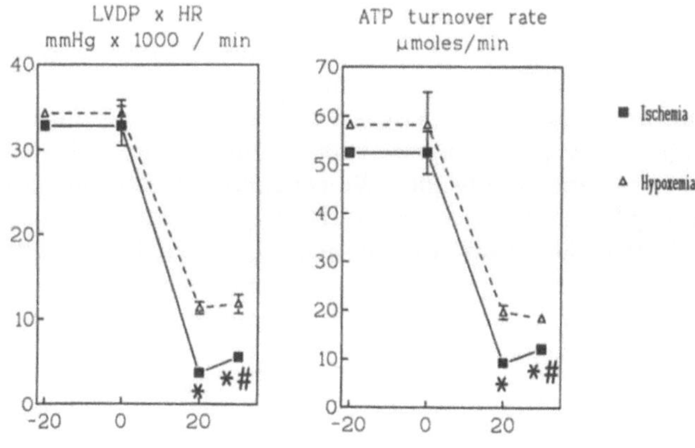

Figure 4. Myocardial contractile work (LVDP×HR) and ATP turnover rate (J_{ATP}) during baseline (t=0 min), dysoxia (t=20 min) and in dysoxic paced hearts (t=30 min). Other information in Fig.1.

lower glycolytic rate in ischemic than anoxic working rat hearts (Rovetto *et al.*, 1973; Kobayashi and Neely, 1979). However, these data also show that the depression of anaerobic glycolysis in ischemic hearts can be relieved following the increased demand of energy, while this mechanism does not operate during hypoxemia.

Venous [lactate] was higher in ischemic than in hypoxemic hearts reflecting impaired washout in ischemia, consistently with earlier observations (Tani and Neely, 1990). Lactate-induced acidosis is known to inhibit glycolysis (Kobayashi and Neely, 1979; Zhou *et al.*, 1991; Rovetto *et al.*, 1975). An important consequence of the associated functional depression is enhanced post-ischemic metabolic and functional recovery (Currin *et al.*, 1991; Bing *et al.*, 1973; Schaefer *et al.*, 1990). The protective effect of lactate accumulation during ischemia is consistent with our previous reports that hearts recovered from hypoxemia are more injured than those recovered from ischemia, when the two conditions are matched for O_2 supply, duration of the insult and temperature (Corno *et al.*, 1993; Samaja *et al.*, 1994).

In conclusion, ischemic hearts are downregulated. In contrast, hypoxemic hearts are not. Most likely, this process is modulated by lactate: low CFR during ischemia induces high intracellular lactate that depresses glycolysis and myocardial function; high CFR during hypoxemia prevents lactate accumulation releasing the inhibition. The former feature may be considered advantageous in terms of myocardial protection because it prevents energy wasting and allows better recovery.

SUMMARY

Aim of this study was to assess the role of O_2, lactate and energy demand in the regulation of myocardial work during severe dysoxia. For this purpose, we measured function and metabolism in isolated Langendorff-perfused rat hearts exposed to either ischemia or hypoxemia (matched for the O_2 supply, 10% of baseline) with/out electrical stimulation. When hearts could adjust their HR, hypoxemia demanded more energy than ischemia ($p < 0.05$) despite same O_2 supply. Venous PO_2 was 12 ± 2 or 139 ± 20 mmHg ($p < 0.0001$), respectively, but VO_2 was the same. After 10 min at HR=300 min^{-1}, myocardial performance increased in ischemic but not in hypoxemic hearts. P_vO_2 and VO_2 were not affected by pacing. In contrast, both venous [lactate] and lactate production rate increased, but in ischemic hearts only. We conclude that ischemic hearts were downregulated while hypoxemic hearts were not. Likely, depressed washout of lactate during ischemia could offset the effects of O_2 in severely dysoxic hearts. Anaerobic glycolysis provided the energy necessary to meet increased energy demand in ischemic hearts, but could not exploit this action in hypoxemic hearts probably because in these hearts it was already working near maximum.

ACKNOWLEDGEMENTS

We thank Drs. Galavotti and Menghini, Dideco, Italy, for help to design the oxygenator used in this study. We also thank Dr.Sanna for help in building the monitoring system. This work was supported by grants from the Fondazione San Romanello del Monte Tabor, Milan, and the Target Project BTBS, Rome, Italy.

REFERENCES

Bing, O.H.L., Brooks, W.W., and Messer, J.V., 1973, Heart muscle viability following hypoxia: protective effect of acidosis. *Science* 180:1297.

Connett, R.J., Honig, C.R., Gayeski, T.E.J., and Brooks, G.A., 1990, Defining hypoxia: a systems view of VO_2, glycolysis, energetics, and intracellular pO_2. *J. Appl. Physiol.* 68:833.

Corno, A.F., Motterlini, R., Brenna, L., Santoro, F., and Samaja, M., 1993, Ischaemia/reperfusion in the posthypoxaemic re-oxygenated myocardium: haemodynamic study in the isolated perfused rat heart. *Perfusion* 8:113.

Currin, R.T., Gores, G.J., Thurman, R.G., and Lemasters, J.J., 1991, Protection by acidotic pH against cell killing in perfused rat liver: evidence for pH paradox. *FASEB J.* 5:207.

Dodd, S.L., Powers, S.K., Brooks, E., and Crawford, M.P., 1993, Effects of reduced O_2 delivery with anemia, hypoxia, or ischemia on peak VO_2 and force in skeletal muscle. *J. Appl. Physiol.* 74:186.

Hogan, M.C., Nioka, S., Brechue, W.F., and Chance, B., 1992, A ^{31}P NMR study of tissue respiration in working dog muscle during reduced O_2 delivery conditions. *J. Appl. Physiol.* 73:1662.

Hogan, M.C., Bebout, D.E., and Wagner, P.D., 1993, Effect of blood flow reduction on maximal O_2 uptake in canine gastrocnemius muscle in situ. *J. Appl. Physiol.* 74:1742.

Kobayashi, K. and Neely, J.R., 1979, Control of maximum rate of glycolysis in rat cardiac muscle. *Circ. Res.* 44:166.

Paul, R.J., 1980, Chemical energetics of vascular smooth muscle. In Handbook of Physiology. The Cardiovascular System. Vascular Smooth Muscle, p. 201, Am.Physiol.Soc., Bethesda, MD.

Roughton, F.J.W. and Severinghaus, J.W., 1973, Accurate determination of O_2 dissociation curve of human above 98.7% saturation with data on O_2 solubility in unmodified human blood from 0 ° to 37 °C. *J. Appl. Physiol.* 35:861.

Rovetto, M.J., Whitmer, J.T., and Neely, J.R., 1973, Comparison of the effects of anoxia and whole heart ischemia on carbohydrate utilization in isolated working rat hearts. *Circ. Res.* 32:699.

Rovetto, M.J., Lamberton, W.F., and Neely, J.R., 1975, Mechanisms of glycolytic inhibition in ischemic rat hearts. *Circ. Res.* 37:742.

Samaja, M., Motterlini, R., Santoro, F., Dell'Antonio, G., and Corno, A.F., 1994, Oxidative injury in reoxygenated and reperfused hearts. *Free Rad. Biol. Med.* 16:255.

Schaefer, S., Schwartz, G.G., Gober, J.R., Wong, A.K., Camacho, S.A., Massie, B., and Weiner, M.W., 1990, Relationship between myocardial metabolites and contractile abnormalities during graded regional ischemia. *J. Clin. Invest.* 85:706.

Stainsby, W.N., Brechue, W.F., O'Drobinak, D.M., and Barclay, J.K., 1990, Effects of ischemic and hypoxic hypoxia on VO_2 and lactic acid output during tetanic contractions. *J. Appl. Physiol.* 68:574.

Tani, M. and Neely, J.R., 1990, Vascular washout reduces Ca^{2+} overload and improves function of reperfused ischemic hearts. *Am. J. Physiol.* 258:H354.

Zhou, H.Z., Malhotra, D., and Shapiro, J.I., 1991, Contractile dysfunction during metabolic acidosis: role of impaired energy metabolism. *Am. J. Physiol.* 261:H1481.

EFFECT OF PERFUSION PRESSURE ON REGIONAL MYOCARDIAL OXYGEN CONSUMPTION AND END-DIASTOLIC SEGMENT LENGTH IN SWINE MYOCARDIUM

Brian A. Cason, Igor Shubayev, Robert F. Hickey

University of California, San Francisco, CA 94143
Veterans Affairs Medical Center of San Francisco, CA 94121

INTRODUCTION

Increases in coronary perfusion pressure cause both myocardial oxygen consumption and force of contraction to increase (Gregg's phenomenon) (Gregg 1963). Although the underlying mechanism of this phenomenon is unclear, several have been proposed. One proposal is that Gregg's phenomenon is largely artifactual, and is found primarily when isolated coronary perfusion causes mild ischemia (Schulz, Guth et al. 1991). Increasing flow then relieves ischemia and improves contractile function. Another mechanism which has been proposed to explain the Gregg phenomenon is that increased coronary flow increases the delivery of oxygen, calcium, or metabolic substrates, which are rate-limiting under normal conditions, even though there is no resting ischemia. Increased substrate delivery improves contractility. A third hypothesized mechanism is the "garden hose effect": increased perfusion pressure may distend arterioles and increase resting stretch to adjacent sarcomeres, thus increasing the force of contraction by the Frank-Starling mechanism (Arnold, Kosche et al. 1969; Arnold, Morgenstern et al. 1970; Scharf and Bromberger-Barnea 1973). Although work in isolated papillary muscles suggests that the "garden hose effect" does not play a significant role in the Gregg phenomenon (Schouten, Allaart et al. 1992), this has not been studied in vivo. In order to determine whether the "garden-hose effect" plays a significant role in Gregg's phenomenon in vivo, we used a swine model of isolated coronary perfusion to study the relationship between perfusion pressure, myocardial oxygen consumption and end-diastolic segment length.

METHODS

General Methods

Anesthesia. This experimental protocol was approved by our Animal Welfare Committee, and follows the guidelines for animal use provided by the American Physiological Society. Studies were performed under general anesthesia in 8 open-chest domestic swine weighing 40-50 kg. Swine were premedicated with ketamine (10 mg·kg^{-1} s.c.), then anesthesia was induced by mask using oxygen and isoflurane (1-4%). A tracheostomy was performed under deep general anesthesia, ventilation was controlled, and anesthesia was maintained with isoflurane 0.8 - 1.5%. Ventilation was controlled to keep P_aCO_2 at 35-40 mmHg. After completion of the surgical preparation, isoflurane was discontinued, anesthesia was converted to a high-dose narcotic technique to avoid the potentially confounding coronary vasodilator effects of isoflurane anesthesia (Cason, Verrier et al. 1987). Each animal received 25 $mg \cdot kg^{-1}$ pentobarbital over a 30-minute period, and a loading dose of fentanyl (50 $\mu g \cdot kg^{-1}$) followed by a continuous fentanyl infusion (0.5 $\mu g \cdot kg^{-1} \cdot min^{-1}$). Temperature was maintained at 36.5 - 37°C by use of a heating blanket and by warming humidified inhaled gases. Inspired gas concentration was measured by mass spectrometry. Arterial blood gases were measured using a Radiometer ABL-II blood gas laboratory. Hemoglobin and oxyhemoglobin saturation were measured using a Radiometer OSM-2 hemoximeter with electronic correction made for swine hemoglobin absorption characteristics.

Surgery and Hemodynamic Instrumentation. Through a neck dissection, 16-gauge catheters were inserted into the internal jugular vein and carotid artery. A median sternotomy was performed and a pressure transducer-tip (Millar) catheter was inserted through the left atrium into the left ventricle for measurement of left ventricular pressure and its first derivative with respect to time (dP/dt).

Epicardial pacing electrodes were attached to the right atrium, and pacing was instituted at a rate 20% higher than the intrinsic heart rate in order to maintain constant heart rate throughout the experiment.

After all surgery and instrumentation was completed, the animal was heparinized systemically (10,000 u heparin i.v. bolus, and 5,000 $u \cdot hr^{-1}$ continuous infusion).

Sonomicrometry. End-diastolic length and myocardial contractile function were quantified using a Triton sonomicrometer. A small epicardial incision was made and (2 mm) lensed piezoelectric crystals were inserted, using a Teflon guide tube, to a position within 3 mm of the subendocardium. These crystals were inserted 10-15 mm apart, facing each other, and oriented parallel to the minor axis of the heart. Crystal position was confirmed by direct inspection at dissection of the heart, and function data was used only if crystals are confirmed to be within 3 mm of the endocardium, and properly oriented facing each other.

Systolic segment shortening was calculated as segment shortening during systole, averaged over at least 5 heartbeats:

$$\text{Systolic Shortening (\%)} = \frac{\text{(end-diastolic length - end-systolic length)}}{\text{end-diastolic length}} \times 100$$

End-diastole was defined as the onset of positive left ventricular dP/dt; end-systole was defined as the time of peak negative dP/dt (Abel 1981).

LAD coronary cannulation and perfusion. Initial measurements of segmental function and coronary pressure were made, then the LAD coronary artery was cannulated proximally using a plastic cannula manufactured in our laboratory (3mm o.d). Mean coronary artery pressure (CAP)was measured just distal to the tip of this cannula by a 25-gauge catheter which passed through the cannula. Oxygenated blood was withdrawn from a carotid artery and pumped through a windkessel into the LAD coronary artery, using a Masterflex digital roller pump (Cole-Parmer). Coronary blood flow (CBF) was measured by an in-line ultrasonic flowmeter (Transonic). Coronary flow was initially set to provide a mean intracoronary pressure equal to mean aortic pressure. Adequacy of perfusion was assessed by the quick return of segmental function to pre-cannulation values. If function did not return to pre-cannulation levels within 3 minutes the animal was excluded from study.

Determination of normal or "control" coronary flow. After return of segmental function to pre-cannulation values and a 20 minute stabilization period, "control flow" was defined as that coronary flow at which mean coronary pressure was equal to mean aortic pressure.

Cardiac dissection. At the end of each experiment, the area of myocardium perfused by the cannulated LAD artery was defined by a dye infusion technique: blood stained with Evans blue dye was infused into the LAD perfusion circuit at normal aortic pressures. The stained myocardial area was sharply demarcated in swine, and represented the area of LAD perfusion.

Regional Myocardial Oxygen Consumption. MVO_2 was calculated for the LAD-perfused zone, using the Fick Principle, as the product of LAD blood flow and the (coronary arterial - coronary venous) oxygen content difference.

Quantification of Gregg Effect and effects of end-diastolic length. To determine the relationship between the Gregg Effect and end-diastolic length (EDL), associated values of EDL, CBF and CAP were obtained over the pressure range of 60 to 100 mm Hg. To determine each point, CBF was altered slightly until coronary pressure stabilized at one of the target values (60, 70, 80, 90, or 100 mmHg). When both coronary flow and coronary pressure remained stable for one minute, measurements were taken. The average time required to obtain each point was approximately 3 minutes. Then, coronary flow was altered again, and a new pressure-flow-EDL point was obtained.

In each experiment, 5 measurements of regional MVO_2 and EDL were made at intervals over the range of coronary pressures from 60 to 100 mmHg, and the relationships between coronary pressure, MVO_2 and EDL were determined by linear regression.

Data Analysis: Data are expressed as mean ± standard deviation. Repeated-measures ANOVA was used to test for differences in serial hemodynamic measurements, and the Newman-Keuls test was used for multiple comparisons, when indicated by ANOVA. To determine the relationships between MVO_2, CAP and EDL, linear regression equations were fit to the experimental data in the forms shown in Table 2. To summarize these results, mean values (± s.d.) for the coefficients and constants in the regression equations are expressed in Table 2. Relationships between CAP, EDL and MVO_2 are also depicted in Figure 1.

RESULTS

Hemodynamics were normal for swine anesthetized with this anesthetic regimen, and did not change significantly during the experiment (Table 1). Systolic shortening was normal in all swine at the onset of each experiment, or they were excluded from study, as described above.

Table 1. Hemodynamic measurements

	Mean Arterial Pressure (mm Hg)	Heart Rate (beats · min^{-1})	LV dP/dt (mm Hg·sec^{-1})	Systolic Shortening (%)
Control Period	61.4 ± 7.7	123.6 ± 10.2	1122 ± 136	25.8 ± 5.3
Coronary Hyperperfusion, at CAP = 100 mm Hg	61.0 ± 8.0	124.4 ± 11.8	1076 ± 136	23.8 ± 5.6

We found that MVO_2 was directly and linearly correlated with coronary artery pressure, according to the regression equation (Table 2, Fig. 1):

$$MVO_2 = 0.024\ (CAP) + 6.05$$

We found, however, no significant correlation between changes in EDL and changes in MVO_2. Furthermore, changes in EDL which were observed in this study were not correlated with changes in MVO_2.

Table 2: Summary of regression results.

Relation	Regression form	a	b	r^2	p
MVO_2 vs CAP	$MVO_2 = a\,(CAP) + b$	$0.024 \pm .008$	6.05 ± 2.03	.87	.01
MVO_2 vs. EDL	$MVO_2 = a\,(EDL) + b$	0.88 ± 1.97	18.35 ± 27.29	.47	ns
EDL vs. CAP	$EDL = a\,(CAP) + b$	$0.0018 \pm .009$	14.6 ± 2.04	.45	ns

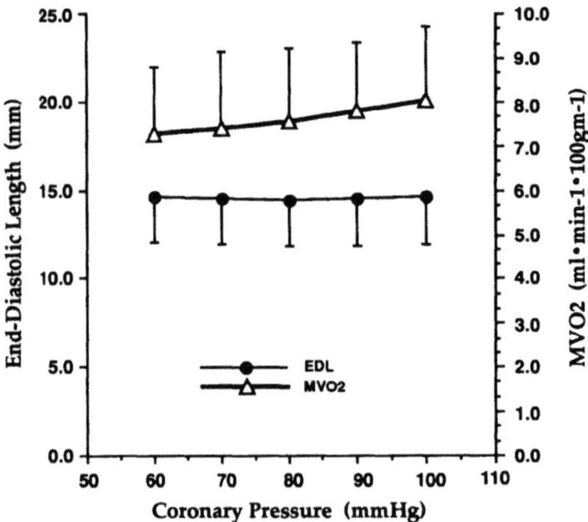

Figure 1. Effects of coronary artery pressure on end-diastolic length and MVO_2. End-diastolic length was unaffected by coronary pressure over the range tested. Increases in coronary pressure within the autoregulatory range of 60 - 100 mm Hg caused significant increases in MVO_2 (P = .01, mean r^2 = .87).

DISCUSSION

Over a pressure range of 50 mmHg within the autoregulatory range, we found a significant Gregg effect. Increasing coronary pressure from 60 mm Hg to 100 mm Hg increased regional myocardial oxygen consumption by approximately 12%, even though there was no evidence of deficient blood flow or oxygen supply in the control state. Over this pressure range, regional myocardial oxygen consumption was highly correlated with coronary perfusion pressure, but not with end-diastolic length. Furhermore, EDL was not significantly affected by perfusion pressure changes in this range, and the small changes in EDL which were observed were not correlated with changes in regional MVO_2. These results suggest that the mechanism of the Gregg phenomenon in this model is not via increased stretch at end-diastole.

Our findings are in agreement with data from the group of Schouten et al. (Schouten, Allaart et al. 1992). Using an isolated, perfused, rat papillary muscle model, this group showed that a Gregg effect can be demonstrated even in vitro. Perfusion-induced changes in segment length were minimal, and did not cause the Gregg effect. Data from the current experiments extend the findings of Schouten et al., by demonstrating that the in-vivo Gregg effect is not dependent on perfusion-induced changes in segment length.

In contrast, other recent investigations have suggested that supranormal coronary blood flow does not increase myocardial contractile function or oxygen consumption. Schulz et al. (Schulz, Guth et al. 1991) used a pump-perfused swine coronary artery model and found that pressure-induced and adenosine-induced increases in coronary blood flow did not increase oxygen consumption or regional myocardial function. Schwartz, et al. used an autoperfused swine coronary artery model, and found that adenosine-induced increases in coronary flow did not change regional contractile function, oxygen consumption, or regional ATP, as measured by nuclear magnetic resonance spectroscopy. The discrepancies between the findings of the current study and the two studies mentioned above, which found no Gregg effect, could be best explained by one of two possible explanations: a) either the Gregg effect is an experimental artifact caused by perfusion-induced relief of baseline ischemia, or b) the Gregg effect is model-dependent for other reasons.

In the current series of experiments, we can reject the possibility that either baseline ischemia or myocardial "stunning" was present. Regional subendocardial segment shortening was measured before coronary artery cannulation, remained normal at the initiation of the experimental protocol, and was not changed by increasing regional perfusion. Other studies have confirmed, however, that the Gregg effect can be model-dependent. Miller et al. (Miller, Nellis et al. 1990) found no Gregg effect in normally-ejecting swine hearts, but found that a significant Gregg effect was present in isovolumically-contracting hearts. In the current series of experiments, we studied normally-ejecting swine hearts, yet still found a significant Gregg effect. We are therefore left to conclude that the Gregg effect can be model-dependent for reasons which are incompletely determined.

In summary, we found that in this swine model of isolated coronary perfusion, increases in regional coronary perfusion and perfusion pressure significantly increased regional myocardial oxygen consumption. This Gregg phenomenon was not dependent on changes in regional end-diastolic segment length.

REFERENCES

Abel, F. (1982). "Maximal negative dP/dt as an indicator of end of systole." AM J Physiol. 240: H676-H679.

Arnold, G., Kosche, F., Miessner, A., Neitzert, A. and Lochner, W. (1969). "The importance of the perfusion pressure in the coronary arteries for the contractility and the oxygen consumption of the heart." Pflugers Arch. 299: 339-356.

Arnold, G., Morgenstern, C. and Lochner, W. (1970). "The autoregulation of the heart work by the coronary perfusion pressure." Pflugers Arch. 321: 34-55.

Cason, B., Verrier, E., London, M., Mangano, D. and Hickey, R. (1987). "Effects of isoflurane and halothane on coronary vascular resistance and collateral myocardial blood flow: Their capacity to induce coronary steal." Anesthesiology. 67: 665-675.

Gregg, D. E. (1963). "Effect of coronary perfusion pressure or coronary flow on oxygen usage of the myocardium." Circ Res. XIII(6): 497.

Miller, W., Nellis, S., Liedtke, A., Whitesell, L. and Effero, B. (1990). "Coronary hyperperfusion and ventricular function in intact and isovolumic pig hearts." Am J Physiol. 258: H500-H507.

Scharf, S. and Bromberger-Barnea, B. (1973). "Influence of coronary flow and pressure on cardiac function and coronary vascular volume." Am J Physiol. 224:918-925.

Schouten, V., Allaart, C. and Westerhof, N. (1992). "Effect of perfusion pressure on force of contraction in thin papillary muscles and trabeculae from rat heart." Journal of Physiology. 451: 585-604.

Schulz, R., Guth, B.D., and Heusch, G. (1991). "No Effect of Coronary Perfusion on Regional Myocardial Function Within the Autoregulatory Range in Pigs." Circulation. 83(4): 1390-1403.

Schwartz, G., Schaefer, S., Trocha, S., Garcia, J., Steinman, S., Massie. B. and Weiner, M. (1992). "Effect of supranormal coronary blood flow on energy metabolism and systolic function of porcine left ventricle." Cardiovascular Research. 26: 1001-1006.

A NEW APPROACH FOR QUANTITATIVE EVALUATION OF CORONARY CAPILLARIES IN LONGITUDINAL SECTIONS

Karel Rakusan, Sanjay Batra, and Marcia I. Heron

Department of Physiology
University of Ottawa
Ottawa, Ontario   Canada K1H 8M5

INTRODUCTION

A better knowledge of oxygen transport to tissue requires accurate, reliable quantification and modelling of a tissue's vascular supply.   Understanding the geometry of the capillary network is of particular importance, since capillaries are the regions primarily involved in the exchange of oxygen between red blood cells and working muscle.

The traditional approach for evaluation of myocardial capillarization is based on measuring the density of coronary capillaries from muscle cross-sections, and then calculating derived parameters such as capillary domain area and the heterogeneity of capillary spacing (e.g. Hoofd et al., 1985; Rakusan and Turek, 1985; Turek, et al., 1987).   More recently, a new dimension was added to this evaluation method, by the reconstruction of longitudinal capillary sets based on the differential staining of capillary walls (Lojda, 1979; Batra et al., 1991).   A capillary set was defined as all capillaries originating from a single arteriole and draining into a single venule.   From these sets, parameters such as capillary set length, minimum capillary length, and capillary segment lengths could be identified, measured and analyzed based on their histochemical properties (Batra and Rakusan, 1991; Batra et al., 1991).   To our knowledge, this approach provides the only available systematic measurements of longitudinal capillary parameters in heart muscle.

The technique of complete capillary set reconstruction however, has its limitations. First of all, complete set reconstruction is very time consuming, as many sections have to be examined in order to find a sufficient number of reliable capillary sets. Secondly, due to the tortuosity and intermingling of capillaries which may in fact be supplied by separate arterioles, some capillaries observed within each section are not always continuous from arteriole to venule, and may not necessarily belong to the same capillary set.   Finally, capillary networks oriented at oblique angles to the cutting plane are only partially captured within that section, and thus are excluded from our analysis.   In fact, only a few sets may be safely recognized within the sections and used for morphometric analysis.

Oxygen Transport to Tissue XVI
Edited by M.C. Hogan *et al.*, Plenum Press, New York, 1994

Therefore, we decided to test a short-cut method, where instead of reconstructing whole sets from arteriole to venule, we measured the length of individual capillary segments, as they were randomly observed within longitudinal sections. A capillary segment was defined as the portion of a capillary located between two clearly visible bifurcations. Intercapillary distances between capillary segments were also measured from the longitudinal sections. It is important to note however, that the short-cut method does not examine all of the longitudinal parameters measured in the capillary set reconstruction method which are of possible importance, such as capillary set length, minimum capillary length, the number of segments per path, the length and width of the capillary set, the capillary set area, and the capillary set length-to-area ratio (Batra and Rakusan, 1992). Consequently, the method of capillary set reconstruction can not be replaced by the short-cut method for the thorough analysis of longitudinal parameters.

In the second part, we compared data obtained from the longitudinal sections, i.e. segment lengths and intercapillary distances, with the most prominent parameter derived from the cross-sections, namely the capillary domain radius.

We propose that this short-cut approach would be less time-consuming than complete capillary set reconstruction, while at the same time being as reliable for estimating capillary segment length and intercapillary distance. In addition, when this capillary segment length data is taken together with data from cross-sectional analysis, a relatively accurate three-dimensional modelling of the myocardial capillary supply would be possible.

METHOD

This new approach has been used by two independent observers, in two different projects, based on relatively large numbers of hearts having varying degrees of capillarization.

In the first project, 26 hearts from eu-, hypo-, and hyperthyroid adult male Sprague Dawley rats (300 - 500 gm) were studied. All hearts were quick frozen in liquid nitrogen, subsequent to their removal from the anaesthetized animals. Longitudinal sections (16 μm) of left ventricular mid-myocardium were cut and stained using the Alkaline Phosphatase (AP)/Dipeptidyl Peptidase IV (DPP) staining technique (Lojda, 1979; Batra et al., 1991), which enzymatically differentiates between the arterial and venous portions of capillaries.

For each heart, 150 capillary segments were chosen at random from longitudinal sections. These segments were measured from branch point to branch point, using a digitizing tablet linked to an IBM computer-based image analysis program (Bioquant, R&M Biometrics, Inc.). Only entire capillary segments located between two clearly visible, successive branching points were measured from all capillaries randomly distributed throughout the longitudinal sections. Measured segments were recorded in one of three categories, based on the segment colour: (1) AC (arteriole; blue) segment, (2) VC (venule; red) segment, (3) Mixed (transitional; blue and red) segment. "H" segments, which were defined as short perpendicular segments joining two adjacent, parallel capillaries, were not measured in either of the two projects.

These results were compared to capillary segment length measurements made from completely reconstructed capillary sets. Three or four capillary sets were reconstructed for each heart, from AP/DPP stained mid-myocardial longitudinal sections (16 μm), prepared from the same 26 eu-, hypo-, and hyperthyroid hearts. In this case, all capillary segments which could be identified within the capillary sets, were measured from branch point to branch point, and were recorded in one of the three categories described previously (AC, VC, or Mixed segments). The results of the segment length measurements, obtained by using the two different approaches, were compared and statistically analyzed.

In the second project, 32 hearts from spontaneously hypertensive rats and their Wistar-Kyoto control were studied. Cryostat sections of both longitudinal (16 μm) and cross-sections (16 μm) of left ventricular mid-myocardium were prepared and stained using the AP/DPP staining method.

As in the first project, 150 capillary segments were chosen at random from the longitudinal sections, measured, and recorded in one of the previously defined three categories. Capillary set reconstructions were not obtained for these hearts. In addition to the segment measurements, approximately 150 intercapillary distances (ICD) were measured per heart. In order to make objective, systematic measurements, a grid overlay consisting of a series of parallel lines, spaced 50 μm apart, was used. The grid was oriented perpendicular to the long axis of the capillaries. ICD were measured from the middle of one capillary to the middle of an immediately adjacent capillary, at the point of intersection between the capillaries and the grid lines. The ICD were categorized based on the colour of the measured capillaries: AC - AC (blue - blue), VC - VC (red - red), AC - VC (blue - red).

From the AP/DPP stained cross-sections, capillary profile drawings were made with the use of a drawing tube attached to an Olympus microscope. Five drawings were made per heart, each drawing containing approximately 75 to 125 capillary profiles. These drawings were then digitized using a Summagraphics digitizing tablet linked to an IBM computer. From the digitized data, arteriolar and venular domain areas and their radii, were calculated with the aid of our capillary domain area program (Hoofd et al., 1985). The results of these cross-sectional measurements, were compared to the results of the segment length and intercapillary distance measurements.

The Pearson product-moment correlation coefficient, was used as an index of the similarity between the two different approaches for measuring segment lengths. This correlation coefficient was also used as an index of the direction and degree of relationship between capillary segment lengths and capillary domain radii, and intercapillary distances and capillary domain radii. Linear regression lines were computed for all comparisons.

RESULTS

Capillary segment length measurements for arteriolar, venular, and mixed capillaries, made by using the two different approaches in various hearts from the first project are shown in Table 1.

TABLE 1. Length data for arteriolar, venular, and mixed capillary segments, as measured by complete set reconstruction and the short-cut method.

| | Segment Lengths (μm) | | | | | |
| | Capillary set reconstruction | | | Short-cut method | | |
	Arteriolar	Venular	Mixed	Arteriolar	Venular	Mixed
Control	78 ± 3	73 ± 4	135 ± 5	94 ± 1	87 ± 2	137 ± 2
Hyperth.	61 ± 3	55 ± 4	126 ± 5	79 ± 1	68 ± 3	120 ± 4
Hypoth.	66 ± 3	56 ± 3	120 ± 4	79 ± 1	61 ± 2	125 ± 2

Using the two methods independently, the same trend was noted in both sets of results. Namely, arteriolar segment lengths were longer than venular segment lengths, and in all cases, mixed segment lengths were the longest. It is especially interesting to note that the average arteriolar and venular segment lengths, as determined from capillary sets, were comparably lower in all three groups, than the segment lengths measured using the short-cut method. The average mixed segment length measurements however, appeared to be more consistent within each group, despite the fact that two different methods were used. The average segment lengths for each capillary portion (arteriolar, venular, and mixed) and each heart are presented in the scatterplot of Figure 1.

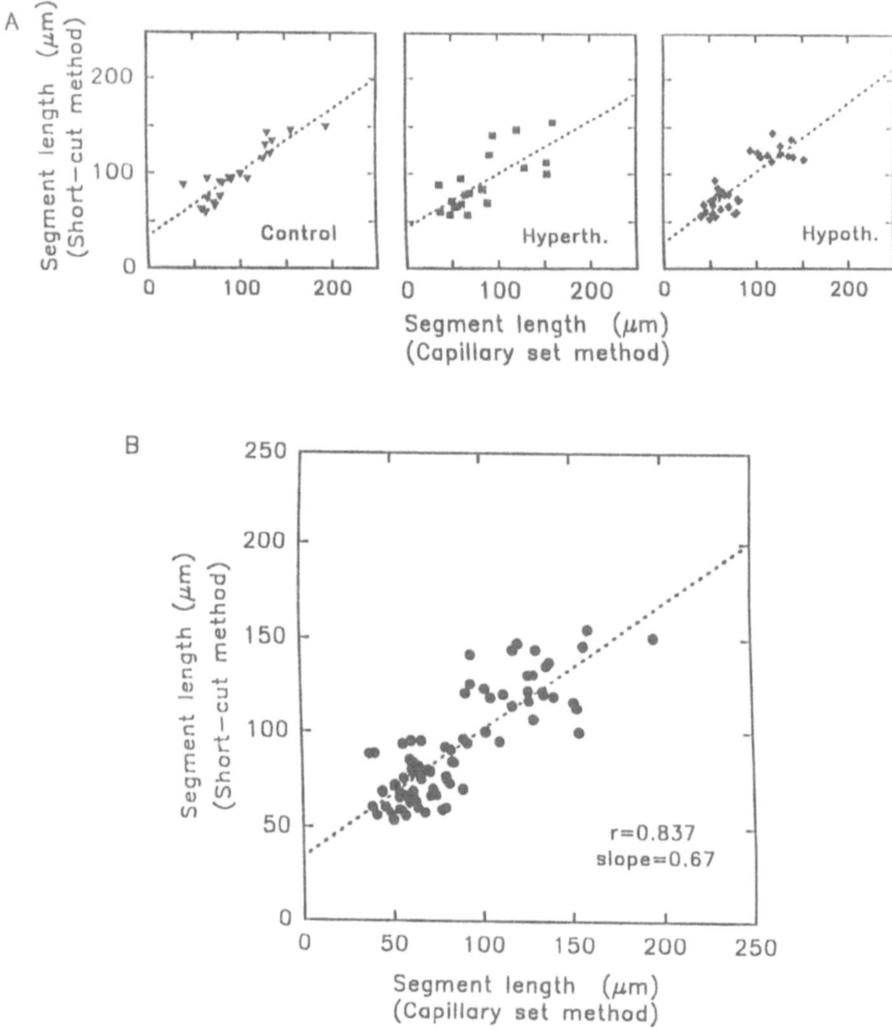

Figure 1. Panel A: Illustrates, for each of the three groups, the linear relationship between segment lengths measured by the capillary reconstruction and short-cut method. Linear regression line (-----) is shown. (Control: N=21, r=0.896, slope=0.67; Hyperthyroid: N=21, r=0.732, slope=0.87; Hypothyroid: N=36, r=0.863, slope=0.74). Panel B: Depicts the existing linear relationship between segment lengths measured by the two different approaches, when all hearts from the three groups are combined. Linear regression line (-----) is shown. (N=78).

Comparison of the average segment length estimations for each capillary portion and each heart, within the individual groups, resulted in a highly significant linear correlation (p<0.001, for each group). Not surprisingly, when all of the average segment length measurements, from all of the hearts within each group, were combined and analyzed, a highly significant correlation (p<0.001) was also observed. These significant correlations provide strong evidence in support of the idea that the two different approaches for measuring capillary segments, result in very similar capillary segment length estimations. In addition, the significant positive correlation indicates at least a general tendency for the results of these two methods to vary in the same direction. Thus, the use of the short-cut segment method alone, to determine capillary segment length, would be a valid approach.

From the second project, the average intercapillary distance measurements for each category (AC-AC; VC-VC) in each heart and the corresponding capillary domain radius (arteriolar, venular) are presented in panel A of Figure 2, while the average capillary segment length for each capillary portion (arteriolar, venular), in each heart, and the corresponding capillary domain radius are presented in the scatterplot in panel B of Figure 2.

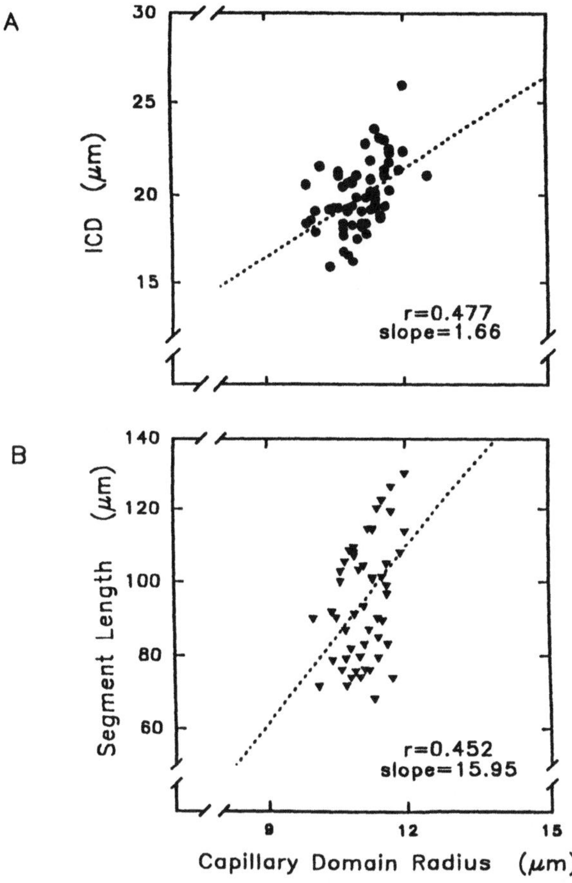

Figure 2. Panel A: Illustrates the linear relationship between the intercapillary distance measured by the short-cut method and capillary domain radius. (N=64). Panel B: Illustrates the linear relationship between the capillary segment lengths, measured by the short-cut method, and the capillary domain radius. (N=54).

The intercapillary distance measured on longitudinal sections showed a significant positive correlation with the radius of capillary domains (p<0.001). This finding was not surprising, due to the relationship which exists between domain radius and intercapillary distance - in theory, the capillary domain radius, as measured from cross-sections, should equal approximately half of the ICD value measured from longitudinal sections. There was also a significant positive correlation between the segment length and the capillary domain radius (p<0.001). These results suggest that larger capillary supply areas in the cross-section tend to be associated with longer capillary segment lengths in longitudinal sections.

DISCUSSION

As more research becomes directed towards assessing and characterizing stimulated cardiac growth of the heart and of its vascular supply, the importance of using adequate, reliable quantitative techniques increases. The introduction of the AP/DPP staining technique, which differentiates arteriolar from venular capillary portions, in combination with morphometric techniques for measuring certain longitudinal parameters from completely reconstructed capillary sets, provided a great deal of information regarding the capillary supply of the myocardium under different experimental conditions. The method of capillary set reconstruction however, has certain limitations which made it cumbersome and laborious to use. In contrast, the new proposed short-cut approach for measuring and examining capillary segment lengths does not have these limitations, and can therefore be considered an easier, and reliable method to use.

The short-cut method was found to be less time-consuming than the capillary set approach, as fewer sections had to be examined in order to find sufficient, acceptable segments. One important distinction between the two approaches was that the short-cut method minimized the subjectivity inherent in the capillary set approach, since segments were chosen randomly as they were observed within longitudinal sections. In addition, small sample sizes were not a problem in the short method, since one was not restricted to only measuring segments which were part of a capillary set. In fact, any complete capillary segment appearing within the longitudinal sections had as equal a probability of being measured as any other capillary segment. This meant that sufficient measurements could be made, in order to obtain a representative estimation of the average capillary segment length. In this study, it was determined that 150 measurements per heart was a sufficient number for accurately estimating capillary segment length. It is well known that adequate sampling and a sufficient number of measurements are two necessary and important criteria for any morphometric technique (Rakusan et al., 1986). The short-cut method was shown to be appropriate for determining capillary segment length as accurately as segment length estimations from reconstructed capillary sets. In fact, this method proved to be consistent, even when it was applied to hearts having different degrees of capillarization, namely the hyperthyroid and hypothyroid groups. The control group results obtained by using this new approach were in accordance with those documented in the literature (Batra and Rakusan, 1991; Batra et al., 1991).

The discrepancy noted in Table 1, between the two methods, namely the higher values in the short-cut method for arteriolar and venular capillary segments, could be a result of the ease with which "shorter" capillary sets, rather than "longer" sets, are found in longitudinal sections. In turn, these shorter sets may be comprised, on average, of shorter capillary segments, thus leading to a lowering of the mean segment length values. On the other hand, when one is not limited to only evaluating

412

segments from identifiable capillary sets, the probability of predominantly measuring "shorter" segments is reduced, since the population from which one is sampling is larger and not as restrictive. For similar reasons, it was decided that "H" segments, as defined in the method section, would not be measured. These segments are, on average very short and therefore more likely than longer, tortuous segments, to be completely captured within the plane of a 16μm sections. Consequently, entire "H" segments are more easily found, compared to entire longer segments, and can thus produce a distortion of the capillary segment length average.

In combination with the capillary domain method and intercapillary distance measurements, this proposed short-cut method can be very illustrative in examining the changes in myocardial capillary geometry, which occur during growth or under pathological conditions. The results of the second project lend support to this assertion. It is important to keep in mind that the primary objective of this portion of the study, was to determine the relationship which existed between capillary domain radius, capillary segment length, and intercapillary distance.

Examination of the relationship between longitudinal and cross-sectional capillary data is quite relevant, for it enables one to conceptualize the capillary supply in a three dimensional model. We have defined the resulting product of the average capillary domain area and average capillary segment length, as the Capillary Supply Unit (Fig. 3).

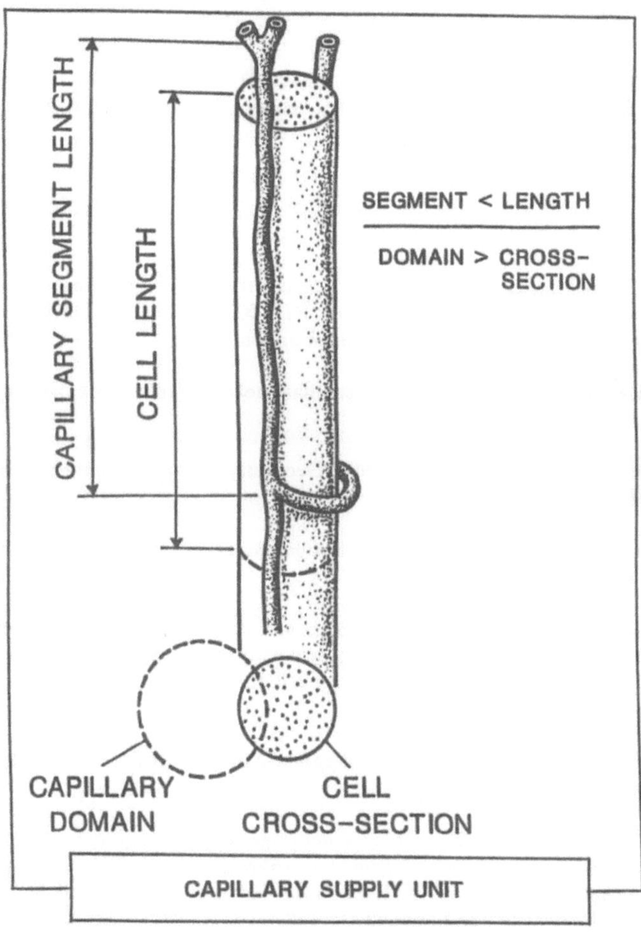

Figure 3. A schematic representation of the Capillary Supply Unit and its possible relationship to a myocardial cell.

To our knowledge, this capillary supply unit is possibly the smallest tissue supply volume which can be modelled in three dimensions, with some degree of accuracy. Use of the AP/DPP histological technique enables this three-dimensional modelling as a function of arterial and venous capillary portions. The ability to quantify and model a three dimensional volume supplied by a capillary segment, not only increases our knowledge of capillary geometry, but also provokes certain questions regarding, for instance, the relationship of this supply unit to myocyte volume. In fact, we believe that the capillary supply volume is closely related to individual myocyte volume.

In summary, we have demonstrated that the short-cut method is a fast and accurate method for estimating capillary segment lengths and intercapillary distances. Because coronary capillaries are organized in an unbounded network, the capillary segment length may be regarded as the most informative structural parameter of this system. It is important to note however, that this short-cut method in no way replaces the method of capillary set reconstruction, but is a method which can be used for quick, accurate capillary segment length estimations. Furthermore, we have shown that when the short-cut method is used in concert with the capillary domain method, a three dimensional volume of tissue, supplied by a coronary capillary - the capillary supply unit - can be estimated. Consequently, a greater knowledge of capillary geometry is achieved.

Acknowledgement : This study was supported by the Medical Research Council of Canada. The authors which to thank Mrs. Barbara Hebert and Mrs. Ching Kuo for their expert technical assistance.

REFERENCES

Batra, S., Kuo, C., and Rakusan, K., 1991, Spatial distribution of coronary capillaries: A-V segment staggering, in: "Oxygen Transport to Tissue - XI," K. Rakusan, G.P. Biro, T.K. Goldstick, and Z. Turek, Plenum Press, New York and London.

Batra, S. and Rakusan, K., 1991, Geometry of capillary networks in volume overload rat heart, *Microvasc. Res.* 42:39.

Batra, S., Rakusan, K., and Campbell, S.E., 1991, Geometry of capillary networks in hypertrophied rat heart, *Microvasc. Res.* 41:29.

Batra, S. and Rakusan, K., 1992, Capillary network geometry during postnatal growth in rat hearts, *Am. J. Physiol.* 262:H635

Hoofd, L., Turek, Z., Kubat, K., Ringnalda, B.E.M., and Kazda, S., 1985, Variability of intercapillary distance estimated on histological sections of rat heart, in: "Oxygen Transport to Tissue - VII," F. Kreuzer, S.M. Cain, Z. Turek, T.K. Goldstick, Plenum Press, New York and London.

Lojda, Z., 1979, Studies on Dipeptidyl(Amino)Peptidase IV (Glycyl-Proline Naphthylamidase), *Histochemistry*. 59:153.

Rakusan, K., and Turek, Z., 1985, The effect of heterogeneity of capillary spacing and O_2 consumption blood flow mismatching on myocardial oxygenation of cardiac muscle, in:" Oxygen Transport to Tissue - VII," F. Kreuzer, S.M. Cain, Z. Turek, T.K. Goldstick, Plenum Press, New York and London.

Rakusan, K., Korecky, B., Sarkar, K., and Turek, Z., 1986, Merits and pitfalls in morphological assessment of cardiac growth, *Federation Proc.* 45:2580.

Turek, Z., Hoofd, L., Rakusan K., 1987, A comparison of the methods for assessment of the heterogeneity of myocardial capillary spacing, *in*:"Oxygen Transport to Tissue - IX," I.A. Silver and A. Silver, Plenum Press, New York and London.

MISINTERPRETATION OF CORONARY CHOLESTEROL ATHEROMATA IN CHOLESTEROL-FED RABBITS AS SUITABLE MODEL FOR CONVENTIONAL HUMAN CORONARY PLAQUES

F. Thimm, M. Frey, G. Fleckenstein-Grün

Study Group for Calcium Antagonism
Physiological Institute
University of Freiburg
Hermann-Herder-Str. 7
D-79104 Freiburg, FRG

INTRODUCTION

Occlusion of coronary arteries following arteriosclerotic plaque formation with thrombotic complications is the main etiological factor in the development of myocardial ischemia with subsequent infarction and cell necrosis. Originally, in the 16th century, arteriosclerosis was considered to be an "ossification" of arteries[1]. Hodgson[2] was the first to report in 1815 chemical analyses of arterial plaques, which contained 65% calcium (Ca) salts and 35% organic matter. However, current concepts of the development of arteriosclerosis are overshadowed by the assumption, that the main etiological factor is an enormous accumulation of lipids, particularly cholesterol, in the arterial walls brought about by an increase in serum cholesterol[3]. Consequently, arterial cholesterol atheromata of cholesterol-fed rabbits are widely considered to be suitable experimental models for conventional human arteriosclerosis[4]. Moreover, standard anti-arteriosclerotic therapy particularly aims at the so-called "normalization" of serum lipids. Our present data, obtained from microchemical analyses of coronary artery plaques of cholesterol-fed rabbits, and humans respectively, enhance the doubts about an exclusive cholesterol hypothesis of conventional atherogenesis.

EXPERIMENTAL PROTOCOL

Rabbit coronary artery segments were taken from male New Zealand rabbits, fed a 2% cholesterol-rich diet for 14 weeks, and compared with those from New Zealand rabbits fed a normal chow. Segments were divided in healthy segments (stage 0), segments with early lesions (not altered in stereomicroscopic inspection = stage I), and segments with visible atheromatotic alterations (stenosing stage). Healthy and pathologically altered human coronary artery specimens of 41-90 year-old patients originated from autopsies, carried out in the Freiburg University Institutes for Pathology and Forensic Medicine. Coronary artery plaques were classified according to the WHO classification scheme: Stage I = fatty streaks; stage II = fibrous plaques; stage III = stenosing bulky plaques, often complicated by ulceration, necrosis, hemorrhage, and thrombosis. The tissue Ca content was measured by atomic absorption spectroscopy, using the analytical procedure of flame and graphite cell technique (Perkin-Elmer Type 3030). Coronary artery segments of 30-80 mg fresh weight were blotted on filter paper and dried overnight at 95°C in teflon tubes. After determination of dry weight, the specimens were dissolved in a mixture (2:1) of nitric acid (65%) and perchloric acid (70% suprapur quality), at 150°C for 6 hours. Ca was analyzed after the addition of Lanthanum oxide to a final concentration of 0.5% to reduce interference with other minerals. Serum cholesterol concentrations and their subfractions (VLDL/chylomicrones, LDL and HDL) were separated by sequential ultracentrifugation technique and determined by standard enzymatic procedures. Coronary arterial wall cholesterol was extracted in single-use, teflon-lined, screw-cap glas vials with chloroform (2.25 ml/vial) for 1 hour. Thereafter a mixture of acetone, methanol and ether (ratio 1:2:1) was added to a total volume of 5 ml and incubated over night. The lipid extract was used for the determination of free and total cholesterol after evaporation and silylation with 300 µl MSTFA (n-methyl-n-trimethylsilyl-trifluoroaceta-mide) using a Varian gas chromatograph (Varian Type 3400).

RESULTS

Figure 1 demonstrates a cross section of a stenosing human coronary plaque which had led to a lethal coronary infarction. Large amounts of Ca in the arterial wall are visualized as black material using H. E. staining. Most of the Ca salt has been broken away during the cutting procedure, because the histological cross section was not decalcificated as normally done. In sharp contrast, the cross section of the descending branch of the left coronary artery of a cholesterol-fed rabbit (Fig. 2), shows luminal occlusion by large amounts of accumulated cholesterol. This experimentally induced coronary atheromatosis in rabbits was characterized by the following peculiarities (Fig. 3): (i) significant incorporation of cholesterol as early lesions and massive accumulation of free and total cholesterol in the stenosing plaques up to 19.9 and 43.3 times above normal; (ii) very modest increase in mural calcium content by a factor of 4.2; and (iii) soft consistency of cholesterol plaques in contrast with the stiffness of calcified arteriosclerotic lesions of humans.

Under the cholesterol-rich regimen, serum cholesterol concentrations were dramatically increased, total cholesterol amounting to approx. 2,000 mg/dl (Fig. 4).

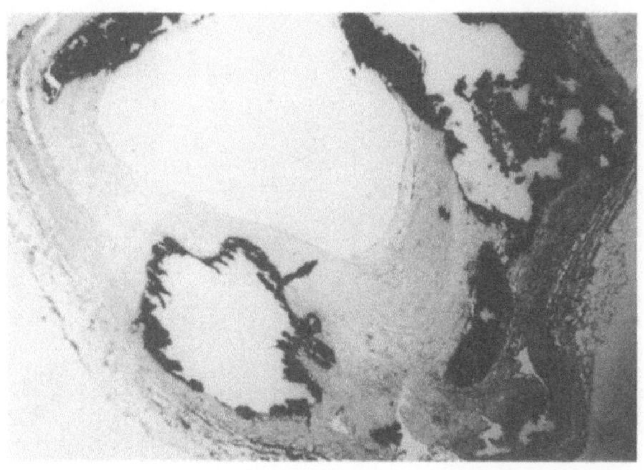

Figure 1. Cross section of a lethal human coronary artery plaque (stage III, WHO classification; H. E. staining). Severe Ca overload is visualized as black material, partially broken out (Prof. Dr. E. Herbst and Prof. Dr. H.-E .Schaefer, Head of the Pathological Institute of the University of Freiburg).

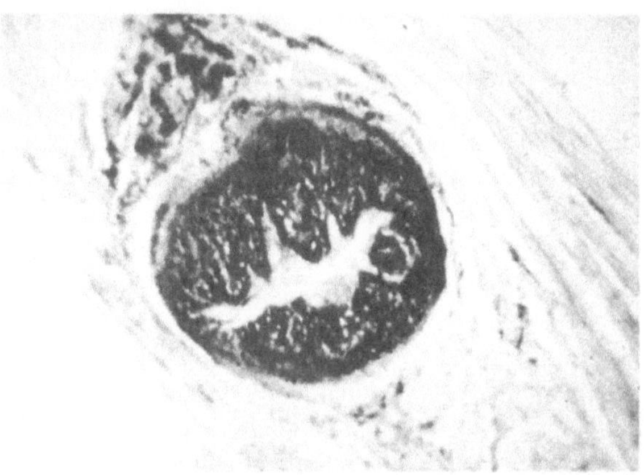

Figure 2. Cross section of a stenosing coronary artery plaque of a cholesterol-fed New Zealand rabbit. Massive atheromatous occlusions after 16 weeks of 2% cholesterol diet (Oil red O staining).

Consequently, a generalized lipid storage disease developed in New Zealand rabbits under the cholesterol-rich regimen. Therefore, as reflected in Fig. 5, massive cholesterol accumulations were observed - not only in coronary arteries - but also in aortae, liver, kidney, and myocardium.

The chemical analysis of conventional human coronary plaques revealed fundamental differences in comparison to the experimental atheromatosis of cholestrol-fed rabbits (Fig. 6). In healthy coronary artery segments (age group: 41-90 years), the content of free (9.4 ± 1.2 g/kg dry weight (n=55)) and total (16.5 ± 1.8 g/kg dry weight (n=55)) cholesterol, always surpassed that of Ca (2.2 ± 0.3 g/kg dry weight (n=113)). In fatty streaks (stage I), free and total cholesterol rose by a factor of 2.9 and 4.2 respectively, whereas the Ca content was increased by more than 12 times above normal. In fibrous plaques (stage II), Ca was accumulated 23-fold, but free and total cholesterol concentrations even declined below the level in fatty streaks (2.8- and 3.4-fold increase above normal). In complicated stenosing lesions

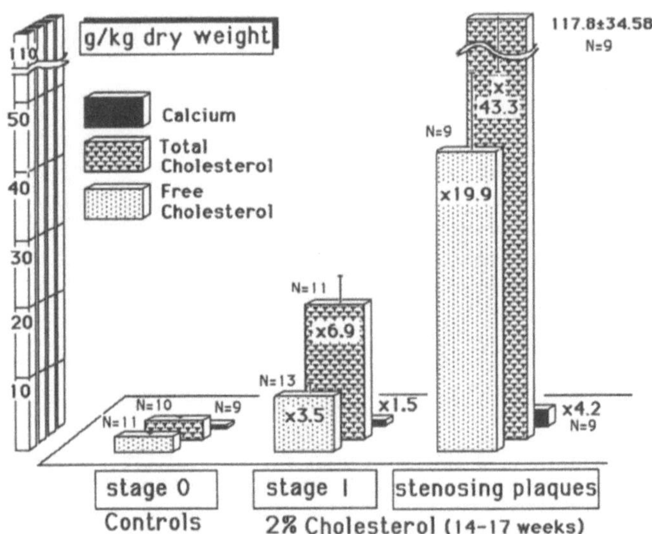

Figure 3. Microchemical analysis of coronary artery segments of New Zealand rabbits. Stage 0: Healthy controls; stage I: Early lesions after 16 weeks of cholesterol-rich diet (2% cholesterol), not yet visible with stereomicroscopic inspection; stenosing plaques: Fully developed cholesterol atheromata after 16 weeks of cholesterol-rich diet (2% cholesterol).

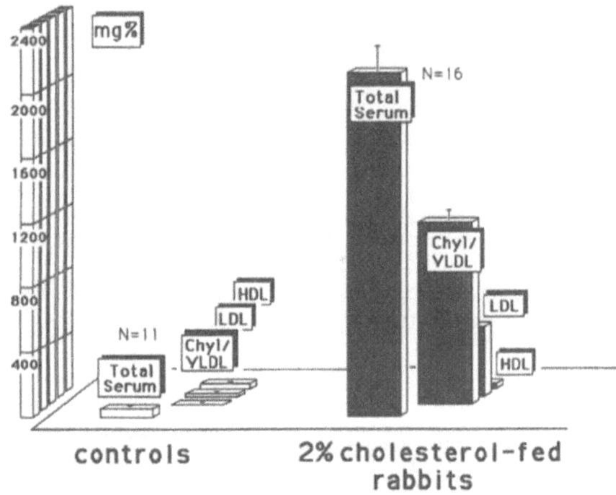

Figure 4. Massive increase of cholesterol concentrations (total cholesterol and the lipoprotein fractions, chylomicrones/VLDL and LDL) in the serum of New Zealand rabbits fed a 2% cholesterol diet for 14 weeks, versus New Zealand rabbits without cholesterol diet (controls).

(stage III), Ca rose by a factor of 83 (184.5 ± 14.8 g/kg dry weight (n=44)). In contrast, free and total cholesterol only amounted to 17.0 ± 2.3 g/kg dry weight (n=51) and 25.9 ± 2.6 g/kg dry weight (n=46) respectively, i.e. the cholesterol content of fully developed stenosing plaques was less than doubled, compared to healthy artery segments of the same age groups.

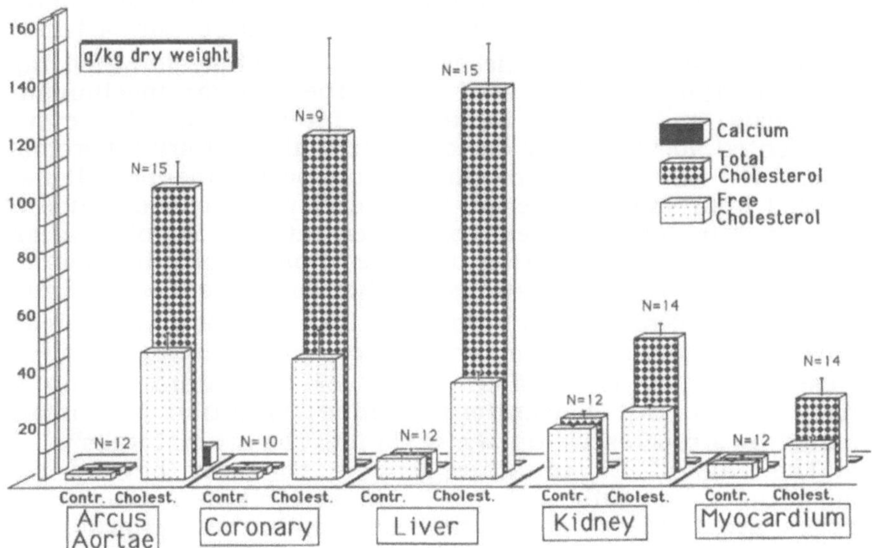

Figure 5. Generalized lipid storage disease concerning aorta, coronary artery, liver, kidney, and myocardium in New Zealand rabbits, fed a cholesterol-rich diet (2% cholesterol) for 14-17 weeks.

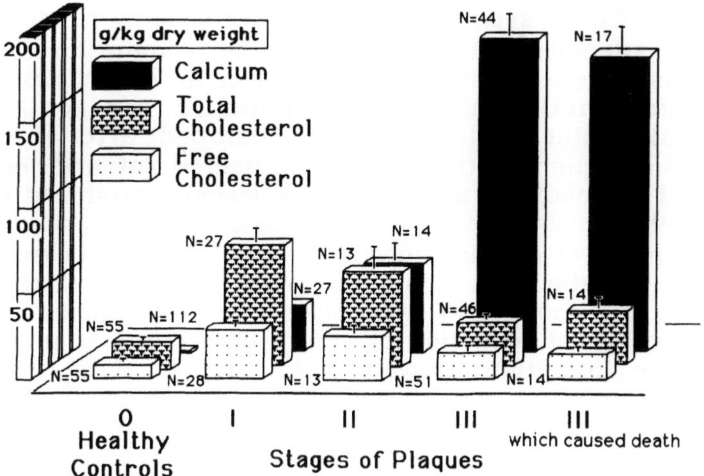

Figure 6. Positive correlation between the degree of Ca accumulation and the severity of plaque formations (stage I-III, WHO classification) in human coronary arteries. Inverse correlation between aggravation of coronary sclerosis and the cholesterol contents.

DISCUSSION

In 1913 Anitschkow introduced the aortic atheromata of cholesterol-fed New Zealand rabbits as suitable models for conventional human arteriosclerosis and laid down the dictum that increased cholesterol in the blood is the decisive cause for arteriosclerosis[4]. Since those days, aortic atheromata of cholesterol-fed rabbits are experimentally used to study both the cellular mechanisms of atherogenesis, and possible therapeutic interventions. However, our data presented above, clearly demonstrate that coronary atheromata of cholesterol-fed New Zealand rabbits are unspecific manifestations of a generalized lipid storage disease, that develops in consequence of an excessive rise in serum cholesterol concentrations. Histological aspect and microchemical composition of these cholesterol atheromata may correspond to arterial alterations in the very rare cases of human familial hypercholesterolemia[3]. However, they are not at all comparable to conventional human coronary plaques, that develop slowly, over decades[5]. Here, by no means, "clots" of cholesterol occlude the coronary lumen. But conventional plaque formation in humans is dominated by a progressive uptake of Ca into the arterial walls[6].

The role of Ca in atherogenesis goes far beyond simple "calcification" of necrotic cell debris. In fact, various Ca-consuming processes are involved, from the very beginning onwards, in the arteriosclerotic process[7]: (i) Ca ions serve as messengers, notably in vascular smooth muscle cells, for migration, proliferation, and matrix production; (ii) Ca "killer" effects induce cell death and necrosis if Ca is excessively accumulated in the intracellular space; (iii) specific binding of Ca to matrix molecules and calcification of elastic fibres are typical criteria of most types of experimental and human arterioscleroses. Obviously, risk factors such as nicotine, hypertension, diabetes, oxidatively modified lipoproteins, oxygen radicals etc., lastly make use of Ca messenger and killer functions to display their atherogenic potencies[8]. Thus a considerable demand for Ca exists in arterial walls, where Ca-dominated types of arteriosclerosis develop. Under experimental conditions, specific Ca antagonists of the verapamil/nifedipine /diltiazem type, damped arterial Ca incorporation[9]. Subsequently, these drugs protected rat arteries from Ca overload and Ca-mediated arteriosclerotic lesions, produced by various risk factors[10,11].

Prevention of coronary occlusion to ensure myocardial oxygen supply, is one of the most significant goals of cardiovascular therapy. According to the cholesterol hypothesis of atherogenesis, alimentary and pharmacological reduction of serum cholesterol is propagated to inhibit plaque formation. However, our present data indicate that conventional human coronary plaques represent a Ca-dominated type of arteriosclerosis, that may be accessible to a therapeutic intervention with suitable Ca antagonists. Positive clinical results are already available in that the Ca antagonist nifedipine significantly reduced the number of early coronary lesions (INTACT study[12]). Simultaneously, the INTACT study demonstrated that there was no statistical correlation between coronary plaque development and serum cholesterol levels in the untreated group of patients. Further comparative trials have to consider the benefit of both antiatherogenic principles in the prevention of conventional human plaque formation: Reduction of serum cholesterol versus normalization of arterial Ca homeostasis by Ca antagonists.

CONCLUSION

We conclude that coronary atheromata of cholesterol-fed New Zealand rabbits are unspecific manifestations of a generalized lipid storage disease following excess increase in serum cholesterol levels. Consequently, they do not represent a suitable model for conventional human coronary plaque formation which frequently occurs in persons with "normal" cholesterol levels, and is dominated by a progressive Ca uptake into arterial walls. Interestingly, coronary plaques of completely different etiology (conventional human Ca-dominated arteriosclerosis and rabbit cholesterol atheromata) lastly lead to the same diminution of oxygen supply, with the fatal consequence of myocardial infarction. However, precise knowledge of the underlying pathogenetic processes is the prerequisite for a suitable therapy. Our data indicate that specific Ca antagonists may optimize oxygen transport to human tissues, not only by their pronounced vasodilator potencies, but also by the prevention of Ca-mediated arteriosclerotic plaque formation.

REFERENCES

1. E.R. Long, Development of our knowledge of arteriosclerosis,*in*: "Cowdy's arteriosclerosis. A survey of the problem," H.T. Blumenthal, ed., First Edition., Charles C Thomas, Springfield/Ill USA (1933).
2. J. Hodgson. "Treatise on the diseases of arteries and veins," (1815).
3. M.S. Brown, and J.L. Goldstein, How LDL receptors influence cholesterol and atherosclerosis. *Sci Am*. 251(5):58 (1984).
4. N. Anitschkow, and S. Chalatow, Über experimentelle Cholesterinsteatose und ihre Bedeutung für die Entstehung einiger pathologischer Prozesse. *Zentralbl. Allgemeine Pathol u Anat*. 14:1 (1913).
5. G. Fleckenstein-Grün, M. Frey, F. Thimm, C. Luley, A. Czirfusz, and A. Fleckenstein, Differentiation between calcium- and cholesterol-dominated types of arteriosclerotic lesions: Antiarteriosclerotic aspects of calcium antagonists. *J Cardiovasc Pharmacol*.18(Suppl 6): S1 (1991).
6. A. Fleckenstein, M. Frey, F. Thimm, and G. Fleckenstein-Grün, Excessive mural calcium overload - A predominant causal factor in the development of stenosing coronary plaques in humans. *Cardiovasc Drugs and Therapy*. 4:1005 (1990).
7. G. Fleckenstein-Grün, and A. Fleckenstein, Calcium - A neglected key factor in arteriosclerosis. The pathogenetic role of arterial calcium overload and its prevention by calcium antagonists, *Ann Med*. 23:589 (1991).
8. G. Fleckenstein-Grün, F. Thimm, M. Frey, and A. Czirfusz. Role of calcium in arteriosclerosis. Experimental evaluation of antiarteriosclerotic potencies of Ca antagonists, *Basic Res Cardiol*, accepted, in press (1993).
9. A. Fleckenstein, M. Frey, J. Zorn, and G. Fleckenstein-Grün, Calcium, a neglected key factor in hypertension and arteriosclerosis. Experimental vasoprotection with calcium antagonists or ACE inhibitors, *in*: "Hypertension: Pathophysiology, Diagnosis, and Mannagement." J.H. Laragh and B.M. Brenner eds.,Raven Press, New York, 471 (1990).

10. G. Fleckenstein-Grün, M. Frey, F. Thimm, and A. Fleckenstein, Protective effects of various calcium antagonists against experimental arteriosclerosis, *J Human Hypertension* 6 (Suppl 1) : S13

11. G. Fleckenstein-Grün, F. Thimm, A. Czirfusz, S. Matyas, and M. Frey, Experimental vasoprotection by calcium antagonists against Ca-mediated arteriosclerotic alterations. *J Cardiovasc Pharmacol*, accepted, in press (1993).
12. P.R. Lichtlen, P.G. Hugenholtz, W. Rafflenbeul, H. Hecker, S. Jost, P. Nikutta, J.W. Deckers, Retardation of coronary artery disease in in humans by the calcium channel blocker nifedipine: Results of the INTACT study (International Nifedipine Trial on Antiatherosclerotic Therapy). *Cardiovasc Drugs and Therapy* . 4 : 1047 (1990).

FUNCTIONAL HETEROGENEITY OF THE HEART

James B. Bassingthwaighte, Joseph I. S. Chan
Center for Bioengineering, University of Washington, Seattle WA 98195

The variation in regional flows in the heart is NOT random, but rather it is a fractal, which implies there is organization and correlation. Any mild asymmetry of flows at branch points in a bifurcating network results in varied regional flows and when the asymmetry increases slightly with each generation the result is a fractal. The fractal relationships can only hold over the 12 to 16 generations of the arterial tree, so these are not classic mathematical fractals which scale to infinitely small regions.

The data showing a range of flows from 20% to 200% of the mean in the normal heart posed a conundrum. The heart is a syncytium and is virtually monofunctional: it serves by contracting throughout its entirety each beat; since the cells contract in synchrony, one might think that force development is the same throughout the ventricular wall. (This premise is somewhat questionable, for observations of Prinzen et al. (*Am. J. Physiol.* 259:300, 1990) show that early activation in the region of a pacing catheter leads to reduced local flow at that site, presumably because this region contracts early and more isotonically than regions activated later which must contract somewhat more isometrically.) If local work were uniform there should be uniform oxygen consumption. When the mean oxygen extraction for the whole heart is 50% (or more, as in exercise), it simply is not possible for the low flow regions to combust as much oxygen as the average region, since there is not high enough O_2 delivery at the observed flows. Thus these regions *must* have lower oxygen requirements and *must* do less work per gram of myocardium. (Even so, we know from the work of Weiss and Sinha (*Circ. Res.* 42:119, 1978) showing variation in hemoglobin saturation in regional veins, that flow and metabolism are not perfectly matched.)

A generic hypothesis is that work and oxygen metabolism are matched by the local flow under optimal circumstances. The direct test is to estimate both flow and metabolism locally, by independent measures. Lacking a way to measure oxygen consumption or ATP turnover regionally, secondary measures, the concentrations of ATP or metabolic enzymes have been made, but because they measure a capacity rather than a flux, are found inadequate to answer the question (Griggs et al., *Am. J. Physiol.* 222:705, 1972; Franzen et al., *Am. J. Physiol.* 254:H344, 1988).

In the heart fatty acid is by far the preferred substrate, and the intracellular pool for fatty acid, as acyl-CoA and triglyceride, is so great that the local retention is long, so the rate of uptake into the cell is not a good measure of metabolism (Bassingthwaighte et al., *Mol. Cell Biochem.* 88:51, 1989; van der Vusse et al., *Physical Rev.* 72:881, 1992). Uptake does provide a good measure of the local transport capacity, the permeability-surface area products, of the membranes of the endothelial cells and myocytes in series, which Caldwell et al. (*Am. J. Physiol.*, in revision) found to be proportional to the local flows in running dogs. This supported the idea that there might be local impedance matching of flows, transport, metabolism and work, but evidence on local metabolism and work is still lacking.

Insight into the degree of workable mismatching is provided through an understanding of the interactions between substrate delivery by flow and by diffusion. Oxygen diffusion is rapid, so allowing supplying relatively large regions, more than a single capillary tissue unit; fatty acid and glucose diffusion and permeation are much slower, but they are supplied in excess to the extracellular fluid and so are adequately supplied over a few intercapillary distances also. When flow distributions are fractal, then the diffusion of oxygen spreads its delivery into adjacent lower flow regions, smoothing the spatial oxygen profiles. The result, demonstrable by the modeling of spatial profiles of flow, diffusional exchanges and consumption, implies that modest degrees of mismatching of flow to metabolism are adequately handled by the tissues. (Supported by NIH grants HL50238 and RR1243.)

RESPIRATORY GAS EXCHANGE AND INERT GAS RETENTION DURING PARTIAL LIQUID VENTILATION

E.A. Mates[1], J.C. Jackson[2], J. Hildebrandt[1,3], W.E. Truog[2],
T.A. Standaert[2], and M.P. Hlastala[1,3]

[1]Department of Physiology and Biophysics, SJ-40
[2]Department of Pediatrics, RD-20
[3]Department of Medicine, RM-12
University of Washington
Seattle, Washington 98195

INTRODUCTION

Perfluorocarbon (PFC) fluids have been used as ventilatory media due to their unique combination of low toxicity, low solubility in body fluids, ability to lower interfacial tensions, and high oxygen and carbon dioxide carrying capacities[12]. Cardiorespiratory support has been achieved during liquid ventilation with a variety of perfluorinated chemicals in a range of experimental animals[1,2,7,8,9,18]. In animal models of respiratory distress syndrome, introduction of PFC to the lungs has been shown to improve both compliance and gas exchange[11,13,14,15,16]. However, the presence of fluid in the gas exchange regions of the lung should impede convective and diffusive mass transport and have a deleterious effect on gas exchange. Indeed, while it is possible to achieve adequate oxygenation during liquid ventilation, CO_2 retention and acidosis can be problematic[7]. In order to maximize the benefits of this mode of ventilation while minimizing the side effects of CO_2 retention, acidosis, and O_2 toxicity due to high F_IO_2, a clearer understanding of the determinants of gas exchange through a fluorocarbon medium is necessary.

Once liquid is instilled in the lung, O_2 and CO_2 exchange with the environment are typically accomplished in one of two ways. The first is full tidal liquid ventilation (FTLV) in which tidal volumes of oxygenated PFC are cycled in and out of the airless lung, mixing with resident fluid and exchanging respiratory gases[10]. The second method, referred to as partial liquid ventilation (PLV), involves filling only the functional residual capacity (FRC) with PFC, then delivering tidal volumes of oxygen in order to exchange O_2 and CO_2[3]. PLV was used in this study to investigate the effects of varying volumes of liquid in the lung on gas exchange.

It is the purpose of this study to quantitate gas exchange limitation in the healthy piglet whose lungs are filled with graded amounts of liquid perflubron ($C_8F_{17}Br$). We hypothesized that PLV would increase intrapulmonary shunt and alveolar-arterial differences ((A-a)D's) of respiratory gases in normal lungs. We further speculated that the magnitude of shunt and (A-a)D's would be proportional to the volume of fluid in the lungs. To test these hypotheses, we measured shunt during PLV in two ways: by oxygen alveolar-arterial difference according to the Berggren shunt equation, and by inert gas retention using the multiple inert gas elimination technique (MIGET)[6]. In gas-ventilated lungs, the inert gas shunt is normally near zero whereas the O_2 shunt is 2-5% reflecting post-pulmonary "shunts" from bronchial and coronary circulations. The two methods should correlate exactly for changes of intrapulmonary shunting. It turned out that these relationships did not hold during PLV. We also measured alveolar-arterial differences of CO_2 calculated from end-tidal and arterial CO_2.

MATERIALS AND METHODS

Nine healthy, 7-14 day old piglets weighing 2.7 ± 0.6 (SD) kg were anesthetized by induction with ketamine/xylazine (24 and 2.75 mg/kg IM, respectively) and maintenance with pentobarbital (1 mg/kg/min IV, supplemented with 15-30 mg/kg hourly). Pancuronium bromide was administered in 0.2 mg/kg IV doses as needed to prevent respiratory efforts. Carotid and jugular catheters as well as a pulmonary artery thermodilution catheter were placed. A tracheotomy was performed and a metal endotracheal tube was secured in the airway.

A Harvard single piston animal ventilator was used to deliver tidal breaths of 100% O_2 to animals in the supine position. Tidal volume (17.2 ± 2.9 ml/kg) and frequency (21±4 bpm) were set to maintain arterial PCO_2 less than 40 mmHg before perflubron was added to the lungs (baseline condition). Ventilatory parameters were unchanged thereafter. Airway pressure was continuously monitored and expiratory flow and volume assessed using a pneumotachometer and signal integrator, respectively. Positive end-expiratory pressure (PEEP) was applied by immersion of the distal expiratory port in 5 cm water. Carotid and pulmonary arterial pressures were recorded and cardiac outputs determined in triplicate via thermodilution.

A dilute solution of six inert gases (sulfur hexafluoride (SF_6), ethane, halothane, cyclopropane, ether, and acetone) in 5% dextrose solution was continuously infused in the jugular vein according to the standard MIGET technique[5,20]. The MIGET technique allows one to distinguish shunt, dead space, and the general pattern of the $\dot{V}_A/\dot{Q}$ distribution from mixed venous, arterial, and mixed expired concentrations of the six inert gases. Expiratory gas was collected in a heated one liter flow-through glass chamber by separating inspiratory and expiratory pathways with a 3-way solenoid valve triggered by the ventilator. Mixed venous and arterial blood, and mixed expired gas samples were collected simultaneously for each condition. One hour of equilibration was allowed between manipulating the animal and drawing MIGET samples, a time period normally thought to be sufficient for establishing steady state gas exchange (a fundamental assumption of the MIGET analysis).

Blood gases were assessed using a Corning model 170 pH/blood gas analyzer which showed linear PO_2 response to tonometered blood up to 700 mmHg with and without perflubron vapor present. Arterial and mixed venous blood samples were drawn after acquiring MIGET samples for each condition. Venous samples were drawn in triplicate and blood O_2 content values averaged for greater accuracy in calculating O_2 shunt using the Berggren equation:

$$\frac{\dot{Q}_s}{\dot{Q}_t} = \frac{(C_{c'} - C_a)_{O_2}}{(C_{c'} - C_{\bar{v}})_{O_2}}$$

with the assumption that end-capillary (c') PO_2 is equivalent to alveolar PO_2. Alveolar PO_2 during 100% O_2 breathing was calculated as $P_B - P_{H_2O} - P_{CO_2} - P_{PFB}$. "Alveolar" refers to gas adjacent to the fluid layer and does not necessarily represent gas tensions in resident

PFC. (a-A)DCO$_2$ was determined from arterial and end-tidal PCO$_2$'s. Exhaled CO$_2$ was continuously monitored using a Novametrix model 7000 infrared analyzer situated between the piglet and solenoid valve.

Warmed, non-preoxygenated perflubron (Alliance Pharmaceutical Corp, San Diego, CA) was delivered in 10 ml/kg doses via a sideport of the endotracheal tube during gas ventilation with 100% O$_2$. Small aliquots of PFC were deposited in the tube at end-expiration over 3-4 minutes until the entire dose was administered. During dosing, the animal was rotated to improve distribution such that approximately 1/3 of the dose was delivered in each of supine, left lateral, and right lateral positions.

Arterial and mixed venous blood gases and MIGET samples were obtained for baseline gas exchange analysis. The first dose of 10 ml/kg perflubron was then administered. After equilibrating for one hour, blood gases and MIGET samples were drawn and a second dose of 10 ml/kg was administered. A final dose of 10 ml/kg perflubron was added at three hours (the volume of fluid in the lung now approximating FRC[3]). Finally, 10 ml/kg fluid was drained out and blood gas and MIGET samples were repeated at the end of the final equilibration period.

RESULTS

One of the nine animals studied was excluded due to the presence of a large shunt prior to beginning liquid ventilation and data are reported from the remaining eight. There were no significant changes in cardiac output, mean arterial pressure, PA wedge pressure, or peak airway pressures between baseline and post-PFC conditions (Table 1). If respiratory efforts were noted, additional pancuronium was given. Table 2 summarizes alveolar and blood gas data for this study.

Figure 1 shows O$_2$ shunt (Q$_s$/Q$_t$) and (a-A)DCO$_2$ increase with progressive addition of PFC to the lung. Linear regression analysis of O$_2$ shunt vs. perflubron (ml/kg) for each animal produced a group average slope of 0.002$\pm$0.001 (SD) which was significantly different from zero (p = 0.0002). Similar analysis of (a-A)DCO$_2$ vs. perflubron (ml/kg) produced a mean slope of 0.32$\pm$0.20 (SD) also significantly different from zero (p=0.0014). MIGET shunt, determined primarily from retention of SF$_6$ in arterial blood, did not show a significant trend with fluid volume (mean slope 0.001$\pm$0.002 (SD), p=0.15).

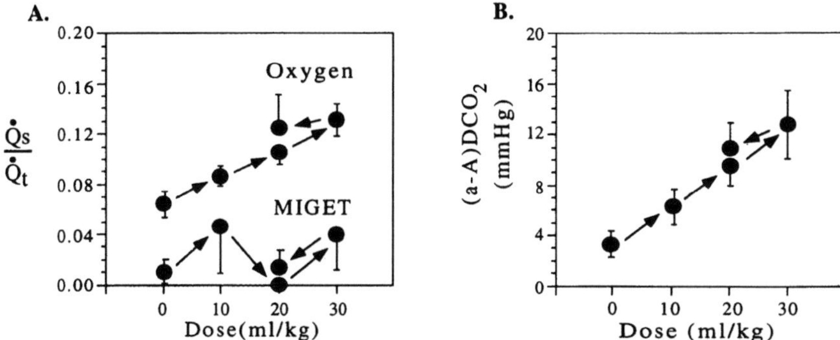

Figure 1. Gas exchange analysis in 8 animals with graded addition and removal of PFC. **A.** Oxygen and MIGET shunts **B.** Arterial-alveolar CO$_2$ difference ((a-A)DCO$_2$). All values (mean ± SEM) vs. PFC dose (ml/kg).

Table 1. Mean arterial pressure, cardiac output, mean pulmonary artery wedge pressure and peak airway pressure (n=8; mean ± SEM)

Condition	Pa, mmHg	CO, ml/min	PA wedge, cmH_2O	Paw, cmH_2O
baseline	71.0 ± 7.2	439 ± 29	11.0 ± 1.8	20.0 ± 0.96
10 ml/kg	74.5 ± 5.4	476 ± 52	10.3 ± 1.5	20.4 ± 1.40
20 ml/kg	62.0 ± 3.3	445 ± 49	10.5 ± 1.9	19.8 ± 1.10
30 ml/kg	56.1 ± 4.5	470 ± 55	10.8 ± 2.1	19.9 ± 0.97
20 ml/kg	54.4 ± 4.5	491 ± 53	9.5 ± 1.8	19.9 ± 1.30

Table 2. Summary of alveolar and blood gas data (n=8; mean ± SEM)

Condition	$P_{A_{O_2}}$, mmHg	$P_{a_{O_2}}$, mmHg	$P_{A_{CO_2}}$, mmHg	$P_{a_{CO_2}}$, mmHg	pHa
baseline	669 ± 2.97	522 ± 15	33 ± 1.2	35.1 ± 1.8	7.45 ± .016
10 ml/kg	664 ± 3.38	456 ± 20	34 ± 1.4	40.5 ± 2.5	7.40 ± .022
20 ml/kg	663 ± 2.47	408 ± 11	32 ± 1.3	41.8 ± 2.3	7.38 ± .026
30 ml/kg	660 ± 3.19	345 ± 21	31 ± 1.9	44.4 ± 3.4	7.34 ± .026
20 ml/kg	662 ± 3.17	419 ± 34	32 ± 1.5	43.4 ± 2.8	7.37 ± .023

Table 3. Inert and respiratory gas solubilities (ß, ml gas•100 ml solvent^{-1}•mmHg^{-1}) and perflubron-to-blood partition coefficients (λ).

Gas	β_{blood}	β_{PFC}	$\lambda_{PFC/blood}$
O_2	0.15 to 0.24	0.0658	0.27 to .44
CO_2	0.58	0.276	0.47
SF_6	0.000974	0.410	421.
ethane	0.0116	0.234	20.2
cyclopropane	0.0749	0.791	10.6
halothane	0.396	0.826	2.09
ether	1.34	5.09	3.80
acetone	38.4	3.86	0.101

Solubilities of MIGET gases in perflubron were determined by the double extraction technique described by Wagner et.al.[21]. Following this method, a volume of perflubron containing each of the six gases was equilibrated with a volume of tracer-free air. MIGET gas content in the volume of air was determined via gas chromatography. A second extraction into air was performed and a set of peak heights describing gas content was similarly obtained. From mass balance it can be shown that the solubility of each of the six gases in perflubron may be determined from the ratio of peak heights (equivalent to content ratios) between the extractions. Table 3 lists solubility coefficients (ß) for the six inert gases in blood and in perflubron as well as the ratio of the two ß's, i.e., the perflubron-to-blood partition coefficients (λ). Effective O_2 and CO_2 blood solubilities were calculated as slopes of arterial-venous chords of content vs. partial pressure curves. The values reported for O_2 bracket the normal range of possible slopes on the nonlinear hemoglobin-oxygen dissociation curve assuming normal hemoglobin with concentration 15 g/dl. O_2 and CO_2 solubilities in perflubron were provided by Alliance Pharmaceutical Corp., San Diego CA.

DISCUSSION

The primary goal of this study was to describe alterations in gas exchange expressed as shunt and (a-A)DCO_2 in normal lungs filled with graded amounts of perfluorocarbon fluid. A second objective was to determine the mechanisms responsible for impaired gas exchange during fluid breathing, i.e. shunt, diffusion limitation, and/or ventilation-to-perfusion (V_A/Q)

mismatch. We chose the normal animal model uncomplicated by disturbances at the alveolar-capillary membrane due to injury or disease to determine the effects of PFC alone. The description of gas exchange during PLV in the normal animal may be thought of as a "best case" against which lung injury models can be compared.

We found that O_2 shunt and (a-A)DCO_2 increase linearly with volume of PFC in the lung. MIGET shunt, however, showed no significant increase with fluid volume. There are several potential explanations why MIGET and O_2 shunts do not match. It has been reported[4] that O_2 shunt calculated using the Berggren equation and MIGET shunt differ systematically over the range 0-20%. MIGET measured a smaller shunt compared to the O_2 method and the discrepancy was thought to be due to the fact that MIGET does not detect fixed extra-pulmonary shunts that form part of the O_2 shunt. This may explain some of the difference seen here, although the magnitude of the discrepancy between techniques is much larger in our study and furthermore the differences increase with volume of perflubron in the lungs. The more likely explanation is that MIGET gases, some of which are highly soluble in the perflubron lining alveoli, are not achieving steady state exchange between blood and gas phases, a requirement which must be met in order for MIGET modeling to accurately predict shunt.

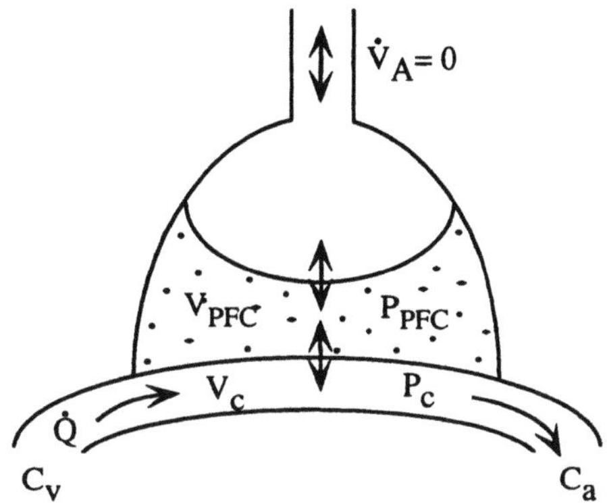

Figure 2. Schematic diagram illustrating a model of alveolar gas exchange during PLV. See text for explanation.

To evaluate this hypothesis, we developed a simple mathematical model consisting of two well mixed compartments (Figure 2) representing gas exchange in the fluid filled alveolus. The model is described by a system of two linear first order differential equations expressing mass balance in each compartment (see Appendix). The capillary compartment, V_c, receives systemic mixed venous blood with gas content, C_v (ml gas•100 ml blood^{-1}), and is drained by pulmonary venous blood with content C_a. It exchanges tracer gas with the PFC compartment, V_{PFC}. Diffusion of gases between PFC and alveolar air compartments was not represented (i.e., alveolar ventilation, $\dot{V}_A$, was set to zero) because we were only interested in the rate of equilibration between blood and PFC, the limiting factor in gas transport in regions of shunt. The blood and PFC compartments (V_c and V_{PFC}) were each assumed to be well-mixed with a single partial pressure of dissolved gases throughout. In our case, membrane conductance between compartments was assumed to be infinite such that gas exchange dynamics are not affected by membrane diffusion limitation. Gas exchange between V_c and V_{PFC} is dependent on rate of delivery of gas to the capillary compartment, i.e. transport is flow-limited. With these assumptions, the rate of equilibration between blood and PFC was found to be described by the time constant, $\tau=[V_c+\lambda_{PFC/c}*V_{PFC}]/\dot{Q}$

(see Appendix). Figure 3 shows model solutions of tracer gas accumulation in PFC for each of the six MIGET gases after a step change in P_c from Pv_0 to Pv_1. Model parameter values were chosen as follows: V_{PFC} was 90ml (30ml/kg in a 3kg animal), V_c was 5ml, $\dot{Q}$ (cardiac output) was 200 ml/kg/min, and β_{PFC} and β_c were gas-dependent and are listed in Table 3. Comparison of gas partition coefficients ($\lambda=\beta_{PFC}/\beta_c$) vs. equilibration times illustrated in Figure 3 shows the dependence of mass transport kinetics on $\lambda_{PFC/blood}$. SF_6, with the largest partition coefficient (421), has an equilibration time (to 95%) of more than three hours. The implications of this are that SF_6 will not be retained in the arterial blood (hence no shunt) if regions of shunt contain PFC and are not stable for roughly three hours or more. PFC acts as a "sump", absorbing much of the SF_6 delivered by mixed venous blood despite lack of ventilation. From these model predictions we concluded that SF_6 retention will not accurately represent shunt during PLV because it does not reach equilibrium within the time frame of usual physiologic manipulations.

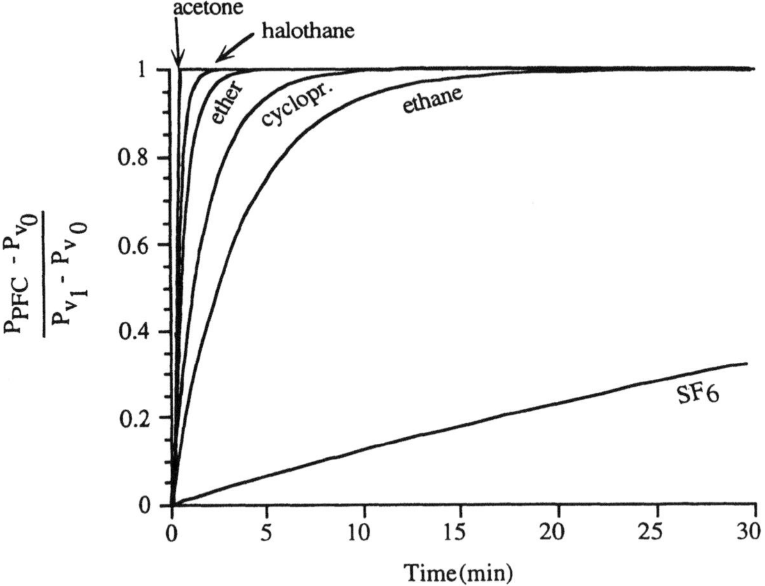

Figure 3. Solutions of two compartment model showing time course of PFC saturation with each of six MIGET gases after step change in blood partial pressure from Pv_0 to Pv_1. See text and Appendix for details of the model and calculations.

The second objective of this study was to characterize the type of mass transport limitation being imposed by the fluid. It is possible to rule out ventilation/perfusion ($\dot{V}_A/\dot{Q}$) mismatch as a cause of increased $(A\text{-}a)DO_2$ during 100% O_2 breathing because effects of $\dot{V}_A$ are made small when PA_{O_2} is high and essentially homogeneous throughout the lung. However, in this study it was not possible to distinguish between the other two causes of $(A\text{-}a)DO_2$, i.e., shunt and diffusion-limited transport. In fact, one of the main assumptions underlying the Berggren shunt equation is that there is no diffusion limitation to O_2 transport between alveolar gas and capillary blood. In a fluid-lined alveolus, however, it is likely that diffusion gradients and hence diffusion-limited transport exist. Therefore, our calculated oxygen "shunt" is probably a combination of anatomic shunt and diffusion limitation. Creation of anatomic shunt could be imagined in the traditional sense as due to regions that become flooded with PFC and receive infrequent or no ventilation. Shunt of this type may occur in areas where the distribution of fluid is uneven or in areas of high surface tension adjacent to those of low surface tension resulting in alveolar instability and collapse. An alternative mechanism creating apparent shunt would be increased times to diffusional equilibrium for O_2 and CO_2, i.e., longer than pulmonary capillary transit times. One may calculate, using the Einstein diffusion equation and measured diffusion coefficients of O_2 and CO_2 in fluorocarbon fluids[19], the average distance traveled by an oxygen molecule in one second is about 100μ. If the mean diameter of a fluid filled alveolus is 200μ and a significant proportion of alveoli were flooded with PFC, it would take about 2 sec for gas molecules to traverse the fluid layer in either direction. This is equivalent to the time to complete a single breath and about 3 times the transit time of a red cell through a capillary. This mechanism would contribute to lowered PaO_2 even in steady-states. The final mechanism, described by the model above, involves the "sump" effect of PFC on certain gases due to their high solubilities in perflubron. This property should be manifested mainly in transient situations where λ is extremely high, as it is for SF_6 but not for O_2.

Our (a-A)DCO$_2$ results also create interesting problems in interpretation of possible mechanisms. It is not traditionally thought that shunt, or $\dot{V}_A/\dot{Q}$ mismatch, can cause the differences seen here (mean (a–A)DCO$_2$ of 12 torr at 30 ml/kg of PFC). Shunt and $\dot{V}_A/\dot{Q}$ mismatch alone may account for a maximum of 2-4 mmHg difference. Diffusion limitation is not typically thought to cause widening (a–A)DCO$_2$ in the gas-ventilated lung because CO_2 is very soluble in tissue and diffusion distances are relatively short (a few μ at most). Until a more comprehensive mathematical model is developed, it is not possible to quantitatively discern the components creating the CO_2 differences seen in our experiments but it would seem that diffusion limitation and slight sump effects of PFC are at least as important as they are for O_2.

In this experiment, V_E was set during conventional gas breathing and was unchanged thereafter. This was done so that changes in gas exchange status could be attributed to effects of the liquid and not to other ventilatory parameters.   Hypoventilation (elevated $PaCO_2$) was evident during PLV but could easily have been corrected by altering PEEP or V_E. These results should therefore not be interpreted as indicating that PLV causes CO_2 retention and respiratory acidosis.

We conclude from this study in normal piglets that O_2 and CO_2 exchange in the lung is impeded roughly in proportion to the volume of added fluorocarbon fluid. For certain gases such as SF_6, equilibration times between blood and perflubron pools are long (up to 3 hours) due to high $\lambda_{PFC/blood}$.   For this reason, it was not possible to quantitate shunt using the MIGET gas SF_6. It is also not possible to determine the exact nature of the O_2 shunt calculated using the Berggren equation. The gas exchange impairment is most likely a combination of diffusion limitation in the fluid-containing alveoli and creation of true shunt regions consisting of flooded unventilated alveoli.

APPENDIX

Gas transport between capillary (c) and perflubron (PFC) compartments can be described by the following mass balance equation:

$$\frac{d(M_c + M_{PFC})}{dt} = \dot{Q} \cdot (C_{\bar{v}} - C_a) \qquad (1)$$

where M_c and M_{PFC} represent the mass of tracer gas (ml gas) in capillary and perflubron compartments, respectively, Q is flow (ml blood•min^{-1}) into the capillary compartment, $C_{\bar{v}}$ and C_a are mixed venous and arterial gas contents respectively (ml gas•100 ml blood^{-1}). Rewriting in terms of gas contents ($C=M/V$):

$$\frac{d}{dt}(C_c \cdot V_c + C_{PFC} \cdot V_{PFC}) = \dot{Q} \cdot (C_{\bar{v}} - C_a) \tag{2}$$

where C_c and C_{PFC} are capillary and PFC contents, respectively (ml gas•100 ml blood^{-1}), and V_c and V_{PFC} are blood and PFC volumes (ml). Using the Henry's law for gases in physical solution: $P=C/ß$, and assuming capillary and perflubron compartments are in equilibrium with respect to partial pressure $(P_{PFC}=P_c)$, we substitute $C_c=C_a=C_{PFC} \cdot (ß_c/ß_{PFC})$ into 2:

$$\frac{dC_{PFC}}{dt} \cdot (\frac{ß_c}{ß_{PFC}} \cdot V_c + V_{PFC}) + \dot{Q} \cdot \frac{ß_c}{ß_{PFC}} \cdot C_{PFC} = \dot{Q} \cdot C_{\bar{v}} \tag{3}$$

Solving for $C_{PFC}(t)$ with the boundary condition $C_{PFC}(0)=0$:

$$C_{PFC}(t) = C_{\bar{v}} \cdot \frac{ß_{PFC}}{ß_c}(1 - \exp[\frac{-\dot{Q}}{V_c + \frac{ß_{PFC}}{ß_c} \cdot V_{PFC}} \cdot t]) \tag{4}$$

Finally, the time constant for equilibration given the assumptions stated above is equivalent to the inverse of the coefficient of t in the exponent:

$$\tau = \frac{V_c + \frac{ß_{PFC}}{ß_c} \cdot V_{PFC}}{\dot{Q}} = \frac{V_c + \lambda_{PFC/c} \cdot V_{PFC}}{\dot{Q}} \tag{5}$$

Since V_c is typically much smaller than $\lambda \cdot V_{PFC}$, Eq. 5 becomes $\tau = \lambda \cdot V_{PFC}/\dot{Q}$.

ACKNOWLEDGMENTS

The expert technical assistance of Wayne Lamm and Mical Middaugh was invaluable in successfully completing these experiments. Alliance Pharmaceutical Corp (San Diego, CA) provided the perflubron used in this study. This research was supported in part by NIH grant R37HL12174.

REFERENCES

1. Clark, L. C. and F. Gollan. Survival of mammals breathing organic liquids equilibrated with oxygen at atmospheric pressure. Science 152: 1755-1756, 1966.
2. Curtis, S. E., B. P. Fuhrman and D. F. Howland. Airway and alveolar pressures during perfluorocarbon breathing in infant lambs. JAP 68: 2322-2328, 1990.
3. Fuhrman, B. P., P. R. Paczan and M. DeFrancisis. Perfluorocarbon-associated gas exchange. Crit. Care Med. 19: 712-22, 1991.

4. Hlastala, M. P., P. S. Colley and F. W. Cheney. Pulmonary shunt: a comparison between oxygen and inert gas infusion methods. JAP 39: 1048-1051, 1975.

5. Hlastala, M. P. and H. T. Robertson. Inert gas elimination characteristics of the normal and abnormal lung. JAP 44: 258-266, 1978.

6. Hlastala, M. P. Multiple inert gas elimination technique. JAP 56: 1-7, 1984.

7. Koen, P. A., M. R. Wolfson and T. H. Shaffer. Fluorocarbon ventilation: maximal expiratory flows and CO_2 elimination. Ped. Res. 24: 291-296, 1988.

8. Matthews, W. H., R. H. Balzer, J. D. Shelburne, P. C. Pratt and J. A. Kylstra. Steady-state gas exchange in normothermic, anesthetized, liquid-ventilated dogs. Undersea Biomed. Res. 5: 341-354, 1978.

9. Modell, J. H., H. W. Calderwood, B. C. Ruiz, M. K. Tham and C. I. Hood. Liquid ventilation of primates. Chest 69: 79-81, 1976.

10. Moskowitz, G. D., T. H. Shaffer and S. E. Dubin. Liquid breathing trials and animal studies with a demand-regulated liquid breathing system. Med. Instrum. 9: 28-33, 1973.

11. Richman, P. S., M. R. Wolfson and T. H. Shaffer. Lung lavage with oxygenated perfluorochemical liquid in acute lung injury. Crit. Care Med. 21: 768-774, 1993.

12. Sargent, J. W. and R. J. Seffl. Properties of perfluorinated liquids. Fed. Proc. 29: 1699-1703, 1970.

13. Schwieler, G. H. and B. Robertson. Liquid ventilation in immature newborn rabbits. Biol. Neonate 29: 343-353, 1976.

14. Shaffer, T. H., D. Rubenstein, G. D. Moskowitz and M. Delivoria-Papadopoulos. Gaseous exchange and acid-base balance in premature lambs during liquid ventilation since birth. Pediat. Res. 10: 227-231, 1976.

15. Shaffer, T. H., P. R. Douglas, C. A. Lowe and V. K. Bhutani. The effects of liquid ventilation on cardiopulmonary function in preterm lambs. Pediat. Res. 17: 303-306, 1983.

16. Shaffer, T. H., C. A. Lowe, V. K. Bhutani and P. R. Douglas. Liquid ventilation: effects on pulmonary function in distressed meconium-stained lambs. Ped. Res. 18: 47-52, 1984.

17. Shaffer, T. H., M. R. Wolfson and L. C. Clark Jr. Liquid ventilation. Pediatr. Pulmonol. 14: 102-109, 1992.

18. Sivieri, E. M., G. D. Moskowitz and T. H. Shaffer. Instrumentation for measuring cardiac output by direct Fick method during liquid ventilation. Undersea Biomed. Res. 8: 75-83, 1981.

19. Tham, M. K., R. D. Walker and J. H. Modell. Diffusion coefficients of O_2, N_2, and CO_2 in fluorinated ethers. J. Chem. Eng. Data 18: 411-412, 1973.

20. Wagner, P. D., H. A. Saltzman and J. B. West. Measurement of continuous distributions of ventilation-perfusion ratios: theory. JAP 36: 588-599, 1974.

21. Wagner, P. D., P. F. Naumann and R. B. Laravuso. Simultaneous measurement of eight foreign gases in blood by gas chromatography. JAP 36: 600-605, 1974.

EFFECTS OF DIFFERENT MECHANICAL VENTILATION MODES ON OXYGENATION IN SURFACTANT DEPLETED RABBIT LUNGS

Jozef Kesecioglu[1], Lüfti Telci[2], Ahmet S. Tütüncü[2], Figen Esen, Wilhelm Erdmann[1], and Burkhard Lachmann[1]

[1]Department of Anesthesiology, University Hospital Dijkzigt
Dr. Molewaterplein 40, 3015 GD Rotterdam, The Netherlands
[2]Department of Anesthesiology, University of Istanbul, Faculty of Medicine, Istanbul, Turkey

INTRODUCTION

Application of the conventional volume controlled ventilation (VCV) with positive end-expiratory pressure (PEEP) is usually a successful mode as an immediate therapy to relieve hypoxemia in acute respiratory failure (ARF).[1] However, this form of ventilation is associated with high tidal volumes (V_T) and high peak inspiratory pressures (PIP) which are suggested to cause barotrauma and morphological changes in the lungs.[2-5]

Pressure controlled inverse ratio ventilation (PCIRV) is reported[6,7] as a method providing better oxygenation with lower PIP, compared to VCV with PEEP.

High frequency ventilation (HFV) is another model preventing lung injury due to low V_T and its successful use in providing gas exchange are reported.[8,9] However, HFV has some limitations in application in humans as monitoring of the respiratory variables, alarm systems and humidification of gases are sometimes not suitable for long-term use.

On the other hand, superimposing HFV on the expiratory phase of pressure regulated volume controlled ventilation (PRVCV) with an inspiration/expiration (I/E) ratio 4:1, is expected to provide further increase of mean airway pressure (mPaw) and auto-PEEP leading to an improvement of oxygenation and CO_2 elimination, while increase in PIP levels are avoided.

Therefore, aiming to achieve a PaO_2 level above 300 mmHg by adjusting preset static-PEEP ($PEEP_S$) and PIP levels, and constant expiratory minute volumes (V_E), VCV with PEEP, PRVCV with an I/E ratio 4:1 and superimposed expiratory HFV on PRVCV with an I/E ratio 4:1 (SEHFV-PRVCV) were investigated in this study, concerning their influence on gas exchange and lung mechanics in surfactant depleted rabbit lungs.

METHODS

Six male New Zealand rabbits, 2.8 ± 0.2 kg (range 2.6-3.2 kg), were used in this study. Anesthesia was induced with pentobarbital given intravenously through a 22 G cannula placed into an ear vein. Tracheostomy was performed and a tube ID 3 mm with two additional side entrances for HFV and airway pressure monitoring was inserted. The tube was fixed tightly to the trachea to avoid air leakage. Anesthesia was maintained with infusion of pentobarbital and fentanyl; pancuronium bromide was administered for muscle paralysis. Lungs were ventilated with Servo 900C (Siemens-Elema, Solna, Sweden). SEHFV was performed with a high Frequency Unit 970 (Siemens-Elema, Solna, Sweden) connected to and triggered by the Servo ventilator 900C.

A catheter was placed in the femoral artery for invasive arterial blood pressure monitoring and blood sampling. Cardiovascular monitoring was done by means of a Marquette monitor system 7000™ (Marquette Electronics Inv. Milwaukee, Ws, USA) and printed directly with a Marquette monitor printer 7100. Arterial blood gases were measured with the blood gas analyzer ABL 330 (Radiometer, Copenhagen). Total PEEP [$PEEP_T$ = PEEPs + auto PEEP] in M2 and M3 displayed by pressing the end-expiratory hold button of the ventilator, PIP and Mpaw displayed by the ventilator and intrapulmonary pressure amplitude (ΔP), defined as the difference between PIP and $PEEP_T$, or PEEPs were recorded.

After completing baseline measurements, respiratory failure was induced by surfactant depletion by lung lavage performed four times with 100 ml warm saline[2] to produce a PaO_2 between 40 and 70 mmhg. Before lavage [control mode (CM) 1], during and after lung lavage (CM2) lungs were ventilated with VCV with a preset PEEPs of 4 cm H_2O V_T 12 ml/kg, frequency (f) 30 breaths/min and I/E ratio 1:2 (25% I, 10% pause).

After a stabilization period of 15 min randomly applied modes of ventilation to achieve a PaO_2 above 300 mmhg and constant V_E were as follows:

Mode (M) 1: VCV with a $PEEP_s$ between 8 and 14 cm H_2O to achieve PaO_2 above 300 mm Hg, f 30 breaths/min, V_T 12 ml/kg and I/E ratio of 1:2 (25% I, 10% pause).

M 2: PCVRV with f 30/min and I/E ratio of 4:1 (80% I). 35 cm H_2O PIP was applied during the first 5 min of artificial ventilation for alveolar recruitment. PIP was then reduced to a level of 1-2 cm higher then the opening pressure of the alveoli. PEEPs was adjusted to keep V_T at 12 ml/kg.

M 3: PCVRV with f 30/min and I/E ratio of 4:1 (80% I). 35 cm H_2O PIP was applied during the first 5 min of artificial ventilation for alveolar recruitment. PIP was then reduced to a level of 1-2 cm higher then the opening pressure of the alveoli. PEEPs was adjusted to keep V_T at 12 ml/kg. Additionally an open system was connected to flush the airways with SEHFV with a f 15 Hz, V_T 0.3 ml/kg and pulse duration of 20%.

FiO_2 was 1.0 in all modes. The rabbits were treated 30 min with each mode and the measurements were made thereafter. Before the use of different ventilatory settings CM2 was applied for 30 min to investigate whether any changes in gas exchange occurred over time.

Data were compared by two-way analysis of variance (ANOVA) test. All data are expressed as mean $\pm$ SD. Student's t-test was used for pair-wise comparisons. Significance was considered at $P < 0.05$.

RESULTS

With the same minute ventilation, PRVCV increased oxygenation significantly compared to control modes and M1. M3 also improved PaO_2 values but SEHFV in this mode did not produce significant changes compared to M2 (Figure 1). $PaCO_2$ decreased significantly with the application of M2 and M3. SEHFV did not further effect CO_2 elimination (Figure 2).

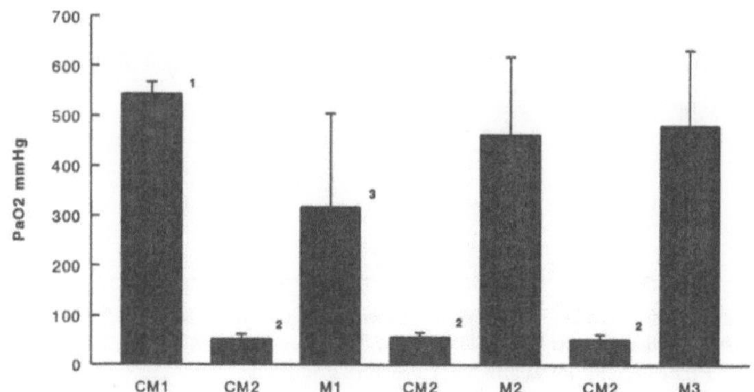

Figure 1. PaO_2 values observed during the study. CM= control mode; M= mode. 1= significantly different from CM2 and M1; 2= significantly different from M1, M2 and M3; 3= significantly different from M2 and M3.

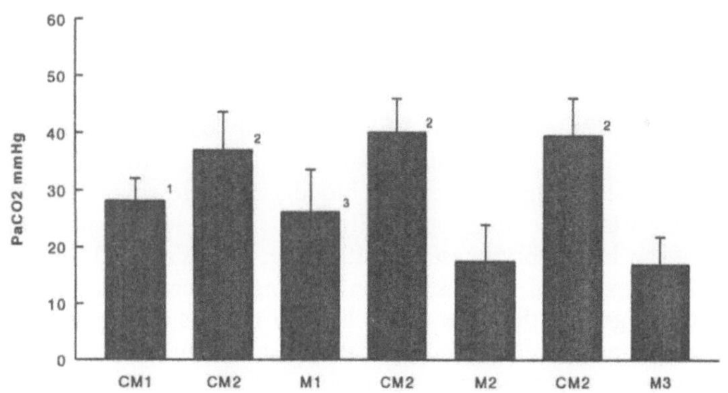

Figure 2. $PaCO_2$ values observed during the study. CM= control mode; M= mode. 1= significantly different from CM2, M2 and M3; 2= significantly different from M1, M2 and M3; 3= significantly different from M2 and M3.

PEEPs in M1 and $PEEP_T$ measured in M2 and M3 were similar and no statistical significance was observed between groups. ΔP was significantly higher M1 compared to M2 and M3. PIP values were significantly lower and mPaw values were significantly higher in M2 and M3 compared to the VCV with PEEP (Table 1).

Table 1. Lung mechanics parameters during the application of ventilatory modes (mean ± SD).

n = 9	PIP (cm H_2O)	Mpaw (cm H_2O)	$PEEP_T$ (cm H_2O)	ΔP (cm H_2O)
CM1	15.8 ± 1.3[a]	6.8 ± 1.0[a]	4[c]	11.2 ± 0.9[a]
CM2	26.3 ± 3.0	8.7 ± 0.8[c]	4[c]	22.8 ± 1.7[d]
M1	37.1 ± 7.0[b]	17.2 ± 4.4[d]	13.2 ± 1.8	24.3 ± 5.4[d]
CM2	24.3 ± 3.2	8.7 ± 0.9[c]	4[c]	20.9 ± 1.8[d]
M2	28.5 ± 2.3	23.8 ± 2.1	13.4 ± 2.1	14.6 ± 2.0
CM2	25.9 ± 2.8	8.0 ± 0.9[c]	4[c]	21.3 ± 2.3[d]
M3	28.7 ± 3.7	24.9 ± 2.1	14.2 ± 2.3	14.7 ± 2.6

CM = control mode; M = mode. PIP = peak inspiratory pressure; Mpaw = mean airway pressure; $PEEP_T$ = total PEEP [values shown in CM1, CM2 and M1 are static PEEP (PEEPs) and in M2 and M3 are PEEPs + auto PEEP]; ΔP = intrapulmonary pressure amplitude.
a = Significantly different from CM2, M1, M2 and M3 ($p < 0.05$).
b = Significantly different from CM2, M2 and M3 ($p < 0.05$).
c = Significantly different from M1, M2 and M3 ($p < 0.05$).
d = Significantly different from M2 and M3 ($p < 0.05$).

DISCUSSION

The results of this study show that best oxygenation was achieved with PRVCV and SEHFV-PRVCV, yet PIP values maintaining constant minute ventilation were lower with these modes. Considerably high PIP was necessary to provide the desired gas exchange in VCV with PEEP compared to M2 and M3, although the PaO_2 achieved was much lower than in the other treatment modes.

In a recent review article Lachmann[10] stated that during the application of VCV with clinically used PEEPs levels the set PEEPs will only balance the retractive forces of parts of the damaged lungs and only these parts of the lung will not collaps during the whole respiratory cycle leading to improved blood gases. However, PEEPs will not be enough to keep all parts of the lungs open. Highly damaged lung regions will be aerated only at the end of the inspiratory phase. Additionally, perfusion will decrease due to the high intra-alveolar pressure of end inspiration limiting the contribution of affected lung regions to gas exchange. On the other hand, healthy regions of the lungs can also have capillary compression and a ventilation/perfusion mismatching due to the applied PEEPs which will be even more prominent during the inspiration phase, causing the over-distension of these parts.

On the other hand at PRVCV if the PIP is adjusted only to compensate for the retractive forces of the whole lung, overdistansion of the alveoli can never occur. An auto-PEEP can be created by either increasing the I/E ratio at constant frequency or increasing the frequency at a constant I/E ratio (or both). This will establish an expiration time too short to allow emptying of the lung. Therefore even the stiffest parts of the lung will have no time for collapse, will be kept open and can be ventilated with a smaller ΔP compared to VCV with PEEP.

In this study SEHFV-PRVCV did not effect oxygenation of CO_2 elimination impressively compared to PRVCV. Earlier reports show improvement of these parameters with HFV combined with conventional ventilation.[11-14] However, in those studies superimposed HFV on conventional ventilation was compared with VCV with

PEEP or continuous positive airway pressure (CPAP). Moreover, from experimental results in pigs with acute respiratory failure, it is known that the reason for these findings may be the increased PIP and mPaw pressures during superimposed HFV. A remarkable improvement in gas exchange is observed with combined ventilation compared to M1 in this study, but when the results of M2 are considered, this change is obviously due to PRVCV rather than HFV. In a similar study Lachmann and colleagues[15] superimposed HFV to the expiratory phase of VCV and achieved improved CO_2 elimination. However, their impression was that this form of ventilation was inferior to PCIRV with a prolonged inspiratory cycle up to 80%. Our results confirm their impression as no additional CO_2 elimination in observed with the application of SEHFV. This lack of further CO_2 elimination was probably due to the adequate CO_2 elimination from the large airways and the tube up to the Ypiece, due to the application of PRVCV.

In conclusion VCV with PEEPs provided adequate oxygenation. However, it resulted in high PIP. Although the PIP levels remained low, SEHFV-PRVCV did not further increase the PaO_2 or $PaCO_2$ elimination in this animal model of acute respiratory failure suggesting no additional advantages with this mode over M2. In this study PRVCV with a prolonged inspiratory cycle of 80% is found to be a self competent ventilatory mode, not necessitating further combinations.

ACKNOWLEDGMENTS

We thank Ms. Sharida Santoe for typing the manuscript and Mr. Ton Muêtgeert for technical assistance.

REFERENCES

1. D.G. Asbaugh, T.L. Petty, D.B. Bigelow, and T.M. Harris, Continuous positive-pressure breathing (CPPB) in adult respiratory distress syndrome. *J Thorax Cardiovasc Surg.* 57:31-41 (1969).
2. B. Lachmann, B. Robertson, and Vogel J, In-vivo lung lavage as an experimental model of the respiratory distress syndrome, *Acta Anaesth Scand.* 24:231-236 (1980).
3. P.P. Hamilton, A. Onayemi, J.A. Smith, J.E. Gillan JE, E. Cutz, A.B. Froese, and Bryan AC, Comparison of conventional and high frequency ventilation: oxygenation and lung pathology, *J Appl Physiol.* 55:131-138 (1983).
4. T. Kolobow, M.P. Moretti, R. Fumagali, D. Mascheroni, P. Prato, V. Chen, and M. Joris, Severe impairment in lung function induced by high peak airway pressure during mechanical ventilation, An experimental study, *Am Rev Respir Dis.* 135:312-315 (1987).
5. D. Dreyfuss, P. Soler, G. Basset, and Saumon G, High inflation pressure pulmonary edema. Respective effects of high airway pressure, high tidal volume and positive end-expiratory pressure, *Am Rev Respir Dis.* 137:1159-1164 (1988).
6. B. Lachmann, E. Danzmann, B. Haendly, and B. Jonson, Ventilator settings and gas exchange in respiratory distress syndrome, *in:* "Applied Physiology in Clinical Respiratory Care," O. Prakash, ed., Martinus Nijhoff Publishers, The Hague, (1982).
7. B. Lachmann, B. Jonson, M. Lindroth, and Robertson, Modes of artificial ventilation in severe respiratory distress syndrome. Lung function and morphology in rabbits after wash-out of alveolar surfactant. *Crit Care Med.* 10:724-732 (1982).

8. G.C. Carlon, W.S. Howland, C. Ray, S. Miadownik, J.P. Griffin, and J.S. Groeger JS, High-frequency jet ventilation, A prospective randomized evaluation, *Chest* 84:551-559 (1983).

9. W.A. Carlo, R.L. Chatburn, and R.J. Marti, Randomized trial of high-frequency jet ventilation versus conventional ventilation in respiratory distress syndrome, *J Pediatr.* 110:275-282 (1987).

10. B. Lachmann, Open the lung and keep the lung open, *Intensive Care Med.* 18:319-321 (1992).

11. J.M. Hurst JM, and C.B. DeHaven, Adult respiratory distress syndrome: Improved oxygenation during high frequency jet ventilation/continuous positive airway pressure, *Surgery* 96:764-769 (1984).

12. N. El-Baz, L.P. Faber, and A. Doolas, Combined high-frequency ventilation for management of terminal respiratory failure: a new technique, *Anesth Analg.* 62:39-49 (1983).

13. E. Barzilay, D. Kessler, and R. Raz, Superimposed high frequency ventilation with conventional mechanical ventilation, *Chest* 95:681-682 (1989).

14. B.R. Boynton, F.L. Mannino, R.F. Davis, R.J. Kopotic, and G. Friederichsen, Combined high-frequency oscillatory ventilation and intermittent mandatory ventilation in critically ill neonates, *J Pediatr.* 105:297-302 (1984).

15. B. Lachmann, W. Schairer, M. Hafner, S. Armbruster, and B. Jonson, Volume-controlled ventilation with superimposed high frequency ventilation during expiration in healthy and surfactant-depleted pig lungs, *Acta Anaesthesiol Scand 33.*, Supp 90:117-119 (1989).

EFFECT OF KETANSERINE ON OXYGENATION AND VENTILATION INHOMOGENEITY IN PIGS WITH ARDS

Jozef Kesecioglu, Can Ince*, Jan C. Pompe*, Ismail Gültuna, Wilhelm Erdmann, and Hajo A. Bruining*

Departments of Anesthesiology and Surgery*
University Hospital Dijkzigt, Dr. Molewaterplein 40
3015 GD Rotterdam, The Netherlands

INTRODUCTION

The lung dynamics in ARDS are characterized by alveolar units with different time constants. This is due to the nonhomogenous distribution of the disease in the alveoli, where the changes in compliance and airways resistance cause an unequal distribution of pressure and volume during mechanical ventilation. This nonhomogenous distribution of time constants results in ventilation inhomogeneity. A multi breath indicator gas washout technique has recently been reported to measure this inhomogeneity.[1]

Contradictory results are obtained in various studies concerning the effects of ketanserine as a serotonin antagonist in animals or patients with ARDS. While some investigators observed no changes in PaO_2 with the administration of ketanserine,[2] others reported improved oxygenation.[3] It is uncertain whether this improved oxygenation was caused by the redistribution of perfusion to well ventilated lung units or improved ventilation to normally perfused lung due to decreased bronchoconstriction in the distal airways.

In this study lung lavage was used to create surfactant depletion in pig lungs, aiming to assess the ventilation inhomogeneity caused by this acute respiratory failure model. It is hypothesized that serotonergic activities might be initiated in the lungs accompanying the hypoxia of acute respiratory failure in this model, leading to vaso- and bronchospastic events such as pulmonary hypertension and ventilation inhomogeneity. Furthermore, the effect of a specific antagonist, ketanserine is investigated in the treatment of pulmonary vasoconstriction and ventilation inhomogeneity of surfactant depleted pig lungs.

METHODS

Seven pigs, (19 - 22 kg) were investigated in this study. Anesthesia was induced with intramuscular ketamine. The trachea was intubated with a portex tube with an internal diameter of 7 mm. The lungs were ventilated with a Servo 900C (Siemens-Elema Solna, Sweden) ventilator thereafter. Anesthesia was maintained by infusion of midazolam and ketamine. Pancuronium bromide infusion was administered after a bolus dose, for muscle relaxation.

A catheter for adequate fluid replacement and a 7F Swan-Ganz thermodilution catheter for hemodynamic monitoring were inserted through the right and left internal jugular veins respectively.

The carotid artery was cannulated for blood sampling and invasive blood pressure monitoring. An 18F Foley urine catheter was placed into the bladder by cystotomy to monitor urine output.

Mean arterial pressure (MAP) and mean pulmonary artery pressure (MPAP), were measured. Cardiac output (CO) was measured by thermodilution technique, using three 10-ml injections saline at room temperature, randomly. Arterial blood gases were determined by AVL 945 automatic bloodgas system.

A multiple breath indicator gas washout test using 2% SF_6 was used to calculate end expiratory volume (EEV) and an index for ventilation inhomogeneity (S) as described by Huygen and colleagues.[1] The indicator and the metabolic gases were continuously measured by a mass-spectrometer (Airspec MGA 3000, UK) and EEV and S were calculated.

Pigs were subjected to lung lavage to establish a model of ARDS.[4] They were ventilated before lung lavage (BLL), after lung lavage (ALL) and after ketanserine administration (K) with volume controlled ventilation (VCV) with PEEP 4 cm H_2O tidal volume (V_T) 8-12 ml/kg, frequency (f) 12 breaths/min, Inspiratory (I)/Expiratory (E) ratio 1:2 and FiO_2 1.0. Ventilation parameters were not changed during the whole course of the study. After completing baseline measurements at BLL lung lavage was performed with 150 ml/kg of warm saline solution to induce severe respiratory failure. PaO_2 < 100 mm Hg at ALL was accepted as ARDS and measurements were repeated after allowing a stabilization period of at least 2 h. Ketanserine was given thereafter as a bolus dose of 0.2 mg/kg. The same set of measurements were done 5 min after the administration of the drug.

All data expressed as mean $\pm$ SEM were compared between groups by pairwise t-tests. Significance was considered at $p \leq 0.05$.

RESULTS

Significant improvement was observed in ventilation inhomogeneity expressed by lowered S values, after the administration of ketanserine (Figure 1). No effect of ketanserine was seen on EEV (Figure 2).

PaO_2 values decreased significantly after surfactant depletion (Table 1). No significant change was observed between ALL and K. $PaCO_2$ increased significantly ALL compared to BLL. No significant change was observed after ketanserine administration (Table 1).

Changes in CO, MPAP and MABP are shown in Table 1. The use of ketanserine did not effect these parameters significantly.

DISCUSSION

Multi-breath indicator gas washout test enabled us to investigate functional

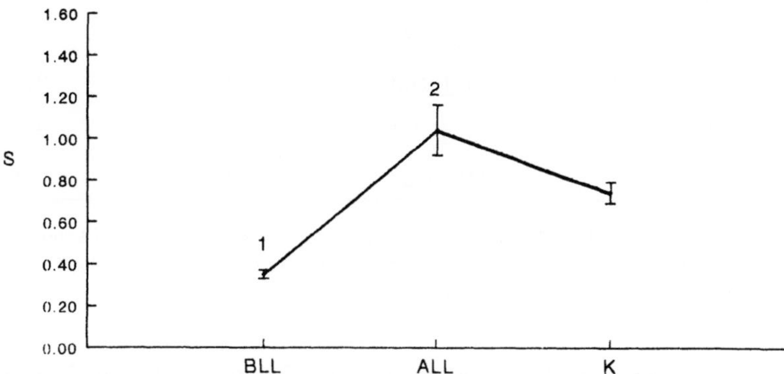

Figure 1. Ventilation inhomogeneity observed in the study. S= ventilation inhomogeneity index; BLL= before lung lavage; ALL= after lung lavage; K= after ketanserine administration. 1= significantly different from ALL and K (P<0.05); 2= significantly different from K (P<0.05).

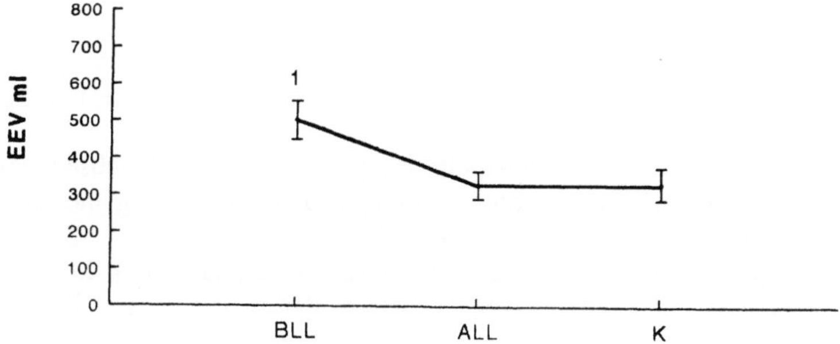

Figure 2. End expiratory volume observed in the study. EEV= end expiratory volume; BLL= before lung lavage; ALL= after lung lavage; K= after ketanserine administration. 1= significantly different from ALL and K (P<0.05).

Table 1. Respiratory and hemodynamic parameters observed during the study (mean ± sem).

n = 7	BLL	ALL	K
PaO_2 (mm Hg)	525 ± 34[1]	70 ± 5	87 ± 6
$PaCO_2$ (mm Hg)	46 ± 0.6[1]	58 ± 4.8	58 ± 3
CO (L/min)	3.6 ± 0.3	3.3 ± 0.5	3.4 ± 0.6
MPAP (mm Hg))	19.9 ± 1.6[1]	37 ± 3.6	36.1 ± 14
MABP (mm Hg)	109 ± 6[1]	85 ± 6	72 ± 6

BLL= before lung lavage; ALL= after lung lavage; K= after the administration of ketanserine; CO= cardiac output; MPAP= mean pulmonary artery pressure; MABP= mean arterial blood pressure. 1= significantly different from ALL and K (p<0.05).

properties of the lung in the acute respiratory failure model used in this study. Additionally, the use of mass-spectrometry provided an accurate and continuous measurement of the indicator gas. Recently Huygen and colleagues[1] have used this washout test in postoperative intensive care patients and stressed the importance of the index S as a sensitive diagnostic parameter for the assessment of ventilation inhomogeneity.

Lung lavage is an acute respiratory failure model which is known to produce surfactant depletion, alveolar collapse and hypoxemia. This model introduced by Lachmann and co-workers[4] imitates the early phase of ARDS and high S values obtained confirms it as a good ARDS model providing inhomogeneous distribution of ventilation leading to the hypoventilation of some alveoli while others are overdistended.

The results of this study show that ketanserine improved the ventilation inhomogeneity without significant changes in EEV. Although some pigs showed improved PaO_2 values, the mean value over the whole group was unchanged. In contrast to previous studies, no changes in MPAP was observed after the administration of ketanserine.

Activation of platelets and their pulmonary entrapment with subsequent release of the smooth muscle constrictor serotonin is thought to occur during the early phase of ARDS resulting in pulmonary hypertension and airway constriction.[2,5] Serotonin is also known to cause peripheral bronchospasm in airways less than 3 mm in diameter.[6] This constriction may play an important role in the ventilation inhomogeneity of the ARDS lungs.

Ketanserine, a specific antagonist of serotonin is reported to decrease pulmonary hypertension when administered in patients with ARDS.[2,7] Moreover, the hemodynamic changes occurred, were not accompanied by a deterioration of oxygenation. Vincent and colleagues[2] observed no changes in PaO_2 with the administration of ketanserine and related this to the comparable increase in ventilation as well as perfusion due to the antagonism of airway constriction caused by ketanserine. Huet and co-workers[3] reported improved oxygenation with ketanserine in pulmonary embolism, and, related this either to the redistribution of perfusion to well ventilated lung units or improved ventilation to normally perfused lung due to decreased bronchoconstriction. Olson[8] has also suggested that ketanserine might relieve airway constriction in hypoventilated lung areas.

The significant reduction of ventilation inhomogeneity and better distribution of ventilation in our study, is probably due to bronchodilatation occurring in some constricted small airways related to ketanserine administration. This allows us to speculate on the partial involvement of serotonergic reflexes in ventilation inhomogeneity caused by lung lavage, as well as surfactant deficiency, and its reversal by a specific antagonist. However, the resulting improvement in ventilation perfusion mismatch is not accompanied with a significant improvement of oxygenation. This might be attributed to the still existing alveolar collapse and low EEV due to surfactant depletion.

On the other hand the lack of antagonist effect of ketanserine on pulmonary hypertension suggests that vasoconstrictor agents other than the serotonergic effect might be responsible in increases MPAP in our study model.

In conclusion, according to the multi-breath indicator gas washout test used in this study lung lavage seems to be a good respiratory failure model to investigate ventilation inhomogeneity. In this model, ventilation inhomogeneity is probably caused partly by constriction of the distal small airways due to serotonergic reflex activity and it is reversed by a specific serotonin antagonist ketanserine, resulting in better distribution of ventilation in the lungs. However, no effect of ketanserine was seen on oxygenation and pulmonary hypertension.

ACKNOWLEDGMENTS

We thank Ms. Sharida Santoe for typing the manuscript and Mr. Ton Muêtgeert for technical assistance.

REFERENCES

1. P.E. Huygen, I. Gültuna, C. Ince, A. Zwart, J.M. Bogaard, B.W. Feenstra, and H.A. Bruining, A new ventilation inhomogeneity index from multiple breath indicator gas wash-out tests in mechanically ventilated patients, *Crit Care Med.* in press (1993).
2. J.L. Vincent, J.P. Degaute, M. Domb, P. Simon, J. Berre and A. Vandesteene, Ketanserine, a serotonin antagonist. Administration in patients with acute respiratory failure, *Chest* 85:510-513 (1984).
3. Y. Huet, C. Brun-Buisson, F. Lemaire, B. Teisseire, F. Lhoste and M. Rapin, Cardiopulmonary effects of ketanserin infusion in human pulmonary embolism, *Am Rev Respir Dis.* 135:114-117 (1987).
4. B. Lachmann, B. Robertson, and J. Vogel, In-vivo lung lavage as an experimental model of the respiratory distress syndrome, *Acta Anaesthesiol Scand.* 24:231-236 (1980).
5. P. Radermacher, Y. Huet, F. Pluskwa, R. Herigault, H. Mal, B. Teisseire, and F. Lemaire, Comparison of ketanserin and sodium nitroprusside in patients with severe ARDS, *Anesthesiology* 68:152-157 (1988).
6. H.J.H. Colebatch, C.R. Olsen, and J.A. Nadal, Effects of histamine, serotonin and acetylcholine in the peripheral airways, *J Appl Physiol.* 21:217-226 (1966).
7. W.V. Huval, S. Lelcuk, D. Sherpo, and H.B. Hechtman, Role of serotonin in patients with acute respiratory failure, *Ann Surg.* 200:166-172 (1984).
8. N.C. Olson, Role of 5-hydroxytryptamine in endotoxin-induced respiratory failure of pigs, *Am Rev Respir.* 135:93-99 (1987).

EFFECT OF CONTINUOUS ROTATION ON THE EFFICACY OF PARTIAL LIQUID (PERFLUBRON) BREATHING IN CANINE ACUTE LUNG INJURY

Scott E. Curtis, Samuel J. Tilden, W. Edward Bradley
and Stephen M. Cain

Departments of Pediatrics and Physiology and Biophysics,
University of Alabama at Birmingham, Birmingham, AL
35294-0005, U.S.A.

INTRODUCTION

Perfluorocarbons are liquids characterized by relatively low surface tension (10 to 20 dyne/cm) and high oxygen solubility (Clark, 1985). Liquid breathing can significantly improve gas exchange in animals with acute lung injury or surfactant deficiency (Calderwood et al., 1973; Shaffer et al., 1976, 1983a, 1983b, 1984). In several of the studies by Shaffer et al. (1976, 1983a, 1984), after drainage of most perfluorocarbon from the lung and conversion to gas breathing, improvements in arterial PO_2 and lung mechanics persisted, suggesting that residual perfluorocarbon in the lung may have functioned as an artificial surfactant. The use of gas breathing combined with partial lung doses of perfluorocarbon has recently received increased attention, with positive results seen both in animal models of restrictive lung disease and in premature human infants (Curtis et al., 1993a; Greenspan et al., 1990; Richman et al., 1993; Tütüncü et al., 1993). We previously tested the effects of sequential doses of perflubron (LiquiVent™, Alliance Pharmaceutical Inc.) from one-sixth of FRC up to full FRC in dogs with severe oleic acid induced lung injury (Curtis et al., 1993b). Moderate improvements in PaO_2 occurred after 50% lung filling, with the largest increases in PaO_2 seen after complete lung filling. We speculated that, similar to the situation seen with bolus surfactant administration (Lewis et al., 1993), perflubron may not be homogeneously distributed in acutely injured lungs. The high density of perflubron (1.9 g/ml) would also favor a selectively dependent distribution. This study tested the hypothesis that continuous physical rotation of the subject would improve distribution of perflubron within the lungs, and produce greater improvements of PaO_2 and compliance at smaller perflubron doses. To achieve this rotation, a mechanical bed commonly used in intensive care settings (to prevent the complications of immobility (Choi, 1992)) was employed. Rotated, oleic-acid injured dogs did (R/PFB) or did not (R/CON) receive serial doses of perflubron and were compared to our two previously studied non-rotated groups (NR/CON and NR/PFB).

METHODS

This study was approved by the UAB Institutional Animal Care and Use Committee. Sixteen adult dogs of mixed breed and either sex with a mean$\pm$SD weight of 17.9$\pm$2.8 kg were studied. We first anesthetized all dogs with 30 mg/kg of pentobarbital iv, intubated them with a cuffed endotracheal tube, and restrained them in the supine position on an Infant Roto Rest™ (Kinetic Concepts, Inc.) bed. We paralyzed the dogs with an im injection of succinylcholine (1 mg/kg) followed by a continuous iv drip (0.1 mg/min) that also provided 60 ml/hr of saline as maintenance fluid. A Siemens Servo 900C was used to ventilate the dogs with oxygen at a rate of 20 breaths/min, a positive end-expiratory pressure (PEEP) of 6 cmH$_2$O, an I:E ratio of 1:1, and a tidal volume initially adjusted to maintain PaCO$_2$ at 35 to 45 Torr. Ventilator settings were not changed after that.

We next catheterized the left carotid artery and external jugular vein (EJV) and placed a thermodilution catheter (Abbott) in the pulmonary artery via the right EJV. Cardiac output determinations were made using 5 ml injections of iced saline. The average of 3 to 5 determinations was divided by dog weight to obtain cardiac index. Mean pulmonary arterial pressure (mPpa) and systemic arterial pressure (mPsa) and heart rate were continuously recorded on a Gould TA4000 thermal array recorder. Pulmonary vascular resistance (PVR) was calculated as the quantity mPpa minus wedge pressure divided by cardiac index (CI) and reported as PRU (Torr/ml$\cdot$min^{-1}kg^{-1}). We measured arterial and mixed venous blood gas tensions with a Radiometer blood gas analyzer (ABL 30) and later corrected the values to the animal's core temperature at the time of sampling. Blood oxygen content was calculated from the hemoglobin content and O$_2$ saturation measured with a CO-oximeter calibrated for dog blood (IL-282, Instrumentation Lab) with dissolved O$_2$ added to the calculation. Oxygen delivery and uptake ($\dot{D}O_2$ and $\dot{V}O_2$) were calculated from CI and carotid and mixed venous O$_2$ contents. An Evans blue dye dilution technique was used to periodically measure plasma volume. We then estimated blood volume from the plasma volume and hematocrit. Static respiratory system compliance (Crs) was measured using the interrupter method (Sly et al., 1987) as follows. A 3-second end-inspiratory hold provided an estimate of elastic recoil pressure (Pel). PEEP, peak and mean inspiratory pressure, and exhaled tidal volume (VT$_{ex}$) were recorded from the ventilator's digital display. Effective tidal volume (VT$_{eff}$) was estimated by correcting VT$_{ex}$ for respiratory circuit compliance. Crs was then calculated as (Pel-PEEP)/VT$_{eff}$, then standardized by dividing by dog weight.

After all lines were placed and the animal appeared stable, baseline data were collected (*minute 0*). We then infused 0.15 ml/kg of oleic acid ((0.89 g/ml, Sigma Chemical Co.) into the left EJV catheter over 20 min. We collected data at *minute 30*, then every 30 minutes after that until *minute 270*. At *minute 90*, we randomized the dogs to two groups (n=8 each) that received continuous rotation with (R/PFB) or without (R/CON) serial intratracheal doses of perflubron (PFB). Rotation was done continuously, $\pm$62 deg side-to-side at a rate of 31 deg/min. In the R/PFB group, a 10 ml/kg bolus of warmed perflubron was instilled via the ET tube every 30 min for 5 doses followed by a 6th dose adjusted to fill the lungs (at *minutes 90, 120, 150, 180, 210,* and *240*). The lungs were judged to be filled (to liquid FRC) when a fluid meniscus in the endotracheal tube parallel to the anterior chest wall was achieved at a PEEP of zero. Each perflubron dose was given in thirds with the dog placed right side down, left side down, and supine while gas ventilation continued. Hemodynamic and gas exchange data were collected at the above times before each perflubron dose was instilled and with rotation temporarily halted. Anteriorposterior and left lateral chest radiographs were taken with a General Electric AMX1 at each data collection to assess distribution of peflubron and development of lung injury. To maintain stability after oleic acid injury, a 10 ml/kg iv bolus of

Dextran 70 was given anytime CI decreased to < 90 ml·min⁻¹·kg⁻¹. Also, sodium bicarbonate (7.5%) was given as needed to keep arterial HCO_3 > 20 meq/dl. After the last data collection at *minute 270*, depth of anesthesia was confirmed and the dog killed with an iv bolus of concentrated KCl.

Statistically significant differences ($p < 0.05$) within a group versus time and significant between group differences were detected by analysis of variance with correction for multiple comparisons done using the conservative Duncan's multiple range test.

RESULTS

Gas exchange

Arterial PO_2 decreased significantly after oleic acid infusion in both R/CON and R/PFB groups, from 518±22 and 486±23 Torr at *minute 0* to 74±11 and 77±10 Torr at *minute 90*, respectively (Fig.1). In R/CON, PaO_2 did not change significantly after *minute 90* and averaged 91±11 Torr. PaO_2 in R/PFB was similar to R/CON until after the fifth (145±26 vs. 88±19 Torr) and sixth (213±33 vs. 88±22 Torr) doses of perflubron, when R/PFB had a significantly higher PaO_2. The volume of perflubron required to obtain liquid FRC was 69±7 ml/kg (mean±SD). Arterial CO_2 increased significantly in both groups from *minute 0* to *minute 90* (R/CON: 37±1 to 45±2 Torr, R/PFB: 39±1 to 44±3 Torr) (Fig.2). $PaCO_2$ continued to rise to the mid 50's after that but did not differ at any time between groups.

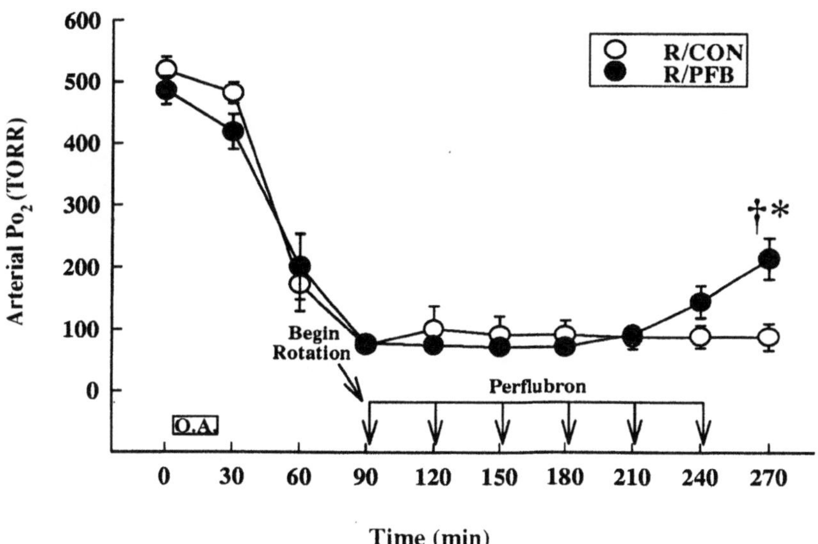

Figure 1. Data are mean±SE. * significantly different from controls at same time, † significantly different from *minute 90*.

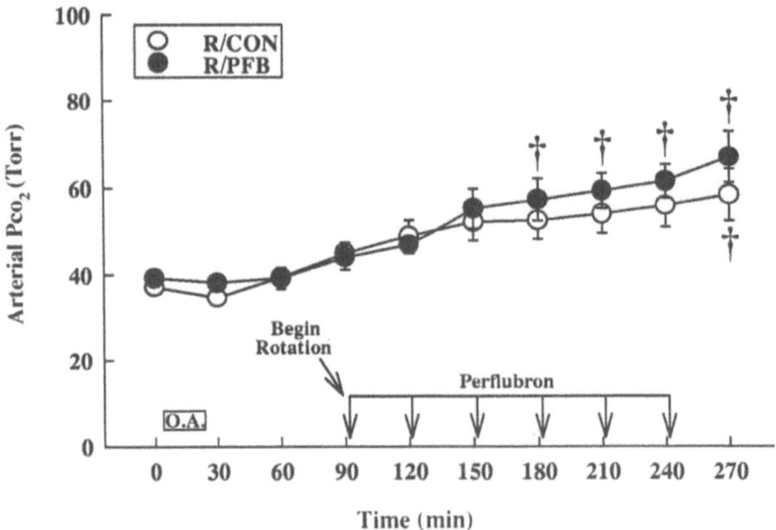

Figure 2. Data are mean±SE. * significantly different from controls at same time, † significantly different from *minute 90*.

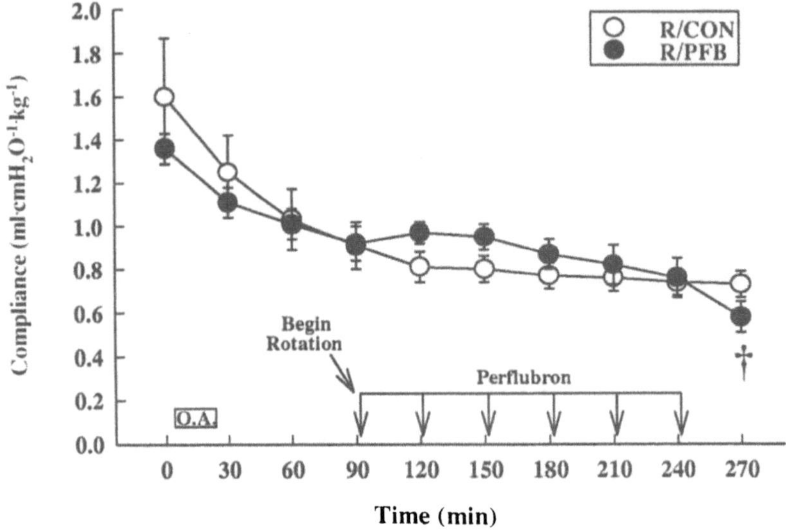

Figure 3. Data are mean±SE. * significantly different from controls at same time, † significanlty different from *minute 90*.

Lung mechanics

Oleic acid infusion caused a significant decrease in static respiratory system compliance (Crs) by *minute 90* in control dogs (1.60±0.27 to 0.91±0.11 ml cmH$_2$O^{-1}kg^{-1}) and was relatively stable thereafter (Fig.3). Crs decreased to a similar degree in perflubron treated dogs (1.36±0.07 to 0.92±0.08 ml·cmH$_2$O^{-1}·kg^{-1}) during the first ninety minutes. Crs appeared to slightly increase, then decrease with serial doses of perflubron, but was never statistically different from Crs in controls. Parallel to the decline in Crs with oleic acid, mean airway pressure (mPaw) increased significantly in both groups from baseline to *minute 90* (8.9±0.4 to 10.5±0.5 cmH$_2$O). mPaw was significantly higher in perflubron treated dogs compared to controls after the fifth (14.2±0.9 vs. 12.3±0.5 cmH$_2$O) and sixth perflubron doses (16.6±1.0 vs. 12.5±0.5 cmH$_2$O).

Hemodynamics

Oleic acid infusion caused a significant decrease in cardiac index from *minute 0* to *minute 30* in the two groups (R/CON: 145±13 to 89±15 ml·min^{-1}·kg^{-1}, R/PFB: 135±10 to 92±3 ml·min^{-1}·kg^{-1}) (Fig.4). R/PFB dogs did have a slightly but significantly lower CI than controls at *minute 270*. R/PFB dogs received an average of 21±11 ml/kg of Dextran 70 and R/CON dogs received 25±15 ml/kg (p=NS). Blood volume, which was measured at *minute 0, 90, 180,* and *270* did not change significantly over time in either group and averaged 102±24 ml/kg (mean±SD). Oxygen uptake was also stable throughout each experiment in the two groups, at 6.3±1.5 ml·min^{-1}·kg^{-1}. Both groups required less than 0.5 meq/kg of exogenous NaHCO$_3$ to maintain levels > 20 meq/dl. Pulmonary vascular resistance increased significantly with oleic acid infusion from baseline (0.06±0.02 PRU) to *minute 30* (0.13±0.06 PRU). PVR gradually rose with time in both groups to a peak of 0.17±0.06 PRU but never differed between groups.

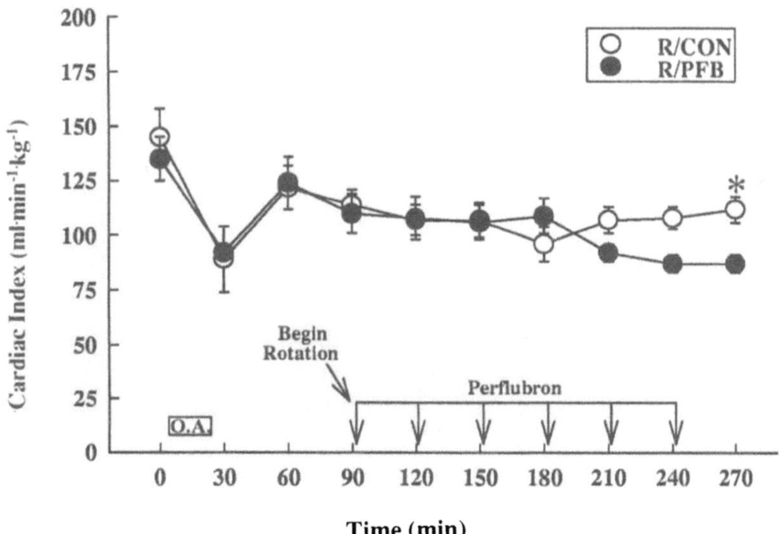

Figure 4. Data are mean±SE. * significantly different from controls at same time, † significanlty different from *minute 90*.

Radiography

All dogs developed diffuse alveolar and patchy infiltrates after receiving oleic acid. The first two perflubron doses always appeared to be concentrated in posterior and inferior lung fields and was virtually absent anteriorly (Fig.5). Perflubron generally appeared anteriorly between the third and fifth dose, but the progressive increase in density seen with the serial dosing always occurred from the back towards the front.

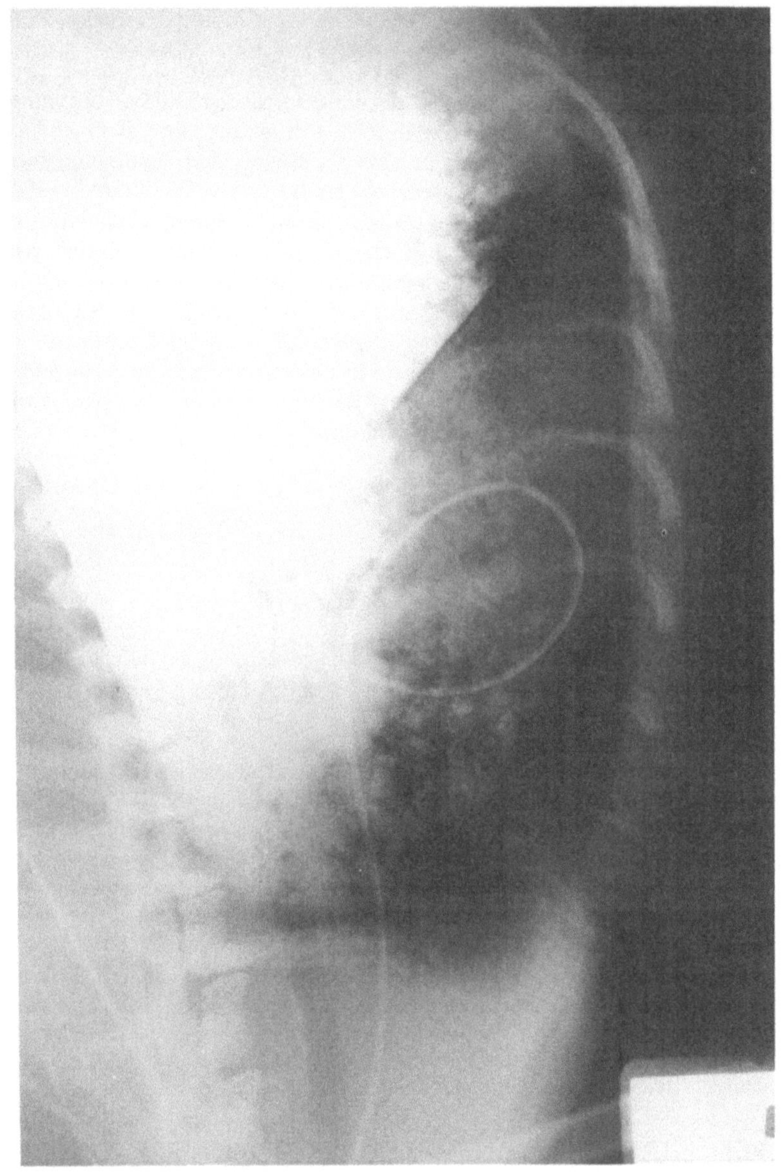

Figure 5. Lateral CXR after instillation of 20 ml/kg LiquiVent™ in a rotated dog. Note that the distribution is largely posterior and inferior.

Several films were taken at end-expiration with no PEEP applied and revealed some gas trapped throughout the lung, despite the presence of a fluid level in the endotracheal tube.

DISCUSSION

Previously we found that doses of perflubron equal to one-half of FRC significantly improved PaO_2 compared to injured controls (108 ± 26 vs. 55 ± 3 Torr) but that the largest effect on PaO_2 only occurred after the lungs were filled to FRC (198 ± 46 vs. 51 ± 3 Torr) (Curtis et al., 1993b). Although no radiographs were taken in that study, we speculated that the high density and spreadability of perflubron may cause it to pool in dependent lung areas. With the assumption that the beneficial effects of perflubron are related to its ability to recruit collapsed or edema-filled alveoli, we further speculated that improved distribution of perflubron might result in improved compliance and gas exchange at lower perflubron doses. This hypothesis was tested in the current study by the use of a continuously rotating bed in oleic acid injured, perflubron treated dogs. Also, chest radiographs were performed in this study to characterize the distribution of perflubron within the lungs. An injured non-perflubron group was included to separate out any effects of rotation alone on this severe animal model of human ARDS.

Mean ($\pm$SEM) PaO_2 after the first four perflubron doses in this study (78 ± 5 Torr) was not significantly different from that of the non-rotated, perflubron treated (NR/PFB) dogs (96 ± 10 Torr) previously studied (Curtis et al., 1993b). Also, the average PaO_2 seen after complete lung filling in R/PFB (213 ± 33 Torr) was not different from NR/PFB dogs (198 ± 46 Torr). PaO_2 did appear somewhat higher in rotated dogs after the fifth perflubron dose (145 ± 26 vs. 101 ± 23 Torr), but this was not statistically significant. Also, R/PFB dogs did not have a significantly higher PaO_2 than controls (R/CON) until the sixth perflubron dose, whereas NR/PFB dogs had a higher PaO_2 than their controls (NR/CON) after the third, fourth, and sixth doses. However, this appears to be due to the fact that PaO_2 in NR/CON was on average about 40 Torr less than in rotated control dogs. This suggests that rotation may have had a beneficial effect upon gas exchange in the injury model, although it failed to amplify the positive effects of perflubron treatment.

Although chest X-rays were not done in the earlier study of non-rotated dogs, the films in this study clearly showed a dependent distribution of early perflubron doses. Rotation was stopped for about 10 minutes during each data sampling period, with the X-rays being taken at the end of that 10 minute period. Thus, they may not have reflected the dynamic distribution of perflubron during active rotation. Also, in both studies perflubron doses were administered in thirds with the animal placed supine and in the left and right lateral decubitus positions. That manipulation itself may have produced an equivalent degree of perflubron dispersion in non-rotated and rotated dogs that overshawdowed any effect of the rotating bed. The attractiveness of testing the mechanical bed lay in its ability to provide continuous rotation, its wide clinical use, and that it has been shown to reduce the incidence of pneumonia and atelectasis in intensive care patients (Choi, 1992). Still, it proved ineffective in ensuring a homogeneous perflubron distribution in this study. It is possible that the degree of rotation used (±62 deg) was insufficient to direct the peflubron into anteriorly oriented airways.

Respiratory system compliance after lung injury in the current two groups (0.91 ± 0.11 $ml \cdot cmH_2O^{-1} \cdot kg^{-1}$) was similar to the injury produced in NR/CON and NR/PFB dogs (1.13 ± 0.07 $ml \cdot cmH_2O^{-1} \cdot kg^{-1}$). Rotation appeared to have no effect on Crs in controls, as Crs did not change further after the development of lung injury. Unlike the previous study, in which the first four doses of perflubron were associated with a Crs greater than in controls, Crs in R/PFB dogs did not differ significantly from R/CON dogs at any time, the opposite result of what was expected. The decrease in Crs with oleic acid injury is due both to surfactant inactivation by edema fluid and also to loss of air space from alveolar flooding, which forces the tidal breath to enter into a smaller lung volume (Grossman et al., 1980). It is not known how perflubron is distributed with respect to edema fluid, but other studies have clearly shown a beneficial effect upon Crs with partial perfluorocarbon dosing (Curtis et al., 1993a; Greenspan et al., 1990; Richman et al., 1993; Tütüncü et al., 1993). Rotation appeared to detract from that benefit, but did not adversely affect PaO_2. Also, hemodynamics were equally stable in perflubron treated and control dogs and did not differ from what was seen in the non-rotated animals.

In summary, we found that an FRC dose of perflubron significantly improved arterial PO_2 in dogs with severe acute lung injury. Continuous rotation using the Infant Roto Rest appeared to improve oxygenation in controls, but did not significantly alter the dose of perflubron required, nor did it ensure homogenous perflubron distribution. Alternative modes of perflubron administration, such as aerosolization, or rotation to more than 90 deg, remain to be studied.

ACKNOWLEDGEMENT

SEC is supported in part by an American Lung Association Trudeau Research Scholar Award and NIH P30 HD28831. PFB was generously supplied by Alliance Pharmaceutical Corp. We thank Siemens-Elema, ventilator division, and Kinetic Concepts, Inc. for the loan of equipment. We also acknowledge the expert technical assistance of Glenda Davis.

REFERENCES

Calderwood, H.W., Modell, J.H., Ruiz, B.C., et al., 1973, Pulmonary lavage with liquid fluorocarbon in a model of pulmonary edema, *Anesthesiology* 38:141.

Choi, S.C. and Nelson, L.D., 1992, Kinetic therapy in critically ill patients: combined results based on meta-analysis, *J. Crit. Care*, 7:57.

Clark, L.C., 1985, Introduction to fluorocarbons, *Inter. Anesth. Clinics* 23(1):1.

Curtis, S.E. and Peek, J.T., 1993a, Partial liquid breathing with perflubron improves arterial oxygenation in acute canine lung injury, In press, *J. Appl. Physiol.*

Curtis, S.E. and Peek, J.T., 1993b, Effects of progressive intratracheal administration of perflubron during conventional gas ventilation in oleic acid lung injury, In press, *Adv. Exp. Med. Biol.*

Greenspan, J.S., Wolfson, M.R., Rubenstein, S.D., and Shaffer, T.H., 1990, Liquid ventilation of human preterm neonates, *J. Pediatr.* 117:106.

Grossman, R.F., Jones, J.F., and Murray, J.F., 1980, Effects of oleic acid-induced pulmonary edema on lung mechanics, *J. Appl. Physiol.* 48:1045.

Lewis, J.F., Tabor, B., Ikegami, M., Jobe, A.H., Joseph, M., and Absolom, D., 1993, Lung function and surfactant distribution in saline-lavaged sheep given instilled vs. nebulized surfactant, *J. Appl. Physiol.* 74:1256.

Richman, P.S., Wolfson, M.R., and Shaffer, T.H., 1993, Lung lavage with oxygenated perfluorochemical liquid in acute lung injury, *Crit. Care Med.* 21:768.

Shaffer, T.H., Rubenstein, D., Moskowitz, G.D., and Delivoria-Papadopoulos, M., 1976, Gaseous exchange and acid-base balance in premature lambs during liquid ventilation since birth, *Pediatr. Res.* 10:227.

Shaffer, T.H., Douglas, P.R., Lowe, C.A., and Bhutani, V.K., 1983a, The effects of liquid ventilation on cardiopulmonary function in preterm lambs, *Pediatr. Res.* 17:303.

Shaffer, T.H., Tran, N., Bhutani, V.K., and Sivieri, E.M., 1983b, Cardiopulmonary function in very preterm lambs during liquid ventilation, *Pediatr. Res.* 17:680.

Shaffer, T.H., Lowe, C.A., Bhutani, V.K., and Douglas, P.R., 1984, Liquid ventilation: effects on pulmonary function in distressed meconium-stained lambs, *Pediatr. Res.* 18:47.

Sly, P.D., Bates, J.H.T., and Milic-Emili, J., 1987, Measurement of respiratory mechanics using the Siemens Servo Ventilator 900c, *Pediatr. Pulm.* 3:400.

Tütüncü, A.S., Faithfull, N.S. and Lachmann, B., 1993, Intratracheal perfluorocarbon administration combined with mechanical ventilation in experimental respiratory distress syndrome: Dose-dependent improvement of gas exchange, *Crit. Care Med.* 21:962.

CONSTRICTION AND DILATATION OF PULMONARY ARTERIAL RING BY HYDROGEN PEROXIDE - IMPORTANCE OF PROSTANOIDS -

Kazuhiro Yamaguchi, Koichiro Asano, Masaaki Mori, Tomoaki Takasugi, Hirofumi Fujita, Yukio Suzuki and Takeo Kawashiro

Department of Medicine School of Medicine, Keio University
Tokyo 160, Japan

INTRODUCTION

Reactive oxygen species (ROS) have been considered as one of the important factors causing acute lung injury associated with alveolar flooding, especially in the case of sepsis, fat embolism, hyperoxic lung damage as well as reperfusion injury (Hefner and Repine, 1989). Among them, hydrogen peroxide (H_2O_2) and its derivatives are relatively stable as compared to other ROS and are taken notice as the substance altering pulmonary hemodynamics accompanied by alveolar flooding (Archer et al., 1989; Gurtner and Burke-Wolin, 1991; Barnard et al., 1992). Effects of H_2O_2 on pulmonary circulation is much complicated, namely vasoactive action of H_2O_2 is dependent on the vascular tone. In other words, H_2O_2 relaxes the pulmonary vessel when the tone is high but constricts it when the tone is low (Burke-Wolin et al., 1991). Gurtner and Burke-Wolin (1991) reported that vasoconstrictive effect of H_2O_2 would be mainly mediated through thromboxane A_2 (TXA_2) generated in the endothelial cells of pulmonary vasculature. However, several authors (Madden et al., 1986; Smith, 1986; Takayasu et al., 1990) showed that arachidonate metabolism might exist not only in endothelial cells but also in smooth muscle cells and fibroblasts in vascular walls, indicating that the conclusion derived by Gurtner and Burke-Wolin (1991) should be reassessed.

Vasodilation caused by H_2O_2 has been considered to break out through the direct stimulation on soluble guanylate cyclase, which augments the concentration of guanosine 3',5'-cyclic monophosphate (cGMP) in the smooth muscle cells (Gurtner and Burke-Wolin, 1991). However, the important role of vasodilator arachidonates such as prostacyclin (PGI_2) for H_2O_2-induced vasodilation has not been decisively excluded. PGI_2 has been recently shown to be yielded in various cells constituting the vascular wall in addition to endothelial cells (Madden et al., 1986; Smith, 1986).

Using isolated pulmonary artery without endothelium obtained from rats, the present study was undertaken to clarify whether or not

Oxygen Transport to Tissue XVI
Edited by M.C. Hogan *et al.*, Plenum Press, New York, 1994

vasoactive arachidonates, especially those mediated by cyclooxygenase, would play a significant role in regulating the H_2O_2-induced vasoconstriction and vasodilation.

MATERIALS AND METHODS

Male Sprague-Dawley rats (weighing 350-500 g) anesthetized with peritoneal injection of pentobarbital sodium (100 mg/kg) were slaughtered by a rapid exsanguination. Main pulmonary arteries were isolated and cut into the rings with a diameter of 1.5-2.0 mm and a length of 3-4 mm. The endothelium was removed by gently rubbing the inside of the arterial ring by means of polyethylene tube. Isolated rings were suspended in an organ bath filled with 10 ml Krebs-Henseleit solution and were equilibrated with a gas mixture containing 95% O_2 and 5% CO_2. Equilibration was performed for 90 min applying the resting tension of 2.5 g, which was preliminarily determined to provoke the maximum tension in response to phenylephrine (see below). The temperature of the organ bath was maintained at 37°C using circulating hot water. Complete removal of the endothelium was functionally confirmed by examining the loss of the dilatation after administration of 1 µM acetylcholine. Maximum contraction of the arterial ring was determined by the addition of 10^{-5} M phenylephrine which was resistible to ROS. Subsequently, H_2O_2 was added to the organ bath in which the ring had been partially contracted by 20 mM potassium chloride.

Following experiments were undertaken to investigate the possible roles of prostaglandins (PGs) for H_2O_2-induced vasoconstriction and vasodilation. 1) Administration of varied concentrations of H_2O_2 ranging from 10^{-6} to 10^{-4} M. 2) Addition of 10^{-4} M H_2O_2 in the presence of 1000 U/ml of catalase (CAT). 3) 10^{-4} M H_2O_2 under the condition where inhibitor for TXA_2 synthase coexists (OKY-046 (Naito et al., 1983), 10^{-5} M). 4) 10^{-4} M H_2O_2 in the presence of TXA_2- and/or PGH_2-receptor blocker (ONO-3708 (Kondo et al., 1989), 10^{-5} M). 5) H_2O_2 at the concentration of 10^{-4} M in the presence of 5 µM indomethacin, a potent inactivator for cyclooxygenase. The effects of varied medicines on the rings responding to H_2O_2 were quantitatively estimated by calculating the relative tension against the maximum contraction generated by phenylephrine (% contraction).

To obtain a direct evidence for the genesis of PGs from the denuded arterial rings, concentration of TXB_2 (stable metabolite of TXA_2) and that of 6-keto prostaglandin $F_{1\alpha}$ ($PGF_{1\alpha}$: metabolite of PGI_2) in the organ bath were examined with radioimmunoassay after exposure to 10^{-4} M H_2O_2 for 15 min. The critical concentration of TXB_2 and of PGF_{1a} was shown to be less than 30 pg/ml.

RESULTS

H_2O_2 caused transient contractile response of the pulmonary arterial ring, sustained for 2 min. Thereafter, the tension of the ring reverted to the basal condition in 10 min after addition of H_2O_2 (Figure 1). The contractile response was enhanced in parallel with the concentration of H_2O_2 (Figure 2), i.e. % contraction yielded by H_2O_2 at 10^{-6}, 10^{-5} and 10^{-4} M was 10%, 40% and 60%, respectively.

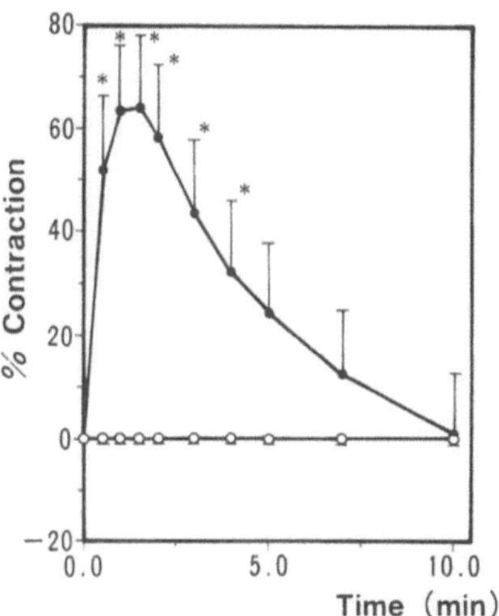

Figure 1. Contraction of the pulmonary arterial rings after addition of 10^{-4} M H_2O_2 (closed circles). In the presence of catalase (1000 U/ml), the contractile response by H_2O_2 completely disappeared (open circles).

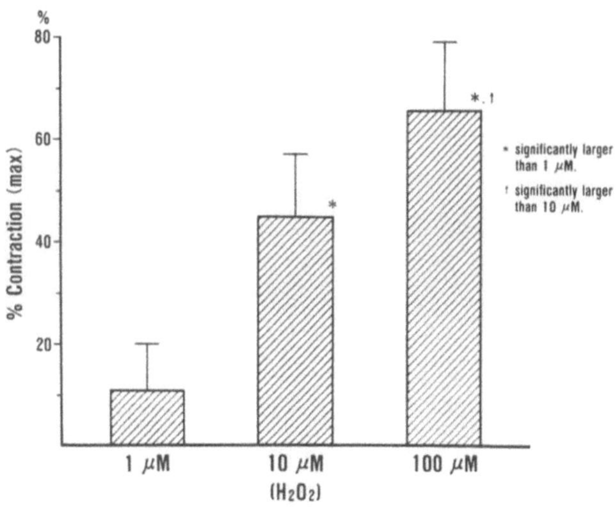

Figure 2. Maximum constriction of the pulmonary arterial rings by H_2O_2 whose concentration ranges from 10^{-6} to 10^{-4} M. H_2O_2 enhanced the contraction of the ring in parallel with the concentration of H_2O_2.

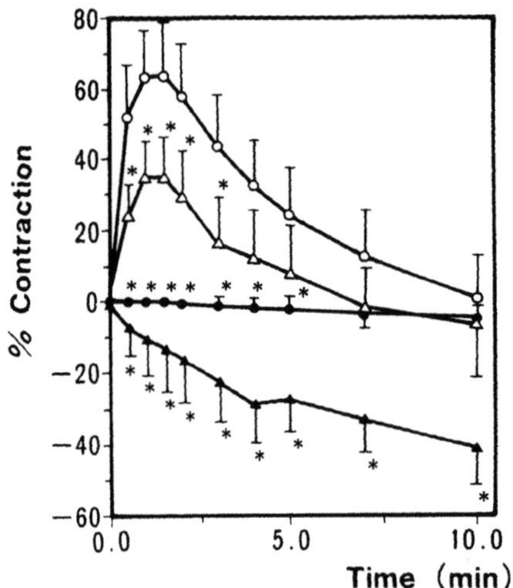

Figure 3. Effects of varied inhibitors on H_2O_2-induced vasoconstriction and vasodilation. Open circles: H_2O_2 (10^{-4} M) alone, open triangles: OKY-046 (10^{-5} M), closed circles: indomethacin (5 µM), closed triangles: ONO-3708 (10^{-5} M). See text for detailed explanation.

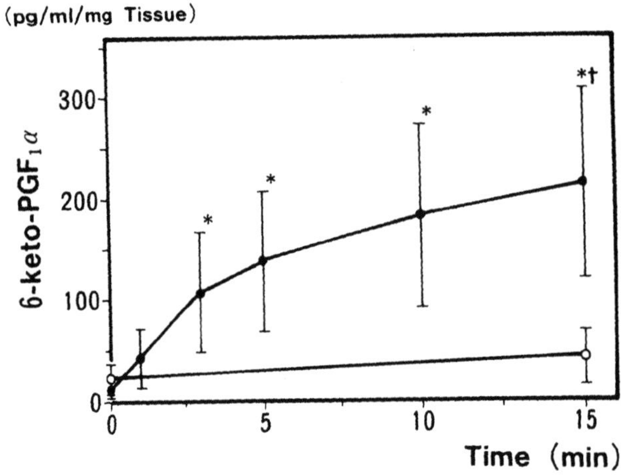

Figure 4. Concentration of 6-keto prostaglandin $F_{1\alpha}$ after addition of H_2O_2 (10^{-4} M). Open circles: without H_2O_2, closed circles: with H_2O_2 (10^{-4} M).

Addition of CAT fully inhibited the contracting as well as dilating response induced by H_2O_2 (Figure 1). TXA_2 synthase inhibitor, OKY-046 slightly reduced the % contraction caused by 10^{-4} M H_2O_2 up to 40% (Figure 3). TXA_2 and/or PGH_2 receptor blocker, ONO-3708 completely eliminated the contractile response and yielded a simple relaxation after addition of H_2O_2 (Figure 3). Cyclooxygenase inactivator, indomethacin successfully inhibited both vasoconstriction and vasodilation induced by H_2O_2 administration, the trend being qualitatively the same as observed in the case of CAT (Figures 1 and 3).

Exposure of the ring to H_2O_2 (10^{-4} M) for 15 min failed to increase TXB_2 concentration in the organ bath, but significantly augmented the production of $PGF_{1\alpha}$ (Figure 4).

DISCUSSION

Vasoconstriction by H_2O_2

several authors reported that administration of H_2O_2 to isolated perfused lungs caused noticeable but transient vasoconstriction accompanied by alveolar flooding (Tate et al., 1986; Seeger et al., 1986; Johnson et al., 1989). These authors considered TXA_2 produced in platelets and/or endothelial cells as a main mediator to modify the pulmonary hemodynamics. Using isolated pulmonary arterial ring without endothelium, we obtained the results qualitatively consistent with the previous studies. The experimental results (Figures 1, 2 and 3) showed that contractile response induced by H_2O_2 was completely eliminated in the presence of H_2O_2 scavenger (CAT), cyclooxygenase inhibitor (indomethacin) and TXA_2 and/or PGH_2 receptor blocker (ONO-3708), but inhibited only a little by TXA_2 synthase blocker. The findings may indicate that H_2O_2 stimulates the metabolism concerning cyclooxygenase pathway even in the absence of endothelium, and that TXA_2 is not a main substance causing H_2O_2-induced vasoconstriction. Our study showed that the most important substance for contractile response of denuded pulmonary artery to H_2O_2 may be PGH_2, a percussor for TXA_2 and various prostaglandins having biological activity. PGH_2 is a potent vasoconstrictive substance and more stable than TXA_2 (Kato et a;., 1990). The administration of H_2O_2 could not enhance the genesis of TXB_2 in the organ bath (Figure 4), being highly consistent with the idea as described above.

Vasodilation by H_2O_2

TXA_2 and/or PGH_2 receptor blocker (ONO-3708) wholly inhibited the vasoconstriction and caused the sustained relaxation of the pulmonary arterial ring without endothelium after administration of H_2O_2 (Figure 3). Indomethacin successfully restrained the vasodilation induced by H_2O_2 (Figure 3). Further, H_2O_2 augmented the release of PGI_2 generated in suspended ring without endothelium (Figure 4). The findings are highly compatible with the idea that H_2O_2 activates PGI_2 production in the pulmonary arterial wall independently of endothelium.

Recently, vasodilation caused by H_2O_2 has been attributed to its direct effect on guanylate cyclase in the vascular smooth muscle cells, i.e. compound of H_2O_2 and catalase (compound I) activates guanylate cyclase, leading to the increase of cGMP in the smooth muscle cells

(Ignarro et al., 1985; Burke and Wolin, 1987; Gurtner and Burke-Wolin, 1991). Although we did not measure cGMP concentration in the rings, the experimental results (Figure 3) might exclude the important role of cGMP for H_2O_2-induced vasodilation. Indomethacin fully inhibited the relaxation of the ring (Figure 3), indicating that H_2O_2-induced vasodilation would be wholly explicable from vasodilator prostaglandin such as PGI_2 and would not require any other factor including cGMP.

In conclusion, 1) H_2O_2 stimulates the metabolism on cyclooxygenase pathway in medial and/or outer layers of the pulmonary arterial wall irrespective of endothelium. 2) PGH_2 but not TXA_2 may be a fundamental substance causing the contraction of the pulmonary artery (without endothelium) responding to H_2O_2. 3) Vasodilator prostaglandin such as PGI_2 may be the most important factor inducing the relaxation of the pulmonary artery (without endothelium) in the presence of H_2O_2.

REFERENCES

Archer, S.L., Peterson, D., Nelson, D.P., DeMaster, E.G., Kerry, B., Eaton, J.W. and Weir, E.K., 1989, Oxygen radicals and antioxidant enzymes alter pulmonary vascular reactivity in the rat lung, J. Appl. Physiol. 66:102.

Bernard, M.L. and Matalon, S., 1992, Mechanisms of extracellular reactive oxygen species injury to the pulmonary microvasculature, J. Appl. Physiol. 72:1724.

Burke, T.M. and Wolin, M.S., 1987, Hydrogen peroxide elicits pulmonary arterial relaxation and guanylate cyclase activation, Am. J. Physiol. 252:H721.

Burke-Wolin, T., Abate, C.J., Wolin, M.S. and Gurtner, G.H., 1991, Hydrogen peroxide-induced pulmonary vasodilation: Role of guanosine 3',5'-cyclic monophosphate, Am. J. Physiol. 261:L393.

Gurtner, G.H. and Burke-Wolin, T., 1991, Interactions of oxidant stress and vascular reactivity, Am. J. Physiol. 260:L207.

Hefner, J.E. and Repine, J.E., 1989, Pulmonary strategies of antioxidant defense, Am. Rev. Respir. Dis. 140:531.

Ignarro, L.J., Harbison, R.G., Wood, K.S., Wolin, M.S., McNamara, D.S., Hyman, A.L. and Kadowitz, P.L., 1985, Differences in responsiveness of intrapulmonary artery and vein to arachidonic acid: mechanism of arterial relaxation involves cyclic guanosine 3',5'-monophosphate and cyclic adenosine 3',5'-monophosphate, J. Pharmacol. Exp. Ther. 233:560.

Johnson, A., Philips, P., Hocking, D., Tsan, M. and Ferro, T., 1989, Protein kinase inhibitor prevents pulmonary edema in response to H_2O_2, Am. J. Physiol. 256:H1012.

Kato, T., Iwama, Y., Okumura, K., Hashimoto, H., Ito, T. and Satake, T., 1990, Prostaglandin H_2 may be the endothelium-derived contracting factor released by acetylcholine in the aorta of the rat, Hypertension. 15:475.

Kondo, K., Seo, R., Omawari, N., Imawaka, H., Wakitani, K., Kira, H., Okegawa, T. and Kawasaki, A., 1989, Effects of ONO-3708, an antagonist of thromboxane A_2/prostaglandin endoperoxide receptor, on blood vessels, Eur. J. Pharmacol. 168:193.

Madden, M.C., Vender, R.L. and Friedman, M., 1986, Effect of hypoxia on prostacyclin production in cultured pulmonary artery endothelium, Prostaglandins. 31:1049.

Naito, J., Komatsu, H., Ujiie, A., Hamano, S., Kubota, T. and Tsuboshima, M., 1983, Effects of thromboxane synthetase inhibitors on aggregation of rabbit platelets, Eur. J. Pharmacol. 91:41.

Seeger, W., Suttorp, N., Schmidt, F. and Neuhof, H., 1986, The glutathione redox cycle as a defense system against hydrogen-peroxide-induced prostanoid formation and vasoconstriction in rabbit lungs, Am. Rev. Respir. Dis. 133:1029.

Smith, W.L., 1986, Prostaglandin biosynthesis and its compartmentation in vascular smooth muscle and endothelial cells, Ann. Rev. Physiol. 48:251.

Takayasu, M., Terashima, Z. and Kondo, K., 1990, Endothelin-1 and platelet activating factor stimulate thromboxane A_2 biosynthesis in rat vascular smooth muscle cells, Biochem. Pharmacol. 40:2713.

Tate, R.M., Morris, H.G., Schroeder, W.R. and Repine, J.E., 1984, Oxygen metabolites stimulate thromboxane production in isolated saline perfused rabbit lungs, J. Clin. Invest. 74:608.

FUNCTIONAL HETEROGENEITY IN THE LUNGS

Peter D. Wagner

Department of Medicine, University of California, San Diego
La Jolla, California, U.S.A.

The purpose of this symposium is to compare amongst several organs the functionally important forms of heterogeneity, their anatomical basis, how they compromize the ability of each organ to perform its duties, how heterogeneity can be measured, and how heterogeneity evolves from health to disease.

In the lungs, the principal forms of functional heterogeneity concern nonuniform distribution of: 1) inspired gas, and 2) mixed venous blood, to the alveoli. While each alveolus can be visualized individually, average alveolar dimensions (~ 300 μ diameter) and their placement in freely communicating groups, results in a functionally homogeneous unit of lung tissue considerably greater in size than a single alveolus. A functional unit is currently considered then as all the alveoli distal to the smallest purely conducting airways, the distal terminal bronchioles. Non-uniform ventilation or blood flow within such a lung unit has no measurable consequence on lung function; differences in ventilation or blood flow between such units will. The importance of non-uniform ventilation or blood flow is the impact on O_2 and CO_2 exchange. Ventilation/perfusion (or $\dot{V}A/\dot{Q}$) inequality, as it is called commonly, causes hypoxemia and can also cause hypercapnia. This is because PO_2 and PCO_2 in each lung unit is a unique function of the ventilation and blood flow of that unit (for given conditions of inspired gas and mixed venous blood). In fact, $\dot{V}A/\dot{Q}$ inequality is the most common consequence of most lung diseases due to the structural changes of disease, be they on airways or blood vessels, and temporary or permanent changes. It therefore becomes important to assess the anatomical nature and functional severity of $\dot{V}A/\dot{Q}$ inequality. Minimally invasive techniques exist employing radioactive tracers that mark both blood flow and inspired gas distribution. These yield anatomical detail, but have limited spatial resolution that generally fails to uncover all of the inequality present. Individual changes in ventilation or blood flow can be visualized. A different approach is to use blood gas measurements to assess the functional severity, employing the respiratory gases O_2 and CO_2 or tracer inert gases (non-radioactive). While these require blood sampling and yield no anatomical information, they do provide, by definition, complete accounting for the principal effect of inequality on pulmonary gas exchange.

INFLUENCE OF INTRATRACHEAL PERFLUBRON VAPOR ON LUNG MECHANICS AND BLOOD GASES

R.J. Houmes*, B. Lachmann*, N.S. Faithfull[#] and W. Erdmann*

*Department of Anesthesiology, Erasmus University, Rotterdam, The Netherlands and [#] Department of Medical Research, Alliance Pharmaceutical Corp., San Diego, USA.

Studies in adult animals with acute respiratory failure have demonstrated that intratracheal administration of perflubron (perfluorooctyl-bromide, Alliance Pharmaceutical Corp., San Diego, CA) improves PaO_2 values [1]. However, data on perflubron administration in healthy animals showed a decrease in PaO_2 values [2]. It is speculated that this decrease could originate from an elevation of surface tension in healthy animals, combined with the mechanical effect of perflubron moving in the lungs during ventilation. The aim of this study was to assess the effect of vaporized perflubron on blood gases and lung mechanics in healthy animals.

Methods. Twelve adult New Zealand rabbits were anaesthetized, tracheotomized, paralyzed and ventilated with a Siemens Servo Ventilator 900C equipped with a Siemens Servo Humidifier 153 at the inspiratory line. After induction of anaesthesia, pressure controlled ventilation was applied. The peak pressure was set to obtain a tidal volume of 10 mL/kg. The other ventilatory parameters were: Frequency=30/min, FiO_2=1.0, PEEP=3 cmH_2O, I/E ratio=1:2. Animals were divided into two groups; Group 1 (n=6) was treated with perflubron, Group 2 (n=6;controls) was only ventilated with the identical ventilatory parameters as Group 1. The perflubron vapor was produced by constant infusion (10 mL/h) of heated (50°C) perflubron into the humidifier. Any surplus of perflubron in the tubings was collected in a reservoir. P-V curves and blood gases were recorded every 30 min for three hours.

Results. The mean PaO_2 in the perflubron group was 565 mmHg vs. 581 mmHg in the control group. Neither the slope of the P-V curves nor any other parameter from the deflation limb (characterizing the surfactant system) were changed due to any effect of the perflubron vapor

Conclusion. From these results it is concluded that administration of perflubron vapor in healthy animals during pressure controlled ventilation has no clinical influence on pulmonary function. These results can form the basis for further studies using perflubron vapor for treatment of acute respiratory failure.

1. Tütüncü AS, Lachmann B, Faithfull NS, Erdmann W. Dose dependent improvement of gas exchange by intratracheal Perflubron (perfluorooctylbromide) instillation in adult animals with acute respiratory failure. Adv Exp Med Biol 1992; 317:397-400.

2. Fuhrman BP, Parczan PR, DeFrancisis M. Perfluorocarbon-associated gas exchange. Crit Care Med 1991; 19:712

INFLUENCE OF PERFLUBRON VAPOR ON PULMONARY GAS TRAPPING AND SURFACE TENSION AFTER INTRAVENOUS PFOB-EMULSION ADMINISTRATION

B. Lachmann*,R.J. Houmes*, N.S. Faithfull[#] and W. Erdmann*

*Department of Anesthesiology, Erasmus University, Rotterdam, The Netherlands and [#]Department of Medical Research, Alliance Pharmaceutical Corp., San Diego, USA

Use of intravenous injection of perfluorocarbon emulsions in rabbits has been reported to result in increased collapsing volume after post-mortem thoracotomy [1]. This effect has been called pulmonary gas trapping (PGT) and has been associated with an increase in intra-alveolar micro-bubbles due to "gas osmosis" [2]. Inhaling the vaporized form of perflubron (perfluorooctyl bromide [PFOB] Alliance Pharmaceutical Corp., San Diego, CA) has been shown to reduce this PGT effect. As histologic evidence for these micro-bubbles is hard to obtain, we postulated that the presence of the micro-bubbles may be related to an increased surfactant activity. The aim of this study was to investigate the hypothesis that the increased collapsing volume is associated with increased surfactant activity.

Methods. Twenty four adult New Zealand rabbits were randomized to four groups (n=6, per group), groups 1 and 2 received 6 ml/kg saline i.v. where as groups 3 and 4 received 6 ml/kg perflubron emulsion iv. Groups 2 and 4 were breathed a fully saturated perflubron vapor for four days. At day four, all animals were anesthetized and tracheotomized, following which the diaphragm was opened through a subcostal incision. A pressure volume-curve was recorded stopping deflation at a pressure of 5 cm H_2O; then the lungs were exposed to ambient pressure for five seconds, representing collapsing volume. The lungs were then removed from the thoracic cavity and the remaining air volume was measured, using the Archimedes principle. The left lung was lavaged twice with saline (37°C, 15 mL/kg) and the recovered lavage fluid was pooled. The surface activity in the pooled lavage fluid was measured using a modified Wilhelmy balance (E. Biegler GmBH, Mauerbach, Austria).

Results. The mean collapsing volumes per kg of bodyweight and mean minimal surface tensions are shown in figs. 1 and 2. The correlation between the collapsing volume per kg bodyweight and the minimal surface tension proved to be statistically significant (p ≤ 0.01).

Conclusion. These results support the hypothesis that the increase in collapsing volume in rabbit lungs may have its origin in overproduction of surfactant, leading to a "bubble lung" similar to the state present immediately after birth.

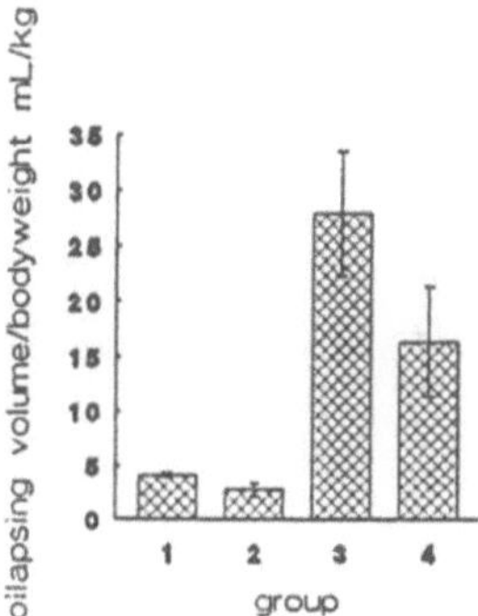

figure 1.

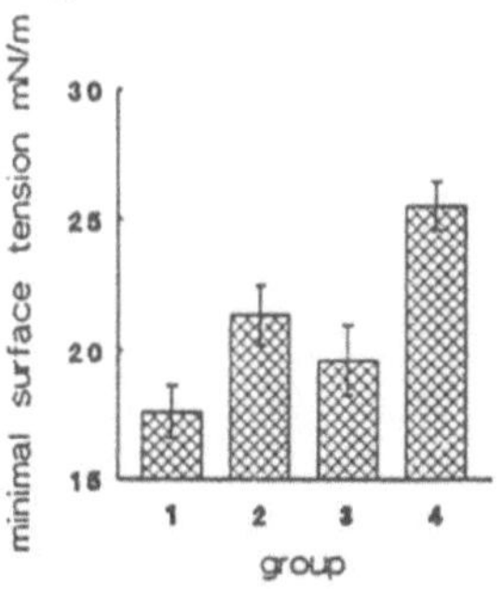

Figure 2.

REFERENCES

1. Clark LC, Response of the rabbit lung as a criterion of safety for perfluorocarbon breathing and blood substitutes. Biomat Art Cells & Immob Biotech 1992; 20:2-4, 1085-1099.
2. Schutt E, Barber P, et al, Proposed mechanism of Pulmonary Gas Trapping (PGT) following intravenous perfluorocarbon emulsion administration. Presented at: V[th] ISBS, San Diego, CA, USA. March 1993.

EFFECT OF MILD HYPOTHERMIA ON ACTIVE AND BASAL CEREBRAL OXYGEN METABOLISM AND BLOOD FLOW

Edwin M. Nemoto[1], Richard Klementavicius[2], John A. Melick[1], and Howard Yonas[2]

[1]Department of Anesthesiology/CCM
[2]Department of Neurological Surgery
 University of Pittsburgh School of Medicine
 Pittsburgh, PA 15261

INTRODUCTION

Compartmentation of Cerebral Metabolic Rate for Oxygen ($CMRO_2$)

The compartmentation of $CMRO_2$ began with Kety's hypothesis that because anesthetics produce anesthesia through the reversible inhibition of synaptic transmission, they should affect $CMRO_2$ only to the extent associated with synaptic transmission (Kety, 1965). Michenfelder (1975) tested this hypothesis and showed that, indeed, massive doses of thiopental infused into dogs on cardiopulmonary bypass attenuated $CMRO_2$ only to the point of cortical electroencephalographic (EEG) silence and about 50% of normal. Continued infusion at doses several fold higher than required to abolish the cortical EEG did not further reduce $CMRO_2$ below 50% of normal. Thus, the concept of active $CMRO_2$ ($ACMRO_2$), i.e., associated with an active EEG, and basal $CMRO_2$ ($BCMRO_2$), i.e., associated with some basal level of cerebral metabolic activity was formulated, each representing about 50% of total $CMRO_2$ ($TCMRO_2$). Astrup and associates (Astrup, Skovsted, et al., 1981; Astrup, Sorensen, et al., 1981) extended this concept and divided the $BCMRO_2$ compartment into that associated with Na^+-K^+ leak flux representing about 40% of $BCMRO_2$, and the other 60% of $BCMRO_2$ attributed to unidentified metabolic processes. Thus, thiopental allows the differentiation of the active and basal compartments by abolishing the active compartment.

Attenuation of Ischemic Brain Damage by Barbiturates and Hypothermia

The efficacy of barbiturate anesthetics in attenuating the severity of ischemic brain damage has long been a subject of intense investigation (Smith et al., 1974); Bleyaert et al., 1978; Moseley et al., 1975). The results show that barbiturates effectively attenuate focal and global ischemic brain injury presumably by reducing $CMRO_2$ (Newberg et al, 1983; Milde et al., 1988). Inhibition of $CMRO_2$ to burst suppression, the usual level of barbiturate coma for brain protection, reduces $CMRO_2$ to 50% of normal (Michenfelder, 1983; Steen et al., 1983).

Oxygen Transport to Tissue XVI
Edited by M.C. Hogan *et al.*, Plenum Press, New York, 1994

However, it has recently been shown that mild to moderate hypothermia which reduces $CMRO_2$ by only 15 to 20%, is even more effective than barbiturates in attenuating ischemic brain injury (Minamisawa et al., 1990). In addition, we note that the inhibition of $CMRO_2$ is almost entirely due to the suppression of active $CMRO_2$ which is associated with EEG activity. Hypothermia, on the other hand, while having a dramatic effect on $CMRO_2$, does not abolish the EEG until $CMRO_2$ has decreased to about 7% of normal (Michenfelder and Milde, 1992)! Thus, we <u>hypothesized</u> that mild hypothermia differentially affects active and basal $CMRO_2$ with a relatively greater effect on basal $CMRO_2$. If the inhibition of basal $CMRO_2$ is more important to cerebral protection, this would explain the greater efficacy of hypothermia in attenuating ischemic brain damage.

METHODS

Whole brain, total cerebral blood flow (TCBF-H_2 clearance, cerebral metabolic rate for oxygen ($TCMRO_2$) was measured in male Wistar albino rats mechanically ventilated on 70% N_2O/30% O_2 at brain temperatures (T_B) of 38°C and 34°C with continuing monitoring of arterial pressure and cortical EEG activity. T_B was reduced with ice packs. Thereafter, at a T_B of 34°C, basal values of CBF (BCBF), $CMRO_2$ ($BCMRO_2$) and CMRG (BCMRG) were measured during mechanical ventilation on 70% N_2/30% O_2 and titrated thiopental infusion to render the EEG isoelectric at a sensitivity of 50 uV/cm. Because of the cardiovascular depression induced by high dose thiopental, steps were taken to maintain arterial pressure during thiopental infusion. In the first series of studies, titrated norepinephrine (NE) was used to maintain arterial pressure. However, we soon learned that the infusion of NE increased $CMRO_2$. CMRG was measured in only the first series of studies. Thus, after completing the studies with NE infusion, a second series was done using donor blood transfusion to maintain arterial pressure during high dose thiopental induction of an isoelectric EEG. Active $CMRO_2$ ($ACMRO_2$) is then calculated as the difference between $TCMRO_2$ and $BCMRO_2$.

Whole brain CBF was measured by equilibration of the brain with 1-5% (inspired) and continuous monitoring of H_2 concentration in the confluence of the sinuses. CBF was calculated by the T1/2 method (Singh et al., 1992). Cerebral venous blood was withdrawn through a catheter in the left transverse sinus inserted via a branch of the internal maxillary artery. Arterial blood samples were withdrawn from a catheter in the femoral artery. Arterial blood gases and pH were measured on a Ciba Corning trielectrode model 178 blood gas analyzer. Arterial and cerebral venous blood oxygen contents were measured using an IL model 482 Co-Oximeter.

Statistical analyses were done by one-way ANOVA and post-hoc analysis using the SNK test and a maximum p value of 0.05 for statistical significance.

RESULTS

In series I studies with NE infusion, physiological variables were maintained within normal limits at both 38 and 34°C (Table 1). Mean arterial pressure (MAP) was relatively constant at about 130 mmHg between 38 and 34°C but fell significantly to 98 mmHg during thiopental infusion despite titrated NE infusion. Nevertheless, MAP was still well above the range of autoregulation. $PaCO_2$, PaO_2 and pHa were well within normal limits and unaltered between 38 and 34°C.

$CMRO_2$ fell by 27.7% from 6.5 to 4.7 ml/100g/min when T_B was reduced from 38 to 34°C while CBF was essentially unchanged (Table 1). Induction of an isoelectric EEG with TP caused a further 21% reduction in $CMRO_2$ to 3.7 ml/100g/min and a 27% reduction in CBF to 81 ml/100g/min. At a T_B of 34°C with NE infusion to sustain mean arterial pressure (MAP) near 100 mmHg, active:basal $CMRO_2$ was 34%:66% or 1/3rd active and 2/3rd basal. The effect on CBF was similar to that of $CMRO_2$.

In series II studies, instead of using NE to maintain MAP during TP infusion to induce an isoelectric EEG, we used freshly drawn, heparinized donor blood transfused prior to TP to expand blood volume. Again, physiological variables were comparable between the different temperatures at both 38 and 34°C (Table 2). However, the % active:basal $CMRO_2$ at T_B of 34°C using whole blood transfusion to maintain MAP was opposite of that obtained with NE infusion, namely, 64%:36% or 2/3rd active and 1/3rd basal. Thus, mild hypothermia to 34°C had a greater effect on basal compared to active $CMRO_2$.

DISCUSSION

The results show that mild hypothermia to 34°C differentially affects the active and basal $CMRO_2$ compartments with a relatively greater depression of basal compared to active $CMRO_2$. They also show that $CMRO_2$ can be influenced by exogenous catecholamine infusion during hypothermia which apparently does not occur at normothermia and that infusion of catecholamines can affect basal $CMRO_2$ in the absence of any cortical EEG activity. These results have important mechanistic and clinical implications.

The observation that mild hypothermia affects basal more than active $CMRO_2$ is consistent with its greater protective effect in ischemic brain damage than the barbiturates (Minamisawa et al., 1990). Barbiturates influence primarily, the $CMRO_2$ associated with EEG activity and therefore, $CMRO_2$ associated with synaptic transmission. This component of $CMRO_2$ may be considered unessential to the survival of the neuron. Indeed, the brain is quite capable of surviving prolonged periods of barbiturate coma without any detrimental effect on its viability. On the other hand, ischemic or hypoxic injury leading to the depression of the basal $CMRO_2$ component, if suppressed for sufficiently long, could result in irreversible brain damage. This component of $CMRO_2$ could be thought of as the component necessary to maintain the basal transmembrane ionic gradients. When this fails, transmembrane equilibration of ions occurs and the cell dies.

The reduction in basal $CMRO_2$ by mild hypothermia must occur through a reduction in the permeability of the membrane to Na^+-K^+ leak flux rather than an inhibition of Na^+-K^+ ATPase activity. The reduction in leak flux would reduce energy demand for this compartment whereas the inhibition of ATPase activity would result in transmembrane ionic equilibration and eventually a worsening of the damage sustained during ischemia. Mild hypothermia may reduce membrane permeability through an effect on the phase transition of membrane lipids which has yet to be determined.

The observation that exogenously infused catecholamines can affect $CMRO_2$ during mild hypothermia which, however, apparently does not occur during normothermia in the normal brain, has important clinical implications in the use of vasopressors in patients under hypothermic therapy. The increased cerebral metabolic demand caused by vasopressors during mild hypothermia may detrimentally increase cerebral oxygen demand in patients with head injury or cerebral ischemic-hypoxic insults that could worsen the outcome.

Finally, the observation that NE infusion can increase basal $CMRO_2$ in the absence of any EEG activity suggests that NE stimulated Na^+-K^+

Table 1. Effects of mild hypothermia on cerebral variables and distribution of active:basal $CMRO_2$ in rats with controlled physiological variables. Mean arterial pressure (MAP) maintained during thiopental (TP) infusion by norepinephrine (NE) infusion. Values are mean $\pm$ SD. N = 7

		T_T (°C)	MAP mmHg	$PaCO_2$ mmHg	PaO_2 mmHg	pH	$CMRO_2$ (ml/100g/min)	CBF
Bsln	X	38.0	130	35	146	7.40	6.5	123
	SD	0.3	19	5	25	0.04	0.8	31
34°C	X	33.5	130	40	156	7.33	4.7*	111
	SD	1.2	19	5	42	0.06	1.0	32
34°C	X	33.4	98	35	159	7.38	3.7**	81
+ TP	SD	1.4	23	8	37	0.07	1.6	55
%Act	X						34.4	35.5
	SD						13.7	17.7
%Bas	X						65.7	62.7
	SD						13.5	16.6

* = P <0.05 ** = P<0.01 compared to 38°C

Table 2. Effects of mild hypothermia on cerebral variables and distribution of active:basal $CMRO_2$ in rats with controlled physiological variables. Mean arterial pressure (MAP) maintained during thiopental (TP) infusion by infusion of donor blood. Values are mean $\pm$ SD. N = 5

		T_B (°C)	MAP mmHg	$PaCO_2$ mmHg	PaO_2 mmHg	pH	$CMRO_2$ (ml/100g/min)	CBF
Bsln	X	38.0	125	28	128	7.50	7.2	89
	SD	0.2	15	4	25	0.04	2.04	31
34°C	X	34.0	105	40	156	7.38	6.3*	90
	SD	0.1	17	5	42	0.05	1.8	28
34°C	X	33.9	76	31	199	7.39	2.1**	42
+ TP	SD	0.1	15	6	126	0.06	0.5	10
%Act	X						64.2	50.0
	SD						11.5	14.0
%Bas	X						35.7	50.3
	SD						11.5	14.0

* = P <0.05 ** = P<0.01 compared to 38°C

leak flux and thereby membrane depolarization. Thus, catecholamine infusion could exacerbate the severity of ischemic brain damage through enhancing the depolarization of injured neurons and transmembrane ionic equilibration in addition to inappropriately increasing O_2 demand in circumstances of compromised O_2 delivery.

REFERENCES

Astrup, J., Skovsted P., Gjerris, F., and Sorensen H.R., 1981, Increase in extracellular potassium in the brain during circulatory arrest: Effects of hypothermia, lidocaine and thiopental, *Anesthesiology* 55:256-62.

Astrup, J., Sorensen, P.M., and Sorensen H.R., 1981, Inhibition of cerebral oxygen and glucose consumption in the dog by hypothermia, pentobarbital, and lidocaine, *Anesthesiology* 55:263-68.

Bleyaert, A.L., Nemoto, E.M., Safar, P., Stezoski, S.W., Mickell, J.J., Moossy, J., and Rao, G., 1978, Thiopental amelioration of ischemic brain damage in monkeys, *Anesthesiology* 49:390-98.

Kety, S.S., 1965, Discussion, *Pharmacol. Rev.* 230-32.

Michenfelder, J.D., 1975, The in vivo effects of massive concentrations of anesthetics on canine cerebral metabolism, *in*: "Molecular Mechanisms of Anesthesia, Progress in Anesthesiology," Vol. 1, B.R. Fink, ed., pp 537-43, Raven Press, New York.

Michenfelder, J.D., 1983, The interdependency of cerebral functional and metabolic effects following massive doses of thiopental in the dog, *Anesthesiology* 41:231-36.

Michenfelder, J.D., and J.H. Milde, 1992, The relationship among canine brain temperature, metabolism and function during hypothermia, *Anesthesiology* 75:130-36.

Milde, L.N., Milde, J.H., Lanier, W.L., and Michenfelder, J.D., 1988, Comparison of the effects of isoflurane and thiopental on neurologic outcome and neuropathology after temporary focal cerebral ischemia in primates, *Anesthesiology* 69:905-13.

Minamisawa, H., Nordstrom, C.H., Smith, M.L., and Siesjo, B.K., 1990, The influence of mild body and brain hypothermia on ischemic brain damage, *J. Cereb. Blood Flow Metab.* 10:365-74.

Moseley, J.I., Laurent, J.P., and Molinari, G.F., 1975, Barbiturate attenuation of the clinical course and pathologic lesions in a primate stroke model, *Neurology* 25:879.

Newberg, L.A., Milde, J.H., and Michenfelder, J.D., 1983, The cerebral metabolic effects of isoflurane at and above concentrations that suppress cortical electrical activity, *Anesthesiology* 59:23-28.

Singh, N.D., Kochanek, P.M., Schiding, J.K., Melick, J.A., and Nemoto, E.M., 1992, Uncoupled cerebral blood flow and metabolism after severe global ischemia in rats, *J. Cereb. Blood Flow Metab.* 12:802-08.

Smith, A.L, Hoff, J.T., Nielsen, S.L., and Larson, C.P., 1974, Barbiturate protection in acute focal cerebral ischemia, *Stroke* 1:1-7.

Steen, P.A., Newberg, L., Milde, J.H., and Michenfelder, J.D., 1983, Hypothermia and barbiturates: Individual and combined effects of canine cerebral oxygen consumption, *Anesthesiology* 58:527-532.

INVESTIGATION OF THE EFFECTS OF HYPOCAPNIA UPON CEREBRAL HAEMODYNAMICS IN NORMAL VOLUNTEERS AND ANAESTHETISED SUBJECTS BY NEAR INFRARED SPECTROSCOPY (NIRS)

H. Owen-Reece[1], C.E. Elwell[2], J. Goldstone[1], M. Smith[4], D.T. Delpy[2], J.S. Wyatt[3]

Departments of Anaesthesia[1], Medical Physics[2] and Paediatrics[3], University College London Medical School, and Anaesthesia[4], The National Hospital for Neurology and Neurosurgery, London, UK

INTRODUCTION

The effect of alterations in arterial carbon dioxide tension ($PaCO_2$) upon the cerebral blood vessels was first described in 1930 by Wolff and Lennox and quantified in 1948 by Kety and Schmidt. Since then many studies have investigated this relationship in terms of cerebral blood flow (CBF) (Novack *et al* 1953, Severinghaus *et al* 1967, Grubb *et al* 1974.) and volume (CBV)(Greenberg *et al* 1978, Artru 1984). CBV has not been as closely studied as CBF because of the paucity of measurement techniques. Since arterial PCO_2 is frequently maintained at an artificially low level during anaesthesia for neurosurgery, to reduce intracranial pressure, it is of interest to measure the changes in cerebral haemodynamics as well as to describe the time course. [133]Xenon washout and positron emission tomography (PET) are relatively invasive and can only provide measurements intermittently. Near infrared spectroscopy (NIRS) has been used to quantify the response of CBV to altered $PaCO_2$ (CBVR) in newborn infants and is capable of measuring changes in CBV at 0.5 second intervals. The purpose of this study was to quantify both the extent and the duration of the haemodynamic response to an alteration in $PaCO_2$ in anaesthetised patients and healthy volunteers.

THEORY

The principles of NIRS were first described in 1977 by Jöbsis. We have developed the technique to allow quantification of CBV, CBF and CBVR. Biological tissue, though

Oxygen Transport to Tissue XVI
Edited by M.C. Hogan *et al.*, Plenum Press, New York, 1994

relatively transparent to near infrared light, contains three chromophores which absorb it - oxyhaemoglobin, deoxyhaemoglobin and cytochrome aa_3. Changes in the absorption of infrared light can be converted to changes in concentration of each chromophore using a modified Lambert-Beer law (Delpy *et al* 1988:

$$\Delta OD = \alpha.\Delta C.L.B. + G$$

Where ΔOD represents absorption, ΔC the change in concentration of the chromophore, α the absorption coefficient, L the distance between the points of light entry and exit, G a constant related to tissue geometry and B a pathlength factor which takes account of the highly scattering nature of biological tissue and the consequent increase in the distance travelled through the tissue by light. This pathlength factor has been measured for adults (van der Zee *et al* 1990) and has a value of 5.9. The NIRS machine (NIRO 500, Hamamatsu Photonics KK, Hamamatsu City, Japan) transilluminates the tissue with four wavelengths of near infrared light generated by laser diodes, measures the change in intensity of the emergent light and displays changes in chromophore concentration <u>uncorrected</u> for pathlength. Dividing these data by the pathlength allows calculation of a real change in concentration in units of micromoles per litre.

The theoretical basis for the calculation of CBV has been described before (Wyatt *et al* 1990). Briefly, a controlled fall in arterial oxygen saturation (SaO_2) of about 5% is induced via a reduction in inspired oxygen tension over a period of several minutes and then reversed. The relationship between change in fractional arterial oxygen saturation $\Delta(SaO_2)$ and the change in oxyhaemoglobin concentration $\Delta[HbO_2]$ during this manoeuvre can be describe by the following equation:-

$$[tHb] \times \Delta SaO_2 = \Delta[HbO_2]$$

It follows that if the change in $[HbO_2]$ is plotted against the change in arterial saturation then the gradient of the resulting line will be proportional to cerebral blood volume. CBV can then be derived arithmetically once the subject's blood haemoglobin concentration, the cerebral to large vessel haematocrit ratio, cerebral tissue density and the molecular mass of haemoglobin are incorporated into the calculation.

METHODS

Subjects Studied

Two groups of subjects were studied: one of six healthy adult volunteers (age range 21-37 years) and one of seven patients scheduled for surgery (age range 28-45 years). Both parts of the study were approved by the University College London Medical School Joint Committee on the Ethics of Clinical Investigations and all subjects gave informed consent.

The NIRS optodes were positioned on the forehead just in front of the hairline avoiding the temporal region and the midline air sinuses. The range of interoptode spacing was 3.4-4.5cm. They were covered with a light occluding material.

A pulse oximeter (model 850, Novametrix, Wallingford, CA.) was applied to the earlobe and end tidal carbon dioxide concentration ($ETCO_2$) was monitored as a guide to $PaCO_2$ (Datex Normocap, Datex Instruments, Finland).

Conscious Subjects. The gas mixture that the subjects breathed was supplied by a gas blender which we have described elsewhere (Elwell *et al* 1993) and was a variable mixture of oxygen and nitrogen. It was supplied through a mouthpiece fitted with a nonreturn expiratory valve. The subjects wore a noseclip. The inspired oxygen concentration was monitored with an oxygen analyser (model 5550, Hudson, CA). After a ten minute baseline period when oxy (HbO_2), deoxy (Hb) and total (Hbsum) haemoglobin concentration changes were recorded the subject was instructed to breathe more deeply but at the same rate until a new level of $ETCO_2$ was reached 1-2kpa lower than the baseline value. By observing the $ETCO_2$ trace it was easy for the subjects to maintain a regular ventilatory pattern. When the changes in Hbsum were complete the experiment was concluded. Finally, the interoptode spacing was measured with a tape measure. A sample of blood was obtained from each subject for an estimation of haemoglobin concentration.

Anaesthetised subjects. For the measurements during anaesthesia, patients undergoing cosmetic plastic surgery or excision of a temporal lobe epileptic focus were studied. None had any significant pathology other than the reason for their operation. Anaesthesia was induced with 100μg fentanyl and 3-5mg.kg^{-1} of thiopentone or propofol. Neuromuscular blockade was achieved with atracurium or vecuronium and was followed by intubation and ventilation with isoflurane 0.5-1.5% in oxygen and nitrous oxide. Analgesia was provided with fentanyl, papaveretum or morphine as indicated. After induction was complete the NIRS equipment was set up as described above. Monitoring was as for the conscious subjects, with the exception that when possible $PaCO_2$ was determined using blood taken from an arterial cannula. The $ETCO_2$ monitor was compensated for nitrous oxide in the sampled gas. After a 10 minute period for baseline data collection the measurements were begun. The CO_2 response was determined by increasing the tidal volume but not the ventilator rate, until an adequate fall in end tidal CO_2 was seen. A sample of blood was taken from each subject for a haemoglobin estimation.

Analysis

The data from the machine were converted from changes in optical density to changes in concentration uncorrected for pathlength using an algorithm which we have previously published. After correcting for pathlength as described above, the changes were expressed in units of micromoles of haemoglobin per 100g of brain tissue per kilopascal, and mL of blood per 100g of brain tissue per kilopascal. In order to calculate the CO_2 drop, mean $ETCO_2$ was calculated over a 2 minute period when HbO_2 and Hbsum were stable before and after the CO_2 change. Hbsum and HbO_2 concentrations were calculated over similar periods. The time for 90% completion of the change was used to define the duration of the change. This was done because the rate of change in Hbsum was not constant but asymptotic. The magnitude and rate of change of CBV in response to a fall in $PaCO_2$ were compared using the Mann-Whitney U test.

RESULTS

Anaesthetised Subjects

The time taken for Hbsum to reach 90% of its new value was a median 10.8 minutes in the anaesthetised subjects. A typical sample of raw data is shown in Figure 1.

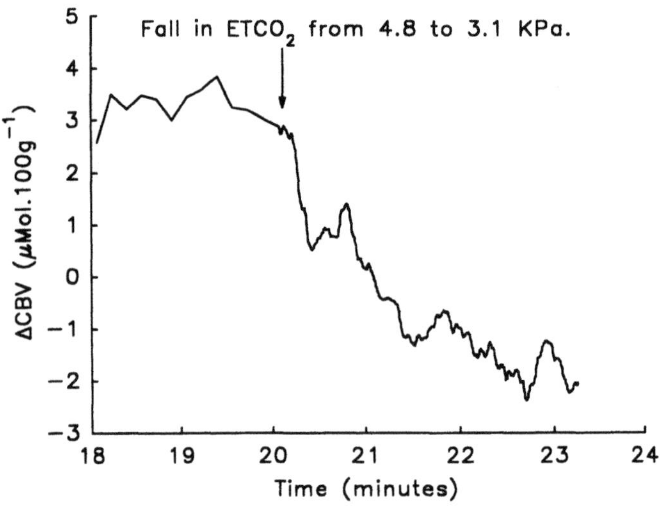

Figure 1: Change in cerebral blood volume with end tidal PCO_2

The response of Hbsum was less than in the awake subjects with a median value of $2.1\mu Mol.kPa^{-1}$ ($0.12mL.kPa^{-1}$). The CBV responses for the anaesthetised patients are shown in Table 1.

Table 1. Cerebrovascular response to altered $PaCO_2$ during anaesthesia.

CBVR ($mL.100g^{-1}.kPa^{-1}$)	CBVR ($\mu Mol.100g^{-1}.kPa^{-1}$)	Time (min)
0.02	0.21	10.1
0.01	0.23	11.8
0.03	0.46	3.3
0.14	2.08	10.6
0.12	1.93	19.8
0.15	2.36	2.3
0.28	3.66	10.8
median 0.12	2.08	10.8

Awake Subjects

The fall in $ETCO_2$ occurred rapidly but the CBVR was considerably slower (median 5.9 minutes). The fall in Hbsum was $3.8\mu Mol.kPa^{-1}$ ($0.19\ mL.kPa^{-1}$); the values for each individual are shown in Table 2.

Table 2. Cerebrovascular response to altered $PaCO_2$ in conscious subjects.

CBVR ($mL.100g^{-1}.kPa^{-1}$)	CBVR ($\mu Mol.100g^{-1}.kPa$)	Time (min)
0.20	2.92	4.2
0.02	0.66	1.6
0.31	4.37	5.8
0.16	5.47	8.1
0.29	4.63	5.9
0.18	3.26	6.4
median 0.19	3.82	5.9

DISCUSSION

We have shown that the fall in CBV following a fall in $PaCO_2$ is slower and smaller during anaesthesia than in conscious patients. We have made several assumptions: It is accepted that $ETCO_2$ approximates to $PaCO_2$. In the absence of lung pathology, $ETCO_2$ may have a value up to 0.3 kpa lower than $PaCO_2$ (Runciman 1990). However, it was not possible to take samples of arterial blood from the volunteers or most of the patients; transcutaneous monitoring would have the drawback of slow response time, concealing the time course of the $PaCO_2$ change; this should ideally be rapid in order that it should not itself limit the rate of decline of Hbsum.

We have also assumed that the response of CBV to an alteration in $PaCO_2$ is linear in the range studied (1.7-8.3kpa). This has been demonstrated by others (Alexander *et al* 1964, Harper *et al* 1965). In view of this underestimate of $PaCO_2$, the lowest value would probably be within the normal range, although the highest could exceed it by 1 kpa. Although the response curve for CBV has not been characterised we expect it to resemble that for CBF (Reivitch 1964) and to have a sigmoid shape due to flattening at the extremes.

We have considered that our method of altering $PaCO_2$ might affect cerebral haemodynamics simply by an increase in intrathoracic pressures causing an alteration in cerebral perfusion. This is an inevitable criticism of any method of lowering $PaCO_2$ which can only be achieved by increasing minute volume. In a subject who is haemodynamically stable and has normal pulmonary and cardiac compliance, alterations in intrathoracic pressure due to changes in tidal volume will be minimal. We have demonstrated cyclical variations in Hbsum and HbO_2 in adult volunteers breathing spontaneously against an end expiratory pressure of 0-20 cm H_2O (Elwell *et al* 1992), which have a magnitude of approximately 1% of CBV. Cycles analogous to these would be expected if intermittent positive pressure ventilation was similarly affecting cerebral haemodynamics but we have never observed any such changes in anaesthetised subjects during IPPV.

Finally we have employed a value of 5.9 for the pathlength factor which has been measured in healthy volunteers(van der Zee 1991). Light enters and leaves through the skull which thus contributes 2-3 cm to the light's path despite being haemodynamically "inert" in terms of NIRS. The haemodynamic changes taking place are only, or almost only, in the brain and thus the pathlength factor is probably an overestimate, causing an underestimate in the calculated value of CBV, CBF and CBVR.

The value for CBVR for the conscious group in this study is about 30% lower than the values quoted by Greenberg *et al* (1978)[6] (using radiolabelled erythrocytes) of 0.053 $mL.100g^{-1}.mmHg^{-1}$ for grey matter and 0.046 $mL.100g^{-1}.mmHg^{-1}$ for white matter (these correspond to $0.35mL.100g^{-1}.kPa^{-1}$ and $0.40 mL.100g^{-1}.kPa^{-1}$). It is notable also that whilst the CBVR of preterm babies is low (Wyatt *et al* 1991), the value rises with gestation until at term it correlates well with Greenberg's values. This value for CBVR ($0.51mL.100g^{-1}.kPa^{-1}$) is also larger than in our conscious adult volunteers and would support our theory of an overestimate in the current pathlength factor since the skull and scalp of a preterm baby are so thin as to make a negligible contribution to the pathlength.

The mean value of CBVR is lower in anaesthetised subjects. This trend is not statistically significant. The time to achieve the lowest value of Hbsum is also greater. This difference does not achieve statistical significance either. There are three possible reasons that the CBVR is lower during anaesthesia. Firstly, anaesthetic drugs and the volatile anaesthetic agents in particular are well known to suppress the cerebral vascular response to alterations in $PaCO_2$ (Artru 1982) although isoflurane is said not to have this effect below 1.3% inspired concentration (Adams 1981). The measurements were made sufficiently long after commencement of anaesthesia for a residual effect of the induction agent to be unlikely.

Secondly hypotension could cause a vasodilatory response in the cerebral circulation overriding the effect of changing $PaCO_2$ and causing an apparently small response CBVR. However, although the mean arterial pressure was lower in the anaesthetised patients than the volunteers it was well within normal limits and thus unlikely to have this effect.

Conclusion

We conclude that CBVR is lower during anaesthesia than in conscious volunteers, although this trend does not reach statistical significance in this small study. The difference is likely to be due to the effect of the anaesthetic agents.

Acknowledgements

We thank the staff of the departments of Medical Physics and Bioengineering, Paediatrics, Anaesthetics and Intensive Care at University College London Hospitals and The National Hospital for Neurology and Neurosurgery for their help with this study. This work was supported by The Wolfson Foundation, The Wellcome Trust, The Medical Research Council, Hamamatsu Photonics KK and The Royal Society.

REFERENCES

Adams R.W., Cucchiara R.F., Gronert G.A. *et al*. 1981. Isoflurane and Cerebrospinal fluid pressure in neurosurgical patients. Anesthesiology **54:** 97-99.

Artru A.A. 1984. Relationship between cerebral blood volume and CSF pressure during anaesthesia with isoflurane and fentanyl in dogs. Anesthesiology **60:** 575.

Artru A.A. 1982. A comparison of the effects of isoflurane, enflurane, halothane and fentanyl on cerebral blood volume and ICP. Anesthesiology **57:** A374.

Alexander S.C., Wollman H., Cohen P.J., Chase P.E., Behar M. 1964. Cerebrovascular response to $PaCO_2$ during halothane anaesthesia in man. J. Appl. Physiol. **19:** 561-565.

Delpy D.T., Cope M., van der Zee P., Arridge S.R., Wray S. and Wyatt J.S. 1988. Estimation of optical pathlength through tissue from direct time of flight measurement. Phys. Med & Biol; **33** (12):1433-1442.

Edwards A.D., Wyatt J.S., Richardson C.E., Delpy D.T., Cope M. and Reynolds E.O.R. 1988. Cotside measurement of cerebral blood flow in ill newborn infants by near infrared spectroscopy. Lancet,*ii*,770-771.

Elwell C.E., Cope M., Edwards A.D., Wyatt J.S., Reynolds E.O.R. and Delpy D.T. Measurement of cerebral blood flow in adult humans using near infrared spectroscopy - methodology and possible errors. 1992. Adv. Exp. Med. Biol. **317**:325-245.

Elwell C.E., Cope M., Kirkby D., Owen-Reece H., Cooper C.E., Reynolds E.O.R., Delpy D.T. An automated system for the measurement of the response of CBV and CBF to changes in arterial CO_2 tension using NIRS. Adv. Exp. Med. Biol 1993 in press.

Elwell C.E., Owen-Reece H., Cope M., Wyatt J.S., Reynolds E.O.R. and Delpy D.T. 1993. Measurement of cerebral haemodynamics during inspiration and expiration using near infrared spectroscopy (NIRS). Adv. Exp. Med. Biol. In press.

Grubb R.L., Raichle M.E., Eichling J.O., Ter-Pogossian M.M. 1974. The effects of changes in $PaCO_2$ on cerebral blood volume, blood flow and vascular mean transit time. Stroke **5**:630.

Greenberg J.H., Alavi A., Reivitch M. *et al* 1978. Local cerebral blood volume response to carbon dioxide in man. Circulation Res. **43**:324.

Harper A.M., Glass H.I. 1965. Effect of alterations in the arterial carbon dioxide tension on the blood flow through the cerebral cortex at normal and low arterial pressures. J. Neurol. Neurosurg. Psychiatry **28**: 449-452

Jöbsis F.F. 1977. Noninvasive infrared monitoring of cerebral and myocardial oxygen sufficiency and circulatory parameters. Science. **198**:1264-1267.

Kety S.S. and Schmidt C.F. 1948. The effect of altered arterial tensions of carbon dioxide and oxygen on cerebral blood flow and cerebral oxygen consumption of normal young men. J.Clin. Invest.**27**:484-491.

Novack P., Shenkin I., Bortin I., Goluboff B., Soffe E.J.1953. The effects of carbon dioxide inhalation upon the cerebral blood flow and cerebral oxygen consumption in vascular disease. J.Clin Invest. **32**: 696-722.

Reinstrup P., Uski T and Messeter K. Modulation by carbon dioxide and pH of the contractile responses to potassium and prostaglandin $F_{2\alpha}$ in isolated human pial arteries. Br. J. Anaes. 1992 **69:615-620**.

Reivitch M., 1964. Arterial PCO_2 and cerebral haemodynamics. Am. J. Physiol. **206** 25-35.

Runciman W.B.1990. Monitoring, *in*: "Anaesthesia", Nimmo W.S. and Smith G. ed., Blackwell Scientific publications, Oxford.

Severinghaus J.W., Lassen N. 1967. Step hypocapnia to separate arterial from tissue PCO_2 in the regulation of cerebral blood flow. Circulation Res. **20**: 272-278.

van der Zee P., Cope M., Arridge S.R., Essenpreiss M., Potter L.A., Edwards A.D., Wyatt J.S., McCormick D.C., Roth S.C., Reynolds E.O.R., Delpy D.T.1992. Experimentally measured optical pathlengths for the adult head, calf and forearm and the head of the newborn infant as a function of inter optode spacing. Adv. Exp. Med. Biol. **416:** 143-153.

Wolff H.G. and Lennox W.G. 1930. The effects on pial vessels of variations in the O_2 and CO_2 content of the blood. *in* The cerebral circulation :XII. Arch. Neurol. Psychiatr. Chicago 23:1097-1120.

Wyatt J.S., Cope M., Delpy D.T., Richardson C.E., Edwards A.D., Wray S. and Reynolds E.O.R. 1990. Quantitation of cerebral blood volume in human infants by near infrared spectroscopy. J. Appl. Physiol. **68**(3): 1086-1091.

Wyatt J.S., Edwards A.D., Cope M., Delpy D.T., McCormick D.C., Potter A. and Reynolds E.O.R. 1991. Response of cerebral blood volume to changes in arterial carbon dioxide tension in preterm and term infants. Pediatr.Res. **29**:553-557.

INTRAOPERATIVE MONITORING OF LOCAL Hb-OXYGENATION IN HUMAN BRAIN CORTEX

Jens Höper[1], Michael R. Gaab[2]

Institut für Physiologie und Kardiologie der Friedrich-Alexander-Universität, Waldstraße 6, D-8520 Erlangen, [2] Klinik und Poliklinik für Neurochirurgie der Ernst-Moritz-Arndt-Universität, Friedrich-Löffler-Straße 23, O-2200 Greifswald

INTRODUCTION

Neurosurgical interventions are frequently performed under conditions of hyperventilation in order to lower intracranial pressure. This treatment is accompanied by a decrease in cerebral blood flow (CBF) and thus a decrease in the oxygen supply rate. Lowering the $PaCO_2$ (arterial PCO_2) from 40 to 30 mm Hg induces a decrease in CBF of 40% (Thoresen et al. 1979). This, in turn, may cause disturbances in local oxygen supply. Because the brain is highly vulnerable to damage from even a brief imbalance of oxygen delivery and demand (O′Sullivan and Cunningham 1989), intraoperative disturbances of local oxygen supply must be avoided. Furthermore, it was shown that anaesthesia can cause a decrease in CBF and in cerebral oxygen uptake rate. Thus it is of interest to know the local oxygen supply to the brain under conditions of anaesthesia and hyperventilation.

METHODS

The investigations were performed in 8 patients (5 male, 3 female). The age ranged from 31 to 67 years (mean age 51.6 years). The study was aproved by the ethical commitee of the Ernst-Moritz-Arndt university, Greifswald. Written consent was obtained from all patients.

In all patients neurosurgical intervention was necessary because of brain tumor. In 4 patients anaesthesia was induced by thiopental (1,5 mg/kg), fentanyl (0,007 mg/kg) and pancuronium (0.06 mg/kg). Anaesthesia was continued by continuous infusion of 100 - 300 mg/h thiopental and 0.2 mg/h fentanyl. Pancuronium was given discontinuously. The inspired gas mixture consisted of N2O, oxygen and air (3:1/2:1).

In the other four patients anaesthesia was induced by propofol (2 mg/kg) and pancuronium (0.06 mg/kg) and was continued with an infusion of propofol (6-12 mg/kg/h), fentanyl (0.2 -0.3 mg/h) and discontinuous injection of pancuronium (0.03 mg/kg). The inspired gas mixture consisted of air and oxygen (FiO$_2$ 0,35).

All measurements were performed in macroscopically normal brain tissue. In 5 patients the first measurements were performed during normocapnia, then the patients were hyperventilated. The other three were first hyperventilated. In these patients the second measurements were done during normocapnia.

The measurements were performed by use of the Erlangen Microlightguide Spectrophotometer (EMPHO).The EMPHO (Diehl GmbH&Co, Nuremberg, FRG) consists of four modules: the light source, the micro-lightguide, the detection device and the computing system. The instrument has been described in detail elsewere (Frank et al. 1989).

RESULTS

Table 1 summarizes blood pressure, F_iO_2 and P_aCO_2. As can be seen, there was no difference in blood pressure or P_aCO_2 between the thiopental and the propofol group.

Table 1 Blood pressure (mm Hg), inspired oxygen concentration (FiO$_2$) and arterial PCO$_2$ during normoventilation and hyperventilation.

	Systolic pressure	Diastolic pressure	FiO$_2$	PaCO$_2$
thiopental	123 $\pm$ 22	75 $\pm$ 11	0.28 $\pm$ 0.04	31 $\pm$ 0.5
propofol	124 $\pm$ 21	66 $\pm$ 7	0.30 $\pm$ 0.1	31 $\pm$ 0.7
thiopental	125 $\pm$ 15	65 $\pm$ 8	0.29 $\pm$ 0.07	37.8 $\pm$ 0.7
propofol	127 $\pm$ 19	69 $\pm$ 10	0.30 $\pm$ 0.09	37.7 $\pm$ 0.8

The results of the local HbO$_2$ measurements in the thiopental group is shown in fig. 1 and 2.

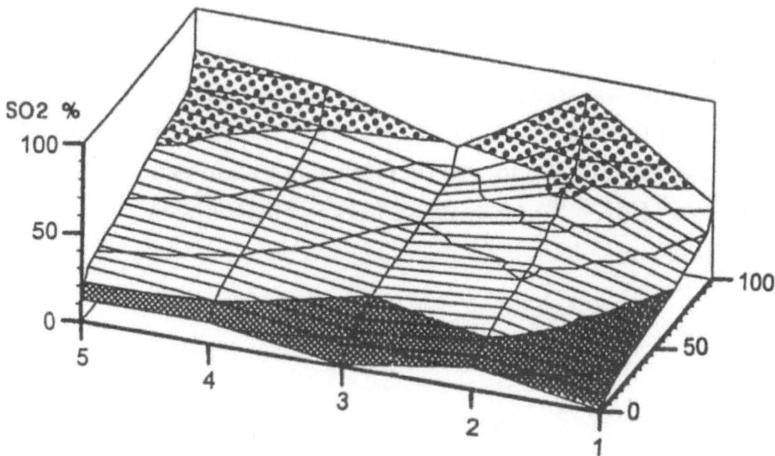

Figure 1 Integrated gradients of intracapillary haemoglobin oxygenation measured in 5 patients (thiopental anaesthesia, hyperventilation). The dotted area represents SO2 values > 50%, the white area values < 50% > 20% oxygenation and the dark area values < 20% oxygenation.

Fig. 1 shows the results obtained during hyperventilation. Only 32% of the measured values are > 50% saturation, 18% of the values are even < 25% Hb-oxygenation. The mean value was 41.4%. An increase in $PaCO_2$ to 37 mm Hg caused an increase in the mean value to 47.8% saturation (Fig.2). The number of values > 50% SO_2 increased to 53%, the number of values < 25% SO_2 was unchanged.

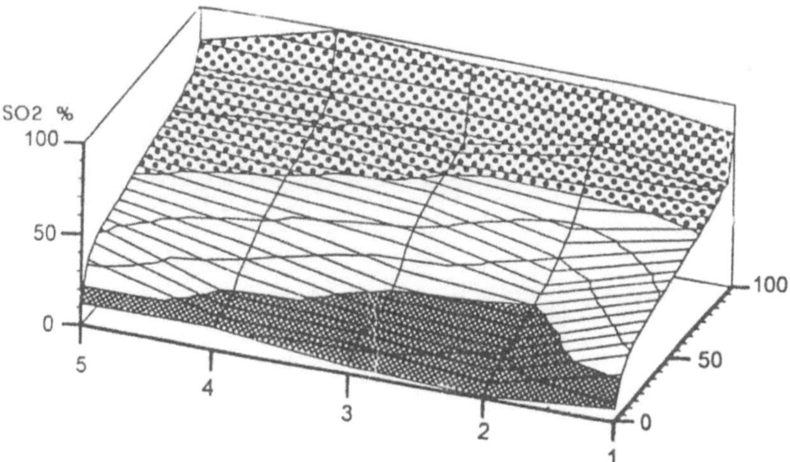

Figure 2 Integrated gradients of intracapillary haemoglobin oxygenation measured during normoventilation in the same patients as shown in fig. 1. For further explanations see fig. 1 and text.

In fig. 3 and 4 the results obtained during propofol anaesthesia are depicted. While in the thiopental group a change in $PaCO_2$ had an effect on the distribution of local haemoglobin oxygenation, no effect was seen in the propofol group. During hypocapnia the mean value was 63.2%, 21% of the values were below 50% SO_2, 5.5% of the measured values were < 25% SO_2. An increase in $PaCO_2$ to mm Hg resulted in a mean value of 61.4% SO_2. 23.5% of the values were <50%, 6.5% of the values were <25% saturation

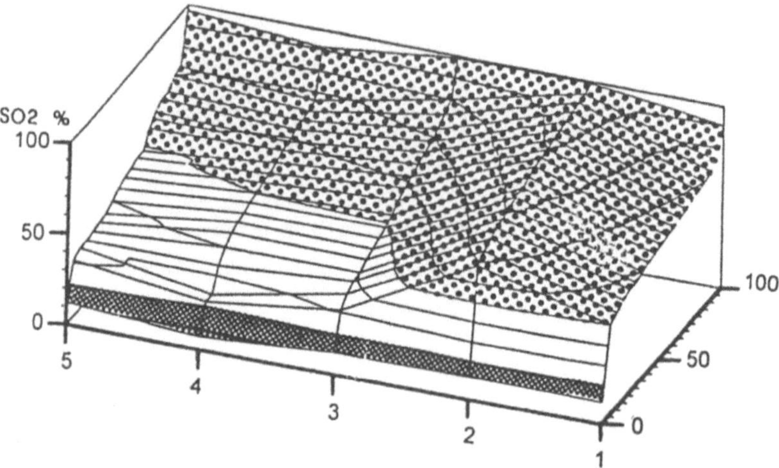

Figure 3 Integrated gradients of intracapillary haemoglobin oxygenation measured in 5 patients during propofol anaesthesia (hyperventilation).

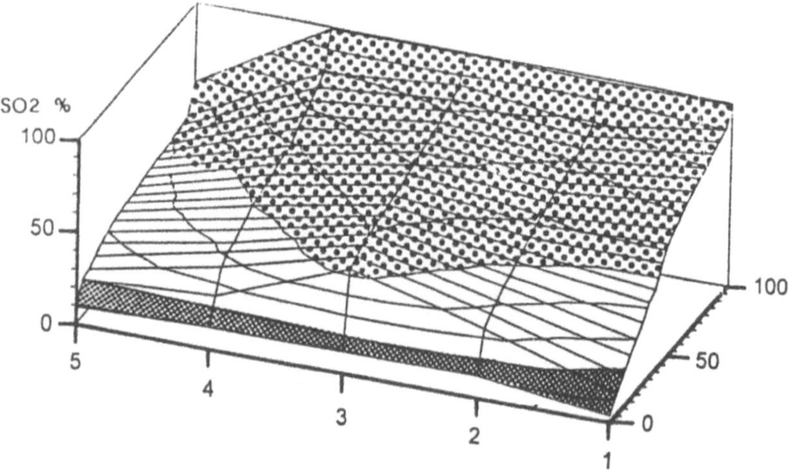

Figure 4 Integrated gradients of intracapillary haemoglobin oxygenation measured in the same patients as shown in fig. 3 (propofol, normoventilation).

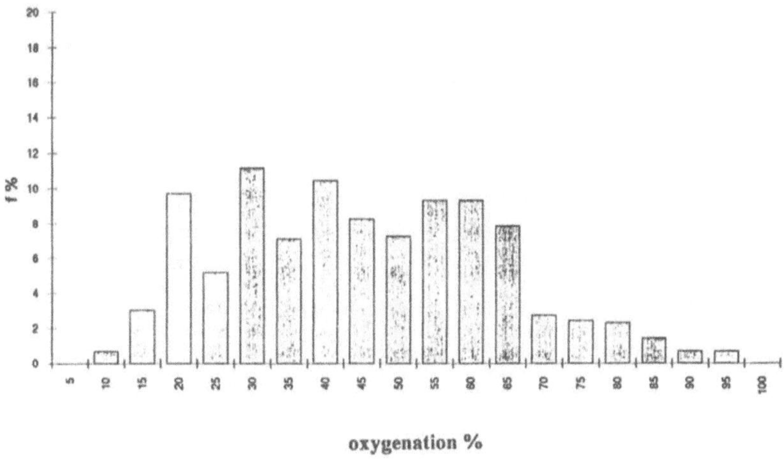

Figure 5: Distribution of local HbO$_2$ values in 2 patients during thiopental anaesthesia

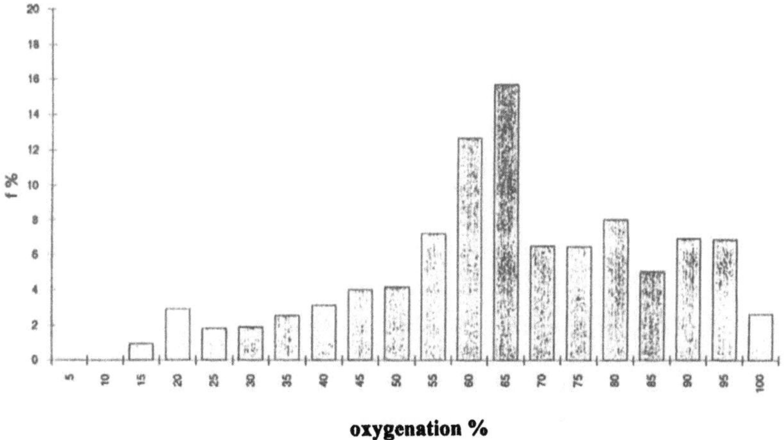

Figure 6: Distribution of local HbO$_2$ values in 2 patients during consequent propofol anaesthesia

Figures 5 and 6 show the results obtained in 2 patients during normocapnia. These patients were ventilated with a gas mixture containing N$_2$O. It is obvious that during propofol anaesthesia higher SO$_2$ values were found despite the presence of N$_2$O.

DISCUSSION

Intraoperative measurements of local oxygen supply to the brain can supply valuable information on pathophysiological events. The EMPHO enables a rapid, non-invasive measurement of local intracapillary haemoglobin oxygenation. Because the measurements are performed with micro-lightguides, a high spatial resolution is attained. Furthermore, direct contact of the end of themicro-lightguide with the brain surface is not necessary. The measurements were performed within 40-60 sec and therefore did not prolong the duration of the neurosurgical operation.

In comparison to other organs low HbO_2 values were found in the brain cortex under the conditions of hyperventilation. 18 % of the values were even below 25 % haemoglobin saturation (SO_2). The lowest value found under these conditions was 1% SO_2. The reason could be a low perfusion rate due to hypocarbia. Thoresen et al. (1979) showed that a reduction of the endexpiratory PCO_2 from 37.5 to 30 mm Hg led to a reduction of cerebral blood flow (CBF) by 40%. Increasing the arterial $PaCO_2$ to 37 mm Hg led not only to an increase in local Hb oxygenation but also in relative Hb concentration indicating the expected increase in brain blood flow. These results indicate that it is possible to determine the CO_2 reactivity of brain blood vessels locally. This could be an important test which can be performed intraoperatively. First measurements in the tumour border zone and in the tumour indicate that the CO_2 reactivity is impaired in these areas (unpublished results).

If the CO_2 reactivity of the brain blood vessels is not attenuated during anaesthesia, the blood flow should increase by almost 40%. It will be necessary to carry out further studies to determine the relation between local blood flow and local haemoglobin concentration. Experiments performed in rats (Höper and Kozniewska, 1992) have shown that an increase in total cerebral blood flow of 215% observed during hypoxia is combined with an increase in cerebrocortical microflow of only 38%. Thus the change in CBF induced by changes in $PaCO_2$ may not be representative of the local changes in the cortex.

Interestingly, low SO_2 values were found in the brain cortex even at normal arterial $PaCO_2$. According to Navari et al (1978), when using an open window preparation one should recognize that progressive hypocapnia of the brain tissue develops owing to the diffusion of CO_2 out of the tissue in contact with the atmospheric air. The resulting low local pCO_2 could induce a consecutive decrease in local blood flow and thus low SO_2 values.

In contrast to the results obtained during thiopental anaesthesia, in propofol anaesthesia not only the mean value of local Hb-saturation was higher (63.2% compared to 41.4%) but also the number of low values (<50% SO_2) was only 21%. Only small changes occurred upon increasing P_aCO_2. Propofol is known to lower oxygen consumption of the brain by 38 %. Concomitantly a decrease in CBF by 51 % was shown to occur (Stephan et al. 1987). From the present results it may be concluded that the oxygenation is better during propofol anaesthesia, at least in the brain cortex. This could be due to the decrease in O_2-uptake rate. The results shown in figures 5 and 6 indicate that the effect is independent of the presence of N_2O.

REFERENCES

Frank, KH, Kessler, M, Appelbaum, K, Dümmler, W (1989) The Erlangen micro-lightguide
spectrophotometer EMPHO I. Physics Med.Biol. 34, 1883-1900.

Höper, J, Kozniewska, E: Attenuation of hypoxic response in cerebral microcirculation following deprenyl.
Int.J.Microcirc.Clin.Exp. 11, 287-295,1992

Navari, RM, Wei, EP, Kontos, HA, Patteron, JL: Comparison of the open skull and cranial window
preparations in the study of the cerebral microcirculation. Microvasc. Res. 16, 304-315, 1978

O´Sullivan, K, Cunningham, AJ (1989) Intraoperative cerebral ischaemia. Br.J.Hosp.Med. 42, 290-296.

Stephan, H, Sonntag, H, Schenk, HD, Kohlhausen, S (1987): Einfluß von Disoprivan (Propofol) auf die
Durchblutung und den Sauerstoffverbrauch des Gehirn und die CO_2-Reaktivität der Hirngefäße
beim Menschen. Anaesthesist, 36, 60-65.

Thoresen, M, Walloe, L: Changes in cerebral blood flow in humans during hyperventilation and CO_2
breathing. J.Physiol. 298, 53-54P, 1979.

SLOW WAVES OF TISSUE PO₂ IN THE BORDER ZONE OF PHOTOTHROMBOTIC BRAIN INFARCTION AND THEIR RELATION TO SPREADING DEPRESSION-LIKE EVENTS

Koen van Rossem, Herman Vermariën and Karin Decuyper

Laboratory of Physiology and Pathophysiology, University of Brussels
VUB Laarbeeklaan 103, B-1090 Brussels, Belgium

INTRODUCTION

Reproducible focal infarction of the cerebral cortex of test animals may be induced photochemically by illuminating a well defined area of the cortex after intravenous injection of the photosensitive dye "Rose Bengal", leading to local photoperoxidative damage of the endothelial cell membranes and subsequent vascular thrombosis and vessel occlusion (Watson et al., 1985). We have modified the original method enabling continuous polarographic monitoring of local tissue PO_2 during and after infarction in awake rabbits. In a previous study (van Rossem et al., 1992c) we documented that iO_2 recorded 1 mm rostral to the illuminated area consistently shows fluctuations of large amplitude and low frequency within the first hour after infarction. Morphologically, brain tissue remains completely normal at this location, the diameter of the lesion only slightly exceeding that of the illuminated area. Interestingly, at the edge of the infarcts a markedly sharp transition from normal to severely damaged tissue is noticed. The aim of the present study was to evaluate the time course of local tissue PO_2 at this edge of the lesion as well as at a larger distance. Therefore, PO_2 electrodes were positioned respectively 0.5 mm and 2.5 mm rostral to the illuminated area. In addition, possible errors due to shifts of the polarising voltage between the monopolar oxygen electrodes and the common reference electrode had to be evaluated. Such shifts might be induced by pathophysiologic depolarisations occurring in the border zone of cerebral infarctions (Nedergaard and Astrup, 1986). Therefore, extracellular DC-potentials were monitored adjacent to the oxygen electrodes in 2 of the 10 investigated animals.

MATERIALS AND METHODS

Electrodes and Measuring Apparatus

Both oxygen and DC-potential measurements were performed with platinum electrodes having a cylindrical measuring tip (length 1 mm, ø 100 µm) covered with a homogenous cellulose acetate membrane (thickness 20 µm ± 2.5 µm). Electrode construction and evaluation were described earlier (van Rossem et al., 1992a). Applying these electrodes, mean values of tissue PO_2 and extracellular DC-potential are measured over 1 mm of cortical thickness. They were fixed into polymethylmetacrylate (PMMA) frames (6 x 8 mm) containing a central shaft in which an optic fiber (ø 3 mm) can be mounted above a perfectly delineated transparent area (ø 3.17 mm) (van Rossem et al., 1992b). Electrodes for PO_2 measurement were fixed respectively in the centre and 0.5 mm and 2.5 mm rostral to the illuminated area. Electrodes for potential measurement were positioned 1 mm medial to the PO_2 electrodes at identical distances relative to the edge of the illumination area. Ag/AgCl reference electrodes for PO_2 and DC-potential measurements were respectively fixed to the left and the right ear after careful removal of the stratum corneum of the skin with a scalpel. Conductive electrode gel was applied between the reference electrodes and the tissue.

Measurements were performed applying a laboratory made 4-channel measuring device (FYSPpO$_2$2) which allows polarographic PO_2 measurement and biopotential recording. Regarding PO_2 measurement a stable internal polarisation voltage (V_{pi} = - 600 mV) or an arbitrary external one (V_{pe}) can be applied between each electrode and the common reference electrode which is connected to ground. The amplifier is essentially a current-to-voltage converter (10^7 VA^{-1}). PO_2 monitoring was performed applying V_{pi}. Polarograms were obtained by sweeping V_{pe} at a low frequency (0.001 Hz) between 100 mV and - 850 mV. Electrodes were calibrated in aerated and deoxygenated Ringer solution at 38 °C in order to check their measuring quality. Since, however, the long-term stability of our electrodes in vitro is only moderate we prefer to mention values of in vivo measured iO_2 instead of PO_2. DC-potentials were measured applying a high input impedance differential amplifier with the - input connected to the reference electrode. The total input resistance of the amplifier and connection cables exceeded 50 x 10^3 MΩ; electrode resistance averaged 30 MΩ.

Animal Preparation and Treatment

Ten male Dutch rabbits (HSD/POC) weighing 1.5 - 2.2 kg were used for the experiments which were performed according to the locally established ethical rules.

Electrode implantation was performed as described earlier (van Rossem et al., 1992b). In brief, rabbits were anaesthetised (Hypnorm® 0.5 ml/kg i.m.) and fixed in a stereotaxic apparatus. After exposure of the skull a hole was drilled according to the contours of the PMMA frame and the dura mater was removed. The centre of the hole was located 1 mm caudal and 3.5 mm lateral (left hemisphere) to the bregma. With the aid of a micro manipulator the frame was placed onto the cerebral cortex so that the entire measuring tip of all electrodes was inserted into the cortex. The frame was fixed to the skull applying cyanoacrylate gel and dental resin.

Ten days after implantation a cortical infarction was induced photochemically. Rose Bengal (7.5 mg/ml in 9 g/l NaCl, subjected to 0.45 µm filtration) was injected intravenously (10 mg/kg) over a 2 min interval via the marginal ear vein. An optic fiber was then mounted into the shaft of the implanted frame and the brain cortex was illuminated with cold green light (spectral width 450 - 600 nm, intensity 110 mW/cm^2) during 20 min. Thirty min before infarction, blood samples for determination of blood gases, pH, haematocrit, platelet count and plasma glucose level were taken from the central artery of the ear. Rectal temperature was monitored during the entire experiment. PO_2 and DC-potentials were monitored continuously from 30 min before until 4 h after onset of illumination. Regarding the animals used for simultaneous measurement of PO_2 and DC-potentials, in vivo polarograms of the oxygen electrodes were recorded before starting the PO_2 measurements.

RESULTS

Physiological variables were normal in all animals (P_{a,O_2} 11.8 ± 0.9 kPa; P_{a,CO_2} 5.0 ± 0.5 kPa; pH 7.43 ± 0.04; haematocrit 39.8 ± 2.2; platelet count 531 x 10^3 ± 85 x 10^3 mm^{-3}; plasma glucose 1.68 ± 0.37 g/l; rectal temperature 38.0 ± 0.3 °C (mean ± SD; n = 10)).

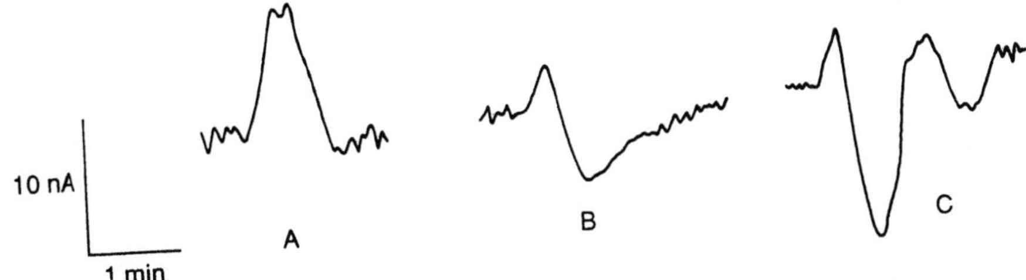

Figure 1. Typical wave forms of SWO_2 recorded in the border zone of a photothrombotic infarction of the brain cortex. A: monophasic increase of iO_2. B: biphasic wave form showing an initial increase. C: multiphasic wave form.

Rectal temperature was stable during the entire experiment: a maximal variation of 0.5 °C was noticed. The in vitro sensitivity and residual current of the PO_2 electrodes respectively were 6.33 ± 0.87 nA/kPa and 15 ± 7 nA (mean $\pm$ SD; n = 30).

Oxygen Measurements

In all animals an infarction was induced as indicated by the time course of iO_2 in the centre of the illuminated area (Table 1). At this location iO_2 significantly decreased after a latency period of 2.9 ± 2.4 min (mean $\pm$ SD) following onset of illumination. Within 10.0 ± 9.4 min iO_2 then dropped gradually to a residual current (assumed to correspond with zero PO_2) which was further maintained. In the border zone recurrent "slow waves" of iO_2 (SWO_2) were recorded at both locations during a "slow variation period" (SVP) of 45 ± 16 min. SVP started after a latency period of 6.6 ± 1.2 min when iO_2 in the centre of the illumination area had decreased to less than 55 % of the initial value minus residual current (range 55 - 0 %). The number of SWO_2's per location per animal varied from 5 to 20. Except for 3 cases, every SWO_2 at the proximal site was coupled with a SWO_2 at the distal site, the total number SWO_2's at 0.5 mm and 2.5 mm respectively amounting to 92 and 89. Most of the SWO_2's at the proximal site were followed within 1 min (30 ± 15 s, n = 77) by a SWO_2 at the distal site. In some animals the time lag between these SWO_2's appeared to be quite variable (maximal range : 6 - 55 s). Exceptionally (n = 5), the distal SWO_2 shortly preceded the proximal one (maximal time interval : 20 s). Only at the onset of SVP, SWO_2's at both sites did sometimes occur simultaneously (n = 4). In two cases the temporal separation could not be quantified due to a very fast succession of subsequent SWO_2's.

Typical wave forms of SWO_2s are presented in Figure 1. In one animal a monophasic increase of iO_2 was recorded. Some wave forms were biphasic, showing an initial rise

Table 1. iO_2 (nA) at different time periods before and after onset of infarction and at different locations relative to the illuminated area.

	Time after onset of illumination					end of SVP
	0 h	1 h	2 h	3 h	4 h	
centre	20.0 ± 5.9	$5.8 \pm 2.0^{*}$	$5.3 \pm 2.0^{*}$	$5.5 \pm 2.0^{*}$	$5.4 \pm 1.9^{*}$	
0.5 mm rostrally	22.2 ± 4.9	23.3 ± 8.3	21.8 ± 6.6	$18.6 \pm 5.6^{*}$	19.3 ± 6.9	23.4 ± 7.4
2.5 mm rostrally	24.5 ± 5.7	30.4 ± 12.7	28.2 ± 10	25.6 ± 6.3	23.8 ± 6.5	29.7 ± 9.2

Values represent mean $\pm$ SD (n = 10). *: significantly lower than initial value (randomisation test for paired data , $p < 0.05$). SVP: slow variation period (45 ± 16 min).

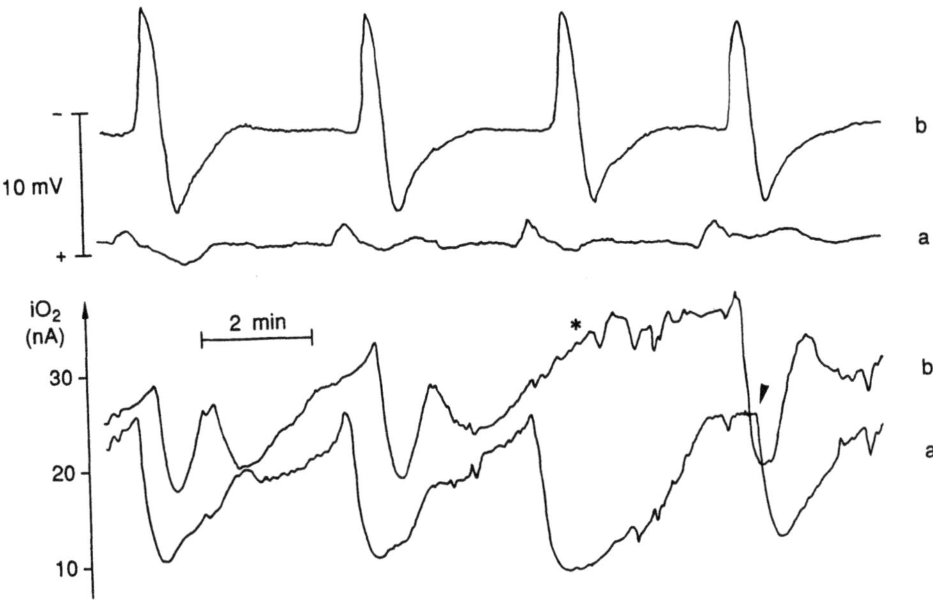

Figure 2. Simultaneous recording of iO$_2$ and DC-potential in the border zone of a photochemically induced infarction of the brain cortex. a: 0.5 mm rostral to the illuminated area; b: 2.5 mm rostral to the illuminated area. Electrodes for DC-potential measurement are positioned 1 mm medial to the oxygen electrodes. Recurrent transients of DC-potential are recorded, showing a temporal separation at the two sites. Except for one time (asterisk), these shifts are accompanied by a SWO$_2$ recorded with the adjacent oxygen electrode. A significantly delayed SWO$_2$ is noticed once (arrow).

of iO$_2$ followed by a decrease, but most of the SWO$_2$'s were multiphasic. For each animal, the pattern of the SWO$_2$'s at a specific site appeared to be constant but differences between the 2 locations were noticed regularly. The amplitudes of the different wave components as well as the total amplitudes of the SWO$_2$'s were variable in each animal but did not show a clear evolution in time. The total amplitudes of the SWO$_2$'s varied from 15 to 200 % of the initial current. Low values of iO$_2$ comparable with residual current were occasionally reached but never maintained. Peak levels of iO$_2$ reached values between 100 % and 215 % of the initial current. The duration of a single SWO$_2$ varied from 2 to 4 min.

Changes of DC-Potential and their Relation to SWO$_2$'s

The resting DC-potential before infarction was 4.5 ± 3.2 mV (n = 6). The maximal drift of this resting potential during 30 min before infarction was 1.7 mV.

In both animals transient shifts of the DC-potential were recorded at both sites of the border zone (Figure 2). Almost all transients at the proximal site were followed by an accompanying change at the distal site after a lag time of about 40 s. In one animal a few solitary transients appeared at the proximal site. In both animals the changes at the distal site were biphasic showing an initial negative wave of respectively 9 and 7.5 mV, lasting 30 - 60 s. This negative deflection was succeeded by a positive wave of respectively 5.5 and 2 mV, lasting 60 -100 s. At the proximal site the positive deflection was not always recorded and total amplitudes (respectively 2.6 mV and 4.0 mV) were significantly lower compared with those at the distal site. Almost every wave of the DC-potential was immediately followed by a SWO$_2$ recorded by the adjacent oxygen electrode. Only exceptionally, the accompanying SWO$_2$ was absent or delayed (Figure 2).

494

Taking into account the slope of the plateau of the in vivo recorded polarograms (range 0.8 - 2.2 nA/100 mV), the variation of iO_2 due to changes of the DC-potential maximally amounted to 0.16 - 4.4 % of the amplitudes of the accompanying SWO_2's.

Following the SVP, DC-potentials remained at resting potential (maximal drift: 0.7 mV).

DISCUSSION

Close to the edge of the photochemically induced infarctions as well as at a larger distance electrode current (iO_2) consistently showed recurrent slow waves (SWO_2), starting when the mean local PO_2 in the centre of the infarction area had dropped to at least 55 % of its initial value and ending within the following hour. In most cases a wave at the proximal site was followed within a minute by a wave at the distal site indicating a phenomenon spreading from the infarct area to the surrounding tissue. In the two animals used for the simultaneous measurement of mean cortical extracellular DC-potential and PO_2, each SWO_2 at a particular site appeared to be accompanied by a slow change of DC-potential recorded with the adjacent electrode. According to the in vivo recorded polarograms the influence of the DC-potential changes on electrode current appeared to be negligible in comparison with the large amplitudes of the SWO_2's. Moreover, in one case a change of DC-potential was recorded without any significant change of iO_2. Hence, we may conclude that the shifts of iO_2 reflect real PO_2 changes.

The shifts in DC-potential resemble cortical spreading depression (CSD)-like slow waves of electrical potential as described by Leão (1951). CSD is a transient profound depression of neural activity which spreads across the cortical surface at a velocity of 2 - 5 mm/min (Leão, 1944), and is characterised by transient increases of extracellular K^+ and intracellular Ca^{2+}. The temporal separation of the DC-potential waves recorded at both locations in our study is in accordance with CSD spreading from the infarction area to the surrounding tissue. The fact that we measured the mean potential over 1 mm of cortical thickness may explain the rather small amplitudes of the shifts in DC-potential. Indeed, CSD preferentially propagates in upper cortical layers and amplitudes of DC-potential waves recorded into the deeper cortical layers are small (Leão, 1951). Yet, the amplitudes recorded at 0.5 mm from the illuminated area are very small. We suggest that at this location the tissue surrounding the electrode is threatened or partially damaged and that ion pump activity is reduced. Moreover, extracellular K^+ at the edge of the infarction may be continuously increased (Nedergaard, 1988). Such increase lowers the treshold for CSD elicitation (Nedergaard and Astrup, 1986) so that the most likely point of origin of the CSD waves is at the margin of the infarction and may therefore be located very close to the proximal electrode.

The recorded CSD-like changes of electric potential always either preceded the accompanying SWO_2 or occurred at the same time. SWO_2's clearly preceding accompanying changes of DC-potential were never noticed. We therefore suggest that the observed SWO_2's are initiated by CSD-like events. Vascular responses might be induced by elevated levels of extracellular K^+ (Hansen et al., 1980; Edvinsson et al., 1992) and intracellular Ca^{2+} during CSD. The latter might for example activate the synthesis of vasoactive substances such as thromboxane and prostacyclin (Siesjö, 1981) and stimulate the nitric oxide synthetase in nitroxidergic neurones (Toda and Okamura, 1990). CSD also alters the local cerebral metabolic rate of oxygen: oxidative phosphorilation increases during repolarisation of the DC-potential shift (Rosenthal and Somjen, 1973) and is associated with increased ion pumping (La Manna and Rosenthal, 1975). As both tissue perfusion and oxygen consumption affect local tissue PO_2, additional monitoring of one of these parameters would be obligatory in order to determine their respective contribution to the observed SWO_2's. Figure 2 clearly shows a CSD-like change of electric potential which is not accompanied by any adjacent SWO_2. A considerable delay between a CSD wave and the accompanying SWO_2 was recorded once. Whatever the underlying mechanisms of these events might be, they may account for the exceptional absence of a distal SWO_2 and the occasional occurrence of a proximal SWO_2 coming after a distal one.

The multiphasic pattern of the SWO_2's might point at a vascular reaction inducing pulsatile flow. The latter could improve both tissue oxygenation (Intaglietta, 1991) and elimination of metabolites (e.g. lactate) thereby acting as a protective mechanism in the border zone of the infarction. As anesthesia may reduce or eliminate such beneficial cerebrovascular response, stroke models using awake animals might reproduce pathophysiologic mechanisms during focal brain infarction more accurately.

ACKNOWLEDGEMENTS

This work was supported by FGWO contract 3.0023.91 (Belgian Fund for Medical Scientific Research) and by the OZR VUB.

REFERENCES

Edvinsson, L., MacKenzie, E.T. and McCullogh, J., 1993, Vascular smooth muscle reactivity in vitro and in situ, *in*: "Cerebral Blood Flow and Metabolism", L. Edvinsson, ed., Raven Press, New York.

Hansen, A.J., Quistorff, B. and Gjedde, A., 1980, Relationship between local changes in cortical blood flow and extracellular K^+ during spreading depression, *Acta Physiol. Scand.*, 109:1-6.

Intaglietta, M., 1991, Arteriolar vasomotion: implications for tissue ischemia, *Blood Vessels*, 28:1-7.

LaManna, J.C. and Rosenthal, M., 1975, Effect of ouabain and phenobarbital on oxidative metabolic activity associated with spreading cortical depression in cats, *Brain Res.*, 88:145-149.

Leão, A.A.P., 1944, Spreading depression of activity in the cerebral cortex, *J. Neurophysiol.*, 7:359-390.

Leão, A.A.P., 1951, The slow voltage variation of cortical spreading depression of activity, *Electroencephalogr. Clin. Neurophysiol.*, 3:315-321.

Nedergaard, M., 1988, Mechanisms of brain damage in focal cerebral ischemia, *Acta Neurol. Scand.*, 77:81-101.

Nedergaard, M. and Astrup, J., 1986, Infarct rim: effect of hyperglycemia on direct current potential and [^{14}C]2-deoxyglucose phosphorilation, *J. of Cerebral Blood Flow and Metabolism*, 6:607-615.

Rosenthal, M. and Somjen, G., 1973, Spreading depression, sustained potential shifts, and metabolic activity of cerebral cortex of cats, *J. Neurophysiol.*, 36:739-749.

Siesjö, B.K., 1981, Cell damage in the brain : a speculative synthesis, *J. of Cerebral Blood Flow and Metabolism*, 1:155-185.

Toda, N. and Okamura, T., 1992, Regulation by nitroxidergic nerve of arterial tone, *News In Physiol. Sci.*, 7:148-152.

van Rossem, K., Vermariën, H. and Bourgain, R., 1992 a, Construction, calibration and evaluation of pO_2 electrodes for chronical implantation in the rabbit brain cortex, *Adv. Exp. Med. Biol.*, 316:85-101.

van Rossem, K., Vermariën, H., Decuyper, K. and Bourgain, R., 1992 b, Photothrombosis in rabbit brain cortex: follow up by continuous pO_2 measurement, *Adv. Exp. Med. Biol.*, 316:103-112.

van Rossem, K., Vermariën, H., Decuyper, K., Van Reempts, J., Laureys, M. and Bourgain, R., 1992 c, Local tissue PO_2 during and after focal brain cortical infarction in rabbits, *Adv. Exp. Med. Biol.*, 317:717-722.

Watson, B.D., Dietrich, W.D., Busto, R., Mitchell, S., Wachtel, B.S. and Ginsberg, M.D., 1985, Induction of reproducible brain infarction by photochemically initiated thrombosis, *Ann. Neurol.*, 17:497-504.

INCREASED BASIC FIBROBLASTIC GROWTH FACTOR mRNA IN THE BRAINS OF RATS EXPOSED TO HYPOBARIC HYPOXIA

Joseph C. LaManna[1], Keith D. Boehm[2], Vladimir Mironov[1], Antal G. Hudetz[3], Martin A. Hritz[1], Jong K. Yun[2], and Sami I. Harik[1]

[1]Department of Neurology, [2]Department of Medicine, Case Western Reserve University, Cleveland, OH 44106, U.S.A.; and,
[3]Department of Physiology, Medical College of Wisconsin, Milwaukee, WI 53226, U.S.A.

INTRODUCTION

Continued exposure of rats to hypobaric hypoxia results in both functional and structural changes in brain. A most striking structural change is the 50-80% increased capillary density in the cerebral cortex and other brain regions after 3 weeks of hypoxic exposure (LaManna et al., 1992; LaManna, 1992). The apparent increased density may be accompanied by a shift in the capillary segment length distribution towards longer capillary segments (LaManna et al., 1993). It is still not clear if the increased capillary density is due to new capillary branching or to elongation of existing capillaries, or both. Nevertheless, it seems likely that the microvascular adaptation to continued hypoxia is controlled by growth factors. The growth factors that control angiogenesis are not yet well understood, but basic fibroblast growth factor (bFGF) is known to activate endothelial cells resulting in capillary growth (de Juan et al., 1990), and this growth factor is increased following metabolic stress such as brain ischemia (Finkelstein et al., 1990; Lyons et al., 1991). In this study, we determined whether bFGF mRNA was increased in brains of rats exposed to 3 weeks of hypoxia.

METHODS

<u>Hypoxic model</u>: Male Wistar rats aged 3 - 6 months (250 - 350 g) were kept for 1 or 3 weeks in hypobaric chambers maintained at 0.5 ATM except for 1 hour per day to feed and water and change the bedding. Littermate controls were kept in similar cages outside the chamber but in the same room (LaManna et al., 1992).

<u>Vascular casts</u>: In the first group, after 1 week of hypoxic exposure, rats were weighed and anesthetized with chloral hydrate (400 mg/kg, i.p.) and cerebrovascular casts made (Lametschwandtner et al., 1990). First, the left carotid artery was cannulated, then the chest cavity was rapidly opened and a cut made in the right atrium. Rats were perfused through the left carotid artery with 20 ml of heparinized physiological buffer solution

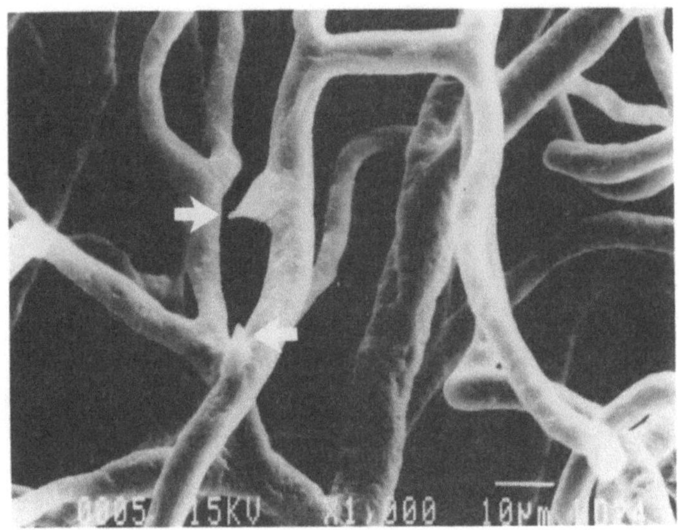

Figure 1

followed by freshly prepared mercox CL-2B (Oken Shoji Company, Japan). At the end of perfusion the neck muscles were clamped and the preparation was left at room temperature for 1 hour for the cast to harden. The head was later removed and incubated in water overnight at 50° C. The next day, the left supratentorial portion of the brain was carefully removed and sectioned in the coronal plane at 1.5-2mm intervals. Sections 5-10 mm

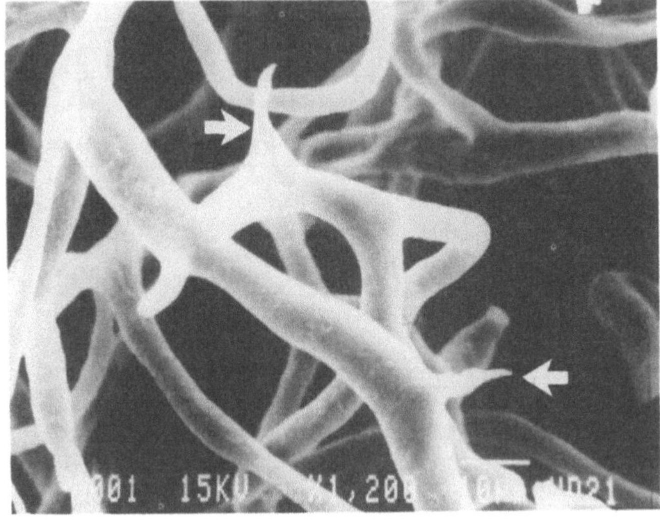

Figure 2

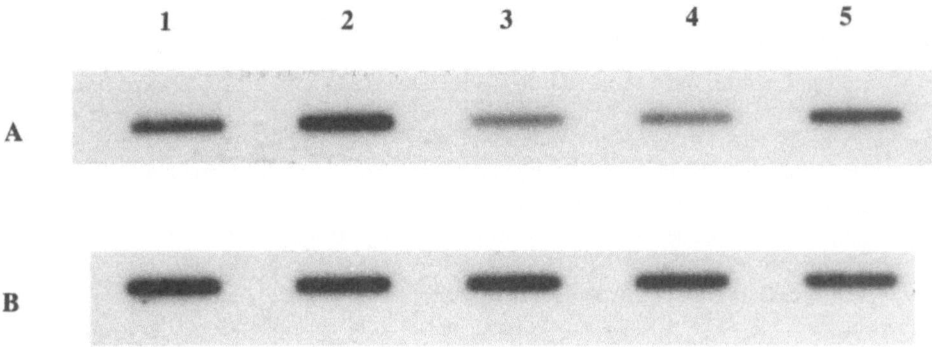

Figure 3

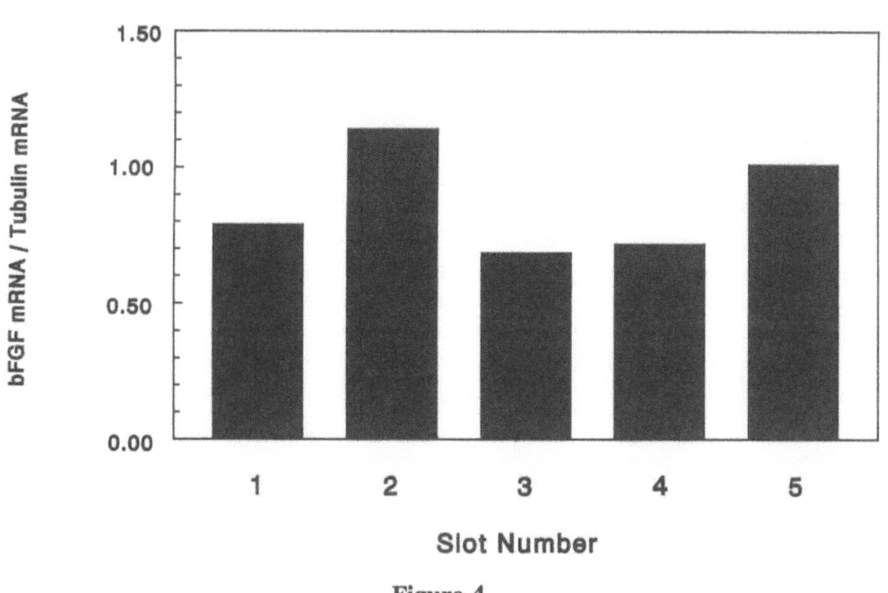

Figure 4

anterior to the interaural line were used for data analysis. The sections were placed in 20% NaOH solution for 2 weeks at 60°C, with frequent changes. Ultrasound agitation was used to speed the process of maceration. The clean corrosion casts were washed in distilled water and air dried. Alternate sections of the dried casts were mounted on stubs with silver conductive glue and coated with gold in a sputtering device (IB3 ion coater, RMC-Eiko Corp.). These casts were examined using a JSM-840A scanning electron microscope (JEOL) at an accelerating voltage of 10-20 kV. Qualitative appraisal of the cerebral cortex was undertaken with special attention to the dorsolateral region of the parietal cerebral cortex.

In situ hybridization: The second group of rats was taken after 3 weeks exposure to hypoxia. The hypoxic rats and their normoxic controls were anesthetized with ether, decapitated, and brains were removed to dry ice and frozen. The brains were sectioned at 5 μm in a cryostat microtome. Serial sections were adhered to poly-L-lysine microscope slides pretreated with Vectabond™ (Vector Labs). A 40-mer conserved coding region specific antisense b-FGF DNA oligonucleotide was used as probe for b-FGF mRNA in situ. Background, non-specific binding controls included: 1) complementary 40-mer sense DNA oligonucleotide probed adjacent sections, and 2) sections pretreated with ribonuclease A. Probes were 5'-end labeled with γ-[^{32}P]ATP to a specific activity of 1 x 10^9 dpm/μg DNA. Hybridization was performed in a solution of 1 M NaCl, 50 mM Tris-HCl, 10% dextran sulfate, 1% SDS, 100 μg/ml sheared, denatured salmon sperm DNA for 18 hrs at 45°C followed by two washes in 0.3 M NaCl/30 mM sodium citrate for 30 min each at 50°C and once for 60 min at 55°C. Slides were then coated with Ilford K.2 nuclear track emulsion and left at -20°C for 3 wks. Autoradiographic grains were developed using Kodak D-19 and the sections were counterstained with hematoxylin and eosin. Some sections were also immunostained for glial fibrillary acidic protein.

mRNA Slot Blot: Poly A$^+$ RNA was isolated from samples of the cerebral cortex taken from the second group of rats that had been kept frozen at -80°C, using a Fasttrack extraction kit (Invitrogen) as described previously (Urata et al., 1993). Ten μg of the resultant poly A$^+$ RNA was applied directly onto nitrocellulose to each slot in a Biorad vacuum apparatus and crosslinked by ultraviolet radiation. Then, γ-[^{32}P]ATP 5'-end labeled oligo 40-mer probe was applied in a solution of 1 M NaCl, 50 mM Tris-HCl (pH 7.5), 10% dextran sulfate, 1% SDS, 100 μg/ml sheared and denatured salmon testes DNA for 18 hrs at 65°C. Following washes in 2xSSC/0.1% SDS at 65°C and then at 25°C, damp nitrocellulose was then exposed to Kodak XAR-B x-ray film for 18 hr in the case of bFGF mRNA probed samples and 6 hr for the α-tubulin mRNA probed samples. Probe specific activities were 1.0 x 10^8 dpm/μg DNA. The blot was bFGF mRNA probed first, then dehybridized and reprobed for α-tubulin mRNA. The data from the blots were analyzed by image analysis using the NIH image software. Images were thresholded and binarized. The area of each spot on the blot was then recorded. The area of the α-tubulin mRNA spot (not expected to increase with hypoxia), was used to normalize the data from the bFGF mRNA spot.

RESULTS AND DISCUSSION

Vascular casts: Scanning electron micrographs of two regions from the superficial parietal cortex (layers 1 - 3) from a rat exposed to 1 week of 0.5 ATM are shown in figures 1 and 2. Evidence for capillary sprouting is indicated by the conus shaped profiles (arrows). Such profiles were neither evident in samples from control rats, nor from rats exposed to 3 weeks of hypoxia, despite the increased capillary density in the cerebral cortex of rats exposed to 3 weeks of hypoxia. This suggests that the increased capillary density seen at 3 weeks in the superficial cerebral cortex is at least partly explained by new capillary branching. Capillary segment length distribution in this region of the cerebral cortex after 3 weeks of hypoxic exposure was unchanged (unpublished observations). Deeper cortical layers of rats exposed to hypobaric hypoxia also show an increased capillary density at 3 weeks, but without evidence of the capillary sprouting that we observed after 1 week. This could mean that branching occurs before 1 week in the deeper cortical layers and is already completed at the time the brain samples were taken, or that the mechanism for increased capillary density in this region of cortex is due to endothelial cell elongation. This latter explanation is more consistent with the finding of increased capillary segment length distributions in deeper cerebral cortex (LaManna et al., 1993).

<u>In situ hybridization</u>: Tissue examination revealed accumulations of autoradiographic grains in brain sections from all rats, normoxic and hypoxic. However, the sections from hypoxic rats had much greater density of grains. No particular pattern of distribution by brain region or by cortical layers was observed, suggesting an overall increase in bFGF message. Thus, we have demonstrated an increase in bFGF mRNA in hypoxia. It remains probable that bFGF is not the only growth factor that responds to hypoxia and it is not definite that bFGF is the angiogenic growth factor, since other growth factors (such as vascular endothelial growth factor) are also stimulated by hypoxia (Shweiki et al., 1992).

<u>mRNA slot blots</u>: The results of the slot blot study are shown in figure 3. Five different rat brains were sampled. Lane 1 represents a control rat brain from an independent group of rats; the other lanes represent brain samples from the same litter. Lanes 2 and 5 represent samples from 2 rats hypoxic for 3 weeks, while lanes 3 and 4 represent samples from 2 normoxic littermate rats. Row A shows the results of the bFGF probe. Even before normalization to α-tubulin (row B), it is apparent that slots 2 and 5 have the highest signal. The result of normalization to α-tubulin is shown in the graph of figure 4. Again, lanes 2 and 5 contain higher ratios of bFGF mRNA to α-tubulin mRNA. These results suggest that there is continued increase in bFGF mRNA levels after a 3 week exposure to hypoxia. We speculate that the increased growth factor message persists for as long as the hypoxic stimulus is present. Furthermore, we hypothesize that the ability of the brain to maintain increased capillary density is dependent upon the continued expression of growth factors.

ACKNOWLEDGEMENT

These studies were supported by PHS grants HL42215 and HL25830.

REFERENCES

de Juan, E., Stefansson, E., and Ohira, A., 1990, Basic fibroblast growth factor stimulates ^{3}H-thymidine uptake in retinal venular and capillary endothelial cells in vivo, <u>Invest. Ophthalmol. Vis. Sci.</u>, 31: 1238-1244.

Finkelstein, S.P., Caday, C.G., Kano, M., Berlove, D.J., Hsu, C.Y., Moskowitz, M., and Klagsbrun, M., 1990, Growth factor expression after stroke, <u>Stroke</u>, 21 (suppl III): III-122-III-124.

LaManna, J.C., 1992, Rat brain adaptation to chronic hypobaric hypoxia, in: "Oxygen Transport to Tissue XIV (Advances in Experimental Medicine and Biology, v.317)," W. Erdmann and D.F. Bruley, eds., pp. 107-114, Plenum Press, New York.

LaManna, J.C., Vendel, L.M., and Farrell, R.M., 1992, Brain adaptation to chronic hypobaric hypoxia in rats, <u>J. Appl. Physiol.</u>, 72: 2238-2243.

LaManna, J.C., Cordisco, B.R., Kneuse, D.E., and Hudetz, A.G., 1993, Increased capillary segment length in cerebral cortical microvessels of rats exposed to 3 weeks of hypobaric hypoxia, in: "Oxygen Transport To Tissue XV (Advances in Experimental Medicine and Biology, v.)," P. Vaupel, ed., pp. (in press)Plenum Publishing Corp., New York.

Lametschwandtner, A., Lametschwandtner, U., and Weiger, T., 1990, Scanning electron microscopy of vascular corrosion casts; technique and applications: Updated review, <u>Scan. Microsc</u>, 4: 889-941.

Lyons, M.K., Anderson, R.E., and Meyer, F.B., 1991, Basic fibroblast growth factor promotes in vivo cerebral angiogenesis in chronic forebrain ischemia, <u>Br. Res.</u>, 558: 315-320.

Shweiki, D., Itin, A., Soffer, D., and Keshet, E., 1992, Vascular endothelial growth factor induced by hypoxia may mediate hypoxia-initiated angiogenesis, <u>Nature</u>, 359: 843-845.

Urata, H., Boehm, K.D., Philip, A., Kimoshita, A., Gabrousek, J., Bumpus, F.M., and Husain, A., 1993, Cellular localization and regional distribution of an angiotensin II-forming chymase in the heart, <u>J. Clin. Invest.</u>, 91: 1269-1281.

OXYGEN SUPPLY TO EXERCISING MUSCLE: ROLES OF DIFFUSION LIMITATION AND HETEROGENEITY OF BLOOD FLOW

Johannes Piiper

Max-Planck-Institut für experimentelle Medizin,
D-37075 Göttingen, Germany

In maximum O_2 uptake exercise, the O_2 uptake is in many cases probably limited by O_2 availability which is determined by blood flow and its distribution, by arterial O_2 content and partial pressure (P_{O_2}), and by blood/tissue diffusion conditions. In this report it is attempted to analyze the role of these factors, on the basis of model calculations and experimental data.

Diffusion limitation

In muscles during maximum O_2 uptake ($\dot{V}_{O_2}$ max), tissue P_{O_2}, at least in the areas least accessible by diffusion, is assumed to be close to zero. The fact that muscle venous P_{O_2} and O_2 content do not approach zero in conditions of increased or maximal O_2 uptake has been attributed to diffusion limitation (Mercker et al., 1949; Stainsby and Otis, 1964; Stainsby et al., 1988; Hogan et al., 1988, 1989; Roca et al., 1989). The Krogh cylinder model and the solid cylinder model have been used calculate critical capillary densities and other diffusion limitation parameters from measured variables.

The Krogh cylinder model may be further simplified for assessment of diffusion limitation. Because of the geometry, most part of the resistance to O_2 uptake resides in the central, pericapillary region. The preponderance of this central resistance is further enhanced by diffusion and reaction resistance within red cells and plasma, whereas the resistance to O_2 transport within the muscle fibers is reduced by facilitated transport by myoglobin (cf. Groebe and Thews, 1990). Thus the radial P_{O_2} gradient for O_2 diffusion may be considered to be mainly located in the intracapillary and pericapillary zone, to be constant along the length of a Krogh cylinder and, in critical O_2 supply conditions, to be equal to the end-capillary or venous P_{O_2} (because the corresponding tissue $P_{O_2} = 0$):

$$\dot{V}_{O_2} \, \text{max} = D_{O_2} \cdot P_{V_{O_2}} \tag{1}$$

The proportionality factor is the effective O_2 conductance or the diffusing capacity for O_2, D_{O_2}.

In rhytmically stimulated dog gastrocnemius muscle, Hogan et al. (1988) found a close to proportional relationship between O_2 uptake and venous P_{O_2} as the O_2 delivery to the muscle stimulated to maximum O_2 uptake was progressively decreased by reduction of arterial P_{O_2} at constant blood flow. There was only a slight increase in the O_2 uptake/$P_{V_{O_2}}$ ratio as hypoxia progressed. A similar result had been obtained by Stainsby and Otis (1964), but with more pronounced increase of the ratio O_2 uptake/venous P_{O_2} with decreasing arterial P_{O_2}. Since these results were expected on the basis of the models, the limitation to O_2 extraction was attributed to diffusion resistance, with diffusion conditions remaining constant or slight improving with progressing hypoxia.

In the homogeneous model, muscle tissue is assumed to have everywhere the same P_{O_2}. This may come about by axial facilitated transport of O_2 and by interaction between neighboring

Oxygen Transport to Tissue XVI
Edited by M.C. Hogan *et al.*, Plenum Press, New York, 1994

capillaries. Indeed, Gayeski and Honig (1986, 1988) found by cryophotometric determination of myoglobin O_2 saturation only small radial and longitudinal P_{O_2} variations in dog gracilis muscle fibers. In this simplified model, blood-capillary O_2 transfer is described by the model conventionally used for analysis of alveolar-capillary diffusion in lungs, with tissue P_{O_2} formally replacing alveolar P_{O_2}. The P_{O_2} difference driving the diffusion is in this case the difference between the mean capillary P_{O_2} and the tissue P_{O_2}. The mean capillary P_{O_2} may be determined by the Bohr integration technique in the same manner as for pulmonary capillaries. The relationship, analogous to eq. (1), is obtained ($P\bar{c}_{O_2}$, mean capillary P_{O_2}):

$$\dot{V}_{O_2} \max = D_{O_2} \cdot P\bar{c}_{O_2} \tag{2}$$

Application of this relationship to the above-mentioned data of Hogan et al. (1988) yields a reasonably proportional relationship also between mean capillary P_{O_2} and O_2 uptake, again showing consistency with the diffusion limitation concept.

Perfusion heterogeneity and shunt: experimental evidence

In a number of studies using various preparations and methods (cf. Piiper, 1990) distribution of blood flow in mammalian skeletal muscle has been found to be unequal. In our laboratory, measurements were made on the in situ isolated dog gastrocnemius muscle at rest and during supramaximal stimulation to rhythmic isotonic tetanic contractions (o.2 sec duration, 30-60 per minute) against varied loads, leading to varied increases in blood flow and O_2 uptake. In the following, the results are summarized.

Washout of inert gases. After equilibration of the muscle preparation with one or several inert gases (He, Ar, CH_4, SF_6), washout was performed by switching the perfusion to blood devoid of the inert gas(es). The decay of the inert gas concentration in venous outflow was markedly non-monoexponential both at rest and during stimulation (Piiper and Meyer, 1984). Similar results, but with even wider scatter of exponential rate constants, have been found in the resting dog gracilis (Grønlund et al., 1989). The simplest explanation is unequal distribution of blood flow to tissue volume. Moreover, a mass balance analysis based on estimated values of the tissue/blood partition coefficient, showed that about 30% of the total blood flow had to be attributed to shunt. A comparison of the washout of test gases with differing diffusivity suggested a role of counter-current veno-arterial back diffusion (Piiper, 1988).

Clearance of locally injected xenon. A small amount of saline containing ^{133}Xe was injected into the muscle at varied sites, and the washout clearance was measured by a counter placed on the injection site. The specific blood flow was calculated from the radioactive decay and the blood/tissue partition coefficient of xenon. The ratio of the blood flow calculated from Xe clearance to the directly measured venous outflow averaged 57% and scattered widely (SD/mean = 33%) (Cerretelli et al., 1984). The large scatter may in part have been due to methodological problems, but mainly it appears to indicate perfusion inhomogeneity. The low blood flow from Xe clearance may be interpreted as due to a large fraction of functional shunt flow (which may have been produced by diffusion limitation local overperfusion or veno-arterial back diffusion).

Microsphere injection. After multiple intraarterial injections of differently labeled microspheres (mean diameter 15 µm) into the arterial inflow, the muscle was cut into 180 to 250 pieces of about 0.75 g each and the number of embolizing microspheres of each label was determined in each piece by differential radioactivity counting techniques. The specific blood flow was assumed to be proportional to the specific radioactivity of the muscle piece. A highly inhomogeneous blood flow distribution was obtained, both at rest and even more so during stimulation (Piiper et al., 1985; Marconi et al., 1988). Similar results were obtained in muscles of intact, resting and running dogs (Pendergast et al., 1985) and on rabbit muscles (Iversen and Nicolaysen, 1989; Iversen et al., 1989).

Unequal distribution of blood flow and shunt: effects on O_2 supply

To estimate the effects of unequal distribution of blood flow on O_2 supply, calculations were performed on simple models with unequal distribution of blood flow and with shunt, but without diffusion limitation, which were roughly matched to the experimental results (Piiper and Haab, 1991).

In the reference model (homogeneous blood flow), as the O_2 delivery (= blood flow x arterial O_2 content) is reduced or the O_2 requirement is increased, the O_2 uptake is equal to O_2 requirement and then starts falling as the ratio O_2 delivery/O_2 requirement ratio drops below unity and venous O_2 reaches zero (critical point). In the model with unequal blood flow distribution, the ratio O_2 uptake/O_2 requirement falls continuously, together with venous O_2 content, as the ratio O_2 delivery/O_2 requirement is reduced. Instead of a critical point, there is a broad critical range.

Of particular interest is the behavior with decreasing arterial O_2 content. With both unequal blood flow and shunt, O_2 uptake and venous O_2 content decrease continuously as the arterial O_2 content is reduced. Similar behavior is found in the model with shunt because with constant shunt, venous O_2 content is proportional to arterial O_2 content when O_2 extraction by tissue is complete. Thus the experimental finding of maximal $\dot{V}_{O_2}$ decreasing with venous P_{O_2} (Hogan et al., 1988) when arterial P_{O_2} was progressively reduced may well have been, at least in part, due to unequal blood flow distribution and/or shunt. For a distinction between the roles of blood flow distribution, shunt and diffusion limitation, accurate measurements by simultaneous application of various methods is required.

Unequal distribution of the metabolic rate

In the above-mentioned studies the reference for blood flow measurement was unit volume (mass) of tissue. However, according to model requirements, the reference should be O_2 requirement.

Therefore, recent experimental data obtained by Iversen and Nicolaysen (1990, 1991a) are of particular interest, in which simultaneously with microsphere distribution, the spartial distribution of dexyglucose uptake, as measure of glucose metabolic rate, was determined. It was found that in rabbit hindlimb muscles, both at rest and during stimulation, deoxyglucose uptake was inhomogeneously distributed, as was blood flow. But, contrary to expectation, there was no correlation between both parameters. Thus the coefficient of variation of the ratio blood flow/metabolic rate appeared to be larger than that of blood flow/tissue volume or of glucose metabolic rate/tissue volume. Although glucose uptake is not necessarily directly related to O_2 uptake, the spartial inhomogeneity of glucose uptake suggests presence of a spatial inhomogeneity of tissue O_2 consumtion.

A similar result was obtained in experiments on cat muscles in which no correlation was found between distributions of blood flow and citric synthase activity, which is assumed to indicate the metabolic turnover rate of the Krebs cycle (Iversen et al., 1992).

Unequal distribution of diffusing capacity

The effects of an unequal distribution of O_2 diffusing capacity (D) to blood flow ($\dot{Q}$) has been studied in lung models (cf. Piiper, 1992a). The general effect is lowering of overall O_2 exchange efficiency of lungs. The factors possibly responsible such inhomogeneity are unequal dimenious of small pulmonary blood vessels. Thus a short capillary would have decreased D and increased $\dot{Q}$ and thus a strongly decreased D/$\dot{Q}$ ratio. Recently, Yamaguchi et al. (1991) have been able to quantify the variation pattern of the D/$\dot{Q}$ ratio in patients' lungs.

In tissues the diffusing capacity per tissue volume may be considered, in first approximation, to be proportional to the capillary surface area per tissue volume, *i. e.* to number of capillaries per volume or number of capillaries per unit area (capillary density).
The effect of unequal capillary density, in terms of variability of Krogh cylinder radii in tissue models, have been investigated by Piiper and Scheid (1991). They showed that upon reduction of diffusive O_2 availability such models with cylinders of varied size display less abrupt changes in diffusion-limited O_2 uptake than cylinders of same mean, but uniform radius. The effect is thus qualitatively analogous to that of specific blood flow variability in perfusion-limited models (Piiper, 1992a, 1992b). Thus the variability of capillary distance may account for part of the parallel fall of O_2 uptake and venous P_{O_2} with falling O_2 delivery in stimulated muscle preparations.

In this connection, the recent experimental findings of Iversen and Nicolaysen (1991 b) are of interest. They found that uptake rates of vitamin B 12 and albumin were unequally distributed within single rabbit skeletal muscles. These uptake rates may be related to capillary surface area, and thus to diffusing capacity for O_2. Again there was no correlation with distribution of blood flow as measured by microsphere embolization.

General Remarks

Accurary of methods. The accuracy of the microsphere method in general, and in particular at high spatial resolution, is still questionable, in spite of the good correlation with distribution of a molecular marker (which is completely extracted from blood in a single tissue passage) found in rabbit heart (Little and Bassingthwaighte, 1983; Bassingthwaighte et al., 1987). But the presence of flow heterogeneity itself, given the large coefficients of variation, is hardly to be doubted, although the quantitative extent may be questionable. .

Sample size. The size of the muscle samples, 0.75 g in experiments of the Göttingen groups, 0.25 g in those of Oslo, must be an important parameter. It was chosen on grounds of feasibility. Such pieces are expected to contain millions of capillaries. Thus comparison with observations of red cell velocities in single muscle capillaries (in hamster tibialis anterior, Damon and Duling, 1985; in rat extensor digitorum longus, Tyml, 1991) appears to be devoid of any justification. However, it is remarkable that the order of magnitude of the coefficient of variation, about 20%-100%, found in these preparations is similar to the heterogeneity of microsphere distribution among 0.25-0.75 g muscle pieces. Application of the fractal approach to the subject, (e. g. Bassingthwaighte, 1988) is expected to provide a basis for description and explanation of heterogeneities in blood flow.

Mechanisms of heterogeneities. The blood flow heterogeneity may be due to anatomical or functional factors. The changes of blood distribution pattern with time were documented, but showed a highly varied extent in individual experiments (Piiper et al., 1985) suggesting involvement of periodic local vasomotion. But extension of microsphere infusion times from 10 sec up to 45 min did not decrease the coefficient of variation of microsphere distribution. This result argues against the role of local blood flow oscillations (Iversen and Nicolaysen, 1990).

Conclusions

Experimental evidence for heterogeneity of resting and stimulated skeletal muscle in terms of blood flow, oxidative metabolism and blood/tissue diffusion is reviewed. This heterogeneity is expected to limit the overall O_2 availability in a manner similar to diffusion limitation. The roles of diffusion limitation and flow heterogeneity are technically difficult to differentiate and discrimination becomes difficult even in theory when spartial inhomogeneity of diffusing conditions is incorporated.

References

Bassingthwaighte, J.B., 1988, Physiological heterogeneity: fractals link determinism and randomness in structures and functions, *News Physiol. Sci.*, 3: 5-10.

Bassingthwaighte, J.B., Malone, M.A., Moffett, T.C., King, R.B., Little, S.E., Link, J.M. and Krohn, K.A., 1987, Validity of microsphere depositions for regional myocardial flows. *Am. J. Physiol.*, 253 (Heart Circ. Physiol. 22): H184-H193.

Cerretelli, P., Marconi, C., Pendergast, D., Meyer, M., Heisler, N. and Piiper, J., 1984, Blood flow in exercising muscles by xenon clearance and by microsphere trapping, *J. Appl. Physiol.*, 56: 24-30.

Damon, D.H. and Duling, B.R., 1985, Evidence that capillary perfusion heterogeneity is not controlled in striated muscle, *Am. J. Physiol.*, 249: (Heart Circ. Physiol. 18): H386-H329.

Gayeski, T.E.J. and Honig, C.R., 1986, O_2 gradients from sarcolemma to cell interior in red muscle at maximal V_{O_2}, *Am. J. Physiol.*, 256 (Heart Circ. Physiol. 20): H789-H799.

Gayeski, T.E.J. and Honig, C.R., 1988, Intracellular P_{O_2} in long axis of individual fibers in working dog gracilis muscle, *Am. J. Physiol.*, 254 (Heart Circ. Physiol. 23): H1179-H1186.

Groebe, K. and Thews, G., 1990, Role of geometry and anisotropic diffusion for modelling P_{O_2} profiles in working red muscle, *Respir. Physiol.*, 79: 225-278.

Gronlund, J., Malvin, G.M., Hlastala, M.P., 1989, Estimation of blood flow distribution in skeletal muscle from inert gas washout, *J. Appl. Physiol.*, 66: 1942-1955.

Hogan, M.C., Roca, J., Wagner, P.D. and West, J.B., 1988, Limitation of maximal O_2 uptake and performance by acute hypoxia in dog muscle in situ, *J. Appl. Physiol.*, 65: 815-821.

Hogan, M.C., Roca, J., West, J.B. and Wagner, P.D., 1989, Dissociation of maximal O_2 uptake from O_2 delivery in canine gastrocnemius in situ, *J. Appl. Physiol.*, 66: 1219-1226.

Iversen, P.O., Standa, M., and Nicolaysen, G., 1989, Marked regional heterogeneity in blood flow within skeletal muscle at rest and during exercise hyperemia in the rabbit, *Acta Physiol. Scand.*, 136: 17-28.

Iversen, P.O. and Nicolaysen, G., 1989a, Heterogeneous blood flow distribution within single skeletal muscles of the rabbit: role of vasomotion, sympathetic nerve activity and effect of vasodilation, *Acta Physiol. Scand.*, 137: 125-133.

Iversen, P.O., and Nicolaysen, G., 1990, The distribution of blood flow and glucose uptake within single skeletal muscles in the awake rabbit, *Acta Physiol. Scand.*, 140: 373-381.

Iversen, P.O. and Nicolaysen, G., 1991a, Local blood flow and glucose uptake within resting and exercising rabbit skeletal muscle, *Am. J. Physiol.*, 260: H1795-H1801.

Iversen, P.O., and Nicolaysen, G., 1991b, Regional distributions of blood flow and tissue uptake rates of vitamin B 12 and albumin within single rabbit skeletal muscles, *Acta Physiol. Scand.*, 143: 311-320.

Iversen, P.O., Flatebø, T., and Nicolaysen, G., 1992, Uneven perfusion within single cat muscles: nitric oxide and citric synthase play no role, *Respir. Physiol.*, 89: 329-339.

Little, J.E. and Bassingthwaighte, J.B., 1983, Plasma-soluble marker for intraorgan regional flows, *Am. J. Physiol.*, 245: H707-H712.

Marconi, C., Heisler, N., Meyer, M., Weitz, H., Pendergast, D.R., Cerretelli, P., and Piiper, J., 1988, Blood flow distribution and its temporal variability in stimulated dog gasrrocnemius muscle, *Respir. Physiol.*, 74: 1-14.

Mercker, H., Ochwadt, B., and Schoedel, W., 1949, Der Einfluss der Erregungsfrequenz und der Belastung auf Durchblutung und Sauerstoffaufnahme des Muskels, *Pflügers Arch.*, 251: 73-82.

Pendergast, D.R., Krasney, J.A., Ellis, A., McDonald, B., Marconi, C., and Cerretelli, P., 1985, Cardiac output and muscle blood flow in exercising dogs, *Respir. Physiol.*, 61: 317-326.

Piiper, J., 1988, Role of Diffusion shunt in transfer of inert gases and O_2 in muscle, *in:* "Oxygen Transport to Tissue X" (Adv. Exp. Med. Biol. 222), M. Mochizuki, C.R. Honig, T. Koyama, T.K. Goldstick and D.F. Bruley, eds. Plenum Press, New York and London, pp. 55-61.

Piiper, J., 1990, Unequal distribution of blood flow in exercising muscle of the dog, *Respir. Physiol.*, 80: 129-136.

Piiper, J., 1992a, Modeling of oxygen transport to skeletal muscle: blood flow distribution, shunt , and diffusion, *in:* "Oxygen Transport to Tissue XIII", T.K. Goldstick et al., eds., Plenum Press, New York, pp. 3-10.

Piiper, J., 1992b, Oxygen supply by perfusion and diffusion in heterogeneous tissue models, *in:* "Oxygen Transport to Tissue XIV", W. Erdmann and D.F. Bruley, eds., Plenum Press, New York, pp. 623-627.

Piiper, J., 1992c, Diffusion-perfusion inhomogeneity and alveolar-arterial O_2 diffusion limitation: theory, *Respir. Physiol.*, 87: 349-356.

Piiper, J., and Haab, P., 1991, Oxygen supply and uptake in tissue models with unequal distribution of blood flow and shunt, *Respir. Physiol.*, 84: 261-271.

Piiper, J., and Meyer, M., 1984, Diffusion-perfusion relationship in skeletal muscle: model and experimental evidence from inert gas washout, *in:* "Oxygen Transport to Tissue V" (Adv. Exp. Med. Biol. 169), D.W. Lübbers, H. Acker, E. Lehniger-Follert and T.K. Goldstick, eds., Plenum Press, New York and London, pp. 457-466.

Piiper, J., and Scheid, P., 1991, Diffusion limitation of O_2 supply to tissue in homogeneous and heterogeneous models, *Respir. Physiol.*, 85: 127-136.

Piiper, J., Pendergast, D.R., Marconi, C., Meyer, M., Heisler, H., and Cerretelli, P., 1985, Blood flow distribution in dog gastrocnemius muscle at rest and during stimulation, *J. Appl. Physiol.*, 64: 241-251.

Roca, J., Hogan, M.C., Story, D., Bebout, D.E., Haab, P., Gonzalez, R., Ueno, O., and Wagner, P.D., 1989, Evidence for tissue diffusion limitation of V_{O_2} max in normal humans, *J. Appl. Physiol.*, 67: 291-299.

Stainsby, W.N. and Otis, A.B., 1964, Blood flow, oxygen tension, oxygen uptake, and oxygen transport in skeletal muscle, *Am. J. Physiol.*, 206: 858-866.

Stainsby, W.N., Snyder, B., and Welch, H.G., 1988, A pictographic essay on blood and tissue oxygen transport, *Med. Sci. Sports Exercise*, 20: 213-221.

Tyml, K., 1991, Heterogeneity of microvascular flow in rat skeletal muscle is reduced by contraction and by hemodilution, *Int. J. Microcirc. Clin. Exp.*, 10: 75-86.

Van Beek, J.H.G.M., Roger, S.A. and Bassingthwaighte, J.B., 1989, Regional myocardial flow heterogeneity explained with fractal networks, *Am. J. Physiol.*, 257 (Heart Circ. Physiol. 26): H1670-H1680.

Yamaguchi, K., Kawai, A., Mori, M., Asano, K., Takasugi, T., Umeda, A., Kawashiro, T. and Yokoyama, T., 1991, Distribution of ventilation and diffusing capacity to ventilation in the lung. *Respir. Physiol.*, 86: 171-187.

DOES ENERGY DEMAND HAVE AN ADDITIONAL CONTROL IN ISCHEMIA OR ARE CURRENT MODELS OF METABOLIC CONTROL ADEQUATE AT EXTREMES?

R.J. Connett, T.E.J. Gayeski, C.R. Honig

Departments of Physiology and Anesthesiology
University of Rochester Medical Center, Rochester, NY 14642

INTRODUCTION

It has been suggested[12] that the response of tissues to falling $[O_2]$ can fall in one of two classes: 1. "regulators", where $\dot{V}_{O_2}$ is constant in the face of falling PO_2 until it is too low to support any ATP production and 2. "conformers", where $\dot{V}_{O_2}$ decreases as $[O_2]$ decreases without a depletion of ATP. A brief review[6] described 3-4 different phases to the metabolic response of a "regulating" cell as $[O_2]$ decreases. Initially there may be no change at all. If oxygen supplies are more than adequate then $[O_2]$ can decrease to some turnover dependent level without affecting the metabolic state or rate of ATP turnover. The next phase shows no change in $\dot{V}_{O_2}$ but there is a change in metabolic state as reflected in decreases in the creatine charge ([PCr]/[total creatine], {PCr}) and changes in the cytosolic redox state as glycolysis is recruited. This phase has been documented in isolated mitochondria and supportive evidence seen in intact tissue, especially working muscle[19,20,21] As the $[O_2]$ decreases further, some critical point (PO_2crit) is reached where the concentration of oxygen limits the ability of the cell to produce ATP aerobically and the cells is dysoxic. The PO_2crit and the PO_2 range of dysoxia depends very much on the ATP turnover rate in the cell, being higher at higher turnover rates[6]. When there is no oxygen available (anoxia) the ATP supply will be severely limited by the capacity to produce ATP from glycolysis and transiently from PCr stores at rates consistent with the ATP consumption. In dysoxia and anoxia the sustainable ATP demand is limited both in rate and in duration.

We generally accept that the regulatory goal of energy metabolism is to maintain the balance between ATP supply and demand so that ATP levels or a closely related parameter is nearly constant. From the standpoint of metabolic regulation, oxygen "conformers" must have an additional control system that decreases the ATP demand, i.e. inhibits normal ATPases, in response to falling oxygen supply. Thus the cell ATP level is maintained not by using other sources but by decreasing demand. From the standpoint of the cell's economy this is not simply a problem of limited oxygen but of survival. If there is insufficient ATP for maintenance functions such as membrane ATPase activity then the cell can suffer irreversible damage. A true regulator thus risks cell death. An alternative strategy is to have controls that close down those demands that are not critical to survival and focus a decreased ATP supply on those that are. This implies a demand control that is not simply a reflection of the oxygen supply. A demand control which responded to PO_2 would be a conformer in the original sense. A demand control could respond to more general signals such as the adenine nucleotide charge, the creatine charge, cell pH, or cytosolic redox. Oxygen "conformers" are identified from observations that $\dot{V}_{O_2}$ of a whole tissue falls when PO_2 in the oxygen supply falls even though the delivered PO_2 is well above that thought to be limiting to mitochondria turnover[16]. This may not reflect an oxygen based control but some other more general demand control evoked by anoxia or dysoxia in the tissue. In other words, the limiting factors are diffusion and other parameters connecting circulatory supply to cell PO_2 rather than oxygen per se. It is necessary to have cell measurements of PO_2 associated with changes in ATP demand to document oxygen conforming behavior.

Working heart and skeletal muscle do appear to have a demand control. When [phosphate] is elevated and pH low due to limited meatbolic capacity either from a substrate limit of the absence of enzymes "fatigue" occurs[7,15,17,18,22]. The actinomyosin ATPase rate declines to a level sustainable by the metabolic ATP supply. The signal on the ATPase is not definitely known but appears to be related to tissue levels of pH and phosphate (Pi). These signals can reflect the state of the glycolytic system and the energy state of the cell. Cytosolic redox potentials, buffered by lactate pyruvate and glycolysis are correlated with $\dot{V}_{O_2}$[3,18]. Thus even the response to falling oxygen supply may be mediated through the glycolytic system.

Contractile activity is a dispensable activity as far as the survival of the muscle cell is concerned and a prime target for a demand control. We approached the question of whether there is conforming activity and a demand control in resting canine muscle. In this case ATPases more fundamental to cell survival must be inhibited. There are reports showing almost no change in phosphocreatine and no increase in cytosolic reduction as measured by the lactate pyruvate ratio after 30 minutes of circulatory occlusion in man[13,14]. This suggests a significant demand control operating in skeletal muscle at rest. Other studies showed a fall in PCr that was linear after ~ 5 minutes of occlusion[1] and elevation of lactate after ~ 10 minutes[11]. Those studies did not include a measurement of cellular PO_2 so it was not clear whether they reached a dysoxic or anoxic range. The studies reported here include both PO_2 and metabolite measurements on the same ischemic muscles.

METHODS

General

The preparation of the vascularly isolated dog gracilis was carried out as described previously[9]. Briefly, hound-type mongrels were anesthetized with pentobarbital sodium, 30 mg/kg body wt, intubated, paralized with pancuronium (2mg/kg) and respirated with room air. The gracilis muscle was vascularly isolated, covered with Saran Wrap$^{@}$ (an O_2 barrier) and maintained at 37°C by external heating. During the control period gracilis blood flow was measured by timed collections. Venous and arterial samples were collected simultaneously and $\dot{V}_{O_2}$ determined by the Fick principle. Ischemia was generated by clamping the femoral artery just above the entry to the gracilis. Muscle was quick frozen

510

in situ after times of occlusion ranging from 2 minutes to 60 minutes as described previously[10].

Myoglobin oxygen saturation was determined by cryomicrospectrophotometry on samples from each muscle. In several muscles hemoglobin cryomicrospectroscopy was also carried out to evaluate the extent at which oxygen from blood vessels was used to support metabolism under these conditions. Enzymatic determinations of PCr, creatine, ATP, phosphate (Pi), lactate and pyruvate were made on samples derived from the surfaces read for spectroscopy as described previously[5]. Sampling was carried out so that both microheterogeneity (within 100 um scale) and macroheterogeneity (over 1-2 cm scale) were included in the sample. The changes due to occlusion must be measured against a control data set from resting muscles that is large enough to give good estimates of population standard deviations for all the data. Rather than expend another set of animals, the control data were taken from those previously published[5]. These were collected in the same labs using the same surgical and analytical techniques and had a large enough sample size (n = 13) to generate reasonable statistical estimates.

RESULTS

Oxidative Stores

Cryospectrophotometric methods for determining the saturation of myoglobin and hemoglobin were applied to samples of muscles subjected to varying periods of circulatory occlusion. Table 1 lists the median, 90th and 10th percentiles of myoglobin saturation for these muscles. Also shown are the median hemoglobin (Hb) saturation in vessels larger than 40 um for selected muscles. In resting muscles with normal blood flow no cells were found with PO_2 near 0 torr. In fact the lowest site found in the muscles used for average resting values was 3.4 torr. During the occlusion, oxygen consumption continued, using oxygen both from the hemoglobin and myoglobin stores in the tissue. The last two muscles in the table show the effect of removing the occlusion after 30 minutes when tissue oxygen contents approach the minimum. After 30 minutes of reperfusion the saturation of myoglobin with oxygen has returned to resting values. PCr stores are also restored in this time period (data not shown). These results show that complete recovery can occur after 30 minutes of occlusion.

Figure 1A illustrates the time course of depletion of oxygen from the tissue as measured by the fraction of the tissue reaching critical PO_2 values. The open circles show the % of the tissue having $PO_2 < 0.6$ torr. This is near the $PcritO_2$ for resting muscle[6]. The closed circles show the fraction of tissue having $PO_2 < 0.3$ torr - the value indistinguishable from 0 by the myoglobin method. The oxygen stores in the tissue were rapidly used in the first 2-5 minutes so that by 5 minutes over 80% of the tissue had $PO_2 < 0.6$ torr. This is clearly into the dysoxic region and oxygen consumption should slow due to rate-limiting concentration of the substrate O_2. The slow approach of the tissue to what amounts to anoxia ($PO_2 < 0.3$ torr) shown in the figure is consistent with this slowing of $\dot{V}_{O_2}$. With one exception, all muscles subjected to 30 minutes or longer occlusion had PO_2 less than or equal to 0.6 torr at over 90% of the sites sampled and thus aerobic ATP production is curtailed.

Table 1. Myoglobin Saturation at various Durations of Ischemia

Muscle	Ischemia (min)	Mb saturation median	10th	90th	Hb Saturation median
				-----percentile-----	
rest	0	0.56	0.47	0.72	
402R	2	0.54	0.39	0.67	
401L	2	0.50	0.32	0.73	
404R	3.5	0.10	0.02	0.18	
400R	5	0.12	0.06	0.30	
399R	5	0.04	0.00	0.13	
399L	10	0.05	0.00	0.12	
400L	10	0.04	0.00	0.12	
397R	15	0.03	0.00	0.11	7
401R	15	0.04	0.00	0.10	
398L	30	0.00	0.00	0.04	5
402L	30	0.06	0.00	0.13	
288L	45	0.06	0.00	0.07	4
398R	45	0.01	0.00	0.07	7
403R	60	0.04	0.00	0.08	
409R	60	0.03	0.00	0.11	
410R	60	0.04	0.00	0.08	
403L	30/recovery	0.43	0.36	0.49	
404L	30/recovery	0.57	0.48	0.72	
409L	30/recovery	0.54	0.43	0.77	

High Energy Phosphate Stores

Phosphocreatine (PCr) and to a limited extent ATP serve as stored sources of energy when the rate of mitochondrial oxidative phosphorylation (and/or glycolysis) cannot match the rate of ATP consumption. Figure 1B illustrates the time course of changes in ATP and phosphocreatine during ischemia. The data are normalized to total creatine for better comparison between muscles. The two dotted lines in the figure show the range of ATP within which there is a 95% probability of finding the resting value, i.e. mean $\pm$ 1.98*SD. Under normal equilibrium conditions of creatine kinase, the ATP value will remain within this range until the creatine pool is almost 95% dephosphorylated[2]. No significant change in the ATP stores were observed through 60 minutes of ischemia. The PCr stores change throughout the early period even though aerobic energy production is occurring. In fact, there appears to be a sharp drop in one muscle during the first 2 minutes when oxygen contents are well above the threshold of dysoxia. PCr is used continuously through the 60 minute period so that after 60 minutes of ischemia the PCr stores are on the order of those found after 3 minutes of stimulation in muscles operating near $\dot{V}_{O_2}max$[5].

Glycolytic Energy Production

Glycolysis is the only relatively unlimited source of ATP in the absence of mitochondrial oxidative phosphorylation. This is quantitated via the accumulation of the

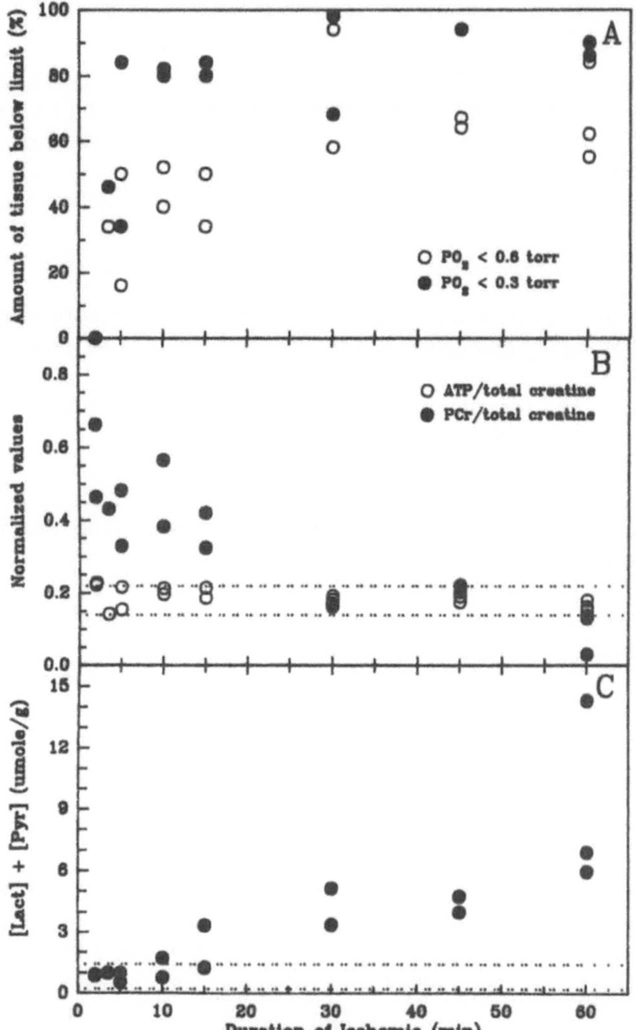

Figure 1. Energy Sources versus Duration of Ischemia
A. % of measured myoglobin saturation sites having PO_2 below the indicated values. 50 total measurements were made on each muscle with controlled sampling as described in Methods.
B. Phosphocreatine [PCr], and [ATP] contents of the each muscle sample. The data are scaled to the total creatine concentration. The dotted lines indicate the mean + or - 1.98*S.D. of normal resting values of [ATP] (see ref 5).
C. The sum of [lactate] & [pyruvate] contents of each muscle sample. The dotted lines indicate the mean + or - 1.98*S.D. of normal resting values (see ref 5).

glycolytic products lactate and pyruvate in the tissue. Figure 1C illustrates the time course of lactate + pyruvate accumulation in these muscles. Again the resting 95% probability range (mean ± 1.98*SD) is shown by the dotted lines. Unlike the changes seen in oxygen and PCr, lactate shows no significant change until about 10 minutes after the start of occlusion which is when 80% of the tissue is dysoxic and ~50% of the tissue is anoxic. Once activated glycolysis continues to supply ATP throughout the rest of the ischemic period.

DISCUSSION

The data in this study contains two kinds of information. One kind is related to the initial question, viz. the question of a control on the ATP demand of a resting tissue. The other is related to the nature of the control signals operating in ischemic muscle. We will discuss these separately.

Energy Balance

In order to evaluate whether there is a regulation on energy demand, the overall energy balance of supply and demand in the muscles needs to be examined.

Energy Supply. The contributions of the various supply sources were estimated as follows:

1. The contribution of the high-energy phosphate store is simply the difference between the total:PCr + ATP measured in control muscles and that measured in the ischemic muscle.

2. The glycolytic contribution was estimated from the increase in lactate + pyruvate over controls, assuming that glycogen was the single source of substrate. Thus 1.5 umole of ATP is produced for each umole of lactate or pyruvate accumulated.

3. To estimate the contribution of aerobic metabolism from oxygen stores in the ischemic muscle we assumed the concentration of myoglobin was 0.5 mM and blood volume in the tissue was 2% of tissue weight. All other data ([Hb], hemoglobin saturation, myoglobin saturation) were directly measured in each muscle. It was assumed that the initial saturation of hemoglobin before occlusion was the average of the measured arterial and venous saturations and initial myoglobin saturation was the same as the average median value given for control muscles in table 1. Since the oxidative contribution is a small fraction of the total in each case an error in these assumptions will have little effect on the final balance shown in table 3.

Energy Demand. Demand was predicted from initial conditions. The resting $\dot{V}_{O_2}$ was measured for each muscle and it was assumed that this reflected all of the resting demand. This rate was converted to high enrgy phosphate (HEP) units by assuming a P/O_2 of 6. It was assumed that this demand was maintained throughout the period of occlusion. Energy balance is expressed as the measured energy consumed minus the predicted demand. If the assumption of constant demand is not true then we should see a deviation from a zero energy balance. A negative balance indicates that less ATP was used than expected from resting turnover. A clear trend in this direction would be support for a control on ATP demand. A positive balance indicates an increased demand during occlusion. This would result from any kind of stimulation of ATPase activity in the muscle. This could be due to anything from marginal K^+ leaks to accidental mechanical stimulation. Figure 2 illustrates the energy balance as a function of the time of ischemia. As in previous figures the dotted lines indicate the 95% confidence interval for the measurements. Any point outside these lines has a $<5\%$ probability of being equal to 0. Most of the muscles show balances that are not significantly different from 0. While there are muscles showing a significant deviation from 0 there is no clear time dependent trend in deviation from 0. At

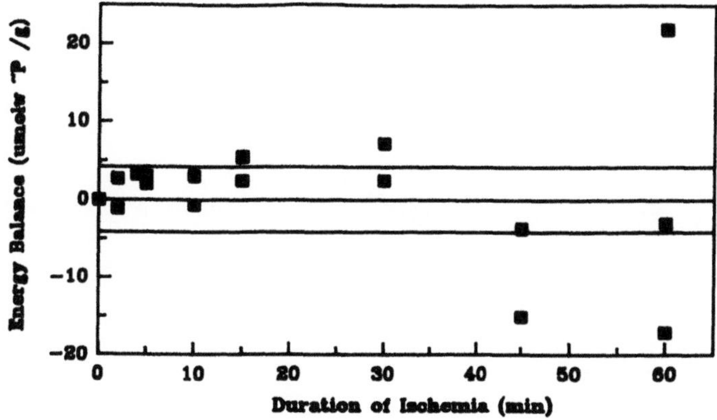

Figure 2. Net Energy Balance versus Duration of Ischemia.
Total energy supplied to the muscle from the sources illustrated in Fig. 1 minus the
expected energy demand based on resting $\dot{V}_{O_2}$. The dotted lines indicate the 95%
confidence interval based on measurements of normal resting muscles.

60 minutes there are deviations in both directions depending on the muscle. Those muscles showing a negative deviation had higher than average resting $\dot{V}_{O_2}$. If they had relaxed to a more normal resting value after occlusion, the predicted ATP demand over the long time interval would be much closer to the measured supply.

The data in this study allow us to ask whether there is an oxygen dependent demand control. The data in figures 1A and 2 do not support such a control in resting muscle. The muscle is anoxic by 15 to 30 minutes and there is no evidence for a decrease in the ATP demand. We therefore conclude that there is no evidence for an oxygen dependent signal on energy demand in resting muscle. The question of oxygen conforming behavior (a fall in $\dot{V}_{O_2}$ as P_{O_2} falls) cannot be answered with this data. A reliable decision in favor of a more general control on demand related to energy state, for example, can only be made if there is overwhelming evidence in favor of an imbalance. There appears to be none in these data.

Control Signals

While changes in PCr and lactate indicate use of stores and glycolysis respectively as sources of ATP, they also reflect changes in the phosphorylation and redox states of the cytosol. These are control signals known to act on both the glycolytic and mitochondrial enzymes. We can evaluate system models of metabolic regulation by comparing the changes observed at constant energy turnover as the system responds to ischemia with observations of the response to aerobically changing energy turnove.

Mitochondria. During the onset of exercise, the activation of mitochondria ATP production can be accounted for by a model that assumes an instantaneous activation of the ATP demand which is supplied by the PCr pool. The change in the phosphorylation state in this pool as reflected in the creatine charge, i.e. [PCr]/[total creatine], then serves as a signal to the mitochondrion to increase ATP production at rates linearly proportional to the creatine charge (4,5). This model accounts for both steady-state and transients in $\dot{V}_{O_2}$[8].

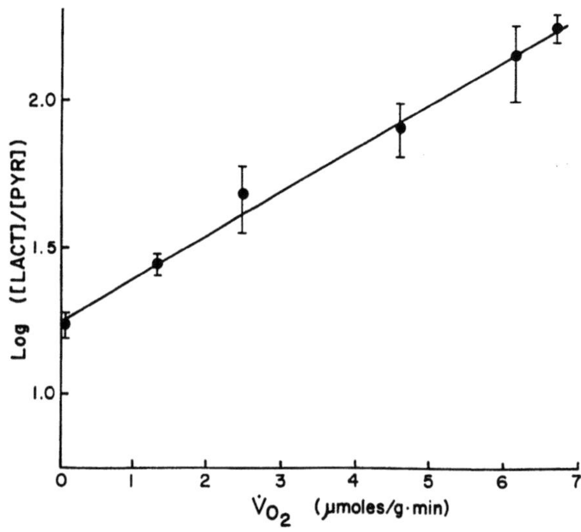

Figure 3. Cytosolic Redox Potential in Stimulated Muscles versus $\dot{V}_{O_2}$. Based on data from ref 5.

The model implies that the phosphorylation state is normally the rate controlling parameter in mitochondrial ATP production and mitochondrial redox is nearly constant. Changes in the cytosolic redox potential, i.e. log (L/P), reflect changes in the mitochondrial membrane potential and are proportional to $\dot{V}_{O_2}$(fig. 3)[3]. In this study there is no change in the rate of ATP consumption, therefore changes in PCr would not be required to increase $\dot{V}_{O_2}$ and the rate of ATP production in the mitochondrion. However, PO_2 falls enough for it also to become a controlling parameter and the same model of mitochondrial control requires changes in the phosphorylation state and redox potential in order to maintain a constant rate of ATP production. These changes would certainly be required to consume oxygen down to anoxic levels. Wilson et al.[21] have suggested that the mitochondria are sensitive to PO_2 throughout the normal range of cell oxygen concentrations. Therefore one would expect changes in the driving function as reflected in the creatine charge as PO_2 decreases even at levels above the dysoxic range. The data in fig. 1 show a decrease in creatine charge in some muscles at 2 minutes when there are no sites in the muscle having a $PO_2 <$ PO_2crit ($\sim$0.5 torr) and 90% of the muscle has a $PO_2 > 2$ torr. This is consistent with Wilson's proposal. By 3.5-5 minutes of ischemia part of the muscle has PO_2 well below the PO_2crit. The creatine charge has fallen further. Although there is no significant increase in the tissue lactate concentration the lactate/pyruvate ratio (L/P) reflecting cytosolic redox as shown in figure 4 has begun to increase. We therefore see the expected changes in control signals in the cell that would be necessary for the mitochondrial to consume the oxygen to anoxic levels and these occur during the period where PO_2 is still measurable and decreasing.

Glycolysis. By 10 minutes of ischemia most of the tissue is dysoxic and PCr stores and glycolysis serve as the sources of energy. If control of glycolysis were simply dependent on the phosphorylation state glycolytic rates should continually increase as the creatine charge falls. In fact, glycolytic rates are fairly constant from 10 minutes to 60 minutes of ischemia. Thus as the stimulating signal increases an inhibitory signal should also be

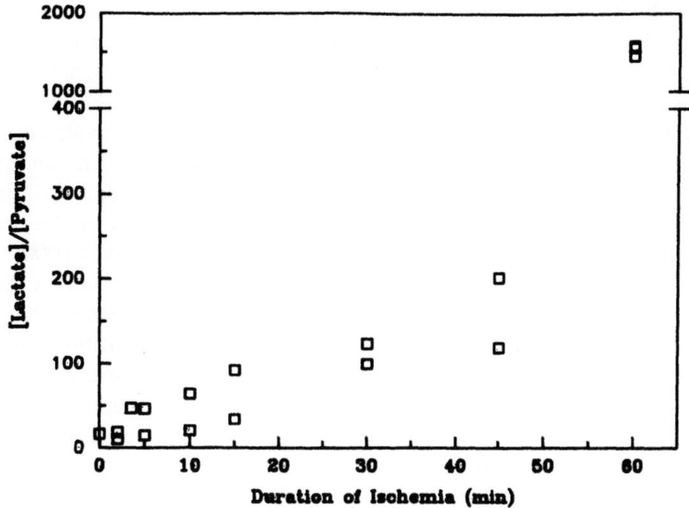

Figure 4. Cytosolic Redox versus the Duration of Ischemia.

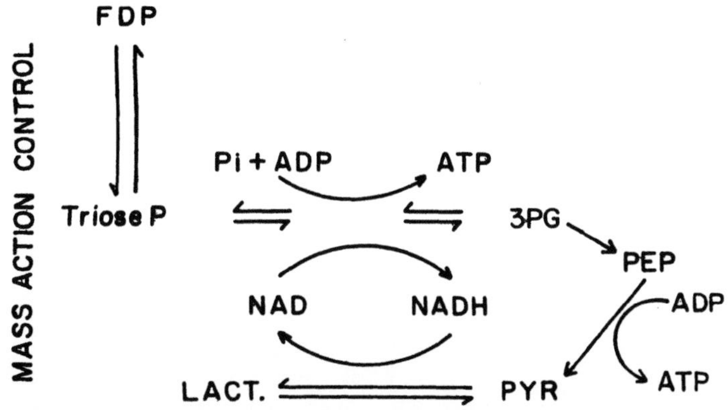

Figure 5. Biochemical Reactions in the Lower Half of Glycolysis.
The steps shown include aldolase, glyceraldehydephosphate dehydrogenase, phosphoglycerate kinase, phosphoglycerate mutase, pyruvate kinase and lactate dehydrogenase.

increasing. Figure 5 illustrates the part of the glycolytic pathway below phosphofructokinase. As the store of PCr is used the concentration of ADP and Pi increase resulting in an increasing driving force on phosphoglycerate kinase (PGK). The flux through PGK and glyceraldehydephosphate dehydrogenase (GAPDH) results in increased reduction of the cytosol and elevation of L/P. If one assumes all these steps to be near equilibrium and a small fixed pool of NAD, then the gradual accumulation of reducing equivalents as ATP is synthesized will result in a continuous rise in L/P and an inhibitory pressure on the forward flux. Thus the forward driving force (phosphorylation state) is balanced by a backward driving force (redox state) and rates are maintained nearly constant. Since the mitochondria can no longer consume reducing equivalents, as the redox state becomes more and more reduced the equilibrium at LDH will be driven toward lactate at the expense of pyruvate and a larger fraction of the limited NAD pool will be in the form of NADH. A rise in L/P will be associated with a fall in pyruvate concentration and ultimately the system will be limited by having almost all the NAD in the form of NADH resulting in a very low concentration of pyruvate and a very high L/P. Figure 6 (solid squares) shows the gradual rise in L/P in the ischemic muscles as the creatine charge declines until the creatine charge is on the order of 0.1 and l/p is over 1000. AT this point the pyruvate levels are very low (< .005 umoles/g). For comparison on the same figure data from aerobically exercising muscles are shown. The relationship between the phosphorylation state and redox are very similar for the two conditions until the limit is reached. In the aerobically working muscles the mitochondria are consuming cytosolic reducing equivalents so the pyruvate levels remain fairly constant (0.05 umole/g). The increase in L/P ratio is associated with increases in lactate rather than small increases in lactate and decreases in pyruvate.

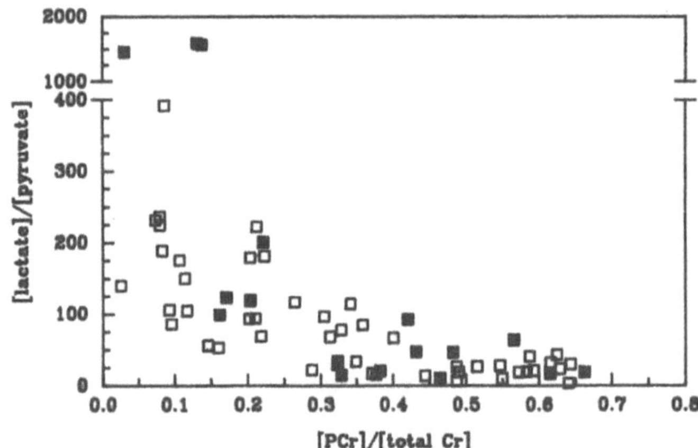

Figure 6. Cytosolic redox versus Creatine Charge
The measured lactate/pyruvate for each muscle is plotted against the creatine charge (phosphocreatine/ total creatine). The open squares show values for normal stimulated muscles covering the range of $\dot{V}_{O_2}$ shown in figure 3. The filled squares are the values measured in the ischemic muscles.

In summary, the changes in control parameters seen in these ischemic muscles are completely consistent with models developed using aerobically working muscles and no sign of an additional control on ATP demand in oxygen limited muscles is observed.

REFERENCES

1. Blei, M. L., K.E. Conley & M.J. Kushmerick. Separate measures of ATP utilization and recovery in human skeletal muscle. J. Physiol. 465:203-222 (1993)

2. Connett, R. J. Analysis of metabolic control: new insights using a scaled creatine kinase model. Am. J. Physiol. 254:R949-R959 (1988)

3. Connett, R.J. The cytosolic redox is coupled to VO2: a working hypothesis. Adv. Exp. Med. Biol. 222:133-142 (1988)

4. Connett, R.J., T.E.J. Gayeski and C.R. Honig. Energy sources in fully aerobic rest-work transitions: A new role for glycolysis. Am.J.Physiol. 248:H922-H929 (1985)

5. Connett, R.J. and C.R. Honig. Regulation of VO2 in red muscle: Do current biochemical hypotheses fit in vivo data?. Am.J.Physiol. 256:R898-R906 (1989).

6. Connett, R.J., C.R. Honig, T.E.J. Gayeski and G.A. Brooks. Defining hypoxia: a systems view of VO2, glycolysis, energetics, and intracellular PO 2. J. Appl. Physiol. 68:833-842.

7. Dawson, M.J. D.G. Gadian, D.R. Wilkie. Muscular fatigue investigated by phosphorus nuclear magnetic resonance. Nature, 274:861-866 (1978)

8. Funk, C.I., A. Clark Jr. and R.J. Connett. A Simple model of aerobic metabolis: applications to work transitions in muscle. Am.J.Physiol. 258:R949-R959 (1990)

9. Gayeski, T.E.J., R.J. Connett and C.R. Honig. Oxygen transport in rest-work transition illustrates new functions for myoglobin. Am.J.Physiol. 248:H914-H921 (1985)

10. Gayeski, T.E.J. and C.R. Honig. Intracellular PO2 in long axis of individual fibers in working dog gracilis muscle. Am.J.Physiol. 254:H1179-H1186 (1988)

11. Harris,R.C., E. Hultman, L. Kaijser & L.-O. Nordesjö. The effect of circulatory occlusion on Isometric Exercise Capacity and energy metabolism of the quadriceps muscle in man. Scand. J. Lab. Invest. 35:87-95 (1975).

12. Hochachka, P.W. Patterns of O_2 dependence of metabolism. Adv. Exp. Med. Biol. 222:143-151 (1988)

13. Katz, A. G-1,6-P2, glycolysis and energy metabolism during circulatory occusion in human skeletal muscle. Am.J.Physiol. 255:C140-C144 (1988)

14. Katz, A., M.K. Spencer and K. Sahlin. Failure of glutamate dehydrogenase system to predict oxygenation state of human skeletal muscle. Am.J.Physiol. 259:C26-C28 (1990)

15. Lewis, S.F., S. Vora and R.G. Haller. Abnormal oxidative metabolism and O2 transport in muscle phosphofructokinase deficiency. J. Appl. Physiol. 70:391-398 (1991)

16. Magnum, C.P. & W. Van Wimkle, Responses of aquatic invertebrates to declining oxygen conditions. Am. Zool. 13:529-541 (1973)

17. Miller, R.G., D. Giannini, H.S. Milner-Brown, R.B. Layzer, A.P. Koretsky, D. Hooper and M.W. Weiner. Effects of fatiguing exercise on high-energy phosphates, force, and EMG: evidence for three phases of recovery. Muscle & Nerve. 10:810-821 (1987)

18. Saltin, B. & J. Karlsson. Muscle ATP, CP and lactate during exercise after

physical conditioning. In: *Muscle Metabolism During Exercise*, edited by B. Pernow and B. Saltin. New York: Plenum, 1971,p.395-399.

19. Sugano, T., N. Oshino and B. Chance. Mitochondrial functions under hypoxic conditions. The steady states of cytochrome c reduction and of energy metabolism. Biochim.Biophys.Acta. $\underline{347}$:340-358 (1974)

20. Wilson, D.F., M. Erecinska, C. Drown and I.A. Silver. Effect of oxygen tension on cellular energetics. Am. J. Physiol. $\underline{233}$:C135-C140 (1977)

21. Wilson, D.F., W.L. Rumsey, T.J. Green and J.M. Vanderkooi. The oxygen dependence of mitochondrial oxidative phosphorylation measured by a new optical method for measuring oxygen concentration. J. Biol. Chem. $\underline{263}$:2712-2718 (1988)

22. Zweier, J. L. W.E. Jacobus. Substrate induced alterations of high energy phosphate metabolism and contractile function in the perfused heart. J. Biol. Chem. $\underline{262}$:8015-8021 (1987)

RED BLOOD CELL TRANSIT TIME IN MAN: THEORETICAL EFFECTS OF CAPILLARY DENSITY

Russell S. Richardson, David C. Poole, Douglas R. Knight, and Peter D. Wagner

Department of Medicine
University of California
La Jolla, Ca 92093-0623

INTRODUCTION

The pioneering development of the knee-extensor exercise model opened a new dimension in the study of human muscle physiology allowing the measurement of blood flow ($\dot{Q}$), oxygen uptake ($\dot{V}o_2$), and work rate in an isolated muscle (Andersen et al., 1985). Original studies using this exercise paradigm reported reduced O_2 extractions presumably as a consequence of the greater maximal muscle $\dot{Q}$ (Andersen & Saltin, 1985). Although this human *in situ* model has elevated the maximum recorded muscle $\dot{Q}$ in man from the previously recorded values of 60-100 ml$\cdot$100g$\cdot$min^{-1} (Mellander, 1981) to values of 250-300 ml$\cdot$100g$\cdot$min^{-1}, it was still not apparent that this was maximal human muscle $\dot{Q}$ (Rowell, 1988). A recent experiment was performed to determine if human muscle $\dot{Q}$ could be elevated above these previously reported values during knee-extensor exercise by recruiting athletic subjects and utilizing a rapidly incremented protocol (Richardson et al., 1993). In this study quadriceps muscle $\dot{Q}$ ($\bar{x} \pm$ SE, 385 $\pm$ 26 ml$\cdot$min$^{-1}\cdot$100g^{-1}), $\dot{V}o_{2max}$ (60.2 $\pm$ 5.8 ml$\cdot$min$^{-1}\cdot$100g^{-1}), and work rate (4.3 $\pm$ 0.3 W$\cdot$100g^{-1}) were by far the highest yet recorded. Despite these being the highest recorded muscle blood flows in man, O_2 extraction at $\dot{V}o_{2max}$ did not appear to be compromised to any greater extent than in conventional cycle exercise (Richardson et al., 1993). Extraction at any given $\dot{V}o_2$ is determined largely by convective O_2 delivery, capillary surface area available for O_2 uptake, and the resultant time available for O_2 release (Honig et al., 1991). It is the objective of this paper to focus on the third of these factors which equates to red blood cell (RBC) transit time during these high human muscle $\dot{Q}$ (Richardson et al., 1993). Previously, it has been demonstrated theoretically that transit time and its variability are major determinants of O_2 extraction from red cells at exercise (Gayeski et al., 1988).

In the most recent knee-extensor studies (Richardson et al., 1993), both muscle $\dot{Q}$ through the quadriceps and O_2 extraction were recorded at five progressive levels of work rate up to maximum. Thus, it is possible to calculate the upper limit of mean red cell transit time (Gayeski et al., 1988; Honig et al., 1991) across a spectrum of potential human capillary densities and compare the predicted O_2 extraction to the measured value.

METHODS

Subjects

Six healthy nonsmoking males participated in these experiments. All were competitive bicycle racers regularly riding 200-400 miles per week. For each subject a health history and physical examination were completed and informed consent was obtained according to the University of California San Diego, Human Subjects Committee requirements. The physical characteristics of the subjects were as follows ($\bar{x} \pm$ SE): age, 23.2 ± 1.9 yr; height 178.3 ± 1.8 cm; and weight 75.1 ± 2.3 kg. Subjects performed an initial incremental cycle ergometry test (100 W initial WR increased by 25 $W \cdot min^{-1}$ until fatigue), and their $\dot{V}o_{2max}$ averaged 4.36 ± 0.06 $l \cdot min^{-1}$ or 58.3 ± 1.4 $ml \cdot kg^{-1} \cdot min^{-1}$ in this test.

Preliminary Tests

All subjects were acclimated to the dynamic knee-extensor apparatus and technique through 5-6 training bouts. The final two practice periods involved a graded maximal exercise test (20 W initial WR increased by 5 $W \cdot min^{-1}$ until the rate could no longer be maintained at 60 rpm) and a simulated final experiment identical to this protocol, but without any vascular catheterization.

Exercise Model

The knee-extensor ergometer was built to replicate the device reported in previous research (Andersen & Saltin, 1985; Andersen et al., 1985; Rowell et al., 1986). The subject was seated on a adjustable chair, with a Monark cycle ergometer placed behind him. The resistance to knee-extension was provided via a metal bar attached to the crank of the ergometer and to a specially designed shoe worn by the subject. Sixty dynamic contractions of the knee-extensor muscles per minute were performed. Contractions of the quadriceps femoris muscle caused the lower part of the leg to extend from approximately 90 to 170° flexion. The momentum of the flywheel assisted in the return of the relaxed leg to the start position. The subjects were constrained in the seat by a safety belt which anchored the angle of the hip to approximately 90°. Both anecdotal reports from the subjects and the examination of tracings from a force transducer placed between the ergometer and the subject allowed the evaluation of subject ability to limit active contraction to the quadriceps (Richardson et al., 1993). In practice it is extremely difficult to perform useful work simultaneously with the hamstrings and quadriceps while maintaining this cadence.

The exercise equipment and methodology was designed based on the recommendations of Andersen and colleagues (1985), with the exception of the inflation of a blood pressure cuff below the knee during measurements. The elimination of the cuffing procedure was to allow a more rapid achievement of physiological steady state at each work rate. Preliminary investigations were conducted on two of our subjects where $\dot{Q}$, leg $\dot{V}o_2$, and pulmonary $\dot{V}o_2$ were measured with no cuff, and with the cuff inflated to 280 mmHg. These measurements revealed no difference in leg $\dot{Q}$, leg $\dot{V}o_2$, or pulmonary $\dot{V}o_2$ with or without cuff inflation (Richardson et al., 1993). These data in conjunction with data from Andersen et al. (1985), who reported a) no difference in pulmonary $\dot{V}o_2$ when data were collected with no cuff and with the cuff inflated and b) EMG analyses of minimal muscle recruitment below the knee during the knee-extensor exercise, indicated that cuff use was not necessary.

Exercise Study With Blood Flow And Blood Gas Measurements

Within 1 wk of the preliminary studies, subjects returned to the laboratory in the morning. After electrocardiographic (ECG) electrode placement, two catheters (radial artery and left femoral vein) and a thermocouple (left femoral vein) were emplaced using sterile technique as reported previously (Poole et al., 1992; Knight et al., 1993; Richardson et al., 1993). During exercise, iced saline was infused through the catheter placed distally in the femoral vein at flow rates sufficient to decrease blood temperature at the thermocouple by $\approx 1\,^{\circ}\mathrm{C}$. Infusions were continued for 15-20 s until femoral vein temperature had stabilized at its new lower value. Saline injection rate was measured by weight change in a reservoir bag suspended from a force transducer which was calibrated before and after each experiment. The calculation of blood flow was performed on thermal balance principles as detailed by Andersen and Saltin (Andersen & Saltin, 1985). O_2 concentration was calculated as 1.39 x [Hb] x measured O_2 saturation + 0.003 x measured Po_2. Arterial-venous $[O_2]$ difference was calculated from the difference in radial artery and femoral venous oxygen concentration. This difference was then divided by arterial concentration to give O_2 extraction. Leg $\dot{V}o_2$ was calculated as the product of arterial-venous O_2 concentration difference and blood flow.

Experimental Protocol

Exercise testing began ≈ 40-50 min Following the completion of catheterization procedures. Each subject initially exercised at a low work rate below 20 W for ≈ 3 min as a warm-up and to check the integrity of all measurement apparatus. After a short break the subject breathed on the mouthpiece, allowing the continuous recording and display of ventilation and gas exchange, and began exercise at the first pre-determined percentage of maximum work rate (WR_{MAX}). After $\approx$ 2-3 minutes and duplicate measurements had been made, work rate was increased from 25 to 50 to 75 and then to 90 and 100 % of WR_{MAX}. At each work rate, measurements were not initiated until 90-95% of the predetermined pulmonary $\dot{V}o_2$ for that WR had been reached (i.e. the rapid phase of the rise in $\dot{V}o_2$ was complete). At the first work rate this was achieved in ≈ 3 min, however at subsequent WR levels this was ≈ 2-3 min (Whipp & Wasserman, 1972; Richardson et al., 1992). Each incremental exercise bout was therefore completed in under 14-16 minutes. The sequence of measurements at each WR was as follows: blood samples (for blood gases) and then femoral vein blood flow. Duplicate measurements were then taken without delay, as an objective of the study was to keep total exercise time to a minimum.

Blood Analyses

Four ml samples of arterial and venous blood were withdrawn from the catheters anaerobically to measure Po_2, Pco_2, pH, O_2 saturation, and hemoglobin concentration ([Hb]). All measurements were made on an IL 1306 blood gas analyzer and IL 282 CO-oximeter (Instrumentation Laboratories, Lexington, MA.) Between each sample, electrodes were calibrated and demonstrated acceptable reproducibility (SD of repeated determinations: Po_2 and Pco_2, 1.5 Torr; pH, 0.003).

Thigh Volume Measurement

The thigh length, three circumference measurements and two skinfold measurements were used to calculate thigh volume as used previously by Andersen and Saltin and others (Jones & Pearson, 1969; Andersen & Saltin, 1985). The quadriceps femoris muscle mass was then calculated with a formula derived from an autopsy study (Andersen & Saltin, 1985).

Calculation Of Red Blood Cell Transit Time

To estimate capillary volume per fiber volume for the spectrum of capillary densities reported in the literature (i.e. 200, 400, 600, and 800 $cap \cdot mm^{-2}$) (Andersen & Henricksson, 1977; Brodal et al., 1977; Coyle et al., 1988), the assumption of a mean capillary diameter of 6 μm was used, which is in the middle of the accepted range for mammalian muscle (1) (Andersen & Saltin, 1985). Based on previous morphometric analyses (Mathieu-Costello et al., 1991) the calculations illustrated below utilize this 6 μm diameter to determine average individual capillary radius (2) and area (3). The product of average capillary area and a given capillary density determines the total capillary area per mm^{-2} (4). Capillary fractional area per fiber area was determined by the division of total capillary area by 1 mm^{-2} (5). As morphometric analyses depend upon an extremely thin slice of muscle, and at this point assume anisotropy, the terms and data for fractional area and fractional volume become synonymous. Additionally, to account for the non-anisotropic components (i.e. capillary tortuosity and branching) a constant of 1.2 was used (6)(1.0 = perfect anisotropy and 2.0 = random orientation; Mathieu-Costello et al., 1991). This is a reasonable estimate based on work by Cutts (1988) who determined theoretical sarcomere length in the human quadriceps muscles to range from 3.1 to 2.0 μm, with the lower leg extended from ≈ 65 to $\approx 167°$, respectively (Mathieu-Costello, 1987; Mathieu-Costello et al., 1991).

(1)   Cap. diameter   =   6.0 μm
(2)   Cap. radius   =   3.0 μm
(3)   Cap. area   =   28.3 μm^2

(4)   Theoretical total   = $(28.3\ \mu m^2)$(number of cap·mm^{-2})
       cap. area per mm^{-2}

(5)   Cap. area. per
       fiber area.   = (capillary area.)/$(1{,}000{,}000\ \mu m^2)$

Cap. fractional area and fractional vol. are considered equal.

(6)   Cap. Vol.
       adjusted for
       tortuosity   = (cap. vol. per fiber vol.)(1.2)

(7)   RBC transit
       time (sec)   = (cap. vol per fiber vol.)/($\dot{Q}$ ml·sec^{-1}·g^{-1})

This final calculation (7), assumes that leg $\dot{Q}$ is homogeneously distributed within and between capillaries in the exercising muscle, and utilizes the relationship between capillaries per fiber volume and the muscle volume normalized leg $\dot{Q}$ to determine red cell transit time. It is recognized that there are both temporal and spatial heterogeneities in flow which are probably greater at the lower work rates. These calculations consequently reflect the longest mean red cell transit times for a given structural capillary density. Transit time would actually be less if functional capillary density was less than total capillary density, which may be the case at low leg $\dot{Q}$.

Prediction Of O_2 Extraction

At WR$_{MAX}$ mean calculated red blood cell transit times, at 200, 400, 600, and 800 cap·mm^{-2}, were extrapolated to predict femoral venous hemoglobin saturation based on the graphical analysis advanced by Gayeski et al. (1988). This algorithm models the time required to release O_2 through the plasma, endothelium, interstitium and sarcolemma (described as the carrier free region) to a myoglobin solution of known saturation (Gayeski et al., 1988). To facilitate this transformation, a 5 Torr myoglobin associated PO$_2$, and a diffusion path length of 1 μm through the carrier free region were assumed (Gayeski et al., 1988; Honig et al., 1991).

RESULTS

Leg $\dot{Q}$ and leg $\dot{V}o_2$ rose in a linear fashion as WR was increased towards WR$_{MAX}$ (Richardson et al., 1993). The average WR$_{MAX}$ was 30-50 W greater than previously reported in knee-extensor exercise (Andersen & Saltin, 1985; Andersen et al., 1985; Bangsbo et al., 1990), and produced greater associated leg $\dot{Q}$ and leg $\dot{V}o_2$ than in these earlier studies. Estimated average quadriceps femoris weight was 2.36 ± 0.09 kg. The resultant maximum mass-specific muscle $\dot{Q}$ and $\dot{V}o_2$ recorded in this study (385 ± 26 and 60.2 ± 5.8 ml·min^{-1}·100g^{-1}, respectively) are the highest to date, in man. Oxygen extraction did not reach an early plateau, but rose from 68 to 76 to 80 to 84 and then to an average maximum of 85% at 25, 50, 75, and 90% of WR$_{MAX}$ (Richardson et al.,

1993). As a function of leg $\dot{V}o_2$, oxygen extraction rose in a hyperbolic fashion similar in nature to previous data from conventional cycle ergometer measurements (Knight et al., 1992; Poole et al., 1992; Richardson et al., 1993). Further detailed results can found in the original account of this research (Richardson et al., 1993).

The calculated capillary transit time varied greatly, dependent upon the measured $\dot{Q}$ and the estimated capillary density used in the computation (Figure 1). The longest red cell transit time was at 25% of WR_{MAX} and an estimated capillary density of 800 cap·mm^{-2} (1.03 sec), at the other end of the spectrum at 100 of WR_{MAX} and an estimated capillary density of 200 cap·mm^{-2} calculated red cell transit time was dramatically reduced to 0.11 sec (Figure 1).

DISCUSSION

WR_{MAX}, Leg $\dot{Q}$, Leg $\dot{V}o_2$, And O_2 Extraction During Knee-extensor Exercise

The group of five subjects studied in this research achieved the highest WR_{MAX} recorded during knee-extensor exercise, the present results extend previous findings to much higher (up to 80% greater) WR's (Andersen & Saltin, 1985; Andersen et al., 1985; Richardson et al., 1993). The progressive increase in WR to this elevated WR_{MAX} was accompanied by a linear increase in leg $\dot{Q}$ and muscle $\dot{V}o_2$, both also rose to the highest levels, as yet, reported in man. Unlike previous studies which reported a maximum O_2 extraction of 70% (Andersen & Saltin, 1985; Andersen et al., 1985), this study demonstrated much more complete extractions, averaging 85% at WR_{MAX}. This high level of O_2 extraction is very similar to values reported for conventional cycle ergometry (Knight et al., 1992; Poole et al., 1992) in which as man approaches maximum exercise, O_2 extraction increases hyperbolically typically achieving values of 80-90% at $\dot{V}o_{2max}$ (Richardson et al., 1993).

There are two potential explanations for this elevated O_2 extraction profile reported herein: a) Earlier knee-extensor research employed lengthy protocols, with each WR lasting between 6 to 10 minutes (Andersen & Saltin, 1985; Andersen et al., 1985). The total exercise time of just 14.5 min $\pm$ 0.8 min in the present study was greatly reduced from previous knee-extensor studies (estimated at 40-60 minutes) permitting the achievement of a 40 to 80% greater WR_{MAX}. The lower WR_{MAX} achieved in previous studies, caused by fatigue, may explain the lower $\dot{V}o_2$ and O_2 extractions reached, and/or b) the difference in subject population of trained in the present study versus relatively untrained in previous studies may have created an intrinsic physiological difference between the groups of subjects studied. Although whole body $\dot{V}o_{2max}$ was not different and there were strong similarities in physiological response at the lower knee-extensor WR, where the only comparisons between subjects within different studies could be made (Richardson et al., 1993), such a difference is still a possibility. As the degree of capillarity may be a decisive factor in determining O_2 extraction, especially at the high muscle $\dot{Q}$ reported here, the effect of varying capillary density is the focus of this discussion.

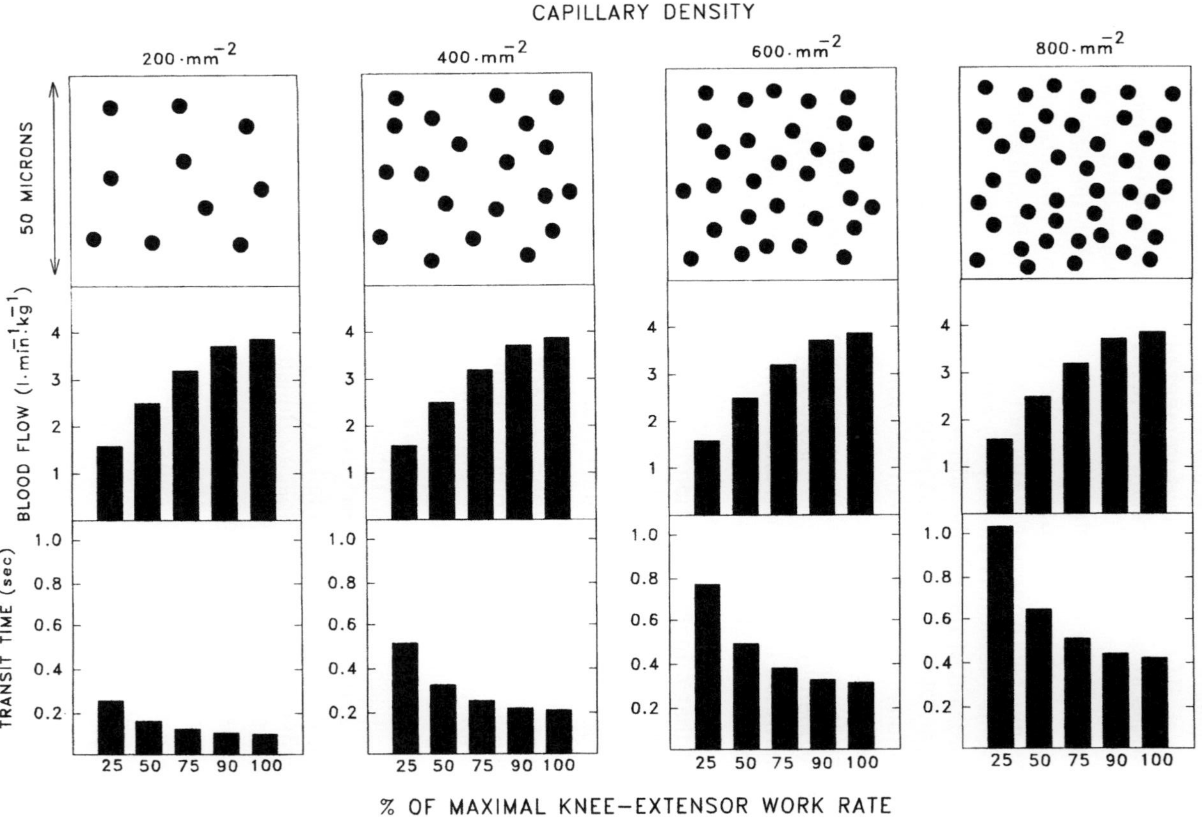

Figure 1. The effect of estimated capillary density on the calculated red blood cell transit time with muscle blood flows measured at 25, 50, 75, 90, and 100% of maximal work rate during knee-extensor exercise. These calculations assume that leg $\dot{Q}$ is homogeneously distributed within and between capillaries in the exercising muscle. Consequently, these data reflect the longest mean red cell transit times for a given capillary density.

Capillary Density

Capillary densities reported for man have varied throughout the literature, ranging from 200 to 800 cap·mm^{-2}, dependent upon the methodology employed and the subject population sampled (Andersen & Henricksson, 1977; Brodal et al., 1977; Coyle et al., 1988). The measurement of capillaries per mm^{-2} is prone to artifactual changes in area dependent upon the preparative technique used (Andersen & Henricksson, 1977). This highly method-dependent variable can account for the large range of capillary densities reported in the same species and muscle group, which may account for the varied data collected from the human quadriceps muscle (Andersen & Henricksson, 1977; Brodal et al., 1977). However, if the same method is employed, capillaries per mm^{-2} can provide a proportional comparison of the relative diffusional distance if the muscle groups examined are similar. Nevertheless, it is recognized that this may not reflect the true *in vivo* conditions (Andersen & Henricksson, 1977).

Early studies revealed no difference in capillary density from bed rest to endurance training (Saltin et al., 1968), or between sedentary and trained men (Hermansen & Watchtlova, 1971). However, more recent studies have repeatedly demonstrated that both man and lower animals respond to endurance training, by increasing capillary density (Mai et al., 1970; Andersen, 1975; Andersen & Henricksson, 1977; Brodal et al., 1977). Additionally, a more recent study determined that a high correlation ($r = 0.74$; $P < 0.002$) existed between capillary density and time to fatigue while cycling at 88% of $\dot{V}o_2$max (Coyle et al., 1988). In the same study, the cyclists were divided into two groups by the intensity of exercise needed to elicit their blood lactate threshold (LT), also an indicator of physical training status (Ivy et al., 1980). The lower LT group had a significantly lower capillary density (327 and 405 cap·mm^{-2}, for the low LT and high LT respectively). Not withstanding methodological problems there is compelling evidence in the literature that increased activity level promotes muscle capillary growth and that capillarity density is greater in endurance trained individuals than in their sedentary counterparts. For example, two groups of subjects, one sedentary and the other endurance trained, had capillary densities of 305 and 425 cap·mm^{-2}, respectively (Brodal et al., 1977). Furthermore, sedentary subjects trained on a cycle ergometer for a period of eight weeks increased their mean capillary density from 329 to 395 cap·mm^{-2} (Andersen & Henricksson, 1977). These data can be interpreted to suggest that trained subjects who are able to perform at high exercise levels have quadriceps capillary densities which may exceed 400 cap·mm^{-2}, whereas untrained subjects may have capillary densities of 300 cap·mm^{-2} or less. With respect to knee-extensor exercise, the limited data available supports this concept. Saltin et al. (Saltin et al., 1986) reported that as capillary density increased across subjects from 280 ($n = 4$), 355 ($n = 4$) to 460 cap·mm^{-2} ($n = 3$), so did their work capacity during knee-extensor exercise (18.2, 20.8, and 24.3 W·kg^{-1}, respectively). Although very simplistic, it is interesting to apply this very linear relationship reported by Saltin et al. (1986) to the data collected in the present knee-extensor study. As the maximal work capacity in this recent study was 42 W·kg^{-1} the simple linear extrapolation of these previous data to the work rates achieved in this study would predict a capillary density of 980 cap·mm^{-2}!

Following this assimilation of previous capillary density measurements in trained and untrained man and in subjects with low and high work capacities it is likely that the subjects in the present study (trained competitive cyclists who produced the highest muscle work capacity to be recorded in man) would have a capillary density of at least 400 cap·mm^{-2} or greater, equating their data more with the right side of Figure 1.

Red Blood Cell Transit Time

Red cell transit time is considered to be a major determinant of O_2 extraction (Gayeski et al., 1988; Honig et al., 1991). Transit time is the ratio of red cell capillary path length to red cell velocity, while velocity is directly proportional to muscle $\dot{Q}$ and varies inversely with capillary bed volume recruited. As relative capillary volume recruitment is less than half that of $\dot{Q}$ increase (Honig et al., 1982), exercise hyperemia must increase mean red cell velocity and shorten transit times (Klitzman & Duling, 1979). Figure 1 illustrates the influence of four different capillary densities on mean red cell transit time using the present measurements of muscle $\dot{Q}$ during knee-extensor exercise. At the blood flows recorded in the present study, and a capillary density of 200 caps·mm^{-2} calculated red cell transit time would be 0.25 sec at only 25 % of WR$_{MAX}$ which would fall to 0.11 sec at WR$_{MAX}$ (Figure 1). At the other extreme, if the subjects had a capillary density of 800 caps·mm^{-2} they would have a calculated red cell transit time of over 1 sec at 25 % of WR$_{MAX}$ and this would fall to 0.42 sec at WR$_{MAX}$ (Figure 1.). It is evident that if endurance training increases capillary density this will have a marked effect on red cell transit time. For example, if a subject with an initial capillary density of 200 cap·mm^{-2} were to double this capillary density through endurance training, and maximal muscle $\dot{Q}$ were unchanged, red cell transit time would also be doubled (Figure 1). Theoretically, this increase in capillary density has the potential to vastly enhance O_2 extraction from ≈ 50 to ≈ 72 % at maximum muscle $\dot{Q}$ of 385 ml·min^{-1}·100g^{-1} as measured in this study (Gayeski et al., 1988) (Figure 2.). The filled bars in Figure 2 illustrate the theoretical fall in femoral vein hemoglobin saturation, as a direct consequence of increased capillarity and therefore the elongated red cell transit time at WR$_{MAX}$. The hatched bars represent the measured femoral vein hemoglobin saturation at $\dot{V}o_{2max}$ at the end of the rapid knee-extensor exercise test (Gayeski et al., 1988). It is interesting to speculate on the meaning of the intersection of these two sets of data. If the capillary density for the present subjects was 560 cap·mm^{-2} then the theoretical calculation of femoral venous hemoglobin saturation would now agree with the measured value of 16 %. This suggests that if the capillary density of the present subjects was indeed approximately 560 cap·mm^{-2} then this large capillary blood volume and consequent elongation of red cell transit time (≈ 0.3 sec) would allow the high O_2 extractions measured, even under such high flow conditions.

SUMMARY

These data indicate that through the reduction in exercise time and recruitment of trained subjects, the exercising muscle in the human dynamic knee-extension model can reach even higher work rates, $\dot{V}o_2$, and $\dot{Q}$ than previously reported (Andersen & Saltin, 1985; Andersen et al., 1985; Rowell et al., 1986). Despite these high muscle $\dot{Q}$, the achievement of high O_2 extractions is possible (Richardson et al., 1993). Previously, this was attributed almost exclusively to the elevated WR_{MAX}, and it was therefore concluded that O_2 extraction is not limited by the high $\dot{Q}$ to any greater extent than in conventional two legged cycle ergometry (Richardson et al., 1993). It is now apparent from the analysis of the data in this paper that it is possible that a difference in capillary density between the subjects in the original studies and the present research may have played a role in the increase in O_2 extraction with increasing muscle $\dot{Q}$. Although, it should be recognized that a) the capillary density necessary to reduce red cell transit time suitably to match the range of measured femoral venous hemoglobin saturation is high·(Gayeski et al., 1988), however it is within the measured values for man (Brodal et al., 1977), and b) where comparable, during low WR knee-extensor exercise and whole body $\dot{V}o_{2max}$, subjects across the studies did not differ. This would be expected if capillary density differed greatly (Brodal et al., 1977). It can therefore be concluded that both the rapid protocol during knee-extensor exercise and the potential increased capillarity of the trained subjects in the present study may have combined to produce the amplified physiologic and WR responses.

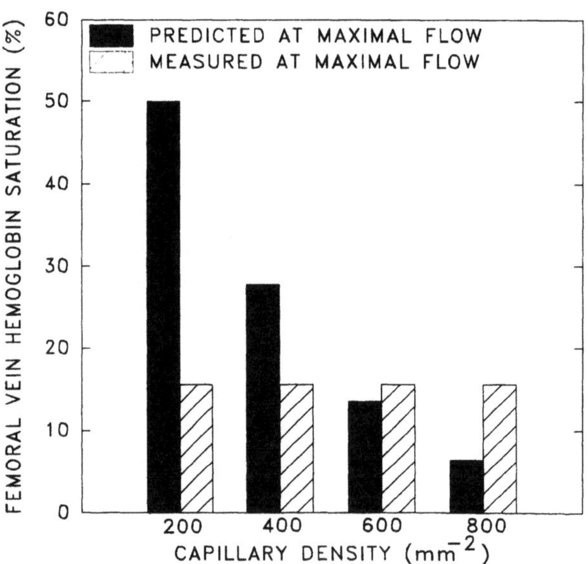

Figure 2. Femoral venous hemoglobin saturation, measured and predicted based on the graphical analysis of Gayeski et al. (1988). Mean calculated red blood transit time, at 200, 400, 600, and 800 cap·mm^{-2} and the maximum measured muscle $\dot{Q}$ were used in conjunction with the assumptions of a 5 Torr myoglobin associated PO_2, and a diffusion path length of 1 μm through plasma, endothelium, interstitium and sarcolemma to calculate the predicted values (Gayeski et al., 1988; Honig et al., 1991).

ACKNOWLEDGEMENTS

The authors would like to express their appreciation to the subjects who participated in this study. The authors are also indebted for the assistance from other investigators in this group, specifically: S. Sadi Kurdak M.D., Michael C. Hogan Ph.D., Bruno Grassi M.D. Ph.D., Emily. C. Johnson Ph.D., and B. Kipp Erickson Ph.D. Additionally, we recognize the expert technical assistance of Harrieth Wagner and Jeffrey Struthers.

This research was supported by funds provided by the National Heart, Lung, and Blood Institute Grant HL-17731 and also by the Cigarette and Tobacco Surtax Fund of the State of California through the Tobacco Related Disease Research Program of the University of California Grant RT 227 and 2KT 0066.

REFERENCES

1.   ANDERSEN, P. Capillary density in skeletal muscle in man. <u>Acta Physio. Scand.</u> 95: 203-205,1975.
2.   ANDERSEN, P., AND J. HENRICKSSON. Capillary supply of the quadriceps femoris muscle in man: Adaptive response to exercise. <u>J. Physiol.</u> 270: 677-690,1977.
3.   ANDERSEN, P., AND B. SALTIN. Maximal perfusion of skeletal muscle in man. <u>J. Physiol.</u> 366: 233-249,1985.
4.   ANDERSEN, P., R. P. ADAMS, G. SJOGAARD, A. THORBE, AND B. SALTIN. Dynamic knee extension as a model for study of isolated exercising muscle in humans. <u>J. Appl. Physiol.</u> 59: 1647-1653,1985.
5.   BANGSBO, J., P. D. GOLLNICK, T. E. GRAHAM, C. JUEL, B. KIENS, M. MIZUNO, AND B. SALTIN. Anaerobic energy-production and O2 deficit-debt relationship during exhaustive exercise in humans. <u>J. Physiol.</u> 422: 539-559,1990.
6.   BRODAL, P., F. INGJER, AND L. HERMANSEN. Capillary supply of skeletal muscle fibers in untrained and endurance-trained men. <u>Am. J. Physiol.</u> 232: H705-H712,1977.
7.   COYLE, E. F., A. R. COGGAN, M. K. HOPPER, AND T. J. WALTER. Determinants of endurance in well-trained cyclists. <u>J. Appl. Physiol.</u> 64: 2622-2630,1988.
8.   CUTTS, A. The range of lengths in the muscles of the human lower limb. <u>J. Anat.</u> 160: 79-88,1988.
9.   GAYESKI, T. E. J., W. J. FEDERSPIEL, AND C. R. HONIG. A graphical analysis of the influence of red cell transit time, carrier free layer thickness, and intracellular PO2 on blood tissue O2 transport. <u>Adv. Exp. Med. Biol.</u> 222: 25-35,1988.
10.   HERMANSEN, L., AND M. WATCHTLOVA. Capillary density of skeletal muscle in well-trained and untrained men. <u>J. Appl. Physiol.</u> 30: 860-863,1971.
11.   HONIG, C. R., C. L. ORDOROFF, AND J. L. FRIERSON. Active and passive capillary, control in red muscle at rest and in exercise. <u>Am. J. Physiol.</u> 243: H196-H206,1982.
12.   HONIG, C. R., T. E. J. GAYESKI, AND K. GROEBE. Myoglobin and Oxygen Gradients. In: *The Lung: Scientific Founditions*, 1st ed., edited by R. G. Crystal, and West, J. B. New York: Raven Press, Vol. 2, 1991, p. 1489-1496.
13.   IVY, J. L., R. T. WITHERS, P. J. VAN HANDEL, D. H. ELGER, AND D. L. COSTILL. Muscle respiratory capacity and fiber type as determinants of lactate threshold. <u>J. Appl. Physiol.</u> 48: 523-527,1980.
14.   JONES, P. R. M., AND J. PEARSON. Anthropometric determination of leg fat and muscle plus bone volumes in young male and female adults. <u>J. Physiol.</u> 294: 63P,1969.
15.   KLITZMAN, B., AND B. R. DULING. Microvascular hematocrit and red cell flow in resting and conrtacting striated muscle. <u>Am. J. Physiol.</u> 237: H481-H490,1979.
16.   KNIGHT, D. R., D. C. POOLE, W. SCHAFFARTZIK, H. J. GUY, R. PREDILETTO, M. C. HOGAN, AND P. D. WAGNER. Relationship between body and leg VO2 during maximal cycle ergometry. <u>J. Appl. Physiol.</u> 73: 1114-1121,1992.

17. KNIGHT, D. R., W. SCHAFFARTZIK, D. C. POOLE, M. C. HOGAN, D. E. BEBOUT, AND P. D. WAGNER. Hyperoxia increases leg maximal oxygen uptake. J. Appl. Physiol. 75: 2586-2594,1993.

18. MAI, J. V., V. R. EDGERTON, AND R. J. BARNARD. Capillarity of red, white and intermediate muscle fibers in trained and untrained guinea-pigs. Experientia 26: 1222-1223,1970.

19. MATHIEU-COSTELLO, O. Capillary tortuosity and degree of contraction or extension of skeletal muscles. Microvasc. Res. 33: 98-117,1987.

20. MATHIEU-COSTELLO, O., C. G. ELLIS, R. F. POTTER, I. C. MACDONALD, AND A. C. GROOM. Muscle capillary-to-fiber ratio: morphometry. Am. J. Physiol. 261: H1617-H1625,1991.

21. MELLANDER, S. Differentiation of fiber composition, circulation, and metabolism in limb muscles of dog, cat, and man. In: *Mechanisms of vasodilation*, 1st ed., edited by P. M. Vanhoutte, and Leusen, I. New York: Raven, 1981, p. 243-254.

22. POOLE, D. C., G. A. GAESSER, M. C. HOGAN, D. R. KNIGHT, AND P. D. WAGNER. Pulmonary and leg VO2 during submaximal exercise: implications for muscular efficiency. J. Appl. Physiol. 72: 805-810,1992.

23. RICHARDSON, R. S., S. C. JOHNSON, AND M. S. WALKER. Heart rate VO2 relationship changes following intense training. Sports Med., Training, and Rehab. 3: 105-111,1992.

24. RICHARDSON, R. S., D. C. POOLE, D. R. KNIGHT, S. S. KURDAK, M. C. HOGAN, B. GRASSI, E. C. JOHNSON, K. KENDRICK, B. K. ERICKSON, AND P. D. WAGNER. High muscle blood flow in man: Is maximal O2 extraction compromised? J. Appl. Physiol. 75: 1911-1916,1993.

25. ROWELL, L. B. Muscle blood flow how high can it go? Med. Sci. Sports. Exerc. 20: S97-S103,1988.

26. ROWELL, L. B., B. SALTIN, B. KIENS, AND N. J. CHRISTENSEN. Is peak quadriceps blood flow in humans even higher during exercise with hypoxemia? Am. J. Physiol. 251: H1038-H1044,1986.

27. SALTIN, B., C. G. BLOMQVIST, J. H. MITCHELL, R. C. Jr. JOHNSON, K. WILDENTHAL, AND C. B. CHAPMAN. Responses to exercise after bed rest and training. Circulation 38: 1-78,1968.

28. SALTIN, B., B. KIENS, G. SAVARD, AND P. K. PEDERSEN. Role of hemoglobin and capillarization for oxygen delivery and extraction in muscular exercise. Acta Physio. Scand. 128(Suppl. 556): 21-32,1986.

29. WHIPP, B. J., AND K. WASSERMAN. Oxygen uptake kinetics for various intensities of constant-load work. J. Appl. Physiol. 33: 351-356,1972.

CAPILLARISATION, FIBRE TYPES AND MYOGLOBIN CONTENT OF THE DOG GRACILIS MUSCLE

H. Degens, B.E.M. Ringnalda, and L.J.C. Hoofd

Department of Physiology
University of Nijmegen
The Netherlands

INTRODUCTION

Heterogeneity of capillary spacing has been shown to play an important role in modelling of pO_2 distribution in rat heart tissue (Turek et al., 1991; Hoofd, 1992; Hoofd and Turek, 1992). Another tissue often considered in modelling of oxygen transport to tissue is the dog gracilis skeletal muscle (Groebe, 1990). However, modelling in this tissue hitherto did not take into account heterogeneity of capillary spacing. In a preliminary study we found that such heterogeneity affects calculated pO_2 considerably (Hoofd and Turek, 1994). However, little is known about the capillarisation in dog gracilis muscle, capillary density values ranging from 510 mm^{-2} (Honig and Odoroff, 1981) to 2580 mm^{-2} (see Plyley and Groom, 1975) and no data of heterogeneity. Myoglobin might play an important role in facilitating oxygen diffusion or as an oxygen buffer (Wittenberg and Wittenberg, 1989). A positive, though weak, correlation between the oxidative capacity of a fibre and its myoglobin concentration was found (Van der Laarse et al., 1985). Since little is known about the capillarisation, fibre type composition and myoglobin content, we measured these parameters in the present study and considered their effect in a model of tissue oxygenation.

METHODS

Beagles (n=3) were obtained from the animal laboratory of the university. The gracilis muscle was excised and muscle blocks were taken from several locations along the length and the width of the muscle. The muscle blocks were frozen in isopentane cooled in liquid nitrogen. Myoglobin content was determined in three blocks of each muscle by the method of Reynafarje (1963). In short, the muscle specimens free of fat and connective tissue were homogenised in phosphate buffer and subsequently centrifuged at 0^0C during 60 min. with 15000 RPM. After centrifugation the supernatant was slowly bubbled with CO during 8 min. Then a pinch of dry Na-dithionite was added to obtain a complete reduction of the myoglobin and the CO was bubbled for another 2 min. Thereafter the absorption spectrum of the carboxymyoglobin was obtained at 538 and 568 nm and the myoglobin content was calculated.

Transverse sections of the other twenty-two muscle blocks were cut at -25^0C and stained for ATP-ase. A photomicrograph was taken from one section of each muscle block. The photomicrographs were analyzed for fibre type composition and capillarisation. Fibres were classified as type I or II based on ATP-ase activity at pH 4.0 and 4.4. In the same sections the capillaries were identified. A consecutive section was stained for succinate dehydrogenase. Since all fibres showed a similar staining intensity for succinate dehydrogenase no further subdivision into oxidative and glycolytic fibres could be made.

Using a digitising tablet, fibre outlines were read into the computer as contour co-ordinates and capillary locations as coordinates of capillary centres. Fibre cross-sectional areas (in μm^2) were derived from complete fibre contours on these photomicrographs. Fibre type composition was assessed as number percentage and as area percentage (area occupied by a certain fibre type divided by the total area occupied by both type I and type II fibres). The amount of non-contractile material was estimated as the area fraction of intercellular tissue of the cross-sections. The capillary density (CD) was defined as the number of capillaries per square millimetre of tissue. Capillary domains were constructed, defined as the area bounded by lines perpendicular to the mid-point of connecting lines to the adjacent capillaries. This method also offers the possibility to estimate the heterogeneity of capillary spacing as the logarithmic standard deviation of the domain areas (Hoofd et al., 1985).

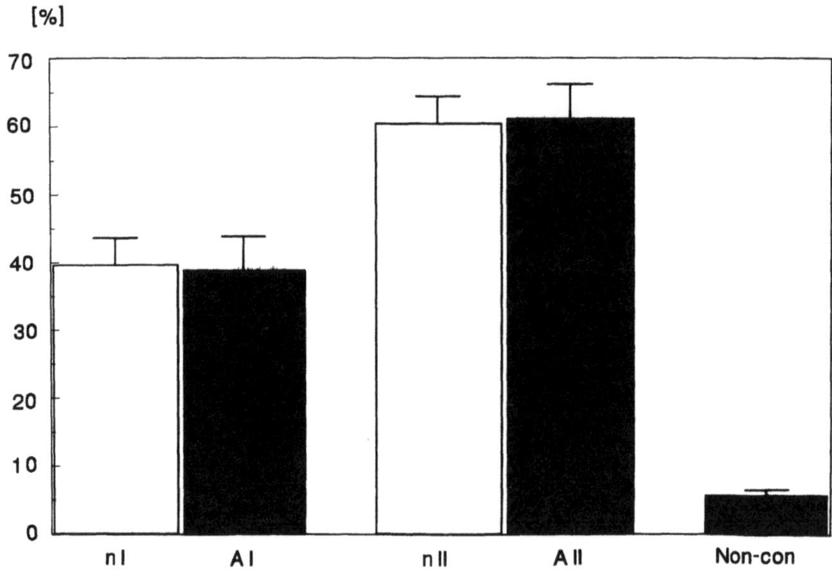

Figure 1. Fibre type composition and amount of non-contractile material (Non-con) in the dog gracilis muscle. nI, nII: number percentage; AI, AII: area percentage of type I or type II fibres.

534

The data were presented as a mean ± SD values unless stated otherwise. Finally the data were fed into a model as described earlier (Hoofd and Turek, 1992; Hoofd and Turek, 1994).

RESULTS

Muscle weights varied between 33 and 39 g. The muscles contained approximately 40% type I and 60% type II fibres. The same distribution holds when the fibre type composition was expressed as area percentages (Fig. 1). This indicates, that type I and type II fibres have similar cross-sectional areas, as confirmed by direct measurements (Table 1). The large variation in fibre cross-sectional areas was similar for both type I and type II fibres (Table 1). Since staining for SDH revealed that all fibres in the muscle were oxidative, no subdivision could be made into oxidative or glycolytic type II fibres. Approximately 6% of the muscle cross-sectional area was occupied by non-contractile material (Fig. 1).

Table 1. Capillarisation and fibre cross-sectional areas in the dog gracilis muscle.

CD	520	FCSA I	2228 (1221)	FCSA II	2233 (1942)
LogSD	0.193	SD I	1076	SD II	1135

CD: capillary density in mm^{-2} estimated from all capillaries as if in one single photomicrograph; LogSD: heterogeneity of capillary spacing; FCSA: Fibre cross-sectional area in μm^2; SD: standard deviation of the FCSA; number of fibres in parentheses.

Table 1 shows overall data for all the muscle samples together. Capillary density (520 mm^{-2}) and its heterogeneity (0.193 = Standard Deviation of logarithm of domain areas) result from combining all photomicrographs into one large area. However, it was obvious from the photomicrographs that there was quite a variation between the samples. Therefore, the following tables show data sorted for the different cases (dog and location along the length of the muscle). For the capillary density, shown in Table 2, averaging these CD's leads to a value of 507 mm^{-2}, only slightly different from the overall value of Table 1. Values of logSD determined in each photomicrograph varied from 0.130 to 0.205 with a mean value of 0.166.

Table 2. Capillary density of the different specimens of the dog gracilis muscle.

	Dog 24/3	Dog 23/4	Dog 10/5	mean ± SD (#)
Proximal	419	603 ± 20 (3)	403 ± 71 (2)	506 ± 114 (6)
Midbelly	551 ± 54 (4)	432 ± 85 (5)	605	497 ± 110 (10)
Distal	497 ± 92 (3)	209	723 ± 224 (2)	524 ± 222 (6)
mean ± SD	514 ± 98 (8)	464 ± 141 (9)	571 ± 199 (5)	507 ± 142 (22)

Capillary density in mm^{-2}; mean value of the photomicrographs ± SD and number of photomicrographs in parentheses when applicable

Table 3. Myoglobin content (mg/g tissue wet weight) of the dog gracilis muscle.

	Dog 24/3	Dog 23/4	Dog 10/5	mean ± SD
Proximal	5.39	9.38	9.15	7.97 ± 2.24
Midbelly	5.04	7.39	7.15	6.53 ± 1.30
Distal	4.69	6.21	5.86	5.59 ± 0.80
mean ± SD	5.04 ± 0.35	7.66 ± 1.60	7.39 ± 1.66	6.70 ± 1.71

Table 4. Area percentage of type I fibres of the dog gracilis muscle.

	Dog 24/3	Dog 23/4	Dog 10/5	mean ± SD (#)
Proximal	73.0	41.7 ± 27.4 (3)	76.1 ± 7.7 (2)	58.4 ± 25.4 (6)
Midbelly	38.2 ± 20.7 (4)	39.7 ± 25.8 (5)	22.4	37.4 ± 20.6 (10)
Distal	15.2 ± 2.9 (3)	29.4	27.4 ± 4.0 (2)	21.6 ± 7.5 (6)
mean ± SD	33.9 ± 23.8 (8)	39.2 ± 22.0 (9)	45.9 ± 28.0 (5)	38.8 ± 23.3 (22)

Values in % ; mean value of the photomicrographs ± SD and number of photomicrographs in parentheses when applicable

Sorted data on myoglobin content are given in Table 3 and for the area percentage of type I fibres in Table 4. Correlations between myoglobin content and area percentage of type I fibres (r=0.38; n=9; p=0.31), capillary density (r=0.01; n=9; p=0.98) or heterogeneity of capillary spacing (r=0.10; n=9; p=0.80) in adjacent muscle blocks were not significant.

For calculations of tissue pO_2 a representative area was selected of one of the photomicrographs with the same capillary density as the overall one (520 mm^{-2}) and a logSD value in the low range (0.14). The area was 500 μm $\times$ 300 μm containing 78 capillaries and was extended to a tissue block of 375 μm thick with two-stage staggered capillaries (Hoofd and Turek, 1992) so as to yield the same venous O_2 values as described by Groebe (1990). Also, all the other parameter values for the calculations were taken from Groebe's publication. The influence of heterogeneity of capillary spacing is best appreciated when comparing with a regular pattern. Therefore, also calculations were done in a tissue block of 264 μm $\times$ 264 μm $\times$ 375 μm containing 36 capillaries in a rectangular pattern (spacing distance 44 μm; CD= 516 mm^{-2}; logSD= 0). For such a regular arrangement, the calculated pO_2 histogram is in agreement with the findings of Groebe (1990) - dark bars in the pO_2 histogram of Fig. 2. However, for realistic capillary spacing a completely different pO_2 histogram results (shaded bars in Fig. 2). This is most obvious for the amount of anoxic tissue calculated (leftmost bars in Fig. 2) - it is insignificant for the regular case (1%) but extremely high for the heterogeneous case (38%).

DISCUSSION

We found that the dog gracilis muscle contained approximately 40% type I and 60% type II fibres both in terms of area and number percentage. In previous studies a similar distribution was found though slightly more type II (70%) and less type I (30%) fibres were reported (Mayne et al. 1991; Amann et al. 1993). However, it is not clear

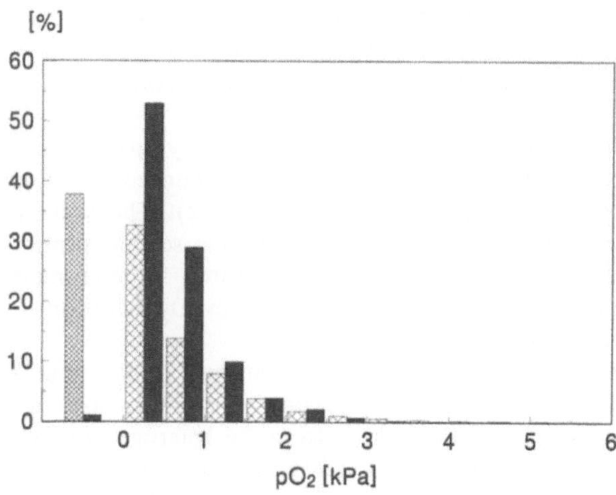

Figure 2. Histogram pO$_2$ calculations for a regular capillary pattern (dark bars) or a typical cross-section containing 78 capillaries (shaded bars) with tissue and blood data according to Groebe (1990).

from these studies where the biopsies were taken. The fibres of both types had a similar size with large variation in their cross-sectional areas as was also found by Mayne et al. (1991). The large variation in fibre size might be the cause of the relatively large heterogeneity of capillary spacing.

We found that the overall capillary density of the gracilis muscle was 520 mm^{-2}. The density is much lower than that estimated as the number of perfused capillaries after contraction (Honig et al. 1982). An explanation might be the use of mongrel dogs in their study whereas we used Beagles. Another possibility might be that the capillarisation in the superficial part of the muscle is dissimilar from that in deeper layers. Concerning the high value of 2580 mm^{-2} reported by Plyley and Groom (1975) it must be mentioned that this comes together with a very small fibre size of 775 μm^2.

For whole dog gracilis muscle we found a myoglobin content of 6.70 mg/g wet weight, which is the same as Reynafarje (1963) reported for dog thigh muscle. Meng et al. (1993), by using the same method, found a content of 4.53 mg/g wet weight for dog gracilis. The value used by Groebe (1990) is about 9 mg/g. The difference might be explained by regional differences in myoglobin content in the dog gracilis muscle, since we found an increase in the myoglobin content of this muscle from distal to proximal (Table 3).

Myoglobin might facilitate oxygen diffusion or serve as a buffer for oxygen (Wittenberg and Wittenberg, 1989). Therefore it might be expected that the more oxidative fibres have a higher myoglobin content. Indeed a positive, though weak correlation was reported between the oxidative capacity of a fibre and its myoglobin concentration (Van der Laarse et al. 1985). The present study could not corroborate this finding. Another explanation for the differences in myoglobin content along the length of the muscle might be related to difference in capillary density as was also suggested by Meng et al.

(1993). This comes as a reasonable assumption, since a reduced capillary density would demand an higher facilitation of oxygen diffusion, when all other factors influencing oxygen demand are similar. However, no significant correlation could be established in our data.

The major influence of heterogeneity of capillary spacing on calculated tissue pO_2 becomes obvious from Fig. 2. It is even more pronounced than found previously for rat heart tissue (Hoofd, 1992; Hoofd and Turek, 1992). The main reason might be that capillary distances are larger in dog gracilis muscle so that similar heterogeneities lead to larger differences in absolute distance measured in micrometers. Thus, the point where distance grows too large to be overcome by O_2 diffusion is reached earlier rendering more tissue anoxic.

Fig. 2 clearly shows that models of O_2 transport in dog gracilis tissue are unrealistic when not accounting for the heterogeneity of capillary spacing. On the other hand, the large anoxic tissue fraction resulting from the heterogeneous model is also unrealistic, even though the actual anoxic fraction will be less because the model overestimates such a fraction (Hoofd, 1992). It must be concluded that present modelling of tissue pO_2 in tetanic dog gracilis muscle is incomplete. Some of the parameter values used in the calculations might be dubious but also modelling of especially O_2 consumption in this tissue (zero-order model) might not be realistic - some of these factors have also been discussed in another model (Turek et al., 1991).

CONCLUSIONS

1. The muscle was homogeneously oxidative as indicated by SDH-stained sections.
2. Type I and II fibres had similar cross-sectional areas.
3. The muscle contained more type II than type I fibres, both in number percentage and area percentage.
4. The amount of non-contractile material was 5.7 ± 0.8 %.
5. The myoglobin content (mean value 6.7 mg/g tissue) decreased from the proximal to the distal part of the muscle.
6. Present modelling of tissue pO_2 in dog gracilis muscle is incomplete.

REFERENCES

Amann, J.F., Wharton, R.E., Madsen, R.W., and Laughlin, M.H., 1993, Comparison of muscle cell fibre types and oxidative capacity in gracilis, rectus femoris, and triceps brachii muscles in the ferret (*Mustela pustorius furo*) and the domestic dog (*Canis familiaris*). *Anat. Rec.* 236: 611-618.

Groebe, K., 1990, A versatile model of steady state O_2 supply to tissue. Application to skeletal muscle. *Biophys. J.* 57: 485-498.

Honig, C.R., and Odoroff C.L., 1981, Calculated dispersion of capillary transit times: significance for oxygen exchange. *Am. J. Physiol.* 240: H199-H208.

Honig, C.R., Odoroff C.L., and Frierson, J.L., 1982, Active and passive capillary control in red muscle at rest and in exercise. *Am. J. Physiol.* 243: H196-H206.

Hoofd, L., 1992, Updating the Krogh model - assumptions and extensions. *In*: Oxygen Transport in Biological Systems. Eds. S. Egginton, H.F. Ross, Cambridge University Press, Cambridge, pp. 197-229.

Hoofd, L., and Turek, Z., 1992, Oxygen pressure histograms calculated in a block of rat heart tissue, *In*: Oxygen Transport to Tissue-XIV. Eds. W. Erdmann, D.F. Bruley, Plenum Press, New York, pp. 561-566.

Hoofd, L., and Turek, Z., 1994, Effect of realistic capillary spacing on pO_2 calculations in dog gracilis muscle. *Med. Biol. Eng. Comp.* in press.

Hoofd, L., Turek, Z., Kubat, K., Ringnalda, B.E.M., and Kazda, S., 1985, Variability of intercapillary distance estimated on histological sections of rat heart. *In*: Oxygen Transport to Tissue-VII. Eds. F. Kreuzer, S.M. Cain, Z. Turek, T.K. Goldstick, Plenum Press, New York, pp. 239-247.

Mayne, C.N., Anderson, W.A., Hammond, R.L., Eisenberg, B.R., Stephenson, L.W., and Salmons, S., 1991, Correlates of fatigue resistance in canine skeletal muscle stimulated electrically for up to one year. *Am. J. Physiol.* 261: C259-C270.

Meng, H., Bentley, T.B., and Pittman, R.N., 1993, Myoglobin content of hamster skeletal muscles. *J. Appl. Physiol.* 74: 2194-2197.

Plyley, M.J., and Groom, A.C., 1975, Geometrical distribution of capillaries in mammalian striated muscle. *Am. J. Physiol.* 228: 1376-1383.

Reynafarje, B., 1963, Simplified method for the determination of myoglobin. *J. Lab. Clin. Med.* 61: 138-145.

Turek, Z. Rakusan, K., Olders, J., Hoofd, L., and Kreuzer, F., 1991, Computed myocardial pO_2 histograms: effects of various geometrical and functional conditions. *J. Appl. Physiol.* 70: 1845-1853.

Van der Laarse , W.J., Maslam S.M., and Diegenbach, P.C., 1985, Relationship between myoglobin and succinate dehydrogenase in mouse soleus and plantaris muscle fibres. *Histochem. J.* 17: 1-11.

Wittenberg, B.A., and Wittenberg, J.B., 1989, Transport of oxygen in muscle. *Annu. Rev. Physiol.* 51: 857-878.

KEY WORDS

Capillarisation, Myoglobin, Gracilis, Fibre type, Skeletal muscle, Tissue oxygenation.

LEUKOCYTE-ENDOTHELIUM INTERACTION

IN THE MICROVASCULATURE OF

POSTISCHEMIC STRIATED MUSCLE

M.D. Menger[1], H. Kerger[2], A. Geisweid[1],
A.J. Leu[3], R. Hecht[1], D. Nolte[1],
and K. Messmer[1]

[1]Institute for Surgical Research, University of Munich, FRG
[2]Department of Anesthesiology University Hospital of Mannheim
 University of Heidelberg, FRG
[3]Angiology Division, Department of Internal Medicine, University Hospital
 Zürich, Switzerland

Ischemia/reperfusion and the microcirculation

Microvascular manifestation of postischemic reperfusion injury is hallmarked by two distinct events, i.e. perfusion failure of nutritive capillaries (*"no-reflow"*) (Majno et al., 1967; Menger et al., 1992a), and the sequelae of *"reflow-paradox"*-associated mechanisms, which include accumulation of leukocytes, leukocyte-endothelium interaction and loss of endothelial integrity (Menger et al., 1992b). Both reperfusion-dependent events contribute to the aggravation of ischemia-induced parenchymal tissue injury, either through prolongation of focal ischemia (*"no-reflow"*), or the action of leukocyte-derived cytotoxic mediators (*"reflow-paradox"*), such as reactive oxygen metabolites and proteolytic enzymes (Granger, 1988; Lehr et al., 1991; Menger et al., 1993a).

Although there is general agreement that leukocytes accumulate in postischemic tissue and contribute to ischemia/reperfusion injury, the anatomical location and the mechanisms elicited by leukocytes are a matter of controversy. During inflammation and ischemia/reperfusion activated leukocytes do not interact with the endothelial lining of terminal (A_2, A_3, A_4) arterioles. In contrast, a variety of in vivo studies have demonstrated that the endothelial lining of postcapillary venules, subjected to inflammatory stimuli (von Andrian et al., 1991) and ischemia/reperfusion (Granger et al., 1989; Lehr et al., 1991; Menger et al., 1992b), represents the primary target for the induction of leukocyte-endothelium interaction. Moreover, activated leukocytes have been proposed to exert detrimental effects to inflamed and ischemic tissue by plugging of nutrtitive capillaries (Engler and Schmid-Schönbein, 1983; Schmid-Schoenbein, 1987; Harris and Skalak, 1993) with the result of prolonged focal ischemia.

Capillary plugging by leukocytes in postischemic *"no-reflow"*

Schmid-Schönbein (1987) has proposed that under conditions of reduced capillary perfusion pressure and/or elevated levels of inflammatory products, granulocytes may become stuck in the capillaries. Despite restoration of perfusion pressure these granulocytes will not be removed from the capillaries due to firm adhesion to the endothelial surface, thus contributing to *"no-reflow"* (Schmid-Schoenbein, 1987). However, this phenomenon has not been documented in capillaries *in vivo*. Bagge et al. (1980) demonstrated leukocyte plugging of capillaries in the low flow shock state, but not following restoration of blood flow (reperfusion). In addition, the authors explicitly state that the leukocytes were not firmely attached to the capillary wall, but often found sliding smoothly back and forth during the low flow state.

In our *in vivo* microscopic analysis of the microcirculation of striated muscle exposed to ischemia/reperfusion, we could rarely document leukocytes, permanently occluding capillaries during postischemic reperfusion (Menger et al., 1992a). Recent reports indicate that inflammation and ischemia/reperfusion may induce reduction of capillary leukocyte velocity (Richardson et al., 1992) and temporary plugging at capillary bifurcations for a time period of seconds (Harris and Skalak, 1993), however, leukocytes have not been shown to obstruct capillaries for prolonged periods of time, resulting in capillary *"no-reflow"*. The notion that plugging of capillaries by activated leukocytes is not the key mechanism of capillary *"no-reflow"* is supported by in vivo analysis of capillary distribution of leukocytes in skeletal muscle microcirculation (Blixt et al., 1987; Ley et al., 1989). Under physiologic conditions leukocytes are known to prefer volume flow dependent pathways to cross the capillary network (Blixt et al., 1987). This shunting of leukocytes is regarded as a mechanism to prevent plugging of nutritional capillaries (Ley et al., 1989). Shunting of leukocytes may even be more pronounced during postischemic reperfusion due to the alteration of the leukocytes visco-elastic properties. However, the fact that leukocyte plugging cannot account for capillary perfusion failure during postischemic reperfusion does not exclude the possibility that leukocytes, which are present within the capillaries during ischemia, may change their visco-elastic properties as consequence of ischemia-induced acidosis, and, therefore, alter microvascular reperfusion by impaired hemorheological conditions.

Apart from plugging of capillaries, several mechanisms have been suggested to promote the degree of postischemic *"no-reflow"*, including (i) thrombosis of microvessels (Quiñones-Baldrich et al., 1991), (ii) swelling of capillary endothelial cells (Gidlöf et al., 1987; Hammersen et al., 1989), (iii) impairment of microvascular blood fluidity (Fischer and Ames, 1972; Menger et al., 1988), and (iv) increase of hydraulic resistance (Gidlöf et al., 1987; Mazzoni et al., 1989). Thrombosis of microvessels does not appear as key mechanism in capillary *"no-reflow"*, since histological studies failed to demonstrate platelet and fibrin thrombi in capillaries (Harman, 1948; Strock and Majno, 1969), and both heparin and prostaglandin E_1 were found ineffective in preventing postischemic breakdown of capillary perfusion (Strock and Majno, 1969; Rosolowsky and Weiss, 1987). Therefore, other mechanisms, such as swelling of capillary endothelial cells, impairment of microvascular blood fluidity, and increased hydraulic resistance, may contribute to ischemia/reperfusion-induced deterioration of capillary perfusion (see Menger et al., 1992a).

In addition to these mechanisms, possibly involved in the pathogenesis of capillary *"no-reflow"*, microhemodynamic changes in postcapillary segments may account for post-ischemic capillary perfusion failure. We have demonstrated a significant decrease of wall shear rate in postcapillary and collecting venules during the initial reperfusion period (Menger et al., 1992a). The decrease of wall shear rate may be caused by increased postcapillary vascular resistance, probably due to leukocyte accumulation and adherence in postcapillary venules (Korthuis et al., 1988; Menger et al., 1993b).

Leukocyte-endothelium interaction in postcapillary venules

It is well established that during postischemic reperfusion leukocyte accumulation within the postcapillary segments of the microvasculature, and their adherence to the endothelial lining are the primary steps in the development of reperfusion injury (Granger et al., 1989; Lehr et al., 1991; Menger et al., 1992b). Leukocyte margination and the interaction with endothelial cells are considered as prerequisites for transendothelial migration and tissue infiltration (von Andrian et al., 1991). The in vivo interaction between leukocytes and endothelial cells has been suggested to follow a multi-step process involving distinct adhesion receptor molecules on the surface of both leukocytes (L-selectin, $ß_2$-integrins) and endothelial cells (P-, and E-selectin, intercellular adhesion molecules) (Springer, 1990; von Andrian et al., 1991). Primary leukocyte-endothelium interaction (rolling), mediated by L-selectin, has been proposed to be the prerequisite for leukocyte adherence (sticking), since monoclonal antibodies directed against this molecule effectively reduce both rolling and sticking in inflammation (von Andrian et al., 1991) and ischemia/reperfusion (Hecht et al., 1993a). Firm attachment of leukocytes to the microvascular endothelium (secondary leukocyte-endothelium interaction) is then mediated by ß2 integrins, involving in particular the CD11b/CD18 glycoprotein complex. This view is supported by experiments, demonstrating that monoclonal antibodies directed against the CD11b or CD18 subunit prevent inflammation- and ischemia/reperfusion-induced leukocyte adherence and emigration into tissue, and, concomitantly, attenuate microvascular injury (Arfors et al., 1987; Hernandez et al., 1987, Hecht et al., 1993b).

Although postischemic reperfusion and reoxygenation is a *'sine qua non'* for the salvage of the ischemic tissue, paradoxically, reflow-mediated events have been shown to contribute to the aggravation of ischemia-induced injury (*"reflow-paradox"*, Menger et al., 1992b). Reactive oxygen metabolites, generated upon reoxygenation, may, besides their toxic action on cellular membranes by lipid peroxidation (McCord, 1985), activate leukocytes (Granger, 1988) and induce expression of adhesion molecules on endothelial cells (Patel et al., 1991), resulting in accumulation and adherence of leukocytes with subsequent transendothelial migration into the tissue (Granger et al., 1989; Menger et al., 1992b). Both, oxygen radicals and activated leukocytes promote the release of potent mediators, such as leukotrienes (Lehr et al., 1991) and platelet-activating factor (Lewis et al., 1988), which in turn have the potential to chemotactically influence leukocytes (Björk et al., 1982; Kubes et al., 1990; Lehr et al., 1991). These pathophysiologic events result in a self-sustaining vitious circle. Leukotrienes (Björk et al., 1982; Lehr et al., 1991) and platelet-activating factor (Björk and Smedegård, 1983; Kubes et al., 1990) cause an increase in microvascular permeability, followed by interstitial edema and elevation of tissue pressure, which may additionally compromise capillary perfusion at the time of reperfusion, and hence increase postischemic tissue damage due to focal ischemia.

CONCLUSION

In conclusion, we propose that accumulation and adherence of leukocytes to microvascular endothelium following ischemia/reperfusion is a characteristic component of the postischemic *"reflow-paradox"*. Leukocyte adherence is observed in postcapillary venules, but not in capillaries and arterioles. Differences in expression of adhesion molecules on the surface of endothelial cells may account for the manifestation of *"reflow-paradox"* in the postcapillary venules. The role of leukocytes in the production and progression of postischemic capillary *"no-reflow"* remains still a matter of controversy, and requires further clarification by experimental *in vivo* studies.

REFERENCES

Andrian, U.H.v., Chambers, J.D., McEvoy, L.M., Bargatze, R.F., Arfors, K.-E., and Butcher, E.C., 1991, Two-step model of leukocyte-endothelial cell interaction: distinct roles for LECAM-1 and the leukocyte ß2 integrins in vivo, *Proc. Natl. Acad. Sci. USA* 88:7538.

Arfors, K.-E., Lundberg, C., Lindbom, L., Lundberg, K., Beatty, B.G., and Harlan, J.M., 1987, A monoclonal antibody to the membrane glycoprotein complex CD18 inhibits polymorphonuclear leukocyte accumulation and plasma leakage *in vivo, Blood* 69:338.

Bagge, U., Amundson, B., and Lauritzen, C., 1980, White blood cell deformability and plugging of skeletal muscle capillaries in hemorrhagic shock, *Acta Physiol. Scand.* 180:159.

Björk, J., Hedqvist, P., and Arfors, K.-E., 1982, Increase in vascular permeability induced by leukoriene B$_4$ and the role of polymorphonuclear leukocytes, *Inflammation* 6:189.

Björk, J., and Smedegård, G., 1983, Acute microvascular effects of PAF-acether, as studied by intravital microscopy, *Eur. J. Pharmacol.* 96:87.

Blixt, Å., Braide, M., Myrhage, R., and Bagge, U., 1987, Vital microscopic studies on the capillary distribution of leukocytes in the rat cremaster muscle, *Int. J. Microcirc.: Clin. Exp.* 6:273.

Engler, R.L., Schmid-Schönbein, G.W., and Pavelec, R.S., 1983, Leukocyte capillary plugging in myocardial ischemia and reperfusion in the dog, *Am. J. Pathol.* 111:98.

Fischer, E.G., and Ames, A., 1972, Studies on mechanisms of impairment of cerebral circulation following ischemia: Effect of hemodilution and perfusion pressure. *Stroke* 3: 538.

Gidlöf, A., Lewis, D.H., and Hammersen, F., 1987, The effect of prolonged total ischemia on the ultrastructure of human skeletal muscle capillaries. A morphometric analysis, *Int. J. Microcirc.: Clin. Exp.* 7:67.

Granger, D.N., 1988, Role of xanthine oxidase and granulocytes in ischemia-reperfusion injury, *Am. J. Physiol.* 255:H1269.

Granger, D.N., Benoit, J.N., Suzuki, M., and Grisham, M.B., 1989, Leukocyte adherence to venular endothelium during ischemia-reperfusion, *Am. J. Physiol.* 257:G683.

Hammersen, F., Barker, J.H., Gidlöf, A., Menger, M.D., Hammersen, E., and Messmer, K., 1989, The ultra structur of microvessels and their contents following ischemia and reperfusion, *Prog. Appl. Microcirc.* 13:1.

Harman, J.W., 1948, The significance of local vascular phenomenona in production of ischemic necrosis in sceletal muscle, *Am. J. Pathol.* 24:625.

Harris, A.G., and Skalak, T.C., 1993, Effects of leukocyte activation on capillary hemodynamics in skeletal muscle, *Am. J. Physiol.* 264:H909.

Hecht, R., Nolte, D., Botzlar, A., Menger, M.D., and Messmer, K., 1993a, Monoclonal antibody to L-selectin prevents postischemic leukocyte-endothelium interaction in postcapillary venules of the mouse, *Eur. Surg. Res.* 25/S1:56.

Hecht, R., Nolte, D., Botzlar, A., Menger, M.D., and Meßmer, K., 1993b, Monoclonal antibody against the leukocyte adhesion molecule MAC-1 (CD11b) prevents postischemic leukocyte adherence in vivo, *Langenbecks Arch. Chir. Forum* 93:333.

Hernandez, L.A., Grisham, M.B., Twohig, B., Arfors, K.E., Harlan, J.M., and Granger, D.N., 1987, Role of neutrophils in ischemia-reperfusion-induced microvascular injury, *Am. J. Physiol.* 253:H699.

Korthuis, R.J., Grisham, M.B., and Granger, D.N., 1988, Leukocyte depletion attenuates vascular injury in postischemic skeletal muscle, *Am. J. Physiol.* 254:H823.

Kubes, P., Suzuki, M., and Granger, D.N., 1990, Modulation of PAF-induced leukocyte adherence and increased microvascular permeability, *Am. J. Physiol.* 259:G859.

Lehr, H.A., Guhlmann, A., Nolte, D., Keppler, D., and Messmer, K., 1991, Leukotrienes as mediators in ischemia-reperfusion injury in a microcirculation model in the hamster, *J. Clin. Invest.* 87:2036.

Lewis, M.S., Whatley, R.E., Cain, P., McIntyre, T.M., Prescott, S.M., and Zimmerman, G.A., 1988, Hydrogen peroxide stimulates the synthesis of platelet-activating factor by endothelium and induces endothelial cell-dependent neutrophil adhesion, *J. Clin. Invest.* 82:2045.

Ley, K., Meyer, J.-U., Intaglietta, M., and Arfors, K.-E., 1989, Shunting of leukocytes in rabbit tenuissimus muscle, *Am. J. Physiol.* 256:H85.

Majno, G., Ames, A., Chiang, J., and Wright, R.L., 1967, No reflow after cerebral ischemia. *Lancet* ii: 569-570.

Mazzoni, M.C., Borgström, P., Intaglietta, M., and Arfors, K.-E., 1989, Lumenal narrowing and endothelial cell swelling in skeletal muscle capillaries during hemorrhagic shock, *Circ. Shock* 29:27.

McCord, J.M., 1985, Oxygen-derived free radicals in postischemic tissue injury, *N. Engl. J. Med.* 312:159.

Menger, M.D., Sack, F.U., Barker, J.H., Feifel, G., and Messmer, K., 1988, Quantitative analysis of microcir culatory disorders after prolonged ischemia in skeletal muscle: Therapeutic effects of prophylactic isovolemic hemodilution, *Res. Exp. Med.* 188:151.

Menger, M.D., Steiner, D., and Messmer, K., 1992a, Microvascular ischemia/reperfusion injury in striated muscle: Significance of *"no-reflow"*, *Am. J. Physiol.* 263:H1892.

Menger, M.D., Pelikan, S., Steiner, D., and Messmer, K., 1992b, Microvascular ischemia/reperfusion injury in striated muscle: Significance of *"reflow-paradox"*, *Am. J. Physiol.* 263:H1901.

Menger, M.D., Vollmar, B., Glasz, J., Post, S., and Messmer, K., 1993a, Microcirculatory manifestations of hepatic ischemia/reperfusion injury, *Prog. Appl. Microcirc.* 19:106.

Menger, M.D., Thierjung, C., Hammersen, F., and Messmer, K., 1993b, Prophylactic isovolemic hemodilution with dextran 60 attenuates postischemic leulocyte adherence in striated muscle, *Circ. Shock* (in press).

Patel, K.D., Zimmerman, G.A., Prescott, S.M., McEver, R.P., and McIntyre, T.M., 1991, Oxygen radicals induce human endothelial cells to express GMP-140 and bind neutrophils, *J. Cell. Biol.* 112:749.

Quiñones-Baldrich, W.J., Chervu, A., Hernandez, J.J., Colburn, M., and Moore, W.S., 1991, Skeletal muscle function after ischemia: "No reflow" versus reperfusion injury, *J. Surg. Res.* 51:5.

Richardson, M., Berker, A., Roberts, A., Guth, P., and Freischlag, J.A., 1992, In vivo microscopy of rat skeletal muscle after ischemia using labeled neutrophils (PMN), *J. Surg. Res.* 53:563.

Rosolowksy, M., and Weiss, H.R., 1987, Effect of blood coagulation and platelet aggregation on perfusable capillaries and arterioles in ischemic and nonischemic myocardium, *Microvasc. Res.* 34:69.

Schmid-Schönbein, G.W., 1987, Capillary plugging by granulocytes and the no-reflow phenomenon in the microcirculation, *Fed. Proc.* 46:2397.

Springer, T.A., 1990, Adhesion receptors of the immune system, *Nature* 346:425.

Strock, P.E., and Majno, G., 1969, Microvascular changes in acutely ischemic rat muscle, *Surg. Gynecol. Obstet.* 129:1213.

FLOW CHARACTERISTICS OF ERYTHROCYTES SUBJECTED TO PROLONGED INCUBATION IN PBS

P. Ram Rao[1], Vandhana Puri[2], Omoefe Abugo[3] and Joseph Rifkind[3]

[1]Department of Mechanical Engineering, UMBC, Baltimore, MD 21228
[2]University of Maryland, College Park, MD 20742
[3]Molecular Dynamics Section, Gerontology Research Center,
National Institutes of Health, Baltimore, MD 21224

INTRODUCTION

In some erythrocyte experimental studies, such as the study of protracted or severe hypoxia, it is necessary to incubate erythrocytes for prolonged periods of time. Therefore, it is relevant to determine the effects of incubation on the flow characteristics and the material properties of erythrocytes. It is known (Richterich, *et al.*, 1981) that incubation of RBC beyond several hours in a medium without special additives to preserve the cellular metabolic processes leads to the depletion of glucose and, eventually, of ATP. Without ATP the sodium-potassium pumps no longer function, and an osmotic perturbation takes place producing an alteration of the cell geometry (viz., discocytes into spheroechinocytes) (Schmid-Schoenbein *et al.*, 1983). These changes may be accompanied by a change in the membrane mechanical properties. Both, the sphericity increase and the perturbation of the mechanical properties of the membrane and the cytoplasm, can potentially lead to a change in the cell deformability. The degree to which these two aspects may contribute, if any, to the altered cell deformability is a matter of speculation. Also, it is unclear if rheological indices present a practical means of detecting biochemical damage to these incubated cells. We will address these questions in this paper. While the present experimental conditions are clearly aphysiological, the results of this study will nevertheless be valuable in that they will provide a baseline data on the effect of incubation by itself.

MATERIAL AND METHODS

The blood suspension was prepared as follows. Each sample of whole blood from normal male and female adult donors was spun at 3000 rpm for ten minutes, and the

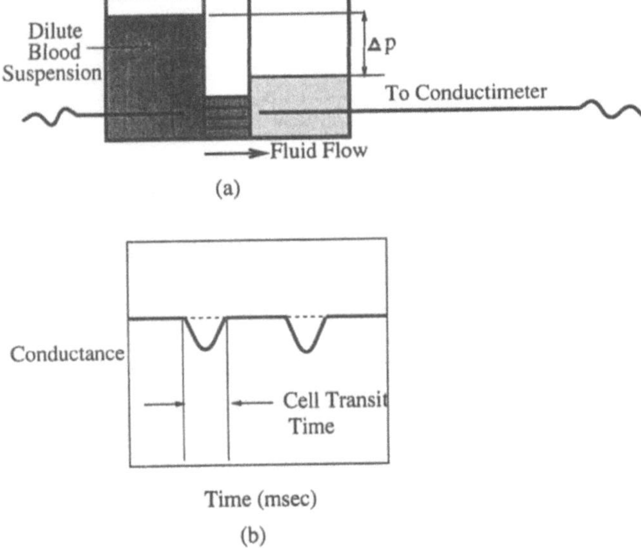

Figure 1. (a) Schematic of the capillary system. The dilute blood sample flows through each of the capillaries due to the applied hydrostatic pressure head, Δp. Due to the low hematocrit (less than 0.1 %) at any given moment only one cell enters the capillary system. The electrodes on either side of the capillaries detect a drop in conductance due to the presence of a cell. (b) The period of drop in the cell conductance is considered the cell transit time. Approximately, two thousand cells are examined per donor to obtain the MCTT.

buffy coat was siphoned off. The sample was then washed thrice by diluting with a phosphate buffered saline (PBS at 290 mOsmol, pH $= 7.4$), spinning at 3000 rpm for five minutes and siphoning off the supernatant. The resultant sample was diluted to 0.1 % hematocrit in the PBS solution, and gently mixed at regular intervals at room temperature. Each day a portion of this sample was removed and studied.

We employed three methods to assess the changes in the characteristics of erythrocytes due to incubation in PBS: a cell transit analyser (CTA), a Coulter counter and an SDS-PAGE apparatus. The purpose of the CTA was to create a capillary flow and to determine the average time (MCTT) taken by the red cells to flow through a fixed length of the capillary and the Filteration Rate (FR). The Coulter Counter provided the mean cell volume data for the same population of erythrocytes. SDS-PAGE apparatus was employed to determine the distribution of the membrane proteins as a function of their molecular weight.

The principle of the CTA is described in Figure 1. A polycarbonate filter with thirty well calibrated pores of diameter $5\,\mu m$ and thickness $15\,\mu m$ is placed between two liquid coloumns. The upstream coloumn contains a dilute suspension (0.1 % hematocrit) of red cells while the second one contains the buffer solution. By varying the heights of the liquid levels in the two coloumns we can develop different known pressure gradients across the parallel capillary system incorporated in the polycarbonate filter. As the red cells flow through these capillaries one at a time (this can be accomplished with sufficient dilution) we can record the duration of the drop in the electric conductance across the filter generated by the presence of the red cell in the capillary. The mean cell transit times, MCTT, for about two thousand cells can thus be recorded in an Apple II GS computer connected to the conductimeter which registers the drop in conductivity. In addition, we can also obtain the Filteration Rate defined as the number of cells per unit time that have flowed through the capillary system.

RESULTS AND DISCUSSION

It was observed that during the initial period of incubation ('Period I') the MCTT decreased till it reached a minimum. Incubation beyond this point ('Intermediate Day') led to a progressive increase in MCTT. (All other variables displayed a monotonic variation with incubation.) The experiment was terminated when FR became too slow to permit more than 500 cells to flow through the capillary system within an arbitrarily specified time of two minutes; this suggested a severe damage to the cell membrane. The durations of Period I and Period II (the remainder of the incubation period after the Intermediate Day) varied with the blood sample (donor). (See Table 1.) In view of this variation so as to be able to make meaningful deductions on the effect of incubation in general, the data for each sample was averged over all the days within a period. Thus, if a sample has three days constituting Period II, then the data for all the three days is averaged and designated as the Period II value for that particular variable. Similarly, if another donor has two days of data prior to the end of Period II, then those two days' of data is averaged to obtain the Period II data for that donor. In this manner, corresponding to each donor, and for each variable we will only have three data points tracing the entire incubation period: data for Period I, for Intermediate Day, and for Period II. The data for Period I for a given donor is *paired* with that for the Intermediate Day, and a Student t-test for paired samples is employed to determine if the changes between the values for Period I and for the Intermediate Day are statistically significant. Similar comparison is made between the Intermediate Day and Period II. We thus determine if there is a trend over the entire incubation period. Although, we need not view the FR and MCV data in the same manner as MCTT, i.e., condense them into three time segments – Period I, Intermediate Day and Period II – it is certainly convenient to do so from the standpoint of recognizing cause and effect.

Table 1. Lengths of 'Period I' and 'Period II', which represent, respectively, the incubation periods prior to and after the Intermediate Day when a minimum in the MCTT occurs.

Donor	A	B	C	D	E	F	G	H
Period I	1	1	2	1	1	1	2	2
Period II	1	1	1	1	1	2	5	2

Figure 2 shows that the FR decreases continuously with incubation. The decrease is statistically significant upto a confidence level of $p < 0.1$ prior to Intermediate Day, and up to $p < 0.05$ after the Intermediate Day. Figure 3 shows that MCTT is apparently not a monotonic function of incubation; it displays a minimum on the Intermediate Day (as mentioned earlier). These two variables, FR and MCTT, appear to behave in conflicting manners during the first period of incubation. While the decrease in FR

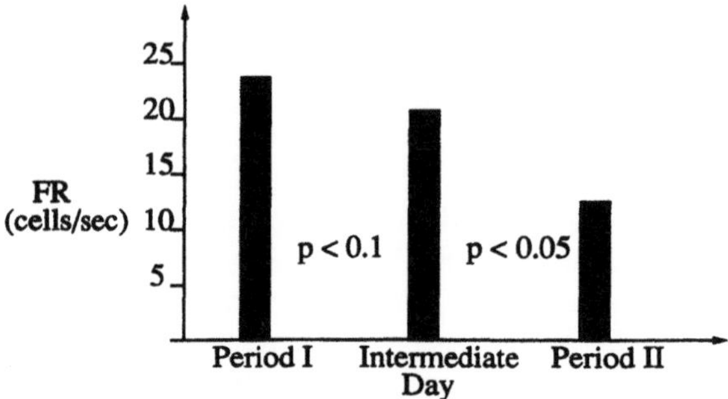

Figure 2. Filteration Rate, or number of cells flowing through the capillary system per second, averaged over 2000 cells per donor and 8 donors. The term 'Intermediate Day' is explained in the text. The monotonic decrease in FR with incubation is statistically significant when the data is viewed as paired samples (i.e., (Period I, Intermediate Day) and as (Intermediate Day, Period II)) and analysed according to paired samples Student t-test.

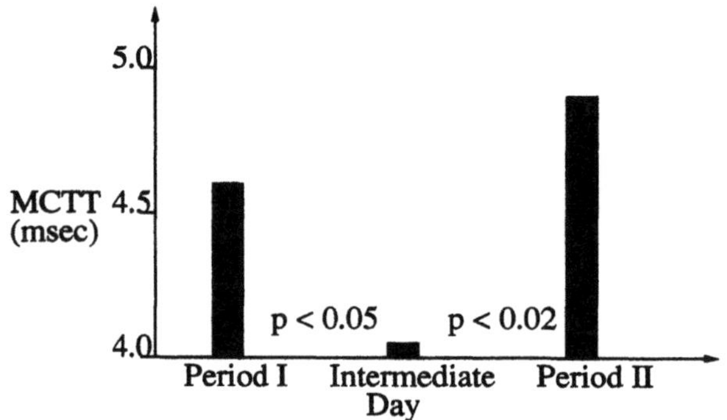

Figure 3. Mean Cell Transit Time (MCTT), or time taken by a single cell to flow through the 15 μm long capillary system averaged over 2000 cells per donor and 8 donors. The term 'Intermediate Day' is explained in the text. The changes in MCTT with incubation are statistically significant according to paired samples Student t-test.

during this period indicates a decrease in the *ease* of the cell flow, the decrease in MCTT
for the same time period indicates that the opposite must be the case. This apparent
contradiction in the trends is resolved if we study the SDS-PAGE data presented in
Figure 4. This figure shows the distribution of the protein components of the membrane
as a function of their molecular weight. In the Spectrin region a new peak is observed
by the Intermediate Day, which by the end of Period II becomes clearly pronounced. The
formation of this new high molecular weight fraction indicates that additional crosslinking in the membrane proteins occurs as incubation progresses. Since additional
membrane cross-linking causes an increase in membrane stiffness (Smith and Hochmuth,
1982), it leads to a reduction in the number of cells that can enter the capillary.

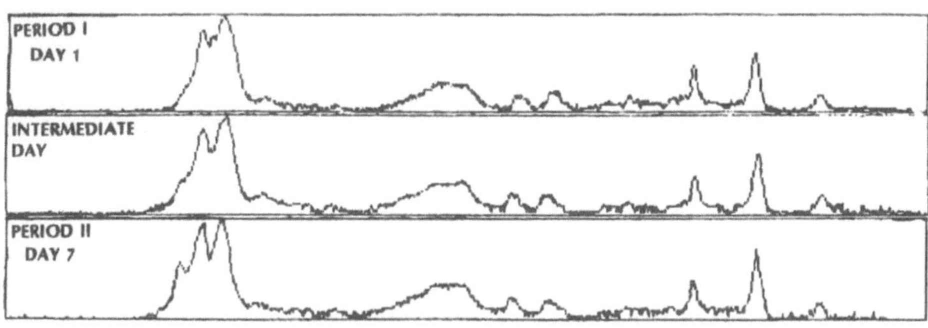

Figure 4. The densitometric trace from SDS-PAGE data for donor G, showing the content
of membrane proteins by molecular weight. Notice the change in Spectrin region on the
'Intermediate Day' which becomes more pronounced by the end of Period II. This indicates
that additional cross-linking of the membrane proteins occurs with incubation, leading to
membrane stiffening with incubation.

Figure 5 shows the effect of incubation on MCV. Although the change in MCV is
not statistically significant ($p < 0.2$) a distinct trend of increasing MCV with incubation appears to exist. Perhaps with a larger sample size this trend would become
statistically significant. This incubation driven increase in MCV will have the effect of
preventing the entry of cells (into the capillary) which were large to begin with.

We thus see that membrane cross-linking/stiffening and increase in MCV accom-
pany a decrease in the Filteration Rate. However, because the increase in MCV is not
found to be statistically significant, or significant only to a low level of confidence for

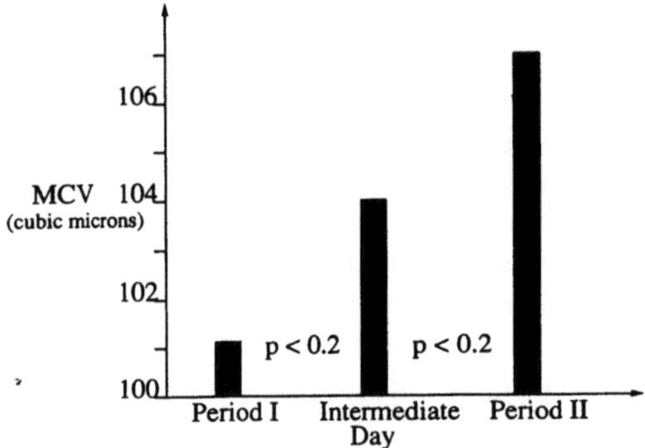

Figure 5. Mean Cell Volume (MCV) averaged over 150,000 cells per donor and 8 donors. The term 'Intermediate Day' is explained in the text. MCV increases monotonically with incubation. This change is, however, not statistically significant according to paired samples Student t-test for the sample size studied.

the given sample size, it may be less of a factor in comparison to membrane stiffening. Also, since the larger or the stiffer members of the cell population are also those which have a greater transit time (Smith and Hochmuth, 1982; Koutsoris et al., 1988; Ducharme et al., 1991), their exclusion from the capillary flow will lead to a lower Mean Cell Transit Time. In other words, the apparent decrease in MCTT in the early stages of incubation is not a reflection of a favorable change in the cell membrane to enable its flow through the capillary. Further incubation will yield progressively larger or stiffer cells, thus lowering the ability of the cells (– those which were able to enter the capillary) to flow through easily. Hence, we see an increase in MCTT in the second phase of incubation.

Why do we see a drastic elimination of a group of cells in the first phase of incubation (causing MCTT to decrease), and only a progressive slowing in the cell flow in the later phase (causing MCTT to increase)? The answer may be that the metabolic breakdown in ATP depleted cells is rapid, and may be complete in several hours. So, in a matter of 24 hours (– the period between two succesive data acquisitions) there must have been a significant deterioration in the cell (membrane) properties. Thus, we see a sizable fraction of the original cell population excluded from capillary entry during the first phase. We find that there is a similar stage of significant alteration of the properties beyond Period II. Indeed, this 'Period III' is so drastic that the cell lysis occurs rapidly. During this stage, we found that no more than 500 cells were able to flow through the capillary system in a period of two minutes which may be compared to an initial rate of 2500 to 5000 cells in less than 2 minutes. To summarize, we find three phases of transformation of the cell membrane during prolonged incubation. There is the initial rapid deterioration followed by a relatively slow one. The transformation, then culminates in the final stage of cell lysis.

CONCLUSION

Prolonged incubation leads to membrane cross-linking, and hence in membrane stiffness. There is also a trend of increasing MCV, but these changes were not found to be statistically significant for the sample size studied. This results in a fractionation of cells wherein the larger and stiffer of the cell population are prevented from entering the capillary. As a consequence, there is an initial apparent drop in MCTT. An important finding of this study is that the membrane damage as observed in the SDS-PAGE plot can also be detected by changes in FR and in MCTT. Thus, the rheological parameters may be valuable as indices of membrane damage in the context of capillary flow.

REFERENCES

Ducharme, R., Kapadia, P., Dowden, J., 1991, A mathematical model of the flow of blood cells in fine capillaries, *J. Biomechanics*, **24**:299.

Koutsoris, D., Guillet, R., Lelievre, J.C., Guillemin, M.T., Bertholom, P., Beuzard, Y. and Boynard, M., 1988, Determination of erythrocyte transit times through micropores, **25**:763.

Richterich, R. and Colombo, J.P., 1981, "Clinical Chemistry", John Wiley & Sons, New York.

Schmid-Schoenbein, H., Grebe, R. and Heidtmann, H., 1983, A new membrane concept for viscous RBC deformation in shear: spectrin oligomer complexes as a Bingham-fluid in shear and a dense periodic colloidal system in bending, *Annals of the New York Academy of Sciences*, 225.

Smith, L. and Hochmuth, R.M., 1982, Effect of wheat germ agglutinin on the viscoelastic properties of erythrocyte membrane, *The Journal of Cell Biology*, **94**:7.

EFFECT OF ELECTROSTATIC FORCE ON
ERYTHROCYTE DEFORMATION IN NARROW CAPILLARIES

P. Ram Rao

Department of Mechanical Engineering
University of Maryland Baltimore County
Baltimore, MD 21228

INTRODUCTION

In the study of oxygen transport, the mechanics of red blood cell (RBC) motion through the blood vessels, especially the narrow capillaries, cannot be overemphasized. A red cell must be able to undergo large deformations so as to be able to flow through capillaries some as narrow as half its diameter. Much has been learnt about the capillary flow through theoretical and experimental models involving flexible capsules and cells. Concise reviews of these studies have been provided by Secomb (1992), and by Skalak (1989) among others.

Essentially, most of the works to date recognize the interaction between loading on the membrane due to the *fluid forces* and the subsequent deformation and motion of the membrane. While it is true that the principal loading on a red cell membrane is hydrodynamic in nature, there is also the possibility of *electrostatic* loading. The latter may arise in narrow capillaries as a consequence of the negative electrostatic charges due to sialic acid residues on the RBC and the endothelial cells. In addition, several types of *in vitro* experimental hemorheologic systems consist of polycarbonate capillaries, and this material has an innate negative electrostatic charge. Thus, in both *in vivo* and *in vitro* capillary flow systems it is reasonable to expect that the electrostatic repulsion (between the capillary wall and the red cell membrane) will induce an enhancement of the lubrication layer, and thus participate in the mechanics and mobility of the capsule. Earlier researchers have reported that the perturbation of surface charge of a red cell by methods such as neuraminidase treatment, can lead to morphological changes in the cell. For example, Schmid-Schoenbein *et al.* (1983)

Oxygen Transport to Tissue XVI
Edited by M.C. Hogan *et al.*, Plenum Press, New York, 1994

found that the neuraminidase treatment led to an enhancement of stomatocyte formation in chlorpromazine treated red cells.

There is, however, no report on the direct impact of surface charge perturbation on the mobility of cells through narrow capillaries. At best, one may conjecture from the work of Reinhart and Chien (1986) that the surface charge perturbation is the explanation for their observation of an increased surface area in salicylate treated erythrocytes which transformed into echinocytes. These cells were observed to flow through micropores more rapidly than the diskocytes. Besides this indirect and speculative evidence, there is no other experimental data on the effect of systematic changes in the surface charges on the cell mobility through

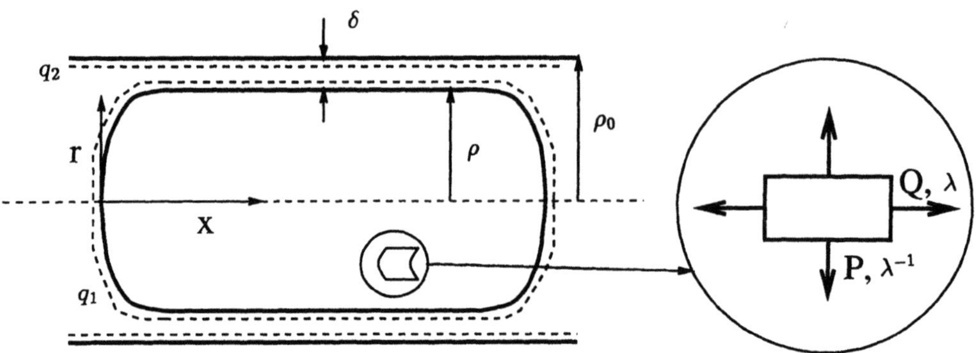

Figure 1. The capsule – capillary system. The capsule and the capillary wall possess negative charges of densities q_1 and q_2, respectively. The lubrication layer thickness is represented by δ. The inset shows the principal stress resultants and the corresponding stretch ratios for an element of the membrane.

narrow vessels. Also, there have been no theoretical analyses presented in literature on the relationship between the flow conditions, the capillary surface charge, and the capsule's surface charge, mechanical properties and mobility.

In this paper, we first present a theoretical analysis based on mechanical equilibrium and minimal potential energy considerations to determine the effect of surface charges on the deformation of the capsule. We will then present an experimental study on the effect of aspirin on the mobility of the RBC, and examine if the above theoretical result can explain the experimental findings.

THE CAPSULE AND THE CAPILLARY

The capsule is assumed to be a two-dimensional area-conserving continuum to render it similar to a red cell membrane. (See Evans and Skalak (1980).) The elastic energy is, therefore associated with only bending and shearing. Upon entering a narrow capillary, the capsule is assumed to acquire an axisymmetric bullet shape with cylindrical body and end-caps which are sections of a sphere (see Figure 1). Assuming the Evans and Skalak model for a red cell membrane, we write the following constitutive relations.

$$Q - P = G(\lambda^2 - \lambda^{-2}) + \eta \frac{1}{\lambda} \frac{d\lambda}{dt}, \tag{1}$$

$$M = B(\kappa - \kappa_0), \tag{2}$$

where P and Q represent the principal stress resultants, λ^{-1} and λ are the corresponding stretch ratios, d/dt is the material derivative, and M is the bending moment resulting from a change in local curvature from κ_0 to κ. The material constants η and G represent, respectively, the moduli of membrane shear viscosity and elasticity, while B is the bending rigidity. Also, we assume the fluid inside the capsule to be incompressible, so that in addition to the surface area, as mentioned above, the volume of the capsule is also a constant. The surface of the capsule as well as that of the capillary wall are assigned distributed surface charges per unit area of q_1 and q_2, respectively. Finally, by virtue of the axisymmetry of the deformed capsule, there is no rotation in the capsule. The only motion associated with the capsule is that of uniform translation. Thus, the fluid internal to the capsule is stationary relative to the wall of the capsule.

THE PROBLEM

The equilibrium of forces on an element of the capsule will require the following.

$$P = \frac{(p_i - p_e - f_{el})}{\rho} \tag{3}$$

$$Q = \mu \left(\frac{\partial V_r}{\partial x} + \frac{\partial V_x}{\partial r} \right), \tag{4}$$

where p_i, p_e, f_{el} and ρ represent the internal pressure of the capsule, the external fluid pressure, the electrostatic force on the capsule per unit area and the radial thickness of the capsule, respectively. V_r and V_x represent the velocity components of the fluid in the radial (r) and the axial (x) directions, respectively. Since our intention is to identify the contribution of the electrostatic portion of the external loading, we will simplify the problem by assuming no flow to begin with. Thus, p_e and the viscous stress terms disappear. The consequence of this simplification will be addressed in later sections.

Next, we can show that the work (W_s) associated with the shear modulus of elasticity, and that (W_b) with the bending rigidity are, respectively:

$$W_s = 2\pi\rho \int_{x=0}^{l} P(\lambda^{-1} - 1) + Q(\lambda - 1)\, dx, \tag{5}$$

$$W_b = 2\pi\rho B \oint \left\{ u_n \frac{d^2\kappa}{ds^2} - u_s \kappa \frac{d\kappa}{ds} \right\} ds, \tag{6}$$

3

where l, s, u_n and u_s represent, respectively, the length of the capsule (excluding the thin end-caps), the arc length of the cross-section, and the normal and the axial displacements. (Due to axisymmetry, there is no angular displacement.) We argue that these two forms of elastic potential energy are approximately proportional to the magnitudes of G and B. If we consider the typical values for a red cell, we have $G/B \approx 10^{10}\ cm^{-2}$. Thus, we may ignore W_b in comparison to W_s. The electrostatic potential energy, W_{el}, for the capillary-capsule charge system can be shown to be equal to

$$W_{el} = (8\pi^2 \ln 2)\,(k\,q_1\,q_2)\,(\rho_0 \rho)l, \tag{7}$$

where ρ_0, q_2 and ρ, q_1 represent the radius and charge density of the capillary and the capsule, respectively. The constant, k, represents the proportionality constant in Coulomb's law, and is equal to $9.00 \times 10^9\ N - m^2/C^2$. In deriving this expression, we have assumed that the annulus thickness between the the capsule and the capillary wall, $(\rho_0 - \rho)$, is small in comparison to the length of the capsule.

We now apply the principle of minimum potential energy, that requires the shape of the capsule such that it minimizes the total potential energy of the membrane. Accordingly, we write

$$\frac{\partial W_s}{\partial \rho} + \frac{\partial W_{el}}{\partial \rho} = 0, \tag{8}$$

and obtain a differential equation relating the capsule stretch ratio with the geometrical and material parameters introduced in the problem.

THEORETICAL RESULTS AND DISCUSSION

After linearizing the strains, we obtain the following closed form analytical solution to the above differential equation:

$$\rho|_{\rho_0}^{\rho} = \exp\{\ln(1 + \epsilon)(1 - \frac{3}{\pi S \ln 2}) + \frac{3\epsilon}{\pi S \ln 2}\}|_{\epsilon_0}^{\epsilon}\,, \tag{9}$$

where ρ is the capsule thickness and ϵ is the corresponding axial strain in the capsule. The subscript '0' refers to the state devoid of electrostatic loading. So, ρ_0 is the capillary radius, ϵ_0 is the strain developed in the capsule due to deformation from a biconcave disk of radius, $4\,\mu m$ and thickness $2\,\mu m$, to a bullet shape tightly fitting into the capillary of radius ρ_0. Finally, S is a dimensionless 'Stiffness Number' defined as follows:

$$S = \frac{k\,q_1\,q_2\,\rho_0}{G}\,. \tag{10}$$

It is proportional to the surface charge densities and the radius of the capillary, and inversely proportional to the capsule shear modulus.

If we write strain ϵ in terms of the capsule length l, and its radius ρ, and enforce volume conservation in the deformed state, then we can reduce equation (9) to the following explicit relation between ρ and S.

$$\ln\left(\frac{\rho}{\rho_0}\right).(6 - 3\,\pi\,S\,\ln 2) - \frac{3V_0}{\pi\,C_0}\left(\frac{1}{\rho^2} - \frac{1}{\rho_0^2}\right) = 0, \tag{11}$$

where C_0 and V_0 are, respectively, the semi-circumference and the volume of the undeformed capsule.

Prior to studying the significance of the above result, we will first present a discussion of the major assumptions made in this analysis. In deriving equation (11) we assumed axisymmetry and the no-flow condition. We also ignored the contribution of the end-caps in the calculation of the electrostatic potential. The precise curvature of the end-caps was ignored in the determination of the axial strain, and instead the circumference of the end-caps was approximated by that of a hemisphere. Further, the possibility of discontinuity of curvature at the interface of the end-caps and the main cylindrical body of the capsule (which can lead to large local bending stresses) has been ignored. We also assumed a cylindrical body with (non-zero) curvature only in the plane perpendicular to axis of the capillary. Finally, we linearized the strains.

Certainly, the above assumptions made it convenient to derive a closed form analytical solution, but it must be recognized that they can be justified to a reasonable extent. This is especially true if the goal of the analysis is to obtain a qualitative insight into the interaction between electrostatic charge and the deformation of the capsule. The assumption of axisymmetry may not hold in the context of a physiological capillary bed, however, in several *in vitro* experimental systems axisymmetry can very well be created. Next, ignoring the end-caps in the electrostatic potential calculations is prompted by the inverse relationship between the potential and the distance of separation of the charged surfaces. As this distance ceases to remain small in comparison to the radius of the capsule (as is in the case of the surfaces of the end-caps), the net contribution of those charges to the potential goes to zero. The assumption of end-cap circumference to be that of a hemisphere introduces an error in the axial strain, in the order of ϵ^2, and is therefore justified for small strains. Also, at the edges of the end-caps we ignored the bending stresses where possibly rapid changes in curvature may occur. This may offset the small value of the bending rigidity to some extent, however, it is not likely that it could eliminate a 10^{10} fold difference between the shear and bending stiffness values. Even if it did, the net contribution of this small region of the membrane to the overall elastic energy integrated over the entire surface will be negligible. The assumption of zero membrane curvature in the plane containing the axis of the cylindrical body may not be rigorously true. However, the radial thickness derived in this analysis may be viewed as an estimate of the mean of the actual location of the lateral surface. Finally, linearization of the membrane strains will restrict the analysis to small strains, perhaps to less than 10 %. This restriction may prohibit a quantitative comparison of this model to the red blood cells where larger strains may be observed. Nevertheless, the qualitative relationships developed here should be equally applicable to such large strain deformations.

The radial thickness of the capsule has been shown to be a function of the Stiffness Number, S, in equation (11). This relationship for different values of the capillary radius, ρ_0, is presented graphically in Figure 2. The figure shows that for a fixed capillary size, as S increases, i.e., as the surface charge increases, the capsule exhibits three phases of deformation. In the first phase, when S begins to rise from zero, the radial thickness of the pellet is not affected, at least not to within graphical accuracy; it remains at ρ_0, the radius of the capillary. This is because, there is already some elastic potential energy stored in the membrane, and the additional deformation caused by the radially inward displacement of the capsule wall does not lead to a minimum potential energy configuration. Therefore, this sluggish response at small values of S is more pronounced in smaller capillaries where the initial elastic strain is much greater. As S increases the electrostatic potential energy begins to outweigh the elastic energy, and so the system minimizes potential energy by moving the capsule farther from the capillary wall. This is the second phase. We observe in this phase a

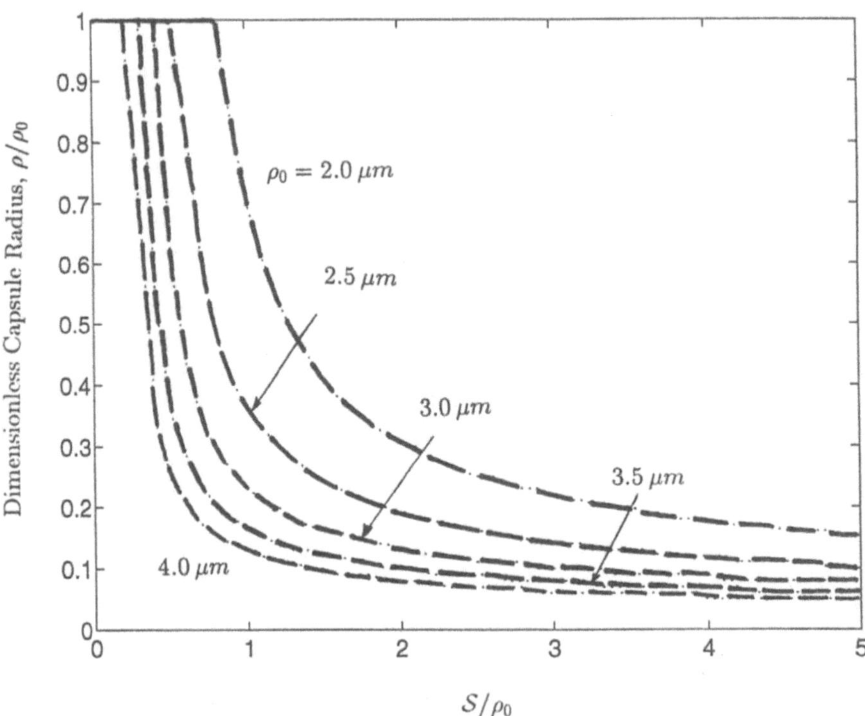

Figure 2. The dependence of the capsule radius on the Stiffness Number, S. For a given capillary radius, ρ_0, as S increases the capsule radius decreases. For smaller capillaries the effect of S is weaker.

rapid decrease in ρ with small changes in S. In the final phase, there is again a diminished effect of S on ρ. This is because as the lubrication layer thickness increases, the elastic potential energy decreases as an inverse function, so that the electrostatic interactions between the capillary wall and the capsule surface no longer are as significant as the elasticity of the capsule. Thus, after attaining a sufficiently large lubrication layer thickness, the system minimizes potential energy by inducing no further elastic strains, that is, by not reducing the radial thickness by any noticeable extent.

The reduction in the capsule radial thickness at moderate to large values of S has significance in the context of capsule mobility when the flow is initiated. From previous works on flow of pellets through narrow capillaries (e.g., Sugihara-Seki and Skalak (1988)) we know that as radial thickness becomes small in relation to the tube diameter, the axial velocity of the pellet becomes larger. Thus, our static problem which yields a decreasing capsule radius with increasing S indicates that when the flow is initiated the capsule will flow more rapidly at the higher S value. Of course, then the capsule will experience a higher level of loading (different from that considered here in the no-flow situation), so that the solution to the pellet thickness presented above will no longer be valid. Nevertheless, the qualitative trend of decreasing capsule thickness with increasing S will still hold, because the net effect of the flow will be only to induce additional membrane strain, ϵ in equation (9).

The significance of the above result is obvious in the context of oxygen transport to tissues. If membrane binding molecules perturb the surface charge on RBC then the transport of oxygen will be affected via perturbation of flow rate of red blood cells. Since the endothelial cells in the human capillaries display a negative surface charge due to the presence of sialic acid residues, the present observations with charged polycarbonate capillaries may well apply in that case as well. We now present below an experiment which we believe suggests that the flow rate of red cells is increased due to perturbation of the surface charge.

INTAKE OF ASPIRIN AND ERYTHROCYTE FLOW RATE

Erythrocytes drawn from normal human donors were washed thrice and resuspended in PBS at less than 1 % hematocrit. The resulting suspension was introduced into a a polycarbonate capillary system and the mean transit time for approximately 2000 cells (MCTT) was determined. The capillaries were $5\,\mu m$ in diameter and $15\,\mu m$ in length. Each donor ingested $650\,mg$ of aspirin immediately after the blood drawing, and 3 hours later donated a second sample of blood. The measurement of the cell transit times was repeated following the same procedure as before. The MCTT showed a highly significant decrease (with $p < 0.01$) after the intake of aspirin. See Figure 3.

We believe that this observation is *not* due to (a) the activity of platelets or (b) changes in the mean cell volume (MCV). Admittedly, the lack of platelet aggregation within the capillaries in the presence of aspirin could have prevented a decrease in the capillary lumen, and therefore might have improved the cell flow, and thus could have yielded a lower MCTT. However, the low initial relative concentration of platelets with respect to RBC ($\approx 1 : 20$), and the repeated washing of the blood sample would have practically eliminated the platelets from the blood sample. Even if residual platelets formed aggregates and occluded a few of the capillaries in the parallel flow system, many of the thirty capillaries would still be open to permit the red cell flow. If a RBC were to flow through a partially occluded capillary (in the absence of aspirin), it would take unusually large cell transit time because the presence of the platelet(s) would reduce an already small lubrication layer (δ, Figure 1). At small values of δ the mobility of a particle decreases in an exponential manner with a decrease in δ (Ducharme *et al.*, 1991). Hence, the computer connected to the conductimeter, which records the electrical conductance, would have ignored this large cell transit time as an outlier. We believe, therefore that platelet activity is not the primary explanation for our observation.

A second possible explanation for observations is that the mean cell volume of the RBC exposed to aspirin decreases, and therefore these cells offer less resistance to squeeze through the narrow capillaries. This is, again, plausible but not applicable to our experiments. The osmolality of the blood sample was essentially the same before and after the aspirin exposure

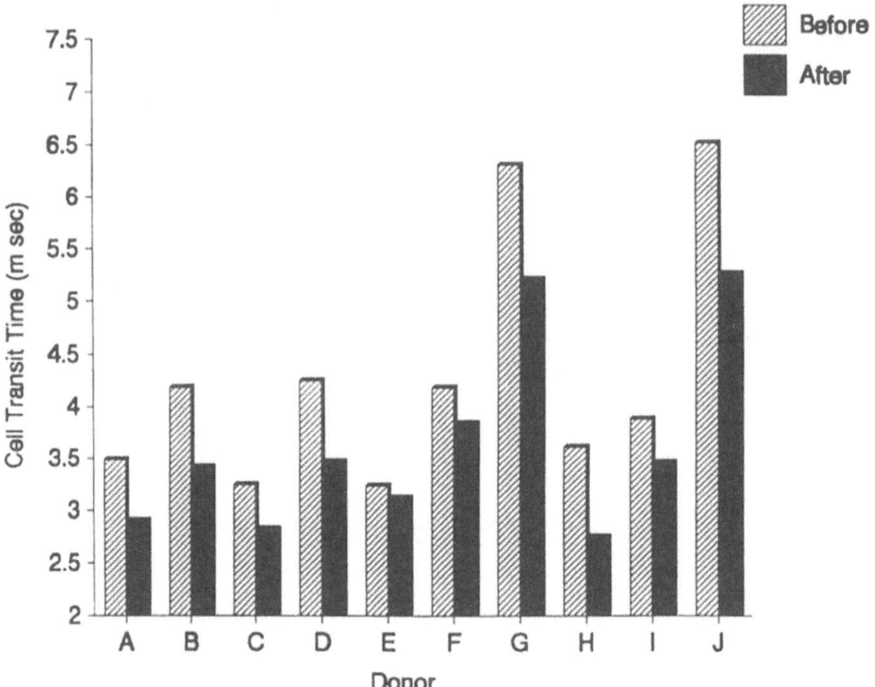

Figure 3. Effect of aspirin on mean cell transit time. Decrease in MCTT is observed in erythrocytes drawn three hours after intake of 650 *mg* of aspirin. The changes are highly significant at 1 % level.

due to the extremely low concentration of cells in the same PBS solution. Therefore, the red cells would have been restored to the same mean cell volume in the two cases.

If the above two phenomena (i.e., platelet deactivation and MCV decrease) are not likely to produce as significant an effect of aspirin as we see here, then we must seek an explanation that is more subtle. We explain this effect of aspirin by citing McLaughlin (1973) who experimentally determined that salicylate adsorbed biomembranes exhibit an enhanced negative surface charge. We believe that the additional negative surface charge leads to an increase in the value of the stiffness number, S, for the red cell-polycarbonate capillary

system, and thus lowers the radial thickness of the red cell in the capillary. This increases
the cell mobility and decreases the MCTT.

CONCLUSION

We have demonstrated that the deformation, and therefore the mobility of an elastic
charged capsule axisymmetrically flowing through a charged capillary is influenced by the
surface charge distribution. Greater the surface charge smaller will be the radial thickness
of the capsule, hence greater will be its mobility. However, capsule thickness is less sensitive
to an increase in S in smaller capillaries.

This result may apply to the experimental data on RBC drawn from donors after intake of
aspirin. If these red cells indeed have adsorbed aspirin into their membranes so as to induce
an additional negative surface charge, then based on our theoretical analysis it is possible
to explain why RBC would flow through polycarbonate capillaries faster after exposure to
aspirin. Since the human capillaries also possess a negative surface charge similar to the polycarbonate model employed here, these results may have physiological relevance. However,
that needs to be confirmed by direct experimentation in human capillaries. Nevertheless,
results of this study are significant, for they have bearing on *in vitro* experimentation of
RBC on which we rely upon to a great extent to understand the mechanics and physiology
of microcirculation and oxygen transport.

ACKNOWLEDGEMENT

The author acknowledges the assistance of Mr. Roopak Manchanda in conducting some
of the experiments.

REFERENCES

Ducharme, R., Kapadia, P., Dowden, J., 1991, A mathematical model of the flow
of blood cells in fine capillaries, *J. Biomechanics*, 24:299.

Evans, E. and Skalak, R., 1980, "Mechanics and Thermodynamics of Biomembranes",
CRC Press, Boca Raton.

Mc Laughlin, S., 1973, Salicylates and phospholipid bilayer membranes, *Nature*, **243**:234.

Reinhart, H.W. and Chien, S., 1986, Red cell rheology in stomatocyte-echinocyte
transformation: roles of cell geometry and cell shape, *Blood*, **67**:1110.

Secomb, T., 1992, Red blood cell mechanics and capillary blood rheology,
Cell Biophysics, **19**:231.

Schmid-Schoenbein, H., Grebe, R. and Heidtmann, H., 1983, A new membrane concept
for viscous RBC deformation in shear, *Annals of the New York Academy of
Sciences*, **1983**:225.

Skalak, R., 1989, Capillary flow, *Biorheology*, **27**:281.

EXERCISE IN PATIENTS WITH INTERMITTENT CLAUDICATION RESULTS IN THE GENERATION OF OXYGEN DERIVED FREE RADICALS AND ENDOTHELIAL DAMAGE

P. Hickman[1], D.K. Harrison[1], A. Hill[2], M. McLaren[2], H. Tamei[3], P.T. McCollum[1] and J.J.F. Belch[2]

[1]Vascular Laboratory Ninewells Hospital and Medical School, Dundee, Scotland,
[2]Section of Vascular Medicine, University Department of Medicine, Ninewells Hospital and Medical School, Dundee, Scotland
and [3]Fuji Chemical Industries Ltd., Takaoka, Toyama, Japan

INTRODUCTION

Peripheral vascular disease is a major cause of morbidity in Britain. Each year approximately 50,000 patients are admitted to hospital in Britain with a principal diagnosis of peripheral vascular disease and of these over 20,000 have major surgery, including amputations, with an operative mortality of about 10% (Department of Health and Social Security Office of Population Censuses and Surveys 1986). Intermittent claudication is the most common symptom of peripheral vascular disease (Dormandy *et al*, 1989) and is a symptom of cramp-like muscle pain, brought on by walking, relieved by rest and reproduced by further exercise. It may affect the calf, thigh or buttock muscle groups depending upon the level and degree of vascular obstruction. Intermittent claudication is caused by an inadequate blood supply to the exercising muscles of the lower limb, although the exact pathophysiological mechanism responsible for the symptom remains unknown (Lorentsen, 1973). Five per cent of men over 50 years suffer from intermittent claudication (Dormandy *et al*, 1989). Claudication itself does not cause death, however, the mortality of claudicants is approximately three times that of age and sex matched individuals (Dormandy *et al*, 1989), being about 50% after ten years (Dormandy *et al*, 1986). Seventy-five per cent of these deaths are due to cardiovascular disease, such as. myocardial infarction, cerebrovascular accidents and aortic aneurysms (Dormandy *et al*, 1989). Paradoxically the disease seems to stabilise symptomatically in seventy-five per cent of patients soon after onset (Dormandy *et al*, 1989) and few claudicants progress to limb threatening ischaemia (Cronenwett *et al*, 1984).

Oxygen Transport to Tissue XVI
Edited by M.C. Hogan *et al.*, Plenum Press, New York, 1994

The only indication for bypass surgery in claudicants is to relieve symptoms and thereby improve quality of life (Creasy *et al*, 1990) and so there is reluctance, amongst vascular surgeons, to offer surgery to most claudicants, because the risks are considered to outweigh the possible benefits (Whyman *et al*, 1991). Exercise is generally believed to be beneficial and indeed increasing physical activity is one of the targets of *The Health of the Nation* (Secretary of State for Health 1992). Supervised exercise has been shown to improve the walking distances of claudicants and may give better results than angioplasty after one year (Creasy *et al*, 1990). Exercise is associated with a reduction in mortality from all causes (Paffenbarger *et al*, 1986) but its effect on the mortality of claudicants is unknown at present. Exercise has been shown to have several other beneficial effects. Exercise in claudicants has been shown to result in the improvement of blood viscosity and blood cell filterability (Ernst *et al*, 1987). It enhances the antioxidant systems of healthy young runners (Robertson *et al*, 1991) but its effect in claudicants remains unknown.

The muscle blood flow of claudicants in the resting state has been found not to differ significantly from that of normal subjects when determined by ^{133}Xe Xenon clearance studies (Tonnensen, 1965). When a patient with a femoropopliteal arterial occlusion walks, the ankle pressure falls as a direct result of shunting of blood to the area of lowest resistance namely the exercising muscle (Yao, 1970). Although this has been attributed to muscular vasodilatation, it would also appear that increased circulation in the thigh muscles during exercise deprives the calf of blood (Angelides *et al*, 1978). The collateral arteries bypassing an occlusion are smaller, high resistance pathways that cannot maintain systolic pressure distal to this area of high resistance and so blood flow distal to this area of high resistance falls (Strandness *et al*, 1964). Pathophysiological considerations suggest that exercise blood flows at comparable work loads must always be reduced in claudicants compared to normal subjects and this has been verified by ^{133}Xe Xenon clearance studies (Tonnensen, 1968). Exercise has been shown to result in a fall in intramuscular oxygen tension in claudicants but a rise in normal subjects (Jussaila *et al*, 1981).

Reoxygenation of hypoxic endothelium (as occurs in the muscles of claudicants after exercise) results in the generation of oxygen derived free radicals within the endothelial cells, which has two effects (Ratych *et al*, 1987). Firstly, it damages endothelial cells and secondly it triggers the arachidonic acid cascade, resulting in the release of inflammatory mediators, which activate neutrophils (Ernster, 1988). Recent evidence suggests that exercise in claudicants results both in generation of free radicals (Ciuffetti *et al*, 1991, Hickman *et al*, 1993) and neutrophil activation (Shearman *et al*, 1988, Hickey *et al*, 1990, Neumann *et al*, 1990). When activated, neutrophils may participate in endothelial injury, both acutely and over time, by adhering to endothelium and causing endothelial damage with toxic oxygen compounds (including free radicals) and proteolytic enzymes (Belch, 1990a). There is now a substantial body of clinical, epidemiological and experimental evidence implicating the neutrophil in the pathogenesis of atherosclerotic disease and its sequelae (Belch, 1990b). Furthermore activated neutrophils or neutrophil-derived mediators may be released from ischaemic tissue and cause adverse effects in remote organs (Anner *et al*, 1987, Schmeling *et al*, 1989, Paterson *et al*, 1980). Endothelial damage and increased vascular permeability are widely believed to be the initiators of atherosclerosis, and are followed by the deposition of lipids and cellular proliferation (Ross, 1986). Free radicals produced in the endothelial cells of claudicants after exercise may activate neutrophils which then cause endothelial damage in remote organs, so contributing to the subsequent mortality and morbidity of claudicants.

AIMS

This study examined the effects of exercise on free radical generation and subsequent endothelial damage in claudicants and normal controls. Free radicals are thought to cause endothelial damage, which may initiate atherosclerotic disease. Whilst previous studies have suggested that exercise in claudicants causes increased free radical production, it is not known whether this leads to significant endothelial damage. Free radical production on exercise in claudicants may therefore cause endothelial damage and contribute to the mortality of claudicants. Supervised exercise training is one potential treatment of claudication, but may, theoretically, have long term deleterious effects.

MATERIALS AND METHODS

Subjects

Fifteen claudicants (11 males & 4 females, aged from 40 - 75 years; median age = 67 years, interquartile range = 61 to 69 years), who had resting ankle brachial pressure indices of less than 0.95 and vascular disease proven by angiography were studied. They were compared to fifteen control subjects (14 males & 1 female aged from 40 - 75 years; median age = 59 years, interquartile range = 50 to 63 years), who had been admitted for minor surgery, had resting ankle brachial pressure indices of more than 0.95 and gave no history of cardiovascular disease.

Clinical Studies

All subjects had a 16 gauge intravenous cannula inserted into an antecubital fossa vein and were rested supine for one hour. Blood samples were then taken for plasma thrombomodulin and plasma malondialdehyde. All subjects then exercised on a treadmill at 3.2 Km/hr with 10% slope until stopped by claudication pain or for a maximum of ten minutes if not stopped by claudication pain. Further blood samples were then taken one minute after exercise.

Free radical activity cannot readily be measured directly but free radicals cause lipid peroxidation, including malondialdehyde formation. The plasma malondialdehyde concentration was measured by Aust's method (Aust, 1987). Thrombomodulin is a membrane glycoprotein on the surface of endothelial cells and is released into the plasma when endothelial cells are damaged (Tomura *et al*, 1990). The plasma thrombomodulin concentration was be measured a one-step sandwich enzyme immunoassay (*Fuji Chemical Industries Ltd., Takaoka, Toyama, Japan*) (Kodama *et al*, 1990).

Statistical Analysis

The results are expressed as medians and interquartile ranges. The results are from paired samples but the data is not normally distributed and so were analysed using a paired Wilcoxon test. The results were analysed on a personal computer using a commercial statistical programme (STATGRAPHICS Statstical Graphics System, *Statistical Graphics Corporation).*

RESULTS

There was a statistically significant rise in malondialdehyde concentration in the plasma of claudicants as measured by Aust's method after exercise. This suggests that exercise results in increased free radical production in claudicants (see **Table 1.**). There was also a statistically significant rise in thrombomodulin in the plasma of claudicants after exercise, suggesting that significant endothelial damage occurs in claudicants after exercise (see **Table 1.**).

Table 1. Changes in Malondialdehyde and Thrombomodulin after exercise in claudicants

	Pre-exercise Median (& interquartile range)	Post-exercise Median (& interquartile range)	Significance of difference (using paired Wilcoxon)
Thrombomodulin (ng/ml)	1.9 (1.1 - 2.6)	2.8 (1.8 - 4.9)	$p < 0.01$
Malondialdehyde (nmol/ml)	6.9 (6.3 - 8.9)	9.3 (7.7 - 11.1)	$p < 0.001$

There was no statistically significant change in malondialdehyde concentration in the plasma of controls after exercise, suggesting that exercise does not cause increased free radical production in normal subjects (see **Table 2.**). There was also no statistically significant rise in thrombomodulin in the plasma of normal subjects after exercise, suggesting that exercise does not cause significant endothelial damage in healthy controls (see **Table 2.**).

Table 2. Changes in Malondialdehyde and Thrombomodulin after exercise in controls

	Pre-exercise Median (& interquartile range)	Post-exercise Median (& interquartile range)	Significance of difference (using paired Wilcoxon)
Thrombomodulin (ng/ml)	1.6 (1.4 - 2.1)	1.4 (1.4 - 2.2)	Not Significant
Malondialdehyde (nmol/ml)	6.9 (5.8 - 7.8)	7.3 (6.6 - 8.4)	Not Significant

DISCUSSION

It is concluded that exercise causes both increased generation of free radicals and endothelial damage in claudicants but not in normal subjects. Endothelial damage is thought to initiate atherosclerosis (Ross, 1986). Although the normal controls were on the whole younger than the claudicants, it is felt that age alone does not account for the differences between claudicants and controls. The generation of oxygen derived free radicals within the endothelial cells triggers the arachidonic acid cascade, resulting in the release of inflammatory mediators, which activate neutrophils (Ernster, 1988). When activated, neutrophils may participate in endothelial injury, both acutely and over time, by adhering to endothelium and causing endothelial damage with toxic oxygen compounds (including free radicals) and proteolytic enzymes (Belch, 1990a). Furthermore, activated neutrophils or neutrophil-derived mediators may be released from ischaemic tissue and cause adverse effects in remote organs (Anner *et al*, 1987, Schmeling *et al*, 1989, Paterson *et al*, 1980). Free radicals produced in the endothelial cells of claudicants after exercise may activate neutrophils and cause endothelial damage in remote organs, so contributing to the subsequent mortality and morbidity of claudicants. Claudication has therefore been shown to initiate a systemic inflammatory response (Ciuffetti *et al*, 1991, Hickman *et al*, 1993, Shearman *et al*, 1988, Hickey *et al*, 1990, Neumann *et al*, 1990). Treatments which reverse the systemic effects of claudication are preferable to those which merely offer symptomatic relief because they may have further benefits in reducing the subsequent morbidity and mortality of claudicants (Hickey *et al*, 1990). Bypass surgery has been shown to ameliorate the deleterious systemic effects in claudicants (Hickey *et al*, 1990) but the effects of other potential treatments, such as angioplasty or supervised exercise training, have not yet been examined and so are unknown at present. Exercise training is a widely used treatment for intermittent claudication and has been shown to improve claudication distance. It is felt, after this work, that controlled trials are now indicated to assess the long term effects of supervised exercise in claudicants. Whilst exercise is generally believed to be beneficial, its benefit in claudicants, other than in terms of symptomatic relief, has largely been unevaluated.

REFERENCES

Angelides N., Nicolaides A., Needham T. *et al.* The mechanism of calf claudication : studies of
    simultaneous clearance of ^{99}Tc from the calf and thigh. *Br. J. Surg.* **65** : 204-209 (1978).
Anner H., Kaufman R.P., Kobzik L. *et al.* Pulmonary hypertension and leukosequestration after lower
    torso ischaemia. *Ann. Surg.* **206** : 642-8 (1987).
Aust S.D. *Lipid peroxidation.* In : *Handbook of Methods for Oxygen, Free Radical Research.*
    CRC Press Inc. Florida : 203-7 (1987).
Belch J.J.F. The role of the white blood cell in arterial disease. *Blood Coag. Fibrinol.* **1** : 183-192.
    (1990a).
Belch J.J.F. The white blood cell as a risk factor for thrombotic vascular disease. *Vascular Medicine*
    Review. **1** : 203-13 (1990b).
Ciuffetti G., Mercuri M., Mannarino E. *et al.* Free radical production in peripheral vascular disease :
    A risk for critical ischaemia ? *Int. Angiol.* **10** : 81-7 (1991).
Creasy T.S., McMillan P.J., Fletcher E.W.L., Collin J. Is percutaneous transluminal angioplasty better than
    exercise ? - Preliminary results from a prospective randomised trial. *Eur. J. Vasc. Surg.*
    4 : 135-40 (1990).
Cronenwett J.L., Warner K.G., Zelenock G.B. *et al.* Intermittent claudication : Current results of
    nonoperative management. *Arch. Surg.* **119** : 430-6. (1984).
Department of Health and Social Security Office of Population Censuses and Surveys.
    Hospital In-Patient Enquiry. Her Majesty's Stationary Office, London (1986).
Dormandy J.A. and Mahir M.S. *The natural history of peripheral atheromatous disease of the leg.*
    In : Greenhalgh R.M., Jamieson C.S., Nicolaides A.N. (eds.) *Vascular Surgery : Issues in Current*
    Practice : 3-18. Grune and Stratton, London (1986).
Dormandy J., Mahir M., Ascady G. *et al.* Fate of the patient with chronic leg ischaemia.
    J. Cardiovascular Surg. **30** : 50-7 (1989).
Ernst E., Matrai A. Intermittent claudication, exercise and blood rheology. *Circulation.* **76** :
    1110-1114. (1987).
Ernster L. Biochemistry of reoxygenation injury. *Critical Care Medicine* **16** : 947-53 (1988).
Hickey N.C., Gosling P., Baar S. *et al.* Effect of surgery on the systemic inflammatory response to
    intermittent claudication. *Br. J. Surg.* **77** : 1121-4 (1990).
Hickman P, Hill A, McLaren M, Belch J.J.F., McCollum PT. Exercise induces neutrophil activation in
    claudicants but not in age matched controls. *Br. J. Surg.* (in press).
Jussaila E.J., Nijnikoski J. Effect of vascular reconstructions on tissue gas tensions in the calf muscles
    of patients with occlusive arterial disease. *Ann. Chir. Gynaecol.* **70** : 56 (1981).
Kodama S., Uchijima E., Nagai M. *et al.* One-step sandwich enzyme immunoassay for soluble human
    thrombomodulin using monoclonal antibodies. *Clinica Chimica Acta* **192** : 191-200 (1990).
Lorentsen E.. Blood pressure and flow in the calf in relation to claudication distance.
    Scand. J. Clin. Lab. Investig. **31** : 141 (1973).
Neumann F.J. Wass, W., Diehm C. *et al.* Activation and decreased defomability
    of netrophils after intermittent claudication. *Circulation* **82** : 922-9 (1990).

Paffenbarger R.S. Jr, Hyde P.H.R.T., Wing A.L., Hsieth C-C. Physical activity, all cause mortality, and longevity of college alumni. *N.Engl. J. Med.* **314** : 605-13 (1986).

Paterson I.S., Klausner J.M., Pugatch R. *et al.* Non-cardiac pulmonary edema after abdominal aortic aneurysm surgery. *Ann. Surg.* **209** : 231-6 (1989).

Ratych R.E., Chuknyischa R.S., Bulkey G.B. The primary localisation of free radical generation after anoxia/reoxygenation in isolated endothelial cells. *Surgery.* **102** : 122-31 (1987).

Robertson J.D., Maughan R.J.,. Duthie G.G, Morrice P.C. Increased blood antioxidant systems of runners in response to training load. *Cin. Sci.* **80** : 611-8. (1991).

Ross R. The pathogenesis of atherosclerosis - an update. *N. Engl. J. Med.* **314** : 488-500 (1986).

Schmeling D.J. Caty, M.G., Oldham K.T. *et al.* Evidence for neutrophil-related acute lung injury after intestinal ischaemia reperfusion. *Surgery* **106** : 195-202 (1989).

Secretary of State for Health. *The Health of the Nation. A strategy for health in England.* Her Majesty's Stationary Office, London : 46-7, 62-4 (1992).

Shearman C.P., Gosling P., Gwynn B.R. *et al.* Systemic effects associated with intermittent claudication : a model to study biochemical aspects of of vascular disease. *Eur. J. Vasc. Surg.* **2** : 401-404 (1988).

Strandness E., Bell J.W. An evaluation of the haemodynamic response of the claudicating extremity to exercise. *Surg. Gynae. Obstet.* **119** : 1237-42 (1964).

Tomura S. Nakamura, Y., Deguchi F. *et al.* Plasma von Willebrand and Thrombomodulin as markers of vascular disorders in patients undergoing regular haemodialysis therapy. *Thromb. Res.* **58** : 413-9. (1990).00

Tonnensen K.H. The blood flow through the calf muscle during rhythmic contraction & at rest in patients with occlusive arterial disease measured by 133 Xenon. *Scand.J.Clin.& Lab.Investig* **17** : 433-446 (1965).

Tonnensen K.H. Muscle blood flow during exercise in intermittentclaudication : Validation of the 133 Xenon clearance technique : Clinical use by comparison to plethysmography & walking distance *Circulation.* **37** : 402-410 (1968).

Whyman M.R., Ruckley C.V., Fowkes F.G.R. Angioplasty for mild intermittent claudication *Br.J.Surg* **78** : 643-645 (1991).

Yao S.T. Haemodynamic studies in peripheral arterial disease *Br. J. Surg.* **57** : 761-66 (1970).

TRAINING, IMMOBILIZATION , AND STRUCTURE-FUNCTION RELATIONSHIPS IN DOG GASTROCNEMIUS MUSCLE. D. E. Bebout, O. Mathieu-Costello, M. C. Hogan, and P. D. Wagner. Department of Medicine, 0623, University of California, San Diego, 9500 Gilman Drive, La Jolla, CA 92093-0623.

To investigate the effects of exercise training and immobilization on structure-function relationships in skeletal muscle, three groups of purpose-bred hounds [control (C), exercised trained (E), and immobilized (I)] were studied. Group E exercised on a treadmill 1 h/day, 5 days/wk for 8 wk, while groups C and I were cage-confined for 8 wk, with group I undergoing left hindlimb immobilization for the last 3 wk. The functional results (Bebout et al. *J Appl. Physiol.* 74(4): 1697-1703, 1993) showed that exercise training increased peak muscle O_2 uptake ($\dot{V}O_2$) by 38 % and estimated diffusive conductance (DO_2) by 71 % in dog gastrocnemius muscle, while immobilization had no effect on peak $\dot{V}O_2$/g muscle or DO_2/g muscle in spite of a 31 % reduction in muscle weight and a 68 % decrease in citrate synthase activity.

To relate these effects of training and immobilization to muscle structure, we investigated the individual roles of capillary surface area, inter-capillary diffusion distance and mitochondrial volume in determining muscle $\dot{V}O_2$ and DO_2 under the conditions of our experiment. Following glutaraldehyde perfusion-fixation, samples were taken from the mid belly of the medial gastrocnemius and capillary-fiber geometry was quantified using standard morphological techniques. Some of the results are summarized in the table below.

Group	capillary-fiber number ratio (n/n)	capillary density (n/mm^2)	diffusion distance (μm)	mitochondrial volume density (%)
C	2.44±0.06	1672±84	12.8±0.3	9.2±1.0
E	3.54±0.11*	1997±81*	11.8±0.2*	12.3±0.5*
I	2.21±0.03*	2452±179*	10.6±0.3*	5.6±0.3*

There were no differences in capillary diameter or tortuosity between groups while fiber area **increased** 22 % after training (P < 0.05) and **decreased** 38 % after immobilization (P < 0.01; not shown). The number of capillaries per fiber **increased** 45 % after training (P < 0.01) and **decreased** 9 % after immobilization (P < 0.05). Capillary density **increased** 19 and 47 % and inter-capillary diffusion distance **decreased** 8 and 17 % after training and immobilization, respectively (P < 0.05). Mitochondrial volume density **increased** 34 % after training (P < 0.05) and **decreased** 39 % after immobilization (P < 0.05).

These results show that muscle $\dot{V}O_2$ and DO_2 are closely related to the number of capillaries per fiber (hence the amount of capillary surface area) and are not related to capillary density, inter-capillary diffusion distance or mitochondrial volume density. This supports the idea that the major resistance to O_2 diffusion in skeletal muscle resides from the capillary wall to the sarcolemma, a very short distance, and is dependent on capillary surface area and not on the longer diffusion distance through the muscle fiber (Gayeski et al. *Am J. Physiol.* 251(20): H789-H799, 1986; Groebe and Thews. *Respir. Physiol.* 79: 255-278, 1990). We conclude that capillary surface area is likely the major determinant of muscle $\dot{V}O_2$ and DO_2 in the dog gastrocnemius, and that the major resistance to O_2 flux lies at the capillary-tissue interface and is not dependent on muscle fiber size. These results are interesting in the context of symmorphosis (the close matching of structure to function, e.g. mitochondrial volume to $\dot{V}O_2$) because mitochondrial volume density decreased 39 % after immobilization with no effect on peak $\dot{V}O_2$ or DO_2, suggesting that the controls had a large excess in mitochondrial volume.

Cryogenic Microspectrophotometry of Myoglobin

Thomas E. J. Gayeski, M.D., Ph.D.

University of Rochester, Rochester, N.Y.

This technique involves the rapid freezing of myoglobin containing tissue through the use of a modified Wollenberger clamp. Myoglobin saturation is estimated by measuring light intensity at three wavelengths and deriving a ratio of these intensities that accounts for light scattering and variations in myoglobin concentration within the cell. In our original system we believed that the spatial resolution was primarily determined by the size of the measuring spot. However, more recent experimental evidence indicates that light scattering plays a more important role than originally appreciated. We estimate that the radius of the catchment volume is approximately that of a dog gracilis myocyte.

Through the use of this technique an appreciation of the spatial distribution of myoglobin saturation of the muscle, a cluster of cells (such as a fascicle within the muscle or a group of cells surrounding a vessel) and along the length of a fascicle or myocyte is obtainable. The principle finding using this technique to date has been the importance of anatomic relationships at the capillary level as they impact on oxygen transport between convective channels and mitochondria.

The tissue oxygen tension of myocytes in both resting and exercising muscles is essentially constant. This constancy of tissue PO_2 over a 100-fold range of oxygen flux prevails in the presence of the large physiologic reserve for diffusive oxygen transport. Hence, tissue oxygen tension appears to be regulated.

We have begun to explore the scale of heterogeneities of oxygen tension and biochemistry within muscle fascicles. Preliminary data suggests that at rest whole muscle distributions of tissue PO_2 are not affected by varying oxygen offered while intrafascicular heterogeneity of PO_2 is inversely related to oxygen offered. Hence, the site of control of tissue regulation of PO_2 may be dependent on the physiologic state of the animal at rest.

In exercising muscle of anemic dogs, there is a large interfascicular variability of median PO_2 as well as intrafascicular heterogeneity. By isolating fascicles with different median PO_2's and distinguishable distributions, we have studied the effect of this difference on cell biochemistry within the same muscle. Our preliminary but reproducible findings are consistent with an interaction between PO_2 and cellular biochemistry in a range of PO_2 above the threshold predicted to affect in vitro mitochondria.

TISSUE OXYGEN TENSION INDICATES TISSUE OXYGEN DEBT DURING PROGRESSIVE ISCHEMIA. AN EXPERIMENTAL STUDY.

Hofer SOP[1], Kleij van der AJ[2], Gründeman PF[1], Scholten EW[3], Klopper PJ[1].
Departments of Experimental Surgery[1], Surgery & Hyperbaric Medicine[2] and Cardiopulmonary Surgery[3], Academic Medical Centre, Meibergdreef 9, 1105 AZ Amsterdam, Holland.

INTRODUCTION. Tissue oxygenation should be monitored to detect a state of malperfusion as soon as possible. Below a critical threshold oxygen delivery (DO_2) an oxygen supply dependency develops, which is generally taken as evidence for a tissue oxygen debt to develop. The aim of this study was to assess a critical threshold tissue oxygen tension (PO_2) value in skeletal muscle below which a tissue oxygen debt develops.

MATERIALS AND METHODS. A continuous PO_2 sensor (Continucath 1000 TM, BMS, Shiley, UK) was used in an isolated hindlimb model in the pig (n=6) to measure skeletal muscle PO_2 (PmO_2) during progressive ischemia. The isolated hindlimb received controlled blood flow by means of pressure regulated extracorporeal circulation (ECC). Blood pressure at the outlet site of the roller-pump was matched with mean arterial blood pressure and provided the baseline blood flow (100% perfusion) offered to the extremity. Baseline recordings were made after 1 hour. After baseline flow, every 20 minutes, flow reductions consisted of 2 steps of 25%, followed by 5 steps of 10% of the initial baseline flow value. Before a change in blood flow samples were obtained for oxygen content determination in an OSM 3 unit (Radiometer, Copenhagen, DK). DO_2 = blood flow x arterial O_2-content; Oxygen consumption (VO_2) = blood flow x arteriovenous O_2-content difference; oxygen extraction (EO_2) = VO_2/DO_2.

RESULTS. Results are given as mean ± SEM. All animals showed an oxygen supply dependency below a critical threshold DO_2. This DO_2 was 4.71 ± 0.21 ml.kg^{-1}.min^{-1} as the average for individual pigs. The critical EO_2 ratio as the average for individual pigs was 0.63 ± 0.04. The critical PmO_2 was 15.2 ± 0.4 mm Hg as the average for individual pigs.

CONCLUSIONS. With this model an oxygen supply dependency could be achieved. PmO_2 offers an absolute value which reflects the adequacy of tissue oxygenation. A critical PmO_2 value could be calculated below which a tissue oxygen debt develops.

INTRAVENOUS POTASSIUM INFUSION REVERSES SKELETAL MUSCLE FATIGUE

R. Kiiski, E. Fernandez, G. Gutierrez; Division of Pulmonary and Critical Care Medicine, University of Texas Health Science Center; Houston, TX 77030

<u>BACKGROUND</u>: Intracellular potassium loss has been implicated as the basis of skeletal muscle fatigue (1). A decrease in the cell membrane K^+ gradient diminishes the action potential. In turn, Ca^{++} release decreases and a smaller tension is developed. The largest K^+ losses have been reported during high-intensity, dynamic exercise. The purpose of this study was to determine the effect of KCl infusion on hindlimb forces in rabbits.

<u>PROTOCOL</u>: A group of New Zealand white rabbits was given an infusion of KCl (POT, n=7) and compared to a control group given saline. (control, n=7). A string was tied around the left paw and attached to a force transducer (Hewlett-Packard, Waltham, MA). Muscle fatigue was induced by low-frequency twitch stimulation of the femoral nerve (40 V, 0.25 Hz) causing a 30 % decrease in the isometric contractions in both groups. An i.v. infusion of 0.22 M KCl or normal saline was started (64 ml/hr for both groups) and stimulation was continued for another two hours. Gas exchange and O_2 transport were assessed hourly, and blood samples were taken for lactate measurement.

<u>STATISTICS</u>: Comparison of groups: analysis of variance for repeated measures; comparison within groups: post-hoc Newman-Keuls method. * = p<0.05; † p<0.01 from baseline.

<u>RESULTS</u>: The infusion of KCl resulted in complete reversal of fatigue as shown in Fig. 1. A shown in Table 1, limb lactate was higher and pH lower in the potassium group.

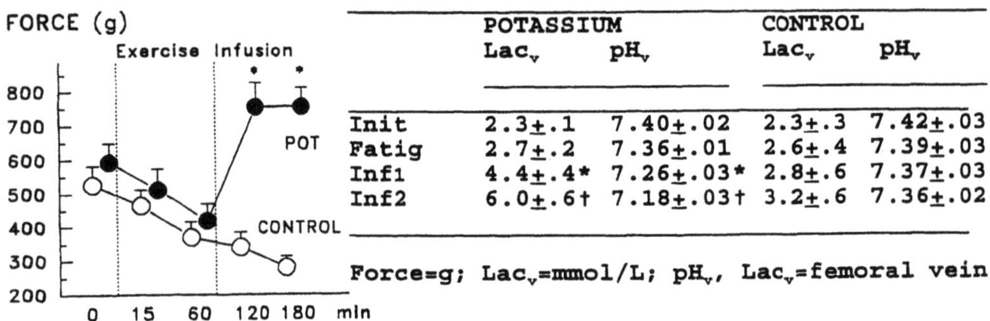

	POTASSIUM		CONTROL	
	Lac_v	pH_v	Lac_v	pH_v
Init	2.3±.1	7.40±.02	2.3±.3	7.42±.03
Fatig	2.7±.2	7.36±.01	2.6±.4	7.39±.03
Inf1	4.4±.4*	7.26±.03*	2.8±.6	7.37±.03
Inf2	6.0±.6†	7.18±.03†	3.2±.6	7.36±.02

Force=g; Lac_v=mmol/L; pH_v, Lac_v=femoral vein

<u>CONCLUSIONS</u>: An intravenous infusion of KCl reverses skeletal muscle fatigue in the rabbit hindlimb perhaps by preventing K^+ efflux from the skeletal muscle cells.

1. Sjogaard G. Exercise-induced muscle fatigue: The significance of potassium. Acta Physiol Scand 1990;suppl 593:1-63

Effect of Increased Ambient Pressure (3 ATA) on Human Skeletal Muscle PO$_2$.

Kleij van der A.J.[1], Vink H.[2], , Bakker D.J.[1], M Günderoth[3]. Departments of Surgery & Hyperbaric Medicine[1], of Medical Physics & Informatics[2], University of Amsterdam & Academic Medical Center. Meibergdreef 9, 1105 AZ Amsterdam, The Netherlands. Eppendorf-Netheler-Hinz GmbH, Hamburg, Germany[3]

Introduction

Increasing FiO$_2$ (90 - 100%) during hyperbaric conditions (2 - 3 ATA) is used to achieve hyperoxygenation (arterial pO$_2$ values ranging between 1200 - 2000 mm Hg) for well-defined clinical conditions which are mainly characterized by tissue hypoxia.

Transcutaneous pO$_2$ (TcpO$_2$) measurements during hyperbaric oxygen administering revealed high pO$_2$ values up to 700 mm Hg (Kleij van et al., 1992). These TcpO$_2$ values are used to predict the final outcome of hyperbaric oxygen therapy for compromized tissues (Mathieu et al., 1990) suggesting an increased oxygen tension in deeper layers. However, the degree to which hyperbaric oxygen may increase tissue pO$_2$ in compromized human skeletal muscle is yet to be determined. As a first step we measured simultaneously TcpO$_2$ and skeletal muscle pO$_2$ (MpO$_2$) in humans during increased ambient pressure (3 ATA, FiO$_2$: 21%).

Material and Methods

Healthy volunteers (n=9, age: 26 - 46 years, 2 females and 7 males) were placed in a Multi-Place (98 m^3) hyperbaric chamber. A transcutaneous pO$_2$ electrode (heated to 44 °C, TINA^{TM}, Radiometer, Kopenhagen) was applied on the mid-thigh region. Skeletal muscle pO$_2$ measurements (MpO$_2$) were performed with a polarographic pO$_2$ needle electrode (SIGMA pO$_2$-Histograph KIMOC, Fa. Eppendorf, Hamburg FRG) according the method as previously described by Fleckenstein et al. (1985). The side of the skeletal muscle pO$_2$ measurement was beneath the transcutaneous measurement. In addition, ambiènt temperature, skin temperature, systemic blood pressure and pulse rate were measured in each volunteer. Measurements were performed during control conditions (1 ATA, FiO$_2$: 21%) and after one hour increased ambient pressure (3 ATA, FiO$_2$: 21%).

Results

During control conditions TcpO$_2$ values and MpO$_2$ were 74,0 $\pm$ 4,6 mm Hg and 21.9 $\pm$ 2.9 mm Hg, respectively. One hour increased ambient pressure (3 ATA) did not significantly affect mean MpO$_2$ (29.4 $\pm$ 3.7 mm Hg) in contrast to a 4-fold increased TcpO$_2$ (310 $\pm$ 14.9 mm Hg). We found considerable variability between interdivual pO$_2$ histograms and no correlation could be found between the behaviour of the initial median pO$_2$-values and the effect of increased ambient pressure.

Conclusions

TcpO$_2$ values measured in human voluteers at 3 ATA (FiO$_2$ 21%) suggest an increased oxygen tension in deeper layers. In contrast mean skeletal muscle pO$_2$ values revealed only a small but not significant increase at 3 ATA. These data suggest that regulatory mechanisms in normal human skeletal muscle are conceived to keep tissue pO$_2$ values within a narrow range.

References

1. Fleckenstein, W., Heinrich R., Kersting Th., Schomerus H., Weiss Ch. A new method for bedside recording of tissue pO$_2$ histograms. Verh. Dtsch Ges Inn Med 90:439, 1984.
2. Kleij van der A. J., Vink H., Henny Ch. P., Bakker D.J., Spaan J.A.E. Red blood cell velocity in nailfold capillaries during hyperbaric oxygenation. 20th ISOTT meeting, Mainz, 1992. Plenum Press(in press).
3. Mathieu, D., Wattel, F., Bouachour, G., Billard, V., Defoin. Post-traumatic limb ischemia: Prediction of final outcome by transcutaneous oxygen measurement in hyperbaric oxygen. J. Trauma. 30:307-314, 1990.

TISSUE MORPHOMETRY: INFORMATION CONTENT AND LIMITATIONS

Odile Mathieu-Costello
Dept. of Medicine, Univ. of California, San Diego, La Jolla CA 92093-0623, USA

The quantitative assessment of the capacity for blood-tissue exchange in muscle depends on the estimation of several interrelated components of capillarity and fiber structure. Capillary number determines the amount of circuitry available for supply. Aggregate surface areas of membranes are important to the distribution of resistances to O_2 delivery along the diffusion path. Functional quantities (blood flow, red cell spacing, velocity patterns and oxygenation) need to be incorporated into structural data in order to understand and model muscle capacity for O_2 flux and utilization.

We briefly review quantitative aspects of capillary-fiber structure and their impact on the understanding of fiber O_2 supply and constraints in muscles at work. Classic approaches to examine muscle capillarization have been to use transverse sections and estimate capillary density (i.e. capillary number per tissue sectional area) or capillary-to-fiber ratio (i.e. capillary number per fiber number) and view those quantities in terms of their effect on intercapillary and diffusion distances. In recent years, efforts were made to quantify capillary geometry and incorporate its contribution into the estimates of muscle capillarization and potential for O_2 flux. Experimental and theoretical data suggested an important role of the capillary-fiber interface, i.e. capillary number rather than diffusion distances, in determining O_2 flux rates in working red muscles (Gayeski and Honig, Am. J. Physiol. 251: H789-H799, 1986). Sullivan and Pittman (Am. J. Physiol. 252: H149-H155, 1987) pointed out that matching O_2 supply and demand in muscles can be achieved by nature via different strategies: it can change fiber size (which affects capillary surface per fiber volume), or capillary-fiber contact area (i.e. the size of the capillary-fiber interface), or both.

We present data from comparative and experimental studies which examined the relationship between vascular arrangement and the size of the mitochondrial compartment in highly aerobic muscles. Comparisons across species and studies of structural plasticities support an important role of the capillary-fiber interface in determining muscle capacity for high O_2 flux rates from capillary to fiber mitochondria. Remarkably similar structural potentials for O_2 flux were found in flight muscles of bird and bat, but they were achieved via different strategies, i.e. different capillary geometry, fiber size and capillary-to-fiber ratio. In red muscle of one of the most athletic fishes (tuna), capillary surface was systematically smaller than in highly aerobic muscles of bird and mammal for the volume of mitochondria to be supplied, suggesting differences in capillary function or mitochondrial properties or both. In all muscles examined so far, aggregate surface areas of mitochondrial membranes relative to capillary surface area were substantially smaller than those considered in the literature to characterize the distribution of resistances to O_2 delivery in working red muscles.

Supported by NIH P01 HL 17331

ASSESSMENT OF METABOLIC STATE IN LOCALIZED REGIONS OF RESTING SKELETAL MUSCLE

Andras Toth, Miklos Pal, Marc E. Tischler and Paul C. Johnson, Departments of Physiology and Biochemistry, Univ of Arizona College of Med., Tucson, AZ, 85724 and 2nd Department of Physiology, Semmelweis University, Budapest, Hungary

Findings from whole organ studies in which blood flow and/or oxygen consumption have been varied have led some investigators to suggest that there are localized regions of resting skeletal muscle in which blood flow is not sufficient to fully support oxidative metabolism. To test this hypothesis we have examined changes in NADH fluorescence of localized tissue areas in the exteriorized sartorius muscle of the anesthetized cat during reduction and stoppage of blood flow, induced by occlusion of arterial inflow or stimulation of the sympathetic nerve supply to the muscle. A special microscope system which measures fluorescence at 450 nm in a 15 to 25 micron area during transillumination at 366 nm was used to monitor tissue fluorescence in small avascular regions of the muscle. The microscope also allows continuous observation and video recording of the adjacent microcirculatory vessels for measurement of red cell velocity. Tissue sites in the arteriolar and venular regions of the capillary network were chosen for study. Under control conditions the fluorescence signal was stable and did not vary during short term periodic variations in blood flow (vasomotion) occasionally seen in adjacent capillaries. With complete occlusion of inflow, fluorescence began to rise 47 $\pm$ 25 sec after flow stopped (range 5-105 sec). Fluorescence rose within 5 sec at only 3% of sites and within 15 sec at 10% of sites. The percentage rise in fluorescence during a 5 min occlusion was not significantly different from that in total tissue NADH as determined from quick frozen tissue samples. The latter observation suggests that the rise in fluorescence is due principally, if not entirely, to an increase in tissue NADH. When capillary blood flow was reduced by sympathetic nerve stimulation (2 to 12 Hz), a 50% reduction in flow, sustained for at least 30 seconds, was necessary to elicit a rise in fluorescence. There was no difference between arteriolar and venular regions in the duration of flow stoppage or degree of flow reduction required to elicit a rise in fluorescence. Assuming that a rise in tissue NADH reflects a shift from aerobic to anaerobic metabolism, these findings suggest that the blood supply to all but a small fraction of local tissue areas is more than sufficient to support oxidative metabolism in resting cat sartorius muscle. Our findings may be reconciled with whole organ studies if oxygen consumption varies with oxygen availability even when the latter is above the "critical" PO2.

(Supported by NIH grants HL 17421 and HL 15390)

SYMMORPHOSIS OR DYSMORPHOSIS?

Peter D. Wagner, M.D.

Department of Medicine, University of California, San Diego
La Jolla, California, U.S.A.

The concept of symmorphosis as applied to oxygen utilization in mammals has received much attention recently. It is based upon the hypothesis that the amount of (mitochondrial) structure required to support a given metabolic rate is well-tuned to the oxygen utilization requirements. In other words, structure and function are well-matched. Drs. Weibel, Taylor and co-workers have extensively examined this hypothesis principally by observations across a scale of size encompassing several orders of magnitude. However, within a single species, there appear to be several lines of evidence that suggest symmorphosis is not obeyed. In fact, structure and function appear to be seriously mismatched, giving rise to a concept of dysmorphosis. The best example appears to be the average but well-trained equine. Dysmorphosis would appear to be evident at least two levels (*i.e.,* in the lungs and in the skeletal muscles).

In the lungs, while the resting arterial PO_2 is approximately 100 Torr and the alveolar-arterial PO_2 difference is generally less than 10 Torr, treadmill exercise at full speed generates significant hypoxemia and increases in the alveolar-arterial difference that have been shown to be based primarily on the development of diffusion limitation of oxygen transport across the blood gas barrier. Thus, the geometry of the lung is insufficient to support the requisite diffusive flux of oxygen for maximum exercise. However, increasing inspired oxygen concentration to 35% overcomes this deficit and restores essentially complete saturation of hemoglobin even during maximal exercise. Dysmorphosis may also be represented in the equine lung by the severe pulmonary hypertension that accompanies heavy exercise, with mean pressures close to 100 Torr. Exercise-induced pulmonary hemorrhage is a very common observation and may well occur on the basis of mechanical disruption of the microvasculature due to high transmural pressures, suggesting an inadequate overall cross-sectional area of the pulmonary vasculature. Finally, CO_2 retention is almost universally seen in the maximally exercising equine, testifying to a ventilatory process in which either the cost of breathing required to maintain eucapnia is prohibitive, or mechanical limits to ventilation are reached.

Dysmorphosis is also evident in the skeletal musculature. Most clearly, maximum $\dot{V}O_2$ can be increased almost 20% as a result of correcting the arterial desaturation by administering 35% oxygen during exercise. This shows that the metabolic machinery at the mitochondrial level is in excess of that required to sustain maximum $\dot{V}O_2$ during normal sea level air breathing. Moreover, not all of the oxygen supplied to the muscle can be extracted during room air breathing, and the likeliest explanation for this is limited diffusional conductance for oxygen out of the muscle microvasculature. With current evidence suggesting the major hindrance to oxygen efflux is between the red cell and the sarcolemma, the observation of incomplete extraction suggests a limited microvascular cross-sectional area in the muscle.

In summary, while it is inevitable that a comparison amongst very small and very large mammals must reveal a correlation between mitochondrial mass and maximum oxygen utilization, and that comparison of active and inactive species of similar mass likely would reveal similar relationships, a closer examination within a species reveals several pieces of evidence suggesting dysmorphosis not only in the lungs, but also in the skeletal muscles. Similar dysmorphosis is evident when elite human athletes are studied.

FILTRATION, REABSORPTION AND OXYGEN IN THE KIDNEY

Roland C. Blantz

Division of Nephrology-Hypertension
University of California and Veterans Affairs Medical Center
San Diego, CA 92161

INTRODUCTION

The rate of glomerular ultrafiltration in the human with normal kidney function is approximately 150 liters per day. Described in other terms, 150 liters of protein free ultrafiltrate passes from the primary capillary bed in the kidney at the glomerulus into the proximal nephron to be acted upon by renal epithelium of the nephron, reabsorb solutes and water and secrete the appropriate solutes into this fluid. Since only one to three liters of fluid are usually excreted as urine each day by normal humans, this demands that 98-99% of the 150 liters is reabsorbed by nephrons and recycled into the circulation. Since the rate of ultrafiltration is several times total body water in a 24 hour period, one could conclude that the process of glomerular filtration must be highly well regulated and in strict coordination with the process of tubular reabsorption. If such coordination were not present, large swings in total body volume status would occur on a daily basis. In fact, total body volume status is very tightly regulated with swings of much than half a liter each day noted in the normal human on a reasonably normal diet.

The high rate of glomerular ultrafiltration is dictated by a high renal blood flow. In fact, 20% of cardiac output traverses the kidneys in a human with normal kidney function[1]. Blood flow rates then must be on the order of 1200-1400 liters per day. Studies in the literature suggest that the regulation of renal blood flow is highly efficient, at least in response to variations in systemic blood pressure. This autoregulation of renal blood flow is more efficient in the kidney than in any other organ in the body. At the same time, the arteriovenous oxygen difference in the kidney is exceedingly small, in part, because of the high blood flow rate. In the nonfiltering, nonreabsorbing kidney, the nutrient requirements necessary to sustain viable cells within the kidney require no more than 5-10% of normal blood flow. The major requirement for oxygen and substrates in the normal kidney is dedicated to tubular reabsorption of sodium chloride and water through active reabsorption of solutes. From the data supplied, one would suggest that the kidney is living in a condition of oxygen excess whereby much more oxygen is supplied than is needed for active transport and

Oxygen Transport to Tissue XVI
Edited by M.C. Hogan *et al.*, Plenum Press, New York, 1994

sustaining nutrient activity of the cells. However, studies from clinical experience suggest that the kidney is, in fact, not protected from hypoxia or ischemia and that acute renal failure does occur as a result of a variety of hemodynamic insults.

Coordination of Glomerular and Tubular Function

In addition to the above considerations, it is obvious that a system which filters 150 liters a day and reabsorbs 148 liters of this material must be highly coordinated. If the reabsorptive process and the filtration process are not coordinated, large swings in extracellular volume will occur. In part, this coordination is mediated by a process called glomerulo-tubular balance whereby increases in flow into each nephron segment, proximal tubule, loop of Henle and distal tubule are accompanied by a near proportional increase or decrease in the rate of tubular reabsorption[2-5]. However, this system alone will not guarantee full coordination between the processes of filtration and reabsorption. In addition, it has been recognized for several decades that there also exists a tubuloglomerular feedback system, an intrinsic mechanism which regulates the filtered load in relationship to the delivery of sodium chloride to the more distal segments of the nephron[6-8]. The afferent signal for this system appears to reside at the macula densa, a specialized distal tubular cell which is in close physical proximity to the glomerulus of the same nephron and to the vascular pole[9]. This system has been demonstrated and examined using microperfusion techniques in which purposely the distal tubule has been isolated from the proximal nephron by blockade of the nephron[7,8]. When the distal nephron is perfused at varying rates with sodium chloride containing solutions, one observes that increased perfusion rate is associated with a reduction in the filtration rate of that nephron unit and, to a lesser extent, when the flow rate is reduced to the distal nephron, a rise in filtration rate occurs, presumably as a result of an intrinsic system that transmits information to the vascular pole and vascular resistances primarily at the afferent arteriole. Studies in the past have examined the tubuloglomerular system in the open loop, whereby the distal tubule is purposely separated from the proximal nephron and the filtering unit. More recent studies in our laboratory have attempted to estimate the gain and efficiency of this system by examining tubuloglomerular feedback activity in a closed loop system in which the proximal and distal portions of each nephron segment are in constant communication[10]. This technique measures the integrated effects of both glomerulo-tubular balance and tubuloglomerular feedback function on flow rate whereas, with the older open loop microperfusion approach, the contribution of glomerulo-tubular balance is eliminated. This on-line technique also permits examination of the behavior of the tubuloglomerular feedback system in and around its ambient flow rate. The techniques that are utilized involve relatively constant or continuous monitoring of flow rate by videometric flow velocitometry utilizing small boluses of artificial tubular fluid containing a fluorescent dye injected serially into the proximal tubules by pressure pulses at a pressure frequency of 0.3-0.6 Hz utilizing a pneumatic microinjection pump. The dye is then excited by a neon green laser. The magnified image was filtered to maximize resolution and the fluorescence was monitored videomicroscopically with a television camera modified to an increased framing rate. A videometric analyzer measured the intensity of the resultant video image at two points separated along a known distance along the nephron immediately distal to the injection pipette. These records were digitized and stored directly on a disk utilizing an IBM compatible computer. The real time video image was recorded on tape for later determination of tubular diameter using an image shearing monitor. Flow was then perturbed in the free flowing nephron at the next downstream loop from the recording windows by aspiration or delivery of artificial tubular fluid into the free flowing lumen utilizing a microperfusion pump. Microperfusion rate V_H was manipulated between -12 and +12 nanoliters per minute in increments of 3-4 nanoliters per minute. V_H was returned to 0 between the perturbations and a period of equilibration allowed at each new flow rate to study the behavior of the TGF system. The efficiency of TGF was examined in at least three different volume conditions, hydropenia, euvolemia and plasma volume expansion. Studies demonstrated that the efficiency of tubuloglomerular feedback was on the order of 70-75% in each nephron and the maximum efficiency was unaffected by volume status[10]. This means that as fluid was added to the nephron, there was a ~75% compensation by a reduction in flow rate proximal to the perturbation. In addition, when fluid was withdrawn, there was a near complete compensation by increases in flow proximal to the perturbation such as to maintain flow rate to the distal nephron nearly constant. We also found that the peak efficiency of this system, although preserved in all volume states, adapts by shifting its efficiency profile such as to protect volume when volume is contracted and to protect GFR when volume is more replete. The range over which efficiency is high is actually enhanced by further volume expansion. Recent studies using electrodes which measure ionic conductivity changes in the early distal tubule fluid near the macula densa suggested that there is a near linear relationship between changes in ionic concentration within the tubular lumen of the distal tubule and the filtration rate response, an inverse relationship (unpublished observation). Studies utilizing these techniques also have suggested that this feedback system exhibits oscillatory behavior at the rate of 2-3 cycles per minute, suggesting a system with a high gain or efficiency and a defined time delay between the regulated flow rates in the proximal tubule and the sensing segment located at the macula densa in the early distal tubule[11].

This tubuloglomerular feedback system has also been demonstrated in vitro. Studies by Ito and colleagues have dissected glomeruli from rabbit kidneys with the attached macula densa segment and afferent and efferent arterioles, arteriolar segments from the glomerulus[12]. When the macula densa segment is perfused with high sodium chloride concentration solution, the afferent arteriolar segment connected to the glomerulus undergoes vasoconstriction. When low sodium chloride solutions are utilized, the preglomerular vessel vasodilates. This relationship is inhibited by administration of high concentrations of furosemide, a diuretic which inhibits Na-Cl-Cl-K cotransport in the distal tubular segment. This finding further suggests that a signal related to the transport of sodium chloride is mediating the afferent signal which eventually communicates to the vascular segments resulting in vasoconstriction or vasodilation. From a teleologic perspective, the tubuloglomerular feedback system functions to maintain distal tubular flow relatively constant despite variations in proximal reabsorption and glomerular filtration rate in an effort to prevent the capacity of distal tubule reabsorption from being overwhelmed and avoid inordinate extracellular volume fluid losses into the urine.

Relationship of Function to Oxygen Utilization in the Kidney

As has been previously stated, oxygen utilization by the kidney is primarily directed toward sodium chloride and water reabsorption. Studies utilizing oxygen electrodes within the kidney have suggested that the oscillatory behavior of pressure and flow which characterizes the tubuloglomerular feedback system are also duplicated by oscillations of similar frequencies in pO_2[13]. This finding suggests that variations in oxygen consumption in pO_2 may dictate a signal which helps elicit an effector response in the vasoconstriction or vasodilation that occurs in the renal vasculature. Variations in flow to the distal nephron may obligate changes in the rate of tubular reabsorption and oxygen utilization. This would imply that the observed variations in pO_2 of a similar frequency may relate to oscillatory variations in the rate of tubular reabsorption by distal tubular epithelial cells and oscillatory variations in oxygen consumption.

Recent studies with the pO_2 electrode have also demonstrated that there is a significant compartmentalization of oxygen within the kidney cortex and medulla which is relatively unaffected by large variations in systemic pO_2[14,15]. Studies by Schurek and coworkers have demonstrated that the normal pO_2 of the kidney cortex is much lower than systemic arteriolar pO_2 in the range of 50-65 mmHg[14]. Medullary pO_2 remains much lower in the range of 25-40 mmHg, significant compartmental heterogeneity within both the cortex and medulla. These relatively low values for pO_2 remain relatively constant even when arterial pO_2 is raised to 550 mmHg, again suggesting significant compartmentalization of gas mixtures within the kidney. It should be recalled that pCO_2 in the kidney cortex is also elevated above systemic CO_2 at ~60-65 mmHg. This finding is a reflection of the high rate of proton secretion by renal epithelia and the acidification of the urine resulting in high rates of CO_2 production. The relatively low cortical and medullary pO_2 is accomplished by a significant preglomerular arteriolar-venous diffusion shunt pathway. The density of arteriolar networks in the kidney is quite high since there are normally several million nephrons within the normal human kidney. It is presumed that as pO_2 is elevated in the systemic circulation, oxygen diffuses into the exiting blood prior to the glomerulus and prior to the cortical filtration process such that, regardless of pO_2 in systemic blood, the oxygen tension remains relatively low in both cortex and medulla[14]. This finding, makes reasonable sense since the site for erythropoietin production is located within the kidney. The site for production of this hormone which regulates red cell production should be sensitive to variations in oxygenation[16]. Since the pO_2 is lower within the kidney than previously predicted, the cortex, and especially the medulla, may be living on the edge of hypoxia and dependent upon the oxygen demands placed upon the tubule for reabsorption of solutes and water.

What then regulates blood flow to the medulla since it is living in an environment of tenuous oxygen supply? Recent studies have demonstrated that inhibitors of nitric oxide synthase, L-NMMA, when infused into the isolated perfused kidney further decrease medullary pO_2, suggesting either that nitric oxide is linked to the regulation of vascular resistance in the medulla or that nitric oxide somehow influences transport rates or NaCl delivery to the medullary thick ascending limb[15]. It is possible that there are also more complex interactions between NO and oxygen since both gases 1) oxygen and 2) the nitric oxide radical bind to various ferroporphyrin enzyme systems and may regulate, in some direct or indirect way, the vascular resistance in medullary vessels.

Therefore, the kidney remains a unique compartment within the body whereby variations in oxygen utilization may trigger or regulate the rate of blood flow and the filtered load, events which eventually dictate transport dependent oxygen utilization. Therefore, in spite of the fact that delivery of oxygen to the kidney appears great related to a high renal blood flow, the kidney remains in a state of borderline hypoxia in both cortex and medulla. This unique compartmentalization of O_2 and CO_2 may explain why acute renal failure occurs in clinical conditions in spite of the fact that renal blood flow rates are high and autoregulation of blood flow is extremely efficient.

ACKNOWLEDGMENTS

These studies were supported in major part by grants from the National Institutes of Health (DK28602, DK40251, DK07671) and funds from the Veterans Affairs Research Service. Faculty Member, Biomedical Sciences Graduate Program of the University of California, San Diego.

REFERENCES

1. R.W. Berliner, Urine formation, *in:* "The Physiological Basis of Medical Practice," C.H. Best and N.B Taylor, eds., The Williams & Wilkins Co., Baltimore, MD (1961).
2. H.W. Smith. "The Physiology of the Kidney," Oxford, New York (1937).
3. M. Imai, D.W. Seldin, and J.P. Kokko, Effect of perfusion on the fluxes of water, sodium, chloride and urea across the proximal convoluted tubule, *Kidney Int.* 11:18(1977).
4. J. Schnermann, M. Wahl, G. Liebau, and H. Fischbach, Balance between tubular flow rate and net fluid reabsorption in the proximal convolution of the rat kidney, *Pfluegers Arch.* 304:90(1968).
5. O.W. Peterson, L.C. Gushwa, and R.C. Blantz, An analysis of glomerular-tubular balance in the rat proximal tubule, *Pfleugers Archiv.* 407:221(1986).
6. K. Thurau and J. Schnermann, Die natrium-konzentration an den macula densa zellen als regulierender faktor fur das glomerulumfiltrat (Mikropunktionsversuche), *Klin. Wochenschr.* 43:410(1965).
7. J. Schnermann, M. Hermle, E. Schmidmeier, and H. Dahlheim, Impaired potency for feedback regulation of glomerular filtration rate in DOCA escaped rats, *Pfleugers Arch.* 358:325(1975).
8. R.C. Blantz and K.S. Konnen, Relation of distal tubular delivery and reabsorptive rate to nephron filtration, *Am. J. Physiol.* 233:F315(1977).
9. N. Goormaghtigh, L'appareil neuro-myoarterial juxtaglomerulaire du rein: ses reactions en pathologie et ses rapports avec le tube urinifere, *C.R. Seances Soc. Biol.* Paris 124:293(1937).
10. S.C. Thomson and R.C. Blantz, Homeostatic efficiency of tubuloglomerular feedback in hydropenia, euvolemia, and acute volume expansion, *Am. J. Physiol.* 264:F930(1993).
11. N.-H. Holstein-Rathlou, A closed-loop analysis of the tubuloglomerular feedback mechanism, *Am. J. Physiol.* 261:F880(1991).
12. S. Ito and O.A. Carretero, An in vivo approach to the study of macula densa mediated glomerular hemodynamics, *Kidney Int.* 38:1206(1990).
13. H.-J. Schurek and O. Johns, Tubulo-glomerular feedback prevents nephron oxygen deficiency via TAL-segments, *J. Am. Soc. Nephrol.* 1:603A(1990).
14. H.-J. Schurek, U Jost, H. Baumgartl, H. Bertram, and U. Heckman, Evidence for

a preglomerular oxygen diffusion shunt in rat renal cortex, *Am. J. Physiol.* 259:F910(1990)

15. M. Brezis, S. N. Heyman, D. Dinour, F.H. Epstein, and S. Rosen, Role of nitric oxide in renal medullary oxygenation. Studies in isolated and intact rat kidneys, *J. Clin. Invest.* 88:390(1991)

16. H. Scholz, H-.J. Schurek, K.U. Eckardt, A. Kurtz, and C. Bauer, Oxygen-dependent erythropoietin production by the isolated perfused kidney, *Pfleugers Archiv.* 418:228(1991).

RESTRICTION OF HYPOXIC MEMBRANE DEFECT BY GLYCINE IMPROVES MITOCHONDRIAL AND CELLULAR FUNCTION IN REOXYGENATED RENAL TUBULES

G. Gronow[1], N. Klause[2], and M. Mályusz[1]

[1]Department of Physiology
[2]Clinic of Nephrology
University of Kiel
24098 Kiel, Germany

INTRODUCTION

The hypoxic tolerance of renal tubular cells may be improved by specific amino acids. In studies on the protective role of glutathione Weinberg and co-workers (1987) observed that not the tripeptide GSH, but one of its components, the amino acid glycine, suppressed hypoxic alterations in isolated tubular segments of rabbit renal cortex. Later experiments confirmed the cytoprotective role of short-chained, non-essential, neutral amino acids such as glycine or alanine (Heyman et al., 1992). The exact mechanism of glycine protection, however, remains unclear. Some authors postulated an unspecific (Baines et al., 1990) or ligand-bound (Weinberg et al., 1990) membrane-stabilizing effect of glycine and related compounds. Recently we reported that glycine supports cellular volume regulation in renal cortical cells at low extracellular oxygen tension (PO_2 < 1 mm Hg) as well as in a subsequent reoxygenation period (Gronow et al., 1990).

Aim of the present investigation was to test whether hypoxic and posthypoxic renal cellular volume regulation may have been supported by different modes of glycine effects: 1) via an altered volume-regulatory potassium pump activity, 2) via provision of metabolic energy, e.g. by glycolysis in hypoxia, or by an improved mitochondrial respiratory function during reoxygenation, and finally via preservation of cellular function by maintaining the integrity of cellular membranes.

METHODS

Details of the preparation of collagenase-isolated tubular segments (ITS) from rat renal cortex have been reported elsewhere (Gronow et al., 1989). Freshly prepared and washed ITS were incubated at 37°C in a modified Krebs-Ringer-bicarbonate (KRB) medium containing 0.5 g/dl bovine albumine and 5 mM glucose. Either 5 mM glycine or the amino acid derivative taurine (= control) was added prior to the experiments into the incubation medium. Extreme hypoxia (PO_2 < 1 mm Hg) was introduced within 3 min by gassing the surface of ITS suspensions with water saturated 95% N_2 : 5% CO_2. Reoxygenation was achieved by gassing the tubule suspension with 95% O_2: 5% CO_2.

Dedicated to Prof. W. Niedermayer in his 60th year

Oxygen Transport to Tissue XVI
Edited by M.C. Hogan *et al.*, Plenum Press, New York, 1994

Mitochondria were isolated from ITS and tested as described by Goldstein (1975). Respiration was measured polarographically in a sealed chamber with a Clark-type electrode. Respiratory control ratio of isolated mitochondria (RCR = state 3 over state 4 respiration) was calculated as the ratio of 0.5 mmol/l ADP-stimulated oxygen consumption (in the presence of 5 mmol/l malate and 5 mmol/l glutamate) over mitochondrial respiration after the ADP-effect had worn off.

Outer tubular diameter (TD) of isolated tubular segments (ITS) was estimated light-microscopically. Tubular and mitochondrial protein was measured colorimetrically by the method of Lowry et al. (1951). Intracellular K^+ was extracted after homogeni-

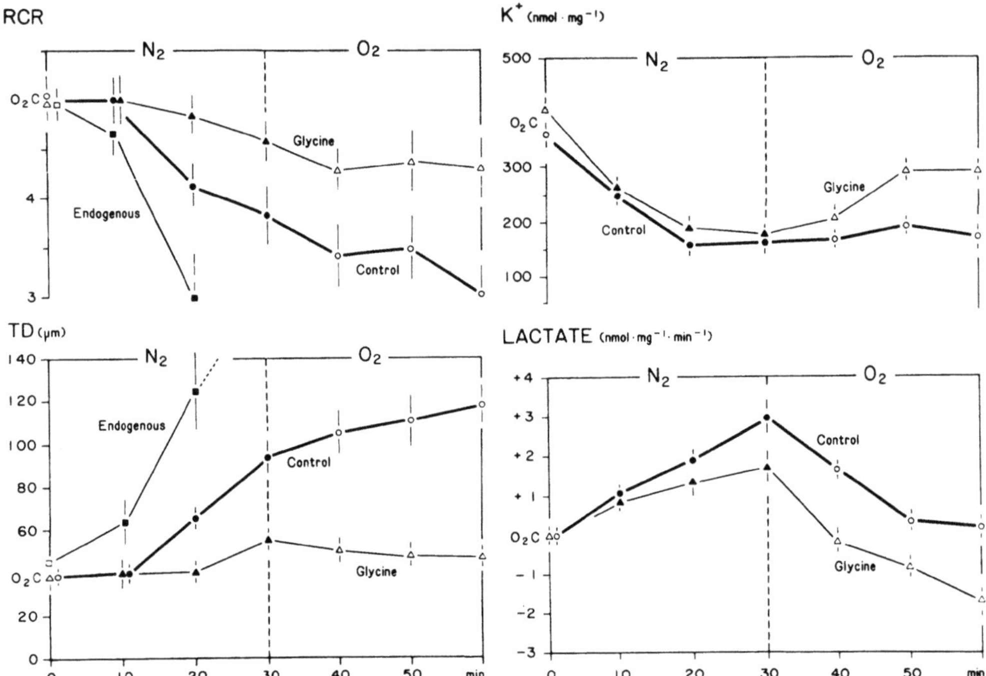

Fig. 1 Functional parameters of renal cortical cells in the time course (abscissa) of extreme hypoxia (N_2 = PO_2 < 1 mm Hg) and subsequent reoxygenation (O_2) in Ringer-Bicarbonate medium containing 5 mM glucose (37°C). Upper left panel: RCR = mitochondrial respiratory control rate (state 3 / state 4). Lower left panel: TD = mean outer tubular diameter. Upper right panel: K^+ = intracellular potassium. Lower right panel: lactate formation (+) and lactate uptake (-). Endogenous= no glucose added, glycine= 5 mM, control= 5 mM taurine. Mean ± SD, n=14

sation of ITS in destilled water and determined by standard flame photometry. Measurements of enzyme activities in the supernatant of at 50 x g in the cold centrifuged ITS suspension (LDH = lactate dehydrogenase, EC 1.1.27; τGT = gamma-glutamyl-transferase, EC 2.6.1.1; NAG = N-acetyl-ß-D-glucosaminidase, EC 3.2.1.30; GlDH = glutamate dehydrogenase, EC 1.4.1.3) were performed according to standard procedures and previously described methods (Gronow et al., 1989). Values are expressed per mg tubular protein and are means ± SD of 14 observations (n = 14). Statistical analysis was employed as ANOVA, paired t-Test, and regression analysis. A P-value of 0.05 or less was assumed to indicate a significant difference.

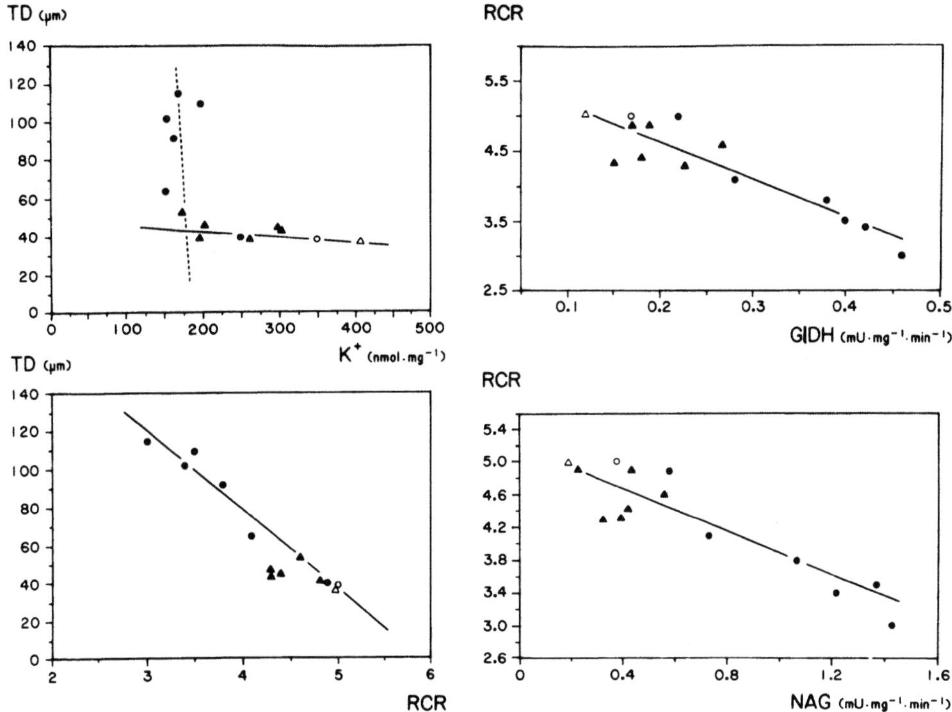

Fig. 2 Losses of marker enzymes from renal cortical cells in the time course (abscissa) of extreme hypoxia (N_2 = PO_2 < 1 mm Hg) and subsequent reoxygenation (O_22) in Ringer-Bicarbonate medium containing 5 mM glucose (37°C). Upper left panel: loss of cytoplasmatic lactate dehydrogenase (LDH). Lower left panel: loss of brush border τ-glutamyltransferase (τGT). Upper right panel: loss of lysosomal N-acetyl-ß-D-glucosaminidase (NAG). Lower right panel: loss ofmitochondrial glutamate dehydrogenase (GlDH). Endogenous = no glucose added, glycine = 5 mM, control = 5 mM taurine. Mean ± SD, n = 14

RESULTS and DISCUSSION

With no glucose, glycine, or exogenous substrates in the incubation medium (= endogenous, Fig. 1, upper left panel), the respiratory control ratio (RCR) in mitochondria isolated from O_2-deprived and subsequently reoxygenated renal tubular cells declined within 20 min rapidly to about 3, and cellular swelling was indicated by an about threefold increase in mean tubular diameter (TD in Fig. 1, lower left panel). 30 min of extreme hypoxia and substrate deprivation induced cell lysis and nearly no ADP-sensitive mitochondrial respiration (not shown). To avoid these starvation effect glucose was added in all further experiments to the incubation medium.

With no glycine (control in Fig. 1, upper left panel), reoxygenation had no beneficial effect on RCR, it declined significantly by about 26% in hypoxia, and up to 40% during reoxygenation. The observed cell swelling and the decrease in RCR was significantly reduced by glycine, after 30 min reoxygenation mean tubular diameter (TD in Fig. 1, lower left panel) was not significantly different from untreated controls (open symbols), and RCR (upper left panel) fell by only 14% in hypoxia and during subsequent reoxygenation.

Extreme hypoxia was accompanied by the loss of nearly 50% of intracellular K^+ (Fig. 1, upper right panel). Addition of glycine had no significant effect, and glycolytic lactate formation (Fig. 1, lower right panel) was markedly suppressed by glycine.

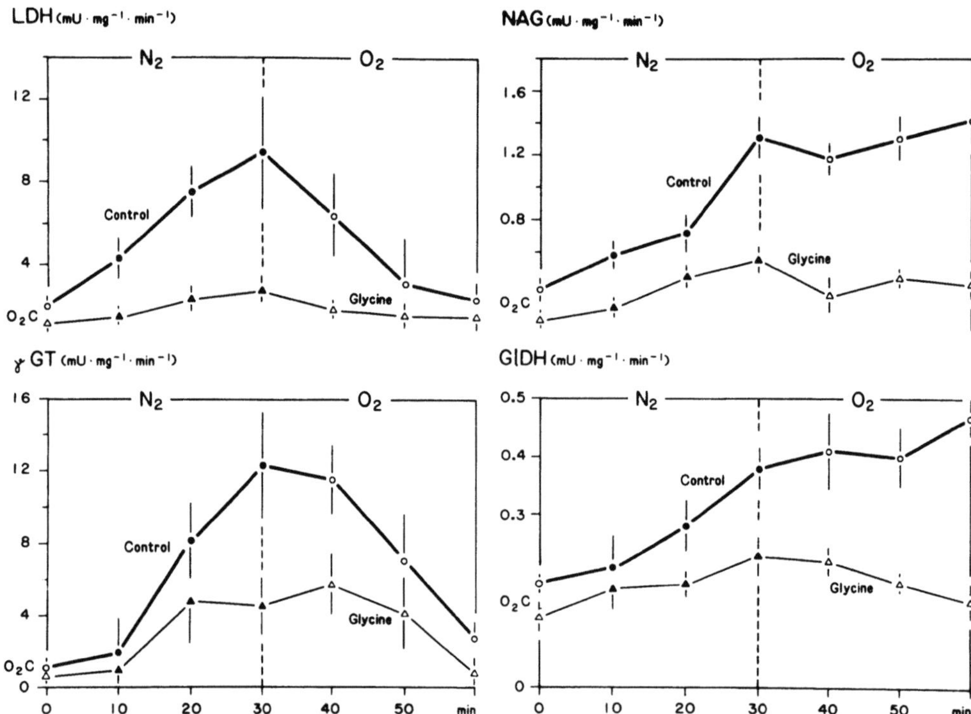

Fig. 3 Outer tubular diameter (TD) and mitochondrial respiratory control rate (RCR) of isolated tubular segments of rat renal cortex incubated at extreme hypoxia (PO_2 < 1 mm Hg) and subsequent reoxygenation (closed symbols). TD is depicted as a function of intracellular potassium (K^+) content (upper left panel) or RCR (lower left panel). RCR is depicted as a function of K^+ (upper right panel) or RCR (lower right panel). Symbols represent mean values of Fig. 1 and Fig. 2. Circles = 5 mM taurine, triangles = 5 mM glycine. Open symbols = untreated controls. For statistical information on regression lines see Tab. 1

Cellular losses of cytoplasmatic lactate dehydrogenase (LDH in Fig. 2, upper left panel), of gamma-glutamyltransferase from renal brush borders (τGT in Fig. 2, lower left panel), of lysosomal N-acetyl-ß-D-Glucosaminidase (NAG in Fig. 2, upper right panel), and of mitochondrial glutamate dehydrogenase (GlDH in Fig. 2, lower right panel) rose nearly linearly in extreme hypoxia. After reoxygenation, LDH and τGT releases declined, whereas NAG and GlDH liberation continued. Glycine significantly suppressed all losses of marker enzymes.

When lack of metabolic energy reduces transmembraneous ion pumping in extreme hypoxia, extracellular fluid is attracted by intracellular colloids and renal tubular cells begin to swell (MacKnight and Leaf, 1977; Völkl et al., 1988). Accordingly, the outer tubular diameter (TD) of ITS increases severalfold at extreme low oxygen tension (Gronow et al., 1990). In the present experiments, alterations of TD did not correlate linearly with changes in cytosolic potassium (K^+ in Fig. 3, upper left panel). Instead, with decreasing intracellular potassium, TD remained nearly constant at about 0.04 mm (straight line). At about half of normal K^+, however, tubular diameter increased sharply and nearly in parallel to the ordinate (dashed line), i. e. marked cell swelling occurred without significant changes in potassium losses. In contrast, increases in TD correlated linearly with decreases in respiratory control rate (RCR in Fig. 3, upper right panel and Tab. 1). Similarly, RCR decreased with increasing losses of mitochondrial GlDH (Fig. 3, upper right panel) as well as with losses of lysosomal NAG (Fig. 3, lower right panel). No significant linear or non-linear correlation could be calculated for changes of TD or RCR in relation to losses of cytoplasmatic LDH or of τGT from renal brush borders.

588

Tab. 1 Changes in mean tubular diameter (TD) in relation to alterations of intracellular potassium content or mitochondrial respiratory control rate (RCR), and of RCR in relation to membrane leaks, indicated by losses of mitochondrial glutamatedehydrogenase (GlDH) and lysosomal N-acetyl-ß-D-glucosa minidase (NAG), n = 14

equation of regression	r	p
$TD = 10445 \times Potassium^{-0.965}$	0.719	< 0.010
$TD = 244.2 - 42.5 \times RCR$	- 0.948	< 0.001
$RCR = 5.19 - 1.3 \times NAG$	- 0.898	< 0.001
$RCR = 5.69 - 5.33 \times GlDH$	- 0.896	< 0.001

The observed reduction of cellular swelling in the presence of glycine maintained the internal milieu and the intactness of cellular membranes in renal tubular cells, as indicated by a minimized loss of cell constituents (Fig. 2) and an improved mitochondrial RCR (Fig. 1, left panels). The latter, in turn, provided obviously more metabolic energy for K^+-reaccumulation during reoxygenation (Fig. 1, upper right panel). The maintenance of renal cellular volume at extreme low oxygen tension, however, was not mediated by a provision of glycolytic energy (Fig. 1, lower right panel) or an alteration of hypoxic K^+-pumping activity (Fig. 3, upper left panel).

The technical assistance of R. Bock is gratefully acknowledged

REFERENCES

Baines,A.D., Shaik,N., and Ho,P., 1990, Mechanism of perfused kidney cytoprotection by alanine and glycine, 1990, *Am. J. Physiol.* 259:F80

Goldstein,L., Glutamine transport by mitochondria isolated from normal and acidotic rats. *Am. J. Physiol.* 229:1027.

Gronow,G., Prechel,P., and Klause,N., 1989, Cytoprotective effect of isotonic mannitol at low oxygen tension. *Adv. Exp. Med. Biol.* 248:755

Gronow.G., Klause,N., and Mályusz,M., 1990, Support of hypoxic renal cell volume regulation by glycine, *Adv. Exp. Med. Biol.* 277:705.

Heyman,S., Spokes,K., Rosen,S., Epstein,F.H., 1992, Mechanism of glycine protection in hypoxic injury: analogies with glycine receptor. *Kidney Int.* 42:41

Lowry,O.H., Rosebrough,N.J., Farr,A.L., and Randall,R.J., 1951, Protein measurement with the folin phenol reagent, *J. Biol. Chem.* 193:265.

MacKnight,A.D.C., and Leaf,A., 1977, Regulation of cellular volume, *Physiol. Rev.* 57:510.

Völkl,H., Paulmichl,M., and Lang,F., 1988, Cell volume regulation in renal cortical cells, *Renal Physiol. Biochem.* 3:158.

Weinberg,J.,M., Abarzua,D.J.A., Rajan,T., 1987, Cytoprotective effect of glycine and gluthathione against hypoxic injury in renal tubules, *J. Clin. Invest.* 80:1446.

Weinberg,J.M., Venkatachalam,M.A., Garza-Quintero,M., Roeser,N.F., and Davis,A., 1990, Structural requirements for protection by small amino acids agaist hypoxic inury in kidney proximal tubules, *FASEB J.* 4:3347

OXYGEN SENSING BY H₂O₂-GENERATING HEME PROTEINS ?

J. Fandrey, S. Frede and W. Jelkmann

Department of Physiology I, University of Bonn
Nussallee 11, D-53115 Bonn, Germany

INTRODUCTION

Erythropoietin (Epo) production in the kidneys and liver will increase up to 100-fold if the Epo-producing tissue becomes hypoxic because of anemia or a decrease in the arterial PO_2 (Jelkmann 1992). Today the only cell culture system in which hypoxia-induced Epo production can be studied are the hepatoma cell lines HepG2 and Hep3B. In these cells a close correlation exists between increasing grades of hypoxia and Epo messenger RNA (mRNA) levels in the cells (Fandrey and Bunn 1993) and secreted Epo protein in the culture medium supernatant (Wolff et al. 1993).

Although much has been recently learned about hypoxia-sensitive promotor- and enhancer-elements in the neighbouring DNA-sequences of the Epo gene (Blanchard et al. 1993, and references therein) the nature of the oxygen sensor is still far from being understood. Goldberg and colleagues (1988) have proposed the involvement of a heme protein in the sensing process. Depending on the oxygen tension this heme protein is thought to change its conformational state as it is well known from hemoglobin. The desoxygenated state will then enhance Epo production whereas the oxy-conformation is inhibitory (Goldberg et al. 1988). Studies from our laboratory have supported the hypothesis that a heme-protein may be involved in the oxygen sensing process. We were able to provide evidence that induction of the cytochrome P_{450} monooxygenase system enhances hypoxia-induced Epo production (Fandrey et al. 1990). Very recently Görlach et al. (1993) have recorded absorption spectra from Epo producing HepG2 cells. They measured spectra typical of cytochromes of the type b heme proteins. These type b-cytochromes change their conformational state in response to the ambient oxygen tension and - like Epo production - are insensitive to cyanide (Görlach et al. 1993).

The group of type b cytochromes includes several heme proteins that produce hydrogen peroxide (H_2O_2). To test whether H_2O_2 might act as a transducing signal between the oxygen sensor (heme protein) and Epo gene expression, the influence of H_2O_2 on the hypoxia-induced Epo formation was investigated.

METHODS

Cell cultures

HepG2 cells from the American Type Culture Collection (ATCC No. HB 8065) were maintained in medium RPMI 1640 (Flow Laboratories, Meckenheim, Germany) supplemented with 10% fetal bovine serum (Gibco, Eggenstein, Germany) and sodium bicarbonate (2.2 g/l) in a humidified atmosphere (5% CO_2 in air) at 37° C (Heraeus incubators, Hanau, Germany). Monolayers of the cells were grown to confluence in 24-well polystyrene dishes (Falcon, Becton Dickinson, Heidelberg, Germany) resulting in a density of 5×10^5 cells/cm^2 at the beginning of the experiments. The medium (0.5 ml/cm^2) was renewed 24 h before the experiment. Under the above described culture conditions HepG2 cells are hypoxic due to diffusion-limited oxygen supply as recently determined by measurements of the pericellular PO_2 (Wolff et al. 1993).

For the determination of Epo protein the medium was collected and frozen at -20° C . The cell layer was washed with phosphate buffered saline and lysed with SDS-NaOH (5 g/l sodium dodecylsulfate in 0.1 mol/l NaOH). Total cellular protein was determined according to the method of Lowry et al. 1951) using a micro determination kit (Sigma Diagnostics, Taufkirchen, Germany).

Cytotoxicity was assessed by the colorimetric tetrazolium salt/formazan method (Hansen et al. 1989). The assay is based on the reduction of (3-(4,5-dimethylthiazol-2-yl)-2,5-diphenyl tetrazolium bromide (MTT; Sigma) to purple formazan by mitochondrial dehydrogenases in living cells. Confluent HepG2 cultures in 96-well dishes were used for these studies.

Quantitation of Epo mRNA by competitive polymerase chain reaction

At the end of the experiment cells were washed with sterile phosphate buffered saline and lysed with 4 mol/l guanidinium isothiocyanate with 0.1 mol/l ß-mercaptoethanol. Total RNA was isolated by CsCl centrifugation as described (Chirgwin et al. 1979), redissolved in water and the concentration determined by absorption at 260 nm.
Competitive polymerase chain reaction (PCR) was performed as recently reported (Fandrey and Bunn 1993). One microgram of total RNA was reverse transcribed into first strand cDNA using oligo dT$_{(15)}$ as a primer for reverse transcriptase (M-MLRV RT Superscript; Gibco, Eggenstein, Germany). The efficiency of reverse transcription was determined as described (Fandrey and Bunn 1993).

Statistics

Results are expressed as the mean ± SD. Dunnett's test was applied to determine the significance of difference ($P < 0.05$) between the mean of a control group and several treatment means.

RESULTS

Exogenous H_2O_2 dose-dependently inhibited hypoxia-induced Epo production in HepG2 cells without being cytotoxic (Fig. 1). Quantitation of cellular Epo mRNA levels by competitive PCR revealed that within 30 min of exposure to H_2O_2 (500 µmol/l) Epo mRNA levels were lowered by more than 50 % (control cells: 1.11 amol Epo mRNA/µg total RNA; H_2O_2 treated cells : 0.69 amol Epo mRNA/µg total RNA; mean of 2 separate experiments). Epo mRNA levels remained at this level in control and H_2O_2-treated cells for 24 h.

When the cellular production of H_2O_2 was increased by incubating HepG2 cells with menadione a dose-dependent inhibition of Epo production was observed (Fig. 2; left panel). Likewise, when the cellular degradation of H_2O_2 by catalase was diminished by the addition of increasing doses of aminotriazole, a known inhibitor of catalase, Epo production decreased dose-dependently (Fig. 2; right panel).

To antagonize the H_2O_2-elicited inhibition of Epo production, HepG2 cells were incubated with 1 mmol/l H_2O_2 and increasing doses of catalase (Fig. 3). Although the concentration of H_2O_2 used was high and almost suppressed hypoxia-induced Epo production completely, the addition of 100 µg/ml catalase restored the full Epo producing capacity of the HepG2 cells.

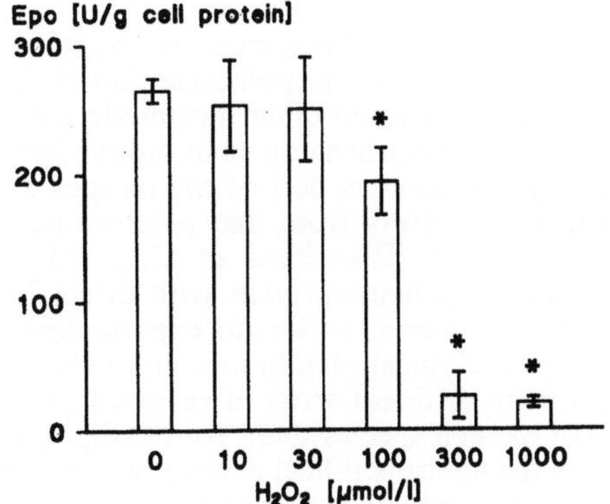

Fig. 1. Inhibition of Epo production in HepG2 cultures by exogenously added H_2O_2. * indicates a statistically significant difference from control; $P < 0.05$; n = 4 - 6 separate cultures.

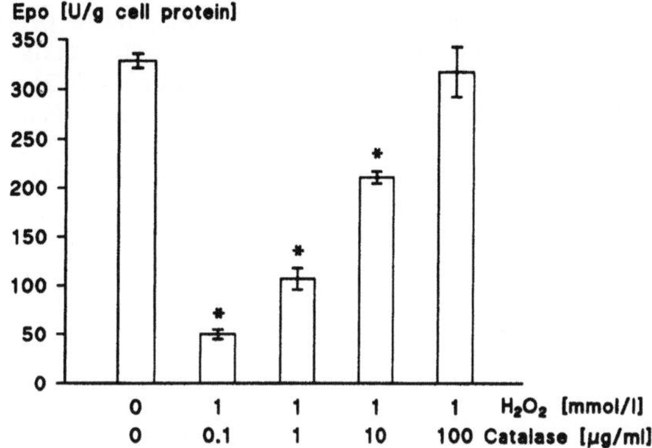

Fig. 2. Inhibition of Epo production in HepG2 cells by menadione (left panel) or aminotriazole (right panel). Data are the means of 4-6 separate cultures per column. A statistically significant difference compared to control cultures is indicated by *; $P < 0.05$; Dunnett's test.

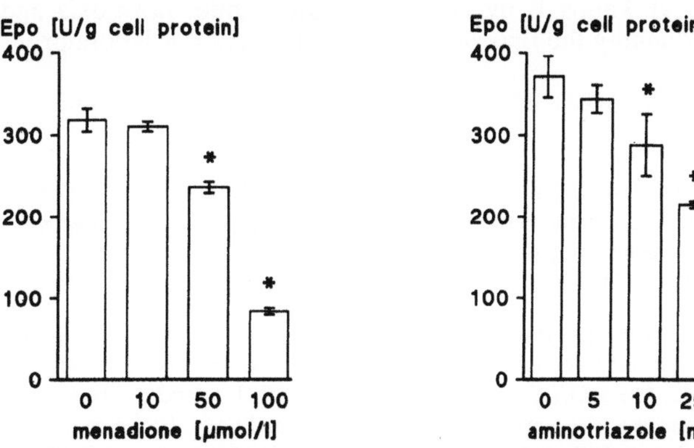

Fig. 3. Prevention of H_2O_2 (1 mmol/l) dependent inhibition of Epo production by coincubation with catalase (increasing concentrations). Data are the means of 4-6 separate cultures per column. A statistically significant difference compared to control cultures is indicated by *; $P < 0.05$; Dunnett's test.

DISCUSSION

In the present study we have provided evidence that hypoxia-induced Epo production in HepG2 cells can be altered by H_2O_2. We hypothesize that H_2O_2 acts as an intracelllular messenger that inhibits Epo gene expression under normoxic conditions. Why is a role for H_2O_2 in the signalling cascade of oxygen sensing an attractive hypothesis ?

Although a role for a heme protein as part of the oxygen sensor is widely accepted, until today no signal tranduction pathway from that putative heme protein to altered Epo gene expression has been proposed (Blanchard et al. 1993). Since oxygen is freely diffusible, one possibility could be a heme-protein with direct DNA binding ability, then acting as a transcription factor. However, so far no experimental evidence of this kind is available. H_2O_2 as well passes biological membranes and cellular compartments (Schreck et al. 1991). Increasing data from different areas of research provide evidence for a much broader role of these reactive oxygen species than, for example, their bactericidal effect as part of the respiratory burst in the neutrophil host defence against infections.

Görlach et al. (1993) have recently demonstrated that HepG2 cells contain cytochrome type b heme proteins with typical spectral changes when switched from normoxia to hypoxia. Among this group of heme proteins is the H_2O_2 generating NADPH-oxidase. Western blots of cellular extracts from HepG2 cells revealed that at least some components of the NADPH-oxidase system are present in these cells. Cross et al. (1990), based on their studies on the oxygen sensor in glomus caroticum cells, have proposed a role for the NADPH-oxidase system in oxygen sensing in chemoreceptor cells. Since the ability to sense the PO_2 may be widespread among cells (Maxwell et al. 1993) the parallel findings in nerve impulse generating glomus cells and Epo producing liver cells are appealing.

The hypothesis that under normoxia higher amounts of H_2O_2 are produced which suppress Epo production is supported by preliminary studies from this laboratory. H_2O_2 production in HepG2 cells can be measured photometrically. A much lower production rate of H_2O_2 was observed when HepG2 cells were exposed to hypoxia. Thus, while the pharmacologically achieved increase in cellular H_2O_2 inhibited Epo production, hypoxia provides a state of lowered cellular H_2O_2 concentration.

Finally, from recent studies on the influence of H_2O_2 on gene expression it has been reported that H_2O_2 can activate NF kappa B (Schreck et al. 1991). Recent studies link NF kappa B to the redox state of the cell which is controlled by the surrounding oxygen tension (Meyer et al. 1993). Sequence analysis of the 5' flanking region of the Epo gene revealed a well conserved consensus sequence for NF kappa B about 300 base pairs upstream of the putative transcriptional start site (Blanchard et al. 1992). It is tempting to speculate whether NF kappa B plays a role as a transcriptional activator or repressor for the Epo gene. Studies on the activation of NF kappa B by H_2O_2 generated under normoxia and its binding to DNA sequences surrounding the Epo gene are in progress in our laboratory.

REFERENCES

Blanchard, K.L., Acquaviva A.M., Galson D.L., and Bunn H.F., 1992, Hypoxic induction of the human erythropoietin gene: Cooperation between the promoter and enhancer, each of which contains steroid receptor response elements. Mol. Cell. Biol. 12: 5373-5385.

Blanchard, K.L., Fandrey J., Goldberg M.A., and Bunn H.F., 1993, Regulation of the erythropoietin gene. Stem cells 11: 1-7

Chirgwin, J.M., Przybyla A.E., Macdonald R.J., and Rutter W.J., 1979, Isolation of biologically active ribonucleic acid from sources enriched with ribonuclease. Biochemistry 18: 5294-5299.

Cross, A.R., Henderson, L., Jones, O.T.G., Delpiano, M.A., Hentschel, J., and Acker, H., 1990, Involvement of an NAD(P)H oxidase as a pO_2 sensor protein in the rat carotid body. Biochem. J. 272: 743-747.

Fandrey, J., Seydel F.P., Siegers C.P., and Jelkmann W., 1990, Role of cytochrome P450 in the control of the production of erythropoietin. Life Sci. 47: 127-134.

Fandrey, J., and Bunn H.F., 1993, In vivo and in vitro regulation of erythropoietin mRNA: Measurement by competitive polymerase chain reaction. Blood 81: 617-623.

Görlach, A., Holtermann G., Jelkmann W., Hancock J.T., Jones S.A., Jones O.T.G., and Acker H., 1993, Photometric characteristics of haem proteins in erythropoietin-producing hepatoma cells (HepG2). Biochem. J. 290: 771-776.

Goldberg, M.A., Dunning S.P., and Bunn H.F., 1988, Regulation of the erythropoietin gene: Evidence that the oxygen sensor is a heme protein. Science 242: 1412-1415.

Jelkmann, W., 1992, Erythropoietin: Structure, control of production, and function. Physiol. Rev. 72: 449-489.

Lowry O.H., Rosenbrough N.J., Farr A.L., and Randall R.J., 1951, Protein measurement with the Folin phenol reagent. J. Biol. Chem. 193:265-272.

Maxwell, P.H., Pugh, C.W., and Ratcliffe, P.J., 1993, Inducible operation of the erythropoietin 3'enhancer in multiple cell lines: Evidence for a widespread oxygen-sensing mechanism. Proc. Natl. Acad. Sci. USA 90: 2423-2427.

Meyer, M., Schreck, R., and Baeuerle, P.A., 1993, H_2O_2 and antioxidants have opposite effects on activation of NF-kappa B and AP-1 in intact cells: AP-1 as secondary antioxidant-responsive factor. EMBO J. 12:2005-2015.

Schreck, R., Rieber P., and Baeuerle P.A., 1991, Reactive oxygen intermediates as apparently widely used messengers in the activation of the NF-kappa B and HIV-1. EMBO J. 10: 2247-2258.

Wolff M., Fandrey J., Jelkmann W., 1993, Microelectrode measurements of pericellular pO_2 in erythropoietin-producing human hepatoma cell cultures. Am. J. Physiol. (in press)

PERFUSED RAT LIVER RESPONSES TO CRUDE VENOMS FROM MARINE SNAILS

David J. Maguire,[1] Jens Höper,[2] Gabi Casel ,[2] and Dagmar Gärtner[2]

[1]Faculty of Science and Technology, Griffith University, Nathan,
 Brisbane, Australia, Q4111
[2]Institut für Physiologie und Kardiologie, Waldstrasse 6, D-8520
 Erlangen, F.R.G.

INTRODUCTION

The venoms of the marine snail (*Conus regius*) contain an astonishing array of peptide toxins. The toxic activities present in any particular species has been shown to be associated with prey category (piscivore, molluscivore or vermivore). The range of activities present in the crude venoms of a number of species has been described by Endean's group. The activities of peptide toxins purified from a small number of piscivores and molluscivores have been described by Olivera's group after testing in laboratory animals such as mice and isolated tissue preparations. The conotoxins that have aroused most interest are compounds which block specific channels involved in nerve or neuromuscular transmission. There are a number of other sites within other tissues which are potential targets for venom components, for example channels located in the vascalature of all organs and peripheral tissues Toxin action on such channels is more difficult to assay than in the case of the neural and neuromuscular channels in mammals, but may be important sites of action in the natural prey and predators of this highly successful species. For a toxin to act most effectively its components should be rapidly distributed to all targets within the the recipient.This is particularly true in the case of slow-moving predators such as marine snails, many of which are particularly challenged in terms of mobility relative to their prey. We describe here the effect of a crude venom preparation upon various parameters of perfused rat liver.

METHODS

Crude venom was collected from sections of duct as described by Maguire and Kwan (1992). This technique involves dissection of the venom duct from the marine snail after it has been removed from its shell. The duct was then laid out on a clean glass surface, irrigated with isotonic saline and cut into sections approximately two to three centimeters in length. Venom was squeezed from each duct section using a spatula. Extracts from duct sections were combined as anterior, mid-region and posterior venom duct pools.

The liver perfusion system was as described by Hoper and Kessler (1988). Livers from rats were maintained at room temperature throughout the duration of each experiment. The total perfusate volume was 120 ml. Spectral measurements were made using the light guide and optical detection system described by Frank et al (1989). Oxygen partial pressures in the

Figure 1. Absorbance at 502 nm recorded for a total of 169 spectral observations taken over the first thirty minutes of five separate perfusions. The absorbance values represent the mean values recorded at each of the 169 time points during that time.

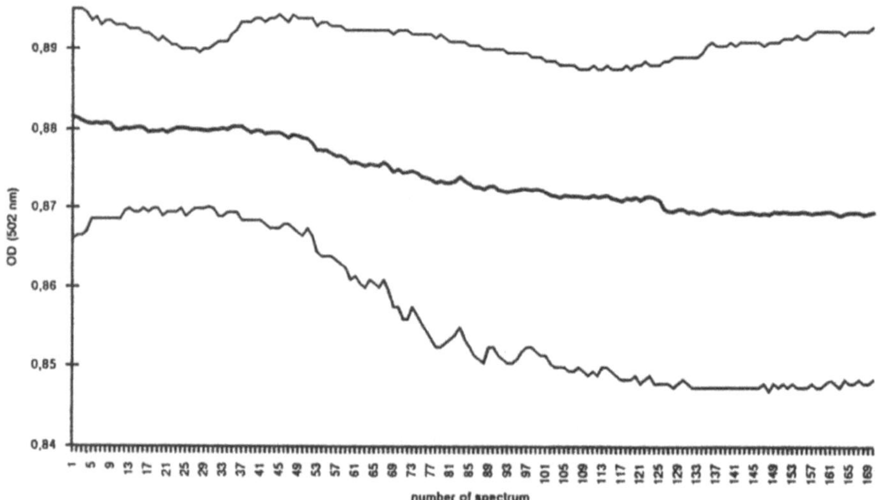

Figure 2. The range of absorbance values recorded for the data presented in Figure 1, showing mean (middle trace), maximal (upper trace) and minimal (lower trace)recordings for five experiments.

perfusate delivery and return circuits were monitored with oxygen electrodes constructed as described by Kessler and Lubbers (1966). All oxygen consumption rates were expressed as micromoles of oxygen consumed per minute per millilitre of perfusate. Perfusate flow rates were monitored continuously.

RESULTS

In Figure 1, absorbance at 502 nm is recorded for a total of 169 spectral observations taken over the first thirty minutes of five separate perfusions. The absorbance values represent the mean values recorded at each of the 169 time points during that time. The range of absorbance values seen over these experiments is recorded in Figure 2, where mean, maximal and minimal recordings for the five experiments are reported. Similar changes were seen in absorbance values at 628 nm (data not shown).

598

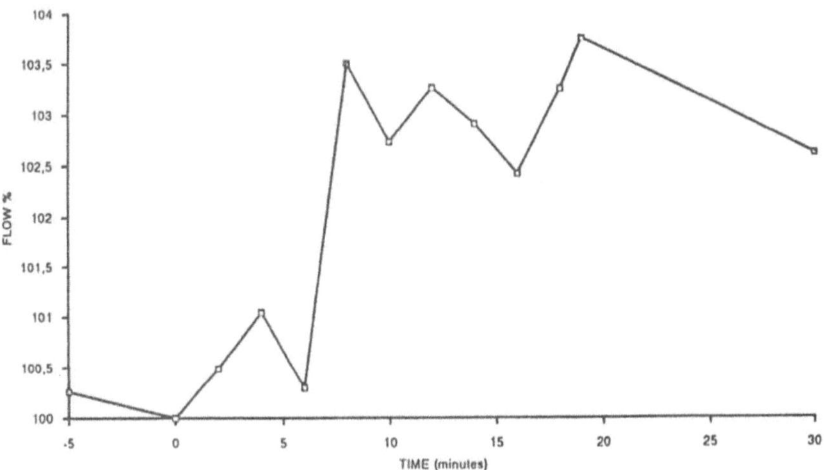

Figure 3. Means of total flow rates at various times during the first thirty minutes of five perfusion experiments, expressed relative to the flow rate at zero time (addition of toxin bolus).

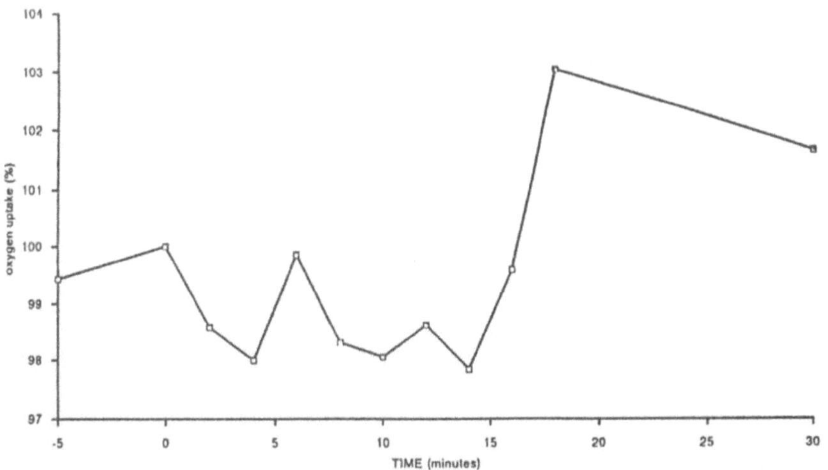

Figure 4. Mean rate of oxygen uptake at various times during the first thirty minutes of five perfusion experiments, expressed relative to the oxygen uptake rate at zero time (addition of toxin bolus).

The means of total flow rates at various times during the first thirty minutes of five perfusion experiments are shown in Figure 3, expressed relative to the flow rate at zero time (addition of toxin bolus).

The mean rate of oxygen uptake at various times during the first thirty minutes of five perfusion experiments are shown in Figure 4, expressed relative to the oxygen uptake rate at zero time (addition of toxin bolus).

The individual and mean (thickened line) spectra for five separate perfusions at zero time and thirty minutes following the addition of venom are shown in Figures 5 and 6 respectively.

In Figure 7, the time course of the spectral changes during one experiment are shown as difference spectra for each tenth spectra. These were derived by subtracting the initial spectrum from each of the other spectra recorded.

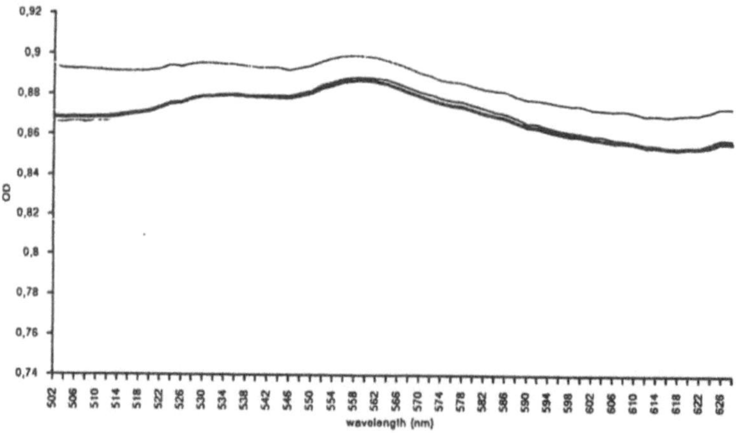

Figure 5. The individual and mean (thickened line) spectra for five separate perfusions at zero time, immediately preceding the addition of venom.

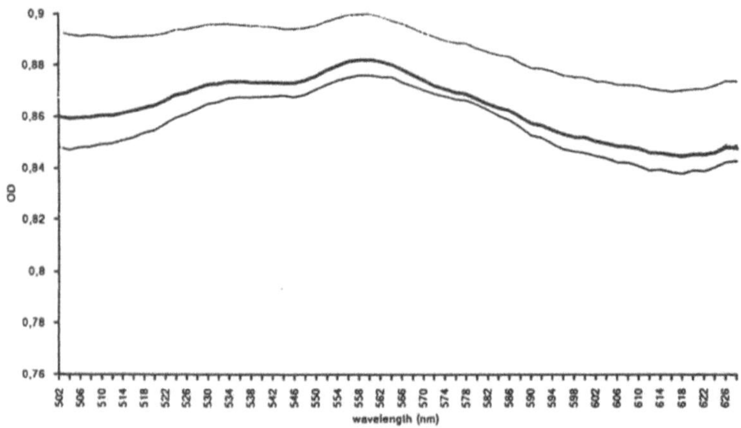

Figure 6. The individual and mean (thickened line) spectra for five separate perfusions at thirty minutes following the addition of venom.

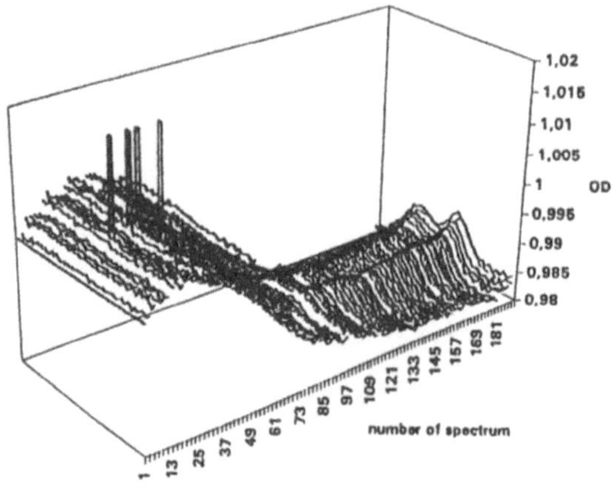

Figure 7. Time course of the spectral changes during one experiment shown as difference spectra for each tenth spectrum. These were derived by dividing the initial spectrum into each of the other spectra recorded.

DISCUSSION

Oxygen consumption and perfusion flow rates before any additions were within the ranges previously reported. Variable responses were recorded for some of the parameters tested after addition of the initial crude venom bolus for some of the snail species tested In the case of C.*regius* , two separate, but not necessarily unrelated series of effects were observed.

Firstly there appeared to be (intra)cellular structural changes, as indicated by decreased optical density readings at both 502 nm and 628 nm for all perfusions. These observations are consistent with an increase in either the number or the volume of small intracellular particles such as mitochondria. These changes occured simultaneously with changes in perfusate flow. Although the mean perfusate flow rate increased marginally at about five minutes, in some experiments flow rates decreased while in others it increased.

Secondly, and apparently secondary to the initial structural and rheological changes, there was a small increase in oxygen uptake at approximately fifteen minutes. This latter change occured in concert with a change in the redox state of cytochromes c and b. Because the changes observed were to cytochromes b and c in the absence of any change to cytochrome aa3, it would appear that the supply of substrate hydrogen was increased.

CONCLUSIONS

In the perfused isolated rat liver preparation, the observed effects of a crude venom from C.*regius* were consistent with a primary effect upon cellular ion channels. The first effects that could be detected after addition of venom were increases in the perfusate flow rate and decreases in total tissue absorbance, this latter effect corresponding to either an increase in tne number of small particles or a decrease in the volume of small particles. Subsequently, and possibly due to changes in intracellular ion activities, an increased oxygen uptake was observed simultaneously with a change in the redox state of cytochromes c and b. The secondary spectral changes observed were consistent with an increase in the supply of substrate hydrogen.

REFERENCES

Frank, K.H., Kessler, M., Appelbaum, K. and Dummler, W., 1989, The Erlangen micro-lightguide spectrophotometer Empho I, Phys. Med. Biol. 34:1883-1900
Hoper, J. and Kessler, M., 1988, Constant-pressure perfusion of the isolated rat liver: local oxygen supply and metabolic funstion, Int. J. Microcirc.: Clin. Exp. 7:155-168
Kessler, M. and Lubbers, D.W.,1966, Structure and possible applications of various pO2 electrodes, Pflugers Arch. 291:R82
Maguire, D., and Kwan, J., 1992, Cone shell venoms - synthesis and packaging, *in* : "Toxins and Targets." Watters, D., Pearn, J., Maguire, D. and Lavin, M., eds, Gordon and Breach, Sydney

GASTRIC INTRAMUCOSAL PH MEASUREMENTS
AS AN INDEX OF TISSUE OXYGENATION
IN PATIENTS WITH SEPSIS SYNDROME

F.Esen, L.Telci, C.Girgin, N.Çakar, K.Pembeci, T.Denkel, K.Akpir

Tissue hypoxia is considered to be the most important factor to the morbidity and mortality in patients with sepsis syndrome. However, the assesment of tissue oxygenation is still contraversial, since direct measurement of the adequacy of tissue oxygenation has not yet been available in the clinical setting.

In this study we aimed to compare tonometric measurements of gastric intramucosal pH with oxygen supply, oxygen consumption and arterial lactic acid values as systemic indices of tissue oxygenation in patients with sepsis syndrome. 18 septic patients meeting the criteria for sepsis syndrome as defined by Bone et al. were included in the study. Pulmonary artery catheters were placed to evaluate hemodynamic variables. Oxygen supply (DO_2) was calculated by the simplified formula $DO_2 = COxHbxSaO_2x1.39$. Oxygen consumption (VO_2) was obtained by two different methods; by direct method of using expired gas (mVO_2) and by calculating the product of cardiac index and arteriovenous oxygen content difference. A gastric tonometer (TRIP, tonometrics, Worchester, MA) was introduced in place of the standard nasogastric tube. Measures of gastric intramucosal PH (PHi) as it was first explained by Fiddian-Green. Dobutamin test was used by infusions starting with $5\mu gr/kg/min$, and increasing to 10 and $15\mu gr/kg/min$ at 60 min intervals while mVO_2, cVO_2, DO_2, lactate levels and PHi values were determined after each dose administration. All patients demonstrated the phenomenon of oxygen supply dependency characterized by a linear function (r = 0.87). Lactate levels were higher than normal before dobutamin infusion while decreases were noted by the increasing levels of DO_2. Gastric PHi measurements showed parallel and significant increase with the increasing levels of oxygen supply.

In summary, our results are of clinical importance that PHi is a valuable adjunct to the oxygen supply, oxygen consumption and lactate measurements as an index of tissue oxygenation.

Dept. of Anesthesiology and Intensive Care, University of Istanbul, Medical Faculty of Istanbul.

INFLUENCE OF NOREPINEPHRINE ON THE OXYGEN SUPPLY,TISSUE PO2 AND LACTATE EXTRACTION RATE OF THE LIVER IN THE SEPTIC PIG

Reiner Schäfer*, Werner Gerling*, Angela Utschakowski+, Dagmar Börner*, Klaus Wagner#

*Klinik f. Anästhesiologie, +Klinik f. Chirurgie, #Inst. f. Physiologie, Medizinische Universität Lübeck, W-2400 Lübeck, Bundesrepublik Deutschland

Background and goal of study: Little is known about oxygen delivery and consumption in the septic state under experimental conditions in animals, if macrocirculation is supported as it is done in human patients, i.e. with therapy like norepinephrin and volume. Therefore this study was performed to investigate the influence of norepinephrine in the treatment of the early septic shock on the oxygen supply , tissue pO2 and lactate extraction rate of the liver.

Materials and methods: After government approval we used 17 female domestic pigs between 20 and 28 kg body wt.; 6 pigs without sepsis served as control group. Anesthesia was induced by ketamin/diazepam and maintained by thiopental, fentanyl and pancuronium. Sepsis was induced by E.coli.-endotoxin (0.25-0.5 µg/kg bw, 0.025µg/kg bw/min; infusion was stopped at PAP>60mmHg) and treatment with norepinephrine (0.5µg/kg bw/min by perfusor) was performed until MAP was >80mmHg. CentPulmonary capillary wedge pressure was held at starting point level by infusing ringer lactate and/or hydroxyethylstarch. The following parameters were measured before and after induction of sepsis: Cardiac output (CO, thermodilution), MAP, PAP, A.hep.flow, V.porta flow (electromagnetic/doppler-ultrasound flow probes), systemic and hepatic oxygen consumption, liver tissue pO_2 (needle electrode, polarographically) and lactate extraction rate. Measuring cues were before (baseline), 20, 40, 60, 90 and 120 min after induction of sepsis.

Results and discussion: Sepsis treated with Norepinephrin increased CO to 137 % of control, SVR decreased to 64 %. Systemic oxygen consumption increased to 126 %, liver oxygen consumption decreased to 65% of control values. The flows of the hepatic artery increased to 146 %, whereas the flows of the portal Venei remained stable at about 95% of control. The ratio Flow hepatic artery/CO increased to 135% , whereas portal venous flow/CO decreased to 65% of its original value. Liver tissue pO2 remained stable at 35 mmHg. Lactate extraction rate remained also stable at 0.45 mmol/kg liver tissue/min.

Our results suggest that treatment of early sepsis by norepinephrine does not influence oxygen delivery and consumption. The ability to extract lactate is also not depressed. The decrease relative portal venous flow perhaps induced by alpha-receptor agonism in the splanchnic circulation is balanced by an increased oxygen extraction and a greater arterial oxygen delivery. The effectiveness of this mechanism is also supported by the stability of the tissue-pO_2-levels. Hepatic arterial vessels seem not to be influenced by norepinephrine.

FUNCTIONAL HETEROGENEITY IN THE GUT: RELEVANCE TO OXYGEN TRANSPORT AND THE MAINTENANCE OF OXYGEN CONSUMPTION

Paul T. Schumacker   Section of Pulmonary and Critical Care Medicine, The University of Chicago, Chicago, IL 60637

The oxygen supply to tissue microvascular units normally exceeds oxygen demand. Under these conditions, cell respiration is set by metabolic activity rather than by O_2 supply. As tissue oxygen delivery is reduced, increases in tissue oxygen extraction are optimized if blood flow among microvascular units is redistributed in proportion to oxygen demand. This redistribution is crucial for the maintenance of cell respiration in tissues with regional heterogeneity in metabolic activity. Heterogeneous tissues that fail to optimally redistribute blood flow during reductions in oxygen supply are more liable to sustain hypoxic tissue injury, because poorly perfused units with high metabolic demands may become O_2 supply-dependent while other units are relatively well perfused. The role of reflex sympathetic vasoconstrictor tone in this redistribution was investigated in vascularly isolated segments of canine small intestine perfused with an occlusive roller pump. Dogs were anesthetized and ventilated with room air. Oxygen delivery to the innervated gut segment was reduced in stages by lowering the speed of the pump. In a normovolemic group, systemic sympathetic tone was minimized during this procedure by intravenous fluid administration. In a hypovolemic group, sympathetic tone was augmented by controlled hemorrhage. The onset of O_2 supply-dependent metabolism in the gut was determined in each experiment from analysis of O_2 delivery-uptake data. Gut critical oxygen extraction in the normovolemic group ($45.6\pm11.5\%$) was significantly poorer than for the hypovolemic group ($68.4\pm3.4\%$). Gut vascular resistance was significantly higher at the critical point in the hypovolemic group. These results suggest that reflex sympathetic vasoconstriction contributes to the increases in gut extraction during progressive ischemia. To clarify the source of this vasoconstrictor tone, α-adrenergic vasoconstriction was inhibited with phenoxybenzamine (3 mg/kg) in another hypovolemic group. Gut critical oxygen extraction after phenoxybenzamine was not less (p=n.s.) than the hypovolemic group. Moreover, phenoxybenzamine did not abolish the progressive increases in gut vascular resistance in response to systemic hypovolemia. These results suggest that non-adrenergic reflex vasoconstriction in the gut mediates the increase in vascular tone, which contributes to the flow redistribution during progressive ischemia. Interestingly, if systemic hemorrhage was initiated in a normovolemic group while the gut was in the middle of the supply-dependent range, progressive increases in gut O_2 extraction were observed at constant gut O_2 delivery. Collectively, these results may reflect a more efficient partitioning of blood flow between mucosa and muscularis, which are characterized by different intrinsic metabolic activities. However, reflex sympathetic vasoconstriction may also act within each of these regions to reduce functional heterogeneity by limiting perfusion to units with low metabolic needs. Conceivably, sympathetic vasoconstriction may also alter perfused capillary surface area in the gut, where capillary recruitment may be actively regulated via pre-capillary sphincters. In either case, these results demonstrate that efficient microvascular regulation in the gut requires extrinsic neural control which is non-adrenergic. In the absence of such tone, local metabolic vasodilation remains intact, but is inadequate to achieve high critical oxygen extractions.

Supported by HL 35440.

DO CHANGES IN TUMOR BLOOD FLOW NECESSARILY LEAD TO CHANGES IN TISSUE OXYGENATION AND IN BIOENERGETIC STATUS?

P. Vaupel, D. K. Kelleher and T. Engel

Institute of Physiology and Pathophysiology
University of Mainz
Duesbergweg 6
D-55099 Mainz
Germany

INTRODUCTION

An increasing number of investigations carried out in recent years provide evidence suggesting that "chronic" decreases in tumor blood flow and/or tissue oxygenation (e.g., during tumor growth) or acute declines in the tissue perfusion (e.g., following therapeutic measures) might be accompanied by significant reductions in the energy status. In several instances, positive correlations between energy status and tumor blood flow or oxygenation have been reported (Lilly et al., 1985; Evelhoch et al., 1986; Tozer et al., 1989; Vaupel et al., 1989a, 1989b; Steen and Graham, 1991), and these investigations have led to the conclusion that blood flow may be the limiting factor in determining the bioenergetic status of tumors during growth. Manipulations of tumor blood flow by vasodilators, hyperthermia, tumor necrosis factor-α (TNF-α), lymphotoxin, interleukin-1, x-irradiation or after i.p. mannitol administration were accompanied by parallel changes in tumor energy status. Only in studies where i.p. or i.v. glucose was administered was energy status found to be stable or even slightly improved despite significant reductions in tumor perfusion (Okunieff et al., 1989; Krüger et al., 1991; Mayer et al., 1992; Schaefer et al., 1993). Similar observations of a dissociation between changes in tumor blood flow, oxygenation and energetic status have been observed in normoglycemic mice during photodynamic therapy (Bremner et al., 1993) or following hydralazine administration in xenografted human and isotransplanted murine tumors (Bremner et al., 1991; Adams et al., 1992).

As a result of these discrepancies, this study was undertaken to attempt to clarify a number of issues regarding the relationship between tumor blood flow and energy status. Under normoglycemic conditions, the bioenergetic status was assessed following both acute and chronic reductions in tumor blood flow to ascertain whether or not alterations in tumor perfusion necessarily lead to changes in bioenergetic status. Furthermore, an attempt was made to delineate critical thresholds and ranges over which such a relationship occurs.

MATERIALS AND METHODS

Animals and Tumors

Sprague Dawley rats (310 ± 5 g) were used in this study. They received a standard diet and water *ad libitum*. Experimental tumors were grown subcutaneously after injection of ascites cells of DS-sarcoma into the hind foot dorsum (for more details see Krüger et al., 1991).

Measurements

Once tumors reached the desired size, animals were anesthetized with sodium pentobarbital (40 mg/kg i.p., Nembutal, Ceva, Paris, France). Polyethylene catheters were surgically placed into the thoracic aorta via the left common carotid artery and connected to a Statham pressure transducer for mean arterial blood pressure measurement, and into the right external jugular vein for drug delivery. Animals breathed room air spontaneously and rectal temperature was maintained at 37.5°C, such that tumor temperature remained within the range 34 - 36°C throughout the experiment.

In all experiments performed in this study, animals were allowed to stabilize following surgical procedures. Relevant parameters assessing tumor perfusion, oxygenation and bioenergetic status were measured as follows either at different stages of growth in the subcutaneously implanted tumors (to provide "chronic" decreases in tumor blood flow), or following acute reductions in tumor perfusion brought about by application of TNF-α:

a) Tumor blood flow was determined using the ^{85}Kr-clearance technique (for more details see Kluge et al., 1992). Measurements were performed on tumors of varying sizes at 20 min intervals before application of TNF-α and at 30 min intervals thereafter over a total time period of 2 h post-treatment.

b) Tumor oxygen tension was determined using polarographic needle electrodes (recessed gold in glass electrode; shaft diameter 250 µm; diameter of the cathode 12 µm) and pO_2 histography (KIMOC-6650, Eppendorf, Hamburg, Germany; for more details see Vaupel et al., 1991). Measurements were made either on tumors of varying sizes or before and 120 min after acute flow declines upon TNF-α application.

c) Metabolic and bioenergetic status: In a first series of experiments, tumors of varying sizes were analysed. In another series, tumors were assayed before or 120 min after administration of TNF-α. Tumors were rapidly frozen in liquid N_2, ground to a fine powder and freeze-dried. After extraction with 0.66 M perchloric acid and neutralization with 2 M potassium hydroxide, ATP, ADP and AMP concentrations were determined using reverse phase HPLC techniques at 254 nm.

Drugs

Recombinant human TNF-α (specific activity: 8.2 x 10^6 U/mg protein; Knoll, Ludwigshafen, Germany) - used to induce acute blood flow reductions in tumor tissue - was diluted in isotonic phosphate-buffered saline solution containing 0.5 wt.% bovine serum albumin. TNF-α was administered via the external jugular vein at a dose of 1 mg/kg over approximately 3 min.

Statistical Analysis

Results are expressed as means ± SEM. Significance was assessed using the paired or unpaired Student's *t*-test, as appropriate. Results were considered to be significant if p-values were less than 5% (p < 0.05)

RESULTS AND DISCUSSION

In the DS-sarcoma, specific tumor blood flow and tissue oxygenation decreased significantly with increasing tumor mass. In the smallest tumors (volumes ≤ 1.0 ml), a mean blood flow of 1.0 ml·g^{-1}·min^{-1} was measured, whereas in the larger tumors, flow had decreased to approximately 50% of this value ($2p < 0.001$). This decrease in tumor blood flow was accompanied by a comparable significant decrease in the mean (and median) tumor pO_2 value ($2p < 0.001$; Fig. 1). In tumors of biologically relevant volumes (i.e., $< 1\%$ of body weight), ATP concentrations and the adenylate energy charge AEC = ([ATP] + 0.5[ADP])/([ATP] + [ADP] + [AMP]) = 0.74 ± 0.03, did not significantly change with increasing tumor volume, despite these significant reductions in tumor perfusion.

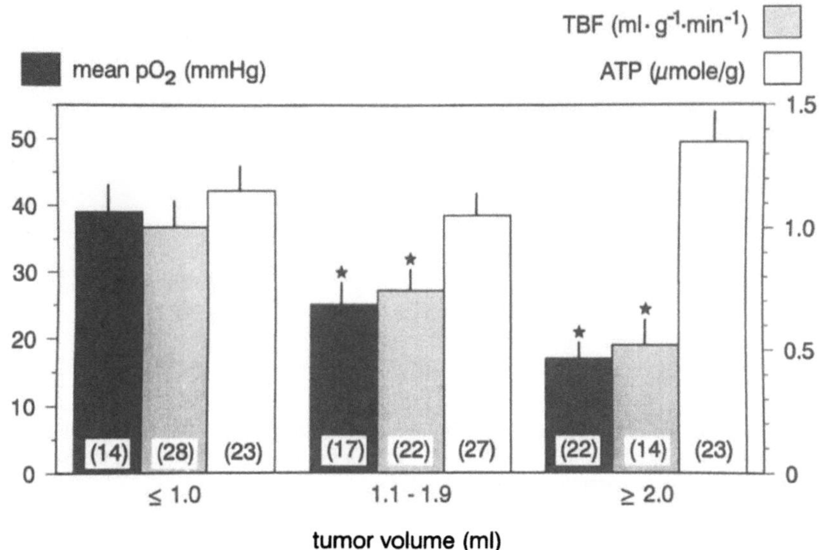

Figure 1. Tumor blood flow (TBF), mean tissue oxygen tension (pO_2) and mean tumor ATP concentrations as a function of tumor size. Values are means ± SEM, with the number of tumors investigated in parentheses ($\star\ 2p < 0.001$)

Over the tumor size range studied (0.7 - 2.5 ml), blood flow and pO_2 values were comparable to those observed in many normal tissues (Vaupel et al., 1989c) and were seen to be accompanied by a stable energy status. While tumor tissue oxygenation can be strongly influenced by changes in tumor blood flow, no such influence of tumor blood flow on ATP concentrations was apparent in this tumor line.

In further experiments in this study, measurements were made following acute reductions of tumor blood flow induced by TNF-α administration. Application of TNF-α resulted in a 50% flow drop within 120 min, a change which was accompanied by a comparable reduction in the mean tumor pO_2 (Fig. 2). However, no concomitant changes were seen in either ATP concentrations or in PCr/P$_i$ and βNTP/P$_i$ ratios (Fig. 3; the latter ratios were estimated using the same experimental protocol and ^{31}P-NMR spectroscopy, B.Elger et al., unpublished data). Under these conditions, energy status remained stable at mean flow values ≥ 0.5 ml·g^{-1}·min^{-1}, mean tumor oxygen tensions ≥ 13 mmHg and mean tumor glucose concentrations ≥ 1.4 µmole/g (Engel and Vaupel, 1993).

It is therefore concluded that, in the rodent tumor system investigated, acute and chronic (growth-related) changes in tumor blood flow are accompanied by comparable alterations in tissue oxygenation. However, providing blood flow values do not fall below a certain threshold level (approximately 0.5 ml·g⁻¹·min⁻¹), the tumor energy status remains stable. Mechanisms which might be responsible for the maintained energy status are:

(i) an intensified glycolytic rate due to deterioration of the oxygenation status, and
(ii) a decreasing number of proliferating cells as the tumor mass increases which would compensate for a deteriorating oxygen supply.

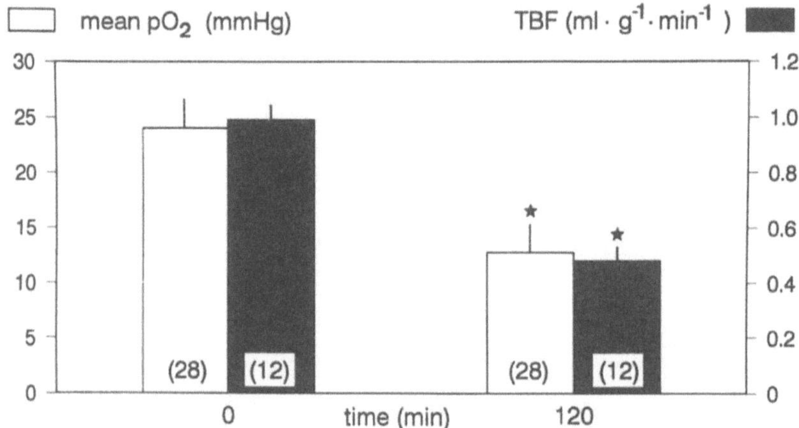

Figure 2. Mean tumor oxygen tension (pO_2) and tumor blood flow (TBF) before and 120 min after administration of TNF-α (1 mg/kg i.v.). Values are means ± SEM, with the number of tumors investigated in parentheses (★ $2p < 0.001$)

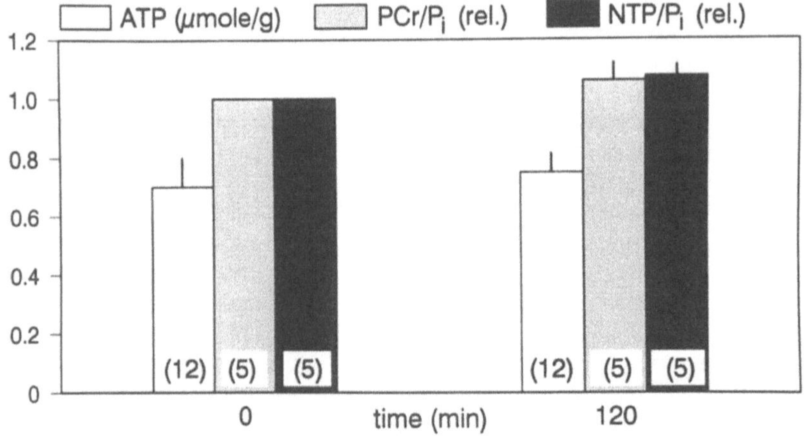

Figure 3. Mean tumor ATP concentration and ³¹P-MRS derived PCr/Pᵢ and βNTP/Pᵢ ratios before and 120 min after administration of TNF-α (1 mg/kg i.v.). Values are means ± SEM, with the number of tumors investigated in parentheses.

ACKNOWLEDGEMENTS

This work was supported by grants from the Deutsche Krebshilfe (Grant M 40/91/Va 1), and from the Vinzenz von Paul Foundation, Basel, Switzerland (Grant 5.1).

DS-sarcoma was kindly provided by Dr. H. Löhrke from the German Cancer Research Center in Heidelberg.

REFERENCES

Adams, G.E., Bremner, J.C.M., Counsell, C.J.R., Stratford, I.J., Thomas, C., and Wood, P.J., 1992, Magnetic resonance spectroscopy studies on experimental murine and human tumors: comparison of changes in phosphorus metabolism with induced changes in vascular volume, *Int. J. Radiat. Oncol. Biol. Phys.* 22:467.

Bremner, J.C.M., Bradley, J.K., Counsell, C.J.R., and Adams, G.E., 1993, Changes in ^{31}P-metabolism and blood flow after photodynamic therapy (PDT): A comparison between a murine sarcoma (RIF-1) and a human xenografted tumour (HT29), *41st Ann. Meeting Radiat. Res. Soc., Dallas, TX, March 20-25*:Abstract P-08-1.

Bremner, J.C.M., Counsell, C.J.R., Adams, G.E., Stratford, I.J., Wood, P.J., Dunn, J.F., and Radda, G.K., 1991, In vivo ^{31}P nuclear magnetic resonance spectroscopy of experimental murine tumours and human tumour xenografts: effects of blood flow modification, *Brit. J. Cancer* 64:862.

Engel, T., and Vaupel, P., 1993, Acute effects of tumor necrosis factor-α or lymphotoxin on oxygenation and bioenergetic status of experimental tumors, *Adv. Exp. Med. Biol.*: in press.

Evelhoch, J.L., Sapareto, S.A., Nussbaum, G.H., and Ackerman, J.H., 1986, Correlations between ^{31}P NMR spectroscopy and ^{15}O perfusion measurements in the RIF-1 murine tumor in vivo, *Radiat. Res.* 106:122.

Kluge, M., Elger, B., Engel, T., Schaefer, C., Seega, J., and Vaupel, P., 1992, Acute effects of tumor necrosis factor α or lymphotoxin on global blood flow, laser Doppler flux, and bioenergetic status of subcutaneous rodent tumors. *Cancer Res.* 52:2167.

Krüger, W., Mayer, W.-K., Schaefer, C., Stohrer, M., and Vaupel, P., 1991, Acute changes of systemic parameters in tumour-bearing rats, and of tumour glucose, lactate, and ATP levels upon local hyperthermia and/or hyperglycaemia, *J. Cancer Res. Clin. Oncol.* 117:409.

Lilly, M.B., Katholi, C.R., and Ng, T.C., 1985, Direct relationship between high-energy phosphate content and blood flow in thermally treated murine tumors, *J. Natl. Cancer Inst.* 75:885.

Mayer, W.-K., Stohrer, M., Krüger, W., and Vaupel, P., 1992, Laser Doppler flux and tissue oxygenation of experimental tumours upon hyperthermia and/or hyperglycaemia, *J. Cancer Res. Clin. Oncol.* 118:523.

Okunieff, P., Vaupel, P., Sedlacek, R., and Neuringer, L.J., 1989, Evaluation of tumor energy metabolism and microvascular blood flow after glucose or mannitol administration using ^{31}P nuclear magnetic resonance spectroscopy and laser Doppler flowmetry, *Int. J. Radiat. Oncol. Biol. Phys.* 16:1493.

Schaefer, C., Mayer, W.-K., Krüger, W., and Vaupel, P., 1993, Microregional distributions of glucose, lactate, ATP and tissue pH in experimental tumours upon local hyperthermia and/or hyperglycaemia, *J. Cancer Res. Clin. Oncol.* 119: 599.

Steen, R.G., and Graham, M.M., 1991, ^{31}P magnetic resonance spectroscopy is sensitive to tumor hypoxia: perfusion and oxygenation of rat 9L gliosarcoma after treatment with BCNU, *NMR Biomed.* 4: 117.

Tozer, G., Suit, H.D., Barlai-Kovach, M., Brunengraber, H., and Biaglow, J., 1989, Energy metabolism and blood perfusion in a mouse mammary adenocarcinoma during growth and following x irradiation, *Radiat. Res.* 109:275.

Vaupel, P., Okunieff, P., Kallinowski, F., and Neuringer, L.J., 1989a, Correlations between ^{31}P-NMR spectroscopy and tissue O_2 tension measurements in a murine fibrosarcoma, *Radiat. Res.* 120:477.

Vaupel, P., Okunieff, P., and Neuringer, L.J., 1989b, Blood flow, tissue oxygenation, pH distribution, and energy metabolism of murine mammary adenocarcinomas during growth, *Adv. Exp. Med. Biol.* 248:835.

Vaupel, P., Kallinowski F., and Okunieff, P., 1989c, Blood flow, oxygen and nutrient supply, and metabolic microenvironment of human tumors: A review, *Cancer Res.* 49:6449.

Vaupel, P., Schlenger, K., Knoop, C., and Höckel, M., 1991, Oxygenation of human tumors: evaluation of tissue oxygen distribution in breast cancers by computerized O_2 tension measurements. *Cancer Res,* 51:3316.

COMPUTERIZED HISTOGRAPHIC CHARACTERIZATION OF CHANGES IN TISSUE pO$_2$ INDUCED BY ERYTHROPOIETIN

D.J. Terris, A.I. Minchinton[†]

Division of Otolaryngology/Head & Neck Surgery
Stanford University Medical Center
Stanford, California USA

[†]British Columbia Cancer Research Centre
Vancouver, British Columbia V5Z 1L3

ABSTRACT

Anemia associated with malignancy is a common clinical problem, and has a negative effect on oxygen delivery and, therefore, response of tumors to radiotherapy. Erythropoietin (EPO) has been shown to increase the hematocrit of rodents when administered subcutaneously.

The objectives of this study were to measure tumor and normal tissue pO$_2$ with computerized pO$_2$ histography, to characterize the change in rodent hematocrit with EPO administration, and to assess the capacity of pO$_2$ histography to measure changes in tumor pO$_2$ during growth, with or without EPO administration.

Ten C$_3$H mice were implanted with SCCVII tumors and after three weeks of tumor growth, EPO (1500 IU/kg/day) was administered for two weeks. Tumor and subcutaneous (SQ) measurements were made weekly with an Eppendorf pO$_2$ histograph and hematocrits were obtained concomitantly. Eight C$_3$H/SCCVII mice, which received no EPO and underwent similar measurements, served as controls.

The hematocrits of the control mice dropped progressively from 42.0 to 23.0% during the two week period. There was a corresponding fall in both the SQ and tumor mean pO$_2$ (55.6 to 40.3 mm Hg, and 19.9 to 10.0 mm Hg, respectively). In the treated group, the hematocrits remained stable (38.0 to 43.1%) as did the mean pO$_2$ of the SQ (46.1 to 54.3 mm Hg) and the tumor (11.1 to 11.3 mm Hg).

These data lend support to the value of EPO in reversing the anemia associated with malignancy and suggest a role of pO$_2$ histography in monitoring the beneficial effects of EPO therapy.

INTRODUCTION

Several authors have presented indirect evidence of the importance of tumor hypoxia and its contribution to radioresistance (Gray et al., 1953; Thomlinson and Gray, 1955; Cater and Silver, 1960). Some reports have correlated hypoxia with attenuated responses to radiation treatments (Gatenby et al., 1988), while others have demonstrated the ease and suitability of using polarographic needle electrodes to measure tumor hypoxia (Kallinowski et al., 1990; Terris et al., 1992; Terris and Dunphy, 1993). These suggest that the ability to accurately detect and then overcome tumor hypoxia could potentially improve outcomes in radiation oncology. Other authors (Bush et al., 1978; Hierlihy et al., 1969; Hill et al., 1971) have correlated poor survival and local control rates following radiotherapy with anemia, and have suggested as its mechanism decreased oxygen delivery to tumor cells.

The beneficial effect of transfusion conferred on anemic patients undergoing radiation treatments for malignancy has been shown previously (Bush et al., 1978), and has stimulated interest in alternative methods for raising the oxygen carrying capacity, such as erythropoietin (EPO). EPO is a relatively newly characterized glycoprotein hormone which is a hematopoietic growth factor, and has demonstrated the capacity to increase hematocrits in rodents and humans (Egrie et al., 1986; Eschbach et al., 1987; Fischl et al., 1990).

We employed a rodent model to evaluate both the ability of erythropoietin to maintain and increase hematocrit levels in malignancy-bearing mice, and the suitability of using polarographic oxygen electrodes to characterize changes in the tissue oxygen tension.

METHODS AND MATERIALS

Tumor Model

Eighteen C_3H mice were implanted intradermally with $2X10^5$ SCCVII carcinoma cells. After two weeks, when tumors had reached a diameter of approximately 10 mm, 10 of these animals began receiving daily subcutaneous injections of recombinant human erythropoietin (r-HuEPO) (1500 IU/kg), while 8 animals served as controls, and received no injections.

Hematocrits and oxygen tension were assessed at weekly intervals. The sites of oxygen tension measurement included both the tumors and areas of normal subcutaneous tissue (SQ) in each animal using anesthesia (50 mg/kg of intraperitoneal pentobarbital). Approximately 40-60 measurements were taken at each site.

Erythropoietin

The r-HuEPO was provided by Ortho Biotech (Raritan, NJ) in 1 ml vials of 4000 IU/ml. It is formulated in a buffered saline solution, and has a specific activity of 174,000 units/$A_{280}\pm5\%$.

Oxygen Tension Measurements

All measurements were performed using the Sigma-Eppendorf pO_2 Histograph Kimoc 6650 (Hamburg, FRG). The tissue oxygen tension is measured polarographically using a fine needle O_2 electrode. The 12 micron glass-insulated gold microcathode is covered with a Teflon membrane and is recessed in a jacket tube of spring steel with a diameter of 0.3 millimeters. The gold cathode is biased with -700 mV toward a Ag/AgCl anode, with a resulting current of 0.01 to 3.0 nA. This current is proportional to the oxygen tension in the connecting elecrolyte (represented by test tissue). The results were analyzed statistically using a Student's t-test.

RESULTS

Hematocrits

The hematocrits of the r-HuEPO-treated animals increased from a mean of $38.0\pm0.9\%$ ($\pm$S.E.) on the 15th post-implant day to $43.1\pm2.4\%$ on the 27th day post-implant (not significantly different), while in the control group, the hematocrit fell from $42.4\pm1.0\%$ to $31.4\pm2.5\%$ ($p<0.01$) (Figure 1).

On the day treatment was instituted, the hematocrits of the control group were significantly higher than the treated group ($p<0.01$), while at 6 and 12 days after treatment, the hematocrits of the treated group were higher ($p<0.01$).

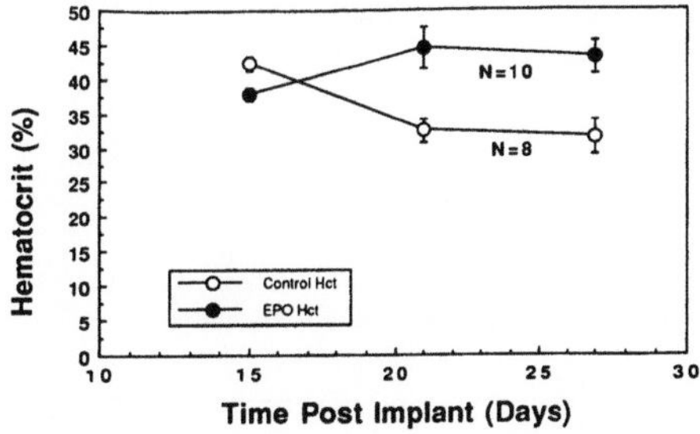

Figure 1:   Graph demonstrating the change in hematocrits (Hct) over time in the erythropoietin (r-HuEPO)-treated (closed circles) and untreated (open circles) animals.

Tissue pO_2

As shown in Figures 2, the mean subcutaneous (SQ) measurement in the r-HuEPO-treated group increased from 46.1 to 54.3 mm Hg (p<0.02). The mean tumor oxygen tension remained stable (11.1 to 11.3 mm Hg). Histograms showing the pooled data for mean tumor oxygen tension in this group over time are pictured in Figure 4. In the control group (Figure 3), the mean SQ oxygen tension fell from 56.6 to 40.3 mm Hg (not significantly different). The mean tumor measurement fell from 19.9 to 10.0 mm Hg (p<0.02).

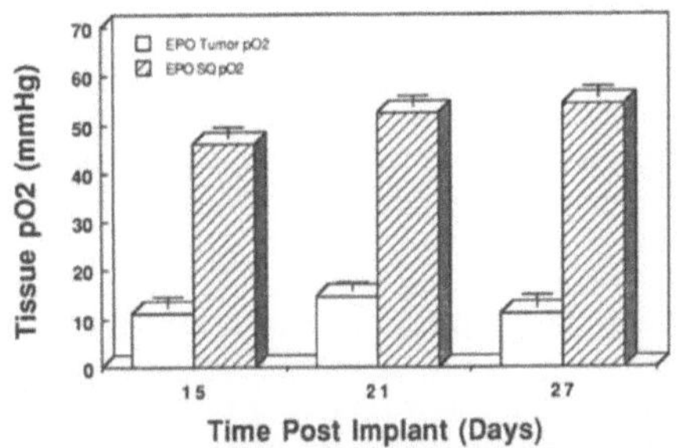

Figure 2: Graph representing the pooled mean tumor and subcutaneous (SQ) tissue oxygen tension, and the change over time, in the erythropoietin (r-HuEPO)-treated animals (N=10).

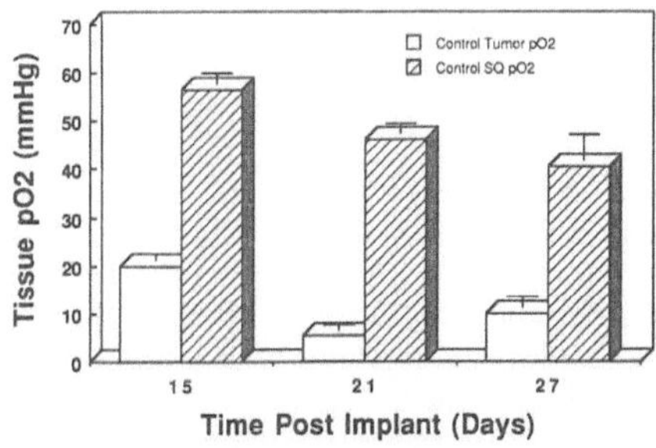

Figure 3: Graph representing the pooled mean tumor and subcutaneous (SQ) tissue oxygen tension (pO_2), and the change over time, in the untreated animals (N=8).

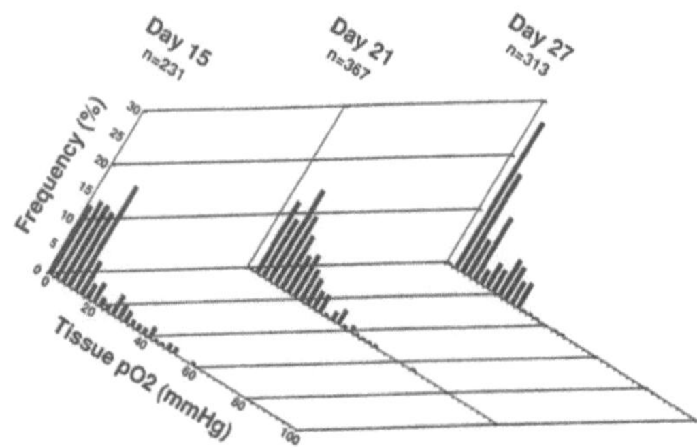

Figure 4: Pooled data for frequency of oxygen tension (pO$_2$) measurements in the tumor tissue of the erythropoietin (r-HuEPO)-treated animals over time (N=10).

DISCUSSION

There are considerable data suggesting that tumor hypoxia has a detrimental effect on the response to radiotherapy (Gray et al., 1953; Thomlinson and Gray, 1955; Gatenby et al., 1988). Much investigation in experimental radiotherapy has included efforts to either reverse this hypoxia (for instance with hyperbaric oxygen (Henk and Smith, 1977; Dische, 1979)), to overcome it chemically (as with radiosensitizers (Brown, 1982)) or to exploit it with bioreductive agents (Zeman et al., 1986).

The ability to quantify and monitor the relative hypoxia of tumors also takes on great importance, and earlier work by Cater and Silver (1960) utilizing oxygen-sensitive probes has been built upon by others using similar instrumentation (Gatenby et al., 1988), and more recently, computerized micro-probes (Kallinowski et al., 1990; Terris et al., 1992; Terris and Dunphy, 1993).

The contribution of hypoxia to radioresistance also led to evaluation of the importance of anemia and decreased oxygen-carrying capacity of blood. Several authors (Hierlihy et al., 1969; Hill et al., 1971; Bush et al., 1978) demonstrated a correlation between anemia and radiation failure both experimentally and clinically. In a prospective investigation, a beneficial effect of transfusion to maintain hemoglobin levels above 12.5 grams% was suggested (Bush et al., 1978).

Subsequent literature has warned extensively of the risks of transfusion, particularly in terms of infectious risks (Bove, 1987). Recent attention has been focused on the nonspecific immunosuppression that heterologous transfusions appear to induce, which carries particular relevance in oncologic patients (Wu and Little, 1988; Jackson and Rice, 1989). Because of these and other disadvantages of transfusion (including iron overload), interest in pharmacologic stimulation of the hematopoietic system has intensified.

In 1977, the purification of human erythropoietin (EPO) was first accomplished (Miyake et al., 1977), and nearly a decade later, the erythropoietin gene was isolated and cloned (Lin et al., 1985). With recombinant DNA techniques, sufficient quantities of EPO became available to perform clinical trials proving its utility in anemic patients (Eschbach et al., 1987; Fischl et al., 1990), and its use is now widespread. A natural next step was to consider EPO administration in anemic patients undergoing radiotherapy.

Experimentally, human recombinant EPO (r-HuEPO) was shown in the current study (and previously by Egrie et al. (1986)) to cause a swift increase in the hematocrit of both anemic and nonanemic rodents (attesting to the highly conserved structure of EPO among mammals, and therefore its crossreactivity (Browne et al., 1986)). Our data demonstrate a statistically significant and sustained rise in the hematocrits of animals treated with EPO, compared with the untreated control animals.

Furthermore, the oxygen tension measurements obtained both in the subcutaneous tissue and in the tumor tissue showed an overall rise or maintenance in pre-treatment oxygen levels, compared with the steady decline in these parameters in untreated control animals. These findings suggest that erythropoietin treatment may enhance the radiosensitivity of the tumors borne in those animals, although proof of this assumption awaits further investigation.

CONCLUSIONS

These studies confirm the swift, predictable rise in hematocrit that is induced by administration of erythropoietin in a murine model. The relatively non-invasive, straight-forward technique of computerized histographic pO_2 measurement was able to quantify and monitor the increase or maintenance in oxygen tension that resulted from the erythropoietin treatment, compared with a progressive decrease in oxygenation in control animals.

REFERENCES

Bove, J.R., 1987, Transfusion-associated hepatitis and AIDS, *N Engl J Med.* 317:242-245.

Brown, J.M., 1982, Clinical perspectives for the use of new hypoxic cell sensitizers, *Int J Rad Oncol Biol Phys.* 8:1491-1497.

Browne, J.K., Cohen, A.M., Egrie, J.C., et al., 1986, Erythropoietin: gene cloning, protein structure, and biological properties. *Quant Biol.* 51:693-702.

Bush, R.S., Jenkin, R.D., Allt, W.E., et al., 1978, Definitive evidence for hypoxic cells influencing cure in cancer therapy, *Br J Cancer.* 37:302-306.

Cater, D.B., and Silver, I.A., 1960, Quantitative measurements of oxygen tension in normal tissues and in the tumours of patients before and after radiotherapy, *Acta Radiol.* 53:233-256.

Dische, S., 1979, Hyperbaric oxygen: the Medical Research Council trials and their clinical significance. *Brit J Radiol.* 51:888-894.

Egrie, J.C., Strickland, T.W., Lane, J., et al., 1986, Characterization and biological effects of recombinant human erythropoietin, *Immunobiol.* 172:213-224.

Eschbach, J.W., Egrie, J.C., Downing, M.R., Browne, J.K., and Adamson, J.W., 1987, Correction of the anemia of end-stage renal disease with recombinant erythropoietin, *N Engl J Med.* 316:73-78.

Fischl, M., Galpin, J.E., Levine, J.D., et al., 1990, Recombinant human erythropoietin for patients with AIDS treated with zidovudine, *N Engl J Med.* 322:1488-1493.

Gatenby, R.A., Kessler, H.B., Rosenblum, J.S., et al., 1988, Oxygen distribution in squamous cell carcinoma metastases and its relationship to outcome of radiation therapy. *Int J Rad Oncol Biol Phys.* 14:831-838.

Gray, L.H., Conger, A.D., Ebert, M., Hornsey, S., and Scott, O.C., 1953, Concentration of oxygen dissolved in tissues at the time of irradiation as a factor in radiotherapy, *Brit J Radiol,* 26:638-648.

Henk, J., and Smith, C., 1977, Radiotherapy and hyperbaric oxygen in head and neck cancer, *Lancet.* 2:104-105.

Hierlihy, P., Jenkin, R.D., and Stryker, J.A., 1969, Anemia as a prognostic factor in cancer of the cervix, *Canad Med Assoc J.* 100:1100-11022.

Hill, R.P., Bush, R.S., and Yeung, P., 1971, The effect of anaemia on the fraction of hypoxic cells in an experimental tumour, *Br J Radiol.* 44:299-304.

Jackson, R.M., and Rice, D.H., 1989, Blood transfusions and recurrence in head and neck cancer, *Ann Otol Rhinol Laryngol.* 98:171-173.

Kallinowski, F., Zander, R., Hoeckel, M., and Vaupel, P., 1990, Tumor tissue oxygenation as evaluated by computerized-pO_2-histography, *Int J Rad Oncol Biol Phys.* 19:953-961.

Lin, F.K., Suggs, S., Lin, et al., 1985, Cloning and expression of the human erythropoietin gene, *Proc Natl Acad Sci.* 82:7580-7584.

Miyake, T., Kung, C.K., and Goldwasser, E., 1977, Purification of human erythropoietin, *J Biol Chem.* 252:5558-5564.

Terris, D.J., Minchinton, A.I., Dunphy, E.P., and Brown, J.M., 1992, Computerized histographic oxygen tension measurements of murine tumors, *Adv Exp Med Biol.* 317:153-159.

Terris, D.J., and Dunphy, E.P., 1994, Oxygen tension measurements of head and neck cancers, *Arch Otolaryngol Head Neck Surg.* In press.

Thomlinson, R.H., and Gray, L.H., 1955, The histological structure of some human lung cancers and the possible implications for radiotherapy, *Brit J Cancer.* 9:539-549.

Wu, J.S., and Little, A.G., 1988, Perioperative blood transfusions and cancer recurrence, *J Clin Oncol.* 6:1348-1354.
Zeman, E.M., Brown, J.M., Lemmon, M.J., Hirst, V.K., and Lee, W.W., 1986, SR-4233: a new bioreductive agent with high selective toxicity for hypoxic mammalian cells, *Int J Rad Oncol Biol Phys.* 12:1239-1242.

GLUCOSE DIFFUSION COEFFICIENTS DETERMINED FROM CONCENTRATION PROFILES IN EMT6 TUMOR SPHEROIDS INCUBATED IN RADIOACTIVELY LABELED L-GLUCOSE

K. Groebe, S. Erz, W. Mueller-Klieser

Institut für Physiologie und Pathophysiologie
Johannes Gutenberg-Universität Mainz
Duesbergweg 6, D-55099 Mainz, Germany

INTRODUCTION

In order to theoretically assess tissue energetic status, conditions for substance exchange need to be known. One group of parameters important in this context are diffusion coefficients of nutrients and metabolic waste products which may be assessed by incubating spheroids in a medium containing tracer amounts of the radioactively labeled substance in question, for a defined period of time. In previous studies, the *overall* amount of ^{14}C-labeled substance taken up by the spheroids was measured by scintillation counters (*e.g.* [1]), or the concentration of ^{3}H-labeled substance *in the spheroid center* was determined by autoradiography and grain counting (*e.g.* [4]). From a number of such measurements, diffusion coefficients may then be obtained using mathematical models. The information content of an autoradiogram can be exploited more efficiently by recording not only *central* grain density but rather grain density *distributions* from each of which a diffusion coefficient may be calculated. This can be done by grain counting on a radial track through an autoradiogram of a central spheroid section for ^{3}H labels or by measuring optical density (densitometry) in autoradiograms of central spheroid sections for ^{14}C labels.

As a first application of this technique glucose diffusivity has been assessed in EMT6/Ro multicellular tumor spheroids by autoradiography of ^{14}C labeled L-glucose which is not taken up by the tumor cells. Glucose has been chosen since it is a particularly important nutrient and since literature values for the glucose diffusion coefficient vary by more than one order of magnitude [7,5,1,2,3,4,11].

MATERIALS AND METHODS

Growth of EMT6/Ro spheroids followed standard procedures [8,9]. After culture in spinner flasks, spheroids were transferred to the P_{O_2} measuring chamber, taken up by suction with a micropipette attached to a 2 *ml* syringe, and incubated in stirred, ^{14}C-L-glucose containing medium. L-glucose was allowed to penetrate into the speroids for incubation times of 4 to 70 *s*. Then spheroids were removed from the medium, rapidly frozen on a precooled brass block, and stored in liquid nitrogen. Before exposure, spheroids were embedded in *TissueTek* precooled to 0 ^{0}C, frozen in liquid nitrogen, and stored at -25 ^{0}C. Central spheroid cryosections of

20 μm thickness were taken up on slides precooled to 0 °C, frozen in liquid N_2, and exposed to radiosensitive *Kodak NMC* film for 15 days at -25 °C. Two-dimensional ^{14}C activity distributions were recorded by densitometry using a *Leitz* microscope system at magnification 1000 ×, monochrome light of wavelength 550 nm, and a spatial resolution of $10 \times 10 \ \mu m^2$. This procedure is superior to using ^3H-labeled glucose in that exposure times with ^3H labels are much longer eye, spatial resolution is lower, and the number of datapoints that can be recorded with reasonable effort is much smaller. Truncated Fast Fourier Transforms were applied to smooth the raw data.

^{14}C activity profiles need to be fitted to solutions of the partial differential equation of transient substance diffusion in spheroids in which a diffusion depleted zone surrounding the spheroid is considered. For a location (x, y) in the profile, a radial coordinate r is defined by $r = \sqrt{(x - x_C)^2 + (y - y_C)^2}$ where x_C and y_C are the center coordinates of the ^{14}C activity distribution, and the dimensionless concentration C at r and time t is given by

$$C(r, t) = v \left(1 - \sum_{\text{ev}\,\mu} \frac{A_\mu}{r} \sin \left(\frac{\mu}{D} r \right) e^{-\frac{\mu^2}{R^2} t}, \right)$$

in which the eigenvalues μ have to solve the equation

$$\tan \sqrt{\frac{\mu}{D}} = \frac{\sqrt{\dfrac{\mu}{D}}}{1 - \dfrac{D_G\, G}{v D (G - R)}}.$$

Coefficients A_μ are defined as

$$A_\mu = \frac{\sin \sqrt{\dfrac{\mu}{D}} \cdot \dfrac{D_G\, G}{v D (G - R)}}{\dfrac{\mu}{D} - \sin^2 \sqrt{\dfrac{\mu}{D}} \cdot \left(1 - \dfrac{D_G\, G}{v D (G - R)} \right)},$$

v is the fraction of extracellular space, and D, R, D_G, G are diffusion coefficients and radii

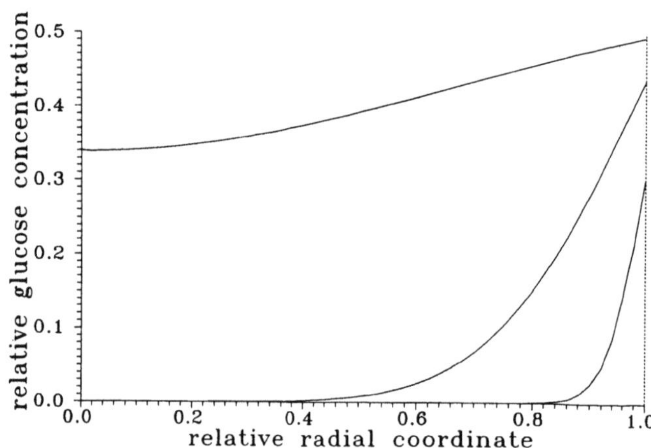

Figure 1. Solutions of the partial differential equation of transient glucose diffusion in spheroids for incubation times of 1, 10, and 100 s (lowest to higest curve). Abscissa: Radial coordinate as fraction of spheroid radius. Ordinate: Normalized concentration.

of spheroid and diffusion depleted zone, respectively. Thus it has been taken into accout that
the autoradiographic measurement integrates over both intra- and extracellular space whereas
L-glucose is present in the extracellular compartment exclusively. In Figure 1, graphs of these
solutions for incubation times of 1, 10, and 100 s are displayed.

From each one of the registered profiles a glucose diffusion coefficient was calculated by
fitting the above analytical solutions to the measured concentration distributions. Diffusion
depleted zone thickness and diffusivity were taken to be 60 μm [7] and $9 \cdot 10^{-6}$ cm^2/s [11],
respectively. Parameters included in this non-linear approximation procedure, comprised diffusion coefficient D, fraction of extracellular space v, and spheroid center coordinates (x_C, y_C).
This procedure differs from that used in former studies in which only one (average or central)
concentration per spheroid had been determined, and this measurement had to be repeated for
a number of different incubation times. In the latter case, a glucose diffusion coefficient could
be computed from the resulting time course only.

RESULTS AND DISCUSSION

Samples of smoothed central profiles in spheroids incubated for 5, 42, 71, and 680 s, respectively, are shown in Figure 2 in 3D-representation. A typical one-dimensional radial section
through such a profile is displayed in Figure 3 (squares). For comparison, also a similar profile
obtained with ^3H labeled glucose is shown (dots).

For each one of the studied spheroids the resulting diffusion coefficients are displayed in
Figure 4. Unexpectedly, diffusion coefficients systematically varied with time of incubation:

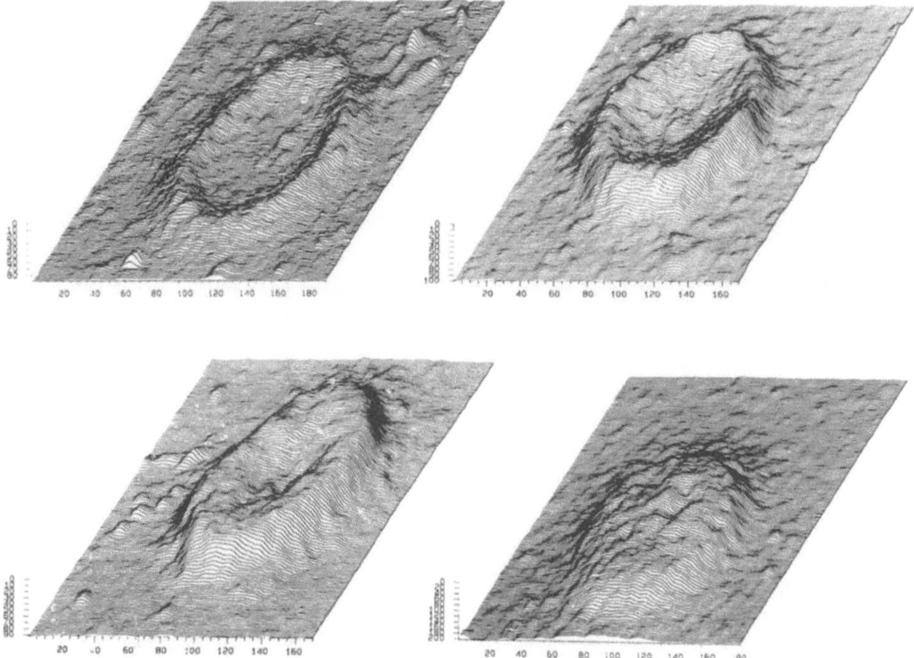

Figure 2. Samples of measured equatorial optical density distributions in spheroids incubated
for 5 s (top left), 42 s (top right), 71 s (bottom left), and 680 s (bottom right), respectively,
after smoothing.

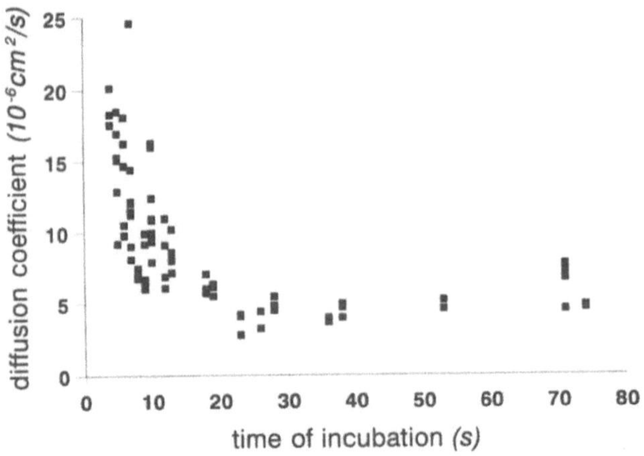

Figure 3. One-dimensional normalized optical density or grain density distributions on radial tracks through spheroids of 1000 μm diameter incubated for 7 s in ^{14}C labeled glucose (squares) or in ^{3}H labeled glucose (dots). Vertical dashed lines indicate edges of spheroid.

Starting out at values higher than physically possible for very short incubation times — glucose diffusion coefficient in water is $9{\cdot}10^{-6}$ cm^2/s —, diffusion coefficients decreased with incubation time to a constant level of about $5{\cdot}10^{-6}$ cm^2/s when incubated for 15 s or longer. Since there is no way of explaining this time course by reasons originating in the process of glucose penetration into the spheroid it must have been introduced artificially during processing or evaluation. Possible sources that may bring about a time dependence of this kind include errors in determining duration of incubation, glucose diffusion during processing caused by

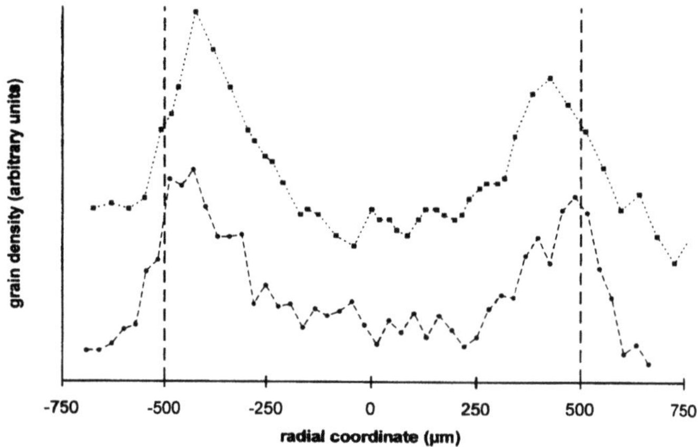

Figure 4. Diffusion coefficients calculated from measured optical density profiles as a function of incubation time.

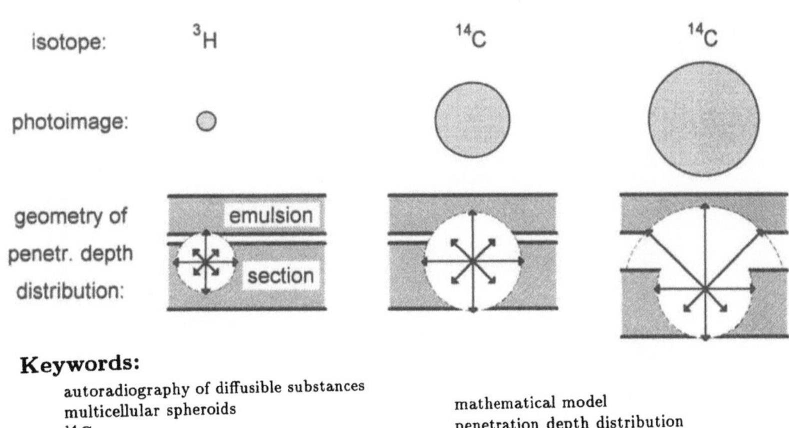

Keywords:
autoradiography of diffusible substances
multicellular spheroids
^{14}C
^{3}H
mathematical model
penetration depth distribution
glucose diffusion coefficient

Figure 5. Schematic presentation of geometries of penetration depth distributions (dashed circles with arrows) with respect to tissue section and photoemulsion, and resulting photoimages for ^3H labels (left), ^{14}C labels (center) and ^{14}C labels with air filled gap between tissue section and photoemulsion.

partial melting of the spheroid, loss of labeled glucose from the outer layers of the spheroid during processing, changes in glucose diffusivity or fraction of extracellular space with depth into the spheroid, etc. A number of such possibilities have been considered and ruled out by appropriate model studies.

There is one more interesting observation that was systematically present and is illustrated in Figure 3: While maximal grain densities in spheroids with ^3H labels are attained very close to the edge of the spheroids (dots), maxima with ^{14}C labels typically are located 70 to 120 μm into the spheroid (squares). This is true even though procedures for incubation and processing were exactly the same in both cases, and suggests that the problems encountered are related to the characteristics of the radioactive label used. The most likely explanation for these phenomena was concluded to lie in the fairly large penetration depth of the ^{14}C β-radiation as illustrated in Figure 5: For very short penetration depths (*e.g.*, the one of ^3H radiation), β-particles emitted by a point source will barely reach the photoemulsion and produce an almost point-like photoimage there (Fig. 5, left). For long penetration depths, larger areas of the photoemulsion get into the reach of the emitted β-radiation, and a larger disk-like image results (Fig. 5, center). This is even aggravated if there is an air-filled gap between tissue section and photoemulsion (Fig. 5, right). Mathematically speaking, a grain density distribution results from a convolution of the radioactivity distribution with the penetration depth distribution. For a narrow penetration depth distribution, distributions of radioactivity and grain density are practically identical. In contrast, a wide penetration depth distribution tends to render recorded grain density profiles more shallow and to smoothen density gradients with respect to the underlying ^{14}C concentration gradients. Based on experimental distributions of penetration depths, it is possible to reverse convolution by deconvoluting measured profiles during the fitting process for determination of glucose diffusion coefficients.

In order to illustrate the effects of large penetration depths, calculated radioactivity profiles (dotted curves) and their convolutions (solid) for incubation times of 1, 10, and 100 s are shown in Figure 6. Since particularly steep gradients are present at short incubation times, these gradients are most severely affected by convolution with wide penetration depth distributions, which would make short time diffusivities to appear far higher than they actually are. On the other hand, for long incubation times and radial coordinates less than 85 % of spheroid radius, distributions of radioactivity and grain density virtually coincide.

Data and algorithms necessary for the procedure of recovering activity distributions from measured optical density distributions are presently being developed. Nevertheless, even

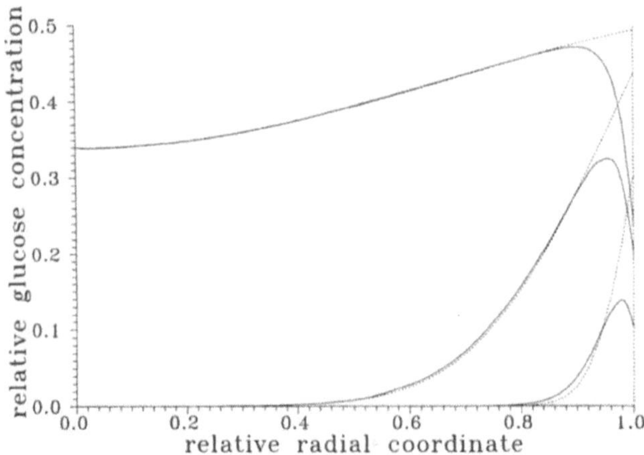

Figure 6. Solutions of the partial differential equation of transient glucose diffusion in spheroids (dotted) and same curves convoluted with estimated penetration depth distribution of ^{14}C radiation under actual experimental conditions (solid) for incubation times of 1, 10, and 100 s. Abscissa: Radial coordinate as fraction of spheroid radius. Ordinate: Normalized concentration.

though consistent evaluations of profiles for short incubation times are at this time not available yet, the various model simulations performed so far indicate that major changes in the resulting diffusion coefficients are to be expected for incubation times below 15 s only, and that a value of $5\cdot10^{-6}$ cm^2/s found for longer incubation times represents a reliable estimate of the glucose diffusion coefficient in EMT6 multicellular tumor spheroids.

SUMMARY AND CONCLUSION

A method for performing and evaluating autoradiography of diffusible ^{14}C labeled substances in multicellular tumor spheroids is presented that allows one to obtain a diffusion coefficient of the substance investigated from each individual spheroid. Application of the method with ^{14}C labeled L-glucose resulted in a glucose diffusion coefficient of $5\cdot10^{-6}$ cm^2/s. It also revealed problems of the method at very short incubation times of about 10 s or less. These problems are most likely caused by the large penetration depth of β particles irradiated by ^{14}C labels (as compared to ^{3}H labels) which tends to transform steep ^{14}C concentration gradients into much more shallow optical density gradients during exposure. This tranformation can be corrected for by deconvolution of the recorded optical density distributions. Basic data and mathematical tools necessary for the process of deconvolution are presently being developed.

It is planned to use this method for determining diffusion coefficients of other substances of interest. One such group of substances are the metabolic waste products, most importantly lactate. Another group consists of larger molecules, *e.g.* peptides and comprises the various growth factors important in tumor biology. Since for members of this latter group little is known about their velocity of penetration into tissue, model calculations may be applied to predict a range of incubation times suitable for determining diffusion coefficients. Moreover,

624

the algorithms for data analysis will have to be modified to allow for receptor binding of the substance under study. Finally, penetration of anticancer drugs and hypoxic markers into tumor spheroids is to be studied. In particular, the method presented allows for directly assessing the effects of chemical modifications of these substances which are intended to improve penetration.

ACKNOWLEDGEMENTS

This research was supported by the Deutsche Forschungsgemeinschaft (Mu 576/2–4).

References

[1] CASCIARI, J. J., S. V. SOTIRCHOS, and R. M. SUTHERLAND, Glucose diffusivity in multicellular tumor spheroids. *Cancer Res.* 48:3905–3909 (1988)

[2] CASCIARI, J. J., S. V. SOTIRCHOS, and R. M. SUTHERLAND, Variations in tumor cell growth rates and metabolism with oxygen concentration, glucose concentration, and extracellular pH. *J. Cell. Physiol.* 151:386–394 (1992)

[3] CASCIARI, J. J., S. V. SOTIRCHOS, and R. M. SUTHERLAND, Mathematical modelling of microenvironment and growth in EMT6/Ro multicellular tumour spheroids. *Cell Prolif.* 25:1–22 (1992)

[4] DOERSCHEL, D., U. KARBACH, GROEBE K., and W. MUELLER-KLIESER, Assessment of glucose penetration into tumor microregions by high resolution autoradiography 37th Meeting of the European Tissue Culture Society, Graz, Austria, 1989. *Cytotechnology* 2 (Suppl.):36 (1989)

[5] FREYER, J. P. and R. M. SUTHERLAND, Determination of diffusion constants for metabolites in multicell tumor spheroids, *Adv.Exp.Med.Biol.* 159:463–475 (1983)

[6] GROEBE, K. and W. MUELLER-KLIESER, Distributions of oxygen, nutrient, and metabolic waste concentrations in multicellular spheroids and their dependence on spheroid parameters, *Eur.Biophys.J.* 19:169–181 (1991)

[7] LI, C. K. N., The glucose distribution in 9L rat brain multicell tumor spheroids and its effect on cell necrosis, *Cancer (Phila.)* 50:2066–2073 (1982)

[8] MUELLER-KLIESER, W., Multicellular spheroids. A review on cellular aggregates in cancer research, *J. Cancer Res. Clin. Oncol.* 113:101–122 (1987)

[9] SUTHERLAND, R. M., Cell and environment interactions in tumor microregions: the multicell spheroid model, *Science* 240:177–184 (1988)

[10] SWABB, E. A., J. WEI, and P. M. GULLINO, Diffusion and convection in normal and neoplastic tissues, *Cancer Res.* 34:2814–2822 (1974)

[11] VAUPEL, P., Oxygenation of human tumors, *Strahlenther. Onkol.* 166:377–386 (1990)

INVESTIGATIONS OF PERFUSION-LIMITED HYPOXIA AND OXYGENATION IN THE KHT SARCOMA

Bruce M. Fenton and Dietmar W. Siemann

Department of Radiation Oncology
University of Rochester Medical Center
Rochester, New York 14642 U.S.A.

INTRODUCTION

A number of previous studies have attempted to manipulate tumor oxygenation and blood flow for the purpose of enhancing either chemo- or radiotherapy. Increased blood flow and oxygen delivery have commonly been used to improve both radioresponse and the delivery of conventional chemotherapeutic agents (Fenton and Sutherland, 1992; Rice et al.,1986). However, decreased oxygen delivery can also be beneficial. If tumor oxygen levels are reduced, the effectiveness of either bioreductive agents or hypoxic radiosensitizers can be enhanced (Bibby et al.,1989; Quinn et al.,1992). To better understand the underlying physiological mechanisms and thereby optimize therapeutic manipulation, a primary goal is to define the basic interrelationships among tumor vascular structure, blood flow, oxygen delivery, and radioresponse.

Over the past few years, we have applied a variety of approaches to the study of a murine tumor model, the KHT sarcoma. Initially, intercapillary distances and blood vessel diameters were measured using i.v. injected colloidal carbon (Fenton and Way, 1993). Hematoxylin and eosin stained histological sections from these tumors, demonstrated that blood vessel diameters tended to increase somewhat with increasing tumor volume, although not significantly. Surprisingly, the mean distance between tumor cells and the nearest blood vessel also remained fairly constant at $\approx 60\text{-}70$ μm for both small and large volume tumors.

In the present study, whole tumor response to radiation therapy was examined using colony survival assays to determine variations in radiobiological hypoxic fraction with tumor growth. This provided a therapeutic basis from which to assess the associated micro-regional variations in oxygenation and intermittent blood flow. To quantitate changes in micro-regional oxygen availability, intravascular HbO_2 saturations were determined in quick-frozen tumors using a cryospectrophotometric technique. Several studies have reported a distinction between chronic (diffusion-limited) and acute (perfusion-limited) hypoxia (Brown, 1979; Chaplin et al.,1986), finding that chronically hypoxic tumor cells may exhibit a significantly different radiation response than cells made acutely hypoxic as a result of temporary cessation of blood flow within the tumor vasculature (Chaplin et al.,1987; Sutherland and Franko, 1980). More recent studies have provided preliminary quantitative information as to the prevalence of transiently perfused vessels in select tumor models and have demonstrated a clear dependence of transient perfusion on tumor size (Trotter et al.,1989). To measure the extent to which individual tumor blood vessels are opening and closing intermittently in the KHT sarcoma, dual fluorescent staining techniques were also included in the present study.

METHODS

Animals and tumors

KHT sarcomas were implanted i.m. by inoculating 2×10^5 cells into the hind legs of 6-8 week-old C3H/HeJ mice and were grown to volumes ranging from 100-1300 mm^3. Tumor volume was calculated as $\pi \cdot d^3/6$, where **d** is the diameter of the tumor.

Colony survival assay and hypoxic fraction determination

Radiobiological hypoxic fraction was determined as described previously (Rofstad et al.,1988). Tumors were irradiated *in vivo* at a dose rate of 5.2 Gy/min using a ^{137}Cs source. Hypoxic conditions were obtained by asphyxiating the mice 15 min before irradiation. Single cell suspensions were prepared from the tumors and cell survival was measured using an *in vivo* to *in vitro* excision assay (Siemann et al.,1985). In brief, KHT sarcomas were excised and single cell suspensions were prepared using a combined mechanical and enzymatic dissociation procedure (Thomson and Rauth, 1974). The cells were counted and various dilutions were plated into 24-well plates with 10^4 lethally irradiated tumor cells in 0.2% agar containing α-minimum essential medium (α-MEM) supplanted with 10% fetal calf serum. Two weeks later, the plates were harvested and colonies larger than 50 cells were scored as surviving with the aid of a dissecting microscope. Survival curves were fitted to the data by linear regression for tumors irradiated in air-breathing and asphyxiated mice. The regression analyses were based on data for doses of 15 Gy and higher (asphyxiated) or 10 Gy and higher (air-breathing), such that only doses that eliminated the oxic cells were considered in the analysis. The fraction of hypoxic cells was determined from the vertical displacement of the two survival curves.

Cryospectrophotometric HbO$_2$ measurements

Cryospectrophotometric procedures have been previously described in detail (Fenton and Gayeski, 1990). Briefly, fur was shaved from the tumors, a depilatory agent applied, and animals sacrificed by cervical dislocation. Tumors were immediately quick-frozen by applying a liquid nitrogen-cooled copper block and stored in liquid nitrogen. Using a cooled scalpel at -73°C, the tumor was cut to expose four cross-sections. From these, a total of 80-120 vessels (≥ 8 μm in diameter) were analyzed cryospectrophotometrically to determine intravascular HbO$_2$ saturations. This technique utilizes reflected microscopy to detect spectral differences between oxy- and deoxy-hemoglobin.

Dual-staining

These methods (Trotter et al.,1989) involved the i.v. injection of one fluorescent marker (Hoechst 33342), followed 20 min later by the i.v. injection of a second marker

(DiOC$_7$(3)). Animals were sacrificed by cervical dislocation 2-5 min following the DiOC$_7$(3) injection, and tumors were excised, frozen in liquid nitrogen-cooled liquid pentane, and stored at -73°C. Hoechst 33342 emits blue fluorescence when excited by ultraviolet light and is dissolved in phosphate-buffered saline and administered i.v. at a dose of either 7.5 or 15 mg/kg. DiOC$_7$(3) exhibits green fluorescence when excited by visible blue light and is dissolved in 75% dimethyl sulfoxide and administered i.v. at a dose of 0.5 mg/kg, either simultaneously or 20 min following Hoechst 33342 injection. Frozen sections, 8 μm in thickness, were cut and immediately examined using a fluorescent microscope (25× obj.). Blood vessels labeled with either or both of the stains were identified (using different filter combinations) and the percent of the vessels containing only one stain (percent flow mismatch) was determined.

RESULTS AND DISCUSSION

Radiobiological hypoxic fraction

From cell survival assays, it was demonstrated that radiobiological hypoxic fraction of KHT tumors increases two- to three-fold with increasing tumor volume. Table 1 summarizes the results as a function of tumor size, including both current results using intramuscular implants and previous findings (Rofstad et al.,1988) using subcutaneous implants. Hypoxic fraction increased with increasing tumor volume and was not substantially different between the two implantation sites.

TABLE 1

IMPLANTATION SITE	TUMOR SIZE	HYPOXIC FRACTION
intramuscular	0.1 - 0.2 g	$\approx$ 10%
intramuscular	0.3 - 0.5 g	11 - 16%
intramuscular	0.7 - 1.0 g	25 - 35%
subcutaneous	200 mm^3	12%
subcutaneous	2000 mm^3	23%

Intravascular HbO$_2$ saturations

Figure 1 summarizes the cryospectrophotometric measurements of HbO$_2$ saturations. Differences between groups were most easily visualized when depicted as percent of the vessels $\geq$ 10% HbO$_2$ saturation (Fenton et al.,1988), which is plotted as a function of distance from the tumor surface. Tumors are divided according to tumor volume as follows: small tumors = 340 $\pm$ 45 mm^3, medium tumors = 730 $\pm$ 32 mm^3, and large tumors = 1230 $\pm$ 34 mm^3. For each volume class, HbO$_2$ levels decreased with increasing distance from the tumor surface and with increasing tumor volume. For the largest tumors, oxygen availability decreased to a minimal level at distances >1 mm from the tumor surface and was not further reduced at greater distances.

Figure 2 presents the percentage of nonfunctional vessels (defined as HbO$_2$ = 0%) versus distance from the tumor surface. The percentage of nonfunctional vessels increased with increasing tumor volume and increasing distance from the tumor surface. For the large tumors, $\approx$35-45% of the vessels were nonfunctional at distances >1 mm from the tumor surface, while even in the smallest tumors, ≈5-25% of the vessel were nonfunctional.

Blood flow intermittency

Most recently, we have been investigating the prevalence of intermittent blood flow in the KHT tumor model using the dual fluorescent staining technique. Figure 3 presents the percentage of the tumor vessels that have either opened or closed over a 20 min time period between stain administrations (% flow mismatch). Since previous studies (Trotter et al.,1990) have reported arterial blood pressure and laser Doppler

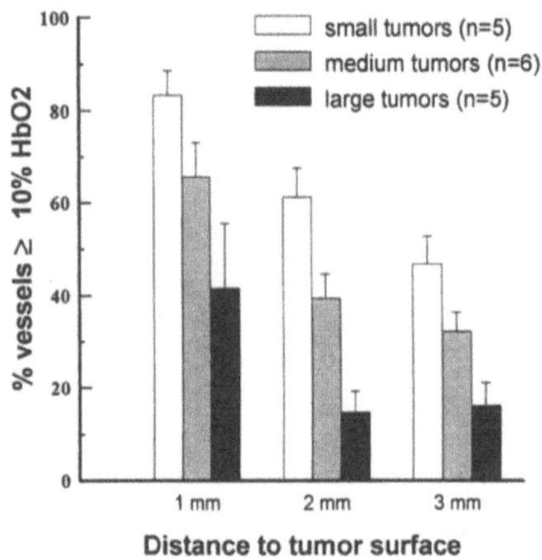

Fig. 1 Oxygen availability vs. distance to the tumor surface.

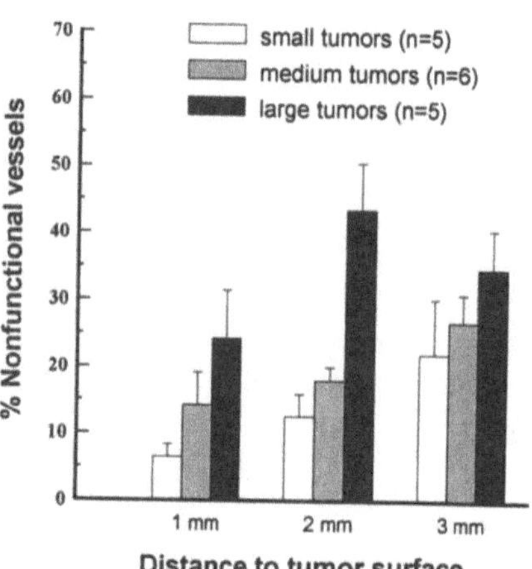

Fig. 2 % nonfunctional vessels vs. distance from the tumor surface.

blood flow variations following Hoechst 33342 administration, two concentrations of the Hoechst 33342 stain were initially included, 7.5 mg/kg and 15 mg/kg.

Small KHT tumors (≈300 mm³) demonstrated a flow mismatch of 1.4%, which was not significantly different from that obtained using simultaneous injections of both stains, as shown by the dashed line in figure 3 and summarized in Table 2. For medium tumors (≈650 mm³) and large tumors (≈1100 mm³), the % flow mismatch ranged from 3.5 to 5%. In alternate tumor models, this percentage has been reported to be as high as 9% and, in agreement with these results, was also found to increase with increasing tumor weight(Trotter et al.,1989).

Tumor volume	[Hoechst 33342]	% flow mismatch	% vessel closures
440 ± 20 mm³	7.5 mg/kg	1.2 ± .13	49 ± 3.0
980 ± 40 mm³	15 mg/kg	1.5 ± .23	59 ± 4.9

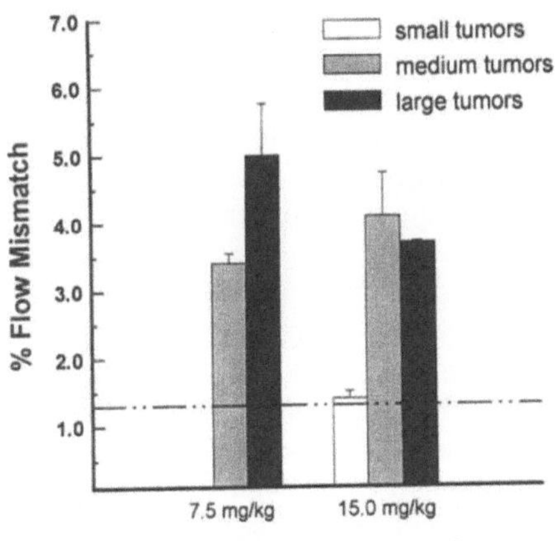

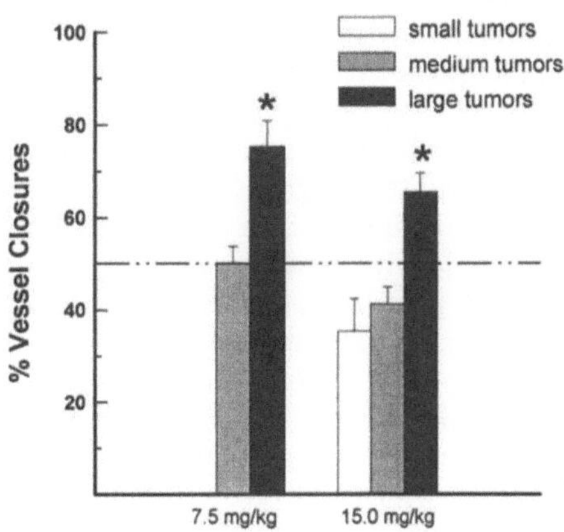

Fig. 3 % flow mismatch vs. Hoechst 33342 concentration

Fig. 4 Percentage vessel closures vs. Hoechst 33342 concentration. *'s denotes significant difference from 50% closures.

For the large volume tumors, the locations of the mismatched vessels were mapped out in relation to the tumor cross-sections (data not shown). Qualitatively, no preferential distribution of the intermittently opening vessels was observed, i.e., the mismatched vessels were equally likely to be found in the tumor center as near the tumor surface. This is in contrast to the findings of Trotter (Trotter, 1990), who reported significantly higher mismatch in the tumor center compared to the periphery.

Assuming that the dual staining measurements are independent of hemodynamic or vasoactive alterations associated with the Hoechst 33342 administration, the percentage of vessel openings over the 20 min period should be approximately equal to the percentage of vessel closures. Figure 4 presents the % vessel closures versus Hoechst 33342 concentration for the three tumor volume classes, with the expected level of 50% closures denoted by the dashed line. Based on the binomial test, neither small nor medium volume tumors were significantly different from the level of 50% vessel closures, indicating that equal numbers of vessels are opening and closing in these tumors. For larger tumors, however, the percentage of vessel closures was significantly higher than the percentage of vessel openings at both the 7.5 and 15 mg/kg concentrations (p=.01 and p=.04, respectively), indicating a likely vasoactive effect due to the Hoechst 33342 administration.

CONCLUSIONS

Blood perfusion in rapidly expanding tumors has generally been reported to decrease, primarily due to blood vessel compression as the tumor cells proliferate in a confined space. This reduction in flow preferentially decreases oxygen delivery to the interior regions, as was observed both with the small volume tumors and, more strikingly, with the medium and large volume tumors. The presence or absence of intermittent flow in a given vessel is most likely dependent on a critical balance between numerous hemodynamic and rheologic factors, including vascular geometry, red cell distribution, and blood viscosity. As the tumor increases in volume, blood vessels dilate, tumor interstitial pressure increases, and tumor blood vessels are compressed, all of which lead to sluggish tumor blood flow. These changes result in a decrease in oxygen availability, which produces the observed corresponding increase in radiobiological hypoxic fraction.

As the growing tumor compresses the interior vessels, these vessels may at first fluctuate on and off until the blood flow is permanently shut down, thus offering a possible explanation for the observed increase in intermittent flow with increasing tumor volume. This increase in intermittent flow is a reasonable expectation, based on the expected corresponding decrease in tumor blood flow and increased red blood cell aggregation with tumor growth. By administering an agent such as nicotinamide that can shift the balance in favor of increased tumor blood flow, intermittent flow can be artificially decreased, as has been demonstrated by Horsman et al. (Horsman et al.,1989).

Hoechst 33342 has been shown to transiently reduce tumor blood flow and to reduce systemic blood pressure for up to 23 min following stain injection (Trotter et al.,1990). In large volume tumors a decrease in systemic blood pressure would be more likely to lead to increased vessel closures, due to an expected closer balance between intravascular and interstitial pressures (Lee et al.,1992). This would not only explain our increased percentage of flow mismatch with increasing tumor volume, but could also answer the question of why Hoechst 33342 produces vasoactive artifacts only in the larger tumors. In view of the reported effects of Hoechst 33342 on systemic blood pressure and blood flow, determinations of flow mismatch must be suspect whenever the % vessel closings differs significantly from 50%.

Estimating the degree of intermittent flow based solely on dual-staining techniques may also tend to underestimate the intermittency of functional flow, as defined by oxygen availability. In the case of the large KHT tumors, up to 42% of the blood vessels contained HbO_2 levels of 0%. But as has been previously reported (Fenton and Boyce, 1993), these 0% HbO_2 vessels are, in fact, predominantly fluorescently labeled. This indicates that stained vessels are not necessarily functional in terms of their capacity for oxygen transport to the surrounding tumor tissue. Instead, the presence of staining can also differentiate vessels having an extremely slow rate of blood flow and no remaining oxygen content.

The 3-5% mismatch observed in the medium and large volume tumors is probably indicative of vessels that are completely closed to flow during the stain administration. This means that the percentage flow mismatch is a conservative estimate of functional mismatch, and, in addition, that flow in individual vessels could vary substantially without a corresponding effect on % mismatch. A similar conclusion was reached by Trotter (Trotter, 1990) in estimates of dual staining using auto-image analysis. Here,

mismatch was scored on the basis of relative stain intensities, instead of based simply
on the presence or absence of each stain. While 5.8% of the vessels opened or closed
completely over the 20 min time period, a further 15.8% of the vessels changed
markedly from the simultaneous intensity ratios, presumably due to a period of transient
flow reduction following one of the stains.

As has previously been pointed out (Trotter et al.,1990), any conclusions as to the
validity of the dual staining techniques are only applicable for the tumor model under
investigation (in this case, the KHT sarcoma). Thus, these dual-staining methods must
be validated independently for each new tumor model. In future studies, it will be
important to not only further characterize both the progression of these blood flow
intermittencies, but also the therapeutic relevance.

ACKNOWLEDGMENTS

The excellent technical assistance of Deborah Boyce and Michelle Meyer is gratefully
acknowledged. We would also like to thank Dr. T.E.J. Gayeski for the use of his
cryospectrophotometer and laboratory. This work was supported by NIH grants
CA52586 and CA55300.

REFERENCES

Bibby, M.C., J.A. Double, P.M. Loadman, and C.V. Duke, Reduction of tumor blood flow by flavone
    acetic acid: A possible component of therapy, *J. Natl. Cancer Inst.* 81:216(1989).
Brown, J.M., Evidence for acutely hypoxic cells in mouse tumours, and a possible mechanism of
    reoxygenation, *Br. J. Radiol.* 52:650(1979).
Chaplin, D.J., R.E. Durand, and P.L. Olive, Acute hypoxia in tumors: implications for modifiers of
    radiation effects, *Int. J. Radiat. Oncol. Biol. Phys.* 12:1279(1986).
Chaplin, D.J., P.L. Olive, and R.E. Durand, Intermittent blood flow in a murine tumor. Radiobiological
    effects, *Cancer Res.* 47:597(1987).
Fenton, B.M. and D.J. Boyce, Micro-regional mapping of HbO_2 saturations and blood flow following
    nicotinamide administration, *Int. J. Radiat. Oncol. Biol. Phys.* (1993).(in press)
Fenton, B.M. and T.E.J. Gayeski, Determination of microvascular oxyhemoglobin saturations using
    cryospectrophotometry, *Am. J. Physiol.* 259:H1912(1990).
Fenton, B.M., E.K. Rofstad, F.L. Degner, and R.M. Sutherland, Cryospectrophotometric determination
    of tumor intravascular oxyhemoglobin saturations: Dependence on vascular geometry and tumor
    growth, *J. Natl. Cancer Inst.* 80:1612(1988).
Fenton, B.M. and R.M. Sutherland, Effect of flunarizine on micro-regional distributions of intravascular
    HbO_2 saturations in RIF-1 and KHT sarcomas, *Int. J. Radiat. Oncol. Biol. Phys.* 22:447(1992).
Fenton, B.M. and B.W. Way, Vascular morphometry of KHT and RIF-1 murine sarcomas, *Radiother.
    Oncol.* (1993).(in press)
Horsman, M.R., J. Overgaard, K.L. Christensen, M.J. Trotter, and D.J. Chaplin, Mechanism for the
    reduction of tumour hypoxia by nicotinamide and the clinical relevance for radiotherapy, *Biomedica
    Biochimica Acta* 48:S251(1989).
Lee, I., Y. Boucher, and R.K. Jain, Nicotinamide can lower tumor interstitial fluid pressure: mechanistic
    and therapeutic implications, *Cancer Res.* 52:3237(1992).
Quinn, P.K., M.C. Bibby, J.A. Cox, and S.M. Crawford, The influence of hydralazine on the
    vasculature, blood perfusion and chemosensitivity of MAC tumours, *Br. J. Cancer* 66:323(1992).
Rice, G.C., C. Hoy, and R.T. Schimke, Transient hypoxia enhances the frequency of dihydrofolate
    reductase gene amplification in chinese hamster ovary cells, *Proc. Natl. Acad. Sci. USA*
    83:5978(1986).
Rofstad, E.K., B.M. Fenton, and R.M. Sutherland, Intracapillary HbO2 saturations in murine tumours
    and human tumour xenografts measured by cryospectrophotometry: Relationship to tumour volume,
    tumour pH and fraction of radiobiologically hypoxic cells, *Br. J. Cancer* 57:494(1988).
Siemann, D.W., K. Maddison, K. Wolf, S.A. Hill, and P.C. Keng, In vivo interaction between radiation
    and 1-(2-chloroethyl)-3-cyclohexyl-1-nitrosourea in the presence or absence of misonidazole in mice,
    Cancer Res. 45:198(1985).

Sutherland, R.M. and A.J. Franko, On the nature of the radiobiologically hypoxic fraction in tumors, *Int. J. Radiat. Oncol. Biol. Phys.* 6:117(1980).

Thomson, J.E. and A.M. Rauth, An in vitro assay to measure the viability of KHT tumor cells not previously exposed to culture conditions, *Radiat. Res.* 58:262(1974).

Trotter, M.J., Intermittent blood flow in the murine SCCVII squamous cell carcinoma (PhD Thesis), University of British Columbia, Vancouver, (1990).

Trotter, M.J., D.J. Chaplin, R.E. Durand, and P.L. Olive, The use of fluorescent probes to identify regions of transient perfusion in murine tumors, *Int. J. Radiat. Oncol. Biol. Phys.* 16:931(1989).

Trotter, M.J., P.L. Olive, and D.J. Chaplin, Effect of vascular marker Hoechst 33342 on tumour perfusion and cardiovascular function in the mouse, *Br. J. Cancer* 62:903(1990).

THE COMBINATION OF NICOTINAMIDE AND CARBOGEN BREATHING TO IMPROVE TUMOUR OXYGENATION PRIOR TO RADIATION TREATMENT

Michael R. Horsman[1], Dietmar W. Siemann[2], Marianne Nordsmark[1], Azza A. Khalil[1], Jens Overgaard[1], and David J. Chaplin[3]

[1] Danish Cancer Society, Department of Experimental Clinical Oncology Nörrebrogade 44, DK-8000 Aarhus C, Denmark
[2] Tumour Biology Division, Rochester Cancer Center, Rochester New York 14642, USA
[3] CRC Gray Laboratory, Mount Vernon Hospital, Northwood Middlesex HA6 2JR, England

INTRODUCTION

Experimental studies have demonstrated the presence of radioresistent hypoxic cells in most animal solid tumours (Guichard et al., 1980; Moulder and Rockwell, 1984). There is also strong evidence to suggest that hypoxic cells can be found in human tumours and that they are probably one of the major reasons for failure to control certain tumour types with conventional radiotherapy (Dische 1989; Overgaard 1989). For many years this hypoxia was generally considered to be chronic in nature, arising as a result of a diffusion limitation of oxygen (Thomlinson and Gray, 1955; Tannock, 1968). More recently, it was postulated that acutely hypoxic cells also existed in tumours (Brown, 1979; Sutherland and Franko, 1980), and this has since been demonstrated experimentally and shown to result from transient fluctuations in tumour blood flow (Chaplin et al., 1986; 1987).

Attempts to eliminate hypoxia in tumours must involve treatments that work against both chronic and acute hypoxia. Diffusion limited chronic hypoxia can be overcome by either increasing oxygen availability by breathing high oxygen content gas, the use of artificial blood substitutes, or agents that increase tumour blood flow; chemically sensitizing these cells to radiation; or preferentially killing them with hypoxic cell cytotoxins (for review, see Horsman, 1993). One of the few agents that can reduce acute hypoxia in tumours is nicotinamide (Chaplin et al., 1990, Horsman et al., 1990). Several studies have now shown that the enhancement of radiation damage achieved using certain agents that attack chronic hypoxia, in particular carbogen or oxygen breathing, perfluorochemical emulsions and hyperthermia, can be increased by including nicotinamide in the treatment schedule (Horsman et al., 1990; Chaplin et al., 1991; 1993; Kjellen et al., 1991).

Since the combination of nicotinamide and carbogen breathing is currently undergoing preliminary clinical testing (Denekamp 1991), we have now investigated the effect of this combination, not only on the radiation response of a C3H mammary carcinoma, but also on the tumour oxygenation status as measured using on Eppendorf oxygen electrode.

MATERIALS AND METHODS

Animal and Tumour Model. All experiments were performed on 10-14 week-old female CDF1 mice. The tumour model used was the C3H/Tif mouse mammary carcinoma. Its derivation and maintenance have been described previously [Overgaard 1980]. Experimental tumours were produced following sterile dissection of large flank tumours. Macroscopically viable tumour tissue was minced with a pair of scissors, and 5-10 μl of this material were injected into the foot of the right hind limb of the experimental animals. This location ensured easy access to the tumour for treatment without involvement of critical normal tissue in the treatment field. In addition, anaesthetization of the animals during treatment could be avoided. Treatments were carried out when tumours had reached a tumour volume of 200 mm^3, which generally occurred within 2-3 weeks after challenge. Tumour size was determined by the formula: D1 x D2 x D3 x $\pi/6$, (where the D values represent three orthogonal diameters).

Drug Preparation. Nicotinamide (Sigma Chemical Co., St. Louis, MO) was dissolved at different concentrations in a sterile saline (0.9% NaCl) solution immediately before each experiment. It was injected intraperitoneally (i.p.) into mice at a constant injection volume of 0.02 ml/g body weight.

Carbogen Breathing. For treatment with carbogen (95% O_2 + 5% CO_2) non-anaesthetized mice were individually restrained in lucite jigs and gassed at a flow rate of 2.5 l/min. Gassing was maintained throughout any subsequent radiation treatment or pO_2 measurement.

Radiation Treatment. Irradiations were given with a conventional therapeutic X-ray machine (250 kV; 10 mA; 2 mm Al filter; 1.1 mm Cu half-value layer; dose rate 2.3 Gy per min). Dosimetry was accomplished by use of an integrating chamber. All treatments to tumour-bearing feet were administered to non-anaesthetized mice restrained in lucite jigs. Their tumour-bearing legs were exposed and loosely attached to the jig with tape, without impairing the blood supply to the foot. Tumours only were irradiated, the remainder of the animal being shielded by 1 cm of lead. To secure homogeneity of the radiation dose, tumours were immersed in a water bath with about 5 cm of water between the X-ray source and the tumour. Tumour response to treatment involved measuring tumour volume 5 times each week following irradiation and calculating the tumour growth time, which was the time taken for tumours to reach 3 times the treatment volume. Mice dying before the tumour growth time was reached were rejected from the analysis and any tumours controlled by the treatments were arbitrarily assigned a tumour growth time of 60 days.

Measurements with the pO_2 Histograph. Unanesthetized mice were restrained in lucite jigs with the tumour-bearing leg exposed and taped to the jig as described earlier. A fine needle autosensitive electrode probe (Eppendorf, Hamburg, Germany) was inserted about 1 mm into the tumour and then moved through the tumour in 0.7 mm increments, followed each time by a 0.3 mm backward step prior to measurement. The response time was 1.4 seconds. Parallel repeated insertions were performed in each tumour until a total of 45 to 100 measurements were made. The relative frequency of the pO_2 measurements was automatically calculated and displayed as a histogram, although the various pO_2 parameters were taken from the original raw data. When making measurements some negative values were observed. These were accepted provided they were small in number and not greater than -1.0 mmHg.

RESULTS

The ability of nicotinamide and/or carbogen to enhance radiation damage in this C3H mammary carcinoma is shown in Figure 1. Nicotinamide or carbogen alone had no effect on tumour growth, but both agents enhanced radiation response. With carbogen this enhancement appeared to be maximal within the first 15 minutes of gas breathing, but slowly decreased as the breathing time increased. The greatest enhancement was observed when nicotinamide and carbogen were combined. Moreover, nicotinamide also appeared to eliminate the carbogen pre-irradiation breathing time dependency effect.

Figure 2 illustrates the influence of nicotinamide dose on tumour radiation response. A constant enhancement of radiation damage by the drug alone was seen with doses from 100 to 1000 mg/kg when given 20 minutes prior to irradiation. Combining nicotinamide with carbogen breathing resulted in an increase in radiation response that also appeared to be independent of nicotinamide dose from 100 to 1000 mg/kg.

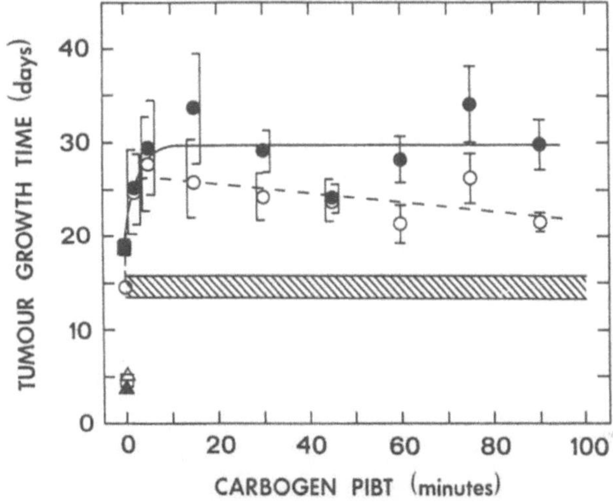

Figure 1. The effect of nicotinamide [1000 mg/kg; i.p.] and different carbogen pre-irradiation breathing times (PIBT) on the response of a C3H mammary carcinoma to radiation (15 Gy). Carbogen breathing was also continued during irradiation. Nicotinamide was injected 20 minutes prior to radiation. Control (△), nicotinamide (▲), carbogen (☐), radiation (☒), nicotinamide + radiation (■), carbogen + radiation (○), nicotinamide + carbogen + radiation (●). Results are means (± 1 S.E.) from 8-12 mice/group, and show the time taken for tumours to regrow to 3 times treatment volume.

The pO$_2$ histograms obtained in this tumour model with low doses of nicotinamide, carbogen breathing, and the combination of these agents are shown in Figure 3, and the various pO$_2$ parameters obtained from this data are summarized in Table 1. Although both 100 and 200 mg/kg doses of nicotinamide appeared to improve tumour oxygenation, the results in general were not significantly different from control values. On the other hand, carbogen breathing, and the combination of nicotinamide and carbogen, did significantly increase oxygenation in this tumour.

DISCUSSION

The radiation response of this C3H mammary carcinoma was increased by either injecting mice with nicotinamide or allowing the animals to breathe carbogen gas. By far the greatest enhancement of radiation damage occurred when nicotinamide and carbogen were combined. These results are entirely consistent with earlier reports for either nicotinamide (Horsman et al., 1989b, 1990, 1993a) or carbogen (Grau et al., 1992) in this C3H tumour, and the combination of these agents in the SCCVII (Chaplin et al., 1993) and CaNT (Kjellén et al., 1991) tumours.

For carbogen, the enhancement of radiation damage is the result of the gas increasing the amount of physically dissolved oxygen in the blood and therapy improving delivery to the tumour (Grau et al., 1992). Such an effect would reduce the level of diffusion limited chronic hypoxia, but be expected to have little or no influence on perfusion limited acute hypoxia. Nicotinamide, on the other hand, has been shown to decrease the transient fluctuations in tumour blood flow (Chaplin et al., 1990; Horsman et al., 1990) which are

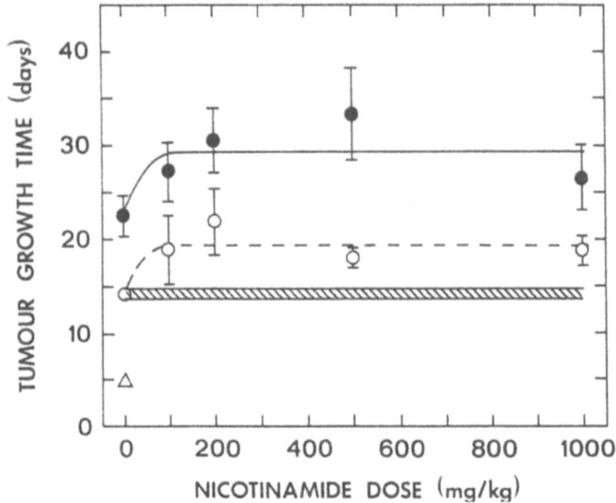

Figure 2. The effect of nicotinamide dose and carbogen breathing on the response of a C3H mammary carcinoma to radiation (15 Gy). Mice were injected i.p. with different nicotinamide doses 20 minutes prior to irradiation. Carbogen breathing was for 10 minutes before and during radiation treatment. Control (△), radiation (▧), nicotinamide + radiation (○), nicotinamide + carbogen + radiation (●). Results are means (± 1 S.E.) from 10-12 mice/group, and show the time taken for tumours to regrow to 3 times treatment volume.

known to result in the development of perfusion limited acute hypoxia (Chaplin et al., 1986; 1987). The finding that the combination of nicotinamide and carbogen enhanced radiation damage to a greater degree than either agent alone is consistent with the agents working on different resistant cell populations. Further evidence that carbogen decreased hypoxia was also shown by the improvement in tumour oxygenation. Although the nicotinamide data also indicated an increase in oxygenation status in the tumour, in general there were no signi-

ficant changes. Similar findings have been reported with high drug doses both in this tumour
(Horsman et al., 1993b) and in a rat DS-sarcoma (Kelleher and Vaupel 1993), although Lee
and Song (1992) did find nicotinamide induced changes in pO_2 in some mouse tumours.
Previous studies on tumour metabolic activity (Wood et al., 1991; Horsman et al., 1993b)
and the binding of ^{14}C-misonidazole (Horsman et al., 1988; 1989a), indirectly indicated that
nicotinamide did improve tumour oxygenation. Since the drug apparently operates at the
microregional level (Chaplin et al., 1990; Horsman et al., 1990) we have suggested
(Horsman et al. 1993b) that such changes may be difficult to detect with pO_2 electrodes, but

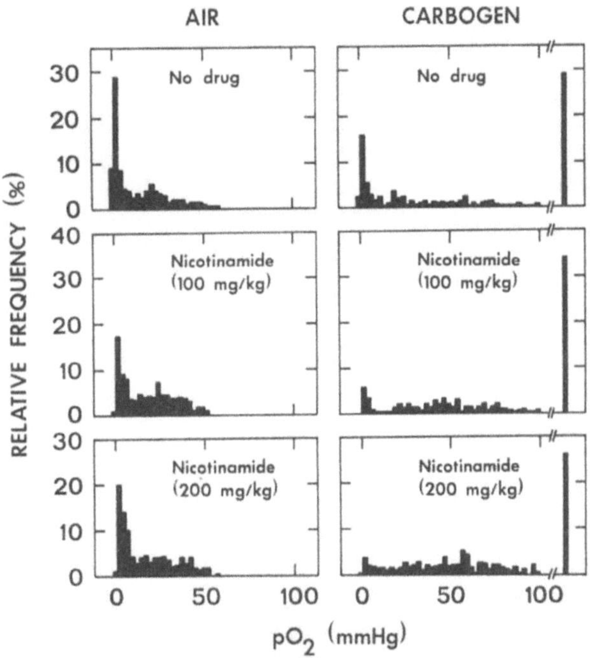

Figure 3. The pO_2 histograms obtained from a C3H mammary carcinoma under different treatment conditions.
Measurements were made in air beathing mice (left panels), or in mice allowed to breathe carbogen for 10
minutes prior to and during measurement (right panels). Nicotinamide (100 or 200 mg/kg; i.p.) was injected
20 minutes prior to the start of the pO_2 measurements. Results show the relative frequency of each pO_2 value
obtained. Values over 100 mmHg have been grouped in one column.

are readily seen with techniques in which the analysis is based on the whole tumour
(Horman et al., 1988; 1989a; 1993b; Wood et al., 1991).

The enhancement of radiation damage in this tumour was somewhat dependent on the
carbogen pre-irradiation breathing time. Such a dependency has been shown before (Siemann
et al., 1977; Chaplin et al., 1993) and may be of critical importance clinically. In our study,

nicotinamide appeared to overcome this breathing time dependency, which is consistent with our results in the SCCVII tumour (Chaplin et al., 1993). That earlier study also showed that although the nicotinamide effect was independent of drug dose above 250 mg/kg, there was a drop-off in sensitization, with and without carbogen, below 250 mg/kg. We found no such loss of sensitization down to 100 mg/kg. Additional studies by us have shown that maximal sensitization is observed at doses as low as 100 mg/kg provided irradiation is given at the time of peak tumour drug concentration, which in this C3H tumour was between 10 to 30 minutes after injection (Horsman et al., 1993a), hence our rational for selecting a 20-minute interval. Loss of sensitization at low doses was seen if radiation was given 60 minutes after drug injection, a time when drug levels in the tumour had started to decrease (Horsman et al., 1993a). In the SCCVII study there was a 1-hour gap between drug administration and irradiation, and it is possible that this time interval was not optimal.

This current study has clearly shown that the combination of nicotinamide and carbogen can substantially enhance radiation damage in a C3H mammary carcinoma. More importantly these effects can occur at nicotinamide doses which are clinically achievable (Horsman et al., 1993a). In fact, the percentage of pO_2 values ≤ 5 mmHg, which probably represents the level of oxygenation of radioresistant hypoxic cells, was decreased from 49% in control tumours to 11% with carbogen + 100 mg/kg nicotinamide, and down to 4% with a 200 mg/kg drug dose. This suggests that such a combination can almost entirely eliminate hypoxia in this tumour. Although additional experimental studies are required, these results strongly support the clinical use of nicotinamide and carbogen in those human tumours where hypoxia may limit radiotherapy.

Table 1. Summary of the pO_2 parameters measured in a C3H mammary carcinoma under different treatment conditions.

Treatment[1]	N (n)[2]	pO$_2$ parameters (mmHg)		
		% $\leq$ 5 mmHg	mean	median
Control	8 (657)	49.1±3.3[3]	13.6±1.4	6.9±1.7
Nicotinamide [100 mg/kg]	5 (291)	27.0±6.0*	18.1±1.8	16.4±3.5*
Nicotinamide [200 mg/kg]	6 (414)	35.3±8.0	15.6±2.5	10.0±3.0
Carbogen	6 (472)	24.1±2.9*	78.4±10.6*	45.1±8.7*
Nicotinamide [100 mg/kg] + Carbogen	6 (365)	11.3±7.8*	97.0±18.9*	75.1±20.2*
Nicotinamide [200 mg/kg] + Carbogen	5 (287)	4.0±2.1**	101.3±23.7*	79.9±24.7*

[1] Results were obtained from control mice under normal air breathing conditions; 20 minutes after i.p. injection with 100 or 200 mg/kg nicotinamide; or 10 minutes after the start of carbogen breathing.

[2] "N" represents the number of mice in each treatment group, while "n" is the total number of pO_2 measurements obtained.

[3] Errors are ± 1 S.E.

* Values significantly different from controls as estimated by a Student's t-test ($p < 0.05$).

** Values significantly different from carbogen alone (($p < 0.05$).

ACKNOWLEDGMENTS

The authors thank Ms. I.M. Johansen and Ms. M.H. Simonsen for excellent technical assistance. This study was supported by grants from the Danish Cancer Society, the American NIH (grant number CA 55300) and the British Cancer Research Campaign.

REFERENCES

Brown, J.M., 1979, Evidence for acutely hypoxic cells in mouse tumours and a possible mechanism of reoxygenation, *Br. J. Radiol.* 52:650.

Chaplin, D.J., Durand, R.E. and Olive, P.L., 1986, Acute hypoxia in tumors: implication for modifiers of radiation effects, *Int. J. Radiat. Oncol. Biol. Phys.* 12:1279.

Chaplin, D.J., Horsman, M.R. and Aoki, D., 1991, Nicotinamide, Fluosol DA and carbogen: a strategy to reoxygenate acutely and chronically hypoxic cells in vivo, *Br. J. Cancer* 63:109.

Chaplin, D.J., Horsman, M.R. and Siemann, D.W., 1993, Further evaluation of nicotinamide and carbogen as a strategy to reoxygenate acutely and chronically hypoxic cells in vivo: importance of nicotinamide dose and pre-irradiation breathing time. *Br. J. Cancer* 68:269

Chaplin, D.J., Horsman, M.R. and Trotter, M.J., 1990, Effect of nicotinamide on the microregional heterogeneity of oxygen delivery within a murine tumour, *J. Natl. Cancer Inst.* 82:672.

Chaplin, D.J., Olive, P.L. and Durand, R.E., 1987, Intermittent blood flow in a murine tumor: radiobiological effects, *Cancer Res.* 47:597.

Denekamp, J. 1991, ARCON: Accelerated radiotherapy with carbogen and nicotinamide. *Eur. Cancer News* 4:3

Dische, S. 1989, The clinical consequences of the oxygen effect, in: "The Biological Basis of Radiotherapy," G.G. Steel, G.E. Adams and A. Horwich, eds., Elsevier Science Publishers, Amsterdam.

Grau, C., Horsman, M.R. and Overgaard, J., 1992, Improving the radiation response in a C3H mouse mammary carcinoma by normobaric oxygen or carbogen breathing, *Int. J. Radiat. Oncol. Biol. Phys.* 22:415.

Guichard, M., Courdi, A. and Malaise, E.P., 1980, Experimental data on the radiobiology of solid tumours. *Eur. J. Radiother.* 1:171.

Horsman, M.R. 1993, Hypoxia in tumours: Its relevance, identification and modification, in: "Current Topics in Clinical Radiobiology of Tumours," H.P. Beck-Bornholdt, ed., Springer-Verlag, Berlin.

Horsman, M.R. Brown, J.M., Hirst, V.K., Lemmon, M.J., Wood, P.J. Dunphy, E.P., and Overgaard, J., 1988, Mechanism of action of the selective tumour radiosensitizer nicotinamide, *Int. J. Radiat. Oncol. Biol. Phys.* 15:685.

Horsman, M.R., Chaplin, D.J. and Brown, J.M., 1989a, Tumor radiosensitization by nicotinamide: a result of improved blood perfusion and oxygenation, *Radiat. Res.* 118:139.

Horsman, M.R., Chaplin, D.J. and Overgaard, J., 1990, Combination of nicotinamide and hyperthermia to eliminate radioresistant chronically and acutely hypoxic tumor cells, *Cancer Res.* 50:7430.

Horsman, M.R., Hansen, P.V. and Overgaard, J., 1989b, Radiosensitization by nicotinamide in tumors and normal tissues: the importance of tissue oxygenation status, *Int. J. Radiat. Oncol. Biol. Phys.* 16:1273.

Horsman, M.R., Høyer, M., Honess, D.J., Dennis, I.F. and Overgaard, J., 1993a, Nicotinamide pharmacokinetics in humans and mice: a comparative assessment and the implications for radiotherapy, *Radiother. Oncol.* 27:131.

Horsman, M.R., Nordsmark, M., Khalil, A., Chaplin, D.J. and Overgaard, J., 1993b, Tumour radiosensitization by nicotinamide: is it the result of an improvement in tumour oxygenation? *Adv. Exper. Med. Biol.* (in press).

Kelleher, D.K. and Vaupel, P.W., 1993, Nicotinamide exerts apparently opposing acute effects on microcirculatory function and tissue oxygenation in rat tumours. *Int. J. Radiat. Oncol. Biol. Phys.* 26:95.

Kjellen, E., Joiner, M.C., Collier, J.M., Johns, H. and Rojas, A., 1991, A therapeutic benefit from combining normobaric carbogen or oxygen with nicotinamide in fractionated X-ray treatments, *Radiother. Oncol.* 22:81.

Lee, I. and Song, C.W., 1992, The oxygenation of murine isografts and human tumour xenografts by nicotinamide, *Radiat. Res.* 130:65.

Moulder, J.E. and Rockwell, S., 1984, Hypoxic fractions of solid tumors, *Int. J. Radiat. Oncol. Biol. Phys.* 10:695.

Overgaard, J., 1980, Simultaneous and sequential hyperthermia and radiation treatment of an experimental tumor and its surrounding normal tissue in vivo, *Int. J. Radiat. Oncol. Biol. Phys.* 6:1507.

Overgaard, J., 1989, Sensitization of hypoxic tumour cells - clinical experience, *Int. J. Radiat. Biol.* 56:801.

Siemann, D.W., Hill, R.P. and Bush, R.A. 1977, The importance of the pre-irradiation breathing times of oxygen and carbogen (5% CO_2: 95% O_2) on the in vivo radiation response of a murine sarcoma. Int. J. Radiat. Oncol. Biol. Phys. 2: 903.

Sutherland, R.M. and Franko, A.J., 1980, On the nature of the radiobiologically hypoxic fraction in tumors, *Int. J. Radiat. Oncol. Biol. Phys.* 6:117.

Tannock, I.F., 1968, The relationship between cell proliferation and the vascular system in a transplanted mouse mammary tumour, *Br. J. Cancer* 22:258.

Thomlinson, R.H. and Gray, L.H., 1955, The histological structure of some human lung cancers and the possible implications for radiotherapy, *Br. J. Cancer* 9:539.

Wood, P.J., Counsell, C.J.R., Bremner, J.C.M., Horsman, M.R. and Adams, G.E., 1991, The measurement of radiosensitizer-induced changes in mouse tumour metabolism by 31-P magnetic resonance spectroscopy, *Int. J. Radiat. Oncol. Biol. Phys.* 20:291.

Assessment of Intracellular Oxygenation of Solid Tumors

Avis L. Sylvia,Frans F.Jöbsis-VanderVliet, Adriaan C. Jöbsis,and
H. Dirk Sostman
Depts. of Cell Biology and Radiology, Duke University Medical
Center, Durham, NC 27710 USA

This study was launched to determine whether oxygenation of tumor
cells, as opposed to intra-tumor red blood cells, could be verified
with a non-invasive methodology. The markers used to differentiate
between the two cell types were hemoglobin for the RBC and
cytochrome c oxidase (cytochrome aa3) for the tumor cells. Near
infrared and visible multiwavelength, differential
spectrophotometry were used to directly assess in vivo kinetic
changes in relative hemoglobin saturation, tissue blood volume, and
cytochrome aa3 redox state in the BA1112 rhabdomyosarcoma and in
skeletal muscle, its derivative tissue. Tissue hemodynamics and
intracellular redox status were monitored both independently in
tumors and concurrently with muscle in anesthetized WAG/Rij/Y rats
breathing normoxic, hyperoxic, hypoxic, and hypercapnic gas
mixtures and also in response to administered vasoactive agents.
Optical signal responses were not necessarily influenced by changes
in arterial perfusion pressure (MAP) as verified by use of
vasoactive agents chosen to increase and decrease MAP. Hypoxia and
hypercapnic acidosis invariably decreased both tumor and muscle
oxygenation as evidenced by hemoglobin deoxygenation, reduced
tissue blood volume, and increased intra-mitochondrial cyt aa3
reduction. Muscle was not responsive to hyperoxia ($F_{I}O_2=1.0$). In
tumors, hyperoxia consistently increased the HbO_2 content even
though tissue blood volume fell. However, the direction and
magnitude of oxidation-reduction changes in cytochrome aa3 varied
among tumors. These responses were not correlated with the size
(volume) of the tumor. We speculate that the underlying cause of
this idiosyncratic behavior was the variable stage and degree of
vascular proliferation/degeneration in the tumors. These findings
emphasize that increased oxygenation of the tumor cells, for
instance prior to radiation treatment, can only be judged by their
intracellular cytochrome response, not by the local blood
oxygenation.

Miwa, M., 171
Mockros, L.F., 382
Mohsenifar, Z., 333
Mori, M., 45
Mortimer, R.H., 389 390
Mueller-Klieser, W., 619

Nakashima, T., 119
Nemoto, E.M., 469
Newton, D.J., 181
Nioka, S., 171
Nolte, D., 541
Nordsmark, M., 635
Nöth, U., 129

O'Hara, J., 119
Opitz, N., 231
Orel, S., 171
Overgaard, J., 635
Owen-Reece, H., 143, 475

Pal, M., 577
Pan, N., 319
Parthasarathi, K., 249
Patel, B., 99
Pawlowski, M., 83, 93
Pembeci, K., 603
Peric, M., 230
Piantadosi, C.A., 157
Piiper, J., 503
Pittman, R.N., 249
Pompe, J.C., 443
Poole, D.C., 235, 521
Popel, A.S., 17
Potter, R.F., 229
Puri, V., 547

Raad, R.A., 181
Rakusan, K., 407
Ram Rao, P., 547 555
Rao, G., 196
Reynolds, E.O.R., 143
Rhoades, G.E., 41
Richardson, R.S., 375 521
Rifkind, J.M., 345, 547
Ringle, A.S., 41
Ringnalda, B.E.M., 533
Rumsey, W.L., 93, 99

Samaja, M., 393
Schäfer, R., 604
Scheid, P., 371
Schmid-Schöenbein, G.W., 385
Scholten, E.W., 573
Schumacker, P.T., 605
Shnall, M., 171
Shubayev, I., 401
Sibbald, W.J., 229

Siemann, D.W., 627 635
Silverton, S.F., 31
Sinaasappel, M., 75, 105
Smith, M., 475
Sostman, H.D., 643
Spilman, S., 206, 215
Standaert, T.A., 427
Stevenson, D.K., 206, 215
Su, E-N., 303
Suematsu, M., 385
Suzuki, Y., 457
Swanson, D., 381
Swartz, H.M., 119
Sylvia, A.L., 643

Takahashi, E., 163
Takasugi, T., 457
Tamei, H., 565
Teicher, B.A., 230
Telci, L., 437 603
Terris, D.J., 613
Thimm, F., 417
Tilden, S.J., 449
Tischler, M.E., 577
Toth, A., 577
Trouwborst, A., 41
Truog, W.E., 427
Turek, Z., 1
Tütüncü, A.S., 437

Ungerleider, R.M., 236
Utschakowski, A., 604

Vallet, B., 295
van der Kleij, A.J., 279, 573, 575
van der Sluijs, J.P., 75, 105
van Dijk, G., 227
Van Houten, J.P., 206, 215
van Rossem, K., 491
Vaupel, P., 607
Vermariën, H., 491
Vink, H., 575
Vinogradov, S.A., 61, 67
Voter, W.A., 313

Wagner, K., 604
Wagner, P.D., 371, 375, 383, 465, 521, 571, 578
Walczak, T., 119
Wang, C.Y., 59
Wartenberg, M., 231
Waters, C.M., 382
Werkmeister, F., 353
Westerkamp, B., 227
Whitaker, E.G., 236
Williams, E.M., 187
Wilson, D.F., 61, 67, 83, 93